Quantitative MRI of the Brain

PRINCIPLES OF PHYSICAL MEASUREMENT

Series in Medical Physics and Biomedical Engineering

Series Editors: John G. Webster, E. Russell Ritenour, Slavik Tabakov and Kwan-Hoong Ng

Recent books in the series:

Quantitative MRI of the Brain: Principles of Physical Measurement, Second edition
Mara Cercignani, Nicholas G. Dowell and Paul S. Tofts (Eds)

A Brief Survey of Quantitative EEG
Kaushik Majumdar

Handbook of X-ray Imaging: Physics and Technology
Paolo Russo (Ed)

Graphics Processing Unit-Based High Performance Computing in Radiation Therapy
Xun Jia and Steve B. Jiang (Eds)

Targeted Muscle Reinnervation: A Neural Interface for Artificial Limbs
Todd A. Kuiken, Aimee E. Schultz Feuser and Ann K. Barlow (Eds)

Emerging Technologies in Brachytherapy
William Y. Song, Kari Tanderup, and Bradley Pieters (Eds)

Environmental Radioactivity and Emergency Preparedness
Mats Isaksson and Christopher L. Rääf

The Practice of Internal Dosimetry in Nuclear Medicine
Michael G. Stabin

Radiation Protection in Medical Imaging and Radiation Oncology
Richard J. Vetter and Magdalena S. Stoeva (Eds)

Statistical Computing in Nuclear Imaging
Arkadiusz Sitek

The Physiological Measurement Handbook
John G. Webster (Ed)

Radiosensitizers and Radiochemotherapy in the Treatment of Cancer
Shirley Lehnert

Diagnostic Endoscopy
Haishan Zeng (Ed)

Medical Equipment Management
Keith Willson, Keith Ison, and Slavik Tabakov

Quantifying Morphology and Physiology of the Human Body Using MRI
L. Tugan Muftuler (Ed)

Quantitative MRI of the Brain

PRINCIPLES OF PHYSICAL MEASUREMENT

Edited by

Mara Cercignani

Nicholas G. Dowell

Paul S. Tofts

CRC Press is an imprint of the
Taylor & Francis Group, an **informa** business

CRC Press
Taylor & Francis Group
6000 Broken Sound Parkway NW, Suite 300
Boca Raton, FL 33487-2742

First issued in paperback 2020

ISBN-13: 978-1-138-03285-9 (hbk)
ISBN-13: 978-0-367-78153-8 (pbk)

Library of Congress Cataloging-in-Publication Data

Names: Cercignani, Mara, editor. | Dowell, Nicholas G., editor. | Tofts, Paul S., editor.
Title: Quantitative MRI of the brain : principles of physical measurement / editors, Mara Cercignani, Nicholas G. Dowell, Paul S. Tofts.
Other titles: Series in medical physics and biomedical engineering.
Description: Second edition. | Boca Raton, FL : CRC Press, Taylor & Francis Group, [2017] | Series: Series in medical physics and biomedical engineering | Includes bibliographical references and index.
Identifiers: LCCN 2017034655| ISBN 9781138032859 (hardback ; alk. paper) | ISBN 1138032859 (hardback ; alk. paper) | ISBN 9781315363578 (e-book) | ISBN 1315363577 (e-book) | ISBN 9781315363554 (e-book) | ISBN 1315363550 (e-book) | ISBN 9781315363561 (e-book) | ISBN 1315363569 (e-book) | ISBN 9781315363547 (e-book) | ISBN 1315363542 (e-book)
Subjects: LCSH: Brain--Magnetic resonance imaging.
Classification: LCC RC386.6.M34 Q365 2017 | DDC 616.8/04754--dc23
LC record available at https://lccn.loc.gov/2017034655

Visit the Taylor & Francis Web site at
http://www.taylorandfrancis.com

and the CRC Press Web site at
http://www.crcpress.com

Contents

Foreword

It is remarkable how much progress has occurred since the publication of Paul S. Tofts' first edition in 2003. Remember that radiology began as a highly qualitative field where shapes and patterns were correlated with diseases. Initially, magnetic resonance imaging thrived because it could produce undistorted views in three planes. It has evolved considerably from beautiful images to being able to truly predict *in vivo* pathology. The previous edition of this book was ahead of its time in articulating the vision of quantitative MRI. How far the field will go in the future depends on the creativity of investigators and their ability to understand and solve technical problems.

To reliably predict biology on combinations of techniques could change approaches to disease. There are indeed many possibilities for the future including accuracy in differentiating aspects of disease such as tumor versus edema, necrosis versus tumor, benign versus malignant, and the grail – biological aggressiveness of lesions. The opportunities are abundant. I will cite two examples. Think about being able to predict aggressiveness and tumor volume of prostate cancer – totally changing how surveillance is carried out. MR may also be able to accurately separate therapeutic tissue changes such as the effects of radiation or immunotherapy from recurrent tumor. This is essential in detecting infiltrating neoplasms such as glioma.

For those engaged in imaging, understanding how physical principles can be applied to produce quantitative results has the prospect of transforming communications, monitoring, and reporting. It is the path where MR and medicine are heading. Quantitation provides objective measures of activity and disease – vital to assessing the extent of pathology and determining the success of particular treatment protocols. The power of quantitative MR is underscored today by its incorporation in clinical trials to demonstrate efficacy.

Quantitative MRI of the Brain begins with basic discussion of data collection and image generation. Chapter 3 discusses quality assurance, precision, and accuracy. These concepts are critical to consistency, an essential element of quantitative imaging. Once grounded in the essentials of reproducibility, the text leads the reader through relaxation times and biophysical principles of susceptibility, diffusion, and magnetisation transfer. The last chapters focus on techniques including functional MRI, spectroscopy, and perfusion.

Paul S. Tofts and his co-editors, Mara Cercignani and Nicholas G. Dowell, have done a magnificent job assembling contributors with appropriate expertise and have been successful in compiling the requisites necessary to achieve a high level of understanding in quantitative MR. This is not a small feat and is necessary for advancing the field. Congratulations on a significant accomplishment!

Robert I. Grossman
NYU Langone Health

Foreword to the First Edition

Paul S. Tofts has succeeded brilliantly in capturing the essence of what needs to become the future of radiology in particular, and medicine in general – quantitative measurements of disease. This is a critical notion. The discipline of radiology started with the ability to discern the shadows that were abnormal. On chest x-rays, one could see the 'white in the right' and that was correlated to the clinical diagnosis of pneumonia. It is truly amazing how long such descriptions were adequate and indeed, the state of the art. This transcended the modern era of cross-sectional imaging. CT and then MR heralded the ability to not only observe pathological states but to specifically define and locate such conditions. Based upon absorption of biophysical parameters combined with position, one could suggest that a particular abnormality was a stroke rather than a tumour or an infection. Make no mistake about it, this was an incredible scientific leap and has totally changed the calculus of medicine. In the twenty-first century radiologists have become both the diagnostician and the arbiter of therapeutic efficacy. In clinical neuroscience the function of the neurologic exam has been diminished by the unbiased, reliable nature of imaging. This has been mirrored throughout the body. The preeminent role of imaging now requires a new level of metric quantitative measurements.

Why is quantitative methodology so vital? First and foremost, it is relatively unbiased compared with qualitative descriptions. Second, it lends itself easily to statistical modelling. Lastly, if performed correctly, the data can be pooled over multiple centres to provide power regarding a clinical trial or longitudinal study. Thus, the natural history of a disease such as multiple sclerosis may be ascertained by a time-dependent study. This was first made apparent when the FDA in the United States approved the use of interferon beta-1b in 1993 based upon MRI data that revealed a decrease in disease activity and lesion burden. The effect of the drug could not be ascertained from the clinical measure of disability, the acknowledged gold standard for multiple sclerosis. Approval of interferon beta changed the course of clinical treatment trials. What has ensued is a discussion of surrogate markers in imaging, sensitivity, specificity, reproducibility, etc. The bottom line is the emergence of the mandatory need to incorporate quantitative imaging techniques into treatment trails.

This book addresses the measurement process, the measures, what the measures mean biologically, and image analysis methodology. Any physician/scientist participating in a scientific study or clinical trial must be familiar with the concepts elucidated in this book. Although the text is focused on the brain, the concepts pertain to any imaging study. How to ensure that your results will stand the test of the critical review is the underlying theme of the first section on the concept of measurement in MR. Thorough knowledge of the principles of accuracy, precision, and quality assurance are essential to the writing of any imaging proposal and the subsequent performance of the study.

The second section is focused on the metrics themselves. Here, there are lucid discussions of MR parameters that are the windows on the pathological processes we wish to study. This is complete and covers the intrinsic MR parameters (T_1, T_2, PD), diffusion, magnetisation, transfer, spectroscopy, dynamic contrast, perfusion, and fMRI. To appreciate the strengths and limitations of these measures enables the reader to identify optimal parameters for particular studies. It also assists in the interpretation of the current literature. The section offers a complete survey of all of the metrics used in clinical MR today.

The chapter on the biological significance of the MR parameters in multiple sclerosis translates the imaging parameters to their biological correlates. This is important, for if the measures are just abstract it is hard to argue for their implementation.

The last major section of the book deals with the topics of registration of images and other measures, including atrophy, texture, and volumetric analysis. Image registration is fundamental when performing any longitudinal analysis. Just think about it. When a radiologist is asked if a lesion has changed on two different studies, one must be careful that the apparent change is not the result of technical differences (slice alignment, slice thickness, etc.).

I was honoured to be asked by Professor Tofts to write this foreword. In my opinion this text is a beautifully executed work, capturing what is essential for radiologists and scientists to understand about quantitative MR measures. There is no more qualified individual up to this task than Dr. Tofts. He is a lucid and most thoughtful scientist. I wish to extend my congratulations to him and the other authors on this effort. This book will become a classic and the first of many on this significant topic.

Robert I. Grossman
New York University School of Medicine

Editors

Mara Cercignani is professor of Medical Physics at the Brighton and Sussex Medical School (BSMS). She has worked in the field of MRI since 1998, and received her Doctorate from University College London in 2007. Before moving to BSMS in 2011, she worked at San Raffaele Hospital in Milan (1998–2002), the Institute of Neurology in London (2002–2007) and Santa Lucia Foundation in Rome (2007–2011). Her main research interests lie with the field of quantitative MRI, spanning from diffusion MRI to quantitative magnetisation transfer imaging.

Nicholas G. Dowell is a lecturer in Imaging Physics at the Brighton and Sussex Medical School (BSMS). He received his Doctorate in solid-state nuclear magnetic resonance from the University of Exeter in 2004 before moving to the Institute of Neurology, University College London, to work on quantitative magnetic resonance imaging. He moved to the newly opened Clinical Imaging Sciences Centre at BSMS in 2007. His research interests lie principally with quantitative magnetisation transfer, diffusion-weighted imaging, data modelling and analysis techniques in the brain.

Paul S. Tofts is emeritus professor at the Brighton and Sussex Medical School (BSMS). After obtaining a BA in Physics from Oxford University in 1970, the new University of Sussex provided an ideal contrasting environment where his D Phil was in experimental NMR studies of helium at low temperature. When biomedical NMR hardly existed, from 1975 he researched radioisotope and CT imaging at the Royal Postgraduate Medical School, London. At University College London, in 1978, he developed a prototype ^{31}P NMR machine for newborn babies and started a career in quantitative MR with the first measurement of absolute metabolite concentrations *in vivo*.

An early MRI machine in 1985 devoted to multiple sclerosis studies at the Institute of Neurology, Queen Square (now part of University College London), enabled Paul to develop a whole range of quantitative imaging techniques and to edit the first book on quantitative MRI. Analysis of dynamic Gd-enhanced image data enabled the quantifying of blood–brain barrier leakage, and his mathematical model is now used extensively.

The new imaging centre at BSMS attracted Paul to return in 2006 to Sussex as the foundation chair of Imaging Physics until 2009. He has 215 publications, 15,000 citations and an h-factor of 62.

Contributors

Daniel C. Alexander
Centre for Medical Image Computing (CMIC)
Department of Computer Science
University College London (UCL)
London, UK

Marco Battiston
Department of Neuroinflammation
UCL Institute of Neurology
University College London (UCL)
London, UK

Sagar Buch
The MRI Institute for Biomedical Research
Waterloo, ON, Canada

David L. Buckley
Division of Biomedical Imaging
University of Leeds
Leeds, UK

Martina F. Callaghan
Wellcome Trust Centre for Neuroimaging
UCL Institute of Neurology
University College London (UCL)
London, UK

Mara Cercignani
Department of Neuroscience
Brighton and Sussex Medical School
Brighton, UK

Yongsheng Chen
Department of Radiology
Wayne State University
Detroit, MI, USA

Chang-Hoon Choi
Institute of Neuroscience and Medicine (INM-4)
Forschungszentrum Jülich
Jülich, Germany

Ralf Deichmann
Brain Imaging Center
Goethe-University
Frankfurt/Main, Germany

Nicholas G. Dowell
Department of Neuroscience
Brighton and Sussex Medical School
Brighton, UK

Audrey P. Fan
Richard M. Lucas Center for Imaging
Stanford University
Stanford, CA, USA

Jörg Felder
Institute of Neuroscience and Medicine (INM-4)
Forschungszentrum Jülich
Jülich, Germany

Shir Filo
Edmond and Lily Safra Center for Brain Sciences (ELSC)
The Hebrew University of Jerusalem
Jerusalem, Israel

Claudia A.M. Gandini Wheeler-Kingshott
Department of Neuroinflammation
UCL Institute of Neurology
University College London (UCL)
London, UK
and
Brain MRI 3T Mondino Research Centre
C. Mondino National Neurological Institute
Pavia, Italy
and
Department of Brain and Behavioural Sciences
University of Pavia
Pavia, Italy

Claudine J. Gauthier
Department of Physics/PERFORM Centre
Concordia University
Montreal, QC, Canada

Leonidas Georgiou
Division of Biomedical Imaging
University of Leeds
Leeds, UK

Kiarash Ghassaban
Magnetic Resonance Innovations Inc.
Detroit, MI, USA

Aurobrata Ghosh
Centre for Medical Image Computing (CMIC)
Department of Computer Science
University College London (UCL)
London, UK

Xavier Golay
UCL Institute of Neurology
University College London (UCL)
London, UK

René-Maxime Gracien
Department of Neurology
University Hospital
Goethe-University
Frankfurt/Main, Germany

Francesco Grussu
Department of Neuroinflammation
UCL Institute of Neurology
University College London (UCL)
London, UK

Ewart Mark Haacke
The MRI Institute for Biomedical Research
Waterloo, ON, Canada
and
Department of Radiology
Wayne State University
Detroit, MI, USA
and
Magnetic Resonance Innovations Inc.
Detroit, MI, USA

Andrada Ianus
Centre for Medical Image Computing (CMIC)
Department of Computer Science
University College London (UCL)
London, UK

Mina Kim
UCL Institute of Neurology
University College London (UCL)
London, UK

Yan Li
Department of Radiology and Biomedical Imaging
University of California San Francisco
San Francisco, CA, USA

Saifeng Liu
The MRI Institute for Biomedical Research
Waterloo, ON, Canada

Aviv A. Mezer
Edmond and Lily Safra Center for Brain Sciences (ELSC)
The Hebrew University of Jerusalem
Jerusalem, Israel

Siawoosh Mohammadi
Medical Center Hamburg-Eppendorf
Department of Systems Neuroscience
Hamburg, Germany

Sarah J. Nelson
Department of Radiology and Biomedical Imaging
University of California San Francisco
San Francisco, CA, USA

Ana-Maria Oros-Peusquens
Institute of Neuroscience and Medicine (INM-4)
Forschungszentrum Jülich
Jülich, Germany

Esben Thade Petersen
Danish Research Centre for Magnetic Resonance
Copenhagen University Hospital Hvidovre,
Denmark

N. Jon Shah
Institute of Neuroscience and Medicine (INM-4)
Forschungszentrum Jülich
Jülich, Germany

Aliaksandra Shymanskaya
Institute of Neuroscience and Medicine (INM-4)
Forschungszentrum Jülich
Jülich, Germany

Stefanie Thust
UCL Institute of Neurology
University College London (UCL)
London, UK

Paul S. Tofts
Brighton and Sussex Medical School
Brighton, UK

Lisa A. van der Kleij
University Medical Center Utrecht
Utrecht, The Netherlands

Tobias C. Wood
Department of Neuroimaging
Institute of Psychiatry, Psychology and Neuroscience
King's College London (KCL)
London, UK

Wieland A. Worthoff
Institute of Neuroscience and Medicine (INM-4)
Forschungszentrum Jülich
Jülich, Germany

Moritz Zaiss
Department of Medical Physics in Radiology
Deutsches Krebsforschungszentrum (DKFZ)
Heidelberg, Germany

Introduction

Quantitative MRI has continued to evolve since the first edition in 2003, *Quantitative MRI of the Brain: Measuring Changes Caused by Disease*. The technological revolution is nowhere near to complete. The need for a new edition was apparent for many years, yet there was no way to manifest it. With retirement came other interests. Then a chance approach from Francesca McGowan at Taylor & Francis, and the realisation that my colleagues at the Brighton and Sussex Medical School could be co-editors, enabled the new edition to be born. The first edition was born 'one balmy March evening on the banks of the river Brisbane'; this edition was conceived on the former Physics Bridge tea room at the University of Sussex and born in the nearby Swan Inn at Falmer.

In this new edition we have placed more attention on the 'how to' aspects and incorporated the profound technological developments that have taken place since the start of this millennium, in both hardware and software. All chapters have been substantially rewritten, mostly with new authors. We have included two new subjects (CEST and multinuclear spectroscopy). The new book is slimmer and, we believe, easier to use.

We appreciate the input from the authors and reviewers, often in difficult circumstances. Bruce Pike, Geoff Parker and Martina Callaghan gave invaluable advice. Robert Grossman again generously agreed to write an introduction. I am grateful to Nicholas G. Dowell and Mara Cercignani for agreeing to undertake this project. Marica Dowell and Claudia Tofts gave key support.

Paul S. Tofts
Brighton and Sussex Medical School

Introduction to the First Edition

This book was conceived one balmy March evening on the banks of the river Brisbane, in Queensland, Australia, where I had just arrived for a sabbatical, and it became clear that the traditions of measurement science and MRI should meet. The notion of a guide, a cookbook, for quantitative MRI (qMR) techniques took seed and attained its own life, insistently telling me, in a variety of auspicious places and times, what had to be included. With the help of a network of enthusiastic colleagues from the International Society for Magnetic Resonance in Medicine, a description of the state of the art has been assembled, which would be impossible for a single author to achieve.

The Muse of qMR visited me in many places: Brighton, Glasgow, Hawaii (Waikiki Beach and Molaka'i), Lewes, London, Oxford and Paris (Jeu de Paume). Others have written in Bordeaux, the Bronx, Chalfont St Peter, Guilford, Leiden, London, Manchester, Nijmegen, Northwood, Nottingham, Oxford, Philadelphia, Utrecht and probably many other places. Jacob Bronowski also inspired me.[1] During the creative part of the process, I have been aware of ideas coming to me in a variety of inspiring places and times, and I am aware of the remark by the composer Stravinsky, whilst writing his *Rite of Spring*: 'I was the vessel through which the Rite passed'. John Cleese's view[2] is that 'The chief [condition in which creativity can thrive] is to give people time and space without pressure, simply to dream. Intelligence increases when you think less. We have too much noise in our heads. We need quiet spaces; it's about allowing something to happen to you'. With the knowledge came the responsibility to make it widely available. Poets speak of 'channelled poetry', where the poet is just told what to write. Gibran[3] (speaking of children) says, 'they come through you, but not from you' and also 'work is love made visible'. Sometimes I seemed to be witnessing the creation of perfection.

At times, it has been a lonely activity; my son Alex regularly brought me back to happiness after bleak days of writing. I was reminded of the composer (possibly Rachmaninov) who worked for 5 days, from 5 o'clock in the morning to 8 o'clock in the evening, then collapsed with tiredness at the end, overcome by the enormity of the work he had just created. The effort to produce this book is equivalent to about three person-years.

Describing the intersection of measurement science and MR imaging has been an international effort by the members of the MRI research community, with much communication by email and rapid access to journal articles online, in a way that would not have been possible a few years ago; some co-authors have not even (yet) met each other. The global village has truly arrived. The overview boxes, and many of the footnotes, are generally my responsibility. The conventions regarding units and abbreviations follow those in the style guide of the journal *Magnetic Resonance in Medicine*, as much as possible.

In the past, physical science has been concerned with our view of the cosmos and of atomic particles. Now we have the chance to see and measure inside our own living brain – to me this is equally profound. Which will history judge as being more important? A decade ago qMR techniques were almost non-existent; in a decade's time they will be routine.

In spite of what science has achieved, I am aware that many people are apparently able to resist disease, and heal themselves and others, in ways that are still mysterious to Western science, using treatments such as acupuncture, body work,[4] homeopathy, reiki and shiatsu. The placebo effect is a phenomenon considered very powerful in medicine, and yet its mechanism of action is not fully understood. With qMR we may be in a position to objectively record responses to such treatments.

Key people have been an inspiration to me at various times in my scientific education: A. Thompson, Eddie Palmer, Donald Edmonds and Michael Richards. Later on, John Clifton, Richard Edwards, Osmund Reynolds and Ian McDonald provided support at crucial early times during my entry into medical physics. I am grateful to my colleagues at the Institute of Neurology for their patience whilst I absented myself, working on this book. The conceptual design of the book had critical input from Clive Baldock, Mark van Buchem and Peter Jezzard. Kate Brunskill, Jackie Cheshire and Jackie Powell gave invaluable help in obtaining references and illustrations.

The contributors and reviewers have put in an enormous amount of work, often working long antisocial hours, and I am most grateful to them. Together they form a body of

[1] Jacob Bronowski's *The Ascent of Man* (reprinted 2002) is a particularly thrilling history of science, giving its cultural and historical contexts.

[2] From the *London Times*, 24 October 2002, based on *Hare Brain, Tortoise Mind* by Guy Claxton.

[3] *The Prophet* by Kahlil Gibran.

[4] For example, Biodanza (see www.biodanza.co.uk).

experts who hold the expertise in the field of qMR. Robert Grossman was extremely generous in his foreword. At Wiley in Chichester, Martin Rothlisberger, Karen Weller, Wendy Pillar and Lynette James steered this project to completion in an enjoyable and professional way. The Multiple Sclerosis Society of Great Britain and Northern Ireland has supported the physics development in the Research Unit at the Institute of Neurology, Queen Square in a very generous way for many years, enabling a broad range of qMR techniques to be built up. Many of the contributors are associated with this unit. Without the support of the Society, this book would not have been possible.

I hope the subject of qMR will become an established subtopic of MRI; the website www.qMRI.org can serve to coordinate activities (and to record errors found in this book). This book can be enjoyed as a view of what is possible in research centres now and what will become increasingly routine in the future.

Paul S. Tofts
Brighton and Sussex Medical School

1

Concepts: Measurement in MRI[1]

Contents

Paul S. Tofts
Brighton and Sussex Medical School

1.1 Introduction

1.1.1 Measurement science and MRI come together

Measurement science has been around a long time; MRI[2] has been around for about 35 years. This book is about the blending of the two paradigms.

We have come to expect to be able to measure certain quantities with great accuracy, precision and convenience. Instruments for mass, length and time are all conveniently available, and we expect the results to be reproducible when measured again and also to be comparable with measurements made by others in other locations. In the human body we expect to measure some parameters (height, weight, blood pressure) ourselves, recognising that some of these parameters may have genuine biological variation with time. More invasive measurements (e.g. blood alcohol level or blood sugar level) are also expected to have a well-defined normal range and to be reproducible. In physics, chemistry, electrical engineering and the manufacturing industry, there is a strong tradition of measurement, international agreements on standards and training courses for laboratory practitioners. International standards of length, mass and time have been in existence for many years. Secondary standards are produced, which can be traced back to the primary standards. National and international bodies provide coordination.

As individual scientists we may have a passionate desire to use our talents for the benefit of mankind, preferring to devote our energy to finding better ways of helping our fellow humans to be healthy than to improving weapons for their destruction. In this context, developing measurement techniques in MRI constitutes a perfect application of traditional scientific skills to a modern problem.

MRI is now widespread and accepted as the imaging method of choice for the brain (and for many body studies). It is generally used in a qualitative way, with the images being reported non-numerically by radiologists. Many MRI machines now have independent workstations, connected to the scanner and the database of MR images, which enable and encourage simple quantitative analysis of the images in their numerical (i.e. digital) form. However, the data collection procedure often prevents proper quantification being carried out; variation in machine parameters such as transmitter output gain, flip angle value (and its spatial variation), receiver gain, and image scaling may all be acceptable for qualitative analysis but cause irreversible confusion in images to be quantified. Researchers may be unaware of good practice in quantification and collect or analyse data in an unsuitable way, even though the MRI machine is capable of more.

[1] Reviewed by Mara Cercignani.

[2] The term *Magnetic resonance imaging* was invented by U.S. radiologists to describe nuclear magnetic resonance (NMR) imaging. The 'nuclear' part was removed from the name NMR to prevent the public being alarmed. NMR spectroscopy (Chapter 12) was originally concerned with identifying chemical compounds, and there was no spatial information contained in the data. It developed separately from imaging, on different machines, and is often referred to as *magnetic resonance spectroscopy* (MRS). Modern MRS is carried out largely on MRI machines and uses the imaging gradients to localise the spectra to particular parts of the body. For these reasons, MRI is now considered to include spectroscopy. *MR*, or *magnetic resonance*, is a more correct term and refers to MRI and MRS together.

The process of quantifying or measuring parameters in the brain necessarily takes more time and effort than a straightforward qualitative study. More MRI scanner time is needed, and considerable physics development effort and computing resources may be needed to set up the procedure. In addition, analysis can be very time-consuming, and support of the procedure is required to measure and maintain its reliability over time. Processes have to be found that are insensitive to operator procedure (whether in the data collection or image analysis) and to scanner imperfections (such as radiofrequency non-uniformity from a particular head coil), that provide good coverage of the brain in a reasonable time and that are stable over study times, which may extend to decades.

The benefits of quantification are that fundamental research into biological changes in disease, and their response to potential treatments, can proceed in a more satisfactory way. Problems of bias, reproducibility and interpretation are substantially reduced. MRI can move from a process of picture-taking, where reports are made on the basis of unusually bright, dark, small or large objects, to a process of measurement, in the tradition of scientific instrumentation, where a whole range of quantities can be tested to see whether they lie in a normal range and whether they have changed from the time of a previous examination.

In this book, the intention is to demonstrate the merging of these two traditions or paradigms, that is measurement and MRI, to form the field of quantitative MRI (qMRI).[3] The MRI measurement process is analysed, often in great detail. Limits to accuracy and precision are identified, as far as possible, with the intention of identifying methods that are reliable and yet practical in a clinical MRI scanning environment. The biological meaning of the many MR parameters that are available is explored, and clinical examples are given where MR parameters are altered in disease. Often these changes have been observed qualitatively, and they serve to encourage us to improve the measurement techniques, in order that more subtle effects of disease can be seen, earlier than currently possible, and in tissue that is currently thought to be normal as judged by conventional MRI. The ideal is to obtain push-button (turnkey) techniques for each of the many MRI parameters in this book, such that an MRI radiographer (technologist) can measure each of these parameters reliably and reproducibly, with a minimum of human training or intervention, in the same way that we can currently step onto a weighing machine and obtain a digital readout of our mass. In the case of qMRI, the output would be considerably richer, perhaps showing images of abnormal areas (computed from large databases of normal image data sets), changes from a previous MRI exam, possible interpretations (diagnoses) and an indication of certainty for each piece of information. The advances in the pre-scan and the spectroscopy MR procedures, which used to be very time-consuming and operator-dependent and are now available as fully automated options, show how this might be possible.

[3] The website www.qmri.org can be used for updates.

Thus, MRI has been undergoing a *paradigm shift*[4] in how it is viewed and used. In the past it was used for forming qualitative images (the 'happy-snappy MRI camera', taking pictures); in the future it may be increasingly used as a scientific instrument to make measurements of clinically relevant quantities. The dichotomy can be seen in the MRI literature; radiological descriptions often speak of signal hyperintensity in a sequence with a particular weighting, whilst studies using physical measurements often report localised concentration values, normal ranges, age and gender effects, and reproducibility. As measurement becomes more precise and analysis enables clinically relevant information to be extracted from a myriad of information, it will become possible, in principle, to make measurements on an individual patient to characterise the state of their tissue, guiding the choice of treatment and measuring its effect. The issues involved in bringing qMRI into the radiological clinic were well summarised in an editorial in the *American Journal of Neuroradiology* (McGowan, 2001; Box 1.1).

As part of this ongoing paradigm shift, our view of what MRI can tell us is changing. When it started, information was largely anatomical (*anatomical MRI*), in the sense that relatively large structures would be observed. Changes in their geometric characteristics (usually size), compared to normal subjects, or to a scan carried out in previous weeks or months, would be noted. Quantitative examples would be volume and atrophy. *Functional MRI* (fMRI) claimed the complementary ground, studying short-term changes in tissue arising from carrying out particular (neural) functions. *Microstructural MRI* occupies a third role, as shown in this book. Many MR parameters (such as diffusion, magnetisation transfer and spectroscopy) show structural changes in tissue arising from damage caused by disease at a microscopic level. To observe these changes directly would require imaging resolution of the order of 1–100 μm,[5] since they generally involve a variety of biological changes at the cellular level. These can be observed by pathologists in post-mortem tissue, using optical or electron microscopy and special staining techniques (histopathology). This resolution is well below the spatial resolution of MRI (which is about 1 mm on clinical scanners). However, changes at the microscopic level (e.g. in cellular structure) give changes in the MR parameters (e.g. in water diffusion) that can be observed at coarser spatial resolution (of about 1 mm). Thus, structural changes of sizes well below those that would be called *anatomical* can be detected. In addition, the concentrations of chemical compounds (metabolites) in cells, and their changes, can be measured with spectroscopy. The physiological permeability of the endothelial membrane around blood vessels can be measured using dynamic imaging of Gd-contrast agent.

[4] Thomas Kuhn, in *The Structure of Scientific Revolutions*, first introduced the idea of paradigm shifts. An example would be the move from a classical physics to a quantum physics view of the world. A paradigm is a pattern or model, a way of viewing the world or part of it, a point of view, or a mindset.

[5] 1 *micron* (μm) is 10^{-3} mm or 10^{-6} m.

Box 1.1 On the use of quantitative MR imaging

There are a growing number of quantitative MR applications that represent evolutionary change in the use of MR imaging. These applications also include magnetisation transfer techniques, absolute T1 and T2 measurements, functional imaging, and a number of spectroscopic techniques.

A significant challenge in the clinical employment of quantitative methods is that underlying physical mechanisms may not yet be fully understood in the context of what can be measured with the MR imaging experiment. For example, one can associate the presence of abnormalities in quantitative measures with the presence of disease, but causality may not be established.

Thus, results are sometimes limited to empirical findings of correlation with some other measure or observable process. Still such results are potentially of great value by providing means of noninvasive disease characterization and, thus, insight into the natural history of disease.

Another substantial benefit is derived from the use of validated methods to study the efficacy of novel therapeutic agents. Coupled with results of other studies, including investigations in animal models in which correlation may be observed between the results of an invasive or destructive test and the results of noninvasive MR imaging, human studies serve to connect clinical observation with imaging findings.

It is statistically advantageous to follow up preliminary studies that use "many" measures with targeted studies that have the power to accept or reject the hypothesis that certain measures are significantly correlated.

Investigators differing from the authors of the original study may do this only when precise and comprehensive data regarding the study methods are provided. However, even when the authors make a good-faith effort to disclose every nuance of the experimental method, it still may be difficult to control for differences in MR hardware and software. This is in part because modern MR system design objectives are focused on obtaining excellent-quality clinical images for conventional, subjective interpretation.

Source: McGowan, J. C., *American Journal of Neuroradiology*, 22(8), 1451–1452, 2001.

These changes may occur both in a so-called lesion, which is tissue seen at post-mortem and in conventional MRI to be visibly different from the surrounding tissue, and in the 'normal-appearing' tissue, which appears normal in conventional MRI. Lesions are usually described as *focal*, meaning that the change is localised to a relatively small area (a few millimetres or centimetres), with a distinct boundary; thus, its differing brightness in an image distinguishes it from the surrounding tissue (considered normal). In contrast, a diffuse change may extend over more area, has no distinct boundary and is harder to detect by simple visual observation of the image. Diffuse changes are often well characterised by quantification, since it is the absolute value of quantities within the area that is measured, without reference to surrounding tissue or the need for a distinct boundary.

1.1.2 Limits to progress

It may appear that qMRI research proceeds under its own impetus. However, the current state and rate of progress in developing reliable qMRI methodology are determined by several factors: MRI manufacturers, research institutions, pharmaceutical companies, computer and electronics technology and publicly funded research councils.

MRI machine manufacturers (vendors) will take on some of the measurement procedures over time, incorporating them into their research and development programs and then offering them as turnkey (push-button) products.[6] The speed of this process is driven by demand from clinical purchasers, by whether competing manufacturers offer such facilities, and by whether public medical funding bodies such as the US Food and Drugs Administration (FDA) are likely to approve reimbursement of the cost of such procedures from medical insurance policies. The existence of a large and growing installed base of high-quality, reliable and ever-improving MRI machines primarily designed for routine clinical use, largely in environments where they can be run as parts of profitable businesses, has enabled and encouraged the development on these machines of qMRI techniques, although they are still of interest to only a (growing) minority of users.[7] As MRI machines evolve, the qMRI techniques usually have to be re-implemented.

Research institutions have particular structural strengths and weaknesses. qMRI needs input from chemists, computer scientists, neurologists, physicists, radiologists and statisticians. There may be good career support for those applying methods to study clinical problems but none for those basic scientists inventing and developing the methods. There may be a clash of paradigms or traditions, between those who have been educated in a hierarchical environment where asking questions is considered to be irrelevant or subversive and those who consider asking questions to be an absolute basic necessity of undertaking modern high-quality scientific research. The availability of talented researchers in turn depends on how much value is placed on science in

[6] These are often sold as extras.

[7] The 'killer app' can sometimes galvanise action around making a qMRI parameter available. This is when an application is found that has a clear clinical importance (e.g. MD in stroke).

society, schools and universities and whether appropriate postgraduate training opportunities exist. The International Society for Magnetic Resonance in Medicine (ISMRM)[8] is a powerful force bringing together researchers from different institutions who are working on similar methodologies, through both its journals and its scientific meetings.

The demand from pharmaceutical companies and neurologists for qMRI measurements to be used in drug trials is large and likely to increase (Filippi and Grossman 2002; Filippi *et al.*, 2002; McFarland *et al.*, 2002; Miller 2002; Sormani *et al.*, 2011; Mallik *et al.*, 2014). The traditional double-blind placebo-controlled Phase III trial involves many patients (typically 100–1000) being studied for several years in order to obtain enough statistical power to determine whether a drug is effective. The large sample size is needed to deal with the variability of disease in the absence of treatment, and the imperfect treatment effect (which may vary according to patient subgroup). Such trials typically cost several hundred million US $. qMRI can potentially shorten the procedure, by identifying treatment failures early on in the testing process, on a smaller sample. If there is no observed biological effect from the treatment, it may be considered unlikely that the drug is working (this will depend on the particular way the drug has been postulated to act). For example, if a potential treatment for multiple sclerosis (MS) showed no effect on all the MR measures that are known to be abnormal in MS, it would probably be dropped in favour of other drugs. With new biotechnology and gene-based treatments being developed, the number of candidate drugs for evaluation will increase by a large factor, and traditional trials will become too expensive and slow to evaluate all of them. Thus, direct *in vivo* qMRI observation of treatment effect could become increasingly valued.

The rapid increase in power and availability of computing technology has also been key in enabling data acquisition and image analysis techniques to be realised. Numerically designed magnets, coils and radiofrequency pulses, digital receivers and rapid image registration and analysis have all changed the way that MRI is carried out.

The resources available from pharmaceutical companies to drive the process of developing and supporting reliable qMRI measures may exceed those available from traditional publicly funded research sources. Traditional research council sources have been willing to support the application of qMRI methods to study particular diseases but often unwilling to support the development of new quantitative methods, sometimes claiming that MRI manufacturers should be doing this.

1.1.3 Using this book

The field of quantitative MRI should include the following four key areas: basic concepts of measurement, how to measure each MR parameter (to include both acquisition and analysis) and the biological significance of each parameter (with input from post-mortem and possibly animal studies). For each MR parameter, the following aspects are important: (1) the biological significance of the MR parameter, (2) how it can be measured accurately and slowly, (3) how it can be measured practically and quickly, (4) examples of clinical applications, (5) what can go wrong in the measurement procedures, (6) QA approaches (controls and phantoms), (7) normal values for tissue, (8) reproducibility performance that can be achieved, (9) multicentre studies and (10) future developments. The editors did their best to get authors to consider all 10 of these aspects in their chapters.

This book is intended to be a repository of qMRI methodology, of particular use to PhD students; hopefully, the methodology will not need to be reinvented by each generation of researchers. The first edition of this book (Tofts 2003) contains some information not present in this edition that may be worth consulting; the chapter authors in this edition have changed and necessarily give a different perspective.

In Chapters 2 and 3, the issues in measurement that occur repeatedly throughout the book, as each MR parameter is considered, are examined in more detail. These are grouped into the processes of data collection, data analysis and quality assurance, all of which crucially affect how well MR quantities can be measured. Units are usually given in SI (System International), and conventions used in this book for physical units and symbols (e.g. *TR*, *TE*, T_1, T_2) are those recommended in the style guide for the journal *Magnetic Resonance in Medicine*, published for the International Society for Magnetic Resonance in Medicine.[9] Most of the focus is on techniques that can be implemented on standard clinical MRI scanners; some techniques (e.g. ^{31}P spectroscopy or ^{23}Na imaging) need non-standard hardware as an add-on.

This book is intended for researchers who already have a basic knowledge of how MRI works and some knowledge of the brain, including the major diseases (cancer, epilepsy, stroke, multiple sclerosis and dementia). Newcomers can find many appropriate books (Table 1.1) and also helpful websites such as ISMRM. A table of common MRI abbreviations is given in Appendix 2.

1.2 History of measurement

1.2.1 Early measurement

Early quantitative techniques focused around the desire to measure distance, mass, monetary value and time. An awareness of these can give us perspective in our own endeavours to quantify!

Developed in about 3000 BC in ancient Egypt, the *cubit* was a ubiquitous standard of linear measurement, equal to 524 mm. It was based on the length of the arm from the elbow to the extended fingertips and was standardised by a royal master cubit of black granite, against which all cubit sticks used in Egypt were to be measured at regular intervals.[10] The precision of the thousands of cubit sticks used in building the great Pyramid of Giza

[8] www.ismrm.org

[9] See http://www.ismrm.org/journals.htm

[10] Much of the historical material in this chapter comes from the *Encyclopaedia Britannica*.

TABLE 1.1 Recommended books for background reading in MRI and neuroanatomy

Title	Authors	Date published	Number of pages	Description
MRI: Physical Principles and Sequence Design	RW Brown, YC Norman Cheng, EM Haacke, MR Thompson, R Venkatesan	2014	976	Thorough exposition of MRI principles; 2nd edition
MRI from Picture to Proton	DW McRobbie, EA Moore, MJ Graves	2017	400	Written by experienced physicists; new edition
Quantitative MRI in Cancer	TE Yankeelov, DR Pickens, RR Price	2011	338	Multi-author book by radiologists and physicists; a 'sister book' to this one
Quantitative MRI of the Spinal Cord	J Cohen-Adad, C Wheeler-Kingshott	2014	330	Multi-author, 'sister' book to this one
Diffusion MRI: Theory, Methods, and Applications	DK Jones	2011	784	Covers much of quantitative brain MRI from a physics point of view
Handbook of MRI Pulse Sequences	M Bernstein, K King, X Zhou	2004	1040	A pulse programmer's friend
MRI in Practice	C Westbrook, CK Roth, J Talbot	2011	456	Established book giving radiographer's viewpoint
Barr's The Human Nervous System: An Anatomical Viewpoint	JA Kiernan, R Rajakumar	2013	448	Includes complete description of the brain

Note: Statistics books are in Chapter 2 (Table 2.2).

is thought to have been very high, given that the sides of the pyramid are identical to within 0.05%.

Early *astronomers* developed remarkably precise measurement methods (as demonstrated at Stonehenge); their ability to guide navigation and predict eclipses brought them fame. In the 16th century, precise calculations of planetary orbits by Copernicus, Kepler and Galileo challenged the intellectual dominance of the Catholic church, bringing an end to the idea that all heavenly bodies rotate around the Earth.

In 1581, the word *quantitative*[11] was first used, meaning 'involving the measurement of quantity or amount'. *Quantity* means 'size, magnitude or dimension' in Middle English. In 1847, *quantitative analysis* was used, meaning 'chemical analysis designed to determine the amounts or proportions of the components of a substance'. In 1878, *quantify* was used to mean 'to determine the quantity of, to measure', and hence *quantification* is 'the operation of quantifying'. In 1927, *quantitate* was used to mean 'to measure or estimate the quantity of, especially to measure or determine precisely'. However, *Webster's dictionary* calls this term a 'back-formation',[12] which is probably as derogatory as a dictionary compiler can be, and this term is not used in this book, nor is it in the *Oxford English Dictionary*.

Francis Bacon (1561–1626) had a great influence on generations of British scientists who followed him (Gribbin, 2003). He stressed collecting as much data as possible, then setting out to explain the observations, instead of dreaming up an idea and then looking for facts to support it. Science must be built on the foundation provided by the facts. What would he say about the modern 'hypothesis-driven' research? In 1662, the Royal Society of London for the Promotion of Natural Knowledge received its charter from King James II, as one of the first (and best known) scientific societies.

[11] See *Webster's Dictionary* and the *Oxford English Dictionary*.

[12] A *back-formation* is a word formed by subtraction of a real or supposed affix from an already existing longer word. Thus, *quantitate* was created from *quantitation*.

1.2.2 The longitude problem: John Harrison

In the 18th century, the problem of navigation around the globe was severe. Although latitude (distance from the equator) could be measured accurately, using the elevation of the sun above the horizon at noon (the time of maximum altitude), longitude (the easterly or westerly distance around the globe, now measured from Greenwich, London, UK) could not be (Sobel, 2005). Samuel Pepys, commenting on the pathetic state of navigation, had written of 'the confusion all these people are in, how to make good their reckonings, even each man's with itself', recognising the distinction between intra- and inter-observer variation. Newton wrote of the sources of error involved in trying to measure time at sea: 'One [method for determining longitude] is by a Watch to keep time exactly. But, by reason of motion of the Ship, the Variation of Heat and Cold, Wet and Dry, and the Difference of Gravity in different Latitudes, such a watch hath not yet been made'.

As a result many lives were lost at sea, through shipwreck and failure of supplies, and navigation was such a sensitive issue that sailors were forbidden to carry out their own calculations, for fear that they would show up errors in those of their superior officers. The growth of vastly profitable world trade was held back. In this context, the Longitude Act of 1714 was passed in the British Parliament, offering a reward of £10,000[13] to anyone who could devise a method of measuring longitude accurately.

The challenge of solving the 'longitude problem', as it came to be known, was taken up by an English clockmaker, John Harrison, who lived near the port of Hull and had heard the stories of souls going to their death and the reward offered. He built four clocks altogether. The first kept good time on land (better than 1 second per month) and in small trips out to sea. The Longitude Board could give incentive awards to help impoverished inventors bring promising ideas to fruition. He succeeded in getting a full trial at sea with the navy, on a voyage to Lisbon in 1736; his clock showed

[13] The sum was graded according to the accuracy that could be achieved.

unexpected error at sea, being susceptible to an artefact caused by accelerations in the motion at sea. His own perfectionism, and obstinacy all round, delayed matters, and the next trial, taking his fourth clock to the West Indies, did not take place for another 25 years. The Longitude Board was dominated by eminent astronomers and others from the naval establishment and repeatedly refused to give Harrison his payment, requiring that the chronometer should first be taken from prototype into mass production. The Board realised that replicate voyages and clocks were needed to establish the reproducibility, without which the accuracy could not be guaranteed. A single measurement could not establish the maximum error. His son William took up his case, and the Royal Society offered him a Fellowship. It was only intervention by King George III, and the passing of a second act by Parliament, that gave him his recognition, at the age of 80, 46 years after he had built his first sea clock.

This story, of finding a scientific solution to a human problem, has all the elements of the struggles that modern scientists may have to develop a technique that they believe will save lives, and many parallels can be seen. Harrison's clocks are preserved in the old Royal Observatory at Greenwich.

1.2.3 Scientific societies

The Lunar Society of Birmingham (England) was a group of forward-thinking scientists who met between 1766 and 1791. They met on the day of the full moon (so that travel would be easier) and flourished independently of the Royal Society (in London). Birmingham was the location of much inventive scientific activity stimulated by the industrial revolution. Both of Charles Darwin's grandfathers (Josiah Wedgewood, the pottery manufacturer, and Erasmus Darwin, the naturalist) were members, as were Matthew Boulton (the manufacturer), Joseph Priestly (who discovered oxygen) and James Watt (who invented the steam engine). The Industrial Revolution in Britain and the rest of Europe gave commercial impetus to the invention of a variety of measuring instruments to be used in the manufacturing process. Lord Kelvin, delivering a lecture on electrical units of measurement in 1883, expressed the desire of his time to quantify:

> *When you can measure what you are speaking about, and express it in numbers, you know something about it; but when you cannot measure it, when you cannot express it in numbers, your knowledge is of a meagre and unsatisfactory kind: it may be the beginning of knowledge, but you have scarcely, in your thoughts, advanced to the stage of science, whatever the matter may be.*

although he might have added a caveat about the danger of numbers giving a pseudo-scientific respectability to some studies.

1.2.4 Units of measurement

In the newly formed United States of America, it was found impossible to reform the archaic system of weights and measures inherited from the British, in spite of the Napoleonic metric system that had recently been adopted in France. The Office of Weights and Standards became the National Bureau of Standards, then the National Institute of Standards and Technology (NIST). In 1960, the 11th General Conference of Weights and Measures, meeting in Paris, established the International System of Units, based on the metre, kilogram, second, ampere, degree Kelvin and candela. These units are often called the *SI units*, after the French expression *Système Internationale*, and are preferred in the scientific community.[14] The kilogram is represented by a cylinder of platinum–iridium alloy kept at the International Bureau of Weights and Measures in France,[15] with a duplicate in the USA; the other units are defined with respect to natural standards (e.g. the metre is defined by the wavelength of a particular visible atomic spectral line). National centres such as the US NIST and the UK National Physical Laboratory are now centres of expertise in measurement science.

1.2.5 Mathematical physics

In parallel with the development of physical instruments had been the discovery of mathematical techniques. Ancient Babylonians, Egyptians, Greeks, Indians (Harappans) and Chinese all had mathematics, originally used for computing areas and volumes of regular objects and also used for handling monetary currency. In the 6th century BC, Pythagoras established the link between the musical note of a string and its length. This bridge between the world of physical experience and that of numerical relationships has been called the birth of mathematical physics, where numbers explain the origin of physical forms and qualities. Newton's differential calculus and Fourier's transform are essential tools used by our current MRI scanners. Early digital computers, most famously used to decipher the Enigma code used by submarines during the Second World War, developed to the stage we take for granted today.

1.2.6 Scientific medicine

In medicine, the concepts of the new scientific methods, including quantification, were applied. William Harvey (1578–1657) was a physician and scientist who studied the blood circulation extensively and was the first to measure the cardiac volume and estimate the total blood volume in the human body. In 1833, William Beaumont, a US army surgeon, published a series of studies[16] on a soldier who had been wounded in the stomach and then developed a flap that could be opened. Beaumont could watch food in the stomach and extract gastric juice. Nowadays we have more convenient ways of making *in vivo* studies.

---

[14] The engineering community in the USA still uses units based on the British imperial system (although these are not used in the UK any more). Incompatibility between imperial and metric units was blamed for a space vehicle failure in the late 1990s.

[15] The BIPM, *Bureau International des Poids et Mesures*; http://www.bipm.org/en/about-us/

[16] From "The Man with a Lid on his Stomach," in the *Faber Book of Science*, edited by John Carey.

Box 1.2 A plea for 'good quality data collection' in MRI[1]

The history of image processing in nuclear medicine shows that collection of good quality image data is at least as important as access to image processing techniques. Even now one could argue that real improvements in the usefulness of image data come from instrumental improvements rather than from more sophisticated ways of image processing. However in the case of large datasets that are already of good quality, the problem is then one of data presentation and reduction, rather than correcting images to compensate for errors in data collection.

With this philosophy we have initially concentrated on collecting good quality data, that are sensitive to the clinical question being studied. For example T2 weighted images of the brain can show Multiple Sclerosis lesions, and one could develop sophisticated algorithms for measuring lesion volume to assess disease and therapies; however the images show oedema and scar tissue, which are secondary to the disease process. Primary visualisation of the disease is shown by the newer technique of GD-DTPA scanning, and therefore we have developed this data collection technique in preference. A second example is the use of expensive classification techniques on image data clearly showing gross nonuniformity which can be removed relatively simply.

Having taken care of the instrumental aspects and obtained good quality data, the processing requirements may become less expensive, and mostly consist of PACS, 3D display, calculation of functional images, and segmentation algorithms. Where sophisticated forms of information processing are required, to make full use of them they must be integrated into a programme that includes aspects of data collection such as sequence design, quality control of instrumental parameters, validation of the quantitative results, and good experimental design. In summary, we believe that data must be appropriate, and of good quality, before undertaking any processing.

Source: Tofts, P.S., *et al.*, *Prog. Clin. Biol. Res.*, 363, 1991a; Tofts, P.S., *et al.*, *Information Processing in Medical Imaging*, Wiley-Liss Inc, 1991b. With permission.

[1] PACS is 'picture archival and computing system' and refers to computer-based systems to store, display and interrogate large quantities of medical images. By 'functional images', it was meant parametric maps of any kind (e.g. permeability).

In the late 1970s, scientists started connecting medical imaging hardware to computers that look extremely basic by modern standards, motivated by the desire to manipulate and interrogate the images. Sophisticated medical imaging instruments were produced in nuclear medicine, ultrasound, X-ray computed tomography and nuclear magnetic resonance.

In about 1978 the annual meetings on Information Processing in Medical Imaging started taking place. In 1989, it was argued that attention to good data collection was at least as important as sophisticated image processing (Box 1.2). The notion that good quantification required attention to both *data collection* and *image analysis* techniques was born, and this complementarity can be seen in the structure of this book. Experience has shown that advances are often made by groups who have access to both data collection (so that the acquisition technique can be optimised for the job in hand) and to advanced analysis techniques (to obtain the most from the data). Computing groups working isolated from the clinical questions and acquisition hardware may produce solutions to non-existent problems, or use data that are degraded by poor acquisition techniques.

1.2.7 Early qMRI

Premature babies were studied with ^{31}P MRS in 1983 (Cady *et al.*, 1983), prompting the measurement of absolute concentration of metabolites in the brain (Wray and Tofts, 1986). In 1985, Bakker completed a PhD thesis, 'Some Exercises in Quantitative NMR Imaging'. The aim was to 'assess the potential of NMR imaging and spectroscopy with respect to tissue characterisation and evaluation of tissue response to radiotherapy and hypothermia' (Bakker *et al.*, 1984). Quantification was recognised by some radiologists as having a potential role in studying disease (Tofts and du Boulay, 1990):

> *Serial measurements in patients and correlation with similar studies in animal models, biopsy results and autopsy material taken together have provided new knowledge about cerebral oedema, water compartmentation, alcoholism and the natural history of multiple sclerosis. There are prospects of using measurement to monitor treatment in other diseases with diffuse brain abnormalities invisible on the usual images.*
>
> *When making quantitative measurements, the physicist can adopt the paradigm of the scientific instrument designer, who is presented with a sample (the patient) about which he or she wishes to make the most careful, detailed measurements possible, in a non-destructive way, using the infinitely adjustable instrument (the imager). The biological question to be answered, and thus the biophysical feature to be measured, needs very careful choice.*

Quantitative magnetic resonance was the subject of a small meeting organised by the UK Institute of Physics and Engineering in Medicine in 1997 at Dundee, Scotland, and it is here that the expression *qMR* was first used. qMRI has now come to denote that part of MR which is concerned with quantitative measurements, in the same way that fMRI, MR angiography (MRA), MRS and quantitative magnetisation transfer (qMT) denote subspecialties of MR.

1.3 Measurement in medical imaging

Physical quantities can be *intensive* or *extensive*, and when we are considering various properties and manipulations to quantities it can be helpful to be aware of these differences. An *intensive quantity*[17] can describe a piece of tissue of any size, and it does not alter as the tissue is subdivided (assuming it is uniform). Examples are density, temperature, colour, concentration, magnetisation, membrane permeability, capillary blood volume and perfusion per unit volume of tissue, texture and the MR parameters proton density, T_1, T_2, the diffusion coefficient of a liquid and magnetisation transfer. An *extensive quantity* refers to a piece of tissue as a whole, and subdivision reduces (or at least changes) the value of the quantity. Examples are mass, volume, shape and total blood supply to an organ.

Some intensive quantities, such as metabolite concentration, local blood flow or local permeability, can be expressed either per unit mass of tissue or per unit volume of tissue. Traditionally, physiologists have used the former system, since the mass of a piece of excised tissue is more easily determined that its volume. In qMRI, where the volume of each voxel is well defined, the latter system is more natural. Conversion from per mass to per volume can be achieved by multiplying by the *density of brain*[18] (1.04 g ml^{-1} or 1040 kg m^{-3} for both white and grey matter) (Whittall *et al.*, 1997).

1.3.1 Images, partial volume and maps

Images and maps are terms used to mean different things. An *image* is produced by the MRI scanner and has an intensity[19] that depends on a variety of parameters, including some that describe the tissue (e.g. PD, T_1, T_2, and combinations of these) and some that are characteristic of the scanner (e.g. the scanner transmit flip angle and receiver gain). The image consists of a two-dimensional (2D) matrix of numbers stored in a computer (often part of a three-dimensional (3D) image data set). Each location in the matrix is called a *pixel* ('picture element'), which is typically square and 1–2 mm wide. The image data come from a slice of brain tissue that has been interrogated, or imaged. This slice has a specified thickness (usually 1–5 mm), and each pixel in the image in fact derives from a cuboidal box-shaped piece of tissue, called a *voxel*. The first and second dimensions of the voxel are those of the image pixel, and the third dimension is the *slice thickness*. Often the image data set is 3D, although we can only see a 2D slice through it at any one time.

The interplay between pixels and voxels is subtle. At times when we are thinking of images, pixels are more natural, and in fact the term originates from the science of interpreting images of 2D surfaces (e.g. in robot vision or remote sensing of the Earth by satellite). Yet when we are thinking of the cuboids of 3D tissue from which the pixel intensities originate, voxels are more natural and serve to remind us to think about the tissue, not the image. Slices of voxels are inside the object, whilst surfaces of pixels are outside the object. Some imaging procedures will use very small pixels ('in-plane' resolution) yet set a large slice thickness (in order to retain signal-to-noise ratio). An extreme example would be a voxel of size 0.7 × 0.7 × 5 mm, which appears to have the ability to resolve small structures, yet any structures that are not oriented nearly perpendicular to the slice plane would be blurred by the large slice thickness. In this case the voxel would be shaped like a matchstick (i.e. have a large *aspect ratio*); a more appropriate voxel size might be 1.5 × 1.5 × 2.2 mm, which has the same volume (and hence signal-to-noise ratio, for a given imaging time) but is more likely to resolve small structures. Three-dimensional imaging sequences can give us voxels which are *isotropic* (i.e. have the same dimensions in all three directions).

Structures in the brain have very fine detail and very often there are two (or more) types of tissue inside the voxel. The resulting NMR signal from this voxel is simply a combination, or weighted average, of what each individual tissue would give if it filled the whole voxel. Thus, if we are trying to measure the T_1 of grey matter, near to cerebrospinal fluid (CSF) the value measured will be somewhere between that of pure grey matter and pure CSF, depending on the relative proportions of brain tissue and CSF in the voxel. This is called the *partial volume effect* and is a major source of error when making measurements in brain tissue at locations near to boundaries with other tissue types. The value measured in the tissue is altered by its proximity to another tissue, and the determination of boundaries and of volumes is brought into error. Partial volume errors can be reduced by using smaller voxels, although the price paid is that of a worsening of the signal-to-noise ratio. An inversion pulse before data collection can remove signal from a tissue with a particular T_1 value (as in the FLAIR and STIR sequences, which null the signal from CSF and fat, respectively).

A parametric *map* can be calculated from two (or more) images of the same piece of tissue. A simple example would be to collect two images with differing amounts of T_2 weighting. The ratio of these two images then only depends on the tissue parameter T_2 and is independent of scanner parameters (such as transmitter or receiver settings). By calculating this ratio for each pixel, a third matrix, or map, can be formed, which has the appearance of an image (brain structures can be identified) but is conceptually different from an image, in that individual pixel values now have a numerical meaning (such as value of T_2, in milliseconds, at each location in the brain), rather than representing signal intensity on an arbitrary scale.

[17] *Intensive* (dictionary definition): of or relating to a physical property, measurement, etc., that is independent of mass; extensive: a property that is dependent on mass.

[18] For example, the normal concentration of water in white matter is about 0.690 g water per g tissue (0.690 kg water per kg tissue), equivalent to 0.718 g water per ml tissue (718 kg water per m^3 tissue) (see first edition, p 91).

[19] Often called the *signal intensity*, since it is proportional to the signal voltage induced in the radiofrequency (RF) coil by precessing magnetisation in that piece of tissue seen in that voxel of the image.

1.3.2 Study design

Many studies set out to compare groups of subjects using the classic double-blind randomised controlled trial design. Typically, a new MR parameter will be evaluated in a particular disease by measuring it in a group of patients and in a group of controls. The controls could be on placebo or another (established) treatment. Other differences between the groups ('confounding variables') should be removed as much as possible, hence the need for age and gender matching. The scanning should be carried out at the same time, using *interleaved controls*, rather than leaving the controls until the end of patient scanning (when a step change in the measurement procedure could produce an artificial group difference). Some patients may be on treatment which alters the MR results. Matching can be improved by *dynamic matching*, carried out as part of subject recruitment as the study proceeds. Thus, if controls are in short supply, but patients plentiful, then each time a control is recruited, a matched patient is selected from the available patients. In placebo-controlled trials, allocation of a patient to the placebo or treated group can be decided at the time of recruitment, to keep the groups matched at all times. Double-blinding[20] is a powerful way of reducing bias in treatment trials. The person giving the treatment, the person making the measurement,[21] and the patient are all blinded to whether they are receiving a genuine treatment or a placebo (Table 1.2).

Inexperienced researchers should beware of 'stamp collecting' when 'interesting patients' are studied, almost at random, with no hypothesis or controls.[22] To design high-quality investigations that will be accepted for publication by the best international journals, the investigator should be aware of what work has already been published or presented at international scientific conferences.[23]

A *literature search*[24] should be carried out. Studies should not be replicated unless there is a case for confirming the results with a different group of patients. Methodological pitfalls, as illustrated by existing published work, should be identified before the study begins. Some errors (e.g. the presence of poor reproducibility, which would be detected with repeated scanning, or scanning controls after an upgrade, not interleaved with the patients) will irreversibly destroy the value of the data.

Selection of MR parameters requires thought. To acquire all the parameters discussed in this book would require more time than can be fitted into one examination (although as scanners get faster and techniques are optimised, acquisition times have come down). Parameters should be selected according to the biological changes that are expected in the particular disease being studied. Measuring several relevant parameters can be powerful (see Chapter 2, Section 2.2.2.5). Mixed-parameters acquisition can address specific questions (e.g. diffusion weighted spectroscopy, or MT prepared multi-echo measurements. Multiparametric studies are addressed further in Chapter 18. The excellent consensus review by O'Connor *et al.*, on imaging biomarkers elegantly summarises many of the issues involved in quantification, and is worth studying in detail (O'Connor *et al.*, 2017).

TABLE 1.2 Good practice in study design and statistics

1.	Optimise the instrumental precision before starting the study.
2.	Talk to a statistician before and after collecting the data.
3.	Collect interleaved control and patient data.
4.	Control for age and gender during subject recruitment.
5.	Inspect the data in scatter plots.
6.	Model the data, including random and systematic error.
7.	Adjust for age and gender during analysis.
8.	Avoid if possible doing t-tests with many comparisons.
9.	Be aware that correlations are hard to interpret.
10.	Give confidence limits on group means and differences.

1.3.3 Usefulness of an MR parameter

From a clinical point of view, a potential new quantity to characterise brain tissue can be evaluated by considering three factors:[25]

> Sensitivity: *does the quantity alter with disease? Is the false negative rate low?*
>
> Validity: *is it relevant to the biological changes that are taking place?*
>
> Reliability: *is it reproducible? Is the false positive rate low?*

Thus, the concept of validity (which is absent from a judgement based merely on accuracy and precision) enables the relevance of a metric to be considered. For example, intracranial volume could be measured very accurately and precisely but would be completely irrelevant in most situations.

[20] The double-blind design is not suitable for treatments where the practitioner plays an essential part in the treatment. This is particularly relevant in so-called alternative therapies (e.g. acupuncture, homeopathy, osteopathy, psychotherapy and reiki). Although a placebo cannot be given, different treatments can be compared. Even in conventional clinical trials, the patient often guesses whether they have a placebo or not from the side effects, and also those with greater side effects may be more likely to drop out. More research on methodology may be needed to find suitable study designs to overcome these problems.

[21] Ideally this includes both the radiographer making the scan and the observer analysing the MR data.

[22] The term *hobby researcher* describes this phenomenon well.

[23] See also Chapter 2, Section 2.2.2.1, on two kinds of studies: 'fishing expedition' and hypothesis-driven.

[24] For example using PubMed, from the US National Library of Medicine, available free of charge online http://www.ncbi.nlm.nih.gov. From here you can download PDF files of papers (provided you are logged on to an academic website, e.g. a university). Usually you can 'search forwards', that is see which papers have cited the paper you are looking at (the 'cited by' list). Thus, a complete picture of publications on a particular topic can be built up quite quickly and conveniently.

[25] See first edition, Chapter 12, discussion by the psychologist N Ramsey in the context of fMRI.

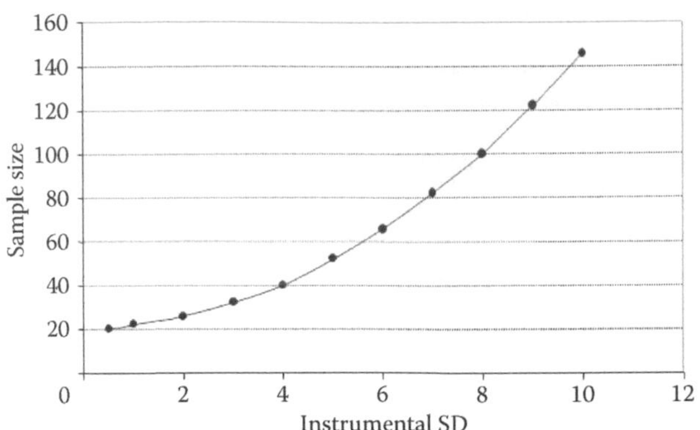

FIGURE 1.1 The effect of instrumental precision (ISD) on the power of a study and the required sample size. By reducing the ISD, the sample size required is dramatically reduced, with a consequent saving in the cost and duration of the study. This is a simulation based on group comparison between controls (parameter value mean = 100, SD = 3) and the same number of patients (effect size = 5, SD = 4.25). Power P = 80%, significance level $\alpha = 0.05$, using G*power3 which is an established software that can be downloaded free of charge.

An alternative viewpoint (closely related) is the set of four psychometric properties often used to assess scores: acceptability, reliability, validity and responsiveness (Hobart *et al.*, 2000). The impact of poor reproducibility on the power of a study can be dramatic (Figure 1.1). Appropriate methods for analysing MR data are still under discussion. The clinical metrics are also being scrutinised and redesigned (Fischer *et al.*, 1999; Hobart *et al.*, 2000). Developments in psychology may be ahead of those used in this field (Krummenauer and Doll, 2000).

1.4 The Future of quantitative MRI

1.4.1 Technology and methodology

Since the first edition of this book, MRI technology has advanced. The standard field strength has moved from 1.5T to 3T, with 4.7T and 7T machines becoming more common. Major manufacturers offer little below 1.5T for brain imaging. Gradients have improved in both strength and speed, enabling fast 3D acquisition to become standard. RF transmit coils have responded to the higher frequencies by including designs to increase uniformity and reduce SAR. RF receive arrays use multiple coils for improved SNR. The only downsides for qMRI are the need to measure the transmit field B_1^+ and the loss of the reciprocity principle (see Chapter 2). Methodologies have continued to advance; this edition has three new topics: advanced diffusion, multinuclear MRS and CEST. Pointers to the future are in Chapter 18.

qMRI has five principal aspects: (1) *concepts* have hardly altered since the first edition of this book, just become more clear; (2) *MRI physics* above 1.5T is more complex, with the loss of reciprocity; (3) *technical advances* continue to alter the environments in which qMRI must be re-implemented; (4) *analysis techniques* make a crucial difference to the value of the MRI data;[26] and (5) the *biological significance*[27] of MR parameters influences what use can be made of qMRI (in particular, can access to the current biology predict the future clinical status?).

Where might these improvements lead? How would we know when no more improvements are worthwhile? We should take time out from the detailed improvements to consider the bigger picture.[28] The concept of the 'perfect qMRI machine' (see Section 1.4.2) might give a clue. A major improvement in qMRI would come about if there were an international certification scheme for qMRI measurements which have reached the level of the perfect machine; a proposal is made below.

1.4.2 International standardisation and certification

Standards already exist for measurements in many physical quantities. Readily available machines to measure voltage, body or food weight, and temperature often come with a certificate conforming to the International Standards Organisation,[29] guaranteeing a particular performance in terms of total error.

The concept of the 'perfect machine' originated in the building of the 200-inch Palomar telescope in the USA in 1933; at the time it was the most perfect telescope that could be built.[30] The concept can usefully be applied to an MRI machine used for quantification. Here it proposed that

> *A Perfect Quantitative MRI machine is one that, in making a measurement, contributes no significant extra variation to that which already exists from biological variation.*

Various grades of performance can be envisaged, depending on the purpose the measurement. Comparison with normal variation will be the most demanding; comparison with variation within a disease might also be appropriate, depending on the context, and would be less demanding. Here a proposal is made for three levels, each with an appropriate medal[31] (see Table 1.3).

---

[26] The first edition contained four chapters on analysis, recognising that it is at least as important as acquisition. Spatial registration, shape, texture, volume, atrophy and histograms were considered.

[27] The first edition had a chapter on the biological significance of MR parameters in multiple sclerosis. Post-mortem studies of tissue can establish a relationship between a qMR parameter (e.g. NAA) and a biological parameter (e.g. neuronal density). Often the relationships are complex and depend on which biological parameters can vary (i.e. the disease context). Spatial registration of the biological specimen and the MR image is crucial.

[28] In Thomas Mann's *Death in Venice*, the writer is on the Venice beach. He sees the detail, in the foreground: children constructing a sand castle. He turns his gaze to the horizon, empty and infinite. What would it be to be a measurement hero?

[29] For example, ISO 17025 is the main ISO standard used by testing and calibration laboratories.

[30] Building the 200-inch telescope at Palomar, California, is described in the book *The Perfect Machine: Building the Palomar Telescope.*

[31] Medals are proposed, inspired by the ISMRM use of medals to acknowledge sponsorship at its annual scientific meeting.

TABLE 1.3 qMRI Medals for perfect machines: a proposal

Medal	Target study	Criterion
Bronze	Group comparison	ISD < 0.3 GSD[a]
Silver	Multicentre study	BCSD < GSD[b]
Gold	Serial study	ISD < 0.3 WSSD[c]

Note: SD = standard deviation; BSD = biological SD; GSD = group SD; ISD = instrumental SD; ICSD = inter-centre SD; BCSD = between-centre SD; WSSD = within-subject SD.

[a] In a group comparison, within-group variation GSD^2 should dominate (i.e. machine variation ISD makes an insignificant contribution to total within-group variation).

[b] The effect of between-centre variation (BCSD) should be less than within-group variation.

[c] In a serial study, total within-subject variation $WSSD^2$ should dominate (i.e. machine variation ISD makes an insignificant contribution to total within-subject variation).

Bronze medal: In a group comparison, the total variance in each group determines the power and sample size needed (see Figure 1.1). This is the sum of the variance from genuine biological spread (characterised by an SD equal to biological SD, BSD) and that given by the imperfect machine (characterised by an SD equal to instrumental SD, ISD), that is total group variance $GSD^2 = BSD^2 + ISD^2$. Thus, if ISD = 0.3 GSD, the contribution of machine variance to the total variance is 9% and may be considered negligible. This concept allows a perfect machine for group comparisons to be specified (Table 1.3). The criteria for each MR parameter would vary; some might be easy to achieve, others might need a long, sustained effort. The value of BSD would depend on the kind of subjects considered; in pooling normal values, a correction for age and gender dependence should be applied (treating them as a confounding variable and standardising all values to a fixed age and gender). The estimates of SD have associated uncertainties, which are significant if the sample size is small (see Chapter 3, Equation 3.2 and Figure 3.6), and these would need to be taken into account when considering if a criterion had been reached.

An example might be the MTR results reported in Chapter 3 (Figure 3.8). The stable scanner gave a normal group SD GSD = 0.4 pu and a measured instrumental SD ISD = 0.15 pu. From these, ISD = 0.375 GSD, and the criterion in Table 1.3 (ISD < 0.3 GSD) is not quite satisfied.

Silver medal: In multicentre studies, inter-centre variation has to be controlled, although some differences can be absorbed by the statistical analysis (provided each subject is always imaged at the same centre). MTR histogram matching using body-coil transmission (Tofts *et al.*, 2006) is probably a perfect silver-medal MTR machine.

Gold medal: In a serial study, instrumental variation can hide subtle within-subject biological changes. The power of a serial study can be limited by such biological variation; often it is small and unknown and may be extremely hard to measure. Gold medals will be the hardest to obtain; for some MR parameters the gold medal may be impossible. An exception is cerebral blood perfusion, measured by arterial spin labelling (ASL; Chapter 16). The natural within-subject variation is large (10%–20%) (Parkes *et al.*, 2004) and it might not be difficult to build a perfect gold-medal ASL machine (i.e. one with ISD < 3%). A second example might be in the context of a serial study in relapsing–remitting MS. The within-subject variation in lesion load is highly variable, and perfect gold-medal machines for lesion volume already exist.

Who might administer such a scheme? Award of medals might be determined by the reviewers of a paper submitted to a journal claiming the status or by an international committee (perhaps sponsored by the ISMRM). Prizes could be awarded (a kind of modern-day John Harrison longitude prize).[32]

The closing words from the first edition are still true:

> *Progress towards such automation [of measurement techniques] will take time, and the persistence of John Harrison the clockmaker may enable us to put our work into its historical perspective. We are present at a true technological revolution which is exposing our inner biological workings in ever increasing detail. A few decades ago this was inconceivable; in a few decades' time the techniques will be as routine as measuring the mass of the body.*

References

Bakker CJ, de Graaf CN, van Dijk P. Derivation of quantitative information in NMR imaging: a phantom study. Phys Med Biol 1984; 29(12): 1511–25.

Cady EB, Costello AM, Dawson MJ, Delpy DT, Hope PL, Reynolds EO, et al. Non-invasive investigation of cerebral metabolism in newborn infants by phosphorus nuclear magnetic resonance spectroscopy. Lancet 1983; 1(8333): 1059–62.

Filippi M, Dousset V, McFarland HF, Miller DH, Grossman RI. Role of magnetic resonance imaging in the diagnosis and monitoring of multiple sclerosis: consensus report of the White Matter Study Group. J Magn Reson Imaging 2002; 15(5): 499–504.

Filippi M, Grossman RI. MRI techniques to monitor MS evolution: the present and the future. Neurology 2002; 58(8): 1147–53.

Fischer JS, Rudick RA, Cutter GR, Reingold SC. The multiple sclerosis functional composite measure (MSFC): an integrated approach to MS clinical outcome assessment. National MS Society Clinical Outcomes Assessment Task Force. Mult Scler 1999; 5(4): 244–50.

Gribbin J. *Science: a history 1543–2001.* London, UK: Penguin; 2003.

Hobart J, Freeman J, Thompson A. Kurtzke scales revisited: the application of psychometric methods to clinical intuition. Brain 2000; 123(Pt 5): 1027–40.

Krummenauer F, Doll G. Statistical methods for the comparison of measurements derived from orthodontic imaging. Eur J Orthod 2000; 22(3): 257–69.

[32] Section 1.2.2.

Mallik S, Samson RS, Wheeler-Kingshott CA, Miller DH. Imaging outcomes for trials of remyelination in multiple sclerosis. J Neurol Neurosurg Psychiatry 2014; 85(12): 1396–404.

McFarland HF, Barkhof F, Antel J, Miller DH. The role of MRI as a surrogate outcome measure in multiple sclerosis. Mult Scler 2002; 8(1): 40–51.

McGowan JC. On the use of quantitative MR imaging. AJNR Am J Neuroradiol 2001; 22(8): 1451–2.

Miller DH. MRI monitoring of MS in clinical trials. Clin Neurol Neurosurg 2002; 104(3): 236–43.

O'Connor JP, Aboagye EO, Adams JE, Aerts HJ, Barrington SF, Beer AJ, et al. Imaging biomarker roadmap for cancer studies. Nat Rev Clin Oncol 2017; 14(3): 169–86.

Parkes LM, Rashid W, Chard DT, Tofts PS. Normal cerebral perfusion measurements using arterial spin labeling: reproducibility, stability, and age and gender effects. Magn Reson Med 2004; 51(4): 736–43.

Sobel D. *Longitude.* New York: Harper Perennial; 2005.

Sormani MP, Bonzano L, Roccatagliata L, De Stefano N. Magnetic resonance imaging as surrogate for clinical endpoints in multiple sclerosis: data on novel oral drugs. Mult Scler 2011; 17(5): 630–3.

Tofts PS. *Quantitative MRI of the brain: measuring changes caused by disease.* New York: Wiley; 2003.

Tofts PS, du Boulay EP. Towards quantitative measurements of relaxation times and other parameters in the brain. Neuroradiology 1990; 32(5): 407–15.

Tofts PS, Steens SC, Cercignani M, Admiraal-Behloul F, Hofman PA, van Osch MJ, et al. Sources of variation in multi-centre brain MTR histogram studies: body-coil transmission eliminates inter-centre differences. Magma 2006; 19(4): 209–22.

Tofts PS, Wicks DA, Barker GJ. The MRI measurement of NMR and physiological parameters in tissue to study disease process. Prog Clin Biol Res 1991a; 363: 313–25.

Tofts PS, Wicks DAG, Barker GJ. The MRI measurement of NMR and physiological parameters in tissue to study disease process. In: Ortendahl DA, Llacer J, editors. *Information Processing in Medical Imaging.* New Jersey, Wiley-Liss Inc, 1991b; p. 313–25.

Whittall KP, MacKay AL, Graeb DA, Nugent RA, Li DK, Paty DW. In vivo measurement of T2 distributions and water contents in normal human brain. Magn Reson Med 1997; 37(1): 34–43.

Wray S, Tofts PS. Direct in vivo measurement of absolute metabolite concentrations using 31P nuclear magnetic resonance spectroscopy. Biochim Biophys Acta 1986; 886(3): 399–405.

2

Measurement Process: MR Data Collection and Image Analysis

Contents

Paul S. Tofts
Brighton and Sussex Medical School

2.1 Magnetic resonance data collection

The process of collecting magnetic resonance (MR) data from a subject, in the form of images, spectra or maps, is analysed in some detail.

2.1.1 Subject positioning and the prescan procedure

The subject is positioned on the scanner couch by the radiographer (technologist). The subject should be comfortable, to reduce movement during the scan as much as possible. The radiographer should use any insight into the subject's emotional state to reduce anxiety if necessary; preparation on a separate couch may be helpful. A cushion under the knees can reduce cramp. Occasionally it is desirable to place the subject prone. *Prone positioning* of the head may be more comfortable if support is provided for the forehead and cheekbones, leaving a gap for the nose, as used in a massage table. *Movement* of the body can cause a head movement; a nasal positioning device (Tofts *et al.*, 1990) can help cooperative subjects to keep still. Some patients will find it hard to keep still because of their disease; researchers involved in the study are usually motivated to keep very still. Some kinds of movement are very common, especially rotation in the sagittal plane ('nodding'). Movement can be monitored by repeated localiser images throughout the study. If spatial registration between different image datasets is used (see Chapter 17), then the amount of movement that took place is available as output from the program. Research on both why some subjects move and on what limits how long a subject can stay in the scanner would improve the quality of MR data that can be obtained. It may become possible to use fast MR (or optical) imaging to dynamically alter the slice positions, tracking the movement of a subject in real time (although movement to a location of different static or RF field value would require some sophisticated correction). If Gd contrast agent is to be injected, a line is placed into the subject's arm, so that injection is carried out without disturbing the positioning of the subject. A power injector is usually used to provide a consistent injection procedure, with synchronisation to the scanner.

After the subject has been placed in the magnet bore, the automatic *prescan* procedure generally includes the following steps, which take account of differences between subjects and are crucial to quantification. The receiver gain is adjusted to use the available dynamic range of the receiver channel, without overloading it. The gain must be fixed for subsequent scans, if image intensity values are to be combined in some way (e.g. for a dynamic Gd scan series, where images are collected at a range of time points after injection of contrast agent – see Chapter 14). The transmitter output is adjusted to give the desired flip angle (FA) in the subject. This can be carried out in a number of ways; ideally, only the signal from the relevant piece of tissue (e.g. a slice or a spectroscopic voxel) is optimised. A multislice or volume acquisition cannot have the correct FA at all locations, because of transmit field non-uniformity.

The pulse sequences, containing long lists of radiofrequency (RF) and magnetic field gradient pulses, are then run. Signals are recorded; localisation of the origin of signal is achieved using a combination of slice or slab selection, frequency encoding and phase encoding gradients. Images can be weighted by various parameters (e.g. T_1, T_2 or D – see the chapters on each MR parameter). Images are reconstructed using Fourier transformation; the magnitude of the complex data is usually calculated (this is not vulnerable to unpredictable phase shifts). Full descriptions of the MR imaging process are available elsewhere (Brown *et al.*, 2014, and also see Table 1.1 in Chapter 1).

2.1.2 The NMR signal

The signal δv from precessing nuclei in a small volume δV_s in the sample is given by the following (Hoult and Richards 1976; Hoult 1978):

$$\delta \upsilon = \omega_0 B_{1xy} M_{xy} \delta V_s \cos(\omega_0 t) \tag{2.1}$$

where ω_0 is the Larmor[1] frequency (in radians s^{-1}); the life of Sir Joseph Larmor, the Irish physicist, is described by Tubridy and McKinstry (2000). B_{1xy} is the component[2] of the RF field B_1 produced in the transverse plane at the location of the sample by unit current in the coil, during transmission.[3] M_{xy} is the transverse component of the magnetisation of the sample.[4] For protons the equilibrium magnetisation M_0 is as follows (Brown *et al.*, 2014):

$$M_0 = \frac{N\gamma\hbar^2 B_0}{4kT} \tag{2.2}$$

where N is the number of protons per unit volume, γ is the *magnetogyric ratio*,[5] $\hbar = h/2\pi$, where h is Planck's constant, B_0 is the magnitude of the main static magnetic field, k is Boltzmann's constant and T is the absolute temperature[6] of the sample.

The proportionality of received signal with the magnitude of the applied field per unit current, shown in Equation 2.1, is called the *principle of reciprocity* (Hoult and Richards 1976), and this has been a key concept in quantitative MR. In simple terms, it says that if we have trouble getting the applied B_1 field into a particular location in the sample, using a particular coil, we will have as much trouble getting the signal out of that location, using the same coil. This is discussed in more detail in Section 2.1.10 below.

The dependence of magnetisation on absolute temperature is relevant when room temperature concentration standards are used (as in measurements of proton density and metabolite concentrations). As a particular concentration of protons is cooled (e.g. from body temperature to room temperature), its magnetisation increases and it can produce more signal (see Chapter 3).

2.1.3 The static magnetic field B_0

In a superconducting magnet, the value of the static field is set at the time of installation by adjusting the amount of circulating current stored in the windings. There may be a very small decay over time, which is compensated for by adjusting the current through room temperature windings or by adjusting the centre frequency of the transmitter.

When the subject is placed in the magnet, the magnetic susceptibility of the tissue alters the field inside the brain slightly. The transmitter centre frequency is adjusted to bring the protons back onto resonance. The shim coil currents are adjusted to obtain a spatially uniform B_0 distribution, as far as possible.

Remaining static field gradients caused by spatially varying tissue susceptibility (particularly near tissue–air interfaces, such as the temporal lobes) can be a problem, particularly for spectroscopy and echo planar imaging, which are very sensitive to such gradients. In spectroscopy the line position will be altered and possibly broadened. In gradient echo and echo planar imaging there may be signal dropout due to intravoxel dephasing.[7] In spin echoes, the dephasing effect of these gradients is corrected provided the spins are stationary; however, in the presence of diffusion, spins moving through a gradient will not be rephased and signal loss will once again be seen. Such signal loss will not normally cause systematic error in quantification, although the lowered signal-to-noise ratio (SNR) will give increased random errors, and in situations where the absolute signal level is important (e.g. PD) there will also be a systematic error.

A further source of degradation is that echo planar images (and to a lesser extent gradient echo images, which have a much shorter echo time than echo planar sequences) will suffer geometric distortion, such that the image is shifted or warped in the locality of susceptibility gradients (Moerland *et al.*, 1995; Hutton *et al.*, 2002; Jezzard 2002). This in turn prevents straightforward spatial registration of such images with those having negligible distortion (principally those that are spin-echo–based, although

1 The *Larmor frequency* is the frequency at which protons precess around the main static field B_0.

2 A linear coil produces two counter-rotating components; one is in the right direction for NMR and is useful; the other is not used but contributes to noise and power requirements. In a circularly polarised coil only the useful component is produced and detected.

3 This equation is valid for the simple case of a single transmit/receive coil at low field. In this case the transmit field B_1^+ equals the receive field B_1^- (i.e. $B_1^+ = B_1^- = B_{1xy}$); see also Section 2.1.10.

4 After a single 90° RF pulse, $M_{xy} = M_0$.

5 $\gamma = \omega_0/B_0$, where B_0 is the static magnetic field strength, in Tesla. For protons, γ = 2.675 10^8 rad s^{-1} T^{-1} (equivalent to 42.57 MHz/T) (Brown *et al.*, 2014). Greek letters are described in Appendix 1.

6 The *absolute temperature* is measured in degrees Kelvin (K) from –273°C, which is called *absolute zero*. Thus the freezing point of water (0°C) is 273K, and body temperature (37°C) is 310K.

7 In *intravoxel dephasing*, the different components of magnetisation in a voxel, experiencing different static fields, become out of phase with each other, and the total transverse magnetisation vector in the voxel is reduced. In a spin echo, this dephasing is corrected by the 180° refocusing pulse; in a gradient echo the uncorrected dephasing leads to signal loss.

gradient echo sequences often also have negligible distortion) and thwarts any attempt at measuring volume. The image intensity is likely to be altered by distortion (since a given amount of signal will be placed into a voxel that is too large or too small). A third degradation is that off-resonance effects in such localities may reduce the apparent FA and distort a 2D slice selection process (see Section 2.1.6).

The static field can be mapped straightforwardly using the phase shift after a gradient echo (Sled and Pike 2000; Hetherington *et al.*, 2006).

2.1.4 Static field gradients

Having taken a lot of care to achieve a uniform static magnetic field, switched field gradients[8] are deliberately introduced as part of the imaging process. The slew rates are very fast, giving typical switching times of <100 μs. Eddy currents can be induced in surrounding conducting structures; these have the effect of producing small transient shifts in B_0, distorting spectra and images. Eddy currents are reduced to low levels by several devices. Actively shielded gradient coils limit the magnetic flux outside the coil; current pre emphasis circuits drive the coils in such a way as to counteract the effects of the eddy currents; and conduction loops are eliminated from the scanner bore construction materials.

The remaining non-idealities are twofold. Firstly, the *gradient amplitudes* may be incorrect, by up to about 1% (depending on the calibration procedure). This gives rise to small errors in the size of objects (since the gradient change corresponds to a change in magnification, or of voxel size) and also to errors in the estimates of diffusion coefficients and tensors. Very precise measurements of voxel dimensions, using image registration, enable the value of small gradient changes to be measured (Lemieux and Barker 1998). Secondly, gradient coils do not produce a completely linear variation of static field with distance (i.e. the gradients are non-uniform); this in turn produces errors in the gradient amplitude (according to the position) and gives rise to spatial distortion (Moerland *et al.*, 1995; Jezzard 2002). *Non-linearity from gradient coils* is minimal in the central (head) region when using body gradient coils. Manufacturers usually measure and correct for the geometric distortion caused by non-linearity; this can be seen by turning the correction off.

2.1.5 Radiofrequency transmit field B_1^+

Usually transmission is by the body coil, which has relatively good uniformity over the head region. Typical pulse amplitudes are 10–20 μT. Sometimes amplitudes are expressed in hz or radians s^{-1}, giving the rate of nutation of magnetisation around a constant RF field of that value.[9] The current in the transmit coil required to achieve this value of RF field depends on the Q[10] of the coil. Q is determined mostly by power losses in the subject (caused mostly by its electrical conductivity) rather than by losses in the coil itself. As the subject is moved into the coil, its conductivity loads the coil, and a greater current, and hence voltage, is required to produce a given B_1 value. The amount of *coil loading* (i.e. the amount of power that is removed from the coil and is deposited in the subject) varies from subject to subject. The prescan procedure sets out to obtain the same value of B_1 (and hence FA), regardless of the loading produced by the particular subject. The *transmitter output* is adjusted, usually automatically, often by adjusting *attenuators* in the amplifier.

Non-linearity in the transmitter output stage may occur, leading to incorrect B_1^+ values and distortion of selective pulses. Such gross non-linearity over the normal range of amplitudes of the selective pulses would probably be picked up as artefacts in the routine imaging (depending on the amplitudes of the selective pulses used). Calibration of the transmitter output stage is usually carried out periodically as part of routine preventive maintenance (Venkatesan *et al.*, 1998). *Transmitter linearity* can be investigated as follows: An oscilloscope can be used to measure the output voltage as a function of software hard pulse amplitude (Alecci *et al.*, 2001); this will be accurate to within a few percent, depending on the oscilloscope. The output can be stepped by the software for convenience. A more accurate method is to use NMR to measure the B_1^+ amplitude, as follows. At each amplitude, observe the signal from a small sample as a function of hard pulse duration. The null duration (i.e. a 180° pulse) gives an accurate and precise measurement of B_1^+. Ensure there is no pulse droop, using an oscilloscope. A plot of B_1^+ versus software pulse amplitude should be linear.

RF non-uniformity is the largest cause of error in qMR.

> *Radiofrequency field inhomogeneities are the most irksome sources of nonidealities (especially as a result of their omnipresence). The spatial variation of the RF field sensitivity of the transmit and receive coils enter the signal expression for any sequence in the form of altering the flip angle at a given spatial location as well as altering the received signal from [the same] spatial location.* (Haacke et al., 1999 p. 661)

At 1.5T the effect is noticeable (Barker *et al.*, 1998); at higher fields the problem becomes worse (see Figure 2.1). An elliptical object (such as the head) in a circularly polarised coil gives a diagonal non-uniformity pattern (Sled and Pike 1998). At 3T using a birdcage coil a 20% reduction in B_1^+ was measured at the

[8] Imaging gradients can be up to about 20 mT m^{-1}; diffusion gradients are often higher (to shorten the echo time), up to 80 mT m^{-1}, and may use a dedicated head coil set. Switching (slew) rates are up to 200 T $m^{-1}s^{-1}$.

[9] The nutation rate is $\omega_1 = \gamma B_1$ rad s^{-1}; thus a RF field of 10 μT corresponds to 2680 rad s^{-1} or 426 hz. The duration of a hard (i.e. non-selective) θ pulse is $\tau_\theta = \theta/(\gamma B_1)$; thus a 90° 10 μT hard pulse lasts 587 us.

[10] Q stands for *quality factor*, denoting how long a coil will ring after being excited. High-Q coils are less damped, lose less energy per cycle, ring for longer, provide a greater B_1 for a given current and provide a greater signal for a given amount of precessing magnetisation.

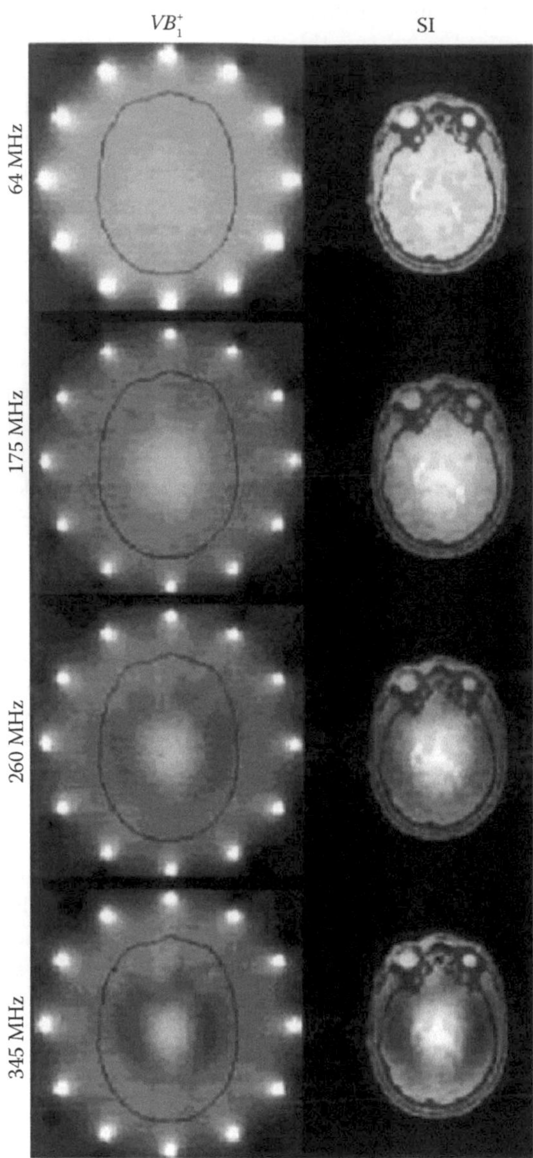

FIGURE 2.1 RF signal non-uniformity in the head. An accurate mathematical model of the head inside a birdcage coil, with flip angle = 90° at the head centre. Frequencies correspond to fields of 1.5T, 4.1T, 6.1T and 8.1T. VB_1^+ is the excitation field (normalised by the factor V), SI is the signal intensity from a gradient echo sequence. The doming effect at the centre of the head becomes increasingly pronounced at higher fields, and an annular region of reduced signal is visible further from the centre. (From Collins, C.M., and Smith, M.B., *Magn. Reson. Med.*, 45(4), 684–691, 2001.)

periphery of the brain, compared to its value at the centre (Alecci *et al.*, 2001). Anatomically accurate models of B_1^+ distribution in the head, using detailed anatomical knowledge (Collins and Smith 2001; Ibrahim *et al.*, 2001) show that at high fields, up to 8T, non-uniformity increases, as dielectric resonance increases the sensitivity near the centre of the head; measurements at 7T confirm this (Collins *et al.*, 2002). Adjustment of the current applied through each port on a multiple transmit coil, and its phase, allows uniformity to be optimised (Ibrahim *et al.*, 2001). This has been termed *RF shimming* (Collins *et al.*, 2005).

In the early days of MRI, it was feared that the reduced RF penetration at high frequencies (the 'skin effect') (Bottomley and Andrew 1978) caused by the electrical conductivity of the tissue would prevent head imaging above 20 MHz. In fact it is more than offset by the amplifying effect of dielectric resonance (which increases B_1^+ in objects whose size is comparable with the half-wavelength of electromagnetic waves at the frequency of observation; see Figure 2.1).

2.1.6 Slice and slab profile

In 2D (slice selective) imaging, slice selection is a key problem. The observed transverse magnetisation is the sum of spins within the voxel that have experienced a variety of histories, according to their location in the slice selective gradient, the local FA and the amount of relaxation (i.e. T_1/TR). Thus the slice profile can be distorted; there is not a single effective FA within a voxel, and signal modelling for quantification is often complex and inaccurate (Parker *et al.*, 2001).[11] Slice selection is often used for an EPI readout and also is needed for some MR parameters (e.g. T_2, T_2*); then such effects must be taken into account. Magnetisation preparation by a hard (non-selective) pulse followed by 2D readout can mitigate the distortion.

Fast and powerful gradients have driven the widespread use of 3D imaging, which has no such problems. Slab selection (in which signal from outside the desired field of view is suppressed) is usually applied in each phase encoding direction, to prevent wrap-around.[12]

2.1.7 B_1^+ transmit field mapping

The determination of the RF 'active' field B_1^+ at each location in space is important for two reasons. First, the local FA can then be found; this is needed for many parameter calculations (e.g. T_1 from VFA Chapter 5[13]). Second, its value is needed for accurate measurement of some MR parameters (e.g. MT).

Early methods of determining FA, or setting it to a required value, were developed in spectroscopy. A sample inside a uniform RF coil gives a maximum signal when the FA is 90°, provided it can relax fully between pulses (i.e. TR >> T1). The pulse amplitude,[14] or duration, is increased from a low value; at first

[11] Slice distortion and correction is described and illustrated in more detail in the first edition.

[12] Slab selection is a more gentle process than slice selection (each voxel has a single well-defined FA) and probably produces little distortion; however, there seems to be no literature on this.

[13] Chapter 5, Section 5.4.4 also contains a discussion of B_1^+ mapping.

[14] A rectangular ('hard') pulse, of amplitude B_1 and duration τ, produces a FA of $\gamma B_1 \tau$. Depending on the spectrometer hardware, either the pulse amplitude or duration is varied to give the required FA. Early spectrometers had fixed amplitude, and the duration was altered. Modern MRI machines often also allow variable amplitude (since a selective ['soft'] pulse must keep its duration fixed).

the signal increases almost linearly, then reaches a maximum, then declines. Further increase in the pulse gives a null signal (corresponding to an FA of 180°), and this condition can often be found more precisely than the maximum at 90°.

Modern methods (summarised in Table 2.1) should meet the following nine criteria: the method should (1) provide an accurate value, to within about 1% (since some measurements are very sensitive to FA errors, e.g. T_1[15]); (2) have good precision;[16] (3) have reasonable imaging time (ideally less than 1 minute; this favours 3D methods); (4) be independent of T_1 effects; (5) be independent of B_0 (off-resonance) effects; (6) be unaffected by 2D slice selection artefacts; (7) have acceptable SAR value; (8) work over a wide FA range if needed;[17] and (9) ideally be capable of implementation using a standard pulse sequence.

TABLE 2.1 B_1^+ mapping methods (a selection)

	Reference	Summary
Magnitude Methods	Yarnykh (2007)	Two pulses, same FA, two TRs (3D)
Actual flip angle imaging (AFI)	Hurley *et al.* (2012)	Combined with T_1 measurement
Dual angle method (DAM)	Stollberger and Wach (1996)	Two pulses, two FAs, long TR (2D)
	Insko and Bolinger (1993)	Original paper
	Cunningham *et al.* (2006)	Saturate for short TR
	Boudreau *et al.* (2017)	EPI-DA; fast uses EPI
180° null	Dowell and Tofts (2007)	Can use short TR
Phase Methods	Sacolick *et al.* (2010)	Phase α B_1^2; 3D
Bloch–Siegert (BS)	Sacolick *et al.* (2011)	3 sec acquisition for prescan
Phase sensitive	Morrell (2008)	Wider range than DAM

Note: FA, flip angle; EPI-DA, EPI-Double Angle.

Early work often used the double-angle method (Stollberger and Wach 1996). Two 2D acquisitions are made, with nominal FA values typically 30° and 60°; the ratio of signals gives the actual FA values but only under the condition of TR >> T1 (complete relaxation). The needs for speed and 3D acquisition may have pushed this method aside. However, an EPI variant provides a 2-minute solution using a standard pulse sequence, and a performance comparable to the AFI and BS methods (Boudreau *et al.*, 2017) (Figure 2.2).

The two principle methods currently in use are actual flip angle imaging (AFI) and Bloch–Siegert (BS) (Sacolick *et al.*, 2010 2011; Whisenant *et al.*, 2016). AFI (Yarnykh 2007) uses two pulses with the same FA, at different (short) TRs, in a 3D sequence. There are residual T_1 effects, which could be measured using VFA (variable flip angle – see Chapter 5). In VAFI (Hurley *et al.*, 2012), VFA and AFI are combined to take proper account of T_1. AFI can be seen as an ingenious variant on the dual angle method. The ratio of signals from two acquisitions is again used; however, it is the TR not the FA that is altered. In the BS approach, an off-resonant pulse is used to produce a phase that is proportional to the square of the local B_1^+ field; this is then read out. It is fast and independent of B_0 error (Sacolick *et al.*, 2010 2011; Duan *et al.*, 2013; Whisenant *et al.*, 2016).

The 180° null method uses a standard 3D sequence to provide a map in 4 minutes – see Figure 2.3 (Dowell and Tofts 2007). The move to higher fields and the consequent increase in B_1^+ non-uniformity has prompted MRI manufacturers to invest in mapping techniques (Sacolick *et al.*, 2010; Nehrke and Börnert 2012). Summaries with discussion, comparisons and

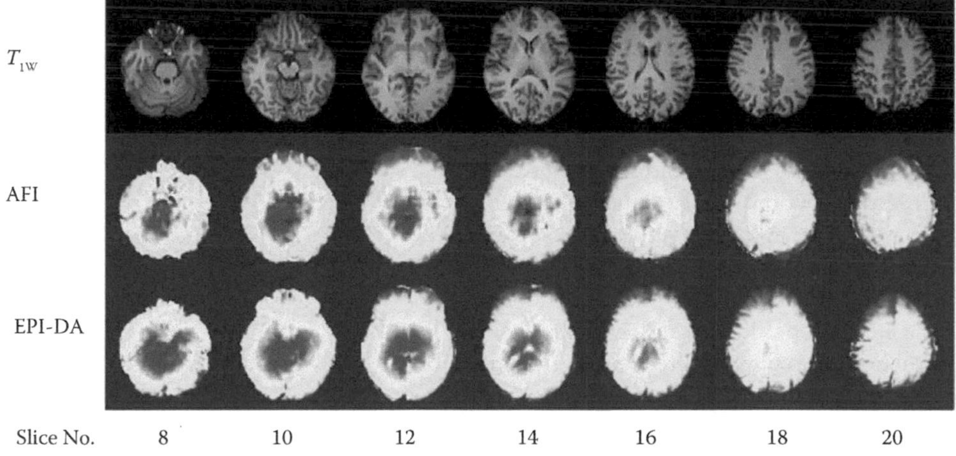

FIGURE 2.2 B_1^+ maps at 3T using the actual flip angle imaging and EPI-DA methods (see Table 2.1) (From Boudreau, M., *et al.*, *J. Magn. Reson. Imaging*, 2017).

[15] A 1% error in FA gives a 2% error in T_1; see Chapter 5, Equation 5.10.

[16] The B_1 function will vary slowly with space, so some spatial smoothing is permissible.

[17] Although some methods presume a small range of FA values, they can usually be adapted to a wider range by adding more measurement pulses.

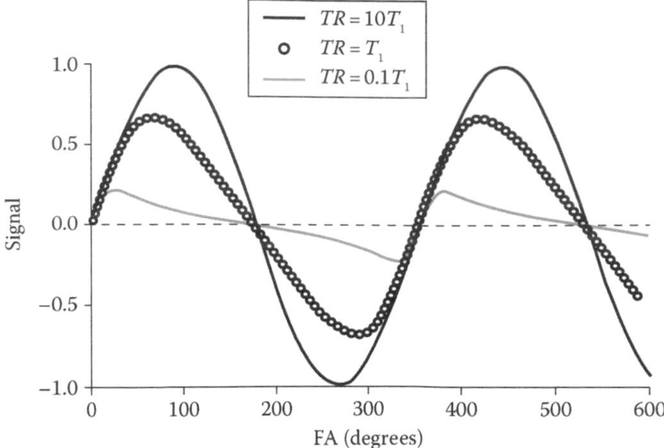

FIGURE 2.3 Simulation showing how the 180° signal null can be used to map B_1^+, regardless of T_1. (From Dowell, N.G., and Tofts, P.S., *Magn. Reson. Med.*, 58(3), 622–630, 2007.)

often optimisation of sequences are given by Lutti *et al.* (2010), Morrell and Schabel (2010), Sacolick *et al.* (2010), Volz *et al.* (2010), Hurley *et al.* (2012), Park *et al.* (2013) and Pohmann and Scheffler (2013).[18]

Inflow effects: In the special case of bulk blood flow, the effective B_1^+ value may be smaller than that in the voxel of interest. ASL[19] uses this effect; blood arrives in the voxel with a different magnetisation from local water that has a recent history in the voxel. In DCE imaging,[20] blood is often imaged with a 2D slice to measure its T_1; however, blood from outside the slice may have a lower FA and hence higher magnetisation than that predicted by a naive use of the FA value in the voxel.[21]

2.1.8 B_1^- receive sensitivity field

Reception is usually by a close-fitting multiple array head coil, for good sensitivity. The receive sensitivity field B_1^-(r) is determined by both the receive coil array geometry and the properties of the object being imaged (whether a head or a phantom). Dielectric resonance tends to give a high B_1^- in the central parts of the object (similar to the B_1^+ distribution), whilst multiple surface coils have greater response from the periphery; these effects can be made to partially cancel (Figure 2.4).

In a multi-array receive system, parallel imaging options (Grappa, Sense, Smash, etc.) can produce image artefacts, arising from the reconstruction of low spatial frequencies, which might be unstable and degrade quantification performance. Thus parallel imaging should be used with caution, with low acceleration factors.

B_1^- cannot be determined directly and with the loss of the reciprocity principle[22] has become one of MRI's 'unsolved problems'. For most MR parameters it is a ratio of signals that is measured, and absolute sensitivity has no influence. However when measuring absolute concentrations of protons (i.e. PD and magnetic resonance spectroscopy (MRS) absolute metabolite concentrations), a map of B_1^-(r) is required, and some attempts have been made to estimate it.

The *bias field* approach uses information that B_1^-(r) varies slowly with position (Volz *et al.*, 2012; see also Watanabe *et al.*, 2011; Jin *et al.*, 2012; Sabati and Maudsley 2013). A PD-weighted image has intensity proportional to B_1^-(r)PD(r); smoothing this (fwhm = 60 mm) gives a map proportional to B_1^-(r) (this assumes PD(r) has little low spatial frequency content). A single reference value for PD (e.g. from cerebrospinal fluid or a mean value for the whole brain) then enables B_1^-(r) to be determined and its effect removed in a PD(r) estimation. The method relies on PD(r) being smooth and there being no large abnormalities. A variation on this is to constrain the possible behaviour of B_1^-(r) by using the Maxwell equations; the method has been successful in the ordered environment of a single transmit/receive coil and a phantom (Sbrizzi *et al.*, 2014; see also Chapter 4, Section 4.3.2).

The *receiver gain* can be altered during the prescan procedure to account for the magnitude of the signal (the gain is much reduced in spectroscopy). Ideally it will be fixed during the acquisition of the image series,[23] at a suitable value that will not overload the receive chain nor introduce extra noise. If it is altered, a correction might be possible, depending on what information on receiver gain is available and whether analogue values of attenuation or gain are accurate.

2.1.9 Image noise

Electrical noise comes from random thermal agitation (Brownian motion) in the subject, the RF coil and possibly the preamplifier. With good design, contributions from the hardware are made insignificant and the dominant source is the subject.

Artefacts (often from movement during the phase encoding process) constitute an additional source of unpredictable error; these can be assessed by viewing the air surrounding the head. The centre of the grey level display window is set to zero. Any visible artefacts are of interest, since they almost certainly extend to the high SNR parts of the image, in the brain, and may exceed the random noise.

2.1.9.1 Optimised sequence parameters

The effect of image noise can be reduced by careful choice of sequence parameters. Increasing the voxel size reduces noise (at the expense of spatial resolution). Increasing the number of

[18] Teaching sessions at the annual International Society for Magnetic Resonance in Medicine meeting by PF van de Moortele (2015) and Lawrence L Wald (2016) were also insightful.

[19] See Chapter 16.

[20] See Chapter 14, Section 14.4.3.

[21] Blood flowing at 1 m s^{-1} into a 5 mm 2D slice will spend only 5 ms in the slice; thus it may experience only one pulse (if TR = 5 ms), insufficient to reach a new equilibrium magnetisation.

[22] See Section 2.1.10.

[23] For example, during acquisition of several TE values to determine T_2 (Chapter 6).

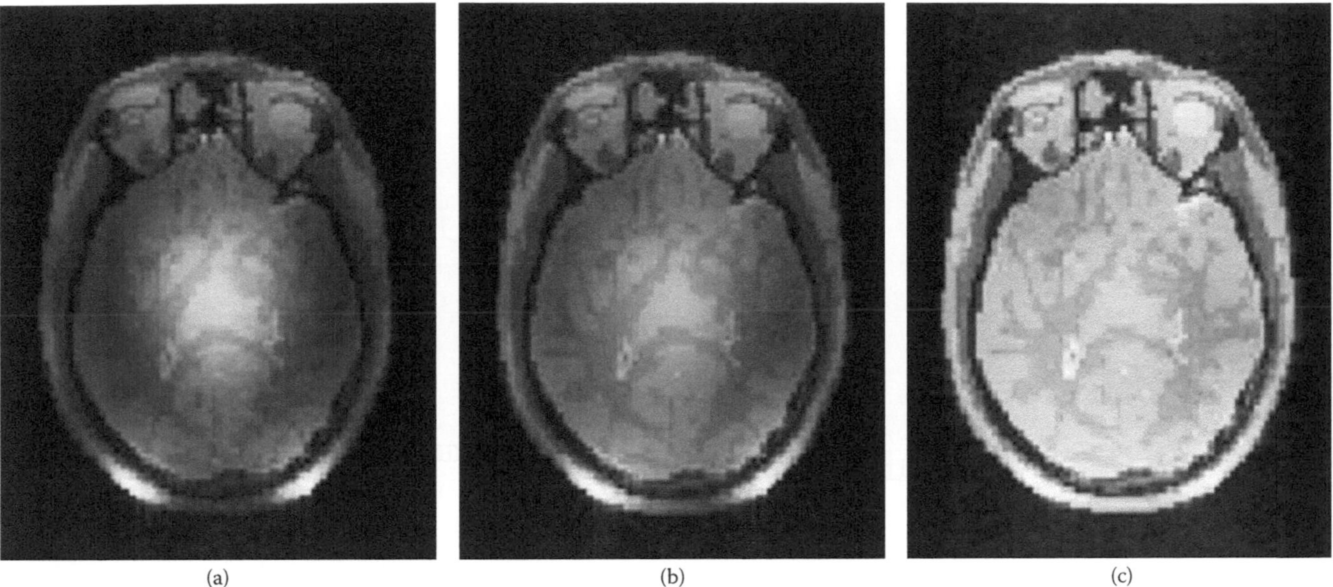

FIGURE 2.4 Simulated magnetic resonance images showing the results of different coil configurations at 300 MHz (7T). Use of a single volume coil in both transmission and reception (a) results in relatively strong B_1^+ and B_1^- fields near the centre (due to constructive interference) surrounded by weaker fields (due to destructive interference) resulting in a centre-bright appearance of the signal intensity distribution. Use of a single volume coil in transmission but an array of decoupled coils in reception with sum-of-magnitude reconstruction (b) results in relatively strong B_1^+ near the centre but relatively strong B_1^- near the periphery, resulting in a more homogeneous signal intensity distribution. Use of a transmit array with RF shimming to lessen the pattern of constructive and destructive interference in transmission and a receive array with sum-of-magnitude reconstruction (c) produces a very homogeneous image, even at this high frequency with only eight elements in transmission and reception. (From Collins, C.M., and Wang, Z., *Magn. Reson. Med.*, 65(5), 1470–1482, 2011.)

averages (NEX – the number of excitations) reduces the noise (SNR α NEX$^{1/2}$). Increasing the TR often also increases the SNR. Changing NEX or TR usually increases the acquisition time, and for a fixed acquisition time an optimal combination of parameters can be found, which maximises SNR. The acquisition of an image dataset from which parameters such as blood–brain permeability are to be estimated can also be optimised (Tofts 1996) (see Figure 2.5 and Chapter 3 section 3.2.3 on modelling error).

2.1.9.2 Rician noise distribution in magnitude image gives systematic error

Most images are constructed from the magnitude of the complex image data, and phase information is discarded. Magnitude data does not have a normal distribution (it cannot be negative). At low values of SNR, the distribution from a single receive coil follows a Rician probability distribution with a non-zero mean value (Henkelman 1985, 1986; Brown *et al.*, 2014). At high values of SNR this approximates to a Gaussian (normal) distribution; at zero SNR it becomes a *Rayleigh distribution* (Figure 2.6).

This effect constitutes a systematic error, or bias, in addition to the random error that noise always contributes. It can be seen most clearly by looking at the mean value in air regions of the image, where a non-zero value will be found. The techniques most affected are those that use low SNR image data, principally T_2 and ADC, where the decay of signal may be followed down into the noise, (Miller and Joseph, 1993; Wheeler-Kingshott *et al.*, 2002), and ASL perfusion, where the signal difference is comparable with the noise (Karlsen *et al.*, 1999). If image averaging is used to improve SNR, this should be carried out on the complex images, before formation of the magnitude.

Correcting single-coil image data is possible; however in multicoil receive systems correction is not straightforward and additionally the noise may vary with position in the image (see Chapter 17).

2.1.9.3 Noise estimation

An estimate of the noise value in high SNR regions is often desirable (e.g. for modelling error propagation). For a single coil system, if the standard deviation in high SNR regions of the image is σ, then sampling the Rician noise in an air region of interest (ROI) will give a mean value of 1.25σ and a standard deviation of 0.66σ (Edelstein *et al.*, 1984; Gudbjartsson and Patz 1995; Andersen 1996), enabling σ to be estimated (see Figure 2.6). Noise has also been estimated by subtracting the squares of images (Sijbers *et al.*, 1998)[24].

In a multicoil receive system, two approaches are possible. (1) Placing an ROI in a uniform region of tissue will give an indication of image noise; however the SD value will also contain a contribution from any tissue non-uniformity in the ROI, and

[24] See also the IPEM QA report (McRobbie 2017), page 21 onwards.

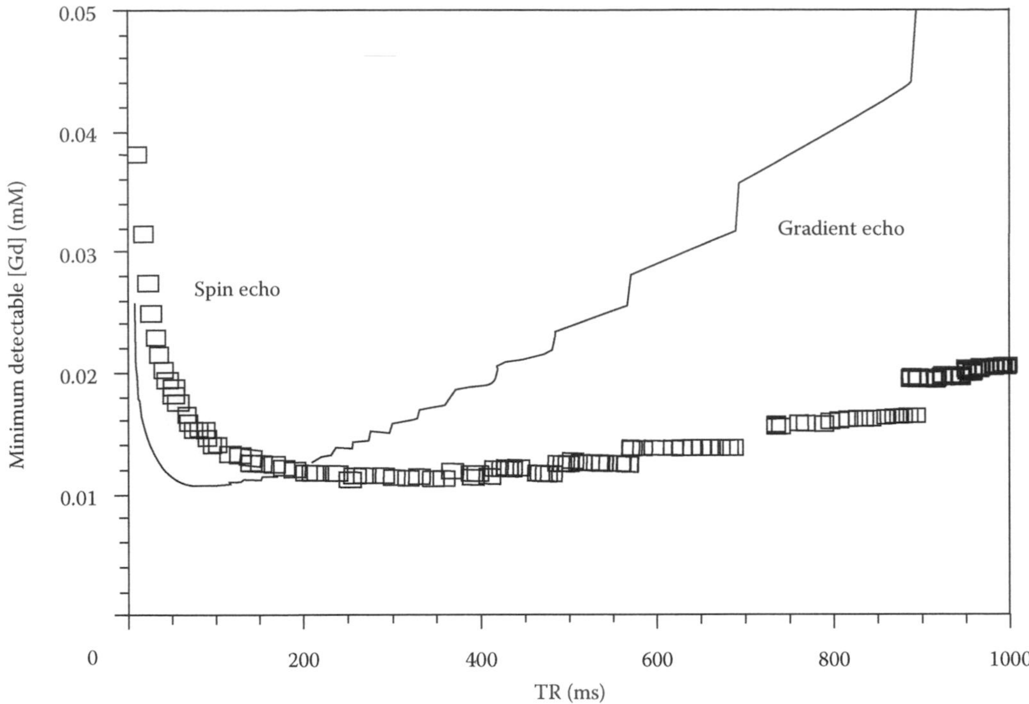

FIGURE 2.5 Sequence optimisation by noise modelling. Mathematical modelling of image noise propagation predicts the minimum amount of Gd contrast agent that can be detected using a T_1-weighted sequence. By optimising the repetition time, *TR*, in a spin echo or gradient echo sequence, its performance in white matter can be optimised. Theory indicates that, for a spin echo, the optimum $TR = T_1/2$ (here it was assumed $T_1 = 600$ ms). The gradient echo (flip angle, FA = 50°) can achieve the same sensitivity, provided the correct *TR* is used. The examination time was fixed at 10 min. (From Tofts, P.S., *Magn. Reson. Imaging*, 14(4), 373–380, 1996.)

the multiple coil system will in any case cause the noise value to vary with position. (2) Subtraction of two repeated images is likely to give a better result. The difference image is expected to have a normal distribution with an SD of $\sqrt{2}\,\sigma$; the distribution of values, and their mean, should be checked for any unexpected effects. This method has been used in phantoms (Murphy *et al.*, 1993; Goerner and Clarke 2011). The noise in brain images is likely to be different (because of different coil loading and B_1^- distribution), and an explicit measurement in brain is preferred. Provided movement is controlled, and measurements are restricted to uniform areas of tissue, then this ought to be possible.

2.1.9.4 Image quantisation errors

Image intensity values are usually stored as integers, typically to 12-bit precision (i.e. 1 in 4096). The floating point values from the image reconstruction process are rounded to the nearest integer.[25] Typical signal values may be 500–1000; thus the electrical noise will be a few units (for SNR = 100), and quantisation noise (maximum value 0.5) should be insignificant compared to electrical noise. If the floating point numbers are truncated (not rounded) a small amount of bias (0.5 image units) will be introduced.

[25] Rounding introduces a maximum error of 0.5, with rms value $1\sqrt{12} = 0.3$

If histograms are to be produced from parameter maps, *image despiking*[26] should be carried out, to prevent the discrete image probability distribution from producing artefacts in the histograms (Tozer and Tofts 2003).

2.1.10 The reciprocity principle and its failure

In the early days of MRI, the reciprocity theorem was key in enabling $B_1^-(r)$ to be found. (Hoult and Richards 1976; Hoult 1978, 2000). In this case $B_1^- = B_1^+ = B_1$, and in this paradigm many papers described 'B_1 mapping' (when in fact they are mapping B_1^+). Reciprocity was demonstrated at fields up to 1.5T (Tofts and Wray 1988; Michaelis *et al.*, 1993; Barker *et al.*, 1998; Fernandez-Seara *et al.*, 2001). Under varying values of coil loading, the product of pulse length[27] or transmitter output voltage and signal is constant. Given B_1^-, absolute measurements of PD and MRS metabolite concentration were possible (Provencher 1993, 2001; Fernandez-Seara *et al.*, 2001).[28] Reciprocity served us well for 40 years.

[26] The integers are converted to floating point numbers, and random noise with maximum magnitude 0.5 is added, forcing the image intensity values to have a continuous distribution, before calculation of the maps.

[27] Data from Tofts and Wray 1988, reanalysed in the first edition, p 304.

[28] The first edition of this book contains much on reciprocity (p 37) and its application in PD (p 95) and MRS (p 304–5).

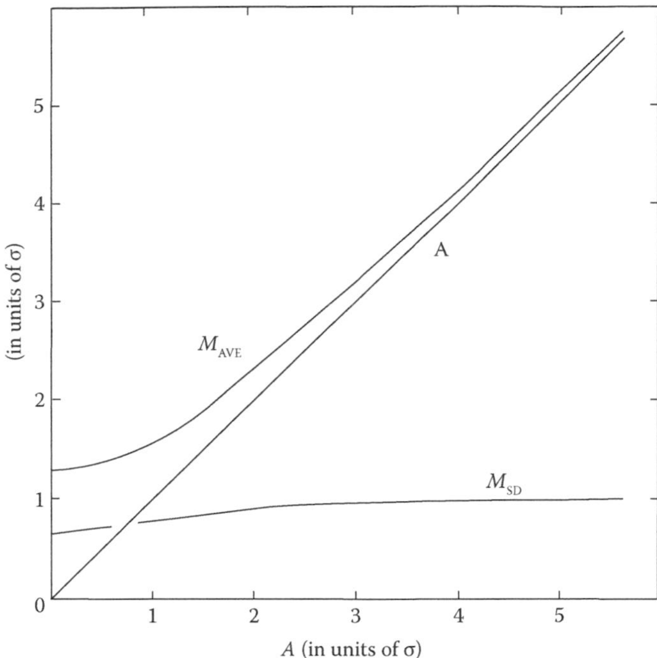

FIGURE 2.6 Rician noise in single-coil magnitude data. For a uniform region of an image with real amplitude A (measured in units of σ, the standard deviation of the noise in each dimension of the normal distribution), the average value M_{AVE} and the standard deviation M_{SD} in the magnitude image are shown, both expressed in units of σ. Thus at high SNR ($A \gg \sigma$), the mean intensity in the magnitude image equals its value in a real image ($M_{AVE} = A$), whilst at low SNR it exceeds the value in a real image, reaching an asymptotic value of 1.253σ in the absence of any signal. At high SNR, the standard deviation equals that in a real image ($M_{SD} = \sigma$), whilst in the absence of any signal it decreases to $M_{SD} =$ 0.655σ. (From Henkelman, R.M., *Med. Phys.*, 12(2), 232–233, 1985.)

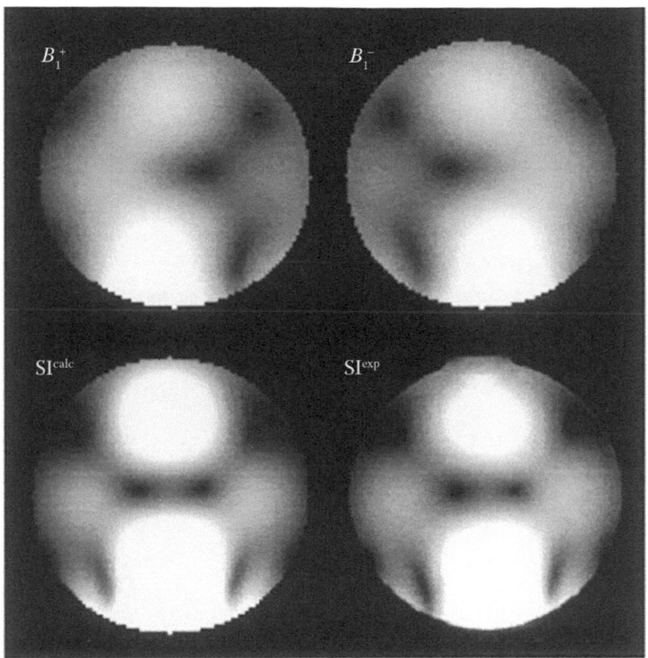

FIGURE 2.7 Reciprocity failure at 7T. Theoretical B_1 distributions in a 16 cm spherical NaCl solution phantom at 7T, imaged with a 10 cm surface coil. B_1^+ is the circularly polarised transmit field, B_1^- is the receive field, and differences between the two can be seen. SI^{calc} and SI^{exp} are the theoretical (calculated) and measured (experimental) signal intensities for a gradient echo image; their similarity gives confidence in the modelling. (From Collins, C.M., *et al.*, *Magn. Reson. Med.*, 47(5), 1026–1028, 2002.)

Later it became clear that the theorem is not valid at higher fields (3T and above),[29] that is $B_1^- \neq B_1^+$ (Sled and Pike 1998; Hoult 2000; Ibrahim 2005; Collins and Wang 2011), and also that as transmit and receive coils improved, single transmit/receive coils would go out of use. Now in the 'brave new technoworld' of higher fields and separate transmit and receive coils, the reciprocity theorem is probably consigned to the past, and we have to travel on without knowledge of B_1^- (Figure 2.7).

PD and MRS measurements have become more difficult with the loss of the principle of reciprocity. It is almost impossible to distinguish whether a change in signal voltage is caused by a change in proton concentration or a change in coil sensitivity (see also Chapter 4, Section 4.3.2 and Chapter 12, Section 12.6.5). The solution may lie in using a dedicated 'reciprocity-friendly scanner' that works at 1.5T or less, with a single transmit/receive coil. Such a PD map could be matched to a higher quality proton-density weighted image from another scanner to provide improved spatial resolution and SNR.

The Distant Dipolar Field method may offer a solution to measuring PD (Gutteridge 2002). This collects a small signal that is proportional to PD^2, and combined with a conventional PD image, the quotient image can give absolute PD without errors from unknown B_1^+ or B_1^-. The 1st edition has more discussion and results on page 96.

2.1.11 Non-Uniformity correction

Early workers found that images were often visibly non-uniform, and much effort was devoted to measuring and correcting non-uniformity.[30] Image non-uniformity (NU) arises from three sources: (1) Transmit field (B_1^+) non-uniformity gives a magnetisation M_{xy} NU that also depends on T_1 (unless a fully relaxed sequence is used). Receive (B_1^-) NU depends on both; (2) the receive coil characteristics; and on (3) the head electromagnetic characteristics. Thus a simple image of a uniform object, or simple smoothing, cannot correct for these three factors in a quantitative way. *Transmit non-uniformity* is usually minimised by using the body coil, and measured if necessary (see Section 2.1.7).

[29] As the wavelength of the B_1 excitation decreases, the approximation of a quasistatic excitation field becomes less valid (from Sled and Pike, 1998). At 3T, the wavelength in water is 260 mm (see Chapter 3, Section 3.5.2).

[30] See first edition, p 37.

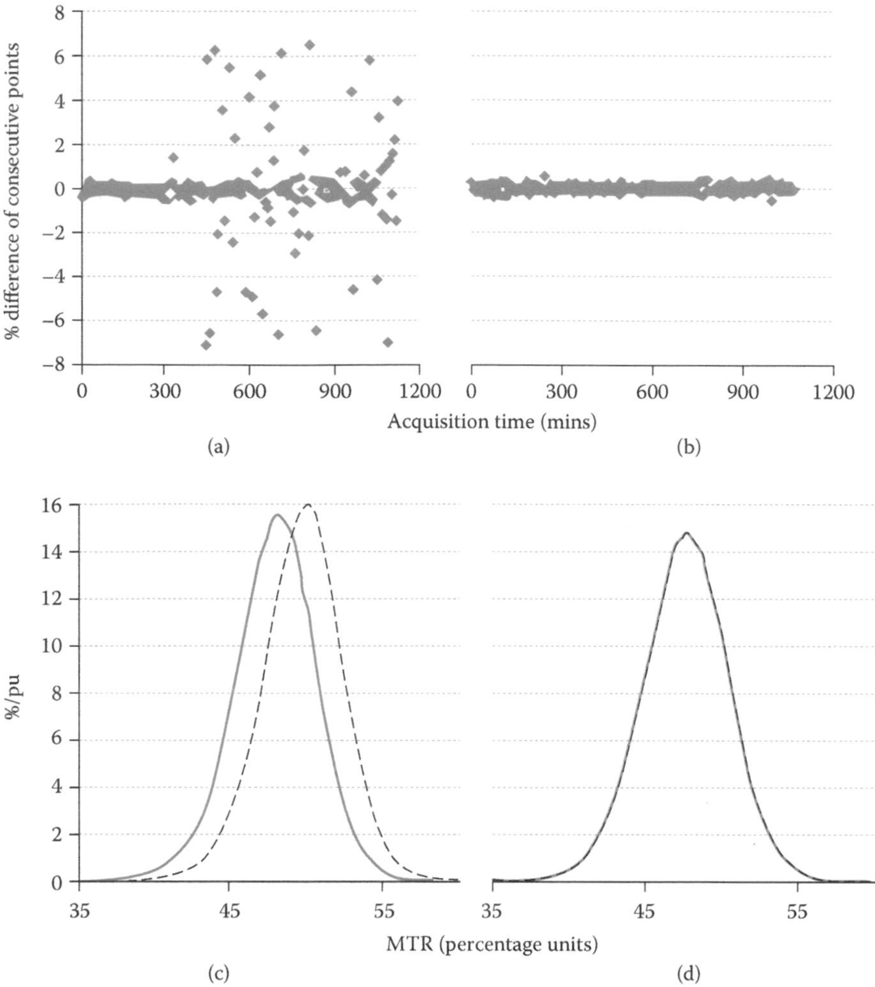

FIGURE 2.8 Unsuspected scanner instability – an invisible problem. MTR histograms showed a large within-subject variation (c). Repeated scanning of a phantom overnight showed large random variation (a). After changing transmitter boards, the scanner was stable (b) and MTR histograms were reproducible (d). (Data from NG Dowell, originally presented in (Haynes *et al.*, 2010) (From Haynes, B.I., *et al.*, Measuring scan-rescan reliability in quantitative brain imaging reveals instability in an apparently healthy imager and improves statistical power in a clinical study. ISMRM annual scientific meeting, Stockholm, p. 2999, 2010.)

2.1.12 Scanner stability

Despite having followed all good practice, image data can be unstable for unknown reasons. Repeated imaging of a phantom may show a short-term variation in signal that is greater than that predicted by image noise (Weisskoff 1996), in addition to long-term drift (see Figure 2.8). Such instability can have two major effects: (1) it degrades reproducibility (i.e. instrumental SD [ISD] – see Chapter 3, Section 3.3.2.1) and (2) in a DCE series acquisition, it introduces time-dependent variation, which may mask a subtle signal enhancement caused by a genuine T_1 change. Thus if a parameter (e.g. T_1) shows an unexpectedly large[31] within-subject variation, it is worth carrying out repeated imaging and looking at the variation of the raw image data. If transmitter instability is suspected, then imaging at the *Ernst angle*[32] ought to produce a stable signal. If receiver instability is suspected, then slowly varying changes would affect the signals at all FAs equally.

2.2 Image analysis, statistics and classification

2.2.1 Types of image analysis

Here an introduction to image analysis concepts is given; a higher-level viewpoint is given in Chapter 17. There are three

[31] The effect of image noise σ on repeated measurements of an ROI mean is to give variation with a standard deviation equal to the standard error of the mean of the ROI (i.e. sem = $\sigma/\sqrt{[\text{no. of pixels}]}$).

[32] In a spoilt gradient echo, the signal is maximal and independent of FA θ at $\cos(\theta_E) = \exp(-TR/T_1)$, where θ_E is the Ernst angle; e.g. for T_1 = 800 ms, TR = 10 ms, then $\theta_E = 9°$.

main ways of extracting relevant image intensities from a set of images that may cover many slices, several tissue parameters and many subjects: ROI's, histograms and group mapping.

Before analysis, the image dataset may be spatially registered. Images from a single subject may be registered, to reduce the effects of movement during the examination. Images from different subjects may be registered to a standard space.[33] Registration can produce subtle changes in image intensity that may thwart attempts at quantification (e.g. in a DCE series, small changes in intensity can sometimes be observed that are caused by the registration process, not by changes in T_1).

2.2.1.1 Region of interest analysis

The study is focused on a particular part or parts of the brain, for example visible lesions or large volumes of normal-appearing white matter, where intensities are to be measured. One or several ROIs are drawn for each subject. Regions can be circular, oval, square, rectangular or irregular. Regions may be defined in a single slice or extend over several slices (then the set is a volume of interest [VOI]). They are often created using a semi-automatic technique, which speeds up the process and improves the reproducibility. ROI size is a compromise between reducing noise (which favours large ROIs) and reducing partial volume error (which favours small ROIs). Alternatively, if the image datasets are in standard (stereotactic) space, then a number of standard VOIs will be available. The process of creating the ROIs takes some time to learn, and different observers will develop different approaches. Inter-observer[34] variation can be reduced by carefully defining the procedure to be used. This may include factors such as how the image is to be displayed, and a detailed description of what anatomical cues are to be used in positioning the ROI. Usually the intra-observer variation is lower, and many studies accept that a single observer should be used to analyse the whole dataset. Even for a single observer, the analysis should be repeated after a few days to ensure the reproducibility is reasonable. A formal measurement of reproducibility can be undertaken (see Chapter 3, Section 3.3).

Unbiased ROI generation: There may be multiple MR images or maps, for example conventional MRI (showing lesions in PD- or T_2-weighted images) and a MTR map. The appearance of lesions may be different on the conventional MRI and the map. If the MTR values of lesions are to be measured and tested, the ROIs should be defined on the conventional MR images (after spatial registration), then transferred to the maps. If the ROIs were defined directly on the maps, the map intensity would influence where the ROI boundary was placed. ROIs tend to be attracted to locations of abnormal intensity (as a result of the process of their creation, where an observer tends to draw around distinct objects). Thus any conclusions about map values in the lesions would be biased, since these values had been used to define which pixels would be included in the region. If serial measurements are made, fixed ROIs should be used for each time point, if conclusions are to be drawn about changes in a parameter value over time.[35] Large regions of most of the normal-appearing white or grey matter have been generated using T_1 maps (Parkes and Tofts 2002) or fractional diffusion anisotropy (Cercignani *et al.*, 2001).

In studies of diffuse disease that affects large parts of the brain, instead of creating large regions to study this, two other approaches are available: histograms and voxel-based group mapping.

2.2.1.2 Histogram analysis

A solution to the problem of ROI placement, and possible bias arising from this process, is to test all of a tissue type (e.g. white matter). This is particularly appropriate for diseases where the biological effects are diffuse and widespread. Histograms do have the disadvantage that localisation information has been lost, and if disease only affects part of the region sensitivity will be reduced by pooling data from the whole region.

Histograms from different centres have sometimes varied; by standardising their generation, multicentre studies are possible (Tofts *et al.*, 2006). The bin width should be chosen to be small enough to capture any fine structure, yet large enough to not display statistical fluctuations caused by a small number of pixels in the bin (typically 5 ms for T_1). The bin should be labelled by its centre value (not the left- or right-hand edge). A normalised bin amplitude should be calculated as follows: find the percentage of the total pixels in that bin, then divide by the bin width. The total area under the histogram curve is then 100%, regardless of bin width (Figure 2.9).[36]

Histogram analysis by Principle Components Analysis (PCA) and linear discriminant analysis can be very powerful; MTR histograms have predicted clinical score and also separated disease subtypes (Dehmeshki *et al.*, 2001, 2002b) – see Figure 2.10.

2.2.1.3 Voxel-based group mapping – beyond rois and histograms

Analysis of group-mapped images provides a way of combining the ability of ROIs to be spatially specific (and thus sensitive) with the ability of histograms to be unbiased. In essence, a complete set of ROIs is automatically generated at all locations across the whole brain, without any bias in where they are located. The image datasets are first spatially normalised to all lie in the same space. Appropriate statistical tests are then carried out on all the ROIs. Further information is in Chapter 17.

[33] The most common is 'MNI-space' (MNI = Montreal Neurological Institute).

[34] *Inter-observer* means 'between observer', that is the difference between measurements made by *different* observers on the same image data. *Intra-observer* means 'within observer', that is the difference if the *same* observer repeats the measurement.

[35] A recent example showed that, after treatment of tumours, regions of Gd enhancement unexpectedly retained the same mean intensity. On being probed, the author disclosed that the ROIs had been drawn and redrawn on the Gd images and had become smaller after treatment. Thus fixed ROIs would have shown a reduction in enhancement.

[36] Examples are shown in the first edition, Chapter 18, along with examples of classification of clinical data.

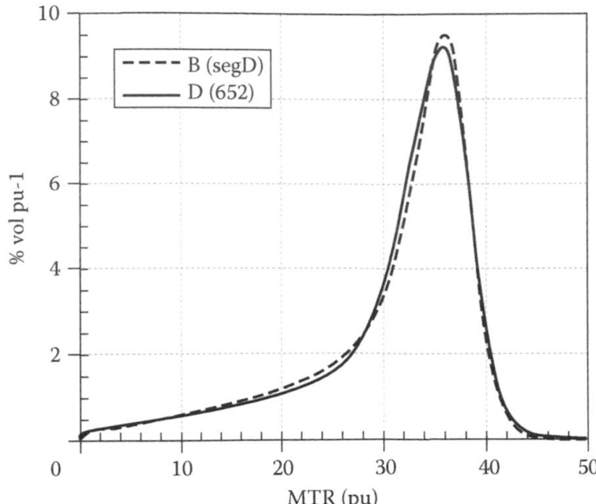

FIGURE 2.9 Matching MTR group histograms from two centres with 1.5T scanners from different manufacturers. By using body coil excitation and standardised histogram generation, inter-centre differences were eliminated. (From Tofts, P.S., *et al.*, *Magma*, 19(4), 209–222, 2006.)

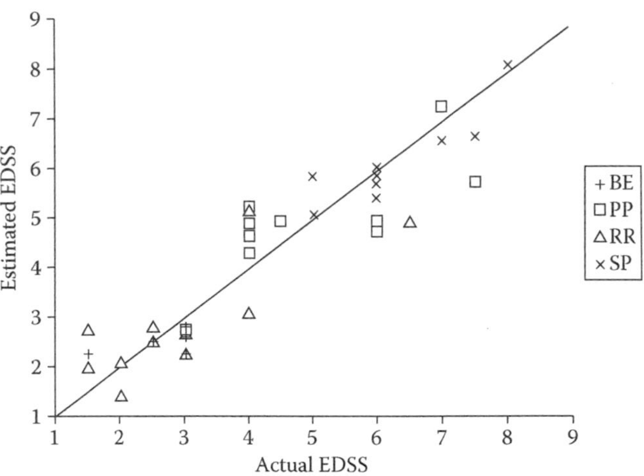

FIGURE 2.10 Correlation of Expanded Disability Status Scale (EDSS, a clinical score used in multiple sclerosis, MS) with principle components enables an estimated EDSS to be calculated for each subgroup of people with MS, solely from the MTR histogram and knowledge of the subgroup. These are usually within one point of the actual clinically measured EDSS value. Compared with conventional features, PCA gives better correlation coefficients. (From Dehmeshki, J., *et al.*, *Magn. Reson. Med.*, 46(3), 600–609, 2001.)

2.2.2 Types of statistical analysis

Statistical analysis of the measurements is an increasingly complex procedure. Involvement with statisticians at an early stage in study design and analysis is advisable. Failure to be aware of statistical pitfalls can lead to embarrassing rejections by journal referees; sometimes re-analysis can address the problems, sometimes data collection is fatally flawed. Clinical datasets can be very expensive to acquire (both financially and in terms of the effort by the patients and controls); this drives an imperative to make the best possible use of clinical data by using appropriate analysis techniques (see Chapter 1, Table 1.2).

In this section the basic concepts relevant to statistical analysis of MR data are summarised; however for full treatment the reader should go to the books (see Table 2.2). Suitable software packages are SPSS (originally 'Statistical Package for the Social Sciences' but now more general), SAS and STATA. These have comprehensive manuals associated with them, and the software is often available through academic sources. With such packages, 'clickity-click' analysis can be a hazard; a few mouse clicks can unleash a sophisticated analysis of a large dataset that the user does not understand and may misinterpret.

2.2.2.1 Group comparisons – t-test

The simplest test of a MR parameter is to compare measurements in two groups (typically diseased and 'normal'). Many MR parameters may have been measured. To test their usefulness, it is tempting to carry out multiple t-tests, for example to see if any of the parameters differ between clinical groups. If 20 tests are carried out at the $p = 0.05$ significance level, on average one test will come out positive by chance (this is known as a *type I error*). Thus the results of multiple comparisons must be treated with caution. The *Bonferroni correction* for multiple comparisons (Bland and Altman 1995) allow for this by suggesting that the appropriate p-value is the value that would have been used for a single test (e.g. $p = 0.05$), divided by the number of comparisons (e.g. 20). A much-reduced p-value (e.g. 0.002) is then used, and the chances of a type I error are reduced (in this example to 0.04). The Bonferroni correction can be unnecessarily cautious, missing an effect that is present (i.e. a *type II error*). If the outcome variables being tested are correlated, then the reduction of the p-value by the number of tests is too extreme. Conversely, if several of the tests show significance (which does imply correlation between the tested variables), then the probability of this occurring by chance are much lower than the probability of just one occurring by chance.

A useful distinction can be made between two kinds of study. A *fishing expedition* looks at many parameters, tests them at $p = 0.05$, accepting that some type I errors will occur, and uses this to gain insight or guide further studies. A strict *hypothesis-driven study* sets up the hypothesis *before-analysis*, makes only one test and is thus able to control type I and type II errors better. Thus a fishing expedition might be used to set up a hypothesis-driven study, which must be carried out on *separate data*. Alternatively, the data could be divided into two parts, the first used for fishing and the second for a strict test (although the power of the study would be reduced by the smaller sample sizes).

Negative results in a group comparison: A negative result has two possible explanations: firstly, the genuine biological spread in each group may be too large to pick out a significant different between them; secondly, the effect of measurement error may have broadened the spread in the groups, beyond its genuine biological value, enough to obscure genuine biological difference (see Chapter 3, Figure 3.5), a kind of 'false-negative'. Thus a negative result can be

TABLE 2.2 Statistics books

Title	Authors	Date Published	Number of Pages	Description
Statistical Methods in Medical Research	Peter Armitage, Geoffrey Berry, JNS Matthews,	2001	832	Classic from statisticians, 4th edition, hardback.
Practical Statistics for Medical Research	Douglas Altman	2000	254	2nd edition, paperback.
An Introduction to Medical Statistics	Martin Bland	2015	448	4th edition, paperback.
Essential Medical Statistics	B Kirkwood, J Sterne	2003	512	
Medical Statistics: A Textbook for the Health Sciences	MJ Campbell, DJ Machin, SJ Walters	2007	344	4th edition, paperback.
Health Measurement Scales: A Practical Guide to Their Development and Use	DL Streiner, GR Norman, J Cairney	2014	416	5th edition, good on clinical scales, paperback.
Analyzing Multivariate Data	J Lattin, JD Carroll, PE Green	2006	556	High level hardback, covers principal components analysis, analysis of variance, clustering, discriminant analysis and much more.
An Introduction to Error Analysis: The Study of Uncertainties in Physical Measurements	JR Taylor	1997	488	A clear description of eternal truths, from a physics point of view.

the consequence of poor measurement technique (high ISD – see Chapter 3, Section 3.3.2.1). Another centre (with better technique) might succeed in showing a group difference in a similar group of subjects. The original centre might succeed with a larger sample (i.e. the study was underpowered). Failure to observe a difference does not mean that none exists.

In the case of a negative result, insight can be obtained in two ways: first, estimates of measurement error and within-group variance should be made (see Chapter 3, Section 3). This enables the intraclass correlation coefficient (ICC) to be calculated; a good ICC means that the groups are genuinely indistinguishable; a poor ICC means that the failure to distinguish may be caused by poor instrumentation. Second, the confidence limits on the group means, and the minimum detectable group difference, should be reported. Other workers can then judge whether improved technique (i.e. reduced measurement error) might enable them to obtain a positive result.

> *The excessive use of hypothesis testing at the expense of more informative approaches ... is an unsatisfactory way of assessing ... findings from medical studies. We prefer the use of confidence intervals ....* (Altman et al., 2008).

Positive results: If a positive result is obtained, the confidence limits on group means and group difference should be given; then other groups can estimate whether their measurement errors are low enough to repeat the positive observation. False positive results can be obtained if there is another (confounding) factor that differs between the groups. This could be the time of scanning (if one group is scanned before a change in measurement procedure, and the other after the change) or other uncontrolled variables, for example age, gender, lifestyle differences or even head size.[37]

[37] A study comparing MS patients and controls found a difference that was actually caused by head size (this gave a different spatial normalisation and hence intensity for the two groups).

2.2.2.2 Correlation with clinical score

In many studies, MR parameters are tested for correlation with a clinical measure (in MS, often the Expanded Disability Status Scale, EDSS), in an attempt to investigate or demonstrate their clinical utility (or lack of it). A parameter with a high correlation coefficient is thought to be a good candidate for a surrogate MR marker of the disease in a clinical trial. In MS, low correlation coefficients r are reported (typically 0.3–0.6, sometimes 0.8; Dehmeshki *et al.*, 2001). Significance values p are also given. Correlation values are attenuated by the imperfect reliability (i.e. scatter) in the MR and clinical scores (thus even if the two measures were intrinsically perfectly correlated, the correlation plot would show scatter about the line describing this relationship). Correlation does not imply causality, only association, and the association may be weakened by the introduction of another causative factor (such as treatment). Thus a good correlation between an MR parameter and a clinical score does not necessarily imply the parameter is a good MR surrogate in a treatment trial; more evidence is needed of a direct causal relationship between the biological changes that happen in the disease and the MR parameter.

An alternative way to think of the linear regression implied in correlation is: how well can the MR parameter predict the current value of the clinical parameter? Thus a high correlation implies that the MR parameter provides a good estimate of the current clinical status (Dehmeshki *et al.*, 2001)– see Figure 2.10. This establishes the relevance of the MR measure, although the ultimate goal is to predict future clinical status. The fraction of the variance in the clinical score explained by the MR parameter is r^2, and this is a useful interpretation of r.

Age and gender correlations with the principle variables can cause problems; they should be included in the correlation, as covariates (Chard *et al.*, 2002).

2.2.2.3 Clinical scores

Clinical status can be quantified using scores. For example in MS the *EDSS* is used to measure disability. In tumours a grading system is used, based on histology of biopsy samples. In psychiatric illness, a battery of psychometric tests, including cognitive and emotional, are in use. Newly introduced MR parameters have often been judged by how well they correlate with existing clinical scores; intensive effort has gone into characterising and improving the performance of the MR parameter. In turn it has been accepted that the same intensive study should go into the clinical scores, and in fact they do have some serious shortcomings. EDSS is non-linear; mixes impairments of ambulation, fine motor skill and cognition; and has limited reproducibility (Hobart *et al.*, 2000).

Scores that are more appropriate are being designed. In MS the *functional composite score* (MSFC score) is increasingly popular (Fischer *et al.*, 1999; Cohen *et al.*, 2000). It consists of three components, which measure different aspects of impairment. Leg function and ambulation are measured by the timed 25-foot walk,[38] arm function by the nine-hole peg test and cognition by the Paced Auditory Serial Addition Test. Correlations of MR parameters with these individual components may be superior, since they reflect different impairments, which may occur at different times in the evolution of the disease and originate from different locations in the CNS. There is still controversy over how reliable these measurements are, since there may be learning effects in the subjects. A correlation plot may show an approximately linear dependence of a MR parameter on EDSS, but a line will often not go through the normal point (normal MR value, EDSS = 0), possibly because subclinical changes happen to the MR parameter, before any clinical disability is apparent.

Tumour grade scoring based on sampling tissue is vulnerable to missing high-grade tissue in a heterogeneous tumour, and multicentre studies can show appreciable differences in grading between pathologists, which can limit how well MRS classifiers can work. The psychological tests are generally better, because there is a longer-standing tradition of test design in that field, although learning, floor and ceiling effects and tiredness are still important limitations.

In view of these shortcomings in clinical scores, and the complex relationship between biological changes and the ensuing clinical changes, a failure of MR parameters to closely predict these scores is hardly surprising. A more realistic test may be to look at how well the MR correlates with the biology and with scores of simple human functions.

2.2.2.4 Classification of individual subjects and receiver operating characteristic curves

If a measurement is performing well in separating two groups of subjects (Section 2.2.2.1), then its performance on individual subjects is worth investigating. In computer science methodology, the MR measurements may be seen as an example of a *classifier*. A classifier is a software tool for deciding which class a number of subjects belong to, based on measurements made on each subject. Classification in its simplest case only attempts to choose between two classes (binary classification). A linear discriminant is often constructed, and a threshold used to assign the class (see e.g. Dehmeshki *et al.*, 2002a). Such binary classification techniques have been used in spectroscopy to classify tumours into several types (Tate *et al.*, 1998). The choice of threshold in a classifier is crucial in balancing false positive and false negative errors, and receiver operating characteristic (ROC) formalism is an ideal way to view and optimise this balance. ROC analysis is also used to characterise how well a radiologist can identify a lesion on a difficult background.

Receiver operating characteristic (ROC) curves (Zweig and Campbell 1993; Altman and Bland 1994b; Armitage *et al.*, 2001; Dendy and Heaton 2002; Huo *et al.*, 2002) originate from studies to characterise screens used by radar operators, and specifically recognised that the number of objects reported would depend on how a particular operator was making their decisions (see Table 2.3). A low threshold of abnormality, that is reporting all objects that could possibly be real, would result in a large number of positive decisions. The proportion of actual objects detected (i.e. true positives) would be large, but at the expense of many false positives (which are in fact noise on the screen). A higher threshold (only reporting objects judged to be certainly real) would result in fewer positive decisions, more missed objects (false negatives) and fewer false positives. Thus the choice of decision threshold allows true positives to be traded off against false positives, and a particular threshold, corresponding to a particular point on the curve, can be chosen according to the relative benefit of true decisions versus the cost of false decisions. For example in a screening program, false negatives are expensive, since missing a tumour may result in death, and false positives are also costly (although less so), since they produce unnecessary worry in the subjects.

In the radar screen context, an ROC curve can be generated by asking the observer to give a score with each object reported – for example 0: object absent (corresponding to a low threshold), 1: object possibly present, 2: may be present, 3: probably present, 4: almost certainly present, 5: certainly present (corresponding to a high threshold). The scores can also be defined by their anticipated probabilities (e.g. 0: <10% of being present, 1: 10%–30% of being present, etc.). The probabilities need not be correct; they just allow the observer to behave consistently. Combinations of scores of True Positive Rate (TPR) and False Positive Rate (FPR) are summed to give points in the ROC space corresponding to different thresholds

TABLE 2.3 Decision matrix, in ROC formalism

	Object reported not to be present (N)	Object reported to be present (P)
Object absent (N)	True negative (TN)	False positive (FP)
Object present (P)	False negative (FN)	True positive (TP)

Note: The test is to report whether an object is present or not; it could equally well be to determine whether a disease or a lesion is present or not.

[38] Often replaced by the 10 m walk in Europe.

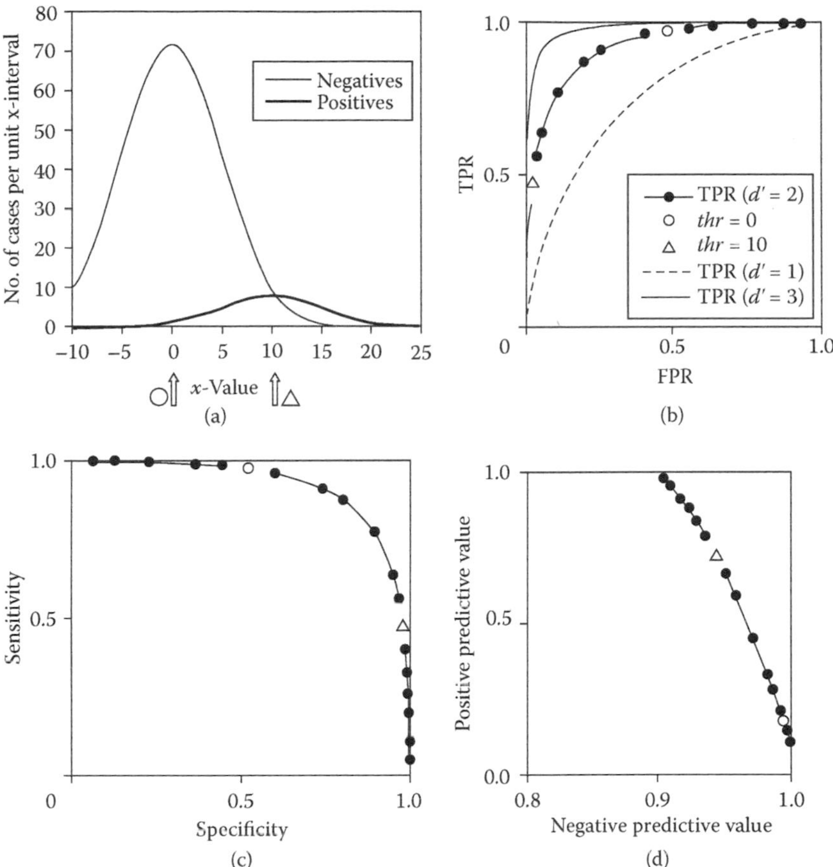

FIGURE 2.11 A receiver operating characteristic (ROC) computer simulation: (a) negative (N) and positive (P) normal distributions. There are 1000 cases, a prevalence of 1 in 10 (so 100 positive cases). Mean values are $x = 0$ and $x = 10$. Standard deviation is 5 for both groups (b) ROC curve (labelled $d' = 2$, meaning that the distributions are separated by 2 SD). Points from thresholds at the centre of each distribution are shown in parts b, c and d (*thr* = 0, centre of negatives; *thr* = 10, centre of positives). (Cases to the right-hand side of the threshold are reported positive. Thus the part of the negative tail to the right of the threshold contributes false positives, which decrease with threshold value. Conversely, the part of the positive tail to the left of the threshold gives false negatives; as the threshold is increased these increase, reducing the true positive rate TPR.) Curves for other separations ($d' = 1$, i.e. 5 units, and $d' = 3$, i.e. 15 units) are shown. The decision criterion to generate the N and P results is that an x-value greater than the threshold indicates that the sample comes from the P distribution. (c) Sensitivity versus specificity plot, which is the left–right reflection of the ROC curve. (d) Positive and negative predictive values both depend on the threshold (and also on the prevalence).

(see Figure 2.11).[39] More generally, an ROC curve can be generated from any model that predicts binary status with a certain probability, depending on one or more predictor variables, such as MR parameters. ROC curves can be fitted to analytic functions, to obtain a measure that enables ROC curves to be compared and to provide an element of smoothing (Constable *et al.*, 1995; Sorenson and Wang 1996).

The terms *sensitivity, specificity, positive predictive value, negative predictive value, accuracy* and *prevalence* are used in the context of how well a test performs (Altman and Bland 1994a, 1994c; Dalton *et al.*, 2002). These can be defined in terms of the number of true and false positive and negative reports (see Table 2.4).

[39] Thus Threshold 1 corresponds to objects detected with scores 1–5 (i.e. all objects), Threshold 2 is object scores 2–5 (i.e. certainty of 'probably' or more), etc., and each threshold gives a point on the ROC curve.

The false positive rate, equal to the fraction of the negative distribution that is to the right of the threshold (see Figure 2.11) is known in statistics as the significance, the *α-value, p-value* or the chance of a *type I error.* The false negative rate, equal to the fraction of the positive distribution that is to the left of the threshold, is the β-value, q-value, or the chance of a *type II error* (Haacke *et al.*, 1999; Armitage *et al.*, 2001). 1-β is the sensitivity or true positive rate (TPR) and is known as the power of the test (the probability of detecting a true association where one exists). Note that parameters such as the sensitivity are all estimated from samples and therefore should be reported with confidence limits.

2.2.2.5 Multiparametric analysis

Multiparametric studies can be powerful. Several MR parameters are measured and combined in such a way to produce a better predictor of biological change or clinical outcome. If several

TABLE 2.4 Radiological terms summarising the performance of a test

Term	Formal Definition	Formula	Comment	Dependent on Prevalence?
Sensitivity	The probability of the test finding the disease amongst those who have the disease	TP/(TP + FN)	Sensitivity characterises the false negatives.	No
Specificity	The probability of the test finding no disease amongst those who do not have the disease	TN/(TN + FP)	Specificity characterises the false positives.	No
Positive predictive value	The fraction of people with a positive test result who do have the disease	TP/(TP + FP)	Fraction of *positive* reports that are correct.	Yes
Negative predictive value	The fraction of people with a negative test result who do not have the disease	TN/(TN + FN)	Fraction of *negative* reports that are correct.	Yes
Accuracy	The fraction of test results that are correct	(TP + TN)/(TP + TN + FP + FN)	Fraction of *all reports* that are correct.	Yes
Prevalence	The fraction of the sample that has the disease	(TP + FN)/(TP + TN + FP + FN)	Distinguish from incidence.[a]	–
True positive rate (TPR)	Fraction of positive cases that are detected	TP/(TP + FN)	Equals sensitivity. The fractional area under the positive curve that is to the right of the threshold.	No
False positive rate (FPR)	Fraction of negative cases that are reported positive	FP/(TN + FP)	Equals 1-specificity. The fractional area under the negative curve that is to the right of the threshold.	No

Note: In this case the test is to find out whether subjects have a disease or not. Negative indicates that they do not, positive that they do. Sensitivity and specificity define the performance of the test (regardless of prevalence), whilst predictive value and accuracy depend on prevalence (i.e. what fraction of the sample has the disease). Thus the latter quantities are very different for an asymptomatic screened population and a symptomatic hospital population.

[a] Thus the prevalence is the fraction of people that have a particular disease at any one time. The *incidence* is the number of *new cases* of disease that arise in a given period (usually 1 year) for a given size of population.

MR parameters all correlate with clinical scores and do not correlate strongly with each other, then there is a case for measuring them all. A plea in the journal *Neurology*, on the subject of MRI techniques to monitor MS evolution (Filippi and Grossman 2002), asked that

1. *Metrics from magnetisation transfer MRI, diffusion-weighted MRI and proton MRS should be implemented to obtain reliable in vivo quantification of MS pathology.*
2. *Multiparametric MRI should be used in all possible clinical circumstances and trials.*
3. *Reproducible quantitative MR measures should ideally be used for the assessment of patients and are essential for trials.*

Diagnosis at an early stage of a disease with a long and variable course (e.g. multiple sclerosis or Alzheimer's disease) and prediction of disease progression can also be viewed as a multiparametric problem. Early symptoms, MRI and other biochemical data are available; their combination can be optimised to provide the best prediction of future clinical status, measured according to various clinical scores. The optimisation can be driven by criteria such as sensitivity, specificity and predictive value (see Table 2.4).

The effects of disease are similarly multidimensional and need several scores to properly characterise them, covering both clinical symptoms and measurable biological changes. To assemble multiparametric MR data, features can be extracted from histograms of each MR parameter and from lesion values. Statistical techniques such as multiple linear discriminant analysis and cluster analysis are appropriate (Tintore *et al.*, 2001). Multiparametric analysis is discussed further in Chapter 18.

References

Alecci M, Collins CM, Smith MB, Jezzard P. Radio frequency magnetic field mapping of a 3 Tesla birdcage coil: experimental and theoretical dependence on sample properties. Magn Reson Med 2001; 46(2): 379–85.

Altman DG, Bland JM. Diagnostic tests 2: predictive values. BMJ 1994a; 309(6947): 102.

Altman DG, Bland JM. Diagnostic tests 3: receiver operating characteristic plots. BMJ 1994b; 309(6948): 188.

Altman DG, Bland JM. Diagnostic tests. 1: sensitivity and specificity. BMJ 1994c; 308(6943): 1552.

Altman DG, Machin D, Bryant TN, Gardner MJ. Statistics with confidence: confidence intervals and statistical guidelines (2nd ed). BMJ books, London, pp. 24; 2000.

Andersen AH. On the Rician distribution of noisy MRI data. Magn Reson Med 1996; 36(2): 331–3.

Armitage P, Matthews JNS, Berry G. *Statistical Methods in Medical Research.* Blackwell; 2001.

Barker GJ, Simmons A, Arridge SR, Tofts PS. A simple method for investigating the effects of non-uniformity of radiofrequency transmission and radiofrequency reception in MRI. Br J Radiol 1998; 71(841): 59–67.

Bland JM, Altman DG. Multiple significance tests: the Bonferroni method. BMJ 1995; 310(6973): 170.

Bottomley PA, Andrew ER. RF magnetic field penetration, phase shift and power dissipation in biological tissue: implications for NMR imaging. Phys Med Biol 1978; 23(4): 630–43.

Boudreau M, Tardif CL, Stikov N, Sled JG, Lee W, Pike GB. B1 mapping for bias-correction in quantitative T1 imaging of the brain at 3T using standard pulse sequences. J Magn Reson Imaging. 2017; 46:1673–1682.

Brown RW, Cheng N, Haacke EM, Thompson MR, Venkatesan R. *Magnetic Resonance Imaging: Physical Principles and Sequence Design* (2nd edition): Wiley-Blackwell, New Jersey; 2014.

Cercignani M, Inglese M, Siger-Zajdel M, Filippi M. Segmenting brain white matter, gray matter and cerebro-spinal fluid using diffusion tensor-MRI derived indices. Magn Reson Imaging 2001; 19(9): 1167–72.

Chard DT, Griffin CM, Parker GJ, Kapoor R, Thompson AJ, Miller DH. Brain atrophy in clinically early relapsing-remitting multiple sclerosis. Brain 2002; 125(Pt 2): 327–37.

Cohen JA, Fischer JS, Bolibrush DM, Jak AJ, Kniker JE, Mertz LA, et al. Intrarater and interrater reliability of the MS functional composite outcome measure. Neurology 2000; 54(4): 802–6.

Collins CM, Liu W, Swift BJ, Smith MB. Combination of optimized transmit arrays and some receive array reconstruction methods can yield homogeneous images at very high frequencies. Magn Reson Med 2005; 54(6): 1327–32.

Collins CM, Smith MB. Signal-to-noise ratio and absorbed power as functions of main magnetic field strength, and definition of "90 degrees" RF pulse for the head in the birdcage coil. Magn Reson Med 2001; 45(4): 684–91.

Collins CM, Wang Z. Calculation of radiofrequency electromagnetic fields and their effects in MRI of human subjects. Magn Reson Med 2011; 65(5): 1470–82.

Collins CM, Yang QX, Wang JH, Zhang X, Liu H, Michaeli S, et al. Different excitation and reception distributions with a single-loop transmit-receive surface coil near a head-sized spherical phantom at 300 MHz. Magn Reson Med 2002; 47(5): 1026–8.

Constable RT, Skudlarski P, Gore JC. An ROC approach for evaluating functional brain MR imaging and postprocessing protocols. Magn Reson Med 1995; 34(1): 57–64.

Cunningham CH, Pauly JM, Nayak KS. Saturated double-angle method for rapid B1+ mapping. Magn Reson Med 2006; 55(6): 1326–33.

Dalton CM, Brex PA, Miszkiel KA, Hickman SJ, MacManus DG, Plant GT, et al. Application of the new McDonald criteria to patients with clinically isolated syndromes suggestive of multiple sclerosis. AnnNeurol 2002; 52(1): 47–53.

Dehmeshki J, Barker GJ, Tofts PS. Classification of disease subgroup and correlation with disease severity using magnetic resonance imaging whole-brain histograms: application to magnetization transfer ratios and multiple sclerosis. IEEE Trans Med Imaging 2002a; 21(4): 320–31.

Dehmeshki J, Ruto AC, Arridge S, Silver NC, Miller DH, Tofts PS. Analysis of MTR histograms in multiple sclerosis using principal components and multiple discriminant analysis. Magn Reson Med 2001; 46(3): 600–9.

Dehmeshki J, Van Buchem MA, Bosma GP, Huizinga TW, Tofts PS. Systemic lupus erythematosus: diagnostic application of magnetization transfer ratio histograms in patients with neuropsychiatric symptoms--initial results. Radiology 2002b; 222(3): 722–8.

Dendy PP, Heaton B. *Physics for Diagnostic Radiology*. Institute of Physics, London; 2002.

Dowell NG, Tofts PS. Fast, accurate, and precise mapping of the RF field in vivo using the 180 degrees signal null. Magn Reson Med 2007; 58(3): 622–30.

Duan Q, van Gelderen P, Duyn J. Improved Bloch-Siegert based B1 mapping by reducing off-resonance shift. NMR Biomed 2013; 26(9): 1070–8.

Edelstein WA, Bottomley PA, Pfeifer LM. A signal-to-noise calibration procedure for NMR imaging systems. Med Phys 1984; 11(2): 180–5.

Fernandez-Seara MA, Song HK, Wehrli FW. Trabecular bone volume fraction mapping by low-resolution MRI. Magn Reson Med 2001; 46(1): 103–13.

Filippi M, Grossman RI. MRI techniques to monitor MS evolution: the present and the future. Neurology 2002; 58(8): 1147–53.

Fischer JS, Rudick RA, Cutter GR, Reingold SC. The Multiple Sclerosis Functional Composite Measure (MSFC): an integrated approach to MS clinical outcome assessment. National MS Society Clinical Outcomes Assessment Task Force. Mult Scler 1999; 5(4): 244–50.

Goerner FL, Clarke GD. Measuring signal-to-noise ratio in partially parallel imaging MRI. Med Phys 2011; 38(9): 5049–57.

Gudbjartsson H, Patz S. The Rician distribution of noisy MRI data. Magn Reson Med 1995; 34(6): 910–4.

Gutteridge S, Ramanathan C, Bowtell R. Mapping the absolute value of M0 using dipolar field effects. Magn Reson Med 2002; 47(5): 871–9.

Haacke EM, Brown RW, Thompson MR, Venkatesan R. *Magnetic Resonance Imaging*; 1999. Physical principles and sequence design.

Haynes BI, Dowell NG, Tofts PS. Measuring scan-rescan reliability in quantitative brain imaging reveals instability in an apparently healthy imager and improves statistical power in a clinical study. ISMRM annual scientific meeting; 2010; Stockholm; 2010. p. 2999.

Henkelman RM. Measurement of signal intensities in the presence of noise in MR images. Med Phys 1985; 12(2): 232–3.

Henkelman RM. Erratum: measurement of signal intensities in the presence of noise [Med. Phys. 12, 232 (1985)]. Med Phys 1986; 13(4): 544.

Hetherington HP, Chu WJ, Gonen O, Pan JW. Robust fully automated shimming of the human brain for high-field ^{1}H spectroscopic imaging. Magn Reson Med 2006; 56(1): 26–33.

Hobart J, Freeman J, Thompson A. Kurtzke scales revisited: the application of psychometric methods to clinical intuition. Brain 2000; 123 (Pt 5): 1027–40.

Hoult DI. The NMR receiver: a description and analysis of design. Progress in NMR Spectroscopy 1978; 12: 41–77.

Hoult DI. The principle of reciprocity in signal strength calculations – a mathematical guide. Concepts in Magnetic Resonance 2000; 12: 173–87.

Hoult DI, Richards RE. The signal-to-noise ratio of the nuclear magnetic resonance experiment. J Magn Reson 1976; 24: 71–85.

Huo Z, Giger ML, Vyborny CJ, Metz CE. Breast cancer: effectiveness of computer-aided diagnosis observer study with independent database of mammograms. Radiology 2002; 224(2): 560–8.

Hurley SA, Yarnykh VL, Johnson KM, Field AS, Alexander AL, Samsonov AA. Simultaneous variable flip angle-actual flip angle imaging method for improved accuracy and precision of three-dimensional T1 and B1 measurements. Magn Reson Med 2012; 68(1): 54–64.

Hutton C, Bork A, Josephs O, Deichmann R, Ashburner J, Turner R. Image distortion correction in fMRI: a quantitative evaluation. NeuroImage 2002; 16(1): 217–40.

Ibrahim TS. Analytical approach to the MR signal. Magn Reson Med 2005; 54(3): 677–82.

Ibrahim TS, Lee R, Baertlein BA, Abduljalil AM, Zhu H, Robitaille PM. Effect of RF coil excitation on field inhomogeneity at ultra high fields: a field optimized TEM resonator. Magn Reson Imaging 2001; 19(10): 1339–47.

Insko EK, Bolinger L. Mapping the radiofrequency field. J Magn Reson series A 1993; 103: 82–5.

Jezzard P. Physical basis of spatial distortions in Magnetic Resonance Images. In: Isaac B, editor. *Handbook of Medical Imaging*: Academic Press, Cambridge, MA; 2002.

Jin J, Liu F, Zuo Z, Xue R, Li M, Li Y, et al. Inverse field-based approach for simultaneous B_1 mapping at high fields - a phantom based study. J Magn Reson 2012; 217: 27–35.

Karlsen OT, Verhagen R, Bovee WM. Parameter estimation from Rician-distributed data sets using a maximum likelihood estimator: application to T1 and perfusion measurements. Magn Reson Med 1999; 41(3): 614–23.

Lemieux L, Barker GJ. Measurement of small inter-scan fluctuations in voxel dimensions in magnetic resonance images using registration. Med Phys 1998; 25(6): 1049–54.

Lutti A, Hutton C, Finsterbusch J, Helms G, Weiskopf N. Optimization and validation of methods for mapping of the radiofrequency transmit field at 3T. Magn Reson Med 2010; 64(1): 229–38.

McRobbie D, Semple S. Quality Control and Artefacts in Magnetic Resonance Imaging (IPEM report 112). York: Institute of Physics and Engineering in Medicine; 2017.

Michaelis T, Merboldt KD, Bruhn H, Hanicke W, Frahm J. Absolute concentrations of metabolites in the adult human brain in vivo: quantification of localized proton MR spectra. Radiology 1993; 187(1): 219–27.

Miller AJ, Joseph PM. The use of power images to perform quantitative analysis on low SNR MR images. Magn Reson Imaging 1993; 11(7): 1051–6.

Moerland MA, Beersma R, Bhagwandien R, Wijrdeman HK, Bakker CJ. Analysis and correction of geometric distortions in 1.5 T magnetic resonance images for use in radiotherapy treatment planning. Phys Med Biol 1995; 40(10): 1651–4.

Morrell GR. A phase-sensitive method of flip angle mapping. Magn Reson Med 2008; 60(4): 889–94.

Morrell GR, Schabel MC. An analysis of the accuracy of magnetic resonance flip angle measurement methods. Phys Med Biol 2010; 55(20): 6157–74.

Murphy BW, Carson PL, Ellis JH, Zhang YT, Hyde RJ, Chenevert TL. Signal-to-noise measures for magnetic resonance imagers. Magn Reson Imaging 1993; 11(3): 425–8.

Nehrke K, Börnert P. DREAM – a novel approach for robust, ultrafast, multislice B_1 mapping. Magn Reson Med 2012; 68(5): 1517–26.

Park DJ, Bangerter NK, Javed A, Kaggie J, Khalighi MM, Morrell GR. A statistical analysis of the Bloch-Siegert B1 mapping technique. Phys Med Biol 2013; 58(16): 5673–91.

Parker GJ, Barker GJ, Tofts PS. Accurate multislice gradient echo T(1) measurement in the presence of non-ideal RF pulse shape and RF field nonuniformity. Magn Reson Med 2001; 45(5): 838–45.

Parkes LM, Tofts PS. Improved accuracy of human cerebral blood perfusion measurements using arterial spin labeling: accounting for capillary water permeability. Magn Reson Med 2002; 48(1): 27–41.

Pohmann R, Scheffler K. A theoretical and experimental comparison of different techniques for B_1 mapping at very high fields. NMR Biomed 2013; 26(3): 265–75.

Provencher SW. Estimation of metabolite concentrations from localized in vivo proton NMR spectra. Magn Reson Med 1993; 30(6): 672–9.

Provencher SW. Automatic quantitation of localized in vivo ^{1}H spectra with LCModel. NMR Biomed 2001; 14(4): 260–4.

Sabati M, Maudsley AA. Fast and high-resolution quantitative mapping of tissue water content with full brain coverage for clinically-driven studies. Magn Reson Imaging 2013; 31(10): 1752–9.

Sacolick LI, Sun L, Vogel MW, Dixon WT, Hancu I. Fast radiofrequency flip angle calibration by Bloch-Siegert shift. Magn Reson Med 2011; 66(5): 1333–8.

Sacolick LI, Wiesinger F, Hancu I, Vogel MW. B1 mapping by Bloch-Siegert shift. Magn Reson Med 2010; 63(5): 1315–22.

Sbrizzi A, Raaijmakers AJ, Hoogduin H, Lagendijk JJ, Luijten PR, van den Berg CA. Transmit and receive RF fields determination from a single low-tip-angle gradient-echo scan by scaling of SVD data. Magn Reson Med 2014; 72(1): 248–59.

Sijbers J, den Dekker AJ, Van Audekerke J, Verhoye M, Van Dyck D. Estimation of the noise in magnitude MR images. Magn Reson Imaging 1998; 16(1): 87–90.

Sled JG, Pike GB. Standing-wave and RF penetration artifacts caused by elliptic geometry: an electrodynamic analysis of MRI. IEEE Trans Med Imaging 1998; 17(4): 653–62.

Sled JG, Pike GB. Correction for B(1) and B(0) variations in quantitative T(2) measurements using MRI. Magn Reson Med 2000; 43(4): 589–93.

Sorenson JA, Wang X. ROC methods for evaluation of fMRI techniques. Magn Reson Med 1996; 36(5): 737–44.

Stollberger R, Wach P. Imaging of the active B1 field in vivo. Magn Reson Med 1996; 35(2): 246–51.

Tate AR, Griffiths JR, Martinez-Perez I, Moreno A, Barba I, Cabanas ME, et al. Towards a method for automated classification of ^{1}H MRS spectra from brain tumours. NMR Biomed 1998; 11(4–5): 177–91.

Tintore M, Rovira A, Brieva L, Grive E, Jardi R, Borras C, et al. Isolated demyelinating syndromes: comparison of CSF oligoclonal bands and different MR imaging criteria to predict conversion to CDMS. Mult Scler 2001; 7(6): 359–63.

Tofts PS. Optimal detection of blood-brain barrier defects with Gd-DTPA MRI-the influences of delayed imaging and optimised repetition time. Magn Reson Imaging 1996; 14(4): 373–80.

Tofts PS, Kermode AG, MacManus DG, Robinson WH. Nasal orientation device to control head movement during CT and MR studies. J Comput Assist Tomogr 1990; 14(1): 163–4.

Tofts PS, Steens SC, Cercignani M, Admiraal-Behloul F, Hofman PA, van Osch MJ, et al. Sources of variation in multi-centre brain MTR histogram studies: body-coil transmission eliminates inter-centre differences. Magma 2006; 19(4): 209–22.

Tofts PS, Wray S. Noninvasive measurement of molar concentrations of 31P metabolites in vivo, using surface coil NMR spectroscopy. Magn Reson Med 1988; 6(1): 84–6.

Tozer DJ, Tofts PS. Removing spikes caused by quantization noise from high-resolution histograms. Magn Reson Med 2003; 50(3): 649–53.

Tubridy N, McKinstry CS. Neuroradiological history: Sir Joseph Larmor and the basis of MRI physics. Neuroradiology 2000; 42(11): 852–5.

Venkatesan R, Lin W, Haacke EM. Accurate determination of spin-density and T1 in the presence of RF-field inhomogeneities and flip-angle miscalibration. Magn Reson Med 1998; 40(4): 592–602.

Volz S, Nöth U, Deichmann R. Correction of systematic errors in quantitative proton density mapping. Magn Reson Med 2012; 68(1): 74–85.

Volz S, Nöth U, Rotarska-Jagiela A, Deichmann R. A fast B1-mapping method for the correction and normalization of magnetization transfer ratio maps at 3 T. NeuroImage 2010; 49(4): 3015–26.

Watanabe H, Takaya N, Mitsumori F. Non-uniformity correction of human brain imaging at high field by RF field mapping of B1+ and B1–. J Magn Reson 2011; 212(2): 426–30.

Weisskoff RM. Simple measurement of scanner stability for functional NMR imaging of activation in the brain. Magn Reson Med 1996; 36(4): 643–5.

Wheeler-Kingshott CA, Parker GJ, Symms MR, Hickman SJ, Tofts PS, Miller DH, et al. ADC mapping of the human optic nerve: increased resolution, coverage, and reliability with CSF-suppressed ZOOM-EPI. Magn Reson Med 2002; 47(1): 24–31.

Whisenant JG, Dortch RD, Grissom W, Kang H, Arlinghaus LR, Yankeelov TE. Bloch-Siegert B1-Mapping improves accuracy and precision of longitudinal relaxation measurements in the breast at 3 T. Tomography 2016; 2(4): 250–9.

Yarnykh VL. Actual flip-angle imaging in the pulsed steady state: a method for rapid three-dimensional mapping of the transmitted radiofrequency field. Magn Reson Med 2007; 57(1): 192–200.

Zweig MH, Campbell G. Receiver-operating characteristic (ROC) plots: a fundamental evaluation tool in clinical medicine. Clin Chem 1993; 39(4): 561–77.

3

Quality Assurance: Accuracy, Precision, Controls and Phantoms[1]

Contents

Paul S. Tofts
Brighton and Sussex Medical School

3.1 Quality assurance

3.1.1 Quality assurance concepts

When an instrument such as an MRI scanner is installed and handed over by the vendor (manufacturer) to the user, a series of acceptance tests is often carried out by the customer (de Wilde *et al.*, 2002; McRobbie and Quest 2002). The vendor's installation engineer will also have carried out extensive testing, according to their own protocols, using phantoms (test objects) to ensure the instrument is operating within the specification of the vendor. For qualitative MRI these may include signal-to noise ratio, spatial resolution and uniformity tests, gradient calibration and ensuring image artefacts are below certain levels.

Quality assurance (QA, sometimes called *quality control*) is used here to denote an *ongoing process* of ensuring the instrument continues to operate satisfactorily (Barker and Tofts 1992; Firbank *et al.*, 2000).

The QA falls into two groups. Firstly, the vendor's ongoing service contract will include some tests, largely to ensure the machine stays within specification. There may be some periodic recalibrations, for example of transmitter output, as components age. The user will not normally be involved in this process.

The second group of QA measurements will be focused on monitoring the quantification performance of the scanner. The quantification methods will often have been implemented in-house, without the explicit support of the vendor, and if they are unreliable the vendor will not be responsible, provided he can ensure the machine is still within the manufacturer's specifications. Thus the user must design, implement and analyse *quantitative quality assurance* (QQA) using appropriate measurements on phantoms and normal subjects (Tofts 1998).

Professional organisations of medical physics sometimes publish material on QA in MRI. The report from the UK Institute of Physics and Engineering in Medicine[2] (McRobbie 2017) Gives a comprehensive description of how to use the Eurospin test objects, and a wealth of detail and insight on other aspects of QA. The American Association of Physicists in Medicine (AAPM) has published some guidance on QA (Price *et al.*, 1990; Och *et al.*, 1992).[3] The American College of Radiology (ACR)[4] has an MRI accreditation scheme and an MRI quality control manual.[5]

1 Reviewed by Mara Cercignani.
2 http://www.ipem.ac.uk
3 *Acceptance Testing and Quality Assurance Procedures for MRI Facilities* free of charge from www.aapm.org
4 http://www.acr.org/
5 Downloadable from the AAPM website.

TABLE 3.1 Relative advantages of phantoms and healthy controls for quantitative quality assurance

	Simple Phantom (Test Object)	Healthy Control Subjects
Availability[a]	Good	Reasonable
Accuracy	Potentially good (e.g. volume)	True value unknown[b]
Uniformity	Poor in gels, good in liquids	Good in white matter
Temperature dependence	D,T_1,T_2 change 2%–3%/°C	Homeostatic temperature control
Stability	Potentially good (e.g. volume) but can be unstable (e.g. gels)	Usually stable
Realism	Generally poor; *in vivo* changes cannot be realistically modelled; B_1 distribution different	Good but no pathology
Standard design for multicentre studies?	Can be made	Use normal range, or travelling subject(s)

[a] Though see institutional constraints (Section 3.5.1).
[b] Although normal values have a narrow range – see Table 3.5.

3.1.2 Quantitative quality assurance

QQA will consume valuable scanner time, yet without it the measurements on research subjects may become valueless. Appropriate QQA provides reassurance that patient data are valid, gives warning if the measurement technique has failed because of a change in equipment or procedure, and may provide some help in rescuing data affected by such a failure. QQA measurements can be carried out in healthy ('normal') controls and in phantoms.

Measurements in *healthy control human subjects* (Section 3.4) are usually completely realistic, provided the parameter is present in normal subjects. Thus brain volume or normal-appearing brain tissue T_1 value could be monitored in this way but lesion volume could not. Increased atrophy or movement in patients might sometimes increase the variability compared to normal control subjects. A few parameters (most notoriously blood perfusion) have large biological intrasubject variation and require special designs for QQA. In addition to long-term monitoring by QQA, short-term reproducibility can be measured in any subject, although there may be ethical issues if Gd contrast agent is to be injected (e.g. for DCE-MRI – see Chapter 14) (Table 3.1).

Phantom measurements (Section 3.5) have the advantages of potentially providing a completely accurate value for the parameter under measurement (e.g. volume or T_1), of potentially being completely stable and of always being available. Often a *loading ring* is inserted into the head coil to provide similar loading to that given by the head. However realism is generally poor, with many potential sources of *in vivo* variation absent (e.g. subject movement, positioning error, partial volume error, variable loading, B_1 variation). Temperature dependence may be a problem (see Section 3.5.5). If a drift is seen in measurements from phantoms, the interpretation is often unclear (was it the scanner or the phantom that was unstable?) (Figure 3.1).

A *short-term test object* may be useful when developing a new measurement technique; this can be made quickly and need not be stable or have good independence of temperature. Later on, as the technique matures and goes into clinical use, full QQA would be needed, using healthy controls or a stable phantom.

A *post-mortem* brain phantom seeks to combine realism with stability and the ability to travel in a multicentre study (Droby *et al.*, 2015).

Frequency: To carry out QQA, controls or phantoms are measured at regular intervals (typically every week or

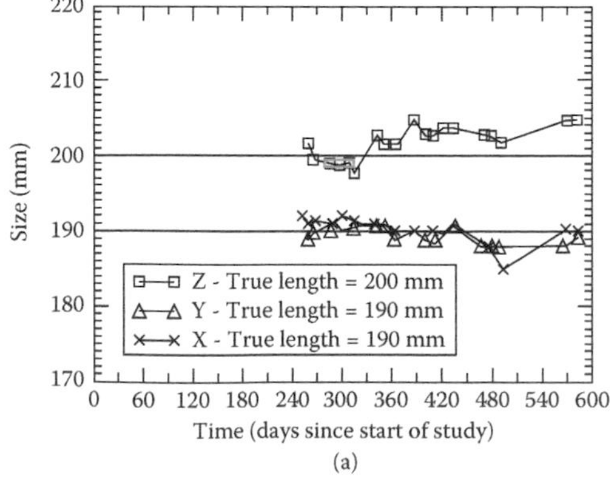

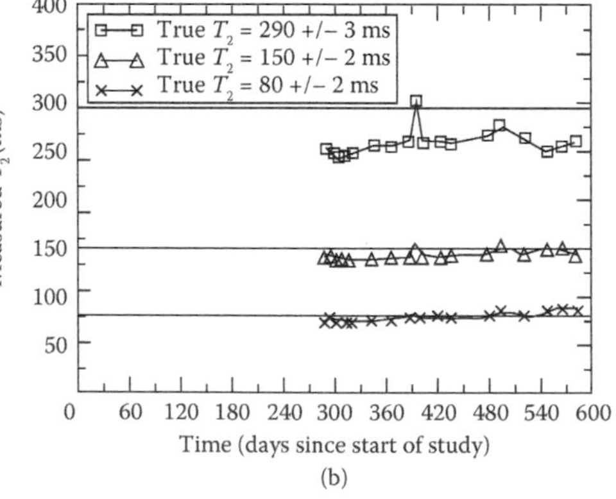

FIGURE 3.1 An early example of quality assurance (QA) measurements of object size (a) and T_2 (b). The apparent size drifts with time, probably because of a fault with gradient calibration. The true size is known accurately and unambiguously. T_2 estimates are inaccurate, particularly for the long-T_2 phantom, and drift with time, suggesting a progressive instrumental error. However inaccuracy and instability in the gel phantoms cannot be ruled out, unless a separate measurement of T_2 is carried out with a procedure known to be reliable. A drift in their temperature is a third possible explanation. (From Barker, G.J. and Tofts, P.S., *Magn. Reson. Imaging*, 10(4), 585–595, 1992.)

TABLE 3.2 Statistical tests used for shewhart charting

Test Number	Name of Test	Description of Test	Action Required
1	Warning	Measure exceeds control limits of mean ± 2 SD of previous measures.	Inspect with Tests 2–6
2	3 SD	Measure exceeds control limits of mean ± 3 SD of previous.	Instrument evaluation
3	2 SD	Two consecutive measures exceed mean ± 2 SD.	Instrument evaluation
4	Range of 4 SD	Difference between two consecutive measures exceeds 4 SD.	Instrument evaluation
5	4 ± 1 SD	Four consecutive measures exceed the same limit (+ 1 SD or – 1 SD).	Instrument evaluation
6	Mean × 10	Ten consecutive measures fall on the same side of the mean.	Instrument evaluation

Source: Adapted from Simmons, A., *et al.*, *Magn. Reson. Med.*, 41(6), 1274–1278, 1999.

month). The frequency has to be a compromise between rapid detection of a change in the instrument and the limited amount of machine time that is available. If an upgrade is planned, *bunched measurements* should be carried out before and after the change. Analysis should be automated as much as possible, both to save human time and to encourage rapid analysis of scan data (Sun *et al.*, 2015). Shewhart charting (Hajek *et al.*, 1999; Simmons *et al.*, 1999) is a set of statistical rules for automatically deciding when a measurement is abnormal enough to warrant human intervention (Table 3.2 and Figure 3.2).

Calibration was sometimes claimed to be a benefit of scanning phantoms with known MR properties. Calibration is measuring the response of the instrument to a stimulus of known value, with the purpose of then being able to apply that knowledge to *in vivo* measurements. For example, it was hoped that by measuring T_1 estimates for phantoms of known T_1 value, the calibration curve between true and estimated T_1 values could be applied to *in vivo* measurements. This concept has limited validity in the context of *in vivo* measurements, because there are many sources of error that are present *in vivo* but not in the phantom or else have different magnitudes in the two cases. Thus T_1 errors arising from incorrect flip angle settings are unlikely to be the same in a phantom and *in vivo*, and in general any systematic errors present in the phantom do not provide a realistic representation of those present *in vivo*. This is true for both a 'same-place' phantom, scanned in the head coil at a different time from the head, and for a 'same-time' phantom, attached to the head but in a different place from the brain.

3.1.3 Multicentre studies

Multicentre studies, where an attempt is made to reproduce the same measurement technique across different centres, or hospitals, often with different kinds of scanners, in different countries, are a challenging test (Podo 1988; Soher *et al.*, 1996; Keevil *et al.*, 1998; Podo *et al.*, 1998; Bauer *et al.*, 2010; Jerome *et al.*, 2016). The European group MAGNiMS[6] has conducted multicentre studies for over 20 years (Filippi *et al.*, 1998; Sormani *et al.*, 2016). The Human Connectome Project seeks to map macroscopic human brain circuits in a large population of healthy adults using DTI and other techniques (Van Essen *et al.*, 2013).

> *Multicentre magnetic resonance imaging (MRI) studies of the human brain enable a more advanced and comprehensive investigation of the disease course of rare and heterogeneous neurological and neuropsychiatric disorders due to increased sample sizes achieved by pooling data from the participating centres. While multi-centre MRI studies allow the acquisition of large amounts of data during a relatively short time period, they are based on the assumption that site-specific differences in MRI equipment do not impose any bias on the data, as this would severely reduce the statistical power of any analysis aimed at detecting differences between groups* (Droby et al., 2015).

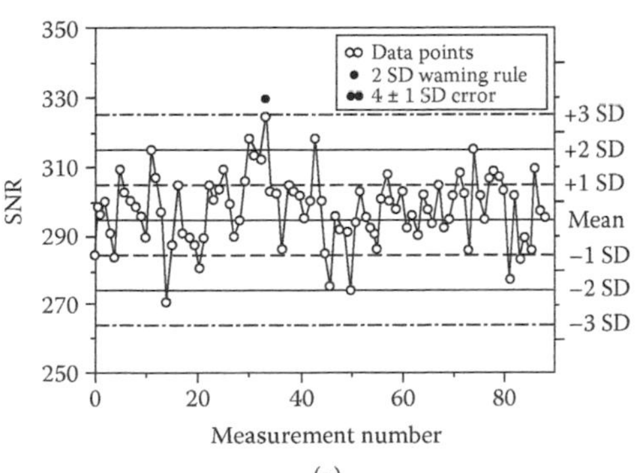

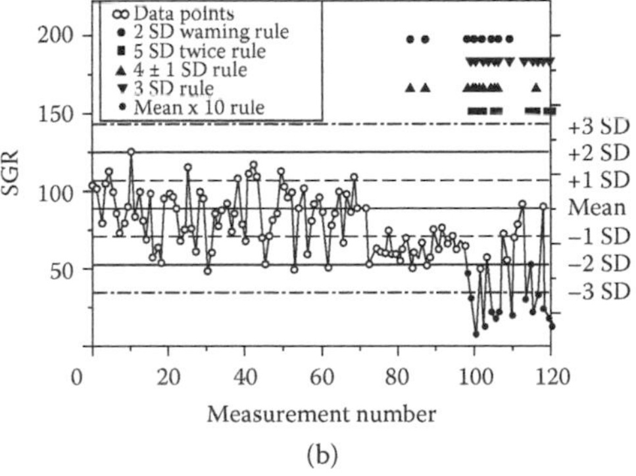

FIGURE 3.2 Shewhart charting of QA parameters. Data points are open symbols; triggering of rules (see Table 3.2) is shown by solid symbols. SNR is signal-to-noise ratio; SGR is signal-to-ghost ratio (used in echoplanar imaging) (From Simmons, A., *et al.*, *Magn. Reson. Med.*, 41(6), 1274–1278, 1999.).

[6] **Magn**etic Resonance **I**maging in **M**ultiple **S**clerosis, https://www.magnims.eu/

Variability among scanners may cancel the benefit of using multiple centers to assess new treatments (Zhou et al., 2017).

Thus the key issue is to minimise contamination of the whole dataset by centres with poor technique.

One approach to minimising between-centre variation is to make the data collection and analysis procedures as identical as possible, so that any systematic errors are replicated across the whole sample of centres. 'Protocol matching' for data collection involves attempting to match scanner type, field strength, sequence timing parameters (TR,TE) and also slice profile and RF non-uniformity (which is often not possible). In pharmaceutical trials there is often a travelling quality control officer, who ensures conformity to the agreed scanning protocol. MT measurements from two centres with different scanners were matched by careful attention to sequences, analysis technique, and by using body coil excitation to reduce B_1^+ differences (Tofts *et al.*, 2006). A second approach is to aim for good accuracy at each centre, measuring the underlying MR or biology parameters independent of the particular measurement procedure, since accurate measurements must necessarily agree with each other. Analysis matching may involve reaching agreement on standardised models, terminology and symbols; this was achieved for the DCE-MRI consensus (Tofts *et al.*, 1999)

Validation can be by measuring healthy controls (which have a narrow spread of values – see Section 3.4 below), measuring travelling controls (which are scanned at each site), measuring a travelling phantom or acquiring a standard phantom at each site.

Thus multicentre studies, although time-consuming and frustrating, are the ultimate test of how good our measurement techniques are. Full discussions of all the issues are available (Padhani *et al.*, 2009; Tofts and Collins 2011; Droby *et al.*, 2015; Jerome *et al.*, 2016). Early identification of outliers may enable problems at particular contaminating centres to be identified (Walker *et al.*, 2013).

Biomarkers: A major driver for developing quantitative MRI is to produce reliable biomarkers, to be used in multicentre treatment trials. Biomarker concepts come from a drug development paradigm; these are well developed and not always aligned with MRI concepts (Padhani *et al.*, 2009; O'Connor *et al.*, 2017).

3.2 Uncertainty, error and accuracy

3.2.1 Concepts

The conventional way to characterise measurement techniques in the physical sciences has been to estimate accuracy and precision (i.e. systematic and random errors). Separating systematic and random error is often helpful, since they occur on different timescales and have different effects on the viability of the measurement. A systematic error, in its ideal form, is one that is constant over the lifetime of the study, whilst a random error is one present in short-term repeated measurements.

A measurement result is complete only when accompanied by a quantitative statement of its uncertainty. The uncertainty is required in order to decide if the result is adequate for its intended purpose and to ascertain if it is consistent with other similar results.[7]

In modern use, *measurement error* is used to mean the difference between the measurement and the true value, whilst *measurement uncertainty* refers to the spread of possible true values that can be inferred from the measurement. Thus, a particular (single) measurement could have zero error but large uncertainty. In psychology and in medicine, the concept of reliability is often used to evaluate the performance of a metric (see Chapter 1, Section 1.3.3).

Accuracy refers to systematic error, the way in which measurements may be consistently different from the truth, or biased. *Precision* refers to random errors, which occur over short time intervals, if the measurement is repeated often. Thus in a determination of T_1, systematic errors could be caused by a consistently wrong B_1^+ value, whilst random errors could be caused by image noise (which is different in each image). However the systematic error could vary over a long period of time (e.g. if the method for setting B_1^+ was improved or a different head coil was installed). Similarly, the precision could be worse if measured from repeat scans over a long period of time, compared with short-term repeats, as additional sources of variation became relevant (e.g. a change of data acquisition technologist) (Figure 3.3).

Thus the differences between *long-term precision* and accuracy become blurred, and the difference is merely one of time scale. Some studies of chronic disease can last for long periods (over a decade in the case of MS, epilepsy, dementia and aging) and considerations of accuracy and its variation over time become increasingly important (see Figure 3.3). Precision can be seen as setting the limits of agreement in a short study on the same machine; accuracy sets the limits of agreement in a long-term or multicentre study, where several machines are to be used, possibly extending over different generations of technology.

3.2.2 Sources of error

Contributions to both inaccuracy and imprecision can arise in both the data collection and the image analysis procedures (see Chapter 2), and both need to be carefully controlled in order to achieve good long-term performance. The major contributors to systematic data collection errors are probably B_1^+ non-uniformity and partial volume errors. Artefacts arising from imperfect slice selection and *k*-space sampling (particularly in fast spin echo and echoplanar imaging) can also give systematic error. Patient positioning and movement contribute to random errors; positioning can be improved with technologist training and liberal use of localiser scans, whilst movement can be reduced by attention

[7] From the US National Institute of Standards and Technology (NIST) website, http://physics.nist.gov/cuu/Uncertainty/index.html. This is a mine of information on constants, units and uncertainty.

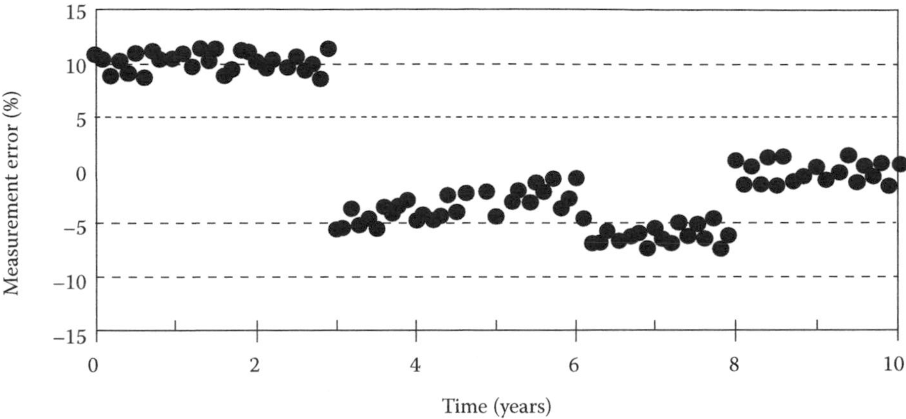

FIGURE 3.3 Long-term precision is dominated by instability in the systematic error. Simulation of fictional change in measurement error over time, during a longitudinal study. Short-term precision is good, and a study completed in the first 3 years is unaffected by the large systematic error (i.e. poor accuracy). A major upgrade at Year 3 dramatically changes the systematic error. A subtle drift in values takes place, followed by two more step changes, at the times of operator change and a minor upgrade. At Year 8 the sources of systematic error are finally identified and removed, giving a system that should provide good accuracy and hence long-term precision for many years.

TABLE 3.3 Potential sources of error in the MRI measurement process[a]

	Random Error	Systematic Error
Biology	Normal variation in physiology	
Data collection	Position of subject in head coil	B_1^+ error
	Coil loading (corrected by prescan?)	Slice profile
	Prescan procedure setting B_1^+	k space sampling (in FSE, EPI)
	Position of slices in head	Partial volume
	Gd injection procedure	Operator training
	Patient movement (cardiac pulsation)	Software upgrade
	Patient movement (macroscopic)	Hardware upgrade
	Image noise	
	Temperature (phantoms only)	
Image analysis	ROI creation and placement	Operator training
		Software upgrade

Note: In their simplest forms, random error is associated with short-term unpredictable variation, whilst systematic error is fixed. However some random processes (e.g. positioning) might only show up over a longer time scale (caused e.g. by change of radiographer [technician]), whilst some sources of systematic error might vary with time (e.g. operator training).

[a] See also Chapter 2.

to patient comfort, feedback devices to assist the subject in keeping still (Tofts *et al.*, 1990) and spatial registration of images (see Chapter 17). Analysis performance can be characterised by repeat analyses, both by the same observer and by different observers. A change of technologist, either for data collection or for analysis, can introduce subtle changes in procedure and hence results. Early work that measured the reproducibility of an analysis procedure has little value without re-scanning the subject, since patient positioning can be a major source of variation (Tofts 1998). In the case of automatic image analysis this is particularly true, since an automatic procedure, being free of a subjective operator, is intrinsically perfectly reproducible (Table 3.3).

The *analysis software* has to be kept stable, and modern software engineering practice[8] defines how to do this. The analysis method should be documented in detail, intra- and interrater differences measured, and software upgrades should be controlled and documented through version control procedures. In long-term studies, some old data should be kept for re-analysis at a later stage, when operators and software may have changed (Tofts and Collins 2011). Alternatively, all the analysis can be carried out at the end of the study, over a relatively short time. However there is often a value in carrying out a preliminary analysis, and in any cases studies are often extended beyond their initially planned duration.

[8] See for example ISO 9001.

3.2.3 Modelling error

3.2.3.1 Error propagation ratio

The error propagation ratio (EPR) is a convenient way of investigating the sensitivity of a parameter estimate to the various assumptions that have gone into the calculation. The EPR is the percent change in a derived parameter arising from a 1% change in one of the model parameters. For example, in a study to measure capillary transfer constant K^{trans} in the breast (Tofts *et al.*, 1995), the estimate is very sensitive to the T_{10} value used (EPR = 1.2) and the relaxivity r_1 (EPR = 1.0) but very insensitive to an error in the echo time (EPR = 0.02). In arterial spin labelling, the sensitivity of the perfusion estimate can similarly be investigated (Parkes and Tofts 2002). Studying error sources in this way immediately brings to light that some errors are truly random, whilst others could be systematic for the same subject in repeated measurements (e.g. a wrongly assumed AIF in T_1w-DCE) but random across other subjects. The effect of random noise on a parameter estimation procedure can also be found by making small changes to the image signal values and measuring the resulting changes in the parameter. Uncertainty budgets and type A and B errors are concepts related to EPR (see Section 3.2.4).

3.2.3.2 Image noise

The contribution of image noise to imprecision in the final parameter can be calculated. If a simple ratio of images is used (e.g. T_1 calculated from images at two different flip angles), then propagation of errors (Taylor 1997) allows the effect of noise in each source image to be calculated. An analytic expression can be derived for the total noise, and this can be minimised as a function of imaging parameters such as TR and the number of averages, keeping the total imaging time fixed (see e.g. Tofts 1996, and Chapter 2, Figure 2.5).

3.2.3.3 Cramer-rao analysis

If least squares curve fitting is used to estimate a parameter from more than two images, simple noise propagation will not work, as the fitted parameter is not a simple function of the source images. However the Cramer-Rao *minimum variance bound* (Cavassila *et al.*, 2001; Brihuega-Moreno *et al.*, 2003) is an analytical method making use of partial derivatives that does calculate the effect of image noise on the fitted parameters. The LC (Linear Combination) model for estimating spectral areas by fitting uses this method to estimate the minimum uncertainty in the metabolite concentration (Provencher 2001). Only uncertainty arising from data noise is included; other factors (both random and systematic) can make the uncertainty higher than this minimum variance bound.

3.2.3.4 Monte carlo

Numerical simulation can simulate the effect of image noise. Noise is added to the source data many times and the effect on the fitted parameter measured.

3.2.4 Uncertainty in measurement: type A and type B errors

The scientific measurement community has moved to refine the traditional concepts of random and systematic error and instead uses a different (though closely related) method of specifying errors.[9] Initiatives have been published from Europe,[10] the USA[11] and UK.[12] Type A errors are those estimated by repeated measurements, whilst type B errors are all others. They are combined into a 'standard uncertainty'. This approach was designed by physical metrologists, primarily for reporting uncertainty in physical measurements. An *uncertainty budget* is drawn up, where error components that are considered important are separately identified, quantified (using propagation of errors), then combined to obtain an overall uncertainty. Thus systematic errors are no longer looked on as being benevolent and unchanging. A simple example of an uncertainty budget is that of measuring diffusion coefficient in a test liquid, where the effects of noise, uncertain temperature and uncertain gradient values were analysed and combined (Tofts *et al.*, 2000)

3.2.5 Accuracy

Accuracy is a measure of systematic error, or bias. It estimates how close to the truth the measurements are, on average. It is intrinsically a long-term measure. Often the truth is unknown in MRI, since the brain tissue is not accessible for detailed exhaustive measurements. Thus the true grey matter volume, or total MS lesion volume, would be extremely hard to measure. A physical model (i.e. a phantom) could never be made realistic enough to simulate all the sources of error present in the actual head.

Yet if accuracy is desired, some basic tests can be applied using simple objects. For the example of measuring lesion volume in MS, simple plastic cylinders immersed in a water bath proved too easy, since the major sources of variation (partial volume and low contrast) were missing. However, by tilting the cylinders (to give realistic partial volume effect), inverting the image contrast (to give bright lesions) and adding noise (to give realistically low contrast-to-noise values for the artificial lesions), images were obtained that gave realistic errors in the reported values of volume (Tofts *et al.*, 1997b). Accuracy (and precision) measured on this phantom represent lower limits to what might be achieved with *in vivo* measurements, since additional sources of error would be present with the latter. Nonetheless, this type of study represents a reasonable test to apply to a measurement technique, since it will identify any major problems (Figure 3.4).

---

[9] The standard work is the *Guidance on the Expression of Uncertainty in Measurement (GUM)*, published by the International Standards Organisation (ISO) in 1995. Available from BIPM (Bureau International des Poids et Mesures; www.bipm.org). There is much commercial activity in this field, as organisations selling measurement services seek ISO accreditation. Many national organisations produce guidance on the expression of uncertainty in measurement, and publish user-friendly versions of GUM. Books are also available.

[10] The European Accreditation group has produced in 2013 *Evaluation of the Uncertainty of Measurement in Calibration*, document EA-4/02. This gives much detail and good examples of uncertainty budgets. See www.european-accreditation.org

[11] The NIST has guidelines from 2000 at http://physics.nist.gov/cuu/Uncertainty/index.html. More recent information is at https://www.nist.gov/

[12] The United Kingdom Accreditation Service has several useful documents; *M3003* and *LAB 12* are concise expositions of the concepts.

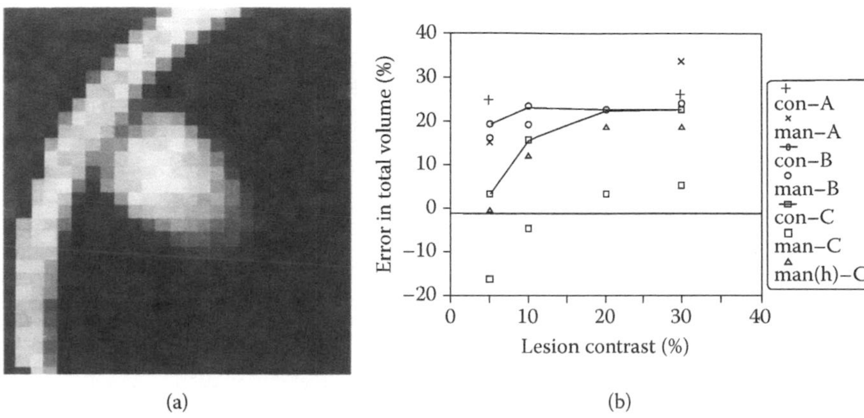

FIGURE 3.4 Lesion volume accuracy measured using an oblique cylinder contrast-adjusted phantom. (a) One small lesion (with a known volume of 0.6 ml), represented as an acrylic cylinder, is mounted on the inside of an acrylic annulus, at an angle to the image slice, giving a realistic partial volume effect. (b) Error in total lesion volume (for nine lesions with volumes 0.3–6.2 ml) showing large variation with lesion contrast, observer (A, B or C) and outlining method (con: semi-automatic contouring; man: manual). (From Tofts, P.S. *et al.*, *Magn. Reson. Imaging*, 15, 183–192, 1997b.)

Importance of accuracy: It has been argued that accuracy is irrelevant in clinical MR measurements, since the systematic error is always present and does not mask group differences. In principle this is true; however actual systematic errors often do not last forever and can change with time (thus forming a contribution to long-term instability or imprecision). An example from spinal cord atrophy measurements shows this (Tofts 1998). The technique (see Figure 3.9) was estimated to have a 6% systematic error, based on scanning a plastic rod immersed in water. The short-term reproducibility was good (0.8% coefficient of variation, CV), and progressive atrophy in MS patients could be seen after about 12 months. After a scanner *software upgrade*, there was an implausible step increase in the normal control values of about 2%. The step change caused by the upgrade prevented atrophy progression through the time of the upgrade from being measured. If the accuracy had been better, and if the sources of systematic error had been understood and controlled, the upgrade would not have been disastrous for this study.

Machine upgrades cannot be avoided; they can only be planned for, and in this context *accuracy provides long-term stability*. As an additional safeguard, if groups of subjects are being compared, subjects from both groups should be collected during the same period, that is 'interleaved controls'. There is a temptation to leave the controls until the end; if there is a step change in the measurement process characteristics after the patients have been measured, but before the controls have been measured, then a group difference cannot be interpreted as caused by disease, since it may have been caused by the change in procedure.

Subtle left–right *asymmetry* or anterior–posterior differences may be seen in a group of subjects. This could be caused by genuine biological difference between the sides or front and back, or by a subtle asymmetry in the head coil. This can be resolved by scanning some subjects relocated with respect to the head coil, for example prone instead of supine.

3.3 Precision

3.3.1 Precision concepts

Precision, *reproducibility*[13] or *repeatability* is concerned with whether a measurement agrees with a second measurement of the same quantity, carried out within a short enough time interval that the underlying quantity is considered to have remained constant. Sometimes this is called the *test–retest* performance in psychology. Good within-subject reproducibility is probably the best indicator of good measurement technique (see Figure 3.5); this is why so much attention is paid here to precision. There is also an ISO definition (Padhani *et al.*, 2009).[14]

Measuring precision: Many studies of reproducibility (precision) have been published, for many MRI parameters. Its value at a particular site depends on the method used to measure the parameter and is often very sensitive to the precise details of the data collection procedure (such as patient positioning and prescan procedure) and data analysis (particularly region of interest placement). The results of a study may not be generalisable – a poor value of reproducibility may be a reflection of poor local technique at a particular site. However a good value gives inspiration to other workers to refine their technique. Detailed studies of the various components in a measuring process can identify the major sources of variation;

[13] A measurement is said to be *reproducible* when it can be repeated (*reproduce*: 'to bring back into existence again, re-create'). However this term is not used by statisticians, who prefer the more precise term *measurement error*. Reproducibility can include factors such as normal short-term biological variation that are not part of measurement error.

[14] According to ISO 5725, *repeatability* refers to test conditions that are as constant as possible, where the same operator using the same equipment within a 'short time interval' obtains independent test results with the same method on identical items in the same laboratory. *Reproducibility* refers to test conditions under which results are obtained with the same method on identical test items but in different laboratories with different operators using equipment.

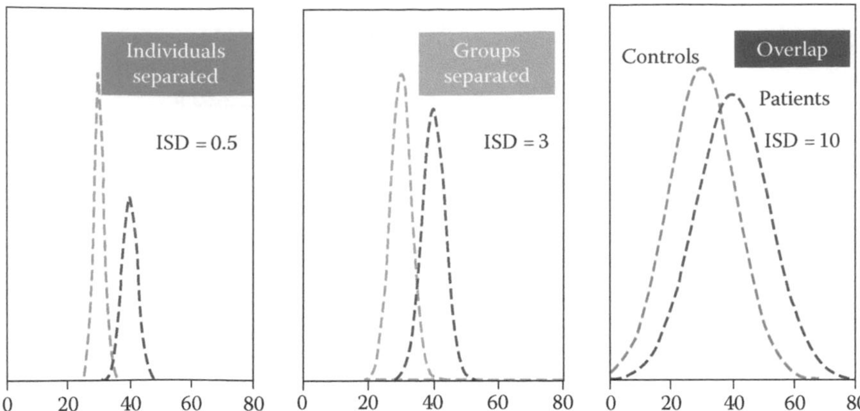

FIGURE 3.5 Simulation showing how magnitude of ISD affects ability to use an MR parameter to separate groups and individuals. Group separation is 10 units. With ISD = 10 (right-hand image), the groups overlap, and considerable statistical power would be needed to separate them (see Chapter 1, Figure 1.3). A reduced ISD = 3 (centre) gives a good group separation. A further reduction to ISD = 0.5 (left-hand image) enables individuals to be accurately classified into their group, ultimately limited only by the inherent biological variation.

for example rescanning without moving the subject will measure effects such as image noise and patient movement, whilst removing and replacing the subject will also include the effect of positioning the subject in the scanner. This knowledge in turn opens the possibility of reducing the magnitude of the variation by various improvements in technique, ranging from more care, training to reduce interobserver effects (Filippi *et al.*, 1998) to formal mathematical optimisation of the free parameters that define the process (Tofts 1996) (see Chapter 2, Figure 2.5). Measuring the reproducibility of various scanner parameters that are thought to have a large effect on the final MR parameter (such as those set during the prescan procedure) may also be of value.

The methods used to report reproducibility are not always standardised – it is hoped that studies will use instrumental standard deviation (SD)*s* and intraclass correlation coefficient (ICC), as described below. Reproducibility may be worse in patients than in normal controls (patients may find it harder to keep still). The reproducibility may depend on the mean value of the parameter (which may be significantly different in patients, e.g. if there is gross atrophy); see also Figure 3.7. Precision may also have a biological component (see Section 3.3.2.3).

3.3.2 Within-subject standard deviation

3.3.2.1 Bland–altman and ISD

The simplest and most useful approach to characterising *measurement error* is that of Bland and Altman, which uses pairs of repeated measurements in a range of subjects; the within-subject standard deviation (SD) *s* of a single measurement, arising primarily from instrumental factors, is estimated (Bland and Altman 1986; Bland and Altman 1996b; Galbraith *et al.*, 2002; Padhani *et al.*, 2002; Wei *et al.*, 2002. The 95% confidence limit on a single measurement is 1.96*s*[15] (Box 3.1).

Box 3.1 Why measure within-subject reproducibility?

1. It tells you *confidence limits on a single measurement.* For example, in measuring the concentration of a compound by MRS, the reproducibility (1 sd) is typically 10%. The 95% confidence limit on a single measurement is then 20% (1.96 sd). This means that there is a 95% chance that the true value lies between these limits, and only a 5% chance that it lies outside this range.
2. It tells you the *repeatability or minimum detectable difference* that can be measured.

In the above MRS example, the concentration might be estimated on two consecutive occasions, perhaps to look for biochemical effects of progressive disease. The sd in difference measurements is 14% (1.4 times the sd in a single measurement), and the 95%CL on a difference measurement is 28% (1.96 times the sd in difference measurements). Thus unless a measured difference is more than 28%, it cannot be ascribed to a biological cause with a confidence exceeding 95%. If the measured difference is less than 28%, it could have arisen by chance.

[15] In a normal distribution with standard deviation *s*, 95% of the area lies between + 1.96*s* of the mean value.

TABLE 3.4 Example of estimating instrumental standard deviation (ISD) using Bland–Altman method

Measurement set number	Replicate 1	Replicate 2	Signed difference
1	107.14	108.12	0.98
2	103.50	98.60	−4.91
3	104.65	104.73	0.08
4	100.97	106.26	5.29
5	96.87	105.76	8.89
6	90.30	98.76	8.46
7	108.97	98.79	−10.19
8	104.55	110.24	5.70
9	99.55	105.13	5.58
10	103.94	99.60	−4.33

SD of differences $sd\Delta$	6.4
mean_difference	1.6
ISD s	4.5
σ_s	1.1
95% CL lower	2.4
95% CL upper	6.6

Note: Ten sets of replicate measurements were simulated, drawn from a random normal distribution with mean = 100, SD = 5 (same data set as Figure 3.6). Signed differences were calculated (left-hand table). From these were calculated (right-hand table) their SD ($sd\Delta$ = 6.4), ISD s = 4.5, the SD of this estimate (σ_s = 1.1) and 95% confidence limits (CL) for s: 2.4–6.6.

For *repeated measurements* on the same subject (who is assumed to be unchanging during this process), the measurement values are samples from a normal distribution with SD s. The signed difference Δ between the repeats in pairs of measurements is also normally distributed, with an SD value of $sd\Delta$:

$$sd\Delta = \sqrt{2}\,s = 1.414s \qquad (3.1)$$

Because of the difficulty in making many measurements on the same subject, and because subjects may in any case vary, pairs of measurements (replicates) are usually made on a number of subjects and the difference calculated for each pair. The SD of this set of differences is then calculated ($sd\Delta$), and from this the SD of the measurements on a single subject (s) (see Table 3.4). At least 10 pairs should be measured (see Figure 3.6).

Mean absolute difference in pairs of replicates: Instead of taking the signed differences (as in Bland and Altman's procedure above), the absolute (unsigned) difference is sometimes taken. Its mean value is 0.80 s and from this the SD can be found.[16]

The *CV* in the measurements is the SD divided by the mean value (i.e. $CV = s/\bar{x}$, where $\bar{x}$ is the mean value) and is usually expressed as a percentage.

When using this technique, consideration should be given to what aspect of the measurement process is to be characterised. To assess the whole process, the subject should be taken out of the scanner between replicates, and it may be desirable to carry out the repeat scan a week later, with a different radiographer (technologist). A separate observer, blinded to the first result, could be used for analysis of the replicate. A Bland–Altman plot should be made to check for dependence on mean value (Figure 3.7).

Estimation of s, also called the *within-subject variability*, in the underlying distribution of measurements (all with the same mean) characterises the measurement process. From this, the coefficient of *repeatability* $\sqrt{2} \times 1.96s = 2.77s$ can be found (assuming there is no bias between the first and the second measurement). The difference between two measurements, for the same subject, is expected to be less than the repeatability for 95% of the pairs of observations. Thus for a biological

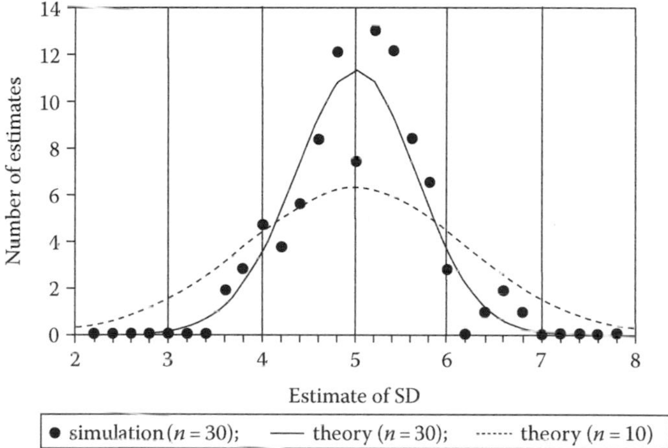

FIGURE 3.6 Simulation of estimation of reproducibility from repeated measurements. Over 8000 samples from a population of random numbers with mean = 100 and SD = 5 were generated. From these, 30 pairs of samples (replicates) were taken, and the differences Δ calculated, retaining the sign of the difference (Δ could be + or −). The SD of the Δ values was found ($sd\Delta$), and from this the SD of the population was estimated (Equation 3.1). Further sets of 30 pairs were taken, to a total of 100 sets, and in each set the population SD estimated. The figure shows the distribution of estimates obtained, showing a mean of 5 (as expected), and clustered mostly between 4 and 6. The theoretical normal distributions are also shown, for 30 and 10 pairs of difference measurements. The theoretical curve for 30 pairs is in agreement with the data. For 30 pairs, an SD of 0.66 was estimated, which gives a 95% CL of ±1.3 (Equation 3.2) in estimating s (i.e. 95% of the estimates will lie in the range 3.7–6.3). On the other hand, with only 10 pairs, this range increases to 2.7–7.3; reducing the number of pairs has reduced the precision with which the SD can be estimated. See also Table 3.4, from 1st edition, pp. 66, Figure 3.7.

[16] See first edition of this book, p. 66.

change to be detected in a single subject with 95% confidence, it must exceed the repeatability (see Box 3.1). These lower and upper limits to differences that can arise from measurement error are sometimes called the *limits of agreement* (Bland and Altman 1986).

Agreement between two instruments has two components: bias (systematic difference) and variability (random differences). Under normal conditions the mean difference between the first and second measurements is expected to be zero, if they come from a set of repeats made under identical conditions. However if two separate occasions, two observers or two scanners are being compared, then a test for bias should be made, using a two-tailed t-test. If the differences are not normally distributed, a Wilcoxon signed rank test is needed.

3.3.2.2 Dependence of SD on mean value

The approach above supposes that the mean value in each pair is similar, so that the differences from paired measurements can be pooled. This assumption can be tested in a *Bland–Altman plot,* where the SD is plotted against mean value (Bland and Altman 1986; Krummenauer and Doll 2000). Any important relationship should be fairly obvious, but an analytic check can be made using a rank correlation coefficient (Kendall's tau) (Bland and Altman 1996a). If SD increases with mean value, (which is often the case) it may need to be transformed in some way to give a quantity that varies less with mean value. For the situation where SD is proportional to the mean, a log transformation is appropriate (Bland and Altman 1996c), although the interpretation of the transformed variable is not so straightforward. An alternative is to use the CV, which is constant under the condition of SD proportional to the mean. For measurements of total lesion volume in MS, the CV is relatively constant over a wide range of volumes (or at least there is no clear evidence of it changing in a systematic way) (see Figure 3.7). In this case the estimates of CV at different volumes can then legitimately be pooled to give a single, more precise value.

In the Bland–Altman approach, the *uncertainty of the estimate of SD* can be found. The uncertainty (one standard deviation) in estimating an SD (s) from n samples is as follows (Taylor 1997; p. 298).

$$\sigma_s = \frac{s}{\sqrt{2(n-1)}} \tag{3.2}$$

See Table 3.4 for an example.

3.3.2.3 Biological variation

Precision may have a significant biological component, in that intrasubject variation may be significant and limit the usefulness of having good machine precision. Thus blood flow varies by about 10% within a day (Parkes *et al.*, 2004), so if a single number is required to characterise the individual, high precision is not required. However if these biological changes are to be studied in detail, for example to find their origin, then a much better instrumental precision would be needed.

Biological variation at time scales longer than a few minutes can be measured using repeated measurements, provided the machine variation (ISD) is known (e.g. from phantoms or fast repeats when the biology is known to be static). Short-term variation might be accessible by the device of *data fractionation*. The data collection procedure is altered, if necessary, to acquire two independent datasets as simultaneously as possible. The easiest way to do this is to use two signal averages for each phase encode and preserve them without addition. Typically the averages are separated by a second or less of time. Two image datasets are then constructed, and differences measured from these, to estimate instrumental precision. These image datasets are statistically completely independent, yet form samples of the biology separated by a second or less.

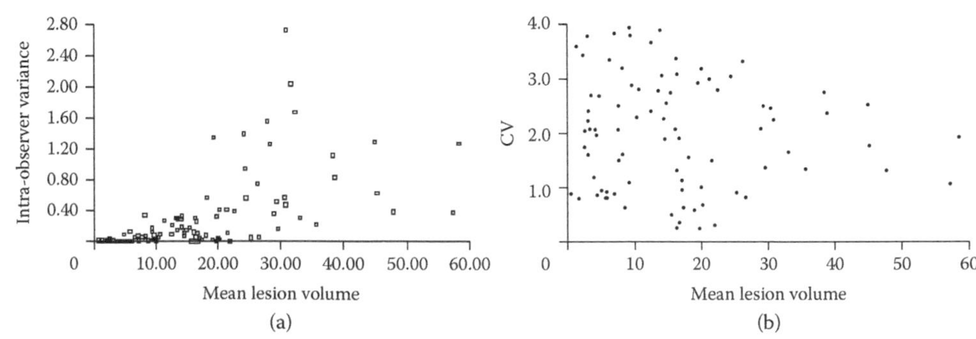

FIGURE 3.7 Bland–Altman plots for estimates of total lesion volume in multiple sclerosis. (a) The variance increases with mean lesion volume (MLV); therefore variance values cannot be pooled. (b) The coefficient of variation (CV) is independent of MLV (i.e. there is no sign of a systematic dependence of CV on MLV); therefore the CV values can be pooled to form a single average value. (From Rovaris, M., *et al.*, *Magn. Reson. Imaging*, 16, 1185–1189, 1998.)

Estimation of biological variation is important in the context of creating a 'perfect quantitative MRI machine', (see Chapter 1, Section 1.4), which contributes no extra variance.

3.3.3 Intraclass correlation coefficient or reliability

This measure considers both the within-subject (intrasubject) variance arising from measurement error (which we have considered in the previous section) and variance arising from the difference between subjects (Cohen *et al.*, 2000; Armitage *et al.*, 2001). If there is a large variance between the subjects (intersubject), measurement variance may be less important, particularly if groups are being compared. The ICC is

$$\text{ICC} = \frac{\text{variance from subjects}}{\text{variance from subjects} + \text{variance from measurement error}} \quad (3.3)$$

The ICC can be thought of being the fraction of the total variance that is attributed to the subjects (rather than measurement error). Thus if measurement error is small compared to the subject variance, ICC approaches 1. Typical values in good studies would be at least 0.9. ICC as a measure has the benefit of placing measurement error in the context of the subjects, and potentially it can stop us being overly concerned about measurement error when subject variance is large.

However ICC has at least two problems. ICC depends on the group of subjects being studied (Bland and Altman 1996c), and a determination in one group does not tell us the value in another group. For example, in normal subjects (who often form a homogeneous group), ICC may be unacceptably low, whilst in patients (who are naturally more heterogeneous) the ICC may be adequate. Secondly, when studying individual patients, and their subtle MR response to treatment, the crucial parameter is the repeatability (or the within-subject standard deviation, from which it is derived), as this is the smallest biological change that can reliably be detected, and ICC has little value.

The ICC is often called the *reliability* (Cohen *et al.*, 2000; Armitage *et al.*, 2001). Reliability is discussed with insight by Streiner and Norman (1995). Although the ICC is not an absolute characteristic of the instrument, it is favoured by many researchers (Chard *et al.*, 2002); see Chapter 1, Section 1.3.3 on psychometric measures. It is probably best to measure both ICC and ISD.

3.3.4 Analysis of variance components

This quite complex analysis is carried out by repeating various parts of the measurement procedure, as well as the whole procedure (see e.g. Chard *et al.*, 2002). The variance arising from different parts of the measurement procedure can be estimated, as well as intersubject and interscanner effects. A model of the variance is first prescribed, with possible interactions, such as allowing some of the variance components to depend on subject or scanner. The measurement can be repeated without removing the subject from the scanner ('within-session variance'), then removed and re-scanned ('intersession variance'). Within-session variance has noise and patient movement (including pulsation); intersession variance also has repositioning (and possibly longer-term biological variation).

3.3.5 Other measures of precision

3.3.5.1 Correlation

In a set of repeated measures, the first result can be correlated with the second one, and high correlation coefficients are usually produced when this is done. However this approach has little value and does not give an indication of agreement between replicated measurements (Bland and Altman 1986). In a trivial example, the measures could differ by large amounts, for example one might be twice the other, and a good correlation could still be produced. A large intersubject variation will also increase its value (Bland and Altman 1996c). Good correlation does not imply good agreement.

3.3.5.2 Kappa coefficient

This is used for categorical or ordinal data (Armitage *et al.*, 2001), where there are few possible outcomes and is not appropriate for continuous quantitative data.

3.4 Healthy controls for QA

The range of values measured in healthy controls ('normals') can be quite small for some parameters, notably T_1, ADC and MTR (Table 3.5). Within-centre CVs of 3%–5% have been achieved for T_1 and ADC, and under 2% for MTR. Between-centre differences are larger (see Section 3.1.3). Values usually depend on location in the brain and age (Silver *et al.*, 1997).

The measured normal range at a centre is influenced by the centre's ISD (measured from repeats – see Section 3.3.2.1). Broadly speaking, the measured spread of values is a convolution of the actual biological spread and that introduced by the instrument. A reduction in ISD can make a dramatic reduction in measured normal range (see Figure 3.8).

Thus healthy controls can be used for QA both within-centre and between-centre. Within-centre stability can be monitored using a few easily available controls who are likely to remain accessible for a long time (see Figure 3.9). Between-centre differences can be studied and minimised using controls at each centre (Tofts *et al.*, 2006). Although T_1 and ADC are the most explored parameters for QA using healthy controls, other parameters may reach this level of standardisation (e.g. MRS metabolite concentrations).

TABLE 3.5A Normal range of T_1 values at 1.5T in white matter

Study[a]		CV[b] (%)	n^c	Mean (ms)	SD (ms)
Stevenson	2000	5	40	666	36
Rutgers	2002	6	15	681	40
Ethofer	2003	4	8	770[d]	30

Source: Adapted from Tofts, P.S., and Collins, D.J., *Br. J. Radiol.*, 84 Spec No 2, S213–S226, 2011.

Note: See also Chapter 5, Table 5.1, for a fuller list of values; coefficients of variation are about 3%.

[a] References for all studies are given in the original tables (Tofts and Collins 2011).

[b] Coefficient of variation = SD/mean.

[c] Sample size.

[d] Used spectroscopic technique; probably some CSF or grey matter contamination.

TABLE 3.5B Normal Range of Mean Diffusivity Values in White Matter

Study		CV (%)	n	Mean (10^{-9} m^2s^{-1})	SD (10^{-9} m^2s^{-1})
Cercignani	2001	5	20	0.93[a]	0.04
Emmer	2006	4	12	0.84	0.03
Zhang	2007	5	29	0.69	0.04
Welsh	2007	3	21	0.73	0.02

Source: Adapted from Tofts, P.S., and Collins, D.J., *Br. J. Radiol.*, 84 Spec No 2, S213–S226, 2011.

Note: See also Chapter 8.

[a] Some CSF contamination.

TABLE 3.5C Normal Range of MTR Values in White Matter

Study		CV (%)	n	Mean (pu)[a]	SD (pu)
Silver	1997	1.9	41	39.5	0.76[b]
Davies	2005	1.0	19	38.4	0.4
Tofts	2006	1.6	10	37.3[c]	0.6

Source: Adapted from Tofts, P.S., and Collins, D.J., *Br. J. Radiol.*, 84 Spec No 2, S213–S226, 2011.

Note: See also Figure 3.8, which shows SD values of 0.5–1.0 pu.

[a] MTR values not comparable between studies (different sequences).

[b] SEM = 0.17 pu; 4 samples each n = 20 or 21; estimated SD = 0.76 pu.

[c] Peak location values in white matter histograms.

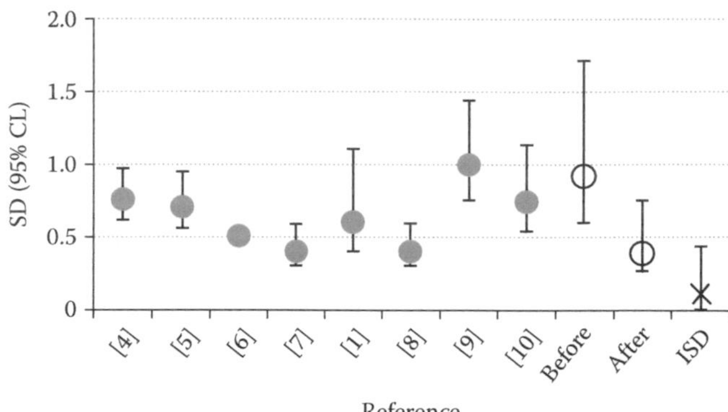

FIGURE 3.8 Normal variation for white matter MTR, and influence of ISD. Blue circles are published values of SD (units for MTR are pu; mean was 38–40 pu) from eight centres; error bars show uncertainty in SD estimate (Equation 3.2). *Before* is authors' first value, almost the highest value of nine centres. After solving a scanner instability problem (Figure 2.8 in Chapter 2), ISD was low (≃0.2 pu) and the re-measured normal range (*after*) dropped to the lowest value of nine centres. (Adapted from Haynes, B.I., *et al.*, *Measuring scan-rescan reliability in quantitative brain imaging reveals instability in an apparently healthy imager and improves statistical power in a clinical study, ISMRM Annual Scientific Meeting, Stockholm*, p. 2999, 2010.)

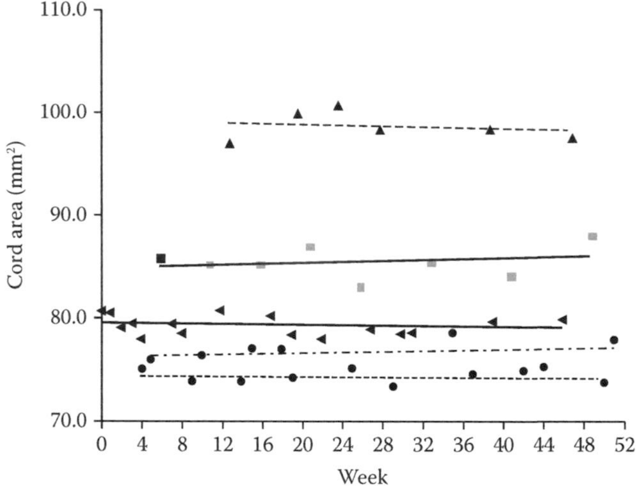

FIGURE 3.9 Early example of quantitative QA in the spinal cord. Data on spinal cord cross-sectional area for five normal controls, which has a short-term precision of 0.8% (CV). The lines are linear regressions. (From Leary, S.M., *et al.*, *Magn. Reson. Imaging*, 17, 773–776, 1999.)

3.5 Phantoms (test objects)

3.5.1 Phantom concepts

Phantom designs for T_1, T_2, ADC and PD are the most developed; these can be made from a single component, or from mixtures. Geometric objects, used for size or volume standards, are often made of *acrylic*.[17] These are immersed in water (doped to reduce its T_1 and T_2 values). Objects with a specified T_1, T_2 or diffusion value can be made from a container filled with liquid or gel, often with various salts added to reduce the relaxation times. Chemical compounds are available from suppliers such as Sigma-Aldrich.

[17] Major manufacturers are Perspex in the United Kingdom and Plexiglas in North America.

Phantoms should ideally be stable with known properties. If a design is to be made into identical phantoms at several centres, as part of a multicentre study, then care is needed on selecting and measuring out the components used in the construction.

Institutional constraints: Those wishing to provide quantitative techniques for clinical studies should be warned that some institutional representatives, operating in a paradigm of health and safety, or ethics, can object to the use of phantoms and volunteers, and slow down the progress of clinical studies.

Phantoms might leak or be damaged, toxic substances might be ingested; ready-made phantoms overcome this objection, though often at considerable cost. Volunteers from the scientist's institution might feel pressurised to volunteer; those from outside might not be covered by insurance. Sometimes a qualitative risk assessment is sufficient to allow progress. Objections might be countered by quoting ethics norms from the paradigm of a chemistry laboratory, or considering the Health and Safety of the patient group whom the clinical study seeks to aid.

3.5.2 Single component liquids

These may be water, oils or organic liquids such as alkanes. They all have the advantage of being readily available either in the laboratory, from laboratory suppliers or from the supermarket, at reasonable prices. No mixing, preparation, weighing or cookery is required. The only equipment needed is a supply of suitable containers. Handling the alkanes should be carried out in accordance with national health and safety regulations.[18]

Water has the advantage of being easily available and of a standard composition. Its intrinsic $T_1 \simeq 3.3$s, $T_2 \simeq 2.5$s at room temperature (see Table 3.8), and in its pure form these long relaxation times usually cause problems. The long T_1 can lead to incomplete relaxation with sequences that may allow full relaxation with normal brain tissue ($T_1 \simeq 600–800$ ms for normal white matter at 1.5T and 3T – see Chapter 5, Table 5.1). The long T_2 can cause transverse magnetisation coherences that would be absent in normal brain tissue ($T \simeq 90–100$ ms). Doped water overcomes these problems (see Section 3.5.3). The low viscosity can also cause problems, with internal movement continuing for some time after a phantom has been moved, giving an artificial and variable loss of transverse magnetisation in spin echo sequences used for T_2 or diffusion.

Water has another particular disadvantage when used in large volumes. Its high *dielectric constant* ($\varepsilon = 80$) leads to the presence of *radiofrequency standing waves* (*dielectric resonance*), where B_1 is enhanced, giving an artificially high flip angle and signal (see Figure 3.10). The high dielectric constant reduces the wavelength of electromagnetic radiation, compared to its value in free space, by a factor $\sqrt{\varepsilon}$; at 3T the wavelength is 260 mm, comparable with the dimensions of a head phantom (Glover *et al.*, 1985; Tofts 1994; Hoult 2000). Standing waves are also present in the head, particularly at high field (see Chapter 2, Figure 2.1), but to a much less extent, because electrical conductivity in the brain tissues damps the resonance. Even at 1.5T this effect is significant, and early attempts to measure head coil *non-uniformity* using large aqueous phantoms are now seen as fatally flawed (see Table 3.6).

Iced water has been used as a diffusion standard (Malyarenko *et al.*, 2013) (Table 3.6).

TABLE 3.6 RF nonuniformity in a uniform phantom – maximum phantom diameter

Field B_0	Water ($\varepsilon = 80$)	Oil ($\varepsilon = 5$)
0.5T	138 mm	551 mm
1.5T	46 mm	184 mm
4.7T	15 mm	59 mm

Note: The maximum diameter of a long cylinder phantom for assessing coil uniformity is given, under the condition that the signal is not to increase by more than 2% as a result of dielectric resonance in the cylinder. A circularly polarised RF coil is assumed. Filling with a low dielectric constant oil ($\varepsilon = 5$) allows larger phantoms to be used. (Adapted from Tofts 1994.)

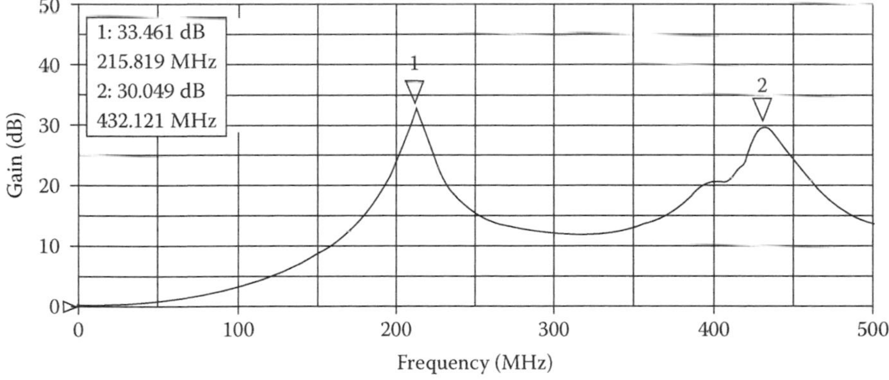

FIGURE 3.10 Dielectric resonance in a spherical flask of water. One small radiofrequency coil was placed inside the 2 litre flask (diameter 156 mm), and one outside. The graph shows the transmission between one coil and the other. The plot is the same regardless of whether the inner coil or the outer coil transmits (an example of the principle of reciprocity). The resonances correspond to wavelengths in water of one diameter and half a diameter. Without water the plot is flat. Adding salt to the water damps the resonances. The lower resonance, at 216 MHz, corresponds to 5.1T for protons. (From Hoult, D.I., *Concepts Magn. Reson.*, 12, 173–187, 2000.)

[18] In the United Kingdom this involves registering the project with a safety representative, using basic protective clothing and carrying out the pouring operation in a fume cupboard.

Oil has a low dielectric constant (ε = 2–3) and has been used for non-uniformity phantoms (Tofts *et al.*, 1997a). Several kinds are available, from various sources, with differing properties. It is stable and cheap; cooking oil is a convenient source. Some are too flammable to use in large quantities. Sources with good long-term reproducibility between samples may be hard to find. T_1 and T_2 values may be closer to *in vivo* values (T_2 values are convenient, at 33–110 ms, whilst T_1 values are generally too low, at 100–190 ms, although some flammable oils have higher values).

Silicone oils of different molecular sizes have been used to obtain a range of T_1 and T_2 values (Leach *et al.*, 1995); pure 66.9 Pa s viscosity polydimethylsiloxane gave $T_1 \simeq 800$ ms, $T_2 \simeq 100$ ms at 1.5T.

Organic liquids such as *alkanes* have been used for diffusion standards (Holz *et al.*, 2000; Tofts *et al.*, 2000). Cyclic alkanes C_nH_{2n} (n = 6–8) are the simplest possible set of organic liquids, with a single proton spectroscopic line. There are only three easily available, and they are toxic. Linear alkanes C_nH_{2n+2} (n = 6–16) are the next simplest set; 11 are readily available, ranging from hexane (which is very volatile, and inflammable) through octane (a major constituent of petrol [gasoline]), to hexadecane (which freezes at 15°C). Their T_1 values are realistic (670–1900 ms), but the T_2 values are rather long (140–200 ms), and currently it is not possible to dope them to reduce the relaxation times. Their diffusion values are ideal, covering the range found in human tissue. Dodecane (n = 12) has a diffusion coefficient of 0.8 10^{-9} m^2s^{-1}, close to the mean diffusivity of normal white matter. Their viscosity is higher than that of water, forcing bulk liquid motion to be rapidly damped. The liquids are anhydrous, so they either should be sealed well or be replaced regularly.

3.5.3 Multiple component mixtures for T_1 and T_2

Doped water has reduced T_1 and T_2, giving a material with more realistic values of relaxation times. Doping compounds are characterised by their relaxivities r_1 and r_2, which describe how much the relaxation rate $R_{1,2}$ ($R_{1,2} = 1/T_{1,2}$) is increased by adding a particular amount of the compound. In aqueous solution:

$$\frac{1}{T_1} = R_1 = R_{10} + r_1 c; \quad \frac{1}{T_2} = R_2 = R_{20} + r_2 c \tag{3.4}$$

R_{10} and R_{20} are the relaxation rates of pure water; c is the concentration of the doping compound, and the increase in relaxation rate is proportional to the concentration (Figure 3.11).

The classic compounds used for doping have been copper sulphate $CuSO_4$ and manganese chloride $MnCl_2$; nickel Ni^{++} has the advantage of a low T_1 temperature coefficient (see Section 3.5.5.4). Gd-DPTA is widely available. Agarose is good for reducing T_2 whilst hardly affecting T_1. $MnCl_2$ is a convenient way of reducing T_2 without the complexity of gel manufacture (Table 3.7).

T_1 of water: The value of this is needed to make up mixtures. (T_2 is less important, because tissue-like phantoms have a much lower T_2 than T_1, and therefore water has less effect on the final T_2 value.) Water T_1 depends on the amount of dissolved oxygen. It is independent of frequency (Krynicki 1966) (Table 3.8).

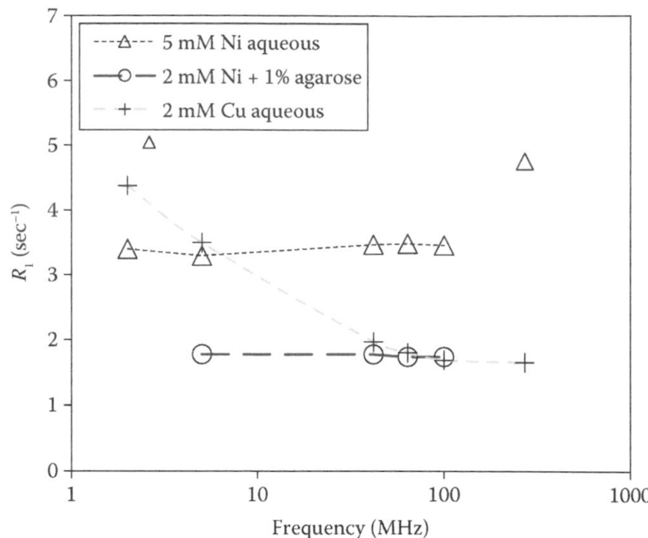

FIGURE 3.11 Field dependence of proton relaxation rate $R_1 = 1/T_1$ for Ni^{++} in aqueous solution and agarose gel and for Cu^{++} in aqueous solution. Cu^{++} has a large frequency dependence. Ni^{++} is independent of frequency up to at least 100 MHz (one point at 270 MHz clearly has a higher R_1 value). Frequency values include 42 MHz (1.0T), 64 MHz (1.5T), 100 MHz (2.4T) and 270 MHz (6.3T). (Re-drawn from Kraft, K.A., *et al.*, *Magn. Reson. Med.*, 5, 555–562, 1987.)

Dissolved oxygen: Water used for making phantoms is likely to have some oxygen in it (depending on whether it was recently boiled). The relaxivity for oxygen is approximately 1.8 ± 0.3 10^{-4} s^{-1} $(mmHg)^{-1}$ (measured in plasma at 4.7T by Meyer *et al.*, 1995). Assuming this value still holds at 3T, fully oxygenated water at 23°C (pO_2 = 150 mmHg) would then have its T_1 reduced from 3.40s to 3.11s, a reduction of 8%. A modern systematic measurement of water T_1 values, under varying conditions of temperature and pO_2, would be valuable, particularly if accompanied by T_2 values (high quality measurements of water T_2 seem to be completely lacking).

Doped agarose gels can be made up in a similar way to doped water (Mitchell *et al.*, 1986; Walker *et al.*, 1988, 1989; Christoffersson *et al.*, 1991; Tofts *et al.*, 1993). There is more control over the values of T_1 and T_2 that can be obtained, since agarose has a high r_2 and low r_1 (see Table 3.7). Agarose flakes are dissolved in hot water, up to concentrations of about 6%, in a similar way to making fruit jelly. A hotplate (Mitchell *et al.*, 1986) or a microwave oven (Tofts *et al.*, 1993) can be used. Stirring is necessary, and care must be taken not to overheat the gel. Fungicide can be added to improve stability. Agarose is relatively expensive if large volumes are to be made up; cooking it is a relatively complex process, and obtaining a uniform gel on cooling also requires skill. Commercially available doped gels with a wide range of T_1 and T_2 values are obtainable (see Section 3.5.6.4); however for many applications single liquids or aqueous solutions will suffice.

TABLE 3.7 Values of relaxivity at 1.5T[a] and room temperature

Relaxation Agent[b]		Source	r_1 (s^{-1} mM^{-1})	r_2 (s^{-1} mM^{-1})
T_1				
	Ni^{++}	Morgan and Nolle (1959)[c]	0.70 ± 0.06	0.70 ± 0.06
		Kraft *et al.* (1987)[d,e]	0.64	–
		Jones (1997)[f]	0.644 ± 0.002	0.698 ± 0.005
	Gd-DTPA	Tofts *et al.* (1993)[g]	4.50 ± 0.04	5.49 ± 0.06
T_2				
	Mn^{++}	Morgan and Nolle (1959)[c]	7.0 ± 0.4	70 ± 4
		Bloembergen and Morgan (1961)[h]	8.0 ± 0.4	80 ± 7
	Agarose	Mitchell *et al.* (1986)[i]	0.05	10
		Tofts *et al.* (1993)	0.01 ± 0.01	9.7 ± 0.2
		Jones (1997)[f]	0.04 ± 0.01	8.80 ± 0.04

[a] There are very few published data at 3T and above; relaxivities for these four agents are similar to values at 1.5T.
[b] More data are shown in first edition, Table 3.5.
[c] At 60MHz, 27°C, calculated by the author from data points on the published figures; 95% confidence limits estimated from scatter in the plots.
[d] See Figure 3.11.
[e] Estimated from published T_1 value.
[f] Estimated from data at 1.5T in the MSc thesis of Craig K Jones (University of British Columbia 1997) (for more details see first edition); 2mM Ni^{++} in 1% agarose gives T_1 = 573 ms, T_2 = 95 ms.
[g] Gd-DTPA r_1 is independent of field up to 4.7T (data at 37°C; Rohrer *et al.*, 2005).
[h] At 60 MHz, 23°C, calculated by the author from data points on the published figures; 95% confidence limits estimated from scatter in the plots.
[i] Data at 5 and 60 MHz.

TABLE 3.8 T_1 values for pure water

Temperature (°C)	T_1 (s)
0	1.73
5	2.07
10	2.39
15	2.76
20	3.15
21*	3.23
22*	3.32
23*	3.40
24*	3.49
25	3.57
37*	4.70

Note: Measurements were made at 28 MHz using a continuous-wave saturation-recovery technique; estimated 95% confidence limits were ±3%. Values marked (*) are linearly interpolated (from data in Krynicki 1966; see also Tofts *et al.*, 2008). Note values are expected to be independent of field strength.

By using a *mixture of two compounds*, a range of T_1 and T_2 values can be obtained, intermediate between those that would be obtained with only one of the compounds. (Mitchell *et al.*, 1986; Schneiders 1988; Tofts *et al.*, 1993). It is important to establish that the two components do not interact; this can be done by plotting relaxation rates versus concentration for the individual components (to establish their relaxivities) and then for mixtures (to show that the individual relaxivities are unaffected). The most useful combinations are pairs where one has high r_2 (much greater than r_1, i.e. $MnCl_2$ or agarose) and the other has low r_2 (about the same as r_1). Thus suitable mixtures are Ni^{++} and Mn^{++} in aqueous solution (Schneiders 1988), Gd-DTPA[19] and agarose (Walker *et al.*, 1989), Ni^{++} and agarose (Kraft *et al.*, 1987), and Ni-DTPA and agarose (Tofts *et al.*, 1993). Linear equations can be produced giving the concentrations of each compound required, for a target T_1 and T_2 value, given the relaxivities of each component, and the T_1 and T_2 of pure water (Tofts *et al.*, 1993) (Figure 3.12).

A mixture of Ni^{++} in agarose provides reduced temperature dependence for T_1. For example a phantom with T_1 = 600 ms, T_2 = 100 ms at 1.5T is produced by mixing 1.77 mM Ni^{++} in 0.96% agarose.[20] Relaxation in Ni^{++} is dominated by fast electron interactions, which are independent of temperature; this also increases the frequency up to which relaxation is almost independent of frequency, although above 4T other relaxation mechanisms come into play (Kraft *et al.*, 1987).

The process of making up the mixtures can be simplified by making up concentrated stock solutions of the components. The required T_1 and T_2 values can be entered into a spreadsheet, along with the relaxivities and stock solution concentrations, to give a simple list of how much stock solution must be added to a particular volume of water to give the required relaxation times.

3.5.4 Other materials

Aqueous *sucrose solutions* have been used for diffusion standards; these are easily made up, and T_1 and T_2 can be

[19] This is preferred to $GdCl_3$, which interacts with the agarose.
[20] See first edition, p. 73.

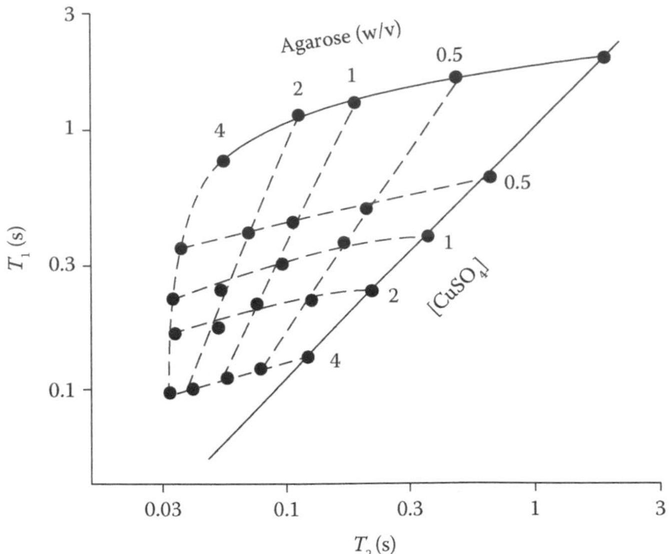

FIGURE 3.12 T_1 and T_2 values for aqueous solutions of agarose and $CuSO_4$. Agarose concentrations are from 0% to 4% weight/volume, Cu is from 0 to 4 mM. Note agarose decreases T_2 but hardly affects T_1, whilst Cu decreases T_1 and T_2 about equally. The agarose line (curved, where [Cu] = 0) and the Cu line (straight, where [agarose] = 0) bound the possible values that the mixture can achieve. Dotted lines connect points of equal agarose or Cu concentration. (Data at 5 MHz from Mitchell, M.D., *et al.*, *Magn. Reson. Imaging*, 4, 263–266, 1986).

controlled by doping (Laubach *et al.*, 1998; Delakis *et al.*, 2004; Lavdas *et al.*, 2013).

PVP (Polyvinylpyrrolidone) is biologically benign (used in postage stamp adhesive and shampoo) and stable. A 25% solution in water at 0°C has T_1 = 533 ms, T_2 = 519 ms at 1.5T and T_1 = 610 ms, T_2 = 500 ms at 3T. ADC is 0.49 10^{-9} m^2s^{-1} (Jerome *et al.*, 2016; Pullens *et al.*, 2017). PVP was first suggested by Pierpaoli *et al.*, in 2009.

Various gels have been used as *MRI radiation dosimeters* (Lepage *et al.*, 2001); the T_2 decreases with dose and is read out after irradiation. In this context, there has been much attention devoted to designing stable gels (De Deene *et al.*, 2000), and this work may result in new designs for MRI QA materials.

3.5.5 Temperature dependence and control

3.5.5.1 Temperature dependence

Temperature dependence of phantom parameter values can be a major problem. The scanner room environment may vary by 1° or 2°C, unless special precautions are taken. The magnet bore may have a colder supply of air blown in to assist the breathing of the MRI subject. A refrigerated phantom could be as cold as 5°C. A phantom positioned next to the subject could warm above room temperature.

T_1, T_2 and ADC all vary by about 1–3%/°C, corresponding to errors of about 5% in the parameter value in an uncontrolled environment. The Eurospin gels (Lerski and de Certaines 1993) have a T_1 temperature coefficient of +2.6%/°C.[21] In alkane phantoms, the diffusion coefficient changes by 2%–3%/°C (Tofts *et al.*, 2000). Agarose has a T_2 coefficient of about –1.25%/°C (Tofts *et al.*, 1993; McRobbie 2017).[22]

PD and MRS concentration measurements are also vulnerable to temperature change; both the density and the magnetic susceptibility vary considerably between refrigerator, room and body temperature. An accurate correction is possible (Section 3.5.5.5).

The effects of temperature dependence can be mitigated by controlling or correcting for the environmental temperature or by using compounds (principally Ni) with reduced temperature dependence.

3.5.5.2 Controlling environmental temperature

Phantoms should be stored in the scanner room at room temperature (not refrigerated). The phantoms should be thermally insulating during their time in the magnet bore (which may have a different and varying temperature), and their temperature should be measured (ideally whilst in the bore, using a thermocouple[23] or a liquid-in-glass thermometer; Tofts and Collins 2011). Temperatures should be known to within better than 1°C, and ideally 0.2°C, in order to allow MR measurements to within 1%. Temperature gradients within the phantoms can be minimised by avoiding both rapid changes in temperature and the presence of any electrical conductors, which might attract induced RF currents and consequent heating. Thermal insulation foam is effective and widely available for house insulation.

Iced water can also be used as a bath for temperature control (Chenevert *et al.*, 2011; Malyarenko *et al.*, 2013; Jerome *et al.*, 2016). Organic liquids freezing above 0°C can also be used (Vescovo 2012).

3.5.5.3 Correcting for phantom temperature

If the phantom temperature varies with each measurement occasion, a correction procedure may still be possible. The temperature coefficient has to be known, the phantom must have a well-defined single temperature (without any temperature gradients), and the temperature must be measured on each occasion. The measured parameter value (e.g. T_1) can then be converted to its estimated value at a standard temperature (e.g. 20°C) (Vassiliou *et al.*, 2016).

[21] This coefficient, calculated from values given in the manual, is approximately independent of T_1 and T_2, since the T_1 behaviour is almost completely determined by the Gd-DTPA.

[22] Walker *et al.* (1988) gives a theoretical temperature coefficient in 2% agarose (T_2 = 60ms) at 20MHz of –1.7%/°C. The Eurospin gels closest to brain tissue have a T_2 coefficient of about –1.5%/°C; most of this originates in the agarose, but it may be attenuated slightly by the positive coefficient of the Gd-DTPA: see also the discussion in McRobbie 2017.

[23] A thin T-type thermocouple has signal dropout limited to within about 3 mm of the tip.

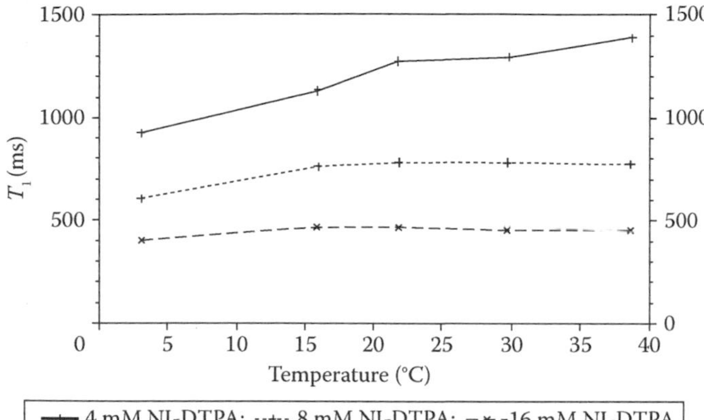

FIGURE 3.13 Temperature dependence of T_1 in a Ni-doped gel at 1.5T. In the tissue-equivalent material (Ni-DTPA in 2% agarose), T_1 is dominated by the relaxation from Ni-DTPA, particularly at the lower T_1 values, and therefore has little dependence on temperature. At room temperature the 8 mM and 16 mM data are flat. Materials with these concentrations of Ni-DTPA, in 1% agarose, have (T_1 = 909 ms, T_2 = 99 ms) and (T_1 = 510 ms, T_2 = 89 ms), covering the range of normal brain tissue. (From Tofts, P.S., *et al.*, *Magn. Reson. Imaging*, 11, 125–133, 1993.)

3.5.5.4 Compounds with reduced temperature dependence for T_1

Ni has a minimum in its T_1 relaxation rate, fortuitously at room temperature (Kraft *et al.*, 1987; Tofts *et al.*, 1993), allowing a brain-equivalent Ni-DTPA agarose gel to have a flat temperature response (Figure 3.13). At 1.5T and 37°C, Ni^{++} agarose phantoms had temperature coefficients of +0.05% K^{-1} (for 530 ms) and +0.7 K^{-1} (for 900 ms) (Vassiliou *et al.*, 2016).

A more general solution is to use a Gd polymer pair (Kellar and Briley-Saebo 1998).[24] One component (NC663868) has a zero temperature coefficient for T_1. The second component (NC22181) has a negative coefficient (about –1.2%/°C) and can be used to neutralise the small effect arising from the positive coefficient of the host material (water and/or agarose). Thus the pair, used in agarose solution, can give zero temperature coefficients for a range of T_1 values.

3.5.5.5 PD and MRS concentrations: correction for temperature

When measuring proton density (Chapter 4) or metabolite concentration by spectroscopy (Chapters 12 and 13), the signal from a concentration standard is often used to measure the absolute gain of the MR system.

The signal from a test object or standard at *room temperature* differs from the same object at *body temperature* for two reasons: first the magnetisation M_0 of a given number of protons is inversely proportional to absolute temperature (Chapter 2, Equation 2.2). This corresponds to a reduction of 0.34%/K at 20°C. Thus the signal at room temperature (about 20°C) will be approximately 5.5% higher than at body temperature (37°C). This was confirmed by phantom measurements of effective spin density versus temperature, over the range 17–36°C, which did indeed show a decrease of 0.32%/K (Venkatesan *et al.*, 2000).

Second, the density of water at room temperature is about 0.5% higher than at body temperature; this small increase in the number of protons will increase the signal at room temperature by this amount. These two factors reinforce.

Thus the signal from a standard, S_{Ts}, measured at room temperature T_s°C, should be converted to the equivalent (lower) value (S_{37}) that it would have at body temperature (37.0°C = 310.2 K):

$$S_{37} = S_{Ts}\frac{\rho_{37}}{\rho_{Ts}}\frac{273.2+T_s}{310.2} \tag{3.5}$$

For a room temperature of 20°C, the correction factor is 0.9406.[25] If the phantom has been kept refrigerated before being imaged, the correction factor will be even larger (up to 11%). Thus for high accuracy the phantom temperature should be recorded, and the signal from the standard corrected to obtain the body temperature value.

3.5.6 Phantom design

3.5.6.1 Phantoms for all quantitative MR parameters

Phantoms for T_1, T_2 and ADC are well developed and have been described above. Phantoms for other parameters are less developed; these may use doped agarose as a host matrix, to obtain realistic T_1 and T_2 values; for example an R_2* phantom uses Ultrasmall Superparamagnetic Iron Oxide (USPIO) particles in agarose doped with Gd-DTPA (Brown *et al.*, 2017). The chapters on each MR parameter will give information on any available phantoms. Desirable qualities are summarised in Table 3.1.

3.5.6.2 Phantom containers

Aqueous and gel-based materials can be conveniently contained in cylindrical polythene containers, about 20–25 mm in diameter. These have plastic screw tops. Foil inserts should probably be avoided. Organic liquids need to be in glass, and polypropylene snap tops are available, although in the author's experience they do allow significant evaporation. For some applications (particularly spectroscopy) spherical containers may be advised, to eliminate the internal susceptibility field gradient. A long cylinder can also give a uniform internal field. Glass spheres with a neck attached for filling are available. Larger objects can be machined from acrylic, although this can be time-consuming and expensive. Convenient airtight polythene containers are often available sold as food containers (lunch boxes).

A matrix of small cylindrical battles can conveniently be supported in a block of expanded polystyrene, with holes drilled

[24] These compounds have to be made up; they are not, to the author's knowledge, available commercially.

[25] Correction factors for signal measured from a standard at a range of temperatures are given in first edition, p 98.

in it. Slabs of polystyrene, about 50 mm thick, are available from builders' merchants for use as wall cavity insulation material. Drilling is a messy operation and should be carried out with a bit that has a tangential blade that rotates around the circumference of the hole. The polystyrene slab can be cut to a circular shape that is a tight fit inside the head coil. Tool blades should be new, with no history of cutting ferrous materials.

EPI sequences probably need the phantoms to be in a water bath (to reduce the susceptibility effects). This can be done by placing bottles in as large a food container as can be fitted into the head coil. Alternatively, a close-fitting water bath, with holes for bottles to be slid in, can be made from acrylic.

3.5.6.3 Stability of phantom materials

The stability of agarose gels is still under discussion.[26] Although there is evidence of stability (Mitchell *et al.*, 1986; Walker *et al.*, 1988; Christoffersson *et al.*, 1991), other workers have reported changes over time, possibly related to how well the containers are sealed or to contamination of the gel. The temperatures involved in melting the gel should sterilise the container; alternatively, a fungicide can be added. Care should be taken to avoid the entrance of air into the container during the gel cooling process (Vassiliou *et al.*, 2016). A glass container with a narrow neck that can be melted to provide a permanent seal is ideal; if the air is pumped out, then as the neck melts, air pressure forces it to narrow and seal.[27] An alternative is to use a cylindrical glass bottle, pour melted wax over the solid gel, then glue the lid on.[28] Evaporation of water (from an aqueous solution or a gel), or water entering and mixing with anhydrous liquids, can be detected by regular weighing of the test objects. However instability of the gel could not be detected by a weight change.

Nonetheless, a slow change in parameter value may be measured, and it is almost impossible to be sure whether this is caused by scanner or phantom change over time (see Figure 3.1) (Vassiliou *et al.*, 2016). The only reliable way to use liquid- or gel-based test objects is to regularly calibrate them (i.e. measure their true parameter value) or else, in the case of single-component liquids, to replace them regularly.

3.5.6.4 Ready-made phantoms and designs

The Eurospin set of test objects (Lerski 1993; Lerski and de Certaines 1993) from Diagnostic Sonar Ltd[29] is comprehensive.[30] The ACR phantom (aqueous $NiCl_2$ + $NaCl_2$) is widely used for evaluation of geometric parameters, weighted images and even diffusion (Ihalainen *et al.*, 2011; Panych *et al.*, 2016; Wang *et al.*, 2016). The Alzheimer's Disease Neuroimaging Initiative (ADNI) phantom is for multicentre geometric measurements (Gunter *et al.*, 2009). The ISMRM/NIST[31] phantom contains multiple compartments with standardised PD, T_1 and T_2 values; aqueous solutions of $NiCl_2$ and $MnCl_2$ are used for T_1 and T_2, respectively (Jiang *et al.*, 2016).

[26] Commercial fruit jams are stable for many years, and these may inspire suitable gel design.

[27] A chemistry glassmaker can often make such a container.

[28] This approach appears to have been used with the Eurospin gels.

[29] http://www.diagnosticsonar.com

[30] Commercial suppliers of various test objects include High Precision Devices, www.hpd-online.com.

[31] ISMRM, International Society for Magnetic Resonance in Medicine; NIST, US National Institute of Standards and Technology.

References

Armitage P, Matthews JNS, Berry G. *Statistical Methods in Medical Research*. Blackwell, New Jersey; 2001.

Barker GJ, Tofts PS. Semiautomated quality assurance for quantitative magnetic resonance imaging. Magn Reson Imaging 1992; 10(4): 585–95.

Bauer CM, Jara H, Killiany R, Initiative AsDN. Whole brain quantitative T2 MRI across multiple scanners with dual echo FSE: applications to AD, MCI, and normal aging. NeuroImage 2010; 52(2): 508–14.

Bland JM, Altman DG. Statistical methods for assessing agreement between two methods of clinical measurement. Lancet 1986; 1(8476): 307–10.

Bland JM, Altman DG. Measurement error. BMJ 1996a; 313(7059): 744.

Bland JM, Altman DG. Measurement error. BMJ 1996b; 312(7047): 1654.

Bland JM, Altman DG. Measurement error and correlation coefficients. BMJ 1996c; 313(7048): 41–2.

Bloembergen N, Morgan LO. Proton relaxation times in paramagnetic solutions. Effects of electron spin relaxation. J Chem Phys 1961; 34(3): 842–50.

Brihuega-Moreno O, Heese FP, Hall LD. Optimization of diffusion measurements using Cramer-Rao lower bound theory and its application to articular cartilage. Magn Reson Med 2003; 50(5): 1069–76.

Brown GC, Cowin GJ, Galloway GJ. A USPIO doped gel phantom for R2* relaxometry. MAGMA 2017; 30(1): 15–27.

Cavassila S, Deval S, Huegen C, van Ormondt D, Graveron-Demilly D. Cramer-Rao bounds: an evaluation tool for quantitation. NMR Biomed 2001; 14(4): 278–83.

Chard DT, McLean MA, Parker GJ, MacManus DG, Miller DH. Reproducibility of in vivo metabolite quantification with proton magnetic resonance spectroscopic imaging. J Magn Reson Imaging 2002; 15(2): 219–25.

Chenevert TL, Galbán CJ, Ivancevic MK, Rohrer SE, Londy FJ, Kwee TC, *et al.* Diffusion coefficient measurement using a temperature-controlled fluid for quality control in multicenter studies. J Magn Reson Imaging 2011; 34(4): 983–7.

Christoffersson JO, Olsson LE, Sjoberg S. Nickel-doped agarose gel phantoms in MR imaging. Acta Radiol 1991; 32(5): 426–31.

Cohen JA, Fischer JS, Bolibrush DM, Jak AJ, Kniker JE, Mertz LA, et al. Intrarater and interrater reliability of the MS functional composite outcome measure. Neurology 2000; 54(4): 802–6.

De Deene Y, Hanselaer P, De Wagter C, Achten E, De Neve W. An investigation of the chemical stability of a monomer/polymer gel dosimeter. Phys Med Biol 2000; 45(4): 859–78.

Delakis I, Moore EM, Leach MO, De Wilde JP. Developing a quality control protocol for diffusion imaging on a clinical MRI system. Phys Med Biol 2004; 49(8): 1409–22.

de Wilde J, Price D, Curran J, Williams J, Kitney R. Standardization of performance evaluation in MRI: 13 Years' experience of intersystem comparison. Concepts Magn Reson Part A 2002; 15(1): 111–6.

Droby A, Lukas C, Schänzer A, Spiwoks-Becker I, Giorgio A, Gold R, et al. A human post-mortem brain model for the standardization of multi-centre MRI studies. NeuroImage 2015; 110: 11–21.

Filippi M, Gawne-Cain ML, Gasperini C, van Waesberghe JH, Grimaud J, Barkhof F, et al. Effect of training and different measurement strategies on the reproducibility of brain MRI lesion load measurements in multiple sclerosis. Neurology 1998; 50(1): 238–44.

Firbank MJ, Harrison RM, Williams ED, Coulthard A. Quality assurance for MRI: practical experience. Br J Radiol 2000; 73(868): 376–83.

Galbraith SM, Lodge MA, Taylor NJ, Rustin GJ, Bentzen S, Stirling JJ, et al. Reproducibility of dynamic contrast-enhanced MRI in human muscle and tumours: comparison of quantitative and semi-quantitative analysis. NMR Biomed 2002; 15(2): 132–42.

Glover GH, Hayes CE, Pelc NJ, Edelstein WA, Mueller OM, Hart HR, et al. Comparison of linear and circular polarization for magnetic resonance imaging. J Magn Reson 1985; 64: 255–70.

Gunter JL, Bernstein MA, Borowski BJ, Ward CP, Britson PJ, Felmlee JP, et al. Measurement of MRI scanner performance with the ADNI phantom. Med Phys 2009; 36(6): 2193–205.

Hajek M, Babis M, Herynek V. MR relaxometry on a whole-body imager: quality control. Magn Reson Imaging 1999; 17(7): 1087–92.

Haynes BI, Dowell NG, Tofts PS. Measuring scan-rescan reliability in quantitative brain imaging reveals instability in an apparently healthy imager and improves statistical power in a clinical study. ISMRM annual scientific meeting; Stockholm; 2010. 2999.

Holz M, Heil SR, Sacco A. Temperature-dependent self-diffusion coefficients of water and six selected molecular liquids for calibration in accurate H-1 NMR PFG measurements. Phys Chem Chem Phys 2000; 2(20): 4740–2.

Hoult DI. The principle of reciprocity in signal strength calculations - a mathematical guide. Concepts Magn Reson 2000; 12: 173–87.

Ihalainen TM, Lönnroth NT, Peltonen JI, Uusi-Simola JK, Timonen MH, Kuusela LJ, et al. MRI quality assurance using the ACR phantom in a multi-unit imaging center. Acta Oncol 2011; 50(6): 966–72.

Jerome NP, Papoutsaki MV, Orton MR, Parkes HG, Winfield JM, Boss MA, et al. Development of a temperature-controlled phantom for magnetic resonance quality assurance of diffusion, dynamic, and relaxometry measurements. Med Phys 2016; 43(6): 2998.

Jiang Y, Ma D, Keenan KE, Stupic KF, Gulani V, Griswold MA. Repeatability of magnetic resonance fingerprinting T1 and T2 estimates assessed using the ISMRM/NIST MRI system phantom. Magn Reson Med 2017; 78(4):1452–57.

Keevil SF, Barbiroli B, Brooks JC, Cady EB, Canese R, Carlier P, et al. Absolute metabolite quantification by in vivo NMR spectroscopy: II. A multicentre trial of protocols for in vivo localised proton studies of human brain. Magn Reson Imaging 1998; 16(9): 1093–106.

Kellar KE, Briley-Saebo K. Phantom standards with temperature- and field-independent relaxation rates for magnetic resonance imaging 1998; 33(8): 472–9.

Kraft KA, Fatouros PP, Clarke GD, Kishore PR. An MRI phantom material for quantitative relaxometry. Magn Reson Med 1987; 5(6): 555–62.

Krummenauer F, Doll G. Statistical methods for the comparison of measurements derived from orthodontic imaging. Eur J Orthod 2000; 22(3): 257–69.

Krynicki K. Proton spin lattice relaxation in pure water between 0°C and 100°C. Physica 1966; 32: 167–78.

Laubach HJ, Jakob PM, Loevblad KO, Baird AE, Bovo MP, Edelman RR, et al. A phantom for diffusion-weighted imaging of acute stroke. J Magn Reson Imaging 1998; 8(6): 1349–54.

Lavdas I, Behan KC, Papadaki A, McRobbie DW, Aboagye EO. A phantom for diffusion-weighted MRI (DW-MRI). J Magn Reson Imaging 2013; 38(1): 173–9.

Leach MO, Collins DJ, Keevil S, Rowland I, Smith MA, Henriksen O, et al. Quality assessment in in vivo NMR spectroscopy: III. Clinical test objects: design, construction, and solutions. Magn Reson Imaging 1995; 13(1): 131–7.

Leary SM, Parker GJ, Stevenson VL, Barker GJ, Miller DH, Thompson AJ. Reproducibility of magnetic resonance imaging measurements of spinal cord atrophy: the role of quality assurance. Magn Reson Imaging 1999; 17(5): 773–6.

Lepage M, Whittaker AK, Rintoul L, Back SA, Baldock C. The relationship between radiation-induced chemical processes and transverse relaxation times in polymer gel dosimeters. Phys Med Biol 2001; 46(4): 1061–74.

Lerski RA. Trial of modifications to Eurospin MRI test objects. Magn Reson Imaging 1993; 11(6): 835–9.

Lerski RA, de Certaines JD. Performance assessment and quality control in MRI by Eurospin test objects and protocols. Magn Reson Imaging 1993; 11(6): 817–33.

Malyarenko D, Galbán CJ, Londy FJ, Meyer CR, Johnson TD, Rehemtulla A, et al. Multi-system repeatability and reproducibility of apparent diffusion coefficient measurement using an ice-water phantom. J Magn Reson Imaging 2013; 37(5): 1238–46.

McRobbie DW, Quest RA. Effectiveness and relevance of MR acceptance testing: results of an 8 year audit. Br J Radiol 2002; 75(894): 523–31.

McRobbie D, Semple S. Quality Control and Artefacts in Magnetic Resonance Imaging (IPEM report 112). York: Institute of Physics and Engineering in Medicine; 2017.

Meyer ME, Yu O, Eclancher B, Grucker D, Chambron J. NMR relaxation rates and blood oxygenation level. Magn Reson Med 1995; 34(2): 234–41.

Mitchell MD, Kundel HL, Axel L, Joseph PM. Agarose as a tissue equivalent phantom material for NMR imaging. Magn Reson Imaging 1986; 4(3): 263–6.

Morgan LO, Nolle AW. Proton spin relaxation in aqueous solutions of paramagnetic ions II Cr +++, Mn ++, Ni ++, Cu ++ and Gd +++. J Chem Phys 1959; 31: 365.

O'Connor JP, Aboagye EO, Adams JE, Aerts HJ, Barrington SF, Beer AJ, et al. Imaging biomarker roadmap for cancer studies. Nat Rev Clin Oncol 2017; 14(3): 169–86.

Och JG, Clarke GD, Sobol WT, Rosen CW, Mun SK. Acceptance testing of magnetic resonance imaging systems: report of AAPM Nuclear Magnetic Resonance Task Group No. 6. Med Phys 1992; 19(1): 217–29.

Padhani AR, Hayes C, Landau S, Leach MO. Reproducibility of quantitative dynamic MRI of normal human tissues. NMR Biomed 2002; 15(2): 143–53.

Padhani AR, Liu G, Koh DM, Chenevert TL, Thoeny HC, Takahara T, et al. Diffusion-weighted magnetic resonance imaging as a cancer biomarker: consensus and recommendations. Neoplasia 2009; 11(2): 102–25.

Panych LP, Chiou JY, Qin L, Kimbrell VL, Bussolari L, Mulkern RV. On replacing the manual measurement of ACR phantom images performed by MRI technologists with an automated measurement approach. J Magn Reson Imaging 2016; 43(4): 843–52.

Parkes LM, Rashid W, Chard DT, Tofts PS. Normal cerebral perfusion measurements using arterial spin labeling: reproducibility, stability, and age and gender effects. Magn Reson Med 2004; 51(4): 736–43.

Parkes LM, Tofts PS. Improved accuracy of human cerebral blood perfusion measurements using arterial spin labeling: accounting for capillary water permeability. Magn Reson Med 2002; 48(1): 27–41.

Podo F. Tissue characterization by MRI: a multidisciplinary and multi-centre challenge today. Magn Reson Imaging 1988; 6(2): 173–4.

Podo F, Henriksen O, Bovee WM, Leach MO, Leibfritz D, de Certaines JD. Absolute metabolite quantification by in vivo NMR spectroscopy: I. Introduction, objectives and activities of a concerted action in biomedical research. Magn Reson Imaging 1998; 16(9): 1085–92.

Price RR, Axel L, Morgan T, Newman R, Perman W, Schneiders N, et al. Quality assurance methods and phantoms for magnetic resonance imaging: report of AAPM nuclear magnetic resonance Task Group No. 1. Med Phys 1990; 17(2): 287–95.

Provencher SW. Automatic quantitation of localized in vivo 1H spectra with LCModel. NMR Biomed 2001; 14(4): 260–4.

Pullens P, Bladt P, Sijbers J, Maas AI, Parizel PM. Technical note: a safe, cheap, and easy-to-use isotropic diffusion MRI phantom for clinical and multicenter studies. Med Phys 2017; 44(3): 1063–70.

Rohrer M, Bauer H, Mintorovitch J, Requardt M, Weinmann HJ. Comparison of magnetic properties of MRI contrast media solutions at different magnetic field strengths. Invest Radiol 2005; 40(11): 715–24.

Rovaris M, Mastronardo G, Sormani MP, Iannucci G, Rodegher M, Comi G, et al. Brain MRI lesion volume measurement reproducibility is not dependent on the disease burden in patients with multiple sclerosis. Magn Reson Imaging 1998; 16(10): 1185–9.

Schneiders NJ. Solutions of two paramagnetic ions for use in nuclear magnetic resonance phantoms. Med Phys 1988; 15(1): 12–16.

Silver NC, Barker GJ, MacManus DG, Tofts PS, Miller DH. Magnetisation transfer ratio of normal brain white matter: a normative database spanning four decades of life. J Neurol Neurosurg Psychiatry 1997; 62(3): 223–8.

Simmons A, Moore E, Williams SC. Quality control for functional magnetic resonance imaging using automated data analysis and Shewhart charting. Magn Reson Med 1999; 41(6): 1274–8.

Soher BJ, Hurd RE, Sailasuta N, Barker PB. Quantitation of automated single-voxel proton MRS using cerebral water as an internal reference. Magn Reson Med 1996; 36(3): 335–9.

Sormani MP, Gasperini C, Romeo M, Rio J, Calabrese M, Cocco E, et al. Assessing response to interferon-β in a multicenter dataset of patients with MS. Neurology 2016; 87(2): 134–40.

Streiner DL, Norman GR. Health Measurement Scales: a practical guide to their development and use. Oxford University Press, Oxford; 1995.

Sun J, Barnes M, Dowling J, Menk F, Stanwell P, Greer PB. An open source automatic quality assurance (OSAQA) tool for the ACR MRI phantom. Australas Phys Eng Sci Med 2015; 38(1): 39–46.

Taylor JR. An introduction to error analysis: the study of uncertainties in physical measurements. Sausalito, CA, USA: University Science Books; 1997.

Tofts PS. Standing waves in uniform water phantoms. J Magn Reson series B 1994; 104: 143–7.

Tofts PS. Optimal detection of blood-brain barrier defects with Gd-DTPA MRI-the influences of delayed imaging and optimised repetition time. Magn Reson Imaging 1996; 14(4): 373–80.

Tofts PS. Standardisation and optimisation of magnetic resonance techniques for multicentre studies. J Neurol Neurosurg Psychiatry 1998; 64 Suppl 1: S37–43.

Tofts PS, Barker GJ, Dean TL, Gallagher H, Gregory AP, Clarke RN. A low dielectric constant customized phantom design to measure RF coil nonuniformity. Magn Reson Imaging 1997a; 15(1): 69–75.

Tofts PS, Barker GJ, Filippi M, Gawne-Cain M, Lai M. An oblique cylinder contrast-adjusted (OCCA) phantom to measure the accuracy of MRI brain lesion volume estimation schemes in multiple sclerosis. Magn Reson Imaging 1997b; 15(2): 183–92.

Tofts PS, Berkowitz B, Schnall MD. Quantitative analysis of dynamic Gd-DTPA enhancement in breast tumors using a permeability model. Magn Reson Med 1995; 33(4): 564–8.

Tofts PS, Brix G, Buckley DL, Evelhoch JL, Henderson E, Knopp MV, et al. Estimating kinetic parameters from dynamic contrast-enhanced T(1)-weighted MRI of a diffusable tracer: standardized quantities and symbols. J Magn Reson Imaging 1999; 10(3): 223–32.

Tofts PS, Collins DJ. Multicentre imaging measurements for oncology and in the brain. Br J Radiol 2011; 84 Spec No 2: S213–26.

Tofts PS, Jackson JS, Tozer DJ, Cercignani M, Keir G, MacManus DG, et al. Imaging cadavers: cold FLAIR and noninvasive brain thermometry using CSF diffusion. Magn Reson Med 2008; 59(1): 190–5.

Tofts PS, Kermode AG, MacManus DG, Robinson WH. Nasal orientation device to control head movement during CT and MR studies. J Comput Assist Tomogr 1990; 14(1): 163–4.

Tofts PS, Lloyd D, Clark CA, Barker GJ, Parker GJ, McConville P, et al. Test liquids for quantitative MRI measurements of self-diffusion coefficient in vivo. Magn Reson Med 2000; 43(3): 368–74.

Tofts PS, Shuter B, Pope JM. Ni-DTPA doped agarose gel - a phantom material for Gd-DTPA enhancement measurements. Magn Reson Imaging 1993; 11(1): 125–33.

Tofts PS, Steens SC, Cercignani M, Admiraal-Behloul F, Hofman PA, van Osch MJ, et al. Sources of variation in multi-centre brain MTR histogram studies: body-coil transmission eliminates inter-centre differences. Magma 2006; 19(4): 209–22.

Van Essen DC, Smith SM, Barch DM, Behrens TE, Yacoub E, Ugurbil K, et al. The WU-Minn human connectome project: an overview. NeuroImage 2013; 80: 62–79.

Vassiliou VS, Heng EL, Gatehouse PD, Donovan J, Raphael CE, Giri S, et al. Magnetic resonance imaging phantoms for quality-control of myocardial T1 and ECV mapping: specific formulation, long-term stability and variation with heart rate and temperature. J Cardiovasc Magn Reson 2016; 18(1): 62.

Venkatesan R, Lin W, Gurleyik K, He YY, Paczynski RP, Powers WJ, et al. Absolute measurements of water content using magnetic resonance imaging: preliminary findings in an in vivo focal ischemic rat model. Magn Reson Med 2000; 43(1): 146–50.

Vescovo E, Levick A, Childs C, Machin G, Zhao S, Williams SR. High-precision calibration of MRS thermometry using validated temperature standards: effects of ionic strength and protein content on the calibration. NMR Biomed 2013; 26(2): 213–23.

Walker L, Curry M, Nayak A, Lange N, Pierpaoli C, Group BDC. A framework for the analysis of phantom data in multicenter diffusion tensor imaging studies. Hum Brain Mapp 2013; 34(10): 2439–54.

Walker P, Lerski RA, Mathur-De Vre R, Binet J, Yane F. Preparation of agarose gels as reference substances for NMR relaxation time measurement. EEC Concerted Action Program. Magn Reson Imaging 1988; 6(2): 215–22.

Walker PM, Balmer C, Ablett S, Lerski RA. A test material for tissue characterisation and system calibration in MRI. Phys Med Biol 1989; 34(1): 5–22.

Wang ZJ, Seo Y, Babcock E, Huang H, Bluml S, Wisnowski J, et al. Assessment of diffusion tensor image quality across sites and vendors using the American College of Radiology head phantom. J Appl Clin Med Phys 2016; 17(3): 442–51.

Wei X, Warfield SK, Zou KH, Wu Y, Li X, Guimond A, et al. Quantitative analysis of MRI signal abnormalities of brain white matter with high reproducibility and accuracy. J Magn Reson Imaging 2002; 15(2): 203–9.

Zhou X, Sakaie KE, Debbins JP, Kirsch JE, Tatsuoka C, Fox RJ, et al. Quantitative quality assurance in a multicenter HARDI clinical trial at 3T. Magn Reson Imaging 2017; 35: 81–90.

4

PD: Proton Density of Tissue Water[1]

Contents

Shir Filo and
Aviv A Mezer
The Hebrew University of Jerusalem

4.1 Introduction

Proton density (PD) measurements indicate the amount of magnetic resonance (MR)-visible protons contributing to the MRI signal. In the brain, PD is used to quantify water content (WC).[2]

Water is fundamental for brain function and protection, and different brain areas tend to have distinct WC (see Tables 4.1 and 4.2). Maturation and ageing involve changes in brain WC (Holland *et al.*, 1986; Neeb *et al.*, 2006a) and several neurological disorders exhibit increase in WC due to inflammation or oedema. Multiple sclerosis (MS), brain tumours, stroke, hepatic encephalopathy and head trauma are all characterised by changes in brain WC (Lin *et al.*, 1997; Ayata and Ropper, 2002; Wick and Küker, 2004; Shah *et al.*, 2008; Volz, Nöth, Jurcoane, *et al.*, 2012).

The interpretation of PD as WC provides a unique opportunity to detect various processes in the human brain including changes in WC due to disease and response to treatment. In addition, any biophysical modelling of MR parameters will benefit from knowledge of the WC generating the signal. As water governs MR signal intensity, its influence is always implicitly present. In this sense, all MR images have intrinsic WC weighting.

Unlike other MR parameters, WC measurement is independent of field strength and has a distinct biological interpretation. The measured protons are attributed to mobile water, because water protons that are tightly bound to macromolecules and protons in phospholipids are usually MR-invisible (Fischer *et al.*, 1990; Horch *et al.*, 2011). This allows an association between PD and mobile WC.

Since more MR-visible water protons in a voxel generate a greater signal, the MRI signal intensity is proportional to WC. Therefore, WC mapping is based on the estimation of signal intensity in the absence of signal loss due to relaxation and magnetic field inhomogeneity. Since the removal of signal loss biases is challenging, especially in the case of receive coil sensitivity (please see also Chapter 2, Sections 2.1.8 and 2.1.10), WC mapping was disregarded for many years. Recent development of methods for efficient acquisition and post-processing now provides accurate WC estimations. See Figure 4.1 for an example.

This chapter introduces the biophysical interpretation of PD as WC and outlines the different methods for measuring WC. It then describes the scientific and clinical applications of WC mapping and its relevance to other MR measurements.

4.2 Biophysical interpretation

The protons measured in brain MRI are commonly attributed to mobile water. Both membrane lipid protons and a small fraction of water protons that are tightly bound to macromolecules tend to have a very short T2 (approximately 1 ms). Thus, these non-mobile

[1] Edited by Mara Cercignani.

[2] Both PD and WC are used in previous publications to describe the measurement of water protons in MRI. This chapter will follow the convention of defining PD as the uncalibrated proton density, and WC as the calibrated water fraction.

[3] In case of very short TE (0.05–0.50 ms), non-mobile protons are MR-visible and may contribute to the PD estimation (Holmes and Bydder, 2005; Du *et al.*, 2014).

TABLE 4.1 Post-mortem and biopsy determinations of water content (percentage weight)[a] in normal brain tissue

Reference	White matter	Grey matter
Norton *et al.* (1966)[b]	71.6 (SEM = 2.2)	81.9 (SEM = 0.5)
Tourtellotte and Parker (1968)[c]	70.6 (SEM = 1.2)	Not measured
Schepps and Foster (1980)[d]	69	80
Takagi *et al.* (1981)[e]	70.4 (SEM = 0.2)	84.7 (SEM = 0.2)
Kaneoke *et al.* (1987)[f]	68 (SEM = 3)	Not measured
Bell *et al.* (1987)[g]	69.7 (SEM = 0.7)	80.5 (SEM = 0.8)
Fatouros and Marmarou (1999)[h]	68.7–69.6 (SEM = 0.2)	Not measured

[a] To convert water content values (g water/g tissue) to proton density values, multiply by the specific gravity of brain tissue (1.04 for both white and grey matter; Whittal *et al.*, 1997; Takagi *et al.*, 1981; Torack *et al.*, 1976). To convert to proton concentration (PC), multiply by the specific gravity and by the molarity of water (55.2 M); thus a water content of 71 pu is 40.6 M. To obtain the PC in μmol g^{-1}, multiply the water content by 55.2 (thus a water content of 71 pu is 39.2 μmol g^{-1}).

[b] Evaporation technique in three subjects. Lipid contents were as follows: white matter, 15.6 pu (SEM = 0.3 pu); grey matter, 5.92 pu (SEM = 0.03 pu).

[c] Evaporation technique in 10 subjects. Lipid content of white matter was 18.7 pu (SEM = 0.4 pu). Others have reported lipid contents for white matter of 16.1 pu and grey matter of 6.3 pu (Brooks *et al.*, 1980).

[d] Samples were dried to constant weight at 105°C. Estimated from published percentage volume figures (74% and 84%) and solid fraction density = 1.3 gml^{-1}.

[e] Gravimetric analysis in at least five subjects (37 samples), SD = 0.5 pu, SEM estimated.

[f] Samples dried to 200°C. Differential scanning calorimetry measured the freezable water fraction; a bound (unfreezable) water fraction of 17.5 ± 3% of the total water content was identified.

[g] Gravimetric determination from biopsy samples (12 of grey matter, 9 of white matter).

[h] From T1 values in 27 subjects. The first figure is frontal white matter; the second is posterior white matter.

protons suffer from rapid signal loss and are invisible in standard MR imaging (Fischer *et al.*, 1990; Horch *et al.*, 2011).[3] Protons in mobile lipids (fat) are MR-visible. However, the abundance of those lipids in the brain is low, and their contribution to the signal is negligible (Delikatny *et al.*, 2011).[4] Therefore, it is commonly accepted to regard PD measurement in terms of WC. This parameter signifies the fraction of water in a tissue relative to that of pure water. It was proposed that the complementary 1-WC can serve as a measurement of the macromolecule and lipid tissue volume (MTV) (Mezer *et al.*, 2013).

The relation between proton density and WC can be derived as follows:

The number of protons per unit volume of tissue is defined as the *absolute proton concentration* or PC of a tissue. For a chemical compound of known composition, the absolute proton concentration is

$$PC = \frac{NP\rho}{MWt}\,[M] \tag{4.1}$$

where NP is the number of protons per molecule, ρ is the density (g/L), MWt is the molecular weight (g/mol) and the units of PC are molar (M). Thus, water has a PC of 110.4 M (NP = 2; ρ = 993.4 g/L at 37°C; MWt = 18 g/mol) at body temperature and a slightly higher value at room temperature. The PC of water is twice the concentration or molarity of pure water (55.2 M), since there are two protons in each molecule of water. The PD measured in MRI is proportional to the PC of the tissue but incorporates other factors determined by the measured signal intensity.

Therefore, to translate PD into WC, a calibration standard is needed.

The WC of a tissue is usually calculated as the concentration of water protons in the tissue, calibrated to the PC of pure water in the same volume at the same temperature. Thus:

$$\left(WC = 100\frac{PC}{110.4}\,[pu]\right) \tag{4.2}$$

where PC is in molar, and WC is in pu (percent units). Pure water, and approximately cerebrospinal fluid (CSF), has PC of 110.4 and therefore WC = 100 pu. White matter has WC ≈ 70 pu. Typical values for WC in various brain regions are shown in Table 4.2.

[4] *MR-visible lipids* or *mobile lipids* are defined as lipids that are observable using proton magnetic resonance spectroscopy in cells and in tissues. These MR-visible lipids are composed of triglycerides and cholesterol esters that accumulate in intracellular neutral lipid droplets, where their MR visibility is conferred due to the increased molecular motion available in this unique physical environment. These mobile lipids are found mostly in adipose tissue and not in the membrane bilayers. As most lipids in the normal brain are contained in bilayer membranes, spectroscopy in normal brain reveals miniature lipid peak compared to water (Delikatny *et al.*, 2011). However, bordering fat tissue can affect the signal in cases such as imaging of the optic nerve (Simon *et al.*, 1988). Increase in mobile lipid content is observed in some brain tumours (Howe *et al.*, 2003) and to a lesser extent in MS lesions (Davie *et al.*, 1994). This fat proton concentration is still most likely insignificant compared to water protons (Howe *et al.*, 2003). Note that this is not the case outside the brain. In organs such as the abdomen and the liver, mobile lipid proton concentration is significant and their MRI signal cannot be neglected. Therefore, it is usually safe to assume that brain MRI signals originate exclusively from water protons, but it should be kept in mind that there are cases in which this assumption is not accurate. Alternately, this can be verified with spectroscopy (Jansen *et al.*, 2006).

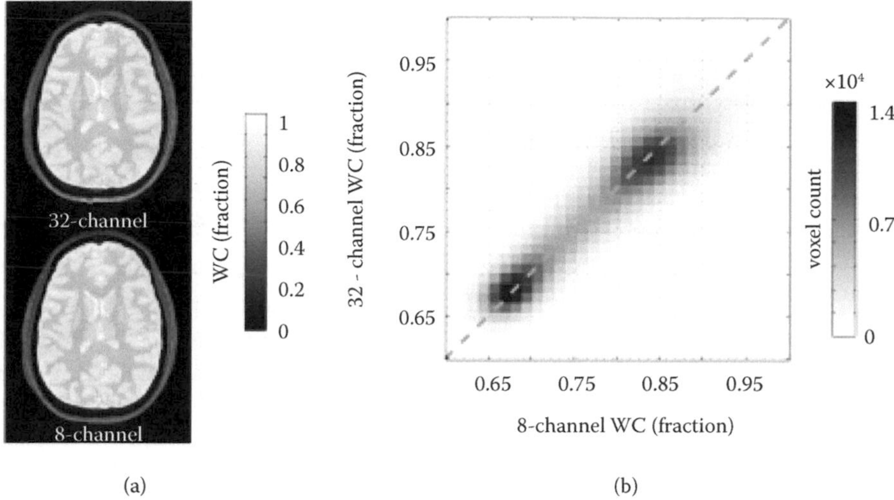

FIGURE 4.1 (a) An axial brain slice showing water content map acquired with 32-channel and 8-channel head coil. (b) 2D histogram of the two maps. (Mezer, A., *et al.*: Evaluating quantitative proton-density-mapping methods. *Hum. Brain Mapp.* 2016. 37. 3623–3635. Copyright Wiley-VCH Verlag GmbH & Co. KGaA. Reproduced with permission.)

TABLE 4.2 Water content for different brain regions as reported in literature

Brain Region		Water content reported in literature (1.5T and 3T)
White Matter	Frontal	68.4 ± 1.9[h]; 68.7 ± 1.0[b]; 69.1 ± 1.7[f]; 66.1 ± 2.9[e]; 69.1 ± 1.1 (left)[g]; 69.7 ± 1.4 (right)[g]
	Occipital	72.7 ± 2.1[h]; 66.9 ± 1.1[f]; 68.4 ± 1.2 (left)[g]; 69.3 ± 0.9 (right)[g]
	Parietal	70.3 ± 1.3[h]
	Temporal	72.5 ± 1.3[h]
	Average WM	69.7 ± 1.3[g]; 70.8[a], 70.9 ± 1.1[d]
Corpus callosum	Genu	69.6 ± 4.6[h]; 67.6 ± 1.2[b]; 69.0 ± 1.3[f]; 68.2 ± 1.4[g]; 71.7 ± 1.0[e]; 72.1 ± 2.9[e]
	Splenium	68.9 ± 1.2[h]; 66.2 ± 1.0[f]; 68.9 ± 1.2[g]; 70.5 ± 4.7[h]
Cortical Gray Matter	Frontal	77.5 ± 3.8[h]; 81.6 ± 1.6 (prefrontal)[f]; 86.2 ± 4.2 (prefrontal)[c]; 80.7 ± 1.1[g]
	Occipital	78.3 ± 2.4[h]; 81.3 ± 1.0[g]
	Parietal	79.1 ± 2.9[h]
	Temporal	82.0 ± 3.1[h]
	Average GM	81.0 ± 1.0[g]; 83.2[a]; 84.6[i], 81.2 ± 1.2[d]
Caudate (head)		83.0 ± 1.3[h]; 80.3 ± 1.1[b]; 81.1 ± 2.5[c]; 80.2 ± 0.7[g]; 81.3 ± 2.2[e]
Globus pallidus		76.8 ± 1.9[h]
Putamen		81.9 ± 1.3[h]; 83.2 ± 1.7[f]; 79.8 ± 1.3[g]; 83.1 ± 0.9[a]; 82.3 ± 2.6[e]
Hippocampus		82.0 ± 1.9[h]
Thalamus		82.5 ± 1.7[h]; 75.8 ± 1.2[b]; 73.4 ± 3.4[c]; 79.8 ± 1.0[a]; 81.0 ± 2.2[e]
Insula		82.9 ± 1.4[h]
Medulla		76.0 ± 1.5[h]
Pons		73.2 ± 2.4[h]
Midbrain		74.2 ± 1.8[h]
Anterior lateral ventricle		99.9 ± 3.7[h]

[a] Whittall *et al.* (1997).
[b] Fatouros and Marmarou (1999).
[c] Gelman *et al.* (2001).
[d] Neeb *et al.* (2006).
[e] Warntjes *et al.* (2007).
[f] Neeb *et al.* (2008).
[g] Volz, Nöth, Jurcoane *et al.* (2012).
[h] Abbas *et al.* (2015).
[i] Berman *et al.* (2017).

4.3 Measurement of WC

WC is measured by mapping and calibrating PD. PD is proportional to the MRI signal intensity, since more MR-visible water protons in a voxel generate a greater signal. However, signal intensity is also affected by relaxation and magnetic field inhomogeneity. Therefore, PD mapping is based on the estimation of the equilibrium magnetisation (M_0), representing the signal intensity in the absence of relaxation mechanisms (Box 4.1).

For example, the signal intensity of a spin-echo sequence is modelled by

$$S_{(TR,TE)} = M_0\left[1-e^{-\frac{TR}{T1}}\right]e^{-\frac{TE}{T2}} \tag{4.3}$$

where TR and TE are repetition time and echo time, respectively, T1 and T2 are the relaxation times, and $S_{(TR,\ TE)}$ is the signal intensity for a given TR and TE. Therefore, PD-weighted images, which are widely used for clinical applications, are typically acquired with short TE to minimise T2 loss and long TR to minimise T1 loss (Nitz and Reimer, 1999).

M_0 depends on PD and the receive coil's sensitivity profile (B_1^-):

$$M_0 = B_1^- \cdot PD \tag{4.4}$$

B_1^- is caused by tissue absorbance and coil spatial inhomogeneity. Thus, extracting PD from M_0 involves a non-trivial estimation of B_1^- imperfection (See also Chapter 2).

PD mapping is challenging and depends on the ability to measure and remove the effects of other MR parameters causing signal loss (e.g. T1 and T2 in the equation above). Any imperfection in the correction for these effects and for the various sources of inhomogeneity will propagate into the PD measurement. Last, translating PD measurements into WC requires additional calibration to a pure water standard.

Validation of WC estimations can be done in phantoms with known water concentration (Mezer *et al.*, 2013; Abbas *et al.*, 2015; Meyers *et al.*, 2016). WC can also be assessed in tissues based on weight loss on evaporation (see Table 4.1).

Box 4.1 Measuring WC step by step

1. **Estimate M_0:** acquisition and analysis of data according to multicompartment T2 model, T2* extrapolation, variable flip angle method or simultaneous fitting combined with relaxation times.
2. **Consider non-M_0 contributions:** correction for signal loss due to relaxation times and inhomogeneities.
3. **Extract PD from M_0:** requires calculation of the receive coil sensitivity profile.
4. **Calibrate PD for WC quantification:** normalise PD values to a pure water standard, either with an external or internal water source.

The following sections describe how to measure WC. The first part specifies methods for acquisition and estimation of the equilibrium magnetisation, M_0. The second part addresses the different corrections that are essential to extract WC from M0 measurement.

4.3.1 Methods for equilibrium magnetisation (M_0) estimation

4.3.1.1 Multicompartment T2

Excited water protons in different brain tissues have different T2 relaxation times (see Figure 4.2). A distribution of those relaxation times in each voxel can be obtained by a multicompartment model. M_0 is equal to the sum over this distribution (Whittall *et al.*, 1997; Laule *et al.*, 2007). In this method, the signal is separated into three components corresponding to different water environments: a long T2 component (~2 s) due to CSF; an intermediate component (~100 ms) arising from intracellular and extracellular water; and a short T2 component (~20 ms) believed to be due to water trapped between the myelin bilayers. The integral over this distribution provides the total contribution of the different water environments to the signal and therefore is proportional to M_0. The major benefit of this method is that it offers additional information about tissue microstructure, due to its ability to probe distinct water environments (Meyers, 2015). It has been shown that accurate M_0 measurement at 3T is feasible using a gradient and spin echo or spin echo sequences (Meyers *et al.*, 2016, 2017) (Figure 4.2).

4.3.1.2 T2* extrapolation

M_0 can also be estimated with T2*-based sequences. In this approach, the T2* relaxation curve is extrapolated to TE = 0. At the first moment of excitation, the maximal available water proton spins are tilted to the transverse plane and produce a signal. At this point, before spin relaxation causes a decay of the signal, the signal is proportional solely to the amount of water protons. Hence, extrapolation of the T2* decay curve to this initial state provides a good estimation for M_0. Neeb *et al.* (Neeb *et al.*, 2006b) map M_0 based on the acquisition of a series of spoiled gradient-echo images with different T2* weightings acquired with the QUTE (quantitative T2* image) sequence. Instead of relying on a specific analytical signal model, the first point on the signal decay curve is acquired with a short TE and a polynomial of third order is used to extrapolate the signal back to TE = 0 (Neeb *et al.*, 2006b). This was shown to be accurate at 1.5T, but for higher field strengths there is a substantial loss of accuracy due to coil sensitivity. To overcome this issue, more careful bias corrections are needed (Volz, Nöth and Deichmann, 2012; Abbas *et al.*, 2014) (see Section 4.2.2).

It is possible to apply variable flip angle of a spoiled gradient echo sequence to quantify T1 and M_0. This method is commonly used (Lin *et al.*, 1997; Volz, Nöth and Deichmann, 2012; Mezer *et al.*, 2013; Sabati and Maudsley, 2013). Estimation of M_0 is based on image intensity. The loss of signal due to the T2* effect can be estimated by acquiring two gradient-echo data sets with different TE (Volz, Nöth and Deichmann, 2012). Alternatively, others assume a minimal T2* effect in a very short TE (~2 ms).

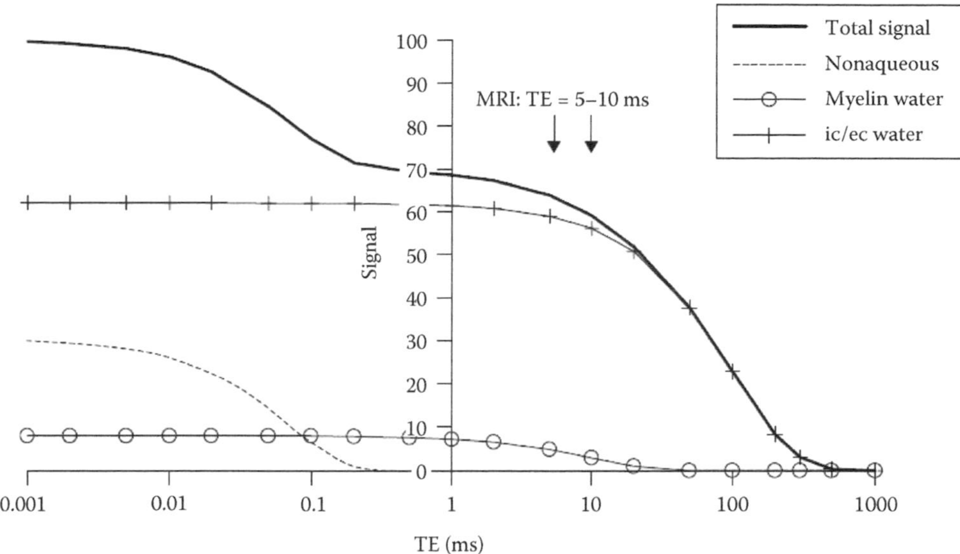

FIGURE 4.2 Symbolic illustration of tri-exponential decay of transverse proton signal in white matter. In this simulation, using realistic pool sizes and T2 values, 30% of the protons are non-aqueous (T2 = 70 μs), 8% are water protons trapped in myelin (T2 = 10 ms) and 62% are in intra- and extracellular (ic/ec) water. The non-aqueous protons are MRI invisible; the myelin water protons are partly MRI visible, for sequences with TE = 5–10 ms. MRI acquisition at several echo times greater enables the full myelin water signal to be recovered, in the presence of decaying ic/ec signal. In reality the non-aqueous proton signal decay is complex and non-exponential (since the line shape does not have fast isotropic motional narrowing). Reproduced with permission from Tofts PS. "PD: proton density of Tissue Water" in Quantitative MRI of the Brain. Wiley, Chichester, England (UK)

Instead of multiple TE extrapolation, a single short TE acquisition is used as an approximation for TE = 0 (Mezer *et al.*, 2013). The variable flip angle method depends largely on correct application of the flip angle. Thus, there is great importance for corrections for field inhomogeneity that may impair flip angle accuracy (B_1^+).

4.3.1.3 Simultaneous fitting of M_0 and relaxation times

Ehses *et al.* (Ehses *et al.*, 2013) proposed using the inversion recovery balanced steady-state free precession (IR-bSSFP) sequence for simultaneous quantification of M_0, T1 and T2 in 1.5T. Sensitivity of this method to magnetisation transfer (MT) effects and slice profile imperfections may impair parameter estimation and require certain corrections.

The QRAPMASTER[5] pulse sequence can also be used for simultaneous measurements of relaxation times and M_0 at 1.5T (Krauss *et al.*, 2015). Multiple TE images are used for determination of T2 relaxation time. T1 and M_0 are estimated based on the images with different TD (delay time). It was found that this method has similar performances to the IR-bSSFP method in terms of accuracy. Overestimation of PD was demonstrated and could be explained by the limitations of this method, such as the number of relaxation points and the mono-exponential fit.

Recently, simultaneous mapping of M_0, relaxation times and mean diffusivity was offered by Gras *et al.* (Gras *et al.*, 2016). They used a diffusion-weighted dual-echo steady-state sequence and optimised it for multiparameter mapping. However, this approach is limited due to its sensitivity to physiological motion.

MR fingerprinting is a novel concept that also aims at providing simultaneous measurements of multiple parameters such as T1, T2 and M_0 (Ma *et al.*, 2013; Cloos *et al.*, 2016). It is based on matching randomly acquired incoherent signals to a dictionary produced by simulations of the time evaluation of the magnetisation. M_0 estimation is done through the ratio between the signal and its matching normalised dictionary entry but may be impaired by B_1 inhomogeneities.

To summarise, there are two main approaches for M_0 estimation: multicompartment T2 and T2* extrapolation. These methods are reliable and more commonly used. Yet they require correction for the loss of signal due to T2 and T1 relaxations. Alternatively, recent approaches utilise simultaneous M_0 and relaxation times acquisition. Although currently suffering from some technical limitations, simultaneous fitting may prove to be an interesting approach for fast acquisition of M_0 with other quantitative MRI parameters.

4.3.2 Bias corrections for total WC estimation

The M_0 estimation can be translated into PD measurement and calibrated to produce WC. In order to extract PD from M_0, all other parameters influencing the signal amplitude should be mapped and removed. The signal is dependent on several parameters such as the relaxation times (T_1, T_2, T_2*), inhomogeneities of the transmit field B_1^+, distortion of the static magnetic field (B_0) and the sensitivity profile of the receiver coil (B_1^-). After

[5] Quantification of relaxation times and proton density by multi-echo acquisition of saturation recovery with TSE read-out.

removing these biases, a calibration standard is needed to translate PD measurement into the WC.

4.3.2.1 Relaxation time contributions

Any MR signal will be affected by processes of transverse (T_2, T_2^*) and longitudinal (T_1) relaxation. To account for this, multiple measurements are acquired and relaxation times are fitted. This way, their influence on the signal equation can be calculated and removed. For example, the spin echo signal equation (Equation 4.3) allows estimation of M_0 from the measured signal ($S_{(TR,\ TE)}$) if T1 and T2 are known:

$$M_0 = \frac{S_{(\mathrm{TR,TE})}}{\left[1-e^{-\frac{\mathrm{TR}}{\mathrm{T1}}}\right]e^{-\frac{\mathrm{TE}}{\mathrm{T2}}}} \tag{4.5}$$

As relaxation times measurement depends largely on correct estimation of B_1^+ and B_0 field, any imperfection in mapping these inhomogeneities impairs the correction for signal loss, and thereby the accuracy of the estimated PD (Volz, Nöth and Deichmann, 2012). While in some methods M_0 is measured based on a T2 decay curve, note that it is still essential to account for T1 effects. In general, this correction should apply to any MR parameter affecting the signal (Neeb *et al.*, 2006b; Ehses *et al.*, 2013; Abbas *et al.*, 2014; Meyers *et al.*, 2016, 2017).

Box 4.2 Receive coil sensitivity: Methods and assumptions

- **Coil-sensitivity functions are smooth over space.** Spatial relation between B_1^- in adjacent voxels is assumed. Calculating a unique PD requires additional assumptions.
- **The biophysical relationship between WC and T1.** WC measurements correlate with T1 in human brain tissue. This relationship can serve as an additional constraint. Other factors influencing T1, such as iron, may cause deviation from the T1–WC relation and impair WC estimation.
- **Multichannel analysis.** Data is acquired from multiple channels; each has a different sensitivity profile. Regularisation is needed to prevent overfitting.
- **Local estimates solution.** Perform smoothing over many small volumes and then smoothly join these estimates. Combining the T1-WC constraint with this method was found to provide particularly accurate WC estimates.
- **Reciprocity theorem.** In theory, the B_1^- field of a receiver coil is completely opposite to the B_1^+ field produced by the same coil if it were to be used for excitation. While this is widely used for 1.5T, for higher fields it was found to be less accurate. See also Chapter 2.

4.3.2.2 Receive coil sensitivity

Removal of the relaxation contributions from the signal leaves only the dependence on M_0. M_0 incorporates the PD and the receive coil sensitivity (B_1^-). The receive coil's ability to pick up the signal is inhomogeneous and depends both on coil size and its distance from the object. Thus, signal intensity has some non-uniformities imposed by spatial variations of the receive coil sensitivity profile (Box 4.2).

M_0 can be presented as the Hadamard product between B_1^- and PD:

$$M_0(x,y,z) = B_1^-(x,y,z) : \mathrm{PD}(x,y,z) \tag{4.6}$$

Therefore, PD can be extracted from M_0 through estimation of the receive coil sensitivity profile (see Figure 4.3).

No simple imaging manipulation can separate PD from B_1^-. It is not possible to fit these two parameters simultaneously, since there are infinite combinations of PD and B_1^- products that give a specific M0. Hence this is an ill-posed problem, and solving it must involve some additional constraints (Mezer *et al.*, 2016). There are several methods for separating PD from B_1^- that can be combined to accurately estimate PD (Figure 4.3).

4.3.2.2.1 Coil-sensitivity functions are smooth over space

In this approach, spatial connection between B_1^- in adjacent voxels is assumed (Noterdaeme *et al.*, 2009; Volz, Nöth and Deichmann, 2012; Volz, Nöth, Jurcoane, *et al.*, 2012; Mezer *et al.*, 2013; Abbas *et al.*, 2015). This constraint is implemented by approximating the coil sensitivity as a smooth function, for

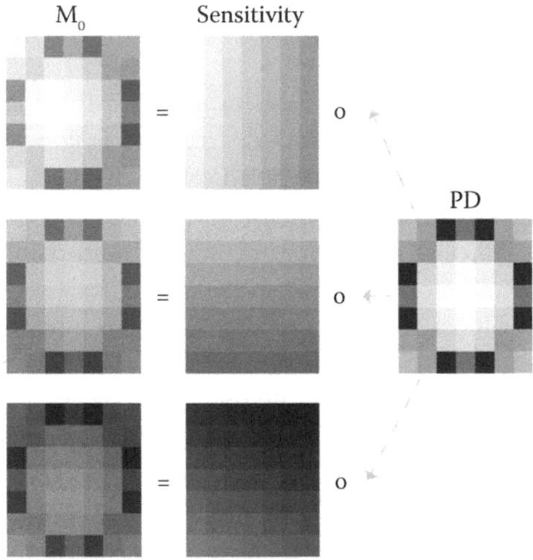

FIGURE 4.3 The relationship between proton density (PD), coil sensitivity (B_1^-) and the equilibrium magnetisation (M_0). The PD values are multiplied, point by point (Hadamard product), by the coil sensitivities, to produce the M_0 images. In this simulation, the coil sensitivities are second-order polynomials. (Mezer, A., *et al.*: Evaluating quantitative proton-density-mapping methods. *Hum. Brain Mapp.* 2016. 37. 3623–3635. Copyright Wiley-VCH Verlag GmbH & Co. KGaA. Reproduced with permission.)

example, a low order 3D-polynomial. Fitting of the PD value of each voxel and the polynomial coefficients for the whole volume is then performed.

The order of the polynomial imposes the number of polynomial coefficients N_p, and the number of voxels N_v is equal to the number of unknown PD values. Since there are more unknown parameters (N_v+N_p) than M0 measurements (N_v), calculating a unique PD requires additional assumptions.

This problem can be solved under the simplifying assumption that the coil sensitivity bias comprises only low frequencies (Noterdaeme *et al.*, 2009; Volz, Nöth and Deichmann, 2012). The risk is that this assumption will smooth physiological variations and thus damage PD estimation. Furthermore, different coils (e.g. 8 channels vs. 32 channels) vary differently in space. Hence coil bias may still exist if the same function is used for different coils.

4.3.2.2.2 The biophysical relationship between WC and T1

It was found that PD and therefore WC measurements correlate with T1 in human brain tissue (Fatouros *et al.*, 1991; Fatouros and Marmarou, 1999; Gelman *et al.*, 2001). Since T1 is largely affected by the fraction of water, a linear relationship between 1/WC and 1/T1 can be stated as follows (Gelman *et al.*, 2001; Volz, Nöth, Jurcoane, *et al.*, 2012; Mezer *et al.*, 2013; Abbas *et al.*, 2015):

$$\frac{1}{WC} = \frac{\gamma}{T1} + \delta \tag{4.7}$$

where γ and δ are arbitrary constants representing the slope and offset of the linear relationship (Figure 4.4). This can serve as an additional constraint when separating PD from B_1^-. However, estimation of WC can be impaired due to additional factors that can change T1 and not WC, such as high iron content (Vymazal *et al.*, 1995, 1999; Rooney *et al.*, 2007; Volz, Nöth, Jurcoane, *et al.*, 2012; Abbas *et al.*, 2014; Stüber *et al.*, 2014). Thus, one should be careful when implanting the linear relationship constraint in abnormal tissues. For the case of MS lesions, it was found that this constraint produces accurate WC estimations (Gracien, Reitz, Wagner, *et al.*, 2016).

In a recent application, the correlation between WC and T1 was combined with the variable flip angle method to estimate both B_1^- and B_1^+ (Baudrexel *et al.*, 2016). This method involves acquisition of gradient-echo datasets at different flip angles. B_1^- and B_1^+ are calculated simultaneously, based on the correlation between PD and T1, combined with the assumption of smoothly varying inhomogeneity fields. Under this assumption, in small volumes B_1^- and B_1^+ are approximately constant.

4.3.2.2.3 Multichannel analysis

Mezer *et al.* (2013) use multichannel data to separate PD from B_1^-. Each channel has a different sensitivity profile, so smoothing over the space is done separately for each channel using polynomial approximation.

Suppose that the number of polynomial coefficients is N_p, then for N_c channels there are N_p*N_c unknowns. In this case, there are more measurements (N_v*N_c) compared to single-channel analysis, and there are still only N_v unknown PD values. Therefore, for relatively low order polynomials the number of measurements exceeds the number of unknown parameters ($N_v* N_c > N_p* N_c + N_c$) and the ill-posed nature of the problem is resolved. However, multichannel measurements significantly increase the size of the acquired data and extend the computation time. In the presence of measurement noise, this method is sensitive to overfitting. To prevent it, a regularisation term is introduced to the PD and B_1^- fitting. Regularisation methods use prior knowledge or assumptions as additional constraint. There are several approaches for regularisation that vary in the constraints they impose. Some use mathematical assumptions, while others exploit the biophysical relationship between T1 and WC. One of the possible mathematical assumptions is ridge regression (Tikhonov regression). This regularisation selects a vector of coil-sensitivity polynomial coefficients (p) with a small vector length. When an orthonormal polynomial basis is chosen, minimising the vector length of p is equivalent to minimising the vector length of the coil-sensitivity coefficients (Bell *et al.*, 1978). Interestingly, the typical ridge regression regulation that is used to estimate the receive coil sensitivity in parallel imaging (Liang *et al.*, 2002; Lin *et al.*, 2004; Hoge *et al.*, 2005) was not sufficient to precisely separate PD from B_1^-. The T1 biophysical prior yields more accurate estimates than any other regularisation (Mezer *et al.*, 2016).

4.3.2.2.4 Local estimates solution

Previously described smoothing methods uses a global estimation of the B_1^- variation in space. The local estimates approach uses a relatively low-order polynomial fitting over many small volumes and then smoothly joins these estimates. Combining the T1–WC constraint with this approach was found to provide particularly accurate WC estimates, even when using conventional single-channel data (Baudrexel *et al.*, 2016; Cordes *et al.*, 2017; Mezer *et al.*, 2016).

4.3.2.2.5 Reciprocity theorem

The reciprocity theorem states that the B_1^- field of the receiver coil is opposite to the B_1^+ field produced by the same coil if it were to be used for excitation (Ibrahim, 2005). This allows correction for both transmit and receive non-uniformity (Hoult, 2000). The reciprocity theorem can still be applied if different coils are used for transmission and reception by incorporating additional acquisition in which the transmit coil is used for receive (Neeb *et al.*, 2008). While this assumption is widely used for 1.5T, for higher fields it was found to be less accurate due to standing wave effects (Volz, Nöth, Jurcoane, *et al.*, 2012; Abbas *et al.*, 2014).

4.3.2.3 Calibration to water standard

After all biases have been removed, an accurate PD estimation is achieved. However, PD is only proportional to the amount of protons in the voxel and does not directly measure it. Thus, a calibration standard is needed to get a quantitative measurement of WC. Since the percentage of protons in pure water is 100 pu, its PD could serve as a good calibration standard. Therefore, WC is a quantitative measurement of the fraction of water protons

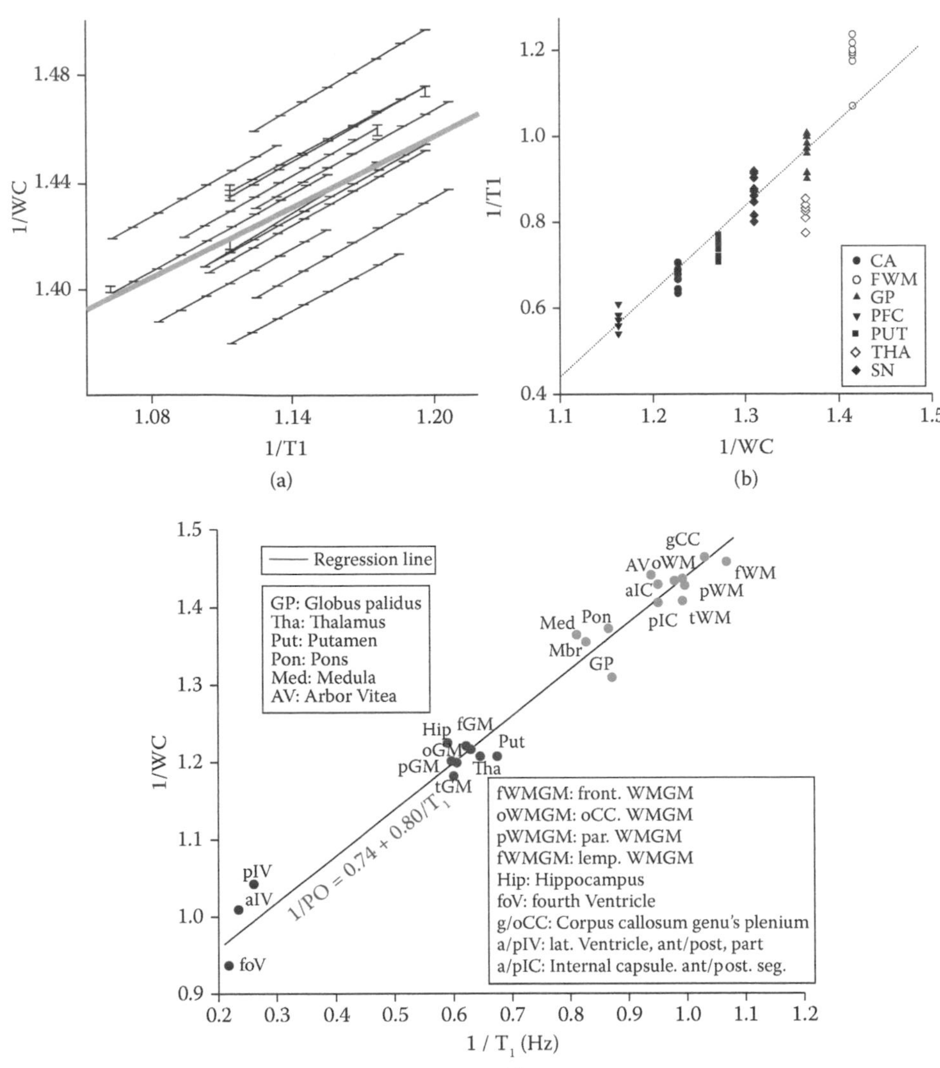

FIGURE 4.4 The linear relationship between 1/T1 and 1/WC as reported in different publications. (a) The white matter 1/T1 plotted against 1/WC across subjects. The 1/T1 values are pooled into bins separated by intervals of 0.05 s^{-1}. The mean number of voxels in each bin is $1.5 \times 105 \pm 5.5 \times 103$. Data from 16 individual subjects are fitted accurately by lines. Data are expressed as means ± s.e.m. The linear fit combining data from all subjects is shown as a thick grey line. (Reproduced with permission from Macmillan Publishers Ltd. *Nat. Med.* Mezer *et al.*, Copyright 2013.) (b) WC and T1 regional analysis for globus pallidus (GP), substantia nigra (SN), putamen (PUT), caudate head (CA), prefrontal cortex (PFC), thalamus (THA) and frontal white matter (FWM). A linear regression (dotted line) was applied to the data from all regions excluding THA and FWM. (Gelman, N., *et al.*: Interregional variation of longitudinal relaxation rates in human brain at 3.0 T: relation to estimated iron and water contents. *Magn. Reson. Med.* 2001. 45. 71–79. Copyright Wiley-VCH Verlag GmbH & Co. KGaA. Reproduced with permission.) (c) WC and T1 regional analysis. Each point corresponds to the inter-subject average of T1 (*x*-axis) and WC (*y*-axis) within a region. For readability, inter-subject standard deviations are not shown in this graphic. Some of the measurements are labelled with the corresponding brain region. (Abbas, Z., *et al.*: Analysis of proton-density bias corrections based on T1 measurement for robust quantification of water content in the brain at 3 Tesla. *Magn. Reson. Med.* 2014. 72. 1735–1745. Copyright Wiley-VCH Verlag GmbH & Co. KGaA. Reproduced with permission.)

in a tissue (PD_{tissue}) relative to the maximum possible number of water protons in a voxel (PD_{water}).

$$WC = \frac{PD_{tissue}}{PD_{water}} \tag{4.8}$$

There are two different approaches for estimating PD_{water}: internal reference or an external reference.

4.3.2.3.1 *Internal reference*

The internal reference method exploits the free water in the ventricles as an approximation of pure water standard. Intra-ventricle voxels with long T1 are used for this calibration. When using variable flip angles, data from those voxels is fitted to the signal equation (Volz, Nöth and Deichmann, 2012; Volz, Nöth, Jurcoane, *et al.*, 2012). Under the assumption of constant T1 in the ventricles, a single PD value for these voxels is found and used as the calibration factor (Mezer *et al.*, 2016). However, large deviations of the CSF signal due to long T1 relaxation and fluid motion may impair the estimation of ventricular PD (Abbas *et al.*, 2014). It is possible to eliminate this effect by applying a T2 constraint that removes voxels exhibiting flow and partial volume effects (Meyers *et al.*, 2017).

4.3.2.3.2 External reference

In the external reference method, a pure water phantom is added to the scan in order to serve as a reference. This approach works well with phantoms but is less accurate in human brain imaging. Since temperature affects the MRI signal, a correction is required for the temperature differences between the reference probe and the subject (Neeb *et al.*, 2008). Moreover, the reference water phantom must be placed alongside the subject and is thereby heavily affected by the B_1^+ inhomogeneity, which tends to be higher closer to the coils. B_1^+ correction is especially problematic in this case. Recently, Meyers *et al.* (Meyers *et al.*, 2017) exploited the reciprocity theorem to calculate the B_1^+ inhomogeneity of an external water source. This application has several limitations; however, it implies that with an accurate coil sensitivity mapping the use of an external water source is feasible.

4.4 Applications

Water constitutes 70%–85% of brain tissue and is critical for brain function and protection (Deoni, 2015; Meyers, 2015). WC varies between brain regions and alters during development and with ageing (Holland *et al.*, 1986; Neeb *et al.*, 2006a, 2006b). In addition, changes in WC are a feature of many neurological diseases, such as MS, brain tumours, hepatic encephalopathy, stroke and head trauma (Ayata and Ropper, 2002; Wick and Küker, 2004; Laule *et al.*, 2006; Shah *et al.*, 2008; Volz, Nöth, Jurcoane, *et al.*, 2012; Mezer *et al.*, 2013). Therefore, mapping WC could serve as a useful tool for research and diagnosis of cerebral diseases and for studying normal ageing and development of the human brain.

A change in brain volume and morphometry is common as a function of age and diseases (Tisserand *et al.*, 2002; Kennedy *et al.*, 2009; Callaert *et al.*, 2014). However, a voxel-based analysis reveals large volumetric changes rather than local, voxel-level effects. Measuring WC would help to differentiate the volumetric changes from the local changes in water and non-WC (Neeb *et al.*, 2006a).

Furthermore, as water is the source of the MRI signal, WC underlies the measurement of other MRI parameters (Gelman *et al.*, 2001; Chen *et al.*, 2005; Vavasour *et al.*, 2011; Mezer *et al.*, 2013; Lorio *et al.*, 2016). Estimation of the correlation between the MR parameter of interest and the WC is essential for the biophysical interpretation of the measurement. WC also serves as a calibration standard for other MR methods, such as spectroscopy (Gasparovic *et al.*, 2009; Lecocq *et al.*, 2013, 2015).

4.4.1 WC in the normal brain

In the normal brain, WC has been shown to differ between white matter and grey matter. Table 4.2 presents WC measurements for different brain regions as reported in the literature. A notable difference between grey and white matter was reported in various studies. The average WC of grey matter was found to be 80%–85%, while in white matter the average WC was approximately 70% (Figure 4.5). These findings are constant through different field strength and acquisition methods. Several studies measured WC in biopsy samples (Table 4.1). In agreement with the MRI WC estimation, the biopsy WC values were found to range from 68.7% to 71.6% for white matter and from 80.5% to 84.6% for grey matter.

4.4.2 WC variation with age

Multiple studies observed WC changes throughout the lifespan.

The alteration in WC with age for different brain structures was tested for a group of 138 volunteers, ranging in age from 19 to 75 years (Callaghan *et al.*, 2014) using voxel-based analysis. WC in the putamen, pallidum, caudate nucleus and red nucleus negatively correlated with age. However, a positive correlation was found between WC and age in the optic radiation and in the superior regions of the white matter. Note that the WC measurements in this study were not corrected for T_2^* weighting and thus may include other contributions such as iron content (given the relatively long TE of 8.5 ms). In addition, this study utilises voxel-based analysis to demonstrate changes in WC. Therefore, the sensitivity of voxel-based analysis to age may account for

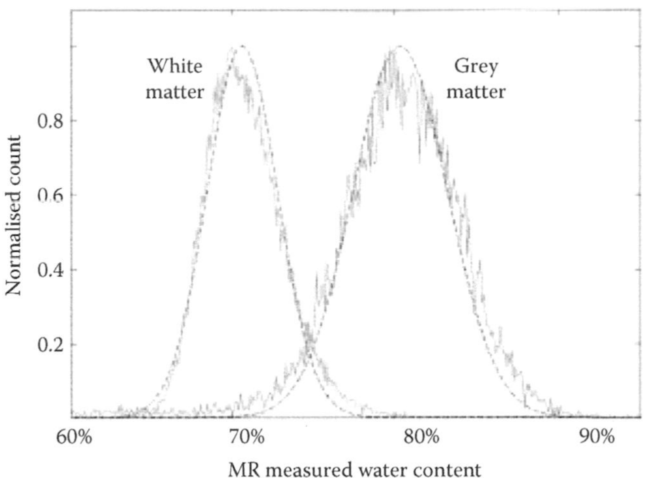

FIGURE 4.5 Histogram with the distribution of WC in white and grey matter. An *in vivo* WC map was segmented to display only white and grey matter based on quantitative T1 information. The segmented WC map was displayed as a one-dimensional histogram and fitted with the sum of two Gaussian functions to extract the mean and standard deviation of WC in both grey and white matter. Fitting the distribution with this function resulted in an estimate of 70.7 ± 2.1% WC for white matter and 80.3 ± 2.9% WC for grey matter. The increased WC in grey matter in comparison to white matter is clearly visible. (Reproduced from *NeuroImage*, 31, Neeb, H., et al., A new method for fast quantitative mapping of absolute water content in vivo, 1156–1168, Copyright 2006, with permission from Elsevier.)

the observed correlation of WC with age (Tisserand *et al.*, 2002; Kennedy *et al.*, 2009; Callaert *et al.*, 2014).

It was shown that grey matter WC decrease with age is sex specific (Neeb *et al.*, 2006a). For females, the grey matter WC decreases at a rate of 0.034% per year in the third and eighth decades of life. For males, a much stronger decrease in grey matter WC is observed after the fifth decade of life. Furthermore, the female average grey matter WC was found to be 1.2% higher than male. No age- or gender-specific changes were observed for white matter WC. It should be noted, however, that the average WC of white and grey matter was estimated on images with slice thickness of 5 mm. Hence, partial volume effects and their interaction with age and sex can influence the results.

Yeatman *et al.* (Yeatman *et al.*, 2014) modelled white matter development and ageing over an 80-year period of the lifespan (*N*=102, ages 7–85). The 1/T1 and MTV (1-WC) growth curves were well fitted by a symmetric curve such as a second-order polynomial over the measured period of the lifespan. This implies that the rate of growth and decline are symmetric. MTV values reach their peak between 30 and 50 years of age and then decline, returning to their 8-year-old levels between age 70 and 80. Interestingly different white matter pathways show different rates of change as a function of age.

A combined quantitative and functional MRI study in children and adults examined the relationship between WC and function (Gomez *et al.*, 2017). Face memory and place recognition were tested in children and adults while scanning relevant brain regions in the ventral temporal cortex. WC measurement was used to extract MTV (1-WC) for these brain regions. It was shown that in the posterior fusiform gyrus, selective for faces, mean MTV increased by 12.6% from childhood to adulthood. Simulations results indicated that increase in the volume of the myelin sheath is not likely to account for the full extent of this observation.

4.4.3 WC in disease

WC changes in disease are not specific to a certain diagnosis but correlate with a wide variety of pathological conditions.

Brain oedema, an accumulation of fluid in the cerebral tissue, is an accompanying feature of many diseases. The rise in brain pressure caused by oedema can be dangerous or even lethal. Evaluating changes in WC is therefore very important in monitoring the extent of the oedema for an accurate prognosis. It was shown that relatively small changes in WC due to oedema reflect much bigger changes in brain swelling (Keep *et al.*, 2012).

Local or global increase in WC due to oedema is associated with common diseases, such as stroke and brain tumours (Badaut *et al.*, 2002; Wick and Küker, 2004). Brain tumour oedema occurs when plasma-like fluid enters the brain extracellular space through impaired capillary endothelial tight junctions in tumours (Papadopoulos *et al.*, 2004). WC mapping of patients with brain tumour revealed global increase in white matter WC (Neeb *et al.*, 2006b). Changes caused by water diffusion from the affected side into the other hemisphere were measurable as well. This was not observable in conventional T1-weighted images.

In patients with liver failure, the presence of brain oedema is a major manifestation of the disease and often determines clinical outcome (Gill and Sterling, 2001). WC mapping provided evidence for the association between the pathophysiology of hepatic encephalopathy and oedema (Shah *et al.*, 2008). Quantification of changes in tissue WC demonstrated that the amount of swelling and the resulting brain oedema correlates with disease grade.

Multiple sclerosis is an autoimmune disease of the central nervous system that is characterised by oedema, inflammation, demyelination and axonal loss (Keegan and Noseworthy, 2002). MS lesions and normal-appearing MS white matter tend to have increased WC (Laule *et al.*, 2004; Volz, Nöth, Jurcoane, *et al.*, 2012; Jurcoane *et al.*, 2013; Mezer *et al.*, 2013; Engstrom *et al.*, 2014; Baudrexel *et al.*, 2016). As an example, cortical WC increase was demonstrated in early relapsing–remitting MS patients (Gracien, Reitz, Hof, *et al.*, 2016) and also at later MS stages (Engstrom *et al.*, 2014). Furthermore, WC measurements were shown to predict gadolinium enhancement in MS (Jurcoane *et al.*, 2013). Gracien *et al.* (Gracien, Reitz, Wagner, *et al.*, 2016) investigated whether different methods for removing the receive coil bias in WC mapping (see Section 4.2.2.2) can be applied to the study of MS patients. In pathologies such as large tumours, WC enhancement can be mistakenly removed by smoothing. In addition, it was not clear whether the linear relationship between T1 and WC used for B_1^- correction still holds in abnormal tissues such as MS lesions. The results of this study show that receiver bias correction via application of the T1 biophysical prior yield reliable WC values in MS patients. In addition, it was demonstrated that discrepancies between the calibrated WC and the predicted WC following from the linear relationship become more prominent in patients with a higher degree of disability (Gracien, Reitz, Wagner, *et al.*, 2016).

Figure 4.6 shows examples of WC measurement in various conditions.

4.4.4 WC effect on other MRI measurements

The MRI signal is generated from water, and thus any MR measurement incorporates WC effects. Accounting for the contribution of WC to the measurement is important, as changes in WC may confound other tissue properties.

T1 is known to correlate with WC. As mentioned earlier, this correlation is exploited to accurately extract PD from M_0. The relation between T1 and WC can also be used to quantify the extent to which changes in T1 are affected by WC.

It was found that differences in WC between grey matter regions account for most of the interregional variation of T1 values (Gelman *et al.*, 2001). However, T1 does not depend solely on WC and is thought to reflect various biophysical properties of the tissue such as iron. Assessment of the T1 variability that is not accounted for by WC variability may capture these biophysical properties. It was shown that the T1 to WC relationship can change between brain regions and as a function of non-water compounds (Kucharczyk *et al.*, 1994; Mezer *et al.*, 2013; Abbas *et al.*, 2015). Several studies (Mezer *et al.*, 2013; Abbas *et al.*, 2015) tested deviations from the

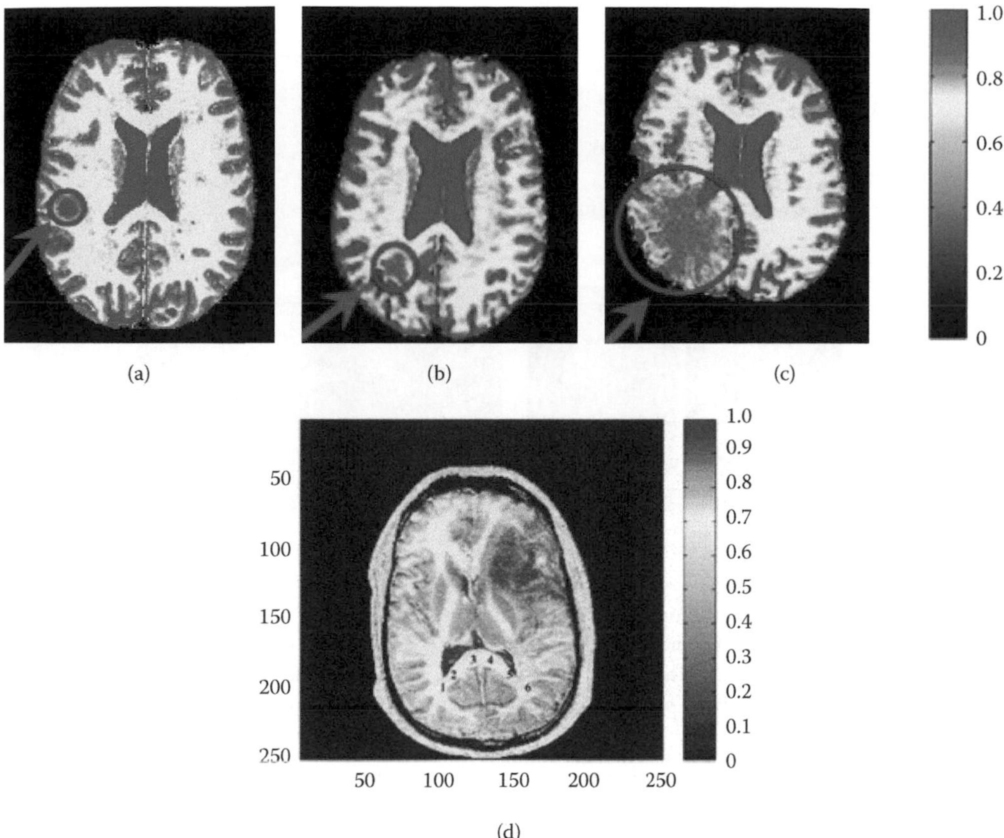

FIGURE 4.6 WC in disease. (a) WC map acquired on an MS patient. The red circle and arrow mark the largest lesion in all images. (b) WC map acquired on an ischemic stroke patient. The red circle and arrow mark the left parietal infarct area in all images. (c) WC map acquired on a patient with recurrent glioblastoma. The red circle and arrow mark the tumour. (d) WC map from a 28-year-old patient with brain tumour. (Reproduced from *NeuroImage*, 63, Volz, S., *et al.*, Quantitative proton density mapping: Correcting the receiver sensitivity bias via pseudo proton densities, 540–552, Copyright 2012, with permission from Elsevier; and from *NeuroImage*, 31, Neeb, H., et al., A new method for fast quantitative mapping of absolute water content in vivo, 1156–1168, Copyright 2006, with permission from Elsevier.)

T1–WC relationship computed for the whole brain. The ability of WC to predict T1 based on this relationship differed between brain regions. For example, WC predicted the white matter T1, but the same prediction did not extend to grey matter and thalamus (Mezer *et al.*, 2013). In addition, the sensitivity of WC and T1 to macromolecule content was evaluated with phantoms containing different lipids compositions. WC was found to be independent of the phantom content and was only sensitive to the fraction of water (Mezer *et al.*, 2013). However, T1 depends on both WC and lipids type (Koenig, 1991; Kucharczyk *et al.*, 1994; Mezer *et al.*, 2013).

Hence, comparing T1 and WC values between tissues may reveal the local physicochemical environment of the macromolecules affecting T1.

An evaluation of the effect WC has on T1 measurement can be achieved by computing a synthetic T1-weighted image. This is an image produced by quantitatively measuring MR parameters and using them to calculate the analytical expression for the signal equation. An inclusion of PD measurement in the synthetic signal calculation was shown to reduce grey matter volume and cortical thickness estimates obtained from T1-weighted MRI images, compared to synthetic images in which PD was not included (Lorio *et al.*, 2016); see Figure 4.7. This highlights the additional value of quantitative MRI for neuroanatomy studies and implies that MRI data used to extract tissue measures should be carefully considered.

Magnetisation transfer is an MRI measurement that examines the interaction between mobile protons in water and bound non-aqueous protons. The amount of magnetisation transferred between these pools can be characterised by the magnetisation transfer ratio (MTR). Changes in this parameter are often attributed to alterations in myelin content (Chen *et al.*, 2005). However, it was found that changes in WC due to oedema and inflammation may also cause changes in MTR unrelated to myelin (see Figure 4.8). It is therefore incorrect to associate decreases in MTR in MS lesions exclusively with demyelination (Vavasour *et al.*, 2011). The combination of quantitative WC mapping in the analysis of MT data enables the separation between the contributions of oedema and demyelination (Levesque *et al.*, 2005; Giacomini *et al.*, 2009). Several quantitative MRI measurements, such as T2 and MT, aim for myelin mapping but are also sensitive to WC (Mezer *et al.*, 2013; West *et al.*, 2016; Berman *et al.*, 2017).

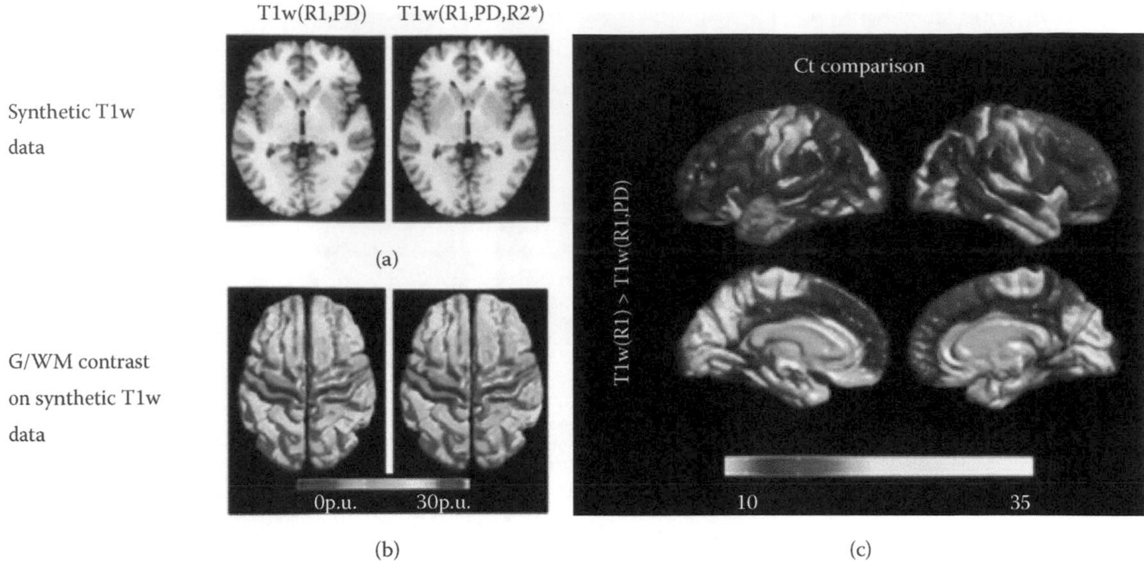

FIGURE 4.7 PD effect on grey matter volume and cortical thickness estimates, obtained from T1-weighted MRI images. (a) Synthetic T1 weighted images with PD incorporation (T1w(R1,PD)) and without PD (T1w(R1)). (b) Grey–white matter contrast in the synthetic T1w images. (c) Statistical comparison of cortical thickness estimates obtained from the different synthetic T1w images. (Reproduced from Lorio, S., *et al.*, *Hum. Brain Mapp.*, 37, 1801–1815, 2016, under the terms of the Creative Commons Attribution License. With permission.)

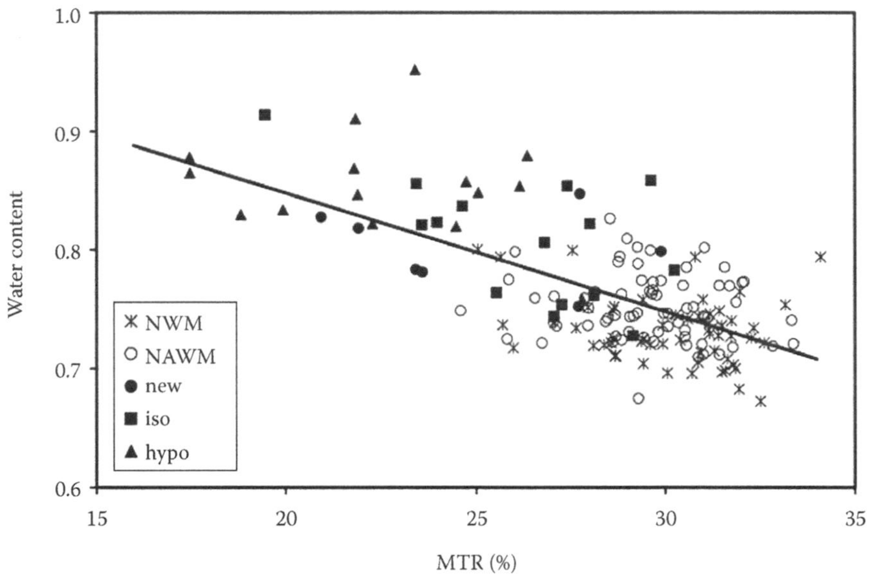

FIGURE 4.8 Correlations between MTR and WC for different tissues. NWM = age- and gender-matched normal white matter (star); NAWM = normal-appearing white matter (open circle); new = lesions less than 2 months old (solid circle); iso = isointense T1 lesions (solid square); and hypo = hypointense T1 lesions (solid triangle). The regression line is for all tissues, R = −0.65. (Vavasour, I.M., *et al.*: Is the magnetisation transfer ratio a marker for myelin in multiple sclerosis? *J. Magn. Reson. Imaging.* 2011. 33. 713–718. Copyrigh Wiley-VCH Verlag GmbH & Co. KGaA. Reproduced with permission.)

Future studies should test how the removal of WC effects can improve their specificity.

Diffusion MRI can also benefit from WC mapping. Mezer *et al.* (Mezer *et al.*, 2013) suggested that combining WC and diffusion measurements may clarify the contributions of different tissue properties. They showed that the fractional anisotropy (FA) value drops substantially in regions where there are known to be many crossing fibres. However, the crossing fibres did not considerably affect the MTV (1-WC). This illustrates that FA and MTV values show complementary aspects of white matter tracts

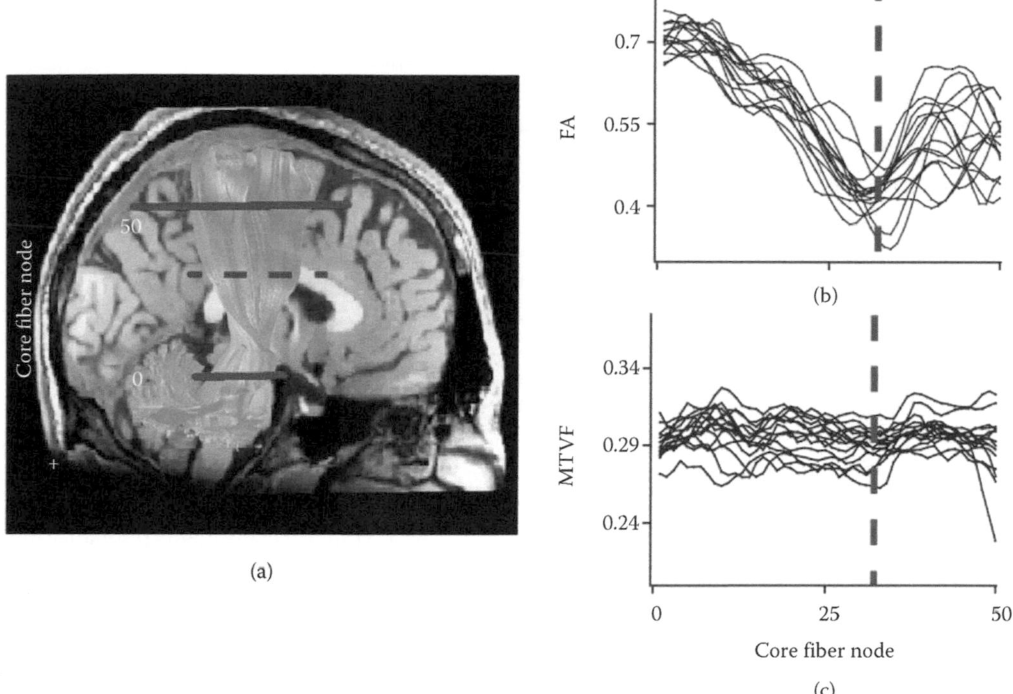

FIGURE 4.9 Diffusion measurements and WC in region of crossing fibre. MTVF (macromolecular and lipid tissue volume fraction, MTVF = 1-WC) and FA (fractional anisotropy) were calculated along the corticospinal tract (CST). The FA value dropped substantially in the region where fibres from the corpus callosum pass through the CST, reducing fibre direction coherence. This region is within the centrum semiovale, where there are known to be many crossing fibres. The MTV values, on the other hand, were constant along the CST. (a) The estimated right CST (blue) is overlaid on a sagittal T1-weighted image. The two solid red lines show axial planes that designate the measurement region; the centroid of the tract (core fibre) is calculated and sampled into 50 nodes. The CST intersection with the callosal fibres is designated by the dashed fuchsia line. (b,c) The curves show FA and MTVF values measured at different nodes along the CST from different control subjects ($n = 15$). The FA value, but not the MTVF value, declines in the region where the CST intersects callosal fibres (dashed fuchsia line). (Reproduced with permission from Macmillan Publishers Ltd. *Nat. Med.* Mezer, A., *et al.*, Copyright 2013.)

(see Figure 4.9). Combining the two measurements can help discriminate between different mechanisms that influence FA. For example, if MTV and FA change together, we would be inclined to explain the change as one caused by an axon-packing difference. If MTV is constant but FA changes, we might explain the change as one caused by mechanisms such as differences in axon coherence.

In addition, diffusion measurement and WC can be combined to produce structural–biophysical properties of white-matter fibres. In recent application, diffusion and WC were used to approximate the ratio between inner and outer axonal diameter (g-ratio), an important structural property of white matter affecting signal conduction (Duval *et al.*, 2016; Berman *et al.*, 2017).

It was suggested to use WC measurement in electrical properties tomography. This technique provides *in vivo* electrical conductivity and permittivity images of biological tissue. Michel *et al.* (Michel *et al.*, 2016) expressed tissue electrical conductivity and permittivity solely in terms of WC by an optimised function. WC was demonstrated to be a reasonable indicator of the tissue's electrical properties at MRI frequencies.

WC mapping is also essential for calibration of different MR measurements. For example, quantification of brain absolute metabolite concentration by spectroscopy imaging requires calibration to water standard, obtained from WC map (Gasparovic *et al.*, 2009; Lecocq *et al.*, 2013, 2015). In addition, WC decreases with maturation was shown to affect parameters of cerebral oxygenation determined by near-infrared spectroscopy. Most commercially available devices do not take these maturational WC changes into account. This may result in a deviation of cerebral oxygenation readings by up to 8% from the correct value (Demel *et al.*, 2014).

Arterial spin labelling, an MRI method for measuring cerebral blood flow, involves calibration to the signal intensity of fully relaxed blood spins. This measurement is also derived from WC map (Alsop *et al.*, 2015; Jen *et al.*, 2016).

4.5 Conclusion

For many years, WC mapping was disregarded. Recent developments in acquisition and post-processing of WC maps, along with enhanced understanding of their biophysical meaning, allows accurate quantification of WC.

WC mapping is now an accessible and reliable MRI measurement, as the influence of biases such as receive coil inhomogeneity is better understood and controlled. However, the ill-posed nature of the separation between coil sensitivity and PD is still not theoretically solved. Moving to higher fields, characterising and removing these biases will become increasingly important. Further study on the implementation of human WC mapping in fields higher than 3T is called for.

Progress has also been made in terms of biophysical interpretation. As it was verified that PD has an explicit association with WC, the clinical implementation of this measurement advanced. It has proven to be accurate across scanners and field strengths, which could have clinical application for standardisation. An increase in WC was shown to characterise several neurological diseases. Thus, the ability to quantify changes in WC has clinical implications. Yet, WC is not specific to a certain pathological condition. Future research may include a multimodality approach of different MR parameters to provide distinct diagnosis, prognosis and treatment monitoring.

All MRI measurements originate from water protons. Thus, WC is considered the most basic MR parameter. Furthermore, WC is unique due to its unambiguous biological interpretation and direct association with water. Hence, the role of WC in biophysical modelling of other MR parameters is likely to increase and reinforce specificity.

References

Abbas Z, Gras V, Möllenhoff K, Keil F, Oros-Peusquens A-M, Shah NJ. Analysis of proton-density bias corrections based on T1 measurement for robust quantification of water content in the brain at 3 Tesla. Magn Reson Med 2014; 72: 1735–45.

Abbas Z, Gras V, Möllenhoff K, Oros-Peusquens A-M, Shah NJ. Quantitative water content mapping at clinically relevant field strengths: a comparative study at 1.5 T and 3 T. NeuroImage 2015; 106: 404–13.

Alsop DC, Detre JA, Golay X, Günther M, Hendrikse J, Hernandez-Garcia L, et al. Recommended implementation of arterial spin-labeled perfusion MRI for clinical applications: a consensus of the ISMRM perfusion study group and the European consortium for ASL in dementia. Magn Reson Med 2015; 73: 102–16.

Ayata C, Ropper AH. Ischaemic brain oedema. J Clin Neurosci 2002; 9: 113–24.

Badaut J, Lasbennes F, Magistretti PJ, Regli L. Aquaporins in brain: distribution, physiology, and pathophysiology. J Cereb Blood Flow Metab 2002; 22: 367–78.

Baudrexel S, Reitz SC, Hof S, Gracien R-M, Fleischer V, Zimmermann H, et al. Quantitative T_1 and proton density mapping with direct calculation of radiofrequency coil transmit and receive profiles from two-point variable flip angle data. NMR Biomed 2016: 29: 349–360.

Bell, B. A., Smith, M. A., Kean, D. M., McGhee, C. N., MacDonald, H. L., Miller, J. D., Barnett, G. H., Tocher, J. L., Douglas, R. H. and Best, J. J. Brain water measured by magnetic resonance imaging. Correlation with direct estimation and changes after mannitol and dexamethasone, Lancet 1987; 1: 66–69.

Bell JB, Tikhonov AN, Arsenin VY. Solutions of Ill-Posed Problems. Math Comput 1978; 32: 1320.

Berman S, West K, Does MD, Yeatman JD, Mezer AA. Evaluating g-ratio weighted changes in the corpus callosum as a function of age and sex. NeuroImage 2017; In Press: 1–10.

Callaert DV, Ribbens A, Maes F, Swinnen SP, Wenderoth N. Assessing age-related gray matter decline with voxel-based morphometry depends significantly on segmentation and normalization procedures. Front Aging Neurosci 2014; 6.

Callaghan MF, Freund P, Draganski B, Anderson E, Cappelletti M, Chowdhury R, et al. Widespread age-related differences in the human brain microstructure revealed by quantitative magnetic resonance imaging. Neurobiol Aging 2014; 35: 1862–72.

Chen JT, Collins DL, Freedman MS, Atkins HL, Arnold DL. Local magnetization transfer ratio signal inhomogeneity is related to subsequent change in MTR in lesions and normal-appearing white-matter of multiple sclerosis patients. NeuroImage 2005; 25: 1272–78.

Cloos MA, Knoll F, Zhao T, Block K, Bruno M, Wiggins C, et al. Multiparamatric imaging with heterogeneous radiofrequency fields. Nat Commun 2016; in press: 1–10.

Cordes D, Yang Z, Zhuang X, Sreenivasan K, Mishra V, Hua LH. A new algebraic method for quantitative proton density mapping using multi-channel coil data. Med. Image Anal. 2017; 40: 154–171.

Davie CA, Hawkins CP, Barker GJ, Brennan A, Tofts PS, Miller DH, et al. Serial proton magnetic resonance spectroscopy in acute multiple sclerosis lesions. Brain 1994; 117 (Pt 1): 49–58.

Delikatny EJ, Chawla S, Leung DJ, Poptani H. MR-visible lipids and the tumor microenvironment. NMR Biomed 2011; 24: 592–611.

Demel A, Wolf M, Poets CF, Franz AR. Effect of different assumptions for brain water content on absolute measures of cerebral oxygenation determined by frequency-domain near-infrared spectroscopy in preterm infants: an observational study. BMC Pediatr 2014; 14: 206.

Deoni SCL, Meyers SM, Kolind SH. Modern methods for accurate T1, T2, and proton density MRI. In Oxford Textbook of Neuroimaging, edited by M. Filippi. Oxford University Press; 2015: 13–26.

Du J, Ma G, Li S, Carl M, Szeverenyi NM, VandenBerg S, et al. Ultrashort echo time (UTE) magnetic resonance imaging of the short T2 components in white matter of the brain using a clinical 3T scanner. NeuroImage 2014; 87: 32–41.

Duval T, Lévy S, Stikov N, Campbell J, Mezer A, Witzel T, et al. g-Ratio weighted imaging of the human spinal cord in vivo. NeuroImage 2016: 145: 11–23.

Ehses P, Seiberlich N, Ma D, Breuer FA, Jakob PM, Griswold MA, et al. IR TrueFISP with a golden-ratio-based radial readout: fast quantification of T1, T2, and proton density. Magn Reson Med 2013; 69: 71–81.

Engstrom M, Warntjes JBM, Tisell A, Landtblom AM, Lundberg P. Multi-parametric representation of voxel-based quantitative magnetic resonance imaging. PLoS One 2014; 9.

Fatouros PP, Marmarou A. Use of magnetic resonance imaging for in vivo measurements of water content in human brain: method and normal values. J Neurosurg 1999; 90: 109–15.

Fatouros PP, Marmarou A, Kraft KA, Inao S, Schwarz FP. In Vivo brain water determination by T1 measurements: Effect of total water content, hydration fraction, and field strength. Magn Reson Med 1991; 17: 402–13.

Fischer HW, Rinck PA, van Haverbeke Y, Muller RN. Nuclear relaxation of human brain gray and white matter: Analysis of field dependence and implications for MRI. Magn Reson Med 1990; 16: 317–34.

Gasparovic C, Neeb H, Feis DL, Damaraju E, Chen H, Doty MJ, et al. Quantitative spectroscopic imaging with in situ measurements of tissue water T1, T2, and density. Magn Reson Med 2009; 62: 583–90.

Gelman N, Ewing JR, Gorell JM, Spickler EM, Solomon EG. Interregional variation of longitudinal relaxation rates in human brain at 3.0 T: relation to estimated iron and water contents. Magn Reson Med 2001; 45: 71–9.

Giacomini PS, Levesque IR, Ribeiro L, Narayanan S, Francis SJ, Pike GB, et al. Measuring demyelination and remyelination in acute multiple sclerosis lesion voxels. Arch Neurol 2009; 66: 375–81.

Gill RQ, Sterling RK. Acute liver failure. J Clin Gastroenterol 2001; 33: 191–8.

Gomez J, Barnett MA, Natu V, Mezer A, Palomero-Gallagher N, Weiner KS, et al. Microstructural proliferation in human cortex is coupled with the development of face processing. Science 2017; 355: 68–71.

Gracien R-M, Reitz SC, Hof SM, Fleischer V, Zimmermann H, Droby A, et al. Changes and variability of proton density and T1 relaxation times in early multiple sclerosis: MRI markers of neuronal damage in the cerebral cortex. Eur Radiol 2016; 26: 2578–86.

Gracien R-M, Reitz SC, Wagner M, Mayer C, Volz S, Hof S-M, et al. Comparison of two quantitative proton density mapping methods in multiple sclerosis. Magn Reson Mater Phys Biol Med 2016; 30: 75–83.

Gras V, Farrher E, Grinberg F, Shah NJ. Diffusion-weighted DESS protocol optimization for simultaneous mapping of the mean diffusivity, proton density and relaxation times at 3 Tesla. Magn Reson Med 2016; 6: 1735–1745.

Hoge WS, Brooks DH, Madore B, Kyriakos WE. A tour of accelerated prallel MR imaging from a linear systems perspective. Concepts Magn Reson 2005; 27A: 17–37.

Holland BA, Haas DK, Norman D, Brant-Zawadzki M, Newton TH. MRI of normal brain maturation. AJNR. Am J Neuroradiol 1986; 7: 201–208.

Holmes JE, Bydder GM. MR imaging with ultrashort TE (UTE) pulse sequences: basic principles. Radiography 2005; 11: 163–74.

Horch RA, Gore JC, Does MD. Origins of the ultrashort-T2 1H NMR signals in myelinated nerve: a direct measure of myelin content? Magn Reson Med 2011; 66: 24–31.

Hoult DI. The principle of reciprocity in signal strength calculations—A mathematical guide. Concepts Magn Reson 2000; 12: 173–87.

Howe FA, Barton SJ, Cudlip SA, Stubbs M, Saunders DE, Murphy M, et al. Metabolic profiles of human brain tumors using quantitative in vivo 1H magnetic resonance spectroscopy. Magn Reson Med 2003; 49: 223–32.

Ibrahim TS. Analytical approach to the MR signal. Magn Reson Med 2005; 54: 677–82.

Jansen JF, Backes WH, Nicolay K, Kooi ME. 1H MR spectroscopy of the brain: absolute quantification of metabolites. Radiology 2006; 240: 318–32.

Jen M, Johnson J, Hou P, Liu H. SU-G-IeP1-07: inaccuracy of lesion blood flow quantification related to the proton density reference image in arterial spin labeling MRI of brain tumors. Med Phys 2016; 43: 3645–45.

Jurcoane A, Wagner M, Schmidt C, Mayer C, Gracien R-M, Hirschmann M, et al. Within-lesion differences in quantitative MRI parameters predict contrast enhancement in multiple sclerosis. J Magn Reson Imaging 2013; 38: 1454–61.

Kaneoke, Y., Furuse, M., Inao, S., Saso, K., Yoshida, K., Motegi, Y., Mizuno, M. and Izawa, A. Spinlattice relaxation times of bound water – its determination and implications for tissue discrimination, Magn. Reson. Imag. 1987; 5: 415–420.

Keegan BM, Noseworthy JH. Multiple sclerosis. Ann Rev Med 2002; 53: 285–302.

Keep RF, Hua Y, Xi G. Brain water content: a misunderstood measurement? Transl Stroke Res 2012; 3: 263–265.

Kennedy KM, Erickson KI, Rodrigue KM, Voss MW, Colcombe SJ, Kramer AF, et al. Age-related differences in regional brain volumes: a comparison of optimized voxel-based morphometry to manual volumetry. Neurobiol Aging 2009; 30: 1657–76.

Koenig SH. Cholesterol of myelin is the determinant of gray-white contrast in MRI of brain. Magn Reson Med 1991; 20: 285–91.

Krauss W, Gunnarsson M, Andersson T, Thunberg P. Accuracy and reproducibility of a quantitative magnetic resonance imaging method for concurrent measurements of tissue relaxation times and proton density. Magn Reson Imaging 2015; 33: 584–91.

Kucharczyk W, Macdonald PM, Stanisz GJ, Henkelman RM. Relaxivity and magnetization transfer of white matter lipids at MR imaging: importance of cerebrosides and pH. Radiology 1994; 192: 521–9.

Laule C, Leung E, Li DKB, Traboulsee AL, Paty DW, MacKay AL, et al. Myelin water imaging in multiple sclerosis: quantitative correlations with histopathology. Mult Scler 2006; 12: 747–53.

Laule C, Vavasour IM, Kolind SH, Li DKB, Traboulsee TL, Moore GRW, et al. Magnetic resonance imaging of myelin. Neurotherapeutics 2007; 4: 460–84.

Laule C, Vavasour IM, Moore GRW, Oger J, Li DKB, Paty DW, et al. Water content and myelin water fraction in multiple sclerosis. A T2 relaxation study. J Neurol 2004; 251: 284–93.

Lecocq A, Le Fur Y, Amadon A, Vignaud A, Bernard M, Guye M, et al. Fast whole brain quantitative proton density mapping to quantify metabolites in tumors. Phys Medica 2013; 29, Supple: e11–e12.

Lecocq A, Le Fur Y, Amadon A, Vignaud A, Cozzone PJ, Guye M, et al. Fast water concentration mapping to normalize (1)H MR spectroscopic imaging. MAGMA 2015; 28: 87–100.

Levesque I, Sled JG, Narayanan S, Santos AC, Brass SD, Francis SJ, et al. The role of edema and demyelination in chronic T1 black holes: a quantitative magnetization transfer study. J Magn Reson Imaging 2005; 21: 103–10.

Liang ZP, Bammer R, Ji J, Pelc NJ, Glover GH. Improved image reconstruction from sensitivity-encoded data by wavelet denoising and Tokhonov regularization. In: *Biomedical Imaging V—Proceedings of the 5th IEEE EMBS International Summer School on Biomedical Imaging*, SSBI, 2002.

Lin F-H, Kwong KK, Belliveau JW, Wald LL. Parallel imaging reconstruction using automatic regularization. Magn Reson Med 2004; 51: 559–67.

Lin W, Paczynski RP, Venkatesan R, He YY, Powers WJ, Hsu CY, et al. Quantitative regional brain water measurement with magnetic resonance imaging in a focal ischemia model. Magn Reson Med 1997; 38: 303–10.

Lorio S, Kherif F, Ruef A, Melie-Garcia L, Frackowiak R, Ashburner J, et al. Neurobiological origin of spurious brain morphological changes: A quantitative MRI study. Hum Brain Mapp 2016; 37: 1801–15.

Ma D, Gulani V, Seiberlich N, Liu K, Sunshine JL, Duerk JL, et al. Magnetic resonance fingerprinting. Nature 2013; 495: 187–92.

Meyers S. Accurate measurement of brain water content by magnetic resonance (Doctoral dissertation). University of British Columbia, 2015.

Meyers SM, Kolind SH, Laule C, MacKay AL. Measuring water content usign T2 relaxation at 3 T: phantom validations and simulations. Magn Reson Imaging 2016; 34: 246–51.

Meyers SM, Kolind SH, MacKay AL. Simultaneous measurement of total water content and myelin water fraction in brain at 3T using a T2 relaxation based method. Magn Reson Imaging 2017; 37: 187–94.

Mezer A, Rokem A, Berman S, Hastie T, Wandell BA. Evaluating quantitative proton-density-mapping methods. Hum Brain Mapp 2016; 37: 3623–35.

Mezer A, Yeatman JD, Stikov N, Kay KN, Cho N-J, Dougherty RF, et al. Quantifying the local tissue volume and composition in individual brains with magnetic resonance imaging. Nat Med 2013; 19: 1667–72.

Michel E, Hernandez D, Lee SY. Electrical conductivity and permittivity maps of brain tissues derived from water content based on T_1 -weighted acquisition. Magn Reson Med; 2017; 77: 1094–1103

Neeb H, Ermer V, Stocker T, Shah NJ. Fast quantitative mapping of absolute water content with full brain coverage. NeuroImage 2008; 42: 1094–109.

Neeb H, Zilles K, Shah NJ. Fully-automated detection of cerebral water content changes: study of age- and gender-related H2O patterns with quantitative MRI. NeuroImage 2006a; 29: 910–22.

Neeb H, Zilles K, Shah NJ. A new method for fast quantitative mapping of absolute water content in vivo. NeuroImage 2006b; 31: 1156–68.

Nitz WR, Reimer P. Contrast mechanisms in MR imaging. Eur Radiol 1999; 9: 1032–46.

Norton, W. T., Poduslo, S. E. and Suzuki, K. Subacute sclerosing leukoencephalitis. Chemical studies including abnormal myelin and an abnormal ganglioside pattern, J. Neuropathol. Exp. Neurol. 1966; 25: 582–97.

Noterdaeme O, Anderson M, Gleeson F, Brady M. Intensity correction with a pair of spoiled gradient recalled echo images. Phys Med Biol 2009; 54: 3473–89.

Papadopoulos MC, Saadoun S, Binder DK, Manley GT, Krishna S, Verkman AS. Molecular mechanisms of brain tumor edema. Neuroscience 2004; 129: 1009–18.

Rooney WD, Johnson G, Li X, Cohen ER, Kim SG, Ugurbil K, et al. Magnetic field and tissue dependencies of human brain longitudinal 1H2O relaxation in vivo. Magn Reson Med 2007; 57: 308–18.

Sabati M, Maudsley AA. Fast and high-resolution quantitative mapping of tissue water content with full brain coverage for clinically-driven studies. Magn Reson Imaging 2013; 31: 1752–9.

Shah NJ, Neeb H, Kircheis G, Engels P, Häussinger D, Zilles K. Quantitative cerebral water content mapping in hepatic encephalopathy. NeuroImage 2008; 41: 706–17.

Schepps, J. L. and Foster, K. R. The UHF and microwave dielectric properties of normal and tumour tissues: variation in dielectric properties with tissue water content, Phys. Med. Biol. 1980; 25: 1149–1159.

Simon J, Szumowski J, Totterman S, Kido D, Ekholm S, Wicks A, et al. Fat-suppression MR imaging of the orbit. Am J Neuroradiol 1988; 9: 961–8.

Stüber C, Morawski M, Schäfer A, Labadie C, Wähnert M, Leuze C, et al. Myelin and iron concentration in the human brain: a quantitative study of MRI contrast. NeuroImage 2014; 93: 95–106.

Takagi, H., Shapiro, K., Marmarou, A. and Wisoff, H. Microgravimetric analysis of human brain tissue: correlation with computerized tomography scanning, J. Neurosurg. 1981; 54: 797–801.

Tourtellotte, W. W. and Parker, J. A. Some spacesand barriers in postmortem multiple sclerosis, Prog. Brain. Res. 1968; 29: 493–525.

Tisserand DJ, Pruessner JC, Sanz Arigita EJ, Van Boxtel MPJ, Evans AC, Jolles J, et al. Regional frontal cortical volumes decrease differentially in aging: an MRI study to compare volumetric approaches and voxel-based morphometry. NeuroImage 2002; 17: 657–69.

Vavasour IM, Laule C, Li DKB, Traboulsee AL, MacKay AL. Is the magnetization transfer ratio a marker for myelin in multiple sclerosis? J Magn Reson Imaging 2011; 33: 713–18.

Volz S, Nöth U, Deichmann R. Correction of systematic errors in quantitative proton density mapping. Magn Reson Med 2012; 68: 74–85.

Volz S, Nöth U, Jurcoane A, Ziemann U, Hattingen E, Deichmann R. Quantitative proton density mapping: Correcting the receiver sensitivity bias via pseudo proton densities. NeuroImage 2012; 63: 540–52.

Vymazal J, Brooks R, Patronas N, Hajek M, Bulte JW, Di Chiro G. Magnetic resonance imaging of brain iron in health and disease. J Neurol Sci 1995; 134 Suppl: 19–26.

Vymazal J, Righini A, Brooks RA, Canesi M, Mariani C, Leonardi M, et al. T1 and T2 in the brain of healthy subjects, patients with Parkinson disease, and patients with multiple system atrophy: relation to iron content. Radiology 1999; 211: 489–95.

West KL, Kelm ND, Carson RP, Gochberg DF, Ess KC, Does MD. Myelin volume fraction imaging with MRI. NeuroImage 2016: In press

Whittall KP, MacKay AL, Graeb DA, Nugent RA, Li DKB, Paty DW. In vivo measurement of T2 distributions and water contents in normal human brain. Magn Reson Med 1997; 37: 34–43.

Wick W, Küker W. Brain edema in neurooncology: radiological assessment and management. Onkologie 2004; 27: 261–6.

Yeatman JD, Wandell BA, Mezer AA. Lifespan maturation and degeneration of human brain white matter. Nat Commun 2014; 5: 4932.

5

T_1: Longitudinal Relaxation Time[1]

Contents

Ralf Deichmann and
René-Maxime Gracien
Goethe University

5.1 Physical basis of T_1

If a hydrogen atom is placed inside a static magnetic field **B**, the *spin* of the hydrogen nucleus can assume two different states, yielding a magnetic moment that is either parallel or antiparallel to the magnetic field. The first state has a slightly higher probability as it is energetically lower. Consequently, under equilibrium conditions, an ensemble of hydrogen atoms inside **B** will produce a macroscopic magnetisation **M** that is parallel to **B**. In general, the vector **M** has two components: the *longitudinal* component, which is parallel to **B**, and the *transverse* component, which is perpendicular to **B**. Under equilibrium conditions, **M** is parallel to **B**, so the transverse component is zero and the longitudinal component assumes the *equilibrium value* M_0.

If the spin system is irradiated by a radiofrequency (RF) pulse with the protons' Larmor frequency, energy is absorbed by the spin system, so a certain number of spins assume the energetically higher state, leaving equilibrium conditions. In the classical view, this corresponds to a rotation of **M** by a certain angle. As a consequence, **M** has now a non-zero transverse component, which rotates around **B** with the Larmor frequency, thus producing the signal that is measured in magnetic resonance (MR) imaging. Furthermore, the longitudinal component of **M** is reduced and assumes a value between $-M_0$ and $+M_0$. Subsequently, if left alone, the spin system approaches again equilibrium conditions. This phenomenon is called *relaxation* and consists of two simultaneous processes, *transverse* and *longitudinal relaxation*. The first process causes an exponential decay of the transverse magnetisation (and thus of the signal; i.e. T_2 decay), while the second process causes a change of the longitudinal magnetisation towards the equilibrium value M_0 (i.e. T_1 relaxation). In this chapter, only the latter process is discussed. During the longitudinal relaxation, the spins release the excess energy, which is absorbed by the surrounding *lattice*, that is by molecules in the neighbourhood. Mathematically, this process is described by the following term in the Bloch equations, assuming that the static magnetic field is applied along the z-axis:

$$\frac{dM_z}{dt} = \frac{M_0 - M_z}{T_1} \tag{5.1}$$

Here, the time constant T_1 is the *longitudinal relaxation time*, sometimes also called the *spin-lattice relaxation time*.

The solution of Equation 5.1 is an exponential change of M_z towards the equilibrium value M_0:

$$M_z(t) = M_0 + \left[M_z(0) - M_0\right]\exp\left(-t/T_1\right) \tag{5.2}$$

[1] Edited by Paul S. Tofts; reviewed by Zaheer Abbas, Nazim Lechea and N. Jon Shah, Institute of Neuroscience and Medicine-4, Forschungszentrum Juelich GmbH, Juelich, Germany

A special case is the *inversion recovery* curve, which describes the time course of M_z after a full spin inversion, so $M_z(0) = -M_0$:

$$M_z(TI) = M_0\left[1 - 2\exp(-\mathrm{TI}/T_1)\right] \quad (5.3)$$

where the *inversion time* TI is the time interval between spin inversion and measurement. As an example, Figure 5.1 shows an inversion recovery curve for a T_1 of 1 s.

5.2 Biological basis of T_1

The T_1 relaxation time depends on the physical properties and microstructural composition of the underlying tissue. In particular, it is related to: (a) the free water content (Fatouros *et al.*, 1991; Gelman *et al.*, 2001), (b) the concentration and types of macromolecules (Rooney *et al.*, 2007) such as myelin (Lutti *et al.*, 2014) and (c) the iron content (Gelman *et al.*, 2001). While increased water content prolongs T_1, increased iron content and myelination reduce T_1. Accordingly, cerebrospinal fluid has a considerably longer T_1 than cerebral white matter and gray matter due to the high water content. Furthermore, T_1 in white matter is shorter than in gray matter, mainly due to the larger proportion of myelin and consequently smaller water fraction in white matter.

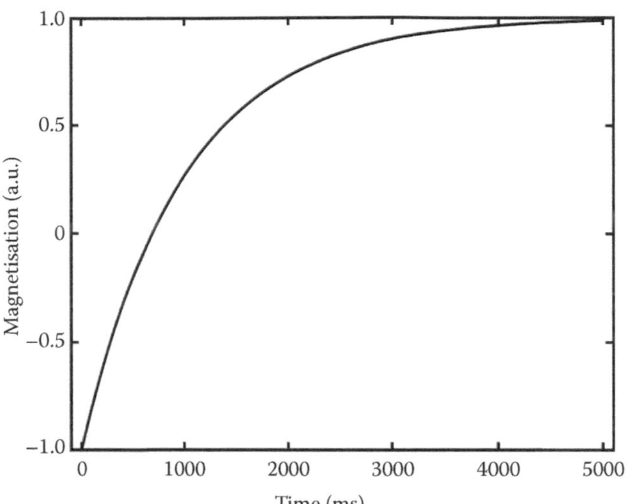

FIGURE 5.1 Inversion recovery curve for $T_1 = 1$ s.

When comparing T_1 values that were measured with different MR systems, for example in multicentre studies, it should be kept in mind that results may be biased by several parameters such as the hardware used or subject age. As an example, T_1 values significantly increase with the magnetic field strength of the respective MR system (Rooney *et al.*, 2007). Furthermore, cerebral T_1 values are known to change over the lifespan (Cho *et al.*, 1997; Gracien *et al.*, 2016c). Table 5.1 shows typical T_1 values of selected brain areas, measured at different field strengths.

5.3 How to measure T_1

5.3.1 Gold standard: The inversion recovery technique

For the sake of simplicity, let us first consider the case of magnetic resonance spectroscopy (MRS), where spectroscopic information is derived from a signal acquired after a single RF excitation pulse (usually 90°). In this case, T_1 quantification via the inversion recovery (IR) technique follows Figure 5.2: several measurements are performed, each of which comprises spin inversion, a subsequent delay TI, spin excitation and signal readout. By varying TI,

TABLE 5.1 T_1 values (in ms) of normal brain tissue at different static magnetic field strengths

Field Strength	Reference	White Matter	Grey Matter	Caudate Nucleus	Putamen	Thalamus
0.2 T	Rooney *et al.* (2007)	361 ± 17	635 ± 54	555 ± 19	524 ± 19	522 ± 44
1.0 T	Rooney *et al.* (2007)	555 ± 20	1036 ± 19	898 ± 45	815 ± 16	807 ± 47
1.5 T	Steen *et al.* (1994)	606 ± 21	1170 ± 43	948 ± 32	834 ± 19	774 ± 16
	Henderson *et al.* (1999)	633 ± 8	1148 ± 24			
	Shah *et al.* (2001)	600 ± 25	1000 ± 90			
	Deoni (2003)	621 ± 61	1060 ± 133	1112 ± 132	1014 ± 101	780 ± 55
	Rooney *et al.* (2007)	656 ± 16	1188 ± 69	1083 ± 52	981 ± 13	972 ± 32
	Warntjes *et al.* (2008)	575 ± 16	1048 ± 61	917 ± 43	832 ±25	738 ± 39
2.0 T	Deichmann *et al.* (1999)	682 ± 4	1268 ± 29			
3.0 T	Clare and Jezzard (2001)	860 ± 20		1310 ± 60	1100 ± 30	1060 ± 40
	Preibisch (2009b)	933 ± 15	1380 ± 59	1450 ± 92	1310 ± 39	
	Marques *et al.* (2010)	810 ± 30	1355 ± 70	1250 ± 70	1130 ± 70	1080 ± 70
	Gras *et al.* (2016)	911 ± 59	1508 ± 208			
4.0 T	Rooney *et al.* (2007)	1010 ± 19	1723 ± 93	1509 ± 53	1446 ± 32	1452 ± 87
7.0 T	Rooney *et al.* (2007)	1220 ± 36	2132 ± 103	1745 ± 64	1700 ± 66	1656 ± 84
	Marques *et al.* (2010)	1150 ± 60	1920 ± 160	1630 ± 90	1520 ± 90	1430 ± 100
	Polders *et al.* (2012)	1085 ± 49	1839 ± 79	1638 ± 73	1477 ± 85	1416 ± 18
9.4 T	Pohmann *et al.* (2016)	1427 ± 52				

Note: If T_1 values were listed for different subareas in the original publications (such as left and right putamen or frontal and occipital white matter), the average value is given in the table. Values are given as mean +/− standard deviation.

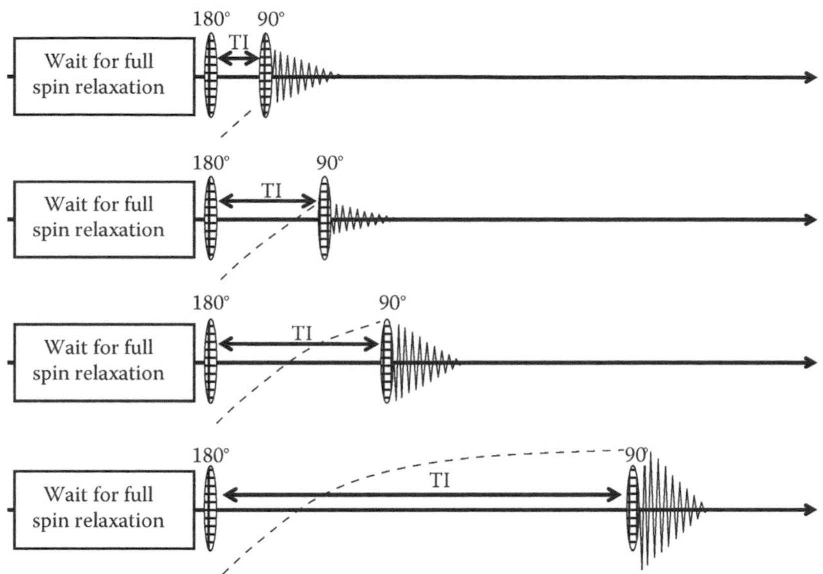

FIGURE 5.2 The inversion recovery (IR) method as gold standard for T_1 quantification, based on several IR measurements with different inversion times (TI). A full spin relaxation is required before each single measurement.

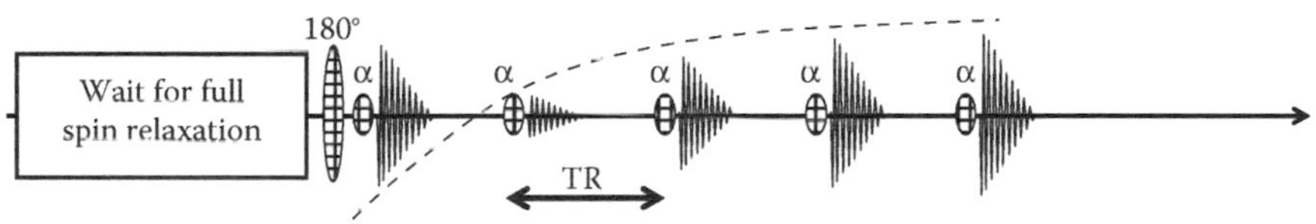

FIGURE 5.3 Principle of the Look-Locker technique: A full T_1 relaxation curve is sampled by sending a series of radiofrequency pulses with a small tip angle and measuring the resulting signals. The temporal resolution for sampling the relaxation curve is the repetition time (TR).

the inversion recovery curve as given in Equation 5.3 is sampled, so T_1 can be obtained via exponential data fitting. The problem is that equilibrium conditions have to be attained before each single experiment, requiring a full spin relaxation before each spin inversion. As can be seen from Figure 5.1 (which refers to a T_1 of 1 s), the equilibrium magnetisation is attained with sufficient accuracy after about five T_1 periods, giving rise to long waiting times between the experiments. As a consequence, even for the relatively simple case of MRS, a full T_1 measurement is time-consuming.

This problem is considerably exacerbated in MRI, where a large number of echoes with different phase encoding have to be sampled to enable image reconstruction. IR-based gold standard techniques for measuring T_1 usually employ spin echo imaging with integrated spin inversion (Stikov *et al.*, 2015). Typical durations are 13 min for a single-slice measurement with an in-plane resolution of 2 mm and a slice thickness of 5 mm, using four different TI values (Stikov *et al.*, 2015). Alternatively, the spectroscopic experiment shown in Figure 5.2 can be converted into an imaging experiment via replacing the spectroscopic signal acquisition by an echo-planar imaging (EPI) module (Preibisch and Deichmann 2009a). In this case, a single-slice measurement with an isotropic resolution of 3 mm, 15 different TI values ranging from 100 ms to 5000 ms and a relaxation delay of 20 s before each inversion has a total duration of about 5:30 min. These relatively long durations stress the need for fast T_1 mapping techniques.

5.3.2 The Look-Locker technique

This technique was originally designed for use in MRS (Look and Locker 1970). The idea is to measure T_1 *during one single T_1 relaxation process*, as shown in Figure 5.3: after inverting the magnetisation, a series of excitation pulses with a small tip angle α and an intermediate *repetition time TR* is sent. Each pulse tilts the magnetisation, creating a transverse magnetisation and thus a signal that is proportional to the current value of the longitudinal magnetisation M_z. Thus, the signal series samples the relaxation curve M_z(t) with a temporal resolution of TR, so T_1 can be obtained via exponential fitting.

The problem is that the excitation pulses distort the free relaxation curve. As an example, Figure 5.4 shows the development of the longitudinal relaxation after spin inversion, assuming T_1 = 1s, α = 30°, *TR* = 250 ms. Clearly, the effective relaxation curve (black) differs considerably from the unperturbed case (red) and has a non-exponential behaviour. However, the measured signal amplitudes represent the values of M_z directly before excitation (circles in Figure 5.4), which show a *modified*

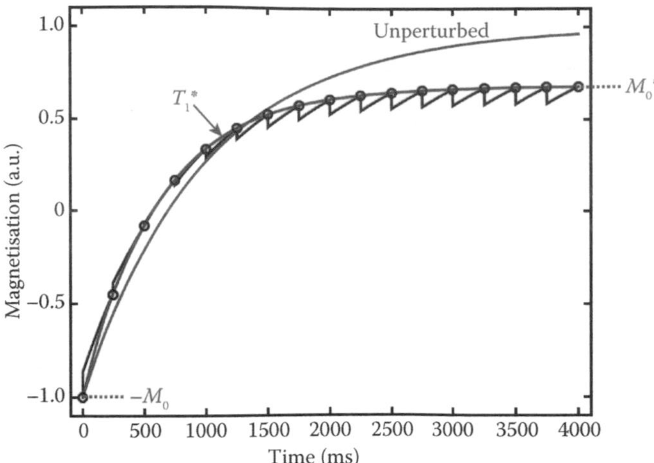

FIGURE 5.4 Look-Locker technique: Development of the longitudinal magnetisation (black) after spin inversion for T_1 = 1 s, α = 30°, TR = 250 ms. The measured signal amplitudes sample the longitudinal magnetisation at the time points of the excitation pulses (circles), showing an exponential behaviour (blue) with a modified time constant T_1^* and approaching the saturation value M_0^*. The red line refers to the case of unperturbed longitudinal relaxation.

exponential behaviour (blue): M_z approaches a *saturation value* $M_0^* < M_0$ with a modified relaxation time $T_1^* < T_1$, where T_1^* and M_0^* are given by the following (Kaptein *et al.*, 1976):

$$\exp(-TR/T_1^*) = \cos(\alpha)\exp(-TR/T_1) \tag{5.4a}$$

or:

$$T_1^* = [1/T_1 - (1/TR)\cdot\ln(\cos(\alpha))]^{-1} \tag{5.4b}$$

and:

$$M_0^* = M_0 \frac{1-\exp(-TR/T_1)}{1-\cos(\alpha)\exp(-TR/T_1)} \tag{5.5}$$

Thus, exponential fitting of the sampled curve yields T_1^*, from which T_1 can be obtained via Equation 5.4b, provided α is known.

In MRI, the Look-Locker (LL) concept is applied by acquiring a series of spoiled gradient echo (GE) images after spin inversion. The idea is that in this way the image amplitudes sample the relaxation process with the spatial resolution of the underlying images, allowing the calculation of a T_1 map. Each GE image acquisition is based on the irradiation of a series of excitation pulses with the repetition time *TR* and the excitation angle α, followed by the acquisition of a gradient echo for each excitation. Thus, the same rules as explained above apply and exponential fitting of the measured relaxation curve yields for each pixel the modified time constant T_1^*, from which T_1 can be calculated according to Equation 5.4b.

It should be noted that the acquisition time for each image must be shorter than T_1, so the relaxation curve can be sampled with sufficient temporal resolution. Thus, *TR* has to be kept relatively short and the number of phase encoding (PE) steps is limited, unless more advanced techniques are used (see below).

The TAPIR sequence (Shah *et al.*, 2001) is based on the LL concept and allows multislice T_1 mapping to be carried out with high spatial and temporal resolutions. The short acquisition time is due to the use of a banded k-space data collection scheme, acquiring three gradient echoes with different PE per excitation pulse. For TAPIR, a duration of 6:44 min has been reported for the acquisition of a T_1 map comprising 32 slices with an in-plane resolution of 1 mm and a slice thickness of 2 mm, sampling the relaxation curve at 20 time points (Möllenhoff 2016).

5.3.3 The variable flip angle technique

This technique is again based on the acquisition of GE data sets. In contrast to the LL technique, acquisition times are considerably longer than T_1^*, due to the use of relatively long *TR* and a large number of PE steps, for example by acquiring three-dimensional (3D) data sets with a high spatial resolution. As a consequence, M_z corresponds to the steady-state value (M_0^*) during the major part of data acquisition, so data are acquired *under steady-state conditions*. The underlying idea is to acquire several data sets with different excitation angles α and to evaluate the signal dependence S(α) for each pixel. As an example, Figure 5.5 shows S(α) for a phantom with an approximate T_1 of 1 s that was scanned with *TR* = 16.4 ms and six different excitation angles. Since the exact shape of this curve depends on T_1, it is possible to derive T_1 from the data (Wang *et al.*, 1987;

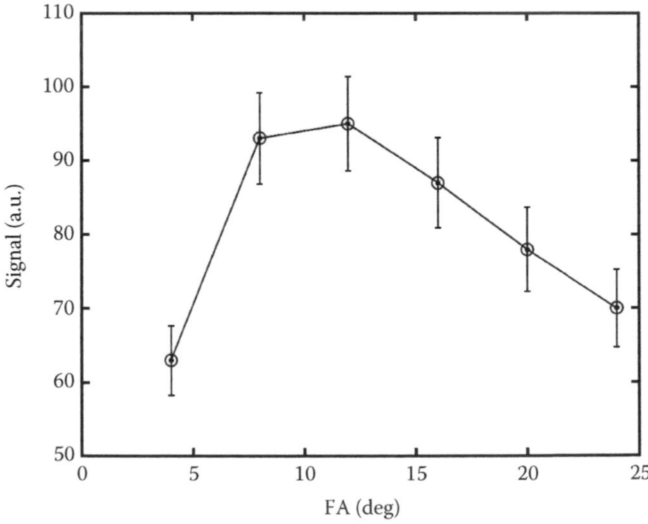

FIGURE 5.5 Variable flip angle technique: Signal dependence on the excitation angle (results of a phantom measurement). The single data points are shown as circles. The error bars denote the standard deviation across the phantom. The data points are connected with lines for visual guidance.

Venkatesan *et al.*, 1998). The signal is given by the longitudinal magnetisation M_z directly before RF excitation, multiplied with the sine of the excitation angle (Figure 5.5). Since in variable flip angle (VFA) data, M_z corresponds to $M_0{}^*$ as defined in Equation 5.5, the signal amplitude follows from

$$S(\alpha) = S_0 \sin(\alpha) \frac{1-\exp(-TR/T_1)}{1-\cos(\alpha)\exp(-TR/T_1)} \tag{5.6}$$

To simplify the analysis, this equation is rewritten

$$S(\alpha)\left[1-\cos(\alpha)\exp(-TR/T_1)\right]/\sin(\alpha) = S_0\left[1-\exp(-TR/T_1)\right] \tag{5.7a}$$

or:

$$S(\alpha)/\sin(\alpha) = \exp(-TR/T_1)S(\alpha)/\tan(\alpha) + S_0\left[1-\exp(-TR/T_1)\right] \tag{5.7b}$$

Thus, if several data sets are acquired with different excitation angles α_i, the different signal amplitudes S_i are determined for a certain pixel and the values $y_i = S_i/\sin(\alpha_i)$ and $x_i = S_i/\tan(\alpha_i)$ are calculated. Equation 5.7 implies that a plot of y_i versus x_i shows a linear dependence with the slope m = $\exp(-TR/T_1)$, from which T_1 can be derived (Wang *et al.*, 1987; Venkatesan *et al.*, 1998). Figure 5.6 shows this linear plot for the phantom data presented in Figure 5.5. There is a clear linear dependence with the slope m = 0.9832, corresponding to a T_1 of about 970 ms for the *TR* chosen.

The advantage of the VFA method is its speed: a full T_1 map can be derived from only two spoiled GE data sets acquired with different excitation angles. Furthermore, a high spatial resolution can be achieved, in particular for 3D data. In the case of a two-point measurement, the two optimum excitation angles can be calculated as follows (Helms *et al.*, 2011): for the *TR* chosen and the approximate target T_1 value, a parameter τ_E is derived

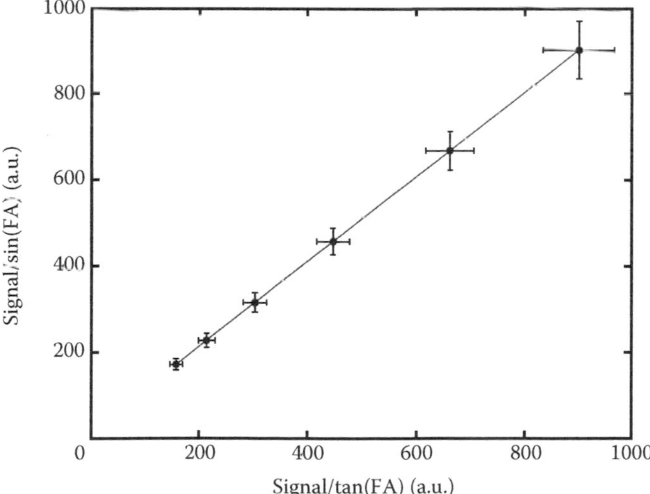

FIGURE 5.6 Variable flip angle technique: Linear plot according to the variable flip angle concept (results of a phantom measurement). The single data points are shown as dots. The error bars denote the standard deviation across the phantom. The straight line represents the linear fit according to Equation 5.7b. *Editor's note*: Although this linearization permits a fast (non-iterative) estimation of T_1, the uncertainty in each point is not equal, and this should ideally be taken into account when making the estimate.

$$\tau_E = 2 \cdot \sqrt{\frac{1-\exp(-TR/T_1)}{1+\exp(-TR/T_1)}} \tag{5.8a}$$

The optimum angles α_1 and α_2 are then given by[2]

$$2 \cdot \tan(\alpha_i/2) = K_i \cdot \tau_E \quad \text{with}: K_1 = 0.4142 \text{ and } K_2 = 2.4142 \tag{5.8b}$$

For the VFA technique, a duration of about 10 min has been reported for the acquisition of a T_1 map with whole brain coverage and an isotropic resolution of 1 mm, based on two GE data sets with different excitation angles (Deoni 2007; Preibisch and Deichmann 2009b).[3] Since VFA requires correction for non-uniformities of the RF transmit profile (see Section 5.4.1), an additional duration of about 1 min for B_1 mapping should be considered when planning the protocol.

5.4 Pitfalls in T_1 measurements

5.4.1 General: B_1 inhomogeneities

Both the LL and VFA techniques require knowledge of the excitation angle for T_1 evaluation. However, the amplitude B_1 of the RF field sent by the transmit coil usually is not uniform, so the local excitation angle can deviate considerably from the nominal value. As an example, Figure 5.7 shows an axial slice of a B_1 map[4] acquired on a healthy subject at a field strength of 3 Tesla (T) (please note that throughout this chapter, B_1 is given in relative units, assuming a value of 1.0 where the actual angle matches the nominal value).

5.4.2 Pitfalls: The IR technique

The analysis of IR data via Equation 5.3 is only warranted if the following conditions are fulfilled: Firstly, there must be a complete spin inversion via a perfect 180° RF pulse. Secondly, there must be a sufficiently long delay after each measurement, allowing full spin relaxation to take place before the next inversion. If perfect spin inversion cannot be guaranteed, data should be analysed via a three-parameter fit. In this case, the factor of two in Equation 5.3 is not fixed but becomes an additional degree of freedom, which is determined during the process of fitting. If the delay between measurements is too short for full spin relaxation (e.g. if *TR* has to be kept short to reduce the experiment duration), a modified equation can be used for fitting (Stikov *et al.*, 2015).

[2] See also the letter by Wood 2015.
[3] A typical sequence uses TR = 16 ms, FA = 4°/25° at 3T.
[4] B_1 mapping is also described in Chapter 2, Section 2.1.7.

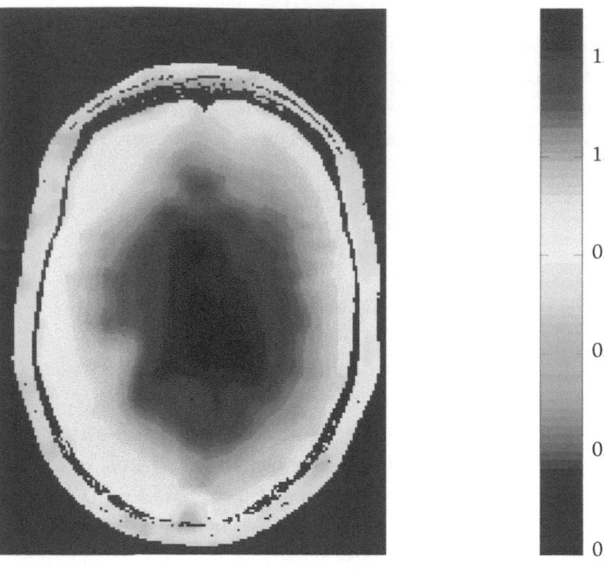

FIGURE 5.7 Axial slice of a B_1 map (isotropic spatial resolution of 4 mm, interpolated to 1 mm), acquired on a healthy subject at a field strength of 3 Tesla, using the method described by Volz *et al.*, (2010).

5.4.3 Pitfalls: The LL technique

Problem 1: The LL technique requires knowledge of the actual excitation angle, which may be difficult to determine in the presence of B_1 inhomogeneities. Furthermore, if two-dimensional (2D) sequences with slice-selective RF pulses are used, the excitation angle varies across the slice in correspondence with the respective slice profile. Fortunately, LL data may be analysed without knowledge of the excitation angle. Since $TR \ll T_1$ usually holds, the term $\exp(-TR/T_1)$ can be approximated as $1-TR/T_1$. A similar approximation holds for $\exp(-TR/T_1^*)$. Inserting Equation 5.4a in Equation 5.5 and using this approximation yields

$$M_0^* = M_0 \frac{T_1^*}{T_1} \tag{5.9}$$

A three-parameter analysis of the relaxation curve as sampled with the LL technique (see Figure 5.4, blue curve) yields the start value ($-M_0$), the asymptotic end value (M_0^*) and the time constant (T_1^*), so T_1 can be calculated from these values via Equation 5.9 (Deichmann and Haase 1992).

Problem 2: In LL, the acquisition time per image must be similar to T_1^* or shorter to sample the relaxation curve with sufficient temporal resolution. This restricts the number of PE steps and therefore the spatial resolution. The TAPIR sequence (Shah *et al.*, 2001) circumvents this problem by repeating the measurement several times, covering different portions of k-space each time. Furthermore, several gradient echoes with different PE are sampled per excitation. As a consequence, TAPIR permits a more detailed sampling of the relaxation curve.

5.4.4 Pitfalls: The VFA technique

Problem 1: If B_1 inhomogeneities are not accounted for in the VFA technique, the analysis yields an *apparent* value T_{1app} given by the following (Helms *et al.*, 2008a; Preibisch and Deichmann 2009a):

$$T_{1app} = T_1 B_1^2 \tag{5.10}$$

Thus, a 5% deviation of B_1 from the ideal value of 1.0 would yield a 10% error in T_1. Consequently, VFA requires additional B_1 mapping, calculation of the actual excitation angle α for each pixel and usage of this angle in Equation 5.7 (Deoni 2007). Several methods for fast B_1 mapping have been reported in the literature (Cunningham *et al.*, 2006; Yarnykh 2007; Helms *et al.* 2008b; Morrell 2008; Volz *et al.*, 2010; Nehrke and Bornert 2012). Furthermore, the B_1 profile can be directly deduced from the VFA data, provided it varies smoothly across space: A method dubbed UNICORT treats reciprocal maps of T_{1app} as anatomical data sets that are affected by a smooth bias given by $1/B_1^2$ (see Equation 5.10), which can be determined via bias field correction (Weiskopf *et al.*, 2011). An algebraic solution to this problem has also been suggested (Baudrexel *et al.*, 2016).

Problem 2: For correct T_1 evaluation via the VFA technique, the exact local excitation angles have to be known. If 3D sequences with non-selective excitation pulses are used, B_1 mapping is required, as explained above. If, however, 2D (multislice) sequences with slice-selective excitation pulses are used, it has to be taken into account that the excitation angle shows a variation across the slice that corresponds to the RF excitation profile. This requires a further correction factor, in addition to the B_1 correction (Gras *et al.*, 2013).

Problem 3: The VFA theory assumes that in GE imaging, residual transverse magnetisation is deleted ('spoiled') after each echo acquisition. However, stimulated echoes may yield considerable deviations of the actual steady-state magnetisation from the theoretical value. A technique dubbed *RF spoiling* (Zur *et al.*, 1991) employs RF pulses that are sent with different pulse phases (i.e. rotation axes), so residual transverse magnetisation components will point in different directions and cancel each other, provided the phase list is chosen appropriately. In detail, the phase of the nth RF pulse should be:

$$\phi_n = \Delta\phi \, \frac{n}{2}(n-1) \tag{5.11}$$

In the original publication on RF spoiling, a 'phase increment' $\Delta\phi$ of 117° was suggested. Figure 5.8 shows the dependence of the actual steady-state magnetisation on $\Delta\phi$ for spoiled GE data acquired with TR = 16.4 ms and α = 20°, assuming T_1 = 1 s and T_2 = 70 ms. Clearly, for most values of $\Delta\phi$ there are considerable deviations from the theoretical value given by Equation 5.5 (shown as a horizontal line). Since this pivotal equation is the basis of the VFA technique, deviations yield erroneous T_1 values, requiring suitable corrections (Preibisch and Deichmann 2009a). Alternatively, it has been proposed to apply very strong

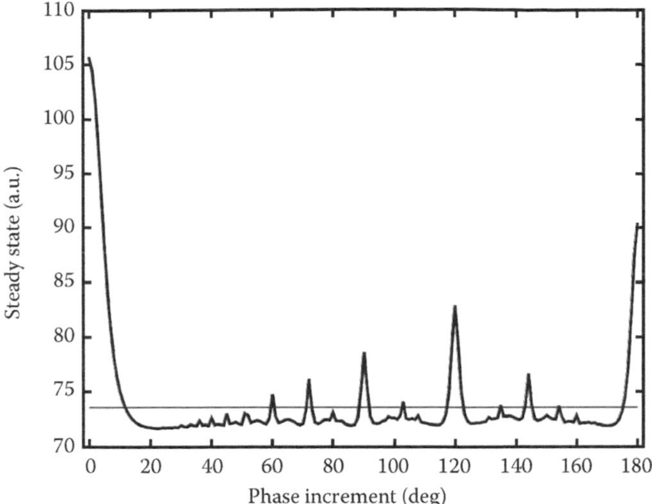

FIGURE 5.8 Steady-state magnetisation independence on the radiofrequency spoiling increment for spoiled gradient echo data acquired with TR = 16.4 ms and α = 20°, assuming T_1 = 1 s and T_2 = 70 ms. The horizontal line shows the value that corresponds to the case of perfect spoiling. The calculation of the steady-state magnetisation was based on a simulation program described in detail in the literature (Preibisch and Deichmann 2009a).

crusher gradients after each echo acquisition, thus giving rise to a faster decay of residual transverse magnetisation components due to diffusion effects (Yarnykh 2010).

5.5 Precision, reproducibility and quality assessment

5.5.1 Precision of Look-Locker method

For a LL protocol sampling the relaxation curve at eight time points with whole brain coverage, in-plane resolution of 1 mm, 30 contiguous slices with a thickness of 4 mm and 9:38 min acquisition time, the measurement was repeated six times on a healthy subject at a field strength of 1.5 Tesla. The standard deviation across measurements was 19 ms in white matter and 33 ms in grey matter, corresponding to a precision of 3.5% and 3.2%, respectively (Deichmann 2005).

5.5.2 Precision of VFA method

For a VFA protocol based on the acquisition of two GE data sets with different excitation angles, whole brain coverage with an isotropic resolution of 1 mm and 10 min acquisition time, the T_1 standard deviation due to background noise has been reported to be 26 ms in white matter and 51 ms in grey matter at a field strength of 3 Tesla (Nöth *et al.*, 2015). These values can be considered as the precision of the measured T_1 value for a single pixel.

5.5.3 Reproducibility of T_1 values in a multicentre study

In a study comparing T_1 data acquired with the VFA method on five healthy subjects and at three different sites operating 3 Tesla MR systems, a high intra-site and inter-site reproducibility of the resulting T_1 maps was reported, with a coefficient of variance of about 5% (Weiskopf *et al.*, 2013). Interestingly, anatomical data sets that were derived from the T_1 maps showed a better intra-site and inter-site reproducibility than conventional T_1-weighted data sets. However, the authors stressed the requirement for accurate B_1 mapping and subsequent data correction (see above) to avoid any hardware and thus site-dependent bias on the results.

5.5.4 Comparison of T_1 mapping methods and quality assessment

In a study comparing three methods (IR, LL, VFA) for T_1 mapping (Stikov *et al.*, 2015), all methods yielded similar T_1 values for a phantom, but considerable discrepancies *in vivo*, with deviations of more than 30% in white matter. The authors observed that in comparison to IR-based techniques, LL and VFA tend to yield shorter and longer T_1 values, respectively. It was suggested that these method-dependent deviations were due to the problems listed above, in particular B_1 inhomogeneities and the effects of insufficient spoiling of transverse magnetisation. The authors therefore recommended suitable quality assessment procedures, comparing results obtained with a certain T_1 mapping protocol with data derived from an IR-based gold standard experiment. In particular, quality assessment should be performed both for a T_1 phantom and under *in vivo* conditions.

5.6 Clinical applications of T_1 quantification

Conventional MRI techniques, as commonly used in the clinical routine, show mixed contrasts. This means that, even though the signal intensity in a conventional T_1-weighted data set is mainly determined by the T_1 value of the investigated tissue, other parameters, such as the relaxation times T_2 or T_2* and the proton density, influence the measured signal. Furthermore, the local intensity in conventional T_1-weighted images also depends on various hardware parameters, such as non-uniformities of the static magnetic field B_0, the transmitted radiofrequency field B_1 and the receive coil sensitivities.

In contrast, quantitative MRI techniques aim to measure actual tissue parameters, thus eliminating any other tissue or hardware-related bias. T_1 relaxometry provides quantitative values for each single voxel, which can be compared between follow-up scans of the same patient and even between different study centres in multicentre-trials. T_1 mapping permits the quantification of tissue properties beyond obvious lesions and, thus, the detection of diffuse or inconspicuous pathologies that are invisible in conventional MRI. T_1 mapping has shown its potential in different medical fields. For example, as the degree of oxygenation influences the T_1 of blood, a comparison of T_1 values in the lung before and after ventilation with 100% O_2 can be used for quantifying changes of blood oxygenation and potentially for the detection of pulmonary pathologies (Jakob et al., 2001).

Particularly in neuroimaging studies, T_1 relaxometry plays an important role, for example for the differentiation of different types of dementia (Besson *et al.*, 1985), for the detection of haemorrhagic transformation in patients with stroke (DeWitt *et al.*, 1987), for the evaluation of cerebral tissue abnormalities in patients with human immunodeficiency virus infection (Wilkinson *et al.*, 1996) or for the detection of tissue changes in patients with temporal lobe epilepsy (Conlon *et al.*, 1988; Cantor-Rivera *et al.*, 2015). Some fields of application will be highlighted more in detail in the following sections.

5.6.1 Multiple sclerosis

Multiple sclerosis (MS) is a chronic inflammatory disease of the central nervous system where focal lesions coexist with global inflammatory and degenerative processes. While many focal lesions are easily visible in clinical routine MRI, quantitative MRI techniques are particularly advantageous for the quantification of pathological tissue changes outside of these macroscopic lesions, allowing the close investigation of normal-appearing tissues and the assessment of diffuse tissue damage. Several authors described increased T_1 values in normal-appearing brain tissues, even at early disease stages (Griffin *et al.*, 2002; Vrenken *et al.*, 2006; Davies *et al.*, 2007). Importantly, a relationship between these changes in tissue composition and the clinical status has been unveiled in a number of studies (Parry *et al.*, 2002; Gracien *et al.*, 2016a), highlighting the clinical relevance of quantitative MRI, especially at chronic disease stages (Gracien *et al.*, 2016b) where global neurodegeneration gains importance.

MR spectroscopic studies suggest that T_1 prolongation might reflect gliosis and axonal loss in MS (Brex *et al.*, 2000). Furthermore, demyelination and oedema are thought to contribute to the increased T_1 values in MS lesions and normal-appearing brain tissue in MS. White matter lesions in conventional MRI are only the tip of the iceberg of tissue pathology in MS (Filippi and Rocca 2005). Accordingly, it seems to be only a matter of time until quantitative MRI methods for T_1 will also be included in clinical therapy studies.

5.6.2 Movement disorders

Parkinson's disease is a progressive neurodegenerative disorder, the underlying biochemical mechanisms of which are still the subject of current research. A microstructural key feature in Parkinson's disease and other extrapyramidal disorders is iron deposition (Dexter *et al.*, 1992).

Studies have used T_1 mapping to investigate disease-related tissue pathology in Parkinson's disease. T_1 decreases were spatially more widespread than T_2* shortening in the brainstem in Parkinson's disease, showing the potential of T_1 relaxometry to assess tissue changes beyond iron deposition (Baudrexel *et al.*, 2010). Furthermore, Vymazal *et al.* reported decreased T_1 values in the frontal cortex, possibly indicating decreased ferritin levels (Vymazal *et al.*, 1999).

Similarly, in multiple system atrophy, a neurodegenerative disease characterised by parkinsonism combined with cerebral ataxia, pyramidal signs and severe autonomic failure, T_1 was shortened in deep grey-matter regions. Interestingly, the estimation of the iron concentration in the globus pallidus with T_1 relaxometry was well in line with values reported in histochemical studies (Vymazal *et al.*, 1999).

These studies suggest that quantitative T_1 mapping has the potential to provide further information that might, in addition to clinical and sonographic data, support the diagnosis of movement disorders and the follow-up of individual patients.

5.6.3 Brain tumours

In patients diagnosed with glioblastoma, malignant cells spread across the whole brain tissue, rather than being restricted to the macroscopic tumour masses. Conventional MRI contrasts fail to visualise the whole extent of the disease. In a preliminary study, the longitudinal comparison of T_1 maps gave an earlier detection of tumour progression than did conventional MRI (Lescher *et al.*, 2015).

Furthermore, quantitative MRI allows the calculation of synthetic anatomies, provided that all contrast relevant physical parameters are measured. These synthetic anatomies can either replicate the typical contrasts of conventional routine data or even provide optimised contrasts. Synthetic anatomies with pure T_1 weighting were shown to provide improved tissue-to-background and tumour-to-background contrasts, thus improving the visibility of brain tumours and oedema (Nöth *et al.*, 2015).

References

Baudrexel S, Nürnberger L, Rüb U, Seifried C, Klein JC, Deller T, et al. Quantitative mapping of T1 and T2* discloses nigral and brainstem pathology in early Parkinson's disease. NeuroImage 2010; 51: 512–20.

Baudrexel S, Reitz SC, Hof S, Gracien R-M, Fleischer V, Zimmermann H, et al. Quantitative T1 and proton density mapping with direct calculation of radiofrequency coil transmit and receive profiles from two-point variable flip angle data. NMR Biomed 2016; 29: 349–60.

Besson JA, Corrigan FM, Foreman EI, Eastwood LM, Smith FW, Ashcroft GW. Nuclear magnetic resonance (NMR). II. Imaging in dementia. Br J Psychiatry 1985; 146: 31–5.

Brex PA, Parker GJ, Leary SM, Molyneux PD, Barker GJ, Davie CA, et al. Lesion heterogeneity in multiple sclerosis: a study of the relations between appearances on T1 weighted images, T1 relaxation times, and metabolite concentrations. J Neurol Neurosurg Psychiatry 2000; 68: 627–32.

Cantor-Rivera D, Khan AR, Goubran M, Mirsattari SM, Peters TM. Detection of temporal lobe epilepsy using support vector machines in multi-parametric quantitative MR imaging. Comput Med Imaging Graph 2015; 41: 14–28.

Cho S, Jones D, Reddick WE, Ogg RJ, Steen RG. Establishing norms for age-related changes in proton T1 of human brain tissue in vivo. Magn Reson Imaging 1997; 15: 1133–43.

Clare S, Jezzard P. Rapid T(1) mapping using multislice echo planar imaging. Magn Reson Med 2001; 45: 630–4.

Conlon P, Trimble M, Rogers D, Callicott C. Magnetic resonance imaging in epilepsy: a controlled study. Epilepsy Res 1988; 2: 37–43.

Cunningham CH, Pauly JM, Nayak KS. Saturated double-angle method for rapid B1+ mapping. Magn Reson Med 2006; 55: 1326–33.

Davies GR, Hadjiprocopis A, Altmann DR, Chard DT, Griffin CM, Rashid W, et al. Normal-appearing grey and white matter T1 abnormality in early relapsing-remitting multiple sclerosis: a longitudinal study. Multiple Sclerosis 2007; 13: 169–77.

Deichmann R. Fast high-resolution T1 mapping of the human brain. Magn Reson Med 2005; 54: 20–7.

Deichmann R, Haase A. Quantification of T1 values by SNAPSHOT-FLASH NMR imaging. J Magn Reson (1969) 1992; 96: 608–12.

Deichmann R, Hahn D, Haase A. Fast T1 mapping on a whole-body scanner. Magn Reson Med 1999; 42: 206–9.

Deoni SCL. High-resolution T1 mapping of the brain at 3T with driven equilibrium single pulse observation of T1 with high-speed incorporation of RF field inhomogeneities (DESPOT1-HIFI). J Magn Reson Imaging: JMRI 2007; 26: 1106–11.

Deoni SCL, Rutt BK, Peters TM. Rapid combined T1 and T2 mapping using gradient recalled acquisition in the steady state. Magn Reson Med 2003; 49: 515–26.

DeWitt LD, Kistler JP, Miller DC, Richardson EP, Buonanno FS. NMR-neuropathologic correlation in stroke. Stroke 1987; 18: 342–51.

Dexter DT, Jenner P, Schapira AH, Marsden CD. Alterations in levels of iron, ferritin, and other trace metals in neurodegenerative diseases affecting the basal ganglia. The Royal Kings and Queens Parkinson's Disease Research Group. Ann Neurol 1992; 32 Suppl: S94–100.

Fatouros PP, Marmarou A, Kraft KA, Inao S, Schwarz FP. In vivo brain water determination by T1 measurements: effect of total water content, hydration fraction, and field strength. Magn Reson Med 1991; 17: 402–13.

Filippi M, Rocca MA. MRI evidence for multiple sclerosis as a diffuse disease of the central nervous system. J Neurol 2005; 252 Suppl 5: v16–24.

Gelman N, Ewing JR, Gorell JM, Spickler EM, Solomon EG. Interregional variation of longitudinal relaxation rates in human brain at 3.0 T: relation to estimated iron and water contents. Magn Reson Med 2001; 45: 71–9.

Gracien R-M, Jurcoane A, Wagner M, Reitz SC, Mayer C, Volz S, et al. Multimodal quantitative MRI assessment of cortical damage in relapsing-remitting multiple sclerosis. J Magn Reson Imaging 2016a; 44: 1600–7.

Gracien R-M, Jurcoane A, Wagner M, Reitz SC, Mayer C, Volz S, et al. The relationship between gray matter quantitative MRI and disability in secondary progressive multiple sclerosis. PLos One 2016b; 11: e0161036.

Gracien R-M, Nürnberger L, Hok P, Hof S-M, Reitz SC, Rüb U, et al. Evaluation of brain ageing. A quantitative longitudinal MRI study over 7 years. Eur Radiol 2016c; 27: 1568–76.

Gras V, Abbas Z, Shah NJ. Spoiled FLASH MRI with slice selective excitation. Signal equation with a correction term. Concepts Magn Reson 2013; 42: 89–100.

Gras V, Farrher E, Grinberg F, Shah NJ. Diffusion-weighted DESS protocol optimization for simultaneous mapping of the mean diffusivity, proton density and relaxation times at 3 Tesla. Magnetic Resonance in Medicine 2016; 78: 130–141.

Griffin CM, Dehmeshki J, Chard DT, Parker, G J M, Barker GJ, Thompson AJ, et al. T1 histograms of normal-appearing brain tissue are abnormal in early relapsing-remitting multiple sclerosis. Multiple Sclerosis 2002; 8: 211–6.

Helms G, Dathe H, Dechent P. Quantitative FLASH MRI at 3T using a rational approximation of the Ernst equation. Magn Reson Med 2008a; 59: 667–72.

Helms G, Dathe H, Weiskopf N, Dechent P. Identification of signal bias in the variable flip angle method by linear display of the algebraic Ernst equation. Magn Reson Med 2011; 66: 669–77.

Helms G, Finsterbusch J, Weiskopf N, Dechent P. Rapid radiofrequency field mapping in vivo using single-shot STEAM MRI. Magn Reson Med 2008b; 60: 739–43.

Henderson E, McKinnon G, Lee T-Y, Rutt BK. A fast 3D Look-Locker method for volumetric T1 mapping. Magn Reson Imaging 1999; 17: 1163–71.

Jakob PM, Hillenbrand CM, Wang T, Schultz G, Hahn D, Haase A. Rapid quantitative lung (1)H T(1) mapping. J Magn Reson Imaging: JMRI 2001; 14: 795–9.

Kaptein R, Dijkstra K, Tarr C. A single-scan fourier transform method for measuring spin-lattice relaxation times. J Magn Reson (1969) 1976; 24: 295–300.

Lescher S, Jurcoane A, Veit A, Bahr O, Deichmann R, Hattingen E. Quantitative T1 and T2 mapping in recurrent glioblastomas under bevacizumab: earlier detection of tumor progression compared to conventional MRI. Neuroradiology 2015; 57: 11–20.

Look D, Locker D. Time saving in measurement of NMR and EPR relaxation times. Rev Sci Instrum 1970: 250–1.

Lutti A, Dick F, Sereno MI, Weiskopf N. Using high-resolution quantitative mapping of R1 as an index of cortical myelination. NeuroImage 2014; 93 Pt 2: 176–88.

Marques JP, Kober T, Krueger G, van der Zwaag W, van de Moortele PF, Gruetter R. MP2RAGE, a self bias-field corrected sequence for improved segmentation and T1-mapping at high field. NeuroImage 2010; 49: 1271–81.

Möllenhoff K. Novel methods for the detection of functional brain activity using 17O MRI; 2016. Université de Liège and Maastricht University. Available at https://cris.maastrichtuniversity.nl/portal/files/2731200/c5362.pdf. [Accessed 14 September 2017]

Morrell GR. A phase-sensitive method of flip angle mapping. Magn Reson Med 2008; 60: 889–94.

Nehrke K, Bornert P. DREAM—a novel approach for robust, ultrafast, multislice B(1) mapping. Magn Reson Med 2012; 68: 1517–26.

Nöth U, Hattingen E, Bähr O, Tichy J, Deichmann R. Improved visibility of brain tumors in synthetic MP-RAGE anatomies with pure T1 weighting. NMR Biomed 2015; 28: 818–30.

Parry A, Clare S, Jenkinson M, Smith S, Palace J, Matthews PM. White matter and lesion T1 relaxation times increase in parallel and correlate with disability in multiple sclerosis. J Neurol 2002; 249: 1279–86.

Pohmann R, Speck O, Scheffler K. Signal-to-noise ratio and MR tissue parameters in human brain imaging at 3, 7, and 9.4 tesla using current receive coil arrays. Magn Reson Med 2016; 75: 801–9.

Polders DL, Leemans A, Luijten PR, Hoogduin H. Uncertainty estimations for quantitative in vivo MRI T1 mapping. J Magn Med Reson 2012; 224: 53–60.

Preibisch C, Deichmann R. Influence of RF spoiling on the stability and accuracy of T1 mapping based on spoiled FLASH with varying flip angles. Magn Reson Med 2009a; 61: 125–35.

Preibisch C, Deichmann R. T1 mapping using spoiled FLASH-EPI hybrid sequences and varying flip angles. Magn Reson Med 2009b; 62: 240–6.

Rooney WD, Johnson G, Li X, Cohen ER, Kim S-G, Ugurbil K, et al. Magnetic field and tissue dependencies of human brain longitudinal 1H2O relaxation in vivo. Magn Reson Med 2007; 57: 308–18.

Shah NJ, Zaitsev M, Steinhoff S, Zilles K. A new method for fast multislice T(1) mapping. NeuroImage 2001; 14: 1175–85.

Steen RG, Gronemeyer SA, Kingsley PB, Reddick WE, Langston JS, Taylor JS. Precise and accurate measurement of proton T1 in human brain in vivo. Validation and preliminary clinical application. J Magn Reson Imaging 1994; 4: 681–91.

Stikov N, Boudreau M, Levesque IR, Tardif CL, Barral JK, Pike GB. On the accuracy of T1 mapping: searching for common ground. Magn Reson Med 2015; 73: 514–22.

Venkatesan R, Lin W, Haacke EM. Accurate determination of spin-density andT1 in the presence of RF-field inhomogeneities and flip-angle miscalibration. Magn Reson Meds 1998; 40: 592–602.

Volz S, Nöth U, Rotarska-Jagiela A, Deichmann R. A fast B1-mapping method for the correction and normalization of magnetization transfer ratio maps at 3 T. NeuroImage 2010; 49: 3015–26.

Vrenken H, Geurts, Jeroen JG, Knol DL, van Dijk, L Noor, Dattola V, Jasperse B, et al. Whole-brain T1 mapping in multiple sclerosis: global changes of normal-appearing gray and white matter. Radiology 2006; 240: 811–20.

Vymazal J, Righini A, Brooks RA, Canesi M, Mariani C, Leonardi M, et al. T1 and T2 in the brain of healthy subjects, patients with Parkinson disease, and patients with multiple system atrophy: relation to iron content. Radiology 1999; 211: 489–95.

Wang HZ, Riederer SJ, Lee JN. Optimizing the precision in T1 relaxation estimation using limited flip angles. Magn Reson Med 1987; 5: 399–416.

Warntjes JBM, Leinhard OD, West J, Lundberg P. Rapid magnetic resonance quantification on the brain: optimization for clinical usage. Magn Reson Med 2008; 60: 320–9.

Weiskopf N, Lutti A, Helms G, Novak M, Ashburner J, Hutton C. Unified segmentation based correction of R1 brain maps for RF transmit field inhomogeneities (UNICORT). NeuroImage 2011; 54: 2116–24.

Weiskopf N, Suckling J, Williams G, Correia MM, Inkster B, Tait R, et al. Quantitative multi-parameter mapping of R1, PD*, MT, and R2* at 3T: a multi-center validation. Front Neurosci 2013; 7: 95.

Wilkinson ID, Paley MN, Hall-Craggs MA, Chinn RJ, Chong WK, Sweeney BJ, et al. Cerebral magnetic resonance relaxometry in HIV infection. Magn Reson Imaging 1996; 14: 365–72.

Wood TC. Improved formulas for the two optimum VFA flip-angles. Magn Reson Med 2015; 74: 1–3.

Yarnykh VL. Actual flip-angle imaging in the pulsed steady state: a method for rapid three-dimensional mapping of the transmitted radiofrequency field. Magn Reson Med 2007; 57: 192–200.

Yarnykh VL. Optimal radiofrequency and gradient spoiling for improved accuracy of T1 and B1 measurements using fast steady-state techniques. Magn Reson Med 2010; 63: 1610–26.

Zur Y, Wood ML, Neuringer LJ. Spoiling of transverse magnetization in steady-state sequences. Magn Reson Med 1991; 21: 251–63.

6

T_2: Transverse Relaxation Time

Contents

Nicholas G. Dowell
Brighton and Sussex Medical School

Tobias C. Wood
King's College London

6.1 Background

6.1.1 Physical basis of transverse relaxation

Following a period of longitudinal relaxation, coherent spin precession may be initiated by the application of a radiofrequency (RF) pulse that tips the net magnetisation into the transverse plane. This net magnetisation does not persist indefinitely. Rather, the spin coherence decays approximately exponentially and is parameterised by the time constant T_2:

$$M_{xy} = M_0 \exp(-t/T_2) \tag{6.1}$$

where M_{xy} is the transverse magnetisation at time t following the pulse and M_0 is equilibrium magnetisation, assuming a perfect RF pulse that fully tips the magnetisation into the transverse plane.

The decay is brought about by a process called *dephasing*. The net magnetisation is composed of individual isochromats (sets of nuclei that experience equal magnetic field and therefore a common resonance frequency) that interact with other nuclei and electrons in their local environment. Dipolar effects between them cause fluctuations in the magnetic field and a shift in Larmor frequency. This makes the isochromats fan out in the transverse plane, and they destructively interfere, reducing the net magnetisation. Since this process involves other nuclei it is often called *spin–spin relaxation*.

MR signals decay more rapidly than suggested in Equation 6.1, because of additional sources of dephasing introduced by the MR scanner hardware or the subject itself. The resonance frequency of isochromats is dependent on the local magnetic field, which is not homogeneous due to imperfections in the magnet, chemical shift effects, macroscopic susceptibility differences or the presence of paramagnetic or ferromagnetic substances. This causes accelerated dephasing, leading to short-lived MR signals parameterised by T_2*. This can be expressed as a combination of irreversible spin–spin relaxation and additional, reversible decay from inhomogeneity, parameterised by T_2':

$$1/T_2^* = 1/T_2 + 1/T_2' \tag{6.2}$$

This equation implies T_2* is always shorter than T_2.

6.1.2 Biological basis of transverse relaxation

Dipolar interactions are dependent on the distance between the coupling nuclei and their orientation with respect to the applied magnetic field of the scanner B_0 (Figure 6.1). The size of the dipolar interaction is proportional to $(3\cos^2\theta-1)/r^3$, where r is the distance between the nuclei and θ is the angle between the dipolar-coupled nuclei and $\boldsymbol{B}_0$. Since the relaxation rate R_2 $(=1/T_2)$ is proportional to the square of the dipolar interaction, relaxation effects are primarily driven by nearby, often

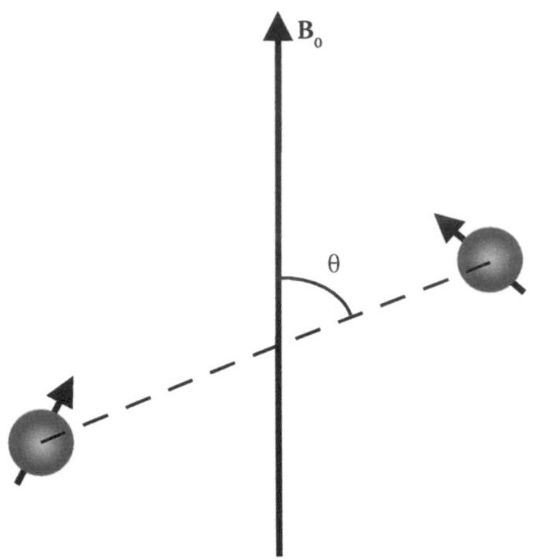

FIGURE 6.1 As two randomly tumbling magnetic dipole moments come into close proximity, they alter the magnetic field the other experiences. This fluctuating magnetic field induces relaxation. The strength of this effect is dependent on the angle θ between $\mathbf{B}_0$ and the spins and the distance between the nuclei (see main text for more details).

intramolecular, nuclei. Where nuclei are free to tumble, such as water in cerebrospinal fluid (CSF), θ may take any number of values so that the dipolar interaction averages out on the MRI timescale. Hence the effect is small and T_2 remains long, in the range of seconds. At the other end of the scale, the biological environment may result in highly restricted motion. This is observed in solid components such as bone, cartilage, ligaments and tendons, which have T_2 of the order of tens of microseconds.

The T_2 of a spin system determines the longevity of the MR signal and hence the detectability with different sequences. The brain possesses several biologically interesting components such as lipids and proteins that are classed as semi-solids that are not able to tumble freely, and so appreciable dipolar coupling reduces T_2 to the sub-millisecond range. These spins are invisible to conventional MR methods, but are visible to ultrashort (UTE) or zero echo-time methods (Waldman *et al.*, 2003). However, most water in brain parenchyma is in the intra- or extracellular space and so is dominated by relatively long T_2 components (T_2 > 10 ms), which can be imaged with standard acquisition approaches.

Assigning a single T_2 value to each voxel in a biological tissue is a simplification of the microstructure, which can be a complex combination of components with differing T_2 values. For example, white matter is commonly quoted as having a T_2 of around 50 ms, but this does not reflect the presence of cell membranes, myelin, proteins, etc., that make up 'white matter'. Each of these has a distinct magnetic local environment and hence T_2, which may differ by orders of magnitude (Vasilescu *et al.*, 1978; Menon and Allen, 1991). A further problem is that measured T_2 values will be weighted both by T_2 and the proportion of nuclei with a particular T_2. Short T_2 components will contribute less to the signal than those with long T_2. Multicomponent models attempt to solve this problem by modelling the signal as a sum of contributions from different spin populations and will be discussed further below.

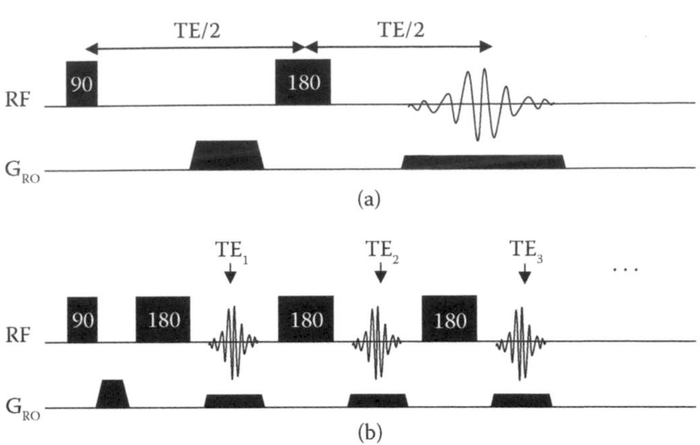

FIGURE 6.2 (a) Pulse sequence diagram for the spin echo. (b) Pulse sequence diagram for the Carr–Purcell–Meiboom–Gill (CPMG) multiple spin echo. Only the readout gradients are shown for clarity.

6.2 Quantification methods

We will now detail the three main families of T_2-mapping methods, starting with the simplest and most accurate, but slowest, spin echo approaches. We will then discuss faster steady-state methods and finish with more recent T_2-prep approaches.

6.2.1 Spin echo

The spin echo refocuses dephasing effects using two RF pulses separated by an evolution period TE/2. The pulse sequence is illustrated in Figure 6.2a, and the isochromat evolution in Figure 6.3. The first pulse tips the net magnetisation so that it lies in the transverse plane, while the refocusing pulse rotates the isochromats by 180° around $\boldsymbol{B}_1$ (the RF pulse axis). This inverts any phase accrued by the isochromats during the first evolution interval. The phase accrued due to T_2' is the same in both evolution intervals and hence cancels out, while T_2 effects remain. The result is an echo collected at time TE following the excitation pulse with the centre of the echo weighted by T_2 only.

Assuming a single T_2 within a voxel, changing the echo time will result in modulation of signal S according to

$$S = S_0 \exp(-\mathrm{TE}/T_2) \tag{6.3}$$

where S_0 is the proton density. T_2 can be calculated from two measurements, S_1 and S_2, with corresponding echo times, TE1 and TE2,

$$T_2 = \frac{\mathrm{TE2} - \mathrm{TE1}}{\log(S_2) - \log(S_1)} \tag{6.4}$$

The echo times are normally chosen so that the expected T_2 lies between them but weighted towards TE2 (Woermann *et al.*, 1998). While signal acquired using two echo times can be used to calculate T_2 for a mono-exponential decay, we discussed above that this is insufficient to describe real tissue. Data collected in this

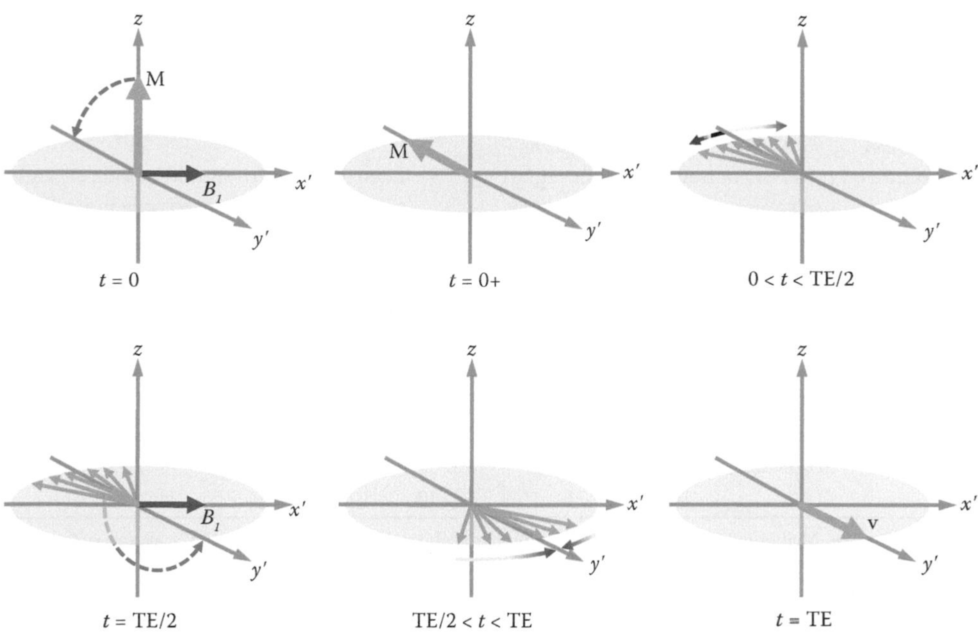

FIGURE 6.3 Vector diagram of the spin echo experiment. Only dephasing due to T_2' effects is shown. Bulk magnetisation vector **M** is initially aligned along the *z*-axis at $t = 0$. After the 90° radiofrequency pulse magnetisation is flipped into the transverse plane. During the first half of the echo time the isochromats fan out until the refocusing pulse at t = TE/2 inverts their phase. During the second half of the echo time the spins rephase until a spin echo is formed at t = TE.

way will suffer partial volume effects, especially at the CSF–grey matter boundary (Woermann *et al.*, 2001; Pell *et al.*, 2004), and is susceptible to noise effects (Whittall *et al.*, 1999). By sampling the signal decay at a range of different echo times it is possible to generate a decay curve with improved accuracy and precision. However, with a simple spin echo this is prohibitively slow due to the need for a long TR to allow full T_1 relaxation. This is required for good SNR but also prevents partial volume effects being exacerbated by differential T_1 weighting. In brain parenchyma at 3T, this would mean a TR of around 5 s. Full brain coverage at 1-mm isotropic resolution using a standard pulse sequence would require around 20 minutes per TE point, which is impractical.

The spin echo can be accelerated by using multiple 180° refocusing pulses and readout periods at increasing TE (Figure 6.2b), referred to as the *multi-echo T_2* (MET$_2$) approach. First introduced in 1954 by Carr and Purcell (Carr and Purcell, 1954), the initial implementation was sensitive to small inaccuracies in the application of the 180° pulses. Meiboom and Gill later improved the method by applying a 180° phase increment to each successive refocussing pulse in the train, creating the Carr–Purcell–Meiboom–Gill (CPMG) sequence (Meiboom and Gill, 1958). Large echo trains can be collected (up to 256 echoes), but the accuracy of T_2 measurements worsens with increasing echoes due to residual inefficiencies in the refocusing pulses (Majumdar *et al.*, 1986). Greater sampling of the T_2 decay curve may result in increased exam time, as both multiple slices and multiple echoes must be accommodated within the repetition time. Mono-exponential approaches require only a few echoes, ranging from 20 to 160 ms. Multi-exponential approaches require more echoes, typically 16 or 32, spanning a wide range of echo times to capture both short and long T_2 components. To accommodate this, investigators must trade off spatial coverage, exam time and accuracy when deciding on the acquisition protocol. SAR is another limitation because of the large number of 180° refocusing pulses, which sets an upper limit on the number of echoes that can be acquired. As a consequence, the number of slices that can be acquired with the CPMG sequence is severely limited.

Important considerations for planning a MET$_2$ study are given in Box 6.1 and typical acquisition parameters in Table 6.1. A Poon–Henkelman sequence is recommended where non-selective composite refocusing pulses ($90_x°$–$180_y°$–$90_x°$) are used to mitigate effects of inhomogeneities in B_0 and B_1 (Poon and Henkelman, 1992). Errors in B_1 result in stimulated echoes that introduce severe errors to T_2 estimation. Slice-selective crusher gradients are applied each side of the refocusing pulse to attenuate flow artefacts and stimulated echoes outside the slice.

The number of acquired slices can be increased by using a gradient and spin echo (GRASE) acquisition (Does and Gore, 2000), which accelerates acquisition of k-space by collecting three or more gradient echoes before and after each spin echo. This results in pure T_2 weighting at the centre of k-space, with additional T_2* weighting towards the edges. 3D GRASE can collect whole-brain MET$_2$ data in 15 minutes (Prasloski *et al.*, 2012b). Data analysis is a challenge due to the presence of a complex combination of primary, stimulated and indirect echoes, but Henning's extended phase graph method (Hennig, 1988) can be used to account for the phase coherence of these different

Box 6.1 MET$_2$ studies

- Use a pulse sequence that mitigates B_1, B_0 errors and stimulated echoes (e.g. Poon and Henkelman, 1992).
- High SNR is required (min. SD of noise 1% of the signal for the shortest TE) – use signal averaging if necessary.
- Echo spacing should be as short as possible (<10 ms *in vivo*).
- Echo train length should be long enough such that the last echo time should return only noise (longest echo >1 s).

TABLE 6.1 Multicomponent Quantitative T_2 Mapping Acquisition Parameters for Sequences Based on CPMG

Reference	Method	TR (ms)	TE1 (ms)	EN/ES (mm)	Voxel size (mm)	Matrix size	NEX	T (min)
Mackay *et al.* (1994)	2D	3000	15	32/15	0.86 × 1.72 × 5	256 × 128 × 1	4	26
Whittall *et al.* (1997)	2D	3000	10	32/10	0.86 × 1.72 × 5	256 × 128 × 1	4	26
Mädler and MacKay (2007)	3D GRASE	1200	10	32/10	0.9 × 1.9 × 5	256 × 108 × 7	1	9[a]
Prasloski *et al.* (2012b)	3D GRASE	1000	10	32/10	1 × 1 × 5	232 × 192 × 20	1	15[b]
Guo *et al.* (2013)	2D	3000	10	32/10	0.85 × 0.85 × 7.5	256 × 256 × 10[c]	1	

Note: CPMG = Carr–Purcell–Meiboom–Gill approach; EN = number of echoes; ES = echo spacing; NEX = averages; T = acquisition time.
[a] EPI factor 3 used.
[b] SENSE (SENSitivity Encoding) factor = 2; slice oversampling factor = 1.3.
[c] GRAPPA (GeneRalized Autocalibrating Partial Parallel Acquisition) factor = 2.

echoes and has been extended to account for imperfect slice profiles (Lebel and Wilman, 2010).

An alternative to GRASE is to use slice-selective refocussing pulses that are much wider than the excited slices, instead of non-selective pulses. This ensures a good slice profile while allowing multislice acquisition but requires a large slice gap to prevent contaminating adjacent slices (Guo *et al.*, 2013).

If more than two echoes have been acquired, then Equation 6.4 is no longer appropriate and instead a least-squares fit of TE against the logarithm of the signal is the most common method of calculating T_2. However, this can lead to noise amplification at the end of the echo-train so an alternative is non-linear fitting to the raw signal values. This requires an initial guess for T_2. If equally spaced echoes have been acquired, then the auto-regression on linear operations method, originally introduced for calculating T_2*, is a fast and robust alternative (Pei *et al.*, 2015). Proponents of the mono-exponential approach claim that, although the T_2 is inaccurate, it is also reproducible and remains sensitive to pathology. However, simulations (Whittall *et al.*, 1999) have demonstrated that fewer points used to characterise the decay curve results in a greater corruptive effect from noise. This effect is reduced by acquiring more data according to the empirical $\sqrt{N}$ law, where N is the number of sampled points along the curve (Whittall *et al.*, 1999).

With sufficient points MET$_2$ data can also be analysed on a multicomponent basis. The measured signal is modelled as a sum of several discrete T_2 decay curves, where the signal at each TE is given by

$$\tilde{S}(\mathrm{TE}_i)=\sum_{j=1}^{M} s_j e^{-\frac{\mathrm{TE}_i}{T_{2,j}}}+\varepsilon_i \tag{6.5}$$

where i is the echo number, a total of M components are considered with relaxation times $T_{2,j}$ and amplitude s_j, and noise is given by ε_i. The precise method used to fit the curve depends on the *a priori* choice of M.

If only a few (typically three or four) compartments are chosen, then non-linear multi-parameter fitting routines can be used. However, almost all MRI studies instead use Lawson and Hanson's non-negative least squares (NNLS) algorithm, with the non-negative constraint placed on s_j (Lawson and Hanson, 1995). This allows M to be set sufficiently high, usually with a logarithmic spacing, that there is no practical limit on the number of components that may be found in the decay curve (Figure 6.4). However, the presence of noise means that many possible T_2 distributions are plausible (Alonso-Ortiz *et al.*, 2015). Regularisation is used to restrict the set of T_2 distributions to the least complex. The favoured solution is found by minimising

$$\Xi=\chi^2+\mu\sum_{j=1}^{M} s_j^2 \tag{6.6}$$

where

$$\chi^2=\sum_{n=1}^{N}\frac{(\mathrm{S}(\mathrm{TE}_i)-\tilde{\mathrm{S}}(\mathrm{TE}_i))^2}{\sigma_i^2} \tag{6.7}$$

where S(TE$_i$) is the measured signal for the ith echo and χ^2 is the misfit between model and measurement. The second term in Equation 6.6 is the energy of the T_2 distribution, weighted by the constant μ. High values of μ increase the constraints, resulting in broad T_2 distributions at the expense of higher χ^2, whilst very low μ results in narrow T_2 distributions. μ is chosen by placing a constraint on the maximum and minimum values of χ^2 (MacKay *et al.*, 2006).

6.2.2 Steady-state

Driven equilibrium single-pulse observation of T_2 (DESPOT2 (Deoni *et al.*, 2003, 2005)) uses a balanced steady-state free precession (bSSFP) pulse sequence, which has a combined T_2/T_1 contrast, to measure T_2 using the same framework as DESPOT1 (see Chapter 5). Neglecting off-resonance effects, the bSSFP signal equation is

$$S(\alpha) = \frac{M_0(1-E_1)\sin(\alpha)}{1-E_1E_2-(E_1-E_2)\cos(\alpha)} \tag{6.8}$$

where $S(\alpha)$ is the signal intensity at flip angle α, $E_1 = \exp(-TR/T_1)$, $E_2 = \exp(-TR/T_2)$ and M_0 is the proton density. This equation can be rearranged into a linear form:

$$\frac{S(\alpha)}{\sin(\alpha)} = \frac{S(\alpha)}{\tan(\alpha)} \times \frac{E_1-E_2}{1-E_1E_2} + \frac{M_0(1-E)}{1-E_1E_2} \tag{6.9}$$

and hence a plot of $S(\alpha)/\sin(\alpha)$ against $S(\alpha)/\tan(\alpha)$ can yield T_2 from the slope m and T_1, which is typically found using DESPOT1:

$$T_2 = -\frac{TR}{\ln\left(\dfrac{m-E_1}{mE_1-1}\right)} \tag{6.10}$$

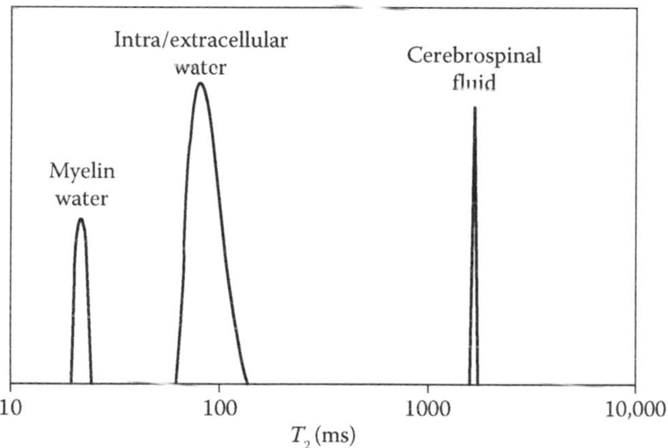

FIGURE 6.4 An example of a T_2 distribution of *in vivo* human brain parenchyma. Regularisation was applied with $1.02\chi^2_{min} \le \chi^2 \le 1.025\chi^2_{min}$. (Modified and reproduced with permission from MacKay, A., *et al.*, *Magn. Reson. Imaging*, 24(4), 515–525, 2006.)

Both DESPOT1 and DESPOT2 rely on an exact knowledge of α, which can be problematic due to B_1 field inhomogeneity. In addition, at high field-strength, off-resonance effects cannot be neglected due to banding artefacts (Zur *et al.*, 1990), but these can be mitigated by acquiring multiple images with different phase increments (also called *phase cycling*). Both effects will be discussed further below.

The DESPOT2 method is particularly attractive because it offers high SNR per unit time and full brain coverage in clinically acceptable durations. Precision of T_1 and T_2 is maximised when the available scanning time is divided in a ratio of 3:1 for T_1 and T_2, respectively (Deoni *et al.*, 2003), but this is difficult to achieve in practice. At 3T, at least six separate images are required, two gradient-echo for DESPOT1 and four bSSFP for DESPOT2 (two flip angles and two phase increments), plus a separate B_1 map.

The SNR of bSSFP is most efficient when the acquisition window fills as much of the TR as possible, so reducing the TR to improve speed implies increasing the receive bandwidth, which in turn increases the readout gradient and will eventually induce peripheral nerve stimulation. In addition, one of the optimum flip angles to use is usually above 60 degrees (Wood, 2015). At 3T and above this will place a limit on TR due to SAR considerations. TRs of around 5 ms are typical, and a split of 1:1 between the time allocated to DESPOT1 and DESPOT2 is common. Example acquisition parameters are given in Table 6.2.

Multicomponent DESPOT (mcDESPOT) assumes separate short and medium T_2 components, and potentially a long T_2 component, are present in each voxel (Deoni *et al.*, 2008). These are thought to correspond to myelin water, intra-/extracellular (IE) water, and CSF. To differentiate the signals from these requires that more than two volumes are acquired with different flip angles. Eight or nine are standard practice, but there has not been a comprehensive assessment of the minimum required. The additional image volumes mean that a reduced voxel-size of around 2 mm isotropic is feasible within a reasonable time (see Table 6.2).

In contrast to MET_2 methods, an additional parameter is introduced to characterise the exchange between the myelin and IE water. The exchange rate is the inverse of the mean residence time of a spin in either compartment. This linking of the water

TABLE 6.2 Acquisition parameters for T_1 and T_2 quantification with DESPOT and mcDESPOT

Reference	Technique	FA	TR (ms)	TE (ms)	Voxel size (mm)	Matrix size	NEX	T (min)
Deoni (2007)	DESPOT1 (SPGR)	4°, 18°	6.2	1.6	1.0 × 1.0 × 1.0	220 × 220 × 140	2	6.5
Deoni (2007)	DESPOT1-HIFI (IR-SPGR)	5°	6.2[a]	1.6	1.0 × 2.0 × 2.0	220 × 110 × 70	2	2
Deoni (2009)	DESPOT2-FM (bSSFP)	15°, 65°	4.2	2.1	1.0 × 1.0 × 1.0	220 × 220 × 140	2[b]	8
Deoni (2011)	mcDESPOT (SPGR)	3°, 4°, 5°, 6°, 7°, 9°, 13°, 18°	5.4	2.4	1.7 × 1.7 × 1.7	128 × 128 × 98	1	9
Deoni (2011)	mcDESPOT (IR-SPGR)	5°	5.4[a]	2.4	1.7 × 1.7 × 1.7	128 × 128 × 98	1	1.5
Deoni (2011)	mcDESPOT (bSSFP)	10°, 13°, 17°, 20°, 23°, 30°, 43°, 60°	4.4	2.2	1.7 × 1.7 × 1.7	128 × 128 × 98	2[b]	15

Note: mcDESPOT = multicomponent DESPOT; bSSFP = balanced steady-state free precession.
[a] Inversion time for DESPOT1-HIFI (IR-SPGR) = 450 ms.
[b] Averages correspond to phase cycling increment 0° and 180°.
[c] Acquisition matrix with 0.75 partial Fourier transform.

pools means that the mcDESPOT signal is not a simple summation of the separate signals, and instead a Bloch–McConnell matrix system with up to 10 free parameters must be solved (Deoni *et al.*, 2013).

There has been some criticism of this model in the literature, as it is difficult to fit with standard non-linear fitting algorithms (Lankford and Does, 2012; Bouhrara and Spencer, 2015; Bouhrara *et al.*, 2015). However, using the stochastic region contraction (SRC) method overcomes some of these issues (Deoni and Kolind, 2014; Hurley and Alexander, 2014) and the addition of extra phase increments appears to stabilise the fitting (Lankford and Does, 2012; Wood *et al.*, 2016). The SRC method is not widespread, although a freely available implementation is available online (https://github.com/spinicist/QUIT). One subtle difficulty with fitting the mcDESPOT model is that the gradient echo and bSSFP data are likely acquired with different receive gains and bandwidths. This means that the two datasets are not scaled by the same amounts, and hence one value of equilibrium magnetisation cannot be fit to them simultaneously. To overcome this, the gradient echo and bSSFP signals are first normalised by their respective means, which still allows the fractional sizes of the T_2 components to be found (Deoni *et al.*, 2008).

6.2.3 T_2 prep

For both MET_2 and DESPOT2, the T_2 weighting of the image is linked to the acquisition sequence. An alternative is to separate the imaging section of the sequence from the T_2 weighting via a preparation module and acquire volumes with different preparation settings. This allows the use of advanced readout methods, which may not be T_2 weighted but are often faster than conventional methods.

A T_2-prep block consists of a 90° pulse, at least one refocusing pulse and a –90° pulse that returns the magnetisation to the longitudinal axis. The refocusing pulse can be a composite hard pulse (Oh *et al.*, 2006) or an adiabatic pulse (Nguyen *et al.*, 2015). At the end of the prep block the T_2 weighting of the longitudinal magnetisation is a function of the number and spacing of the refocusing pulse and can then be sampled with a fast imaging sequence such as 2D spiral (Oh *et al.*, 2006, 2007).

The different T_2 prep images can be fitted to the multicomponent MET_2 model. An important advantage of the T_2 prep method is that the minimum effective TE can be reduced almost to zero by switching off the prep module entirely, permitting the study of short T_2 components. Nguyen *et al.* report an effective minimum echo time of 0.5 ms, while a UTE-type approach can reduce it even further (Waldman *et al.*, 2003).

6.3 Common problems

6.3.1 B_1 inhomogeneity

Owing to limitations of MR hardware and the presence of dielectric resonance effects within the subject (Simmons *et al.*, 1994), it is very difficult to deliver perfect RF pulses and the achieved flip angle across a volume often differs significantly from that prescribed. All methods of measuring T_2 are susceptible to non-uniformity in the applied RF field, but accepted methods to mitigate or correct for B_1 inhomogeneity differ between them.

It is important to distinguish between the profile of the RF field generated by the transmit (generally body) coil and the field of each receive element, which should strictly be referred to as B_1^+ and B_1^-, respectively (although the term B_1 is often used for both in the literature). B_1– is a multiplicative factor on the overall signal scaling that affects acquired points equally and is often ignored, but this simplification is invalid if the subject moves significantly between acquisitions. For a full discussion of this issue, and a potential strategy to solve it, the reader is referred to Papp *et al.* (2015).

The actual flip angle achieved in a voxel is the required parameter for quantitative T_2 analysis. This is the product of B_1^+, which describes only the transmit coil profile, and the slice or slab profile of the RF pulse used. The combined correction factor is usually expressed as a ratio, with values above and below 1 implying that a higher and lower flip angle will be achieved, respectively.

Ideally, the refocusing pulses for MET_2 would result in a perfect rephasing of spins and give a pure T_2 weighting. B_1 inhomogeneities cause some magnetisation to be transferred to the longitudinal axis, where it undergoes T_1 relaxation, before being transferred back to the transverse plane by later pulses. This forms stimulated echoes that corrupt the primary echoes and lead to inaccuracies in T_2 (Pell *et al.*, 2006), and the problem worsens on higher-field systems. By discarding the first echo, which is guaranteed to not contain a stimulated echo, it may be possible to fit the remaining points, but this neglects stimulated echo contributions to later echoes and increases the effective minimum TE (Smith *et al.*, 2001; Maier *et al.*, 2003).

B_1 inhomogeneity can be mitigated to some extent by the Poon–Henkelman sequence (Poon and Henkelman, 1992). This approach has the drawback of incurring increased SAR and effectively limits the acquisition to a single slice. An alternative approach is to use the extended phase graph (EPG) algorithm (Hennig, 1991) to track the progression of the secondary and stimulated echoes and fit for B_1 during the quantification of T_2 (Prasloski *et al.*, 2012a). A major source of stimulated echoes arises from poor (non-rectangular) slice-select profiles that are a feature of 2D multislice imaging. The standard EPG approach assumes perfect slice profiles, but this was addressed by Lebel *et al.*, who provided a more complete EPG-based model (Lebel and Wilman, 2010). A further development is the echo modulation curve algorithm, which uses a Bloch simulation to account for different slice profiles, RF pulse shapes and relaxation during RF pulses (Ben-Eliezer *et al.*, 2015) and has been shown to provide advantages over the EPG (McPhee and Wilman, 2016).

For DESPOT, the accepted approach is to collect a B_1 map to correct the flip angles used in the analysis. A wide variety of sequences have been published in the literature, and a comparison of several can be found in Pohmann and Scheffler (2013), but frustratingly most are not available as manufacturer sequences (Boudreau *et al.*, 2017). The actual flip-angle imaging method has emerged as something of a standard and is available on some platforms (Yarnykh, 2007). It requires strong spoiler gradients and a long TR, which leads to long scan times (Nehrke, 2009; Yarnykh, 2010). A more recent method that strikes a balance between speed, flexibility and accuracy is DREAM (Nehrke *et al.*, 2014). The B_1 field, even at high field strengths, is smooth compared to anatomical features (Brink *et al.*, 2014). Hence B_1 maps are often acquired at low resolution to save acquisition time.

Driven equilibrium single pulse observation of T1 with high-speed incorporation of RF field inhomogeneities (DESPOT1-HIFI) uses an additional inversion-recovery scan (e.g. MP-RAGE), which is widely available, to estimate the B_1 map during T_1 fitting (Deoni, 2007). Implementers should note the TE and TR must be precisely matched to the DESPOT1 sequences and the equation given by Deoni (2007) is incorrect – a correct version is given by Bouhrara and Spencer (2015). It does not produce accurate T_1 or B_1 values in areas with long T_1 (such as the CSF-filled ventricles) without the collection of additional data (Deoni, 2007). Hence the resulting B_1 maps should be smoothed, or fitted to a low-order polynomial, and the T_1 maps recalculated with the resulting B_1 map.

6.3.2 B_0 inhomogeneity

The interaction of the subject with the magnetic field ensures that the magnetic field varies from location to location, particularly in areas with large magnetic susceptibility differences, such as air–tissue interface in the sinuses. In bSSFP, this causes banding artefacts, which destroy signal, leading to nonsense values from DESPOT2. The position of the bands depends on the product of TR and off-resonance frequency, and at 1.5T or less is usually outside the tissues of interest in a well-shimmed magnet, but at 3T or above these effects become significant.

There are several published methods to mitigate the banding artefacts, which require the collection of additional images with different phase increments. A very useful elliptical form of the bSSFP signal equation can be found in Xiang and Hoff (2014).

$$S = M\frac{1-ae^{i\theta}}{1-b\cos\theta} \tag{6.11a}$$

where

$$M = \frac{M_0(1-E_1)\sin(\alpha)}{1-E_1\cos\alpha - E_2^2\left(E_1-\cos(\alpha)\right)} \tag{6.11b}$$

$$a = E_2$$

$$b = \frac{E_2(1-E_1)(1+\cos(\alpha))}{1-E_1\cos\alpha - E_2^2\left(E_1-\cos(\alpha)\right)} \tag{6.11c}$$

are defined as above and

$$\theta = 2\pi\,\Delta\, f_0 T_R + \theta \tag{6.12}$$

is the total phase accrued in one TR, consisting of the off-resonance frequency Δf_0 in Hertz and the scanner-controlled phase increment angle ϕ. From Equation 6.11, it should be clear that low signal occurs when $\theta = 0$ which for the normal value of $\phi = \pi$ implies that $\Delta f_0 = 1/2T_R$. Changing ϕ has the effect of shifting the bands to a different location. Images with different ϕ can then combined to construct band-free images from which T_2 can be calculated (Jutras *et al.*, 2015; Wood *et al.*, 2015). Code is available from https://github.com/jdjutras/QMRI

In contrast, the DESPOT2-FM method fits the bSSFP signal equation directly to the multiple phase increment data and does not calculate band-free images as an intermediate step. This means that it can be used with only two phase increments, compared to four for other methods; however this makes the fitting difficult (Deoni, 2009).

B_0 inhomogeneity also affects MET$_2$ measurements, as later echoes can accrue substantial phase offsets. When combined with noise-floor effects at the end of the echo train, this can lead to artefactual long-T_2 components being recovered in the analysis. Fully modelling the MET$_2$ curve as a complex signal can remove these issues (Bjarnason *et al.*, 2013).

One underexplored effect of B_0 inhomogeneity is that T_2 methods tend to assume that myelin water has the same resonance frequency as other water pools. Evidence from other MR techniques (principally quantitative susceptibility mapping) and from detailed studies of the bSSFP signal profile have shown this to be an oversimplification (Miller *et al.*, 2010).

6.3.3 Magnetisation transfer

During an MET$_2$ acquisition, slice-selective refocussing pulses may act as off-resonance saturation pulses for nearby slices. This will cause magnetisation transfer (MT) effects that result in diminished signal intensity in the free water component and have the potential to cause inaccuracy in T_2 measurements (Forsén and Hoffman, 1963; Maier *et al.*, 2003). The use of 3D acquisition or single-slice acquisition using non-selective pulses are a means of reducing the effects of MT.

For DESPOT2 the bSSFP is inherently susceptible to MT effects due to the short TRs and high flip angles used (Henkelman *et al.*, 2001). One approach to mitigate this is to carefully choose the two flip angles used so that MT effects are similar (Jutras *et al.*, 2015). The MT contribution to mcDESPOT has been explored but requires the introduction of highly complex models that are even more difficult to fit (Liu *et al.*, 2015).

6.3.4 Water exchange

MET$_2$ mapping results are often interpreted with the assumption that there is no water exchange between the different compartments on the T_2 timescale. This assumption has been considered reasonable, but there is growing evidence that water exchange cannot be assumed to be negligible in areas of myelin thinning. Smaller water myelin compartments have a shorter residence time and spins can move between different T_2 environments during the measurement. This leads to an underestimation of the myelin water component and hence lower myelin water fraction (MWF) values. A comparison of the mcDESPOT technique (that includes water exchange in the model) and the 3D GRASE approach has demonstrated that mcDESPOT yields consistently higher estimates of MWF. However, doubts remain whether these differences are wholly due to exchange effects or a result of non-unique solutions in the large mcDESPOT fitting space. (Zhang *et al.*, 2015; Bouhrara and Spencer, 2016, 2017).

6.3.5 Noise

For MET_2 analysis, a common rule of thumb is the shortest echo must have a SNR greater than 100 to avoid any noise effects, which may require signal averaging (MacKay *et al.*, 2006). Image smoothing also has the effect of increasing SNR (Jones *et al.*, 2003). A non-local mean filter can be used to increase the SNR across regions with similar characteristics, but the tuning parameters must be chosen carefully (Guo *et al.*, 2013). Alternatively, voxels can be averaged across a region of interest, but this discards information about the noise that is associated with each constituent component (Bjarnason *et al.*, 2010). This hinders statistical comparison of T_2 or T_2-derived measures across different regions, between subjects or over time. Regularisation is commonly used to improve the resilience of NNLS fits (Groetsch, 1984; Mackay *et al.*, 1994; Graham *et al.*, 1996), but numerical simulations suggest that although the standard errors in T_2 estimates are reduced by this approach, accuracy also suffers (Bjarnason *et al.*, 2010).

6.4 Applications

Single-component T_2 mapping can be used simply as a quantitative indicator of change in biological tissue, without any further biological modelling of why T_2 is changing. This approach has been used to assess tissue viability in knee cartilage (Colotti *et al.*, 2017). However, both single- and multicomponent T_2 mapping can also be used in more advanced biological models of tissue function and structure.

6.4.1 Metabolism and quantitative BOLD

The cerebral metabolic rate of oxygen (CMRO2) is a measure of how fast oxygen is consumed by the brain during mitochondrial energy conversion processes. It is defined as the product of cerebral blood flow and the oxygen extraction fraction (the difference in fractional oxygen saturation in arterial and venous blood). Because of the tight coupling between neural activation and these quantities, there is great interest in quantifying them both in healthy and diseased tissue.

One method to derive a global, whole-brain value of CMRO2 is the T_2 relaxation under spin tagging (TRUST) approach (Lu and Ge, 2008). TRUST uses a T_2-prep–type sequence with the addition of a pulse that labels venous blood (in contrast to arterial spin labelling, which only labels arteries). By acquiring a single slice through the superior sagittal sinus or internal jugular veins in both label and control conditions, the difference in T_2 can be used to derive the venous oxygenation, which can be combined with a phase-contrast velocity measurement in the same vein to derive CMRO2 (Xu *et al.*, 2009).

A related topic is quantitative BOLD, which requires the calculation of the scaling constant M to convert from the relative signal change of standard BOLD to CMRO2 (Blockley *et al.*, 2013). There are multiple methods for deriving M, but a recent approach that does not involve gas administration is to derive $R2'(=1/T_2')$ from the difference of a T_2 and T_2* map and assume that this is purely related to vascular effects (Shu *et al.*, 2016; Stone and Blockley, 2017).

6.4.2 The myelin water fraction

The most developed application for quantitative T_2 is interrogating the fraction of water associated with myelin in the central nervous system. As illustrated in Figure 6.5 myelin is a highly complex structure with layers of lipid and protein interspersed with thin layers of water that are only tens of angstroms thick. This highly restricted environment reduces the amount of motional averaging experienced by the water molecules and hence shortens their T_2s to the region of 10–20 ms.

Mackay *et al.* used the MET_2 approach to separate the T_2 decay curve into three separate components that they assigned to the water trapped within the myelin layers ('myelin water'; T_2 = 5–40 ms), a larger component thought to be the combination of intra- and extra-axonal water (T_2 = 70–100 ms) and a third component with long T_2 (>1 s) that was originally assigned to a CSF fraction (Mackay *et al.*, 1994). Studies in frog nerves, excised cat brain and crayfish nerves have shown up to four components (Menon and Allen 1991; Menon *et al.*, 1992).

Myelin water is considered an indirect but reliable marker of myelin health, as it is an essential component of the myelin sheath. As a consequence, MWF has become an important metric in the study of multiple sclerosis (Laule *et al.*, 2006), Alzheimer's disease (Dean *et al.*, 2014, 2017) and schizophrenia (Flynn *et al.*, 2003) and has been used to study infant myelination over the first few years of life (Leppert *et al.*, 2009; Deoni *et al.*, 2011).

The MWF calculated from MET_2 data has been thoroughly validated against histology (Figure 6.6). mcDESPOT has not been as thoroughly validated, but was recently shown to detect changes in myelin using the established cuprizone mouse model (Wood *et al.*, 2016). A potentially more useful parameter than the MWF is the myelin volume fraction, the fraction of space in a voxel that is occupied by myelin. A mapping between the two was recently introduced in a preclinical model (West *et al.*, 2016). An additional strength of the MET_2 approach compared to mcDESPOT is that it is simple to integrate over the T_2 distribution to derive the total water content in a voxel (Meyers *et al.*, 2016).

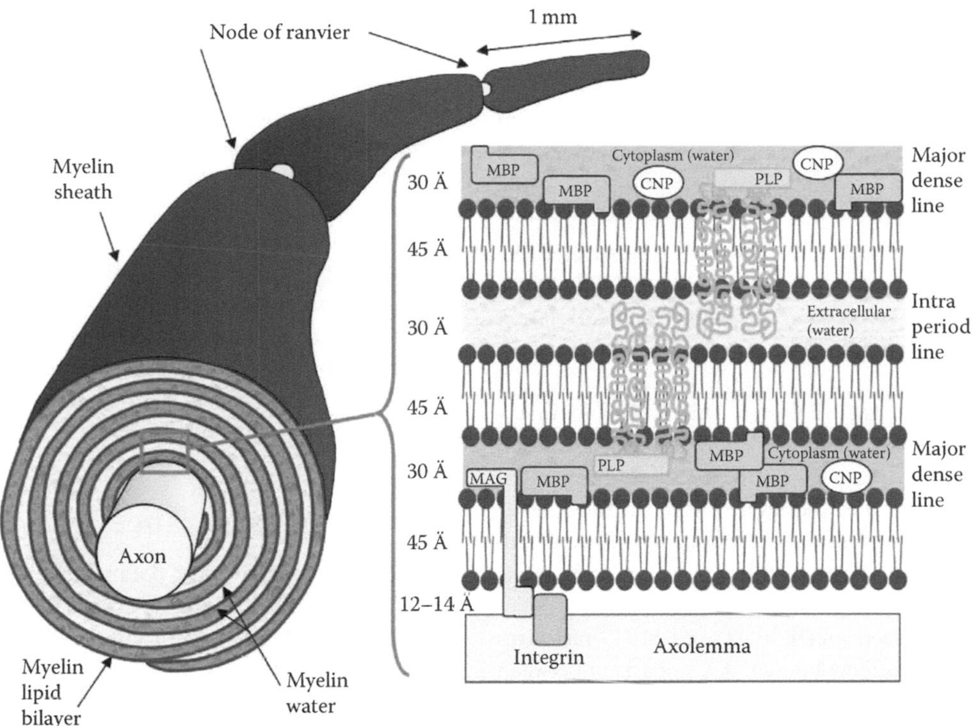

FIGURE 6.5 The CNS myelin sheath surrounding an axon with inset depicting close-up of bilayer, including myelin basic protein (MBP), proteolipid protein (PLP), cyclic nucleotide phosphodiesterase (CNP) and myelin-associated glycoprotein (MAG). (Reproduced with permission from Laule, C., *et al.*, *Neurotherapeutics*, 4, 460–484, 2007.)

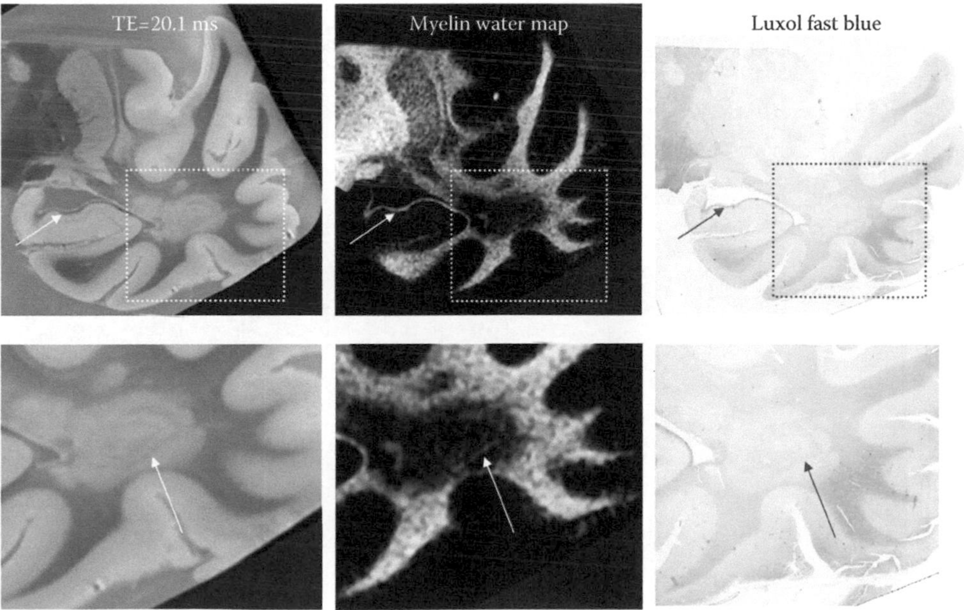

FIGURE 6.6 Example of a 7T TE = 20.1 ms image and myelin water map and corresponding luxol fast blue histology image of the temporal lobe region of an MS patient. The alveus of the hippocampus (upper panel) and faint rings of preserved myelin in the concentric Balo's lesion (lower panel; enlargement of the dashed box in the upper panel) are visible and indicated by arrows. (Reproduced with permission from Laule, C., *et al.*, *NeuroImage*, 40(4), 1575–1580, 2008.)

6.5 Accuracy and reproducibility

6.5.1 Spin echo: The gold standard

The 32-spin echo CPMG T_2-mapping technique is considered by many to be the gold standard for T_2 measurement (Mackay *et al.*, 1994). Histological studies have shown that the size of the myelin water component correlates well with myelin concentration (Laule *et al.*, 2007, 2008). Levesque *et al.* investigated the within-session and longitudinal reproducibility of MET_2 MWF (Levesque *et al.*, 2010); Meyers *et al.* found a strong correlation ($r = 0.98$) between MWF maps collected in a scan–rescan procedure (Meyers *et al.*, 2009).

Similar to the discrepancy between DESPOT1 and inversion-recovery methods, T_2 values are known to differ between DESPOT2 and MET_2. Jutras *et al.* compared single-component measurements using both techniques and reached the conclusion that some disagreement between methods is to be expected, due to the different TRs and TEs sampling different parts of the subvoxel distribution (Figure 6.7).

The discrepancy between mcDESPOT and MET_2 measures of MWF is marked and widely known. A comparison between 3D GRASE-based MET_2 and mcDESPOT demonstrated that mcDESPOT gave 1.5–4 times increased MWF measures compared to those estimated with GRASE (Zhang *et al.*, 2015). Compartmental T_2 times were also different, with mcDESPOT estimating IE water T_2s that were between 1.4 and 1.6 times longer and myelin water having unrealistically short T_2 (< 6 ms). The coefficient of variation (COV) of DESPOT-derived T_2 is good (below 10%), while the COV of MWF is good in white matter but worse in grey (above 10%), and the COV of the exchange parameter is surprisingly high in white matter, at around 30% (De Santis *et al.*, 2014; Wood *et al.*, 2016).

A detailed assessment of the T_2 prep method was conducted by Oh *et al.* (Oh *et al.*, 2006). MWF measures in white matter were found to be between 7% and 9% at 1.5T, in keeping with those reported for the 32-spin echo CPMG approach. The COV was found to be less than 6% for all selected white matter regions. At 3T, MWF was greater by 9.7%–12.3% in the same regions and the COV was as large as 10% for some ROIs.

6.6 Quality assurance

As with all quantitative MR parameters, regular quality assurance (QA) testing is required to ensure that the measurement of T_2 remains reliable over time. Calibrated test objects (phantoms) should be acquired that have a T_2 value that close to *in vivo* values. Normative values from the literature are given in table 6.3. A phantom should also be constructed that is stable and does not decay over time. A common phantom material is agarose gels doped with varying quantities of gadolinium chloride or

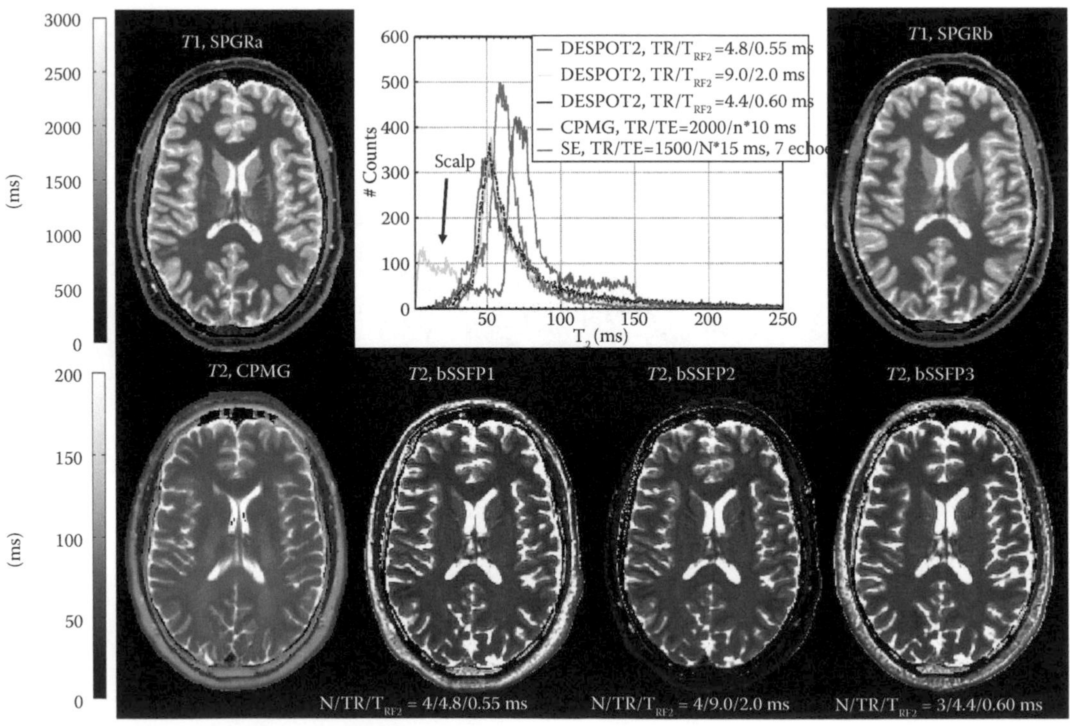

FIGURE 6.7 Axial T_1 maps from DESPOT1 (with SPGRa and SPGRb protocols), and T_2 from DESPOT2 (bSSFP1, bSSFP2 and bSSFP3 protocols in Table 6.1) compared with CPMG (single-component, mono-exponential fit), along with all five T_2 histograms for volunteer v2. Observe how the SE-based T2(WM T2~60 ms) lies between the DESPOT2 (WM T2~50 ms) and CPMG measurements (WM T2~70 ms). Spatial-spectral RF pulse effects in bSSFP2 cause underestimated T_2 in the scalp (arrow). (Reproduced with permission from Jutras, J.-D., *et al.*, *Magn. Reson. Med.*, 76(6), 1790–1804, 2015.)

TABLE 6.3 Normative values of T_2

Parameter	Location	Value	Reference
GMT_2	White matter	78 ms	Levesque *et al.* (2010)
	Grey matter	70–80 ms	Zhang *et al.* (2015)
T_2	Intra-/extracellular white matter	70–90 ms	Zhang *et al.* (2015)
	Myelin water	20–30 ms	Zhang *et al.* (2015)
	CSF	>1000 ms	MacKay *et al.* (2006)
MET_2 MWF	Grey matter	3.6%–4.5%	Oh *et al.* (2006)
	White matter	2%–20%	Levesque *et al.* (2010)
	Frontal white matter	9.4%	Levesque *et al.* (2010)

gadolinium DTPA to generate a range of T_2 (and T_1) values. To improve the stability of the gel, it is common to add a fungicide such as sodium azide. The relaxation values of agarose gels are temperature dependent and while switching the doping agent can almost eliminate this for T_1, this is not the case for T_2, which has a coefficient of about –1.25%/°C (Tofts *et al.*, 1993). Phantoms should be stored in the scanner room to avoid temperature drift during measurement, and temperature should be measured before and after scanning. Repeated QA tests should also be run *in vivo* on a group of five or so normal volunteers selected for their regular availability. Measurements in volunteers have the benefit of being realistic and inherently temperature-controlled by the body. Increased variability from subject movement or physiological noise is likely, but the careful selection of participants may help to mitigate these shortcomings. See Chapter 3 for a more thorough treatment of general QA strategy.

References

Alonso-Ortiz E, Levesque IR, Pike GB. MRI-based myelin water imaging: a technical review. Magn Reson Med 2015; 73(1): 70–81.

Ben-Eliezer N, Sodickson DK, Block KT. Rapid and accurate T2 mapping from multi-spin-echo data using Bloch-simulation-based reconstruction. Magn Reson Med 2015; 73(2): 809–17.

Bjarnason TA, Laule C, Bluman J, Kozlowski P. Temporal phase correction of multiple echo T 2 magnetic resonance images. J Magn Reson 2013; 231: 22–31.

Bjarnason TA, McCreary CR, Dunn JF, Mitchell JR. Quantitative T2 analysis: the effects of noise, regularization, and multivoxel approaches. Magn Reson Med 2010; 63(1): 212–17.

Blockley NP, Griffeth VEM, Simon AB, Buxton RB. A review of calibrated blood oxygenation level-dependent (BOLD) methods for the measurement of task-induced changes in brain oxygen metabolism. NMR Biomed 2013; 26(8): 987–1003.

Boudreau M, Tardif CL, Stikov N, Sled JG, Lee W, Pike GB. B1 mapping for bias-correction in quantitative T1 imaging of the brain at 3T using standard pulse sequences. J Magn Reson Imaging 2017; 46: 1673–1682.

Bouhrara M, Reiter DA, Celik H, Fishbein KW, Kijowski R, Spencer RG. Analysis of mcDESPOT- and CPMG-derived parameter estimates for two-component nonexchanging systems. Magn Reson Med 2015; 75(6): 2406–20.

Bouhrara M, Spencer RG. Incorporation of nonzero echo times in the SPGR and bSSFP signal models used in mcDESPOT. Magn Reson Med 2015; 74(5): 1227–35.

Bouhrara M, Spencer RG. Improved determination of the myelin water fraction in human brain using magnetic resonance imaging through Bayesian analysis of mcDESPOT. NeuroImage 2016; 127: 456–71.

Bouhrara M, Spencer RG. Rapid simultaneous high-resolution mapping of myelin water fraction and relaxation times in human brain using BMC-mcDESPOT. NeuroImage 2017; 147: 800–11.

Brink WM, Börnert P, Nehrke K, Webb AG. Ventricular B1+ perturbation at 7 T—Real effect or measurement artifact? NMR Biomed 2014; 27(6): 617–20.

Carr HY, Purcell EM. Effects of diffusion on free precession in nuclear magnetic resonance experiments. Phys Rev 1954; 94(3): 630.

Colotti R, Omoumi P, Bonanno G, Ledoux J-B, van Heeswijk RB. Isotropic three-dimensional T2 mapping of knee cartilage: development and validation. J Magn Reson Imaging 2017.

Dean DC, Hurley SA, Kecskemeti SR, O'Grady JP, Canda C, Davenport-Sis NJ, et al. Association of amyloid pathology with myelin alteration in preclinical Alzheimer disease. JAMA Neurol 2017; 74(1): 41–9.

Dean DC, Jerskey BA, Chen K, Protas H, Thiyyagura P, Roontiva A, et al. Brain differences in infants at differential genetic risk for late-onset Alzheimer disease: a cross-sectional imaging study. JAMA Neurol 2014; 71(1): 11–22.

Deoni SC. High-resolution T1 mapping of the brain at 3T with driven equilibrium single pulse observation of T1 with high-speed incorporation of RF field inhomogeneities (DESPOT1-HIFI). J Magn Reson Imaging 2007; 26(4): 1106–11.

Deoni SC. Transverse relaxation time (T2) mapping in the brain with off-resonance correction using phase-cycled steady-state free precession imaging. J Magn Reson Imaging 2009; 30(2): 411–17.

Deoni SC. Correction of main and transmit magnetic field (B0 and B1) inhomogeneity effects in multicomponent-driven equilibrium single-pulse observation of T1 and T2. Magn Reson Med 2011; 65(4): 1021–35.

Deoni SC, Matthews L, Kolind SH. One component? Two components? Three? The effect of including a nonexchanging "free" water component in multicomponent driven equilibrium single pulse observation of T1 and T2. Magn Reson Med 2013; 70(1): 147–54.

Deoni SC, Mercure E, Blasi A, Gasston D, Thomson A, Johnson M, et al. Mapping infant brain myelination with magnetic resonance imaging. J Neurosci 2011; 31(2): 784–91.

Deoni SC, Peters TM, Rutt BK. High-resolution T1 and T2 mapping of the brain in a clinically acceptable time with DESPOT1 and DESPOT2. Magn Reson Med 2005; 53(1): 237–41.

Deoni SC, Rutt BK, Arun T, Pierpaoli C, Jones DK. Gleaning multicomponent T1 and T2 information from steady-state imaging data. Magn Reson Med 2008; 60(6): 1372–87.

Deoni SC, Rutt BK, Peters TM. Rapid combined T1 and T2 mapping using gradient recalled acquisition in the steady state. Magn Reson Med 2003; 49(3): 515–26.

Deoni SCL, Kolind SH. Investigating the stability of mcDESPOT myelin water fraction values derived using a stochastic region contraction approach. Magn Reson Med 2015; 73(1): 161–169.

Does MD, Gore JC. Rapid acquisition transverse relaxometric imaging. J Magn Reson 2000; 147(1): 116–20.

Flynn S, Lang D, Mackay A, Goghari V, Vavasour I, Whittall K, et al. Abnormalities of myelination in schizophrenia detected in vivo with MRI, and post-mortem with analysis of oligodendrocyte proteins. Mol Psychiatry 2003; 8(9): 811–20.

Forsén S, Hoffman RA. Study of moderately rapid chemical exchange reactions by means of nuclear magnetic double resonance. J Chem Phys 1963; 39(11): 2892–901.

Graham SJ, Stanchev PL, Bronskill MJ. Criteria for analysis of multicomponent tissue T2 relaxation data. Magn Reson Med 1996; 35(3): 370–8.

Groetsch C. *The theory of Tikhonov Regularization for Fredholm Equations*. Boston, MA: Pitman Publication; 1984, 104p.

Guo J, Ji Q, Reddick WE. Multi-slice myelin water imaging for practical clinical applications at 3.0 T. Magn Reson Med 2013; 70(3): 813–22.

Henkelman R, Stanisz G, Graham S. Magnetization transfer in MRI: a review. NMR Biomed 2001; 14(2): 57–64.

Hennig J. Multiecho imaging sequences with low refocusing flip angles. J Magn Reson (1969) 1988; 78(3): 397–407.

Hennig J. Echoes—How to generate, recognize, use or avoid them in MR-imaging sequences. Part I: fundamental and not so fundamental properties of spin echoes. Conc Magn Reson A 1991; 3(3): 125–43.

Hurley SA, Alexander AL. *Assessment of mcDESPOT Precision Using Constrained Estimation*. ISMRM; Conference Abstract, Proc. Intl. Soc. Mag. Reson. Med. 2014; 22: 3144

Jones CK, Whittall KP, MacKay AL. Robust myelin water quantification: averaging vs. spatial filtering. Magn Reson Med 2003; 50(1): 206–9.

Jutras J-D, Wachowicz K, De Zanche N. Analytical corrections of banding artifacts in driven equilibrium single pulse observation of T2 (DESPOT2). Magn Reson Med 2015; 76(6): 1790–804.

Lankford CL, Does MD. On the inherent precision of mcDESPOT. Magn Reson Med 2013; 69(1): 127–136.

Laule C, Kozlowski P, Leung E, Li DKB, MacKay AL, Moore GRW. Myelin water imaging of multiple sclerosis at 7T: correlations with histopathology. NeuroImage 2008; 40(4): 1575–80.

Laule C, Leung E, Lis DK, Traboulsee AL, Paty DW, MacKay AL, et al. Myelin water imaging in multiple sclerosis: quantitative correlations with histopathology. Mult Scler 2006; 12(6): 747–53.

Laule C, Vavasour IM, Kolind SH, Li DKB, Traboulsee TL, Moore GRW, et al. Magnetic resonance imaging of myelin. Neurotherapeutics 2007; 4: 460–84.

Lawson CL, Hanson RJ. *Solving Least Squares Problems*. Philadelphia, PA: SIAM; Philadelphia; Pennsylvania; 1995.

Lebel RM, Wilman AH. Transverse relaxometry with stimulated echo compensation. Magn Reson Med 2010; 64(4): 1005–14.

Leppert IR, Almli CR, McKinstry RC, Mulkern RV, Pierpaoli C, Rivkin MJ, et al. T(2) relaxometry of normal pediatric brain development. J Magn Reson Imaging 2009; 29(2): 258–67.

Levesque IR, Chia CL, Pike GB. Reproducibility of in vivo magnetic resonance imaging-based measurement of myelin water. J Magn Reson Imaging 2010; 32(1): 60–8.

Liu F, Block WF, Kijowski R, Samsonov A. Rapid multicomponent relaxometry in steady state with correction of magnetization transfer effects. Magn Reson Med 2015; 75(4): 1423–33.

Lu H, Ge Y. Quantitative evaluation of oxygenation in venous vessels using T2-relaxation-under-spin-tagging MRI. Magn Reson Med 2008; 60(2): 357–63.

MacKay A, Laule C, Vavasour I, Bjarnason T, Kolind S, Madler B. Insights into brain microstructure from the T2 distribution. Magn Reson Imaging 2006; 24(4): 515–25.

Mackay A, Whittall K, Adler J, Li D, Paty D, Graeb D. In vivo visualization of myelin water in brain by magnetic resonance. Magn Reson Med 1994; 31(6): 673–7.

Mädler B, MacKay A. Towards whole brain myelin imaging. *International Society for Magnetic Resonance in Medicine*, Berlin, Germany, 2007; 1723.

Maier CF, Tan SG, Hariharan H, Potter HG. T2 quantitation of articular cartilage at 1.5 T. J Magn Reson Imaging 2003; 17(3): 358–64.

Majumdar S, Orphanoudakis S, Gmitro A, O'Donnell M, Gore J. Errors in the measurements of T2 using multiple-echo MRI techniques. II. Effects of static field inhomogeneity. Magn Reson Med 1986; 3(4): 562–74.

McPhee KC, Wilman AH. Transverse relaxation and flip angle mapping: evaluation of simultaneous and independent methods using multiple spin echoes. Magn Reson Med 2016; 77: 2057–65.

Meiboom S, Gill D. Modified spin-echo method for measuring nuclear relaxation times. Rev Sci Instrum 1958; 29(8): 688–91.

Menon R, Allen P. Application of continuous relaxation time distributions to the fitting of data from model systems and excised tissue. Magn Reson Med 1991; 20(2): 214–27.

Menon R, Rusinko M, Allen P. Proton relaxation studies of water compartmentalization in a model neurological system. Magn Reson Med 1992; 28(2): 264–74.

Meyers SM, Kolind SH, Laule C, MacKay AL. Measuring water content using T2 relaxation at 3 T: phantom validations and simulations. Magn Reson Imaging 2016; 34(3): 246–51.

Meyers SM, Laule C, Vavasour IM, Kolind SH, Madler B, Tam R, et al. Reproducibility of myelin water fraction analysis: a comparison of region of interest and voxel-based analysis methods. Magn Reson Imaging 2009; 27(8): 1096–103.

Miller KL, Smith SM, Jezzard P. Asymmetries of the balanced SSFP profile. Part II: white matter. Magn Reson Med 2010; 63(2): 396–406.

Nehrke K. On the steady-state properties of actual flip angle imaging (AFI). Magn Reson Med 2009; 61(1): 84–92.

Nehrke K, Versluis MJ, Webb A, Börnert P. Volumetric B1+ mapping of the brain at 7T using DREAM. Magn Reson Med 2014; 71(1): 246–56.

Nguyen TD, Deh K, Monohan E, Pandya S, Spincemaille P, Raj A, et al. Feasibility and reproducibility of whole brain myelin water mapping in 4 minutes using fast acquisition with spiral trajectory and adiabatic T2prep (FAST-T2) at 3T. Magn Reson Med 2015; 76(2): 456–65.

Oh J, Han ET, Lee MC, Nelson SJ, Pelletier D. Multislice brain myelin water fractions at 3T in multiple sclerosis. J Neuroimaging 2007; 17(2): 156–63.

Oh J, Han ET, Pelletier D, Nelson SJ. Measurement of in vivo multi-component T2 relaxation times for brain tissue using multi-slice T2 prep at 1.5 and 3 T. Magn Reson Imaging 2006; 24(1): 33–43.

Papp D, Callaghan MF, Meyer H, Buckley C, Weiskopf N. Correction of inter-scan motion artifacts in quantitative R1 mapping by accounting for receive coil sensitivity effects. Magn Reson Med 2015; 76(5): 1478–85.

Pei M, Nguyen TD, Thimmappa ND, Salustri C, Dong F, Cooper MA, et al. Algorithm for fast monoexponential fitting based on Auto-Regression on Linear Operations (ARLO) of data. Magn Reson Med 2015; 73(2): 843–50.

Pell GS, Briellmann RS, Waites AB, Abbott DF, Jackson GD. Voxel-based relaxometry: a new approach for analysis of T2 relaxometry changes in epilepsy. NeuroImage 2004; 21(2): 707–13.

Pell GS, Briellmann RS, Waites AB, Abbott DF, Lewis DP, Jackson GD. Optimized clinical T2 relaxometry with a standard CPMG sequence. J Magn Reson Imaging 2006; 23(2): 248–52.

Pohmann R, Scheffler K. A theoretical and experimental comparison of different techniques for B1 mapping at very high fields. NMR in Biomedicine 2013; 26(3): 265–75.

Poon CS, Henkelman RM. Practical T2 quantitation for clinical applications. J Magn Reson Imaging 1992; 2(5): 541–53.

Prasloski T, Madler B, Xiang QS, MacKay A, Jones C. Applications of stimulated echo correction to multicomponent T2 analysis. Magn Reson Med 2012a; 67(6): 1803–14.

Prasloski T, Rauscher A, MacKay AL, Hodgson M, Vavasour IM, Laule C, et al. Rapid whole cerebrum myelin water imaging using a 3D GRASE sequence. NeuroImage 2012b; 63(1): 533–9.

De Santis S, Drakesmith M, Bells S, Assaf Y, Jones DK. Why diffusion tensor MRI does well only some of the time: variance and covariance of white matter tissue microstructure attributes in the living human brain. NeuroImage 2014; 89: 35–44.

Shu CY, Herman P, Coman D, Sanganahalli BG, Wang H, Juchem C, et al. Brain region and activity-dependent properties of M for calibrated fMRI. NeuroImage 2016; 125: 848–56.

Simmons A, Tofts PS, Barker GJ, Arridge SR. Sources of intensity nonuniformity in spin echo images at 1.5 T. Magn Reson Med 1994; 32(1): 121–8.

Smith HE, Mosher TJ, Dardzinski BJ, Collins BG, Collins CM, Yang QX, et al. Spatial variation in cartilage T2 of the knee. J Magn Reson Imaging 2001; 14(1): 50–5.

Stone AJ, Blockley NP. A streamlined acquisition for mapping baseline brain oxygenation using quantitative \{BOLD\}. NeuroImage 2017; 147: 79–88.

Tofts P, Shuter B, Pope J. Ni-DTPA doped agarose gel—a phantom material for Gd-DTPA enhancement measurements. Magn Reson Imaging 1993; 11(1): 125–33.

Vasilescu V, Katona E, Simplaceanu V, Demco D. Water compartments in the myelinated nerve. III. Pulsed NMR result. Cell Mol Life Sci 1978; 34(11): 1443–4.

Waldman A, Rees JH, Brock CS, Robson MD, Gatehouse PD, Bydder GM. MRI of the brain with ultra-short echo-time pulse sequences. Neuroradiology 2003; 45(12): 887–92.

West KL, Kelm ND, Carson RP, Gochberg DF, Ess KC, Does MD. Myelin volume fraction imaging with MRI. NeuroImage 2016; DOI: 10.1016/j.neuroimage/2016/12/067

Whittall KP, Mackay AL, Graeb DA, Nugent RA, Li DK, Paty DW. In vivo measurement of T2 distributions and water contents in normal human brain. Magn Reson Med 1997; 37(1): 34–43.

Whittall KP, MacKay AL, Li DK. Are mono-exponential fits to a few echoes sufficient to determine T2 relaxation for in vivo human brain? Magn Reson Med 1999; 41(6): 1255–7.

Woermann FG, Barker GJ, Birnie KD, Meencke HJ, Duncan JS. Regional changes in hippocampal T2 relaxation and volume: a quantitative magnetic resonance imaging study of hippocampal sclerosis. J Neurol, Neurosurg Psychiatry 1998; 65(5): 656–64.

Woermann FG, Steiner H, Barker GJ, Bartlett PA, Elger CE, Duncan JS, et al. A fast FLAIR dual-echo technique for hippocampal T2 relaxometry: first experiences in patients with temporal lobe epilepsy. J Magn Reson Imaging 2001; 13(4): 547–52.

Wood TC. Improved formulas for the two optimum VFA flip-angles. Magn Reson Med 2015; 74(1): 1–3.

Wood TC, Simmons C, Hurley SA, Vernon AC, Torres J, Dell'Acqua F, et al. Whole-brain ex-vivo quantitative MRI of the cuprizone mouse model. Peer J 2016; 4: e2632.

Wood TC, Wastling SJ, Barker GJ. Removing SSFP Banding Artifacts from DESPOT2 Images Using the Geometric Solution. ISMRM 2015; 23: 3144.

Xiang Q-S, Hoff MN. Banding artifact removal for bSSFP imaging with an elliptical signal model. Magn Reson Med 2014; 71(3): 927–933.

Xu F, Ge Y, Lu H. Noninvasive quantification of whole-brain cerebral metabolic rate of oxygen (CMRO2) by MRI. Magn Reson Med 2009; 62(1): 141–8.

Yarnykh VL. Actual flip-angle imaging in the pulsed steady state: a method for rapid three-dimensional mapping of the transmitted radiofrequency field. Magn Reson Med 2007; 57(1): 192–200.

Yarnykh VL. Optimal radiofrequency and gradient spoiling for improved accuracy of T1 and B1 measurements using fast steady-state techniques. Magn Reson Med 2010; 63(6): 1610–26.

Zhang J, Kolind SH, Laule C, MacKay AL. Comparison of myelin water fraction from multiecho T2 decay curve and steady-state methods. Magn Reson Med 2015; 73(1): 223–32.

Zur Y, Wood M, Neuringer L. Motion-insensitive, steady-state free precession imaging. Magn Reson Med 1990; 16(3): 444–59.

7

T_2*: Susceptibility Weighted Imaging and Quantitative Susceptibility Mapping[1]

Sagar Buch and Saifeng Liu
The MRI Institute for Biomedical Research

Yongsheng Chen
Wayne State University

Kiarash Ghassaban and E. Mark Haacke
Magnetic Resonance Innovations Inc.

Contents

7.1 Introduction

Magnetic resonance imaging (MRI) is a non-invasive modality that provides soft tissue contrast based on properties such as proton density, T_1 and T_2 relaxation times. For gradient echo (GRE) imaging, a broad variety of contrasts are possible by varying the flip angle and echo times (TE) for short repetition time (TR) experiments. However, unlike spin echo sequences, the T_2* relaxation time is affected by local sources of magnetic field inhomogeneities (Brown *et al.*, 2014). These sources of field inhomogeneity, that is the magnetic susceptibility property of the tissues, can also be used to produce another type of contrast. Susceptibility weighted imaging (SWI) utilises this contrast in the form of the phase component of an MRI signal to improve the tissue delineation on magnitude images (Haacke *et al.*, 2004, 2009a; Brown *et al.*, 2014). SWI is a high resolution, 3D, fully flow-compensated GRE method; its initial purpose was the visualisation of cerebral venous vasculature. It has since been applied successfully in imaging stroke, traumatic brain injury (TBI), multiple sclerosis, tumours and haemorrhagic lesions (Haacke *et al.*, 2009b; Park *et al.*, 2009; Wu *et al.*, 2010; Cheng *et al.*, 2013). At this point in time, SWI has become a part of the standard clinical routine (Akter *et al.*, 2007; Goos *et al.*, 2011; Robinson and Bhuta, 2011). As an extension to SWI, more recent applications of phase include quantitative susceptibility mapping (QSM), a method that produces magnetic source images and allows for direct quantification of tissue susceptibility changes (de Rochefort *et al.*, 2008; Shmueli *et al.*, 2009; Haacke *et al.*, 2010b, 2015). One might correctly surmise that whatever applications are available for SWI will also be game for QSM processing (Eskreis-Winkler *et al.*, 2017).

7.2 GRE sequences and T_2* relaxation

GRE-based MRI is considered a conventional technique and is commonly used for nearly every medical application in both 2D and 3D data acquisition (Haacke and Reichenbach 2014).

[1] Edited by Mara Cercignani; reviewed by Richard Bowtell, Sir Peter Mansfield Imaging Centre, Faculty of Science, University of Nottingham, UK.

The magnitude of the MRI signal for a radiofrequency (RF) spoiled, short TR, GRE sequence is given by the following equation (Bernstein *et al.*, 2004; Brown *et al.*, 2014):

$$S(\theta) = \rho_o \sin\theta \cdot \frac{\left[1 - e^{\left(-\frac{TR}{T_1}\right)}\right]}{\left[1 - \cos\theta \cdot e^{\left(-\frac{TR}{T_1}\right)}\right]} \cdot e^{\left(-\frac{TE}{T_2^*}\right)} \tag{7.1}$$

where θ is the flip angle (FA) that is the angle at which the longitudinal magnetisation is tipped (Brown *et al.*, 2014).

Apart from the dephasing due to molecular interactions, the decay of the transverse magnetisation is affected by the local external magnetic field inhomogenities, causing an additional dephasing of the magnetisation (the reversible R_2' component) (Brown *et al.*, 2014). Due to this extra phase dispersion, the decay time of the transverse magnetisation is shortened and is given as T_2*

$$1/T_2^* = 1/T_2 + 1/T_2' \quad \text{or} \quad R_2^* = R_2 + R_2' \tag{7.2}$$

SWI pulse sequences are based on GRE sequences; hence, R_2* or T_2* effects are a key contributor to contrast in SWI magnitude images (Haacke *et al.*, 2004, 2009b). In addition to the usual macroscopic field variations from the main magnet, strong local field inhomogeneities, such as those caused by air–tissue interfaces or iron deposition, can cause significant signal dephasing across a voxel.

7.3 SWI pulse sequence

For SWI, a 3D RF spoiled velocity compensated GRE sequence is used (Figure 7.1) (Haacke and Reichenbach 2014). Gradient moment nulling (GMN) is a method used to modify a gradient waveform in order to suppress the motion sensitivity of a pulse sequence (Bernstein *et al.*, 2004). Gradient moments are values calculated from the integral of a given gradient waveform with time (Bernstein *et al.*, 2004):

$$m_n = \int \left[t^n \cdot G(t) \right] dt \tag{7.3}$$

where m_n is the *n*th gradient moment of the gradient waveform *G*(*t*). The moments of a gradient waveform can be nulled, depending on the application, to various degrees and orders. Signal variations that lead to artefacts in the image are caused by the rapid and pulsatile flow of blood and cerebrospinal fluid. These artefacts include signal loss due to flow-induced dephasing, misregistration artefacts and velocity-induced phase (Bernstein *et al.*, 2004; Brown *et al.*, 2014; Haacke and Reichenbach, 2014). In the presence of a bipolar gradient G_x with duration 2τ, the phase accumulation for a spin moving at a constant velocity (v) is given by

$$\phi = \gamma\, G_x v\, \tau^2 \tag{7.4}$$

Motion or flow with constant velocity is compensated with first-order GMN, by nulling the first moment of a gradient waveform, and is also called *velocity compensation* or *flow compensation* (Brown *et al.*, 2014). For a velocity-compensated pulse sequence, the velocity-induced phase disappears, which leaves the desired susceptibility-induced phase information (Haacke and Reichenbach 2014). However, the background field gradients and the flow acceleration effects may cause errors in flow compensation and induce unreliable phase, especially for arteries with higher velocity (Wu *et al.*, 2016).

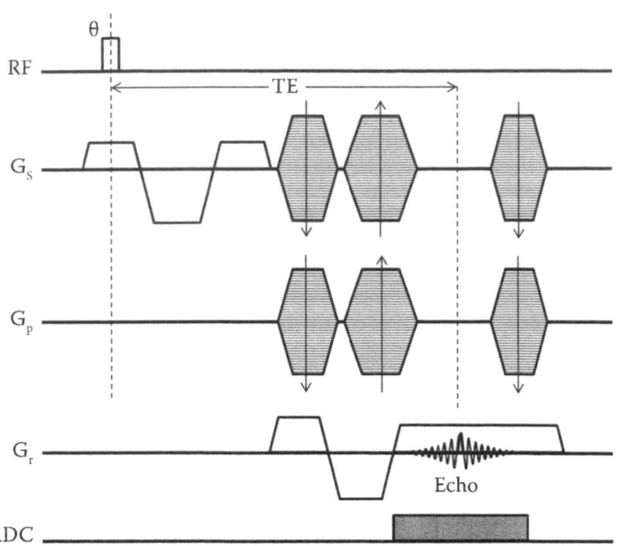

FIGURE 7.1 Pulse sequence diagram for a 3D gradient echo designed for susceptibility weighted imaging acquisition with full flow compensation. The pulse sequence includes first-order gradient moment nulling in slab-select, partition-encoding, phase encoding and readout directions. ADC = analog-to-digital converter.

Figure 7.1 represents the original pulse sequence used for SWI data acquisition. A volume with several centimetres of slab thickness is excited using an RF pulse of low FA, which is then spatially resolved in 3D space by applying the frequency encoding, phase encoding and partition encoding gradients. Velocity compensation is applied in slab-select, partition encoding (G_s), phase-encoding (G_p) and readout (G_r) directions to eliminate oblique flow artefacts (Brown *et al.*, 2014). The gradients in the partition and phase encoding directions are rewound after sampling the echo signal, whereas the readout gradient remains to dephase the spins (Brown *et al.*, 2014).

The data from this sequence can be represented by a complex vector (Brown *et al.*, 2014). Generally, the signal produced by the rotational motion of the spin precession in the presence of a constant magnetic field is deconvolved into two components:

$$S_{xy}(\vec{r},t) = S_x(\vec{r},t) + i\,S_y(\vec{r},t) \quad (7.5)$$

where S_x and S_y represent the real and imaginary channels of the signal. This equation can be rewritten as

$$S_{xy}(\vec{r},t) = |S_{xy}(0)| \cdot e^{i\phi(t)} \quad (7.6)$$

where the phase is dependent on the position of the spin isochromat and time (Brown *et al.*, 2014). For a right-handed system:

$$\phi(\vec{r},t) = -\gamma \Delta B(\vec{r})t \quad (7.7)$$

where γ is the gyromagnetic ratio, and $\Delta B(\vec{r})$ is the variation in the main magnetic field (Haacke and Reichenbach 2014).

7.4 T_2* mapping

T_2* relaxation times can be directly related to the amount of iron present (Gelman *et al.*, 1999; Haacke *et al.*, 2010a; Langkammer *et al.*, 2010; Ghadery *et al.*, 2015) and therefore used to estimate iron content in the brain (Haacke *et al.*, 2010a; Ghadery *et al.*, 2015) and liver (Hankins *et al.*, 2009; McCarville *et al.*, 2010). T_2* relaxation mapping has also been used in blood oxygenation level dependent (BOLD) functional imaging (Ogawa *et al.*, 1993) and the detection and tracking of super-paramagnetic iron oxides (Dahnke and Schaeffter, 2005). Qualitative and quantitative analysis of these decay rates can yield valuable information about the location and severity of the pathology. Accurate quantification of the T_2 and/or T_2* may also make it possible to monitor a treatment plan (Chen *et al.*, 1996). R_2* maps can be generated from two or more echoes, by fitting the magnitude to the exponential decay curve. However, this is a rather simplified model and more sophisticated models include partial volume and background field effects (Yablonskiy and Haacke, 1994; Yablonskiy *et al.*, 2013; Haacke and Reichenbach, 2014). Figure 7.2 shows an example of R_2* mapping results using an 11 echo GRE sequence, with TEs 7.53–32.63 ms (with an echo spacing of 2.51 ms), B_0 = 3 Tesla (T), TR = 70 ms, FA = 15 and bandwidth (BW) = 465 Hz/pixel with a resolution = 0.7 × 0.7 × 1.4 mm^3. In summary, R_2* is still a well-established method for the evaluation of iron load in the basal ganglia during aging (Cherubini *et al.*, 2009; Péran *et al.*, 2009) and in neurological diseases (Lehéricy *et al.*, 2012).

7.5 Magnetic susceptibility

Several sources of magnetic field variation can be found in the body, which can cause signal distortion, signal loss, image artefacts and reductions in T_2*. Two such sources include extracorporeal objects (surgically implanted objects, iron-based tattoos and certain cosmetic products like eyeshadows) and internal magnetic susceptibility differences found between the tissues in the body (Brown *et al.*, 2014). While the extracorporeal objects create distortion artefacts, the internal susceptibility differences can be used to provide a unique contrast in the phase images (Haacke *et al.*, 2009a). This attribute may provide special information about tissues, such as distinguishing lesions from normal tissue and quantifying susceptibility differences between tissues (Haacke and Reichenbach 2014; Haacke *et al.*, 2015).

Magnetic susceptibility can be defined as the property of a substance, when placed within an external uniform magnetic

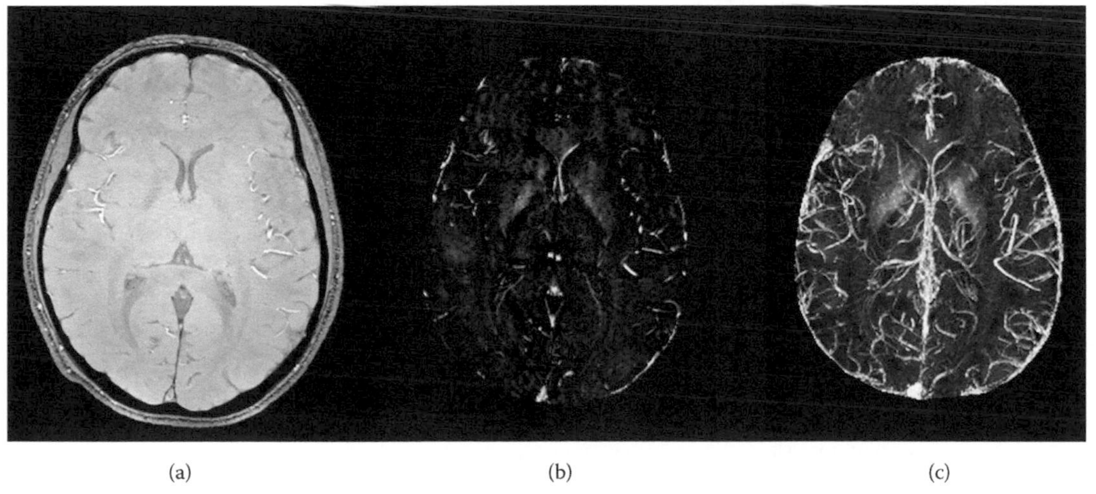

(a) (b) (c)

FIGURE 7.2 (a) Original magnitude at echo time (TE) = 12.55 ms and B0 = 3T, (b) R_2* map generated using TEs: 7.53–32.63 ms (with an interval of 2.51 ms) and (c) maximum intensity projection (MIP) of R_2* maps over 16 slices or an effective slice slab of 22.4 mm. The latter image reveals the presence of veins because of their increased paramagnetic susceptibility caused by their increased level of deoxyhaemoglobin. The edge of the brain shows as a thin bright region because of the rapid changes in field in these regions.

field, that measures its tendency to become magnetised and to alter the magnetic field around it (Brown *et al.*, 2014). The physical magnetic field (measured in Tesla) is given by

$$\vec{B} = \mu \vec{H} \tag{7.8}$$

where μ is the permeability constant of the substance and $\vec{H}$ is measured in Ampere/metre (A/m) (Brown *et al.*, 2014). The induced magnetic field $\vec{B}$ inside a substance is given by

$$\vec{B} = \mu_0 (\vec{H} + \vec{M}) \tag{7.9}$$

where $\vec{M}$ is the induced magnetisation serving as a macroscopic source of internal field contribution of the electron spins inside the substance, and the free space permeability μ_0 is given by $\mu_0 = 4\pi \times 10^{-7}$ Tm/A. The induced magnetisation ($\vec{M}$) can be approximated by (Brown *et al.*, 2014):

$$\vec{M} = \left[\frac{\chi \cdot \vec{B}}{\mu_0 (1+\chi)}\right] \approx (\chi \cdot \vec{B}) / \mu_0 (\text{when } \chi << 1) \tag{7.10}$$

With this approximation, the change in field induced by this magnetisation is given by the following (Salomir *et al.*, 2003; Cheng *et al.*, 2009b; Haacke and Reichenbach, 2014):

$$\Delta B_{dz}(\vec{r}) = B_0 \cdot \text{FT}^{-1}\left[\text{FT}(\chi(\vec{r})) \cdot G(k)\right] \tag{7.11}$$

where FT stands for Fourier transform and FT^{-1} for inverse Fourier transform, and the Green's function $G(k)$ can be approximated by

$$G_{\text{reg}}^{-1}(k) = \begin{cases} \left(\frac{1}{3} - \frac{k_z^2}{k^2}\right)^{-1} & \text{when } \left|\frac{1}{3} - \frac{k_z^2}{k^2}\right| > \delta \\ \text{sign}\left(\frac{1}{3} - \frac{k_z^2}{k^2}\right)\left(\frac{1}{3} - \frac{k_z^2}{k^2}\right)^2 \cdot \delta^{-3} & \text{otherwise} \end{cases} \tag{7.12}$$

to avoid the zeroes when $3k_z^2 = k^2$, in which small unreliable k-space elements below a threshold (δ) are replaced by gradually decreasing values (Shmueli *et al.*, 2009; Haacke *et al.*, 2010b).

7.5.1 Types of magnetic susceptibility

Substances can be classified into diamagnetic, paramagnetic and ferromagnetic materials based on their macroscopic influence over the external magnetic field (Brown *et al.*, 2014). For empty space, the value of χ is zero, whereas a negative value of χ represents a diamagnetic material; if the value of χ is positive the material is paramagnetic (Brown *et al.*, 2014). In imaging human tissue, the terms *paramagnetic* and *diamagnetic* are used relative to the susceptibility of the water rather than vacuum. For ferromagnetic materials, the value of χ is much larger than unity (Brown *et al.*, 2014), whereas, in relation to the susceptibility of air, tissue susceptibility values are generally much less than 1 parts per million (ppm), rather being on the order of parts per billion (Haacke *et al.*, 2009a; Brown *et al.*, 2014; Haacke and Reichenbach, 2014).

7.6 Magnetic field perturbations, phase and data processing steps

To avoid image distortion and the associated changes in signal intensity, the main magnetic field (B_0) should ideally be homogeneous throughout the sample. Practically, there are static field variations, local magnetic field variations, which can be caused by imperfect gradients, eddy currents, air–tissue interfaces and, more interestingly for this discussion, susceptibility differences between tissues (Haacke *et al.*, 2009b; Brown *et al.*, 2014; Haacke and Reichenbach, 2014). The phase can be written as a function of the difference between the uniform field B_0 and the local field variation, $\Delta B(\vec{r})$, at position r. We can rewrite Equation 7.7 as

$$\phi(\vec{r}, \text{TE}) = -\gamma \cdot \left(\Delta B(\vec{r})\right) \cdot \text{TE} \tag{7.13}$$

The MRI signal is acquired in the form of an echo signal centred at the time TE (Bernstein *et al.*, 2004; Brown *et al.*, 2014).

In addition to the phase created by the spatial encoding gradients, there are other unwanted forms of remnant or background phase present (Haacke *et al.*, 2009b; Brown *et al.*, 2014). These global phase effects also need to be understood and dealt with before useful local information can be revealed. The effective phase behaviour can be written as the summation of these fields (Cheng *et al.*, 2009b; Haacke *et al.*, 2009b; Brown *et al.*, 2014):

$$\phi = -\gamma(\Delta B_{\text{main field}} + \Delta B_{cs} + \Delta B_{\text{global geometry}} + \Delta B_{\text{local field}}) \tag{7.14}$$

where ΔB_{cs} represents the field variations due to the chemical shift effects (Haacke *et al.*, 2009b).

7.6.1 Phase aliasing

When phase is measured, it is defined only within the interval $[-\pi, +\pi]$, instead of the full phase evolution of the magnetisation. In order to estimate the actual phase, an integer multiple of 2π must be added or subtracted as appropriate. The phase wraps can be present either spatially or temporally; and the phase change over neighbouring voxels or time points and noise level in the image are the two main factors that govern the accuracy of any method. Spatial unwrapping has been subdivided into path-following unwrapping methods or Laplacian-based unwrapping (Schofield and Zhu 2003). Based on the information travelling through each subsequent voxel, the algorithm can be guided by the presence of change in phase greater than π (Abdul-Rahman *et al.*, 2007; Witoszynskyj *et al.*, 2009) and by avoiding barriers along the path such as singularity/fringe lines (Lu *et al.*, 2005). Temporal phase unwrapping requires data from multiple time points or TEs, utilises the time-dependent effects on phase accumulation such as local field perturbations or blood flow, and offers pixel-wise unwrapping (Feng *et al.*, 2013). The inter-echo difference plays an important role in the effectiveness of unwrapping, where shorter inter-echo differences may result in strong gradient switching rates and dB/dT effects and longer inter-echo differences are not desired due to the presence of phase wraps. This can be overcome by using multiple unequally spaced echoes (Dagher *et al.*, 2014; Robinson *et al.*, 2014). Phase unwrapping algorithms are further considered by Haacke *et al.* (2015) and Robinson *et al.* (2017).

7.6.2 Removal of background fields

The field variation can be written as a combination of the background field ($\Delta B_{bkg}(\vec{r})$) and the local field ($\Delta B_{loc}(\vec{r})$) as

$$\Delta B(\vec{r}) = \Delta B_{bkg}(\vec{r}) + \Delta B_{loc}(\vec{r}) \tag{7.15}$$

In order to remove the field variations due to the inhomogeneities in the main magnetic field and the global geometries, processing techniques such as homodyne high pass filtering (HPF) can be used. HPF image, $\rho'(\vec{r})$, is obtained by complex dividing the original image $\rho(\vec{r})$ by a complex image ($\rho_m(\vec{r})$) generated from truncating the original $n \times n$ pixels to $m \times m$ pixels from the original complex image and zero-filling the elements outside the central $m \times m$ elements to get the same $n \times n$ dimensions as the original image:

$$\rho'(\vec{r}) = \rho(\vec{r}) / \rho_m(\vec{r}) \tag{7.16}$$

HPF images have been successfully used in differentiating tissues with varying susceptibility for small objects such as veins (Haacke and Reichenbach 2014). However, apart from removing the background field effects, a large sized filter will in turn lead to significant signal loss for large structures (Haacke *et al.*, 2009b; Haacke and Reichenbach, 2014). Even though HPF is not the best background field removal method, given the fact that many SWI datasets have already been collected and HPF filter is still routinely used nowadays, it is important to understand the error caused by this filtering process. The level of signal loss is dependent on both the full width at half maximum (FWHM) of the HP filter and the size of the object of interest (Haacke and Reichenbach 2014). Better algorithms that lead to more realistic estimates of the phase inside large objects include (1) *sophisticated harmonic artefact reduction for phase data* (SHARP), which employs a deconvolution process but discards the edge pixels of the field-of-view (FOV) (Li and Leigh 2001; Schweser *et al.*, 2011), except in a recent paper that recovers edge pixels (see Figure 7.3c) (Topfer *et al.*, 2015); (2) a dipole field approach to remove the background fields using the geometry of the brain (Neelavalli *et al.*, 2009); (3) a similar approach using phase inside the brain to estimate the magnetisation sources and avoid the dependence of the final result on the accuracy of the extracted geometry (de Rochefort *et al.*, 2010; Liu *et al.*, 2011); and (4) a recent short TE approach to forward model the fields but also based on the geometry of the brain (see also Section 7.9.3) (Buch *et al.*, 2015).

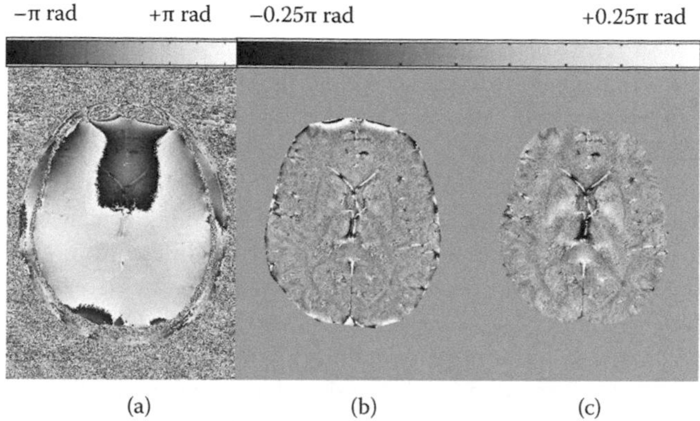

FIGURE 7.3 (a) Original unfiltered phase image with imaging parameters: TE = 17.3 ms, B0 = 3T with 0.5 × 0.5 × 0.5 mm³ resolution; (b) homodyne high pass filtering phase image with filter size of 64 × 64 pixels; and (c) phase images processed with the SHARP algorithm (radius = 6 pixels and th = 0.05). The filtered phase images in Figure 7.3b and c show the underlying tissue information much more clearly due to the reduced background phase.

7.6.3 Susceptibility mapping and the ill-posed inverse problem

The ability to quantify local magnetic susceptibility makes it possible to measure the amount of calcium or iron in the body, whether it is calcium in breast (Fatemi-Ardekani *et al.*, 2009) or iron in the form of non-haem iron (such as ferritin or hemosiderin) or haem iron (deoxy-haemoglobin) (Haacke *et al.*, 2009a; Ropele and Langkammer, 2017). Susceptibility maps are produced using the SWI phase data, where the inverse process utilises the dipoles surrounding a given object to reconstruct the source of phase behaviour, that is susceptibility distribution inside the object (Haacke *et al.*, 2010b).

The expression for reconstructing susceptibility distributions can be derived by rearranging the terms in Equation 7.11 and converting field to phase.

$$\chi(\vec{r}) = \frac{\mathrm{FT}^{-1}\left[v^{-1}(k) \cdot \phi(k)\right]}{\gamma \cdot B_0 \cdot \mathrm{TE}} \tag{7.17}$$

where $\chi(\vec{r})$ is the reconstructed susceptibility map, $\varphi(k)$ is the Fourier transform of phase information (filtered or unfiltered) and $G^{-1}(k)$ is inverse of the Green's function $G(k)$. Using the approximation given in Equation 7.12, the noise amplification in the cone of singularity and the resulting streaking artefacts in the QSM data can be limited by using δ in the range of 0.02–0.3 (Shmueli *et al.*, 2009; Haacke *et al.*, 2010b). However, this leads to a systematic underestimation in susceptibility values with increasing δ. Overall, δ = 0.1–0.2 were found to be optimal values in terms of accurately estimating susceptibility values and reducing streaking to get acceptable QSM data ($\chi(\vec{r})$).

However, the calculated QSM data still lead to underestimated susceptibility and remnant streaking artefacts, especially around structures with strong susceptibility property, such as veins or globus pallidus. These streaking artefacts can lead to errors in quantifying the susceptibilities (Tang *et al.*, 2013). Applying the geometry constrained iterative *susceptibility weighted imaging and mapping* (SWIM) process, a QSM method, where the k-space/image domain approach iteratively fills in the relevant information inside the singularity region of the inverse filter, reduces the external streaking artefacts found in the original susceptibility map, $\chi(\vec{r})$ (Tang *et al.*, 2013).

Another way to compute QSM data is to acquire multiple acquisitions at different orientations with respect to the direction of the main magnetic field. This method is known as *calculation of susceptibility through multiple orientation sampling* (COSMOS) (Liu *et al.*, 2009). Orientation of the singularity cone follows the object orientation in the image domain relative to the magnetic field. Although this approach will provide the missing or unreliable k-space elements defined under the singularity region of the inverse kernel, the clinical applicability is limited due to the prolonged data acquisition time.

7.7 Technical considerations

Modern MRI scanners are readily equipped with some version of a 2D/3D GRE sequence because it is a commonly used fast and robust imaging technique with low specific absorption rate (SAR). At 3T, a brain scan can take around 5–10 minutes, depending on the resolution and FOV. A key requirement for SWI and QSM methodology is to save the complex images (both magnitude and phase) rather than only the magnitude. The SWI sequence can be optimised with respect to the tissue and pathology of interest by modifying TE, FA and TR (see Table 7.1).

7.7.1 Selection of the echo time

According to Equation 7.13, the phase signal is proportional to time. Hence, in order to obtain a stronger phase response, a higher TE should be chosen ideally. Usually the focus of phase imaging is to study veins (Haacke *et al.*, 2010b; Tang *et al.*, 2013) and iron deposition (Liu *et al.*, 2011; Schweser *et al.*, 2011) by acquiring the data at long TE (~20 ms). However, this long TE approach leads to both macroscopic and microscopic (subpixel size) aliasing. For SWI, TE = 20 ms or higher is usually used at 3T, which provides ample contrast to visualise venous vasculature. For QSM, the local phase preservation becomes an important factor and the ability to completely unwrap the pixels inside and around the veins is hindered by high frequency local wraps. Using the analytical model of an infinite cylinder (Brown *et al.*, 2014), we can estimate what TE represents the onset of local phase wraps. For example, the phase for a vein parallel to the main magnetic field is close to π at TE = 26 ms for B_0 = 3T (assuming venous oxygen saturation level (Y_v) of 70%, haematocrit = 44% and difference between fully oxygenated and deoxygenated blood ($\Delta\chi_{do} \approx 3.39$ ppm in SI units)) and will create a phase wrap only inside the vein (Reichenbach *et al.*, 2000; Haacke and Reichenbach, 2014). For a vein perpendicular to the main field, the phase is as high as 3π/2 at the edge of this vein at TE = 26 ms and creates phase wraps over a few pixels even for a TE as short as 18 ms at 3T. In reality, the MRI signal is discretised based on the limits of selected image resolution and, in the presence of

TABLE 7.1 Parameters for brain imaging protocol for the best QSM results using a single orientation, single TE scan

Field strength	Parameters	Microbleeds	Veins	Basal ganglia	Motor cortex
1.5T	Resolution (mm³)	0.7 × 0.7 × 2	0.7 × 0.7 × 2	0.7 × 0.7 × 2	0.7 × 0.7 × 2
	TE (ms)	15	30	40	50
	TR (ms)	20	40	50	60
	FA (degrees)	20	20	20	20
	BW (Hz/pxl)	120	120	120	120
3T	Resolution (mm³)	0.7 × 0.7 × 2	0.7 × 0.7 × 2	0.7 × 0.7 × 2	0.7 × 0.7 × 2
	TE (ms)	7.5	15	20	25
	TR (ms)	20	20	25	35
	FA (degrees)	15	15	15	15
	BW (Hz/pxl)	200	200	200	200
7T	Resolution (mm³)	0.3 × 0.3 × 1	0.3 × 0.3 × 1	0.3 × 0.3 × 1	0.3 × 0.3 × 1
	TE (ms)	3	6.5	8.5	12
	TR (ms)	15	15	15	20
	FA (degrees)	10	10	10	10
	BW (Hz/pxl)	240	240	240	240

Note: Each structure has a different susceptibility property and based on its amplitude, the TE is modified for optimal results. For further details refer to (From Haacke, E.M., *et al.*, Magn. Reson. Imaging, 33, 1–25, 2015.). QSM = quantitative susceptibility mapping; TE = echo time; TR = repetition time; FA = flip angle; BW = bandwidth.

partial volume effects, the phase will get integrated across a voxel leading to T_2* related signal loss. Consequently, the smaller veins will appear thicker (up to the voxel size) in the magnitude image than it really is, and the phase will be an inaccurate estimate of the real phase (Cheng *et al.*, 2007; Schweser *et al.*, 2010). This increase in apparent size leads to a concomitant underestimate of susceptibility. However, the product of susceptibility and the volume of the object (magnetic moment) will remain unaffected (Cheng *et al.*, 2009a, 2015). Hence, for the structures with strong susceptibility, such as the air–bone-tissue interfaces, lower TEs are more suitable (Buch *et al.*, 2015). Detection and quantification analysis for microbleeds is covered by Buch *et al.*, (2017a).

7.7.2 Uncertainty in R_2* and phase

Recent estimates of the error in estimating uncertainty in R_2' from phase images (Haacke *et al.*, 2015) suggests that it is roughly 1/8 of that from magnitude images (i.e. $\sigma_{\Delta R_2',\phi} : \sigma_{\Delta R_2',mag} = \frac{3\lambda}{2\pi} : 1$, where $\lambda \approx 0.26$ is the volume fraction of venous blood). However, these estimates will depend on the extent of local field effects, which in turn are dependent on but not limited to the choice of TE. Rapid phase integration across a voxel due to closely distributed sources at higher TEs will lead to signal loss, but may not manifest as bulk magnetic field effects. For very small levels of iron, such as iron tagged stem cells for example, R_2* maps will be the preferred choice.

One of the advantages of QSM over R_2* mapping is that susceptibility values from QSM are linearly proportional to the concentration of substances that induce a susceptibility effect, such as an iron nanoparticle-based contrast agent. In addition, the diamagnetic effects cause similar signal loss to iron, making it impossible to distinguish the source of increases in R_2*, while QSM provides the discrimination needed through the sign of the susceptibility. Since calcium is diamagnetic it will have a negative value relative to surrounding tissue, making it clear that the source is calcium, not iron (Langkammer *et al.*, 2012). Due to their reproducibility, a combined QSM and R_2* system may be useful to describe and validate iron as a biomarker in neurological diseases (Guan *et al.*, 2017; Santin *et al.*, 2017).

7.8 Applications of QSM

7.8.1 Quantifying oxygen extraction fraction and cerebral activity

Being able to image the haemodynamics of the brain is key to diagnosing and understanding cerebrovascular diseases such as stroke, subarachnoid haemorrhage, vascular dementia and mild TBI. Two key components of brain function are oxygen extraction fraction (OEF) and cerebral metabolic rate of oxygen ($CMRO_2$). One of the missing pieces of this puzzle in MRI today is the ability to measure local oxygen saturation levels.

We can use the local magnetic susceptibility from veins as a means to monitor haem iron (de-oxyhaemoglobin) and therefore predict oxygen saturation (Haacke *et al.*, 2010b). The volume susceptibility of venous blood ($\Delta\chi_v$) can be written as

$$\Delta\chi_v = \Delta\chi_{do} \cdot \text{Hct} \cdot (1 - Y_v) \tag{7.18}$$

where Hct represents haematocrit levels in the blood.

This method provides a quantitative approach to measuring venous oxygen saturation using QSM data. As an example of the potential of QSM data, the change in Y_v before and after the intake of vasodynamic agents with high spatial resolution at 3T can be measured (Figure 7.4) (Buch *et al.*, 2017a). Changes in cerebrovascular reserve can be measured using QSM data, and a recent study shows that the change in Y_v for post-caffeine and post-acetazolamide was measured to be $\Delta Y_{Caffeine} \approx -9\%$ and $\Delta Y_{Acet} \approx 10\%$ inside the major veins (Buch *et al.*, 2017b). In addition, the change in OEF for caffeine and acetazolamide corresponds to the relative changes in cerebral blood flow. Specifically, the relative changes in cerebral blood flow measured using arterial spin labelling in the presence of caffeine and acetazolamide were found to be −30.3% and +31.5%, suggesting that the $CMRO_2$ remains stable between normal and challenged brain states for healthy subjects.

In order to calculate absolute venous oxygen saturation levels, several factors such as haematocrit, arterial oxygen saturation level and $\Delta\chi_{do}$ term are assumed. Normal haematocrit values can vary between 40.7%–50.3% in males and 36.1%–44.1% in females. Estimating the change in OEF values avoids the need to know haematocrit.

7.8.2 Quantifying non-haem iron or ferritin

Iron has a significantly higher abundance within tissues, whereas other metal ions such as copper, manganese and cobalt are present in trace amounts, which makes iron the dominant source of paramagnetic materials in the brain (Beard *et al.*, 1993). Iron is stored as ferritin and hemosiderin, and studies have shown that the amount of this stored iron in deep grey matter (GM) nuclei and cortical areas increases with age (Hallgren and Sourander 1958; Hebbrecht *et al.*, 1999). The iron content rapidly increases from birth till 20 years of age and thereafter the level of iron increases more slowly (Li *et al.*, 2014; Persson *et al.*, 2015). Quantification of iron levels in the brain tissues with QSM has been recently validated by post-mortem studies using alternative iron determination methods, such as inductively coupled plasma mass spectrometry, X-ray fluorescence and Perl's ferric iron staining (Langkammer *et al.*, 2012; Zheng *et al.*, 2013; Sun *et al.*, 2015).

In order to investigate the correlation of iron deposition as a function of age, regional mean susceptibility values, in addition to the global mean, were measured in deep GM nuclei and substantia nigra (Haacke *et al.*, 2010a; Liu *et al.*, 2016). This involved a low-iron (Region I) and high-iron (Region II) content compartmental model, where the significant changes in iron content were shown as a function of age. Liu *et al.*, (2016) extracted region II

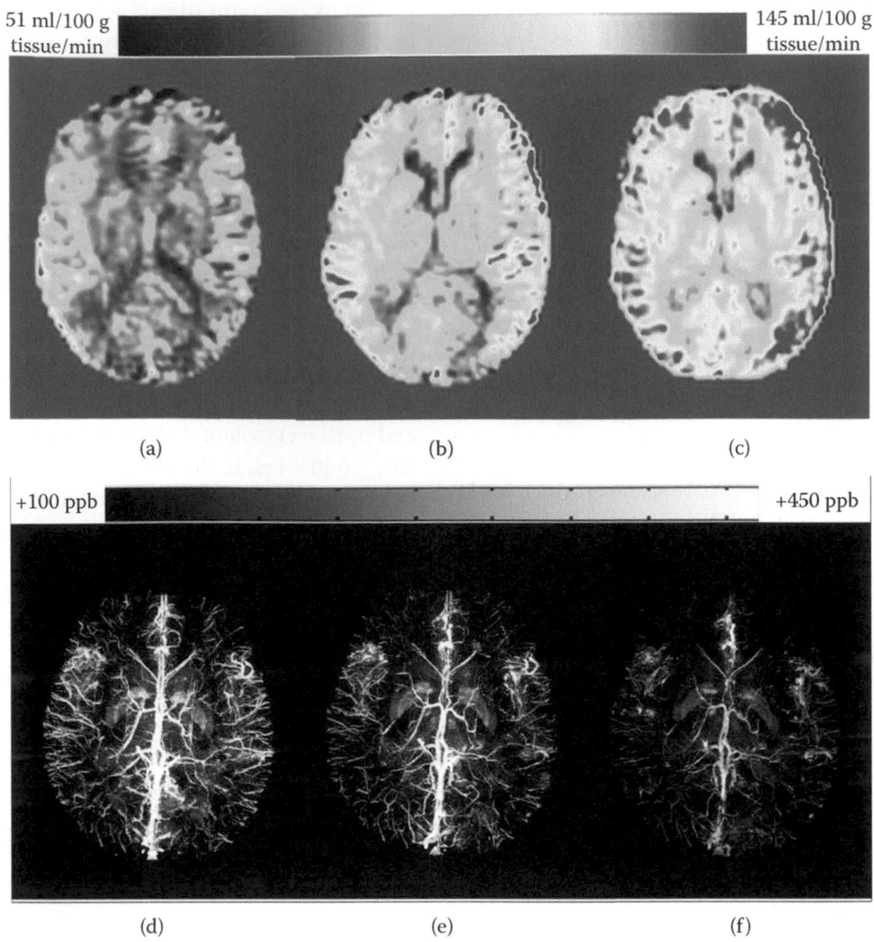

FIGURE 7.4 Cerebral blood flow (CBF) maps: (a) after caffeine intake, (b) normal state and (c) after the intake of acetazolamide. The units of CBF maps are in ml/100 g of tissue/min. MIPs, generated over 32 slices or 16 mm in the slice-select direction, of susceptibility maps (d–f) represent caffeine challenge, normal state and acetazolamide challenged brain states, respectively.

from QSM data by using a threshold of upper 95% interval values of the entire region linear regression analysis. Figure 7.5 demonstrates the change in mean susceptibility inside various structures in the right hemisphere as a function of age using Region II–only analysis, respectively. In this study, 174 normals were scanned by the SWI sequence with the following imaging parameters: TE = 40 ms, TR = 53 ms, FA = 20°, BW = 112 Hz/pixel and voxel resolution = $0.6 \times 0.75 \times 3$ mm^3 at 1.5T. For the high-iron regional analysis, all structures showed a linear increase in mean susceptibility with an increase in age except for the thalamus. Although both global and regional results showed iron increases for putamen, caudate nucleus and red nucleus as a function of age, the slope was significantly higher in the regional analysis (Liu *et al.*, 2016).

7.8.3 QSM of global field sources

In general, the research focus of QSM is on structures within the brain such as the deep GM nuclei, midbrain and veins (Shmueli *et al.*, 2009; Wharton and Bowtell, 2010). In comparison, the sinuses and skull bones have relatively strong susceptibilities (Neelavalli *et al.*, 2009). This creates major non-local phase behaviour, confounding the relevant phase information of the local tissues in the brain (Haacke *et al.*, 2010a). Usually, the phase processing scheme involves reducing artefacts associated with the high susceptibility effects of these air–tissue areas as much as possible using sophisticated methods (Neelavalli *et al.*, 2009; Liu *et al.*, 2011; Schweser *et al.*, 2011).

Susceptibility mapping of tissues with short T_2*, such as bone or teeth, is not feasible with the short TEs because of the severe aliasing at the edge of the brain. Moreover, for regions that have no protons or very short T_2* there will be no discernible signal. To overcome this limitation, a short-TE iterative phase replacement approach is used that combines the inverse computation with the forward modelling approach to image the sinuses, teeth and bones by preserving the phase information in extra-cerebral tissues (Buch *et al.*, 2015), as shown in Figure 7.6.

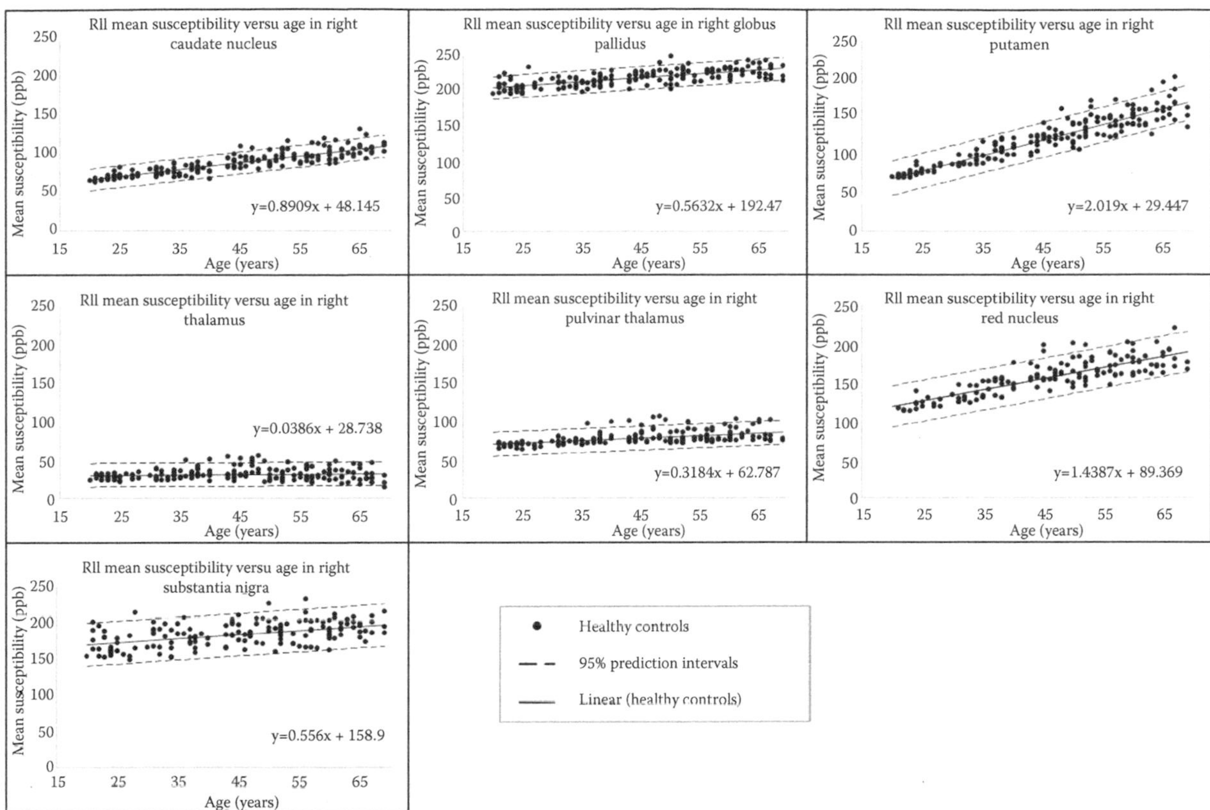

FIGURE 7.5 Correlation of magnetic susceptibility as a function of age from the Region II analysis of deep grey matter nuclei structures. In this study, the quoted values are measured directly from the susceptibility maps to avoid the errors caused by reference selection (Liu, M., *et al.*: Assessing global and regional iron content in deep gray matter as a function of age using susceptibility mapping. *J. Magn. Reson. Imaging.* 2016. 44. 59–71. Copyright Wiley-VCH Verlag GmbH & Co. KGaA. Reproduced with permission.)

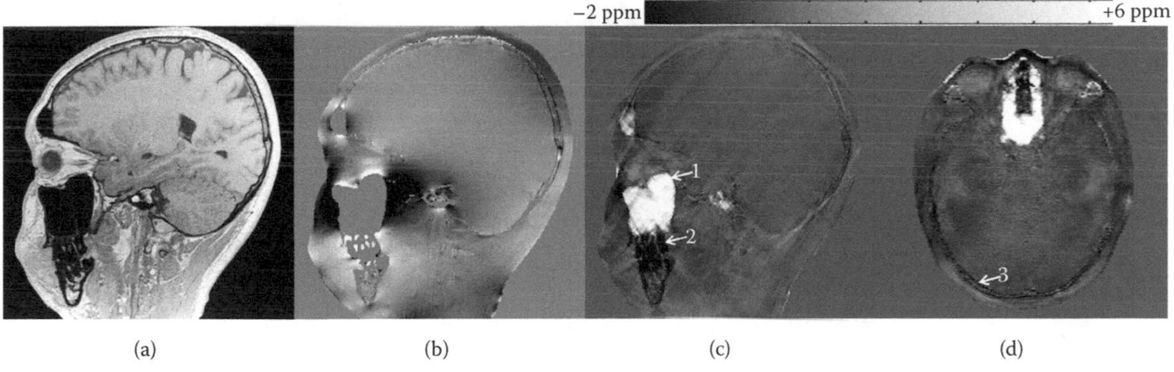

FIGURE 7.6 (a) Original magnitude, (b) phase images at TE = 2.5 ms and B0 = 3T and (c, d) susceptibility maps in sagittal and transverse views, showing spatial differentiation between the sinuses and teeth in terms of their strongly paramagnetic and diamagnetic properties, respectively. Structures are identified by the white arrows: 1 = maxillary sinus, 2 = teeth and 3 = occipital skull bone.

7.8.4 QSM in other parts of the body

Although most of the applications of QSM have been directed towards imaging of the head, magnetic susceptibility mapping outside the brain is an active field of research. QSM methodology is just as applicable to other tissue types; however, the methodology is not easily translated due to several issues that include the presence of fat around relevant structures, rapid signal decay caused by air–tissue interfaces, cardiac and respiratory motions. Recent studies have suggested that QSM data can serve as a robust and direct measure of biomarkers such as measuring iron content in organs such as liver (Sharma *et al.*, 2015), depicting calcifications in prostate cancer patients (Straub *et al.*, 2017) and characterisation of kidneys (Xie *et al.*,

2013), breast tissue (Wang and Liu 2015) and cartilage (Nissi *et al.*, 2015).

7.9 Strategically acquired gradient echo

As shown in many studies, magnetic resonance angiography (MRA) and venography (MRV) as well as T_1, proton density, T_2 and T_2* type contrasts are critical for imaging brain diseases such as stroke, TBI, cerebral microbleeds and many other neurological diseases (Brown *et al.*, 2014; Haacke *et al.*, 2015). Here we introduce a rapid approach named *strategically acquired gradient echo imaging*, or STAGE, that provides the following information with three scans in a few minutes at 3T: T_1 weighted data (T1W), proton density weighted data, T_1 mapping (T_1 MAP), proton density mapping (PD MAP), R_2* mapping (R_2* MAP), SWI, true SWI (tSWI), QSM, magnetic resonance arterio-venous map MRAV and MRA (Figure 7.7).

With STAGE, seven echoes are acquired with different TEs and FAs using three SWI scans. There are two fully flow compensated double echo SWI scans with different FAs of 6° and 24°, which are optimised both for a T1W image with better GM and white matter (WM) contrast and for T1 MAP and PD MAP, and one interleaved scan for the MRAV and MRA with a 12° FA. The short echoes from the two SWI scans can be used for T_1 and PD mapping using the variable FA method (Venkatesan *et al.*, 1998). Meanwhile, all four echoes from the two SWI scans can be used to generate a T1W image with enhanced GM/WM contrast and homogeneity along all directions by using RF field correction derived from T_1 mapping. SWI images can be obtained from all echoes but, in practice, the long echo of 6° is used to derive SWI, tSWI and QSM shown here. Using all echoes would improve the SNR of the QSM images, but the longer echoes may also suffer from reduced susceptibility for the reasons discussed above (Feng *et al.*, 2013).

The interleaved scan includes two flow compensated TEs of 2.5 ms and 12.5 ms and one flow dephased echo of 12.5 ms. The flow rephased–dephased TEs have the same imaging conditions but the gradient preparations for the flowing spins are different. Thus, the subtraction gives us an MRAV image highlighting only the flowing spins (blood vessels) with zero signal from stationary tissues (Ye *et al.*, 2013). We can also use QSM results from the flow rephased TE of 12.5 ms to suppress veins from the

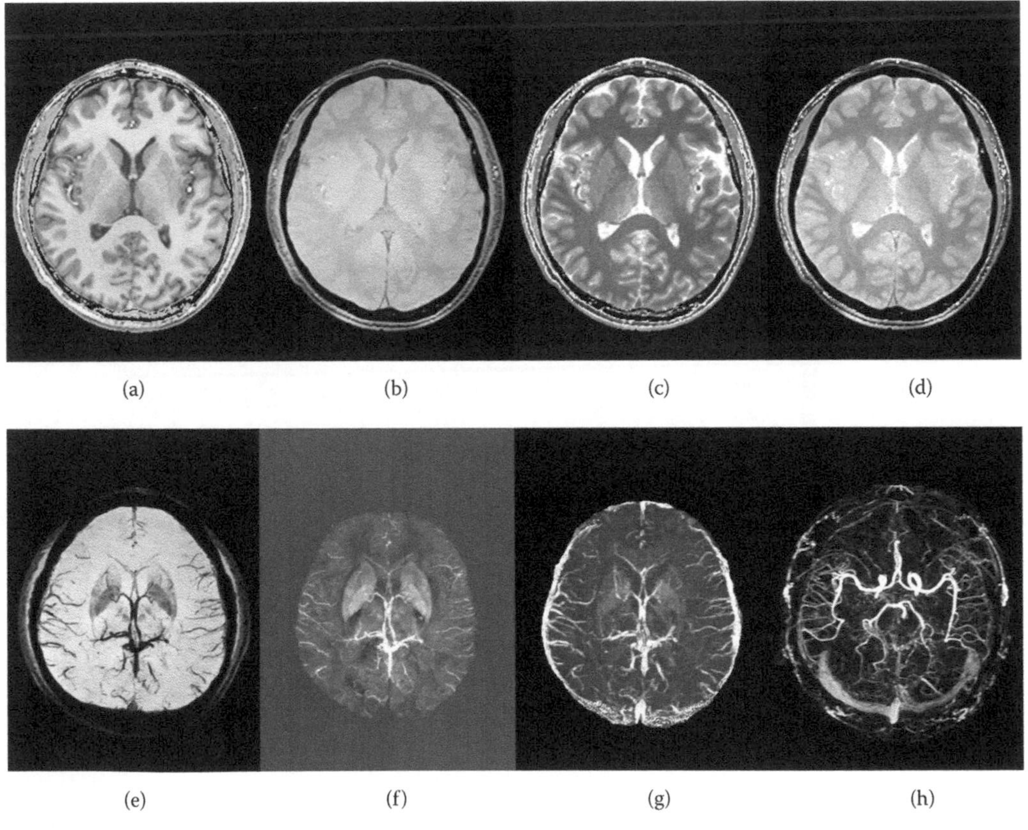

FIGURE 7.7 An example of STAGE from a healthy volunteer at 3T. (a) STAGE T1W, (b) proton density weighted data, (c) T1 mapping (T1 MAP), (d) proton density mapping (PD MAP), (e) minimum intensity projection (mIP) of susceptibility weighted imaging (SWI), (f) MIP of quantitative susceptibility mapping, (g) MIP of R2* MAP, (h) MIP of magnetic resonance angiography (MRA) and venography (MRV). Images in Figure 7.7e through g have an effective slice thickness of 16 mm. Figure 7.7h has an effective slice thickness of 64 mm. The two double echo SWI scans were acquired in 5 minutes and the MRAV in 4 minutes at 3T with a resolution of $0.67 \times 1.33 \times 2$ mm^3 and 64 slices for the whole brain.

MRAV image to get an MRA image. For the R_2* MAP, we can fit two TEs in each scan and average the three R_2* MAPs to increase the SNR of the final R_2* MAP. Therefore, STAGE provides not only the conventional images but also quantitative information for studying the brain.

Conclusions

GRE imaging in the form of SWI, QSM, MRAV and R_2* offers a host of contrasts to help the clinician quantify tissue properties and provide better diagnostic information in the process. Currently, offline processing of QSM can be done for research, and down the line it is anticipated that having QSM networked with the scanner will open the door to more clinical applications of susceptibility mapping.

References

Abdul-Rahman HS, Gdeisat MA, Burton DR, Lalor MJ, Lilley F, Moore CJ. Fast and robust three-dimensional best path phase unwrapping algorithm. Appl Opt 2007; 46: 6623–35.

Akter M, Hirai T, Hiai Y, Kitajima M, Komi M, Murakami R et al. Detection of Hemorrhagic Hypointense Foci in the brain on susceptibility-weighted imaging: clinical and phantom studies. Acad Radiol 2007; 14: 1011–9.

Beard JL, Connor JR, Jones BC. Iron in the brain. Nutr Rev 1993; 51: 157–70.

Bernstein MA, King KF, Zhou XJ. *Handbook of MRI Pulse Sequences*. Burlington, MA: Elsevier Academic Press; 2004.

Brown RW, Cheng Y-CN, Haacke EM, Thompson MR, Venkatesan R. *Magnetic Resonance Imaging: physical Principles and Sequence Design*. 2nd ed. New York: Wiley; 2014.

Buch S, Cheng Y-CN, Hu J, Liu S, Beaver J, Rajagovindan R, et al. Determination of detection sensitivity for cerebral microbleeds using susceptibility-weighted imaging. NMR Biomed 2017; 30(4). doi: 10.1002/nbm.3551.

Buch S, Liu S, Ye Y, Cheng Y-CN, Neelavalli J, Haacke EM. Susceptibility mapping of air, bone, and calcium in the head. Magn Reson Med 2015; 73: 2185–94.

Buch S, Ye Y, Haacke EM. Quantifying the changes in oxygen extraction fraction and cerebral activity caused by caffeine and acetazolamide. J Cereb Blood Flow Metab 2017; 37(3):825–836.

Chen Q, Andersen AH, Zhang Z, Ovadia A, Gash DM, Avison MJ. Mapping drug-induced changes in cerebral R2* by Multiple Gradient Recalled Echo functional MRI. Magn Reson Imaging 1996; 14: 469–76.

Cheng A-L, Batool S, McCreary CR, Lauzon ML, Frayne R, Goyal M, et al. Susceptibility-weighted imaging is more reliable than T2*-weighted gradient-recalled echo MRI for detecting microbleeds. Stroke 2013; 44: 2782–6.

Cheng Y-CN, Hsieh C-Y, Neelavalli J, Haacke EM. Quantifying effective magnetic moments of narrow cylindrical objects in MRI. Phys Med Biol 2009a; 54: 7025–44.

Cheng Y-CN, Hsieh C-Y, Neelavalli J, Liu Q, Dawood MS, Haacke EM. A complex sum method of quantifying susceptibilities in cylindrical objects: the first step toward quantitative diagnosis of small objects in MRI. Magn Reson Imaging 2007; 25: 1171–80.

Cheng Y-CN, Hsieh C-Y, Tackett R, Kokeny P, Regmi RK, Lawes G. Magnetic moment quantifications of small spherical objects in MRI. Magn Reson Imaging 2015; 33: 829–39.

Cheng Y-CN, Neelavalli J, Haacke EM. Limitations of calculating field distributions and magnetic susceptibilities in MRI using a Fourier based method. Phys Med Biol 2009b; 54: 1169–89.

Cherubini A, Péran P, Caltagirone C, Sabatini U, Spalletta G. Aging of subcortical nuclei: microstructural, mineralization and atrophy modifications measured in vivo using MRI. NeuroImage 2009; 48: 29–36.

Dagher J, Reese T, Bilgin A. High-resolution, large dynamic range field map estimation. Magn Reson Med 2014; 71: 105–17.

Dahnke H, Schaeffter T. Limits of detection of SPIO at 3.0 T using T2* relaxometry. Magn Reson Med 2005; 53: 1202–6.

de Rochefort L, Brown R, Prince MR, Wang Y. Quantitative MR susceptibility mapping using piece-wise constant regularized inversion of the magnetic field. Magn Reson Med 2008; 60: 1003–9.

de Rochefort L, Liu T, Kressler B, Liu J, Spincemaille P, Lebon V, et al. Quantitative susceptibility map reconstruction from MR phase data using Bayesian regularization: validation and application to brain imaging. Magn. Reson Med 2010; 63: 194–206.

Eskreis-Winkler S, Zhang Y, Zhang J, Liu Z, Dimov A, Gupta A, et al. The clinical utility of QSM: disease diagnosis, medical management, and surgical planning. NMR Biomed 2017; 30(4). doi: 10.1002/nbm.3668.

Fatemi-Ardekani A, Boylan C, Noseworthy MD. Identification of breast calcification using magnetic resonance imaging. Med Phys 2009; 36: 5429–36.

Feng W, Neelavalli J, Haacke EM. Catalytic multiecho phase unwrapping scheme (CAMPUS) in multiecho gradient echo imaging: removing phase wraps on a voxel-by-voxel basis. Magn Reson Med 2013; 70: 117–26.

Gelman N, Gorell JM, Barker PB, Savage RM, Spickler EM, Windham JP, et al. MR imaging of human brain at 3.0 T: preliminary report on transverse relaxation rates and relation to estimated iron content. Radiology 1999; 210: 759–67.

Ghadery C, Pirpamer L, Hofer E, Langkammer C, Petrovic K, Loitfelder M, et al. R2* mapping for brain iron: associations with cognition in normal aging. Neurobiol Aging 2015; 36: 925–32.

Goos JDC, Flier WM van der, Knol DL, Pouwels PJ, Scheltens P, Barkhof F, et al. Clinical relevance of improved microbleed detection by susceptibility-weighted magnetic resonance imaging. Stroke 2011; 42: 1894–900.

Guan X, Xuan M, Gu Q, Huang P, Liu C, Wang N, et al. Regionally progressive accumulation of iron in Parkinson's disease as measured by quantitative susceptibility mapping. NMR Biomed 2017; 30(4). doi: 10.1002/nbm.3489.

Haacke EM, Liu S, Buch S, Zheng W, Wu D, Ye Y. Quantitative susceptibility mapping: current status and future directions. Magn Reson Imaging 2015; 33: 1–25.

Haacke EM, Makki M, Ge Y, Maheshwari M, Sehgal V, Hu J, et al. Characterizing iron deposition in multiple sclerosis lesions using susceptibility weighted imaging. J Magn Reson Imaging 2009a; 29: 537–44.

Haacke EM, Miao Y, Liu M, Habib CA, Katkuri Y, Liu T, et al. Correlation of putative iron content as represented by changes in R2* and phase with age in deep gray matter of healthy adults. J Magn Reson Imaging 2010a; 32: 561–76.

Haacke EM, Mittal S, Wu Z, Neelavalli J, Cheng Y-CN. Susceptibility-weighted imaging: technical aspects and clinical applications, part 1. Am J Neuroradiol 2009b; 30: 19–30.

Haacke EM, Reichenbach JR. *Susceptibility Weighted Imaging in MRI: Basic Concepts and Clinical Applications.* Hoboken, NJ: Wiley; 2014.

Haacke EM, Tang J, Neelavalli J, Cheng YCN. Susceptibility mapping as a means to visualize veins and quantify oxygen saturation. J Magn Reson Imaging 2010b; 32: 663–76.

Haacke EM, Xu Y, Cheng Y-CN, Reichenbach JR. Susceptibility weighted imaging (SWI). Magn Reson Med 2004; 52: 612–8.

Hallgren B, Sourander P. The effect of age on the non-haemin iron in the human brain. J Neurochem 1958; 3: 41–51.

Hankins JS, McCarville MB, Loeffler RB, Smeltzer MP, Onciu M, Hoffer FA, et al. R2* magnetic resonance imaging of the liver in patients with iron overload. Blood 2009; 113: 4853–55.

Hebbrecht G, Maenhaut W, Reuck JD. Brain trace elements and aging. Nucl Instrum Meth Phys Res Sect B Beam Interact Mater Atmos 1999; 150: 208–13.

Langkammer C, Krebs N, Goessler W, Scheurer E, Ebner F, Yen K, et al. Quantitative MR imaging of brain iron: a postmortem validation study. Radiology 2010; 257: 455–62.

Langkammer C, Schweser F, Krebs N, Deistung A, Goessler W, Scheurer E, et al. Quantitative susceptibility mapping (QSM) as a means to measure brain iron? A post mortem validation study. NeuroImage 2012; 62: 1593–99.

Lehéricy S, Sharman MA, Dos Santos CL, Paquin R, Gallea C. Magnetic resonance imaging of the substantia nigra in Parkinson's disease. Mov Disord 2012; 27: 822–30.

Li L, Leigh JS. High-precision mapping of the magnetic field utilizing the harmonic function mean value property. J Magn Reson 2001; 148: 442–8.

Li W, Wu B, Batrachenko A, Bancroft-Wu V, Morey RA, Shashi V, et al. Differential developmental trajectories of magnetic susceptibility in human brain gray and white matter over the lifespan. Hum Brain Mapp 2014; 35: 2698–713.

Liu M, Liu S, Ghassaban K, Zheng W, Dicicco D, Miao Y, et al. Assessing global and regional iron content in deep gray matter as a function of age using susceptibility mapping. J Magn Reson Imaging 2016; 44: 59–71.

Liu T, Khalidov I, de Rochefort L, Spincemaille P, Liu J, Tsiouris AJ, et al. A novel background field removal method for MRI using projection onto dipole fields (PDF). NMR Biomed 2011; 24: 1129–36.

Liu T, Spincemaille P, de Rochefort L, Kressler B, Wang Y. Calculation of susceptibility through multiple orientation sampling (COSMOS): a method for conditioning the inverse problem from measured magnetic field map to susceptibility source image in MRI. Magn Reson Med 2009; 61: 196–204.

Lu Y, Wang X, He G. Phase unwrapping based on branch cut placing and reliability ordering. Opt Eng 2005; 44: 055601-055601-9.

McCarville MB, Hillenbrand CM, Loeffler RB, Smeltzer MP, Song R, Li CS, et al. Comparison of whole liver and small region-of-interest measurements of MRI liver R2* in children with iron overload. Pediatr Radiol 2010; 40: 1360–7.

Neelavalli J, Cheng Y-CN, Jiang J, Haacke EM. Removing background phase variations in susceptibility-weighted imaging using a fast, forward-field calculation. J Magn Reson Imaging 2009; 29: 937–48.

Nissi MJ, Tóth F, Wang L, Carlson CS, Ellermann JM. Improved visualization of cartilage canals using quantitative susceptibility mapping. PLoS One 2015; 10: e0132167.

Ogawa S, Menon RS, Tank DW, Kim SG, Merkle H, Ellermann JM, et al. Functional brain mapping by blood oxygenation level-dependent contrast magnetic resonance imaging. A comparison of signal characteristics with a biophysical model. Biophys J 1993; 64: 803–12.

Park J-H, Park S-W, Kang S-H, Nam T-K, Min B-K, Hwang S-N. Detection of traumatic cerebral microbleeds by susceptibility-weighted image of MRI. J Kor Neurosurg Soc 2009; 46: 365–9.

Péran P, Cherubini A, Luccichenti G, Hagberg G, Démonet JF, Rascol O, et al. Volume and iron content in basal ganglia and thalamus. Hum Brain Mapp 2009; 30: 2667–75.

Persson N, Wu J, Zhang Q, Liu T, Shen J, Bao R, et al. Age and sex related differences in subcortical brain iron concentrations among healthy adults. NeuroImage 2015; 122: 385–98.

Reichenbach JR, Barth M, Haacke EM, Klarhöfer M, Kaiser WA, Moser E. High-resolution MR venography at 3.0 Tesla. J Comput Assist Tomogr 2000; 24: 949–57.

Robinson RJ, Bhuta S. Susceptibility-weighted imaging of the brain: current utility and potential applications. J. Neuroimaging. 2011; 21: e189–204.

Robinson S, Schödl H, Trattnig S. A method for unwrapping highly wrapped multi-echo phase images at very high field: UMPIRE. Magn Reson Med 2014; 72: 80–92.

Robinson SD, Bredies K, Khabipova D, Dymerska B, Marques JP, Schweser F. An illustrated comparison of processing methods for MR phase imaging and QSM: combining array coil signals and phase unwrapping. NMR Biomed 2017; 30(4). doi: 10.1002/nbm.3601.

Ropele S, Langkammer C. Iron quantification with susceptibility. NMR Biomed 2017; 30(4). doi: 10.1002/nbm.3534.

Salomir R, de Senneville BD, Moonen CT. A fast calculation method for magnetic field inhomogeneity due to an arbitrary distribution of bulk susceptibility. Concepts Magn Reson Part B Magn Reson Eng 2003; 19B: 26–34.

Santin MD, Didier M, Valabrègue R, Yahia Cherif L, García-Lorenzo D, Loureiro de Sousa P, et al. Reproducibility of R2* and quantitative susceptibility mapping (QSM) reconstruction methods in the basal ganglia of healthy subjects. NMR Biomed 2017; 30(4). doi: 10.1002/nbm.3491.

Schofield MA, Zhu Y. Fast phase unwrapping algorithm for interferometric applications. Opt Lett 2003; 28: 1194–6.

Schweser F, Deistung A, Lehr BW, Reichenbach JR. Differentiation between diamagnetic and paramagnetic cerebral lesions based on magnetic susceptibility mapping. Med Phys 2010; 37: 5165–78.

Schweser F, Deistung A, Lehr BW, Reichenbach JR. Quantitative imaging of intrinsic magnetic tissue properties using MRI signal phase: an approach to in vivo brain iron metabolism? NeuroImage 2011; 54: 2789–807.

Sharma SD, Hernando D, Horng DE, Reeder SB. Quantitative susceptibility mapping in the abdomen as an imaging biomarker of hepatic iron overload. Magn Reson Med 2015; 74: 673–83.

Shmueli K, de Zwart JA, van Gelderen P, Li T-Q, Dodd SJ, Duyn JH. Magnetic susceptibility mapping of brain tissue in vivo using MRI phase data. Magn Reson Med 2009; 62: 1510–22.

Straub S, Laun FB, Emmerich J, Jobke B, Hauswald H, Katayama S, et al. Potential of quantitative susceptibility mapping for detection of prostatic calcifications. J Magn Reson Imaging 2017; 45: 889–98.

Sun H, Walsh AJ, Lebel RM, Blevins G, Catz I, Lu JQ, et al. Validation of quantitative susceptibility mapping with Perls' iron staining for subcortical gray matter. NeuroImage 2015; 105: 486–92.

Tang J, Liu S, Neelavalli J, Cheng YCN, Buch S, Haacke EM. Improving susceptibility mapping using a threshold-based K-space/image domain iterative reconstruction approach. Magn Reson Med 2013; 69: 1396–1407.

Topfer R, Schweser F, Deistung A, Reichenbach JR, Wilman AH. SHARP edges: recovering cortical phase contrast through harmonic extension. Magn Reson Med 2015; 73: 851–6.

Venkatesan R, Lin W, Haacke EM. Accurate determination of spin-density and T1 in the presence of RF-field inhomogeneities and flip-angle miscalibration. Magn Reson Med 1998; 40: 592–602.

Wang Y, Liu T. Quantitative susceptibility mapping (QSM): decoding MRI data for a tissue magnetic biomarker. Magn Reson Med 2015; 73: 82–101.

Wharton S, Bowtell R. Whole-brain susceptibility mapping at high field: a comparison of multiple- and single-orientation methods. NeuroImage 2010; 53: 515–25.

Witoszynskyj S, Rauscher A, Reichenbach JR, Barth M. Phase unwrapping of MR images using Phi UN—a fast and robust region growing algorithm. Med Image Anal 2009; 13: 257–68.

Wu D, Liu S, Buch S, Ye Y, Dai Y, Haacke EM. A fully flow-compensated multiecho susceptibility-weighted imaging sequence: the effects of acceleration and background field on flow compensation. Magn Reson Med 2016; 76: 478–89.

Wu Z, Li S, Lei J, An D, Haacke EM. Evaluation of traumatic subarachnoid hemorrhage using susceptibility-weighted imaging. Am J Neuroradiol 2010; 31: 1302–10.

Xie L, Sparks MA, Li W, Qi Y, Liu C, Coffman TM, et al. Quantitative susceptibility mapping of kidney inflammation and fibrosis in type 1 angiotensin receptor-deficient mice. NMR Biomed 2013; 26: 1853–63.

Yablonskiy DA, Haacke EM. Theory of NMR signal behavior in magnetically inhomogeneous tissues: the static dephasing regime. Magn Reson Med 1994; 32: 749–63.

Yablonskiy DA, Sukstanskii AL, Luo J, Wang X. Voxel spread function method for correction of magnetic field inhomogeneity effects in quantitative gradient echo-based MRI. Magn Reson Med 2013; 70: 1283–92.

Ye Y, Hu J, Wu D, Haacke EM. Noncontrast-enhanced magnetic resonance angiography and venography imaging with enhanced angiography. J. Magn Reson Imaging 2013; 38: 1539–48.

Zheng W, Nichol H, Liu S, Cheng Y-CN, Haacke EM. Measuring iron in the brain using quantitative susceptibility mapping and X-ray fluorescence imaging. NeuroImage 2013; 78: 68–74.

8

D: The Diffusion of Water (DTI)[1]

Francesco Grussu
University College London

Claudia A.M. Gandini Wheeler-Kingshott
University College London and University of Pavia

Contents

8.1 Introduction

Every fluid has a characteristic intrinsic self-diffusion constant, *D*, which reflects the mobility of the molecules in their micro-environment (Crank, 1998). Proton nuclear magnetic resonance (NMR, and therefore magnetic resonance imaging, or MRI) can be made sensitive to dynamic displacements of water molecules between 10^{-9} and 10^{-4} m in a timescale of a few milliseconds to a few seconds. Since these displacements are of the same order of magnitude as cellular dimensions within biological tissues, NMR/MRI diffusion measurements can provide a unique insight into tissue structure and organisation, potentially giving information about the size, orientation and tortuosity of both the intra- and extracellular spaces. Moreover, water diffusion *in vivo* is affected by the microdynamics of cellular transport between different compartments of the heterogeneous tissue structure as well as by the presence of non-permeable membranes. Therefore, measurements of *D* can highlight different properties of the system under investigation. The structure of the brain can be highly affected by pathology, even at the early stage, and measurements of the water diffusion coefficient *in vivo* have proven very useful in the study of both normal and pathological brain.

Although introduced by Stejskal and Tanner in 1965 (Stejskal and Tanner, 1965), it was not until the 1990s that *in vivo* water diffusion measurements entered into clinical practice (Wesbey *et al.*, 1984; Le Bihan *et al.*, 1986; Warach *et al.*, 1992). In the last two decades, the hardware developments crucial for MRI diffusion measurements, particularly concerning the strength and performance of field gradients, has led to a larger scale implementation of these techniques. From a clinical perspective, it was in neuroimaging that images made sensitive to molecular water diffusion initially attracted the most attention, because of their excellent capability of showing ischaemic stroke with a high sensitivity and specificity (Moseley *et al.*, 1990a; Warach *et al.*, 1992). Diffusion weighted (DW) imaging is now emerging as a diagnostic and research tool for many other intracranial disease processes, including demyelinating disorders (Filippi *et al.*, 2001), neoplasm (Dzik-Jurasz *et al.*, 2002), intracranial infections (Tsuchiya *et al.*, 2003) and neurodegenerative conditions (Hanyu *et al.*, 1999). While in theory quantification of diffusion is relatively straightforward, the complexity of brain tissue structure means that the results obtained remain pulse-sequence and acquisition-parameter–dependent, making

[1] Edited by Mara Cercignani; reviewed by Christian Beaulieu, Peter S. Allen MR Research Centre, Department of Biomedical Engineering, University of Alberta, Canada.

BOX 8.1 Key points regarding the Apparent Diffusion Coefficient (ADC) and Diffusion Tensor Imaging (DTI)

- The diffusion behaviour of water in the brain is characterised by its apparent diffusion coefficient (ADC).
- Measuring the ADC of brain water gives insight into the confinements (biological barriers) within which water diffuses.
- ADC maps are quantitative and more specific than conventional diffusion weighted images.
- ADC depends on the position of the subject with respect to the diffusion directions.
- Diffusion tensor imaging (DTI) provides rotationally invariant indices that describes the properties of the diffusion profile (amount of diffusion; anisotropy of diffusion).
- DTI metrics are sensitive imaging markers but are characterised by limited specificity.

careful planning (and quality assurance) essential for serial and multicentre studies (Box 8.1).

For further reading on measuring molecular diffusion with MRI, its theory and applications, see Le Bihan (2003), Jones (2010), Johansen-Berg and Behrens (2013); Jones *et al.*, (2013), Bastiani and Roebroeck (2015) and Le Bihan and Iima (2015).

8.2 Physical principles of the phenomenon of diffusion

8.2.1 Diffusion and self-diffusion

The term *diffusion* refers to the general transport of matter whereby molecules or ions mix through normal thermal agitation in a random way. In particular, diffusion is important in the NMR of samples characterised by molecular spins that move in space because of translational Brownian motion,[2] including biological tissues. The following treatment is based on the work of Crank (Crank, 1998), where a more detailed description of the diffusion process can be found.

Each molecule within the sample behaves independently from the others. The collision between molecules provokes a random displacement of each one, without a preferred direction, tracing the path known as the *random walk*. Given a time interval, it is possible to calculate a statistical measure of diffusion distance, averaged over an equilibrium ensemble of molecules, the so-called root mean square displacement, but it is impossible to say in what direction or how far a single given molecule has moved during that time.

Even though the diffusion motion is a random process, there is an underlying driving mechanism. When describing the mixing of two different liquids or gases, diffusion is usually described in terms of the concentration gradient of the diffusing substance. In biological tissues, however, concentration is not the driving force, and the process of interest is instead the motion of water within water driven by thermal agitation and referred to as *self*-diffusion. The following two sections describe molecular diffusion in these two different situations.

8.2.2 Fick's laws and the diffusion tensor – diffusion of a tracer through a medium

In 1855 Fick recognised the analogy between diffusion and transfer of heat by conduction and applied the known equation of heat conduction to quantitatively describe the diffusion process. He supposed that the rate of flow of the tracer in a particular direction would be proportional to the concentration gradient, in that direction, of the tracer. If F is the rate of transfer of the diffusing substance, or tracer, through unit area of each section of the sample studied (i.e. it is the flux), C is the concentration of the diffusing substance and x is the space coordinate measured normal to the section (Figure 8.1), then we can formulate *Fick's first law of diffusion*, which states that the diffusive process drives tracers from areas of higher concentration to areas of lower concentration:

$$F(x,t) = -D\frac{\partial C(x,t)}{\partial x}. \tag{8.1}$$

The negative sign takes into account the fact that the diffusive flow is in the opposite direction to that of increasing concentration and D is the term of proportionality referred to as the *diffusion coefficient* or *diffusivity*. The dimensions of D are $[(\text{length})^2\,(\textit{time})^{-1}]$, giving SI units[3] of $[\text{m}^2\text{s}^{-1}]$, although it is common to also see D reported in $[\text{cm}^2\ \text{s}^{-1}]$, $[\text{mm}^2\ \text{s}^{-1}]$ or $[\mu\text{m}^2\ \text{ms}^{-1}]$.

From Equation 8.1 it is possible to derive *Fick's second law of diffusion*[4] introducing the flux term $F(x, t)$ in a mass balance.[5]

[2] Brownian movement is the random movement of microscopic particles suspended in a fluid, first observed by the Scottish botanist Robert Brown in 1827.

[3] For instance, the diffusion coefficient D measured parallel to white matter fibres is on the order of 2 $\mu\text{m}^2\ \text{ms}^{-1}$.

[4] This should be distinguished from his Second Law, which is on cardiac output.

[5] This can be derived from the equation of conservation of tracer in absence of chemical/nuclear reactions $\left(\frac{\partial C(x,t)}{\partial t} = -\frac{\partial F(x,t)}{\partial x}\right)$, introducing the expression for $F(x,t)$ reported in Equation 8.1. The conservation of tracer states that the local rate of increase in tracer concentration equals the rate at which the tracer flow decreases with distance.

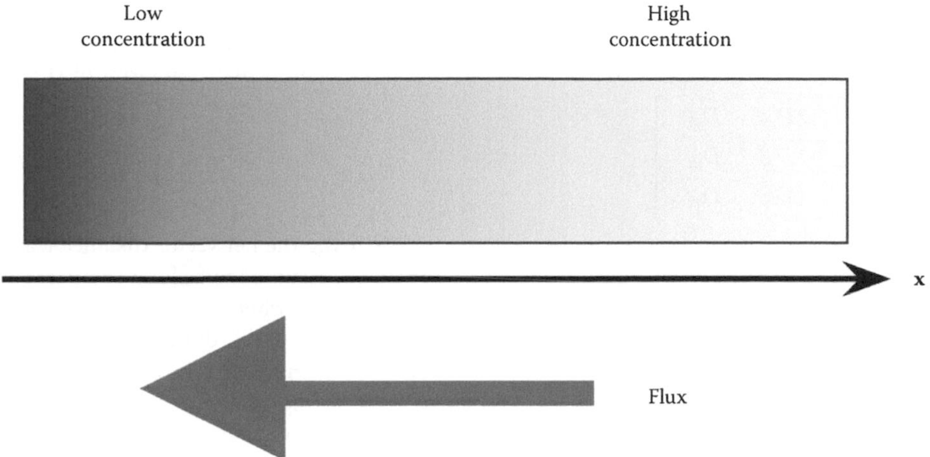

FIGURE 8.1 Representation of a concentration gradient along the x direction. Diffusive flow (or flux) is from left (high concentration) to right (lower concentration).

Specifically, for constant D values and one-dimensional diffusion, this leads to the description of the diffusion process in terms of the temporal and spatial partial derivatives of the concentration $C(x, t)$ as

$$\frac{\partial C(x,t)}{\partial t} = D\,\frac{\partial^2 C(x,t)}{\partial x^2}. \tag{8.2}$$

Integration of Equation 8.2 for an initial concentration $C(x, 0) = \delta(x)$ (i.e. initial point source of diffusing particles located in $x = 0$) and in absence of barriers restricting diffusion leads to the fundamental solution

$$C(x,t) = \frac{1}{\sqrt{4\pi Dt}}\, e^{-\frac{x^2}{4Dt}}. \tag{8.3}$$

Equation 8.3 implies that the spatial distribution of diffusing particles at any time t is Gaussian and that its width increases as t increases (Figure 8.2).

In the case of *isotropic* diffusion, where there is no preferential direction for the diffusive motion, Equation 8.2 can be generalised to the 3D space as

$$\begin{aligned}\frac{\partial C(x,y,z,t)}{\partial t} &= \\ D\left(\frac{\partial^2 C(x,y,z,t)}{\partial x^2} + \frac{\partial^2 C(x,y,z,t)}{\partial y^2} + \frac{\partial^2 C(x,y,z,t)}{\partial z^2}\right) \\ &= D\,\nabla^2 C(x,y,z,t)\end{aligned} \tag{8.4}$$

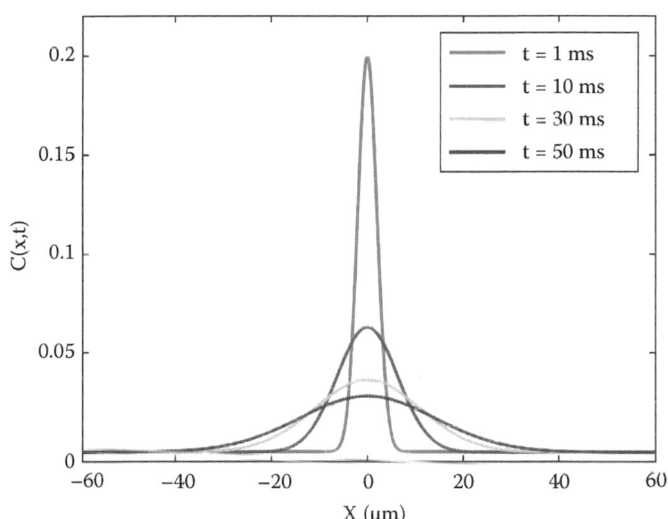

FIGURE 8.2 Plot illustrating the fundamental solution of the mono-dimensional diffusion equation as a function of the spatial position x and for different diffusion times t. The figure illustrates the evolution of the concentration of a tracer initially located in $x = 0$ at $t = 0$. The curves show how the tracer molecules diffuse to locations distant from the initial position as time increases, and that the concentration in the proximity of $x = 0$ decreases over time. In the limit of $t \to \infty$, the concentration of the tracer would be uniform. To generate the figure, an intrinsic diffusion coefficient of $D = 2$ μm^2 ms^{-1} was employed.

The situation becomes more complicated when the value of D is no longer independent of the considered direction. In this case we talk about *anisotropic* media that have different diffusion properties in different directions. The directional dependence of the diffusion properties can be readily characterised in terms of a rank-2 tensor, expressed as a 3×3 matrix, and known as the *diffusion tensor* (DT) **D**. The DT is characterised by nine elements:

$$\mathbf{D} = \begin{bmatrix} D_{xx} & D_{xy} & D_{xz} \\ D_{yx} & D_{yy} & D_{yz} \\ D_{zx} & D_{zy} & D_{zz} \end{bmatrix}. \tag{8.5}$$

D_{xx}, D_{yy} and D_{zz} define the diffusion constants along the main axis of the frame of reference, and the off-diagonal terms, D_{ij}, represent the effect of the concentration gradient along i on the diffusive flow along j, with $i, j = x, y, z$. Assuming that the elements of the DT are constant, it is possible to extend Fick's second law of diffusion to the case of anisotropic diffusion:

$$\frac{\partial C(x,t)}{\partial t} = \sum_{i,j} D_{ij} \frac{\partial^2 C(x,t)}{\partial i \, \partial j}, \tag{8.6}$$

where D_{ij} are the elements of the DT of Equation 8.5. In particular, for uncharged molecules, such as water, the tensor **D** is symmetric, e.g. $D_{ij} = D_{ji}$; hence it is completely defined by six elements: D_{xx}, D_{yy}, D_{zz}, D_{xy}, D_{xz} and D_{yz}. The diffusion coefficient along any direction $\mathbf{n} = [n_x\ n_y\ n_z]^T$ of the 3D space (a 3×1 vector), indicated as $d(\mathbf{n})$, can be obtained from **D** calculating the dot product $d(\mathbf{n}) = \mathbf{n} \cdot (\mathbf{Dn})$ between vectors **n** and (**Dn**).

8.2.3 The random walk description – *Self*-diffusion in a homogeneous fluid

We have seen that that Fick's first law of diffusion (Equation 8.1) connects the particle flux to the particle concentration gradient. In the case of *self*-diffusion, instead of describing the process in terms of the concentration, we use a function that describes the probability of finding a particle in a certain position at a particular time (the *diffusion propagator*). Suppose one wishes to follow the displacements of a molecule that was initially at position **r′** and to estimate the chance of finding it at another position **r″** at time t. The propagator $P(\mathbf{r}''|\mathbf{r}',t)$ describes the probability of this event. In the case of free self-diffusion, the probability is independent of the starting position of the molecule and can be applied to all the molecules in the ensemble; therefore, it is particularly convenient to define a vector for the *relative* dynamic displacement, $\mathbf{R} = \mathbf{r}'' - \mathbf{r}'$, and describe the process with $P(\mathbf{R}|t)$. In analogy with Fick's second law of diffusion (Equations 8.2 through 8.4), the self-diffusion mechanism can be described by

$$\frac{\partial P(\mathbf{R},t)}{\partial t} = D\, \nabla^2 P(\mathbf{R},t) \tag{8.7}$$

In this case D assumes the meaning of *self*-diffusion coefficient or constant.

For the simple case of isotropic unrestricted self-diffusion in three dimensions, the solution to Equation 8.7 again takes the form of a Gaussian function (the three dimensional equivalent of Equation 8.3), and the mean square dynamic displacement, $< R^2 >$, can be calculated in terms of D as

$$< R^2 > \; = \int_0^\infty R^2 P(\mathbf{R},t)\, dR = 6Dt \tag{8.8}$$

with **R** being the net vector distance travelled by a molecule in the time t, assumed to be long compared with the time between collisions, and where $R = \|\mathbf{R}\|$.

The process of self-diffusion can also be modelled (Callaghan 1995) by assuming that the particles of any liquid take a random walk consisting of a succession of n random displacements of constant length ξ, at constant time intervals, τ, over a time $t = n\tau$. After each displacement there is a collision and then a new random orientation for the next displacement. Taking into consideration the random nature of this process, it is possible to determine the mean square displacement, $< R^2 >$, as a function of ξ:

$$< R^2 > \; = n\,\xi^2 = \frac{t}{\tau}\,\xi^2. \tag{8.9}$$

This has the same form of Equation 8.8, i.e. $< R^2 >$ increases proportionally to time (the square displacements are additive). Thus the diffusion coefficient arising from a random walk behaviour is described by the Einstein equation (Einstein, 1905):

$$D = \frac{\xi^2}{6\tau}. \tag{8.10}$$

In a diffusion MRI experiment of the brain, diffusion times t that vary between roughly 10 ms to 100 ms are probed, depending on the MRI sequence used. This corresponds to root mean square distances that vary from a few microns to dozens of microns, given that the neural tissue diffusion coefficient is circa 2 μm² ms⁻¹. It follows that diffusion MRI is a powerful probe of neural tissue microstructure: microstructural structures such as axons and cellular compartments have characteristic dimensions of size similar to the distance travelled by water due to diffusion; hence, water can be used as an endogenous probe.

As the remainder of this chapter will focus on MRI diffusion measurements in the brain, i.e. measurement of the self-diffusion coefficient of water molecules in tissue, for simplicity we drop the *self*- prefix from this point onwards and use the term *diffusion* alone.

8.2.4 How can diffusion affect the MRI signal?

The signal in an MRI image arises from the behaviour of nuclear spins in a magnetic field. To understand how diffusion can affect this signal, we will consider the simple case of the spins of the water molecules in a liquid sample (where there is no motional restriction), although the discussion can be easily extended to the water protons in tissues.

For simplicity we will assume a *spin echo* (SE) sequence (Figure 8.3), but the effects of molecular diffusion will be similar for any other MRI sequence. In the SE case, a 90° radiofrequency (RF) pulse is applied at $t = 0$ to excite the sample and a 180° RF pulse is applied at a time $t = TE/2$ to refocus the magnetisation. This refocusing pulse reverses the phase of the spins and leads to the cancellation of the phase accrual due to magnetic field inhomogeneities and the formation of a SE at exactly $t = TE$ (see Chapter 6 on T_2). This echo formation relies on the fact that the spins do not accumulate a phase distribution that would destroy their coherence (due either to field inhomogeneities or deliberately applied field gradients), as the phase evolution (dephasing) that takes place before the 180° pulse is exactly reversed (rephased) after the 180° pulse. This is possible because the spins experience the same magnetic field throughout the sequence of events (excitation, evolution, echo formation). However, if a magnetic field gradient (i.e. a spatially variant magnetic field $\mathbf{g} = \left[\frac{\partial B_z}{\partial x} \quad \frac{\partial B_z}{\partial y} \quad \frac{\partial B_z}{\partial z} \right]^T = \nabla B_z$, usually called *diffusion encoding* or *diffusion sensitising gradient*) is turned on either side of the refocussing pulse, the refocusing of the spins is only partial, since due to the random walk each spin experiences difference locations (i.e. different magnetic fields) before and after the 180° pulse (i.e. during the dephasing and rephasing). As a consequence, there is an extra damping (loss) term in the transverse magnetisation evolution. The resulting phase distribution is not coherent, which means that it cannot be reversed by the 180° pulse in the SE sequence, resulting in a reduced signal amplitude at $t = TE$ (Figure 8.3).

Indicating with $\varphi(TE)$ the phase accrual of the generic spin due at $t = TE$, the MRI signal intensity is the ensemble average over the whole spin population:

$$S(TE) = S_0 \left| < e^{j\,\varphi(TE)} > \right|. \tag{8.11}$$

Above, j is the imaginary unit (i.e. $j^2 = -1$), S_0 is the signal that would be obtained for the SE sequence when no diffusion encoding gradient is turned on and $< \cdot >$ indicates the ensemble average

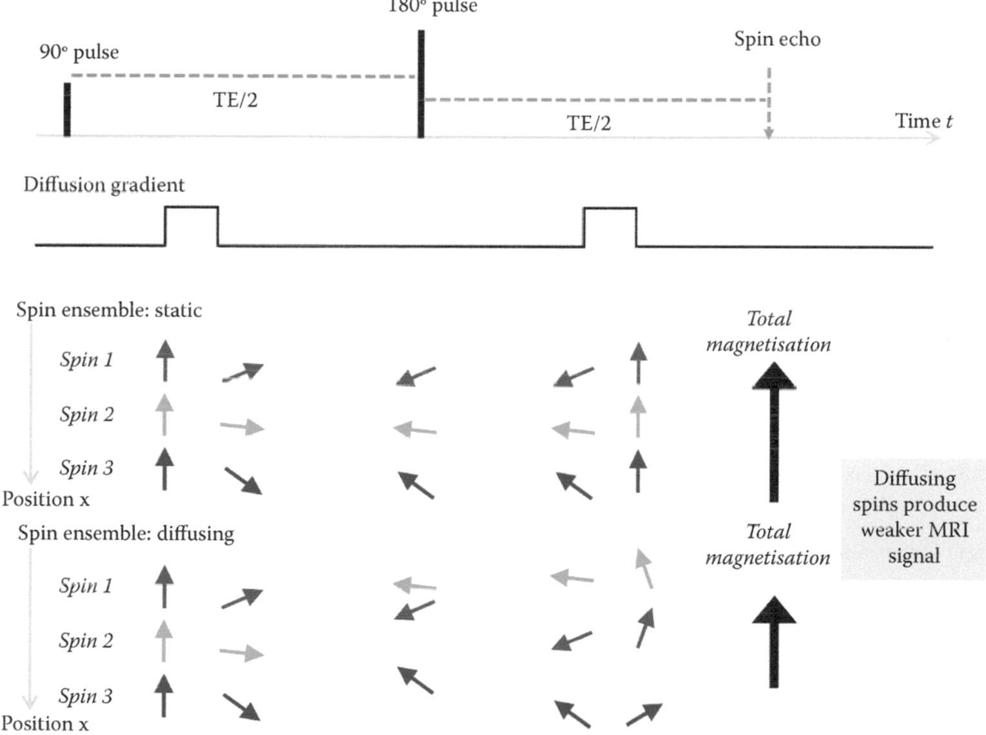

FIGURE 8.3 Illustration of the effect of diffusion weighting gradients on an ensemble for a spin echo (SE) sequence. The figure shows two ensembles of spins: one static and one subject to diffusion (for simplicity, each ensemble is made of three spins, shown in red, green and blue respectively). Static spins are located in the same position when both lobes of the diffusion gradients are played out. Hence, the dephasing induced by the first lobe (which is spatially dependent) is completely reversed by the second, implying that no phase differences among spins other than those occurring for T_2-related phenomena exist. On the other hand, diffusing spins will be located in a different position when the second lobe is played out, as compared to the first. Therefore, they are subject to a different magnetic field, as the magnetic field is spatially dependent in the presence of a gradient. As a result, the phase accrual caused by the first lobe is not nulled by the second lobe, resulting in a distribution of residual phase accruals in the spin ensemble. This leads to an overall attenuation of the transverse component of the magnetisation and consequently of the SE MRI signal. Note that in generating the figure, diffusion occurring when gradients are pulsed is neglected.

over the spin population. The phase accrual $\varphi(TE)$ of each spin for the SE sequence depends on its random walk $\mathbf{r}(t)$ as well as on the temporal pattern of the diffusion encoding gradient $\mathbf{g}(t)$. It follows a relationship obtained from the temporal integration of the Larmor (angular) frequency over the echo time TE (Hall and Alexander, 2009):

$$\varphi(TE) = -\gamma \int_0^{TE/2} \mathbf{g}(t)\cdot\mathbf{r}(t)dt + \gamma \int_{TE/2}^{TE} \mathbf{g}(t)\cdot\mathbf{r}(t)dt. \quad (8.12)$$

In Equation 8.12, γ is the proton gyromagnetic ratio and the different sign of the two integral terms accounts for the change of sign of the phase caused by the refocusing pulse at $t = TE/2$.

8.2.5 Diffusion-Weighted sequences

In presence of diffusion, the differences in phase φ among spins due to their different random walks $\mathbf{r}(t)$ lead to an overall reduction of the coherence of the ensemble, i.e. to a reduction of the measured DW signal as compared to the non-DW signal S_0. Notably, such a signal reduction depends not only on the random walks $\mathbf{r}(t)$ themselves but also on the applied magnetic field gradient $\mathbf{g}(t)$, which can be designed directly to control the amount of diffusion weighting.

A common index employed to characterise the amount of diffusion weighting is the *b*-factor or *b*-value b (measured in units of [s m^{-2}]), which can be defined from the gradient first moment $\mathbf{k}(t) = \gamma \int_0^t \mathbf{g}(t')\,dt'$ as

$$b = \int_0^{TE} \mathbf{k}(t') \cdot \mathbf{k}(t')\, dt', \quad (8.13)$$

having indicated with · the dot product. While theoretically $\mathbf{g}(t)$ can follow any desired waveform, real implementations are limited by hardware specifications (i.e. electronics of the gradient amplifier; heating of gradient coils) or by the fact that excessive gradient slew rate (i.e. $\frac{d}{dt}\left\|\mathbf{g}(t)\right\|$, measured in [T m^{-1} s^{-1}]) can lead to undesirable peripheral nerve stimulation (Setsompop *et al.*, 2013).

At present, the most common SE-based DW sequence is the *pulsed gradient spin echo* (PGSE) method (Stejskal and Tanner, 1965), also known as the *pulsed field gradient* or *single diffusion encoding* method (Shemesh *et al.*, 2016), illustrated in Figures 8.3 and 8.4. During a PGSE experiment, the diffusion encoding gradient $\mathbf{g}$ is linearly polarised and is pulsed on either side of the refocusing pulse. The pulsed gradient has a magnitude G, a duration δ and a separation Δ, which together are known as *Stejskal–Tanner parameters*. The *diffusion time* t_d, which is the time window during which spins probe the local microstructure, is calculated as

$$t_d = \Delta - \delta/3, \quad (8.14)$$

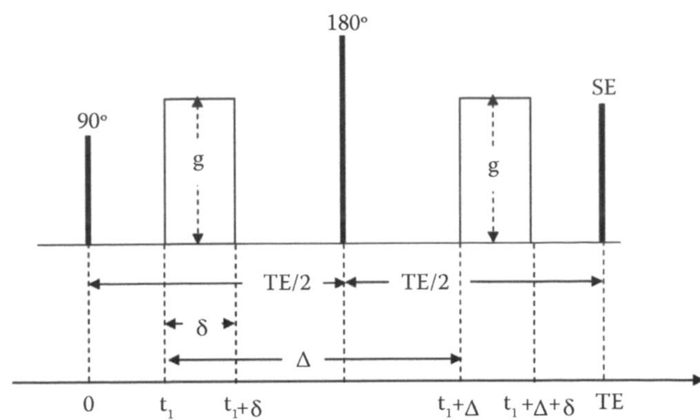

FIGURE 8.4 Stejskal–Tanner pulsed gradient spin echo (PGSE) sequence. A pulsed gradient, **g**, is introduced in a SE sequence, either side of the 180° refocusing pulse. The amplitude, duration and timing of **g** determine the amplitude of the spin echo. The interpulse delay (Δ) and the pulse duration (δ) define the diffusion time ($t_d = \Delta - \delta/3$).

whereas, for negligible background gradients, the strength of the diffusion weighting (i.e. the *b*-factor) is obtained as

$$b = \gamma^2 G^2 \delta^2 (\Delta - \delta/3). \quad (8.15)$$

Regarding a PGSE experiment, the following should be noted.

- If the diffusion coefficient is such that the root mean square displacement occurring when the gradient is pulsed is much smaller than the physical distances that one intends to probe with the PGSE experiment, to a very good approximation only the diffusion taking place between the two gradient lobes contributes to the reduction in signal amplitude and its weight in determining the echo amplitude is independent of when the diffusion occurred within that interval.
- The diffusion gradients have been approximated by rectangular shapes, as clear in Figure 8.4. In reality the gradient lobes usually have a trapezoidal form and Equation 8.15 can be modified to take this into account.[6] Moreover, the presence of the imaging gradients can contribute to the actual effective gradient (Mattiello *et al.*, 1994) and is discussed further in Section 8.4.1.9.
- The PGSE formalism can be applied not only to SE but also to other sequences such as *stimulated echo acquisition*

[6] The full expression for b becomes $b = \gamma^2 G^2 \left(\delta^2 \left(\Delta - \frac{\delta}{3} \right) + \frac{\varepsilon^3}{30} - \frac{\delta \varepsilon^2}{6} \right)$, where δ is measured from the start of the ramp-up to the beginning of the ramp-down. Δ is the time between two identical positions on the two diffusion gradients, e.g. from the start of the ramp-up of the first gradient lobe to the start of the ramp-up of the second gradient lobe. ε is the duration of the rise and fall ramps of the diffusion gradients, assumed to be equal.

mode (STEAM) (Frahm *et al.*, 1985; Merboldt *et al.*, 1985). STEAM is particularly useful when very long diffusion times are required (De Santis *et al.*, 2016).

8.2.6 The Bloch–Torrey equation

The phenomenological Bloch equations that describe NMR/MRI can be extended to include the effect that specific processes have on the transverse magnetisation evolution. The extension of the equation to include the diffusion mechanism was suggested by Torrey; hence the name of the Bloch–Torrey equation (Torrey, 1956). By analogy with Equation 8.7, a diffusion term can be added to the description of the transverse magnetisation (M_{xy}) time evolution after a 90° pulse as follows:

$$\frac{\partial M_{xy}(\mathbf{r},t)}{\partial t} = -j\,\gamma\big(B_0 + \mathbf{g}(t)\cdot\mathbf{r}\big)\,M_{xy}(\mathbf{r},t) - \frac{M_{xy}(\mathbf{r},t)}{T_2} + \nabla\cdot\big(\mathbf{D}(\mathbf{r})\,\nabla M_{xy}(\mathbf{r},t)\big). \tag{8.16}$$

Here, $\mathbf{g}(t)$ is the applied, time-variant magnetic field gradient and $\mathbf{r}$ is the spin position. On the right-hand side of Equation 8.16, the first term represents the nutation in the total applied magnetic field ($B_0 + \mathbf{g}(t)\cdot\mathbf{r}$). The second term describes the rate of change due to T_2 decay, and the third term describes the rate of change of the transverse magnetisation due to the diffusion process, with $\mathbf{D}(\mathbf{r})$ being in the most general case a spatially variant DT, as that introduced in Section 8.2.2.

8.2.7 Diffusion-Weighted signal decay

The behaviour of the magnetisation M_{xy} as a function of time can be derived by solving Equation 8.16 (for analytical and numerical solutions, see Stejskal and Tanner, 1965; Kenkre *et al.*, 1997; Barzykin, 1998; Beltrachini *et al.*, 2015). In particular, it is possible to separate the terms and study the effect of the diffusion mechanisms on the magnetisation evolution. In the case of a magnetic field gradient along a single direction and in the simple case of free (i.e. unrestricted), constant isotropic ($\mathbf{D} = D\mathbf{I}$) diffusion, M_{xy} assumes the following form:

$$M_{xy}(\mathbf{r},t) = M_{xy}(\mathbf{r},0)\,e^{-j\,(\omega_0 t+\,\mathbf{k}(t)\cdot\mathbf{r}\,)}\,e^{-\frac{t}{T_2}}\,e^{-bD}. \tag{8.17}$$

Above, $M_{xy}(\mathbf{r},0)$ is the initial spatial distribution of transverse magnetisation, $\mathbf{k}(t)$ is the gradient first moment $\mathbf{k}(t) = \gamma\int_0^t \mathbf{g}(t')\,dt'$ and b is the b-factor introduced in Equation 8.13.

If the applied gradients during the MR pulse sequence are balanced so that the gradient echo forms at the same time as the spin echo, then $\mathbf{k}(TE) = \mathbf{0}$ and the signal is given by

$$S(TE,b) = S_0(TE)\,e^{-bD}, \tag{8.18}$$

where $S_0(TE) = A\,e^{-\frac{TE}{T_2}}$, with A being proportional to the magnitude of $\int M_{xy}(\mathbf{r},0)d^3\mathbf{r}$ via a spatially variant, instrument-dependent factor.

From Equation 8.18 it is clear that the effect of diffusion on the transverse magnetisation is to add an extra decay term in addition to that seen from normal T_2 decay (i.e. $e^{-\frac{TE}{T_2}}$). This term is equal to e^{-bD} in the simple case of free, isotropic diffusion, where b depends only on the gradient pulse amplitude and timing. $S(TE, b)$ represents the signal from each voxel, which depends on the intrinsic diffusion coefficient, D, and transverse relaxation time, T_2, of the underlying tissues. The echo time, TE, and the b-factor, b, on the other hand, are parameters that can be varied by the user in order to obtain the desired signal attenuation and hence the desired image contrast. Interestingly, Equation 8.18 can be obtained directly from the cumulant expansion of the spin ensemble average in Equation 8.11 (i.e. a series of powers of the b-factor), as long as high order terms carrying the signature of non-Gaussian diffusion are negligible (for detailed information see Chapter 10 of Jones, 2010).

In practice, Equation 8.18 can be generalised to the case when a mixture of water pools with different intrinsic diffusion coefficients coexist within the imaged voxel (Clark and Le Bihan, 2000). Considering the case of N distinct pools and assuming that the pools exchange a negligible amount of water during the timescale of the MRI experiment, then the total signal can be written as a linear combination of the exponential signal decays of each pool, assuming the form of

$$S(TE,b) = S_0(TE)\sum_{n=1}^{N} v_n\,e^{-bD_n}, \tag{8.19}$$

where $S_0(TE) = A\,e^{-\frac{TE}{T_2}}$ and where D_n and v_n indicate respectively the diffusion coefficient and the volume fraction of the nth water pool (subject to $\sum_{n=1}^{N} v_n = 1$). For instance, a model such as that in Equation 8.19 can be used to describe the signal decay in areas of partial volume between the neural tissue and the cerebrospinal fluid (CSF). Importantly, it should be noted that a multi-exponential signal behaviour can be generated also by a single water pool when molecules diffuse within a restricted environment.

Finally, in the case of a continuous distribution of diffusion coefficients D described by a probability density function $P(D)$, Equation 8.18 can be further generalised to

$$S(TE,b) = S_0(TE)\int_0^{\infty} P(D)e^{-bD}dD, \tag{8.20}$$

implying that the signal measured at varying b is the Laplace transform of the diffusivity distribution $P(D)$ (Dhital *et al.*, 2016), a concept used in methods such as diffusion basis spectrum imaging (Wang *et al.*, 2015).

8.2.8 Estimation of the diffusion coefficient from MRI measurements

Let us assume that the aim is to measure the diffusion properties of tissues, as opposed to simply introducing diffusion weighting to the signal amplitude. In that case, multiple DW images have to be acquired so that the signals can be analysed on a voxel-by-voxel basis, providing quantitative parametric maps.

8.2.8.1 Estimation of a single diffusion coefficient from MRI measurements

For the estimation of a diffusion coefficient as that of Equation 8.18, a minimum of two acquisitions with different diffusion weightings are necessary. Independently of the practical issues, Equation 8.18 shows that measuring the signal at a fixed TE while changing b enables the estimation of the diffusion properties (such as the diffusion coefficient D). Where two or more measurements with different b values are available, Equation 8.18 can be fitted using methods such as least squares fitting, enabling the estimation of S_0 and D. In the simplest case, S_0 and D can be estimated directly from two measurements at different b ('two-point estimation'). Suppose that $\tilde{S}(TE, b_1)$ and $\tilde{S}(TE, b_2)$ are the measured signals at a fixed TE and at two different b-factors $b_1 \neq b_2$ (with one of the two commonly set to zero). It is easy to show that the ratio between the two measured signals can be written as

$$\frac{\tilde{S}(TE,b_1)}{\tilde{S}(TE,b_2)} = \frac{A\, e^{-\frac{TE}{T_2}}\, e^{-b_1 D}}{A\, e^{-\frac{TE}{T_2}}\, e^{-b_2 D}} = e^{-(b_1 - b_2)D}, \tag{8.21}$$

enabling the estimation of D as

$$D = \frac{1}{b_1 - b_2} \log\left(\frac{\tilde{S}(TE,b_2)}{\tilde{S}(TE,b_1)}\right). \tag{8.22}$$

While the two-point estimation provides estimates of D with just two images, basing the estimation of D on a higher number of images reduces measurement errors, a valuable option when scan time allows.

From a set of M measurements $\tilde{\mathbf{S}} = \left\{\tilde{S}_m(b_m) \mid m = 1, \ldots, M\right\}$, obtained at M different b-factors and fixed TE, it is possible to estimate diffusion properties such as the diffusion coefficient D fitting the model of the signal decay (Equation 8.18) to the acquired data. In general, such a fitting can be linear or non-linear. In the former case, Equation 8.18 is linearised taking the logarithm of the signal, leading to the linear system of equations

$$\underbrace{\begin{bmatrix} \log\tilde{S}_1 \\ \log\tilde{S}_2 \\ \cdots \\ \log\tilde{S}_M \end{bmatrix}}_{\tilde{\mathbf{m}}} = \underbrace{\begin{bmatrix} 1 & -b_1 \\ 1 & -b_2 \\ \cdots & \cdots \\ 1 & -b_M \end{bmatrix}}_{\mathbf{Q}} \underbrace{\begin{bmatrix} \log S_0 \\ D \end{bmatrix}}_{\mathbf{x}} \tag{8.23}$$

in the unknowns $\log S_0$ and D (where $S_0 = A\, e^{-\frac{TE}{T_2}}$), whose least square solution is obtained as

$$\mathbf{x} = \left(\mathbf{Q}^{\mathrm{T}}\mathbf{Q}\right)^{-1}\mathbf{Q}^{\mathrm{T}}\,\tilde{\mathbf{m}}. \tag{8.24}$$

While linear fitting is fast and easy to implement, it alters the properties of homoscedasticity of the noise, leading to potential inaccuracies of the fitted unknown parameter values. This issue is mitigated by using weighted linear fitting or non-linear fitting, although it should be noted that non-linear fitting can lead to unstable, non-univocal solutions for highly non-linear models and ill-posed model inversions (Jelescu *et al.*, 2016). As an example for the reader, we present below a simple non-linear fitting procedure for the estimation of the parameters (D, S_0) of Equation 8.18 from the set of M measurements $\tilde{\mathbf{S}}$ obtained at M different b-values, under the assumption of additive Gaussian noise.

The measured signal $\tilde{\mathbf{S}}$ can be related to the theoretical value expressed in Equation 8.18 as

$$\tilde{S}(D,S_0;\, b) = S_0\, e^{-bD} + \eta, \tag{8.25}$$

where again $S_0(TE) = A\, e^{-\frac{TE}{T_2}}$ and where η is a stochastic term describing the noise, such that $\eta \sim N(0, \sigma^2)$ (i.e. noise has zero mean and variance equal to σ^2). Equation 8.25 implies that $\tilde{S} \sim N\left(S_0 e^{-bD}, \sigma^2\right)$. This means that $S_0 e^{-bD}$ is the expected value of the measured signal at b-factor b, while the variance of the measured signal equals that of noise. It is important to note that it is common to consider different noise distributions in some application with high diffusion weighting or low *signal-to-noise ratio* (SNR) levels, such as Rician (Gudbjartsson and Patz, 1995) or non-central chi (Koay and Basser, 2006; Sotiropoulos *et al.*, 2013) if multiple coils are used for reception. In those cases, $\tilde{S}$ would be modelled as a Rician/non-central chi distribution, rather than Gaussian.

The parameters D and S_0 can be estimated maximising the posterior probability $P\left(D, S_0 \mid \tilde{\mathbf{S}}\right)$ of the model parameters given the signal. In absence of any prior information about the parameters D and S_0, by means of Bayes' theorem, this is equivalent to maximise the likelihood $P\left(\tilde{\mathbf{S}} \mid D, S_0\right)$ of the measurements given the signal decay model, which here we will indicate as $L\left(\tilde{\mathbf{S}} \mid D, S_0\right)$. Algebraic calculations lead to the following expression for the logarithm $\log L\left(\tilde{\mathbf{S}} \mid D, S_0\right)$:

$$\log L\left(\tilde{\mathbf{S}} \mid D, S_0\right) = -\frac{M}{2}\log\left(2\pi\sigma^2\right) - \frac{1}{2\sigma^2}\sum_{m=1}^{M}\left(\tilde{S}_m - S_0\, e^{-b_m D}\right)^2. \tag{8.26}$$

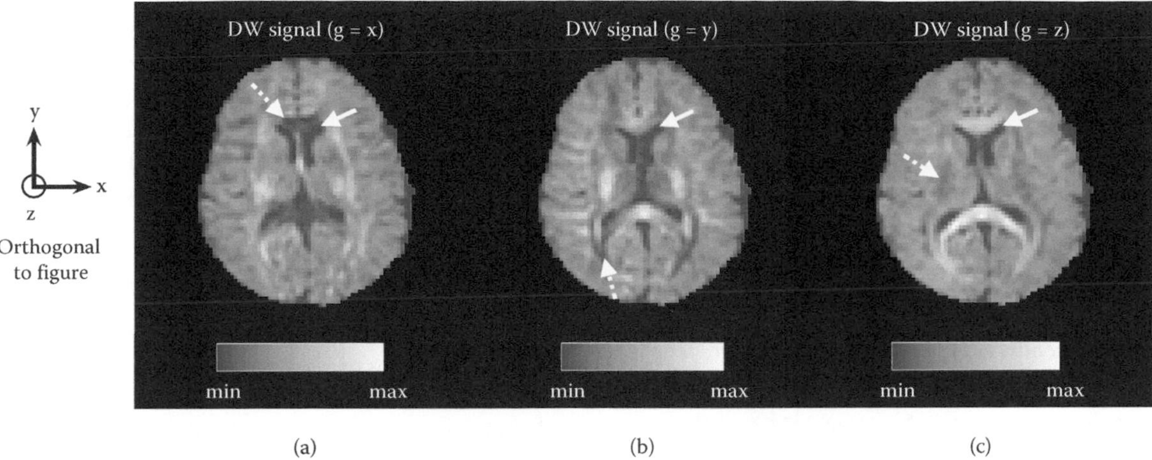

FIGURE 8.5 Examples of axial diffusion weighted images of the human brain. The images were obtained at 3T using a cardiac gated, SE echo planar imaging (EPI) sequence (resolution: 2.5 mm isotropic; *b*-value: 700 s mm^{-2}). The three images were respectively obtained applying gradient directions along the (a) *x*-axis; (b) *y*-axis; (c) *z*-axis (axes shown in the figure). The solid arrows show areas of isotropic signal attenuation in the ventricles. The dashed arrows show areas of high signal attenuation along the specific gradient direction relative to each image (such as direction along the corpus callosum in Figure 8.5a or as direction along the corticospinal tract in Figure 8.5c).

From Equation 8.26, the unknown parameters (D, S_0) can be estimated numerically[7] as:

$$\left(D^{*},S_0{}^{*}\right)= \arg\min_{(D,S_0)}\left(-\log L\left(\tilde{\mathbf{S}} \mid D,S_0\right)\right). \quad (8.27)$$

The solution of Equation 8.27 can be approximated by using gradient descent algorithms and constraints such as $S_0 > 0$ and o $< D < 3\ \mu\text{m}^2\ \text{ms}^{-1}$, with the latter accounting for the fact that the diffusivity of free water at 37°C (body temperature) is circa 3 μm^2 ms^{-1}.

8.2.8.2 Estimation of the multiple diffusion coefficients from MRI measurements

The approach shown in the previous section can be generalised to estimate the parameters of the DW signal model described by Equation 8.19. The parameters (v_1, D_1,…,S_0) can be estimated from a set of measurements $\tilde{\mathbf{S}}$ obtained at various b via minimisation of $-\log L\left(\tilde{\mathbf{S}} \mid v_1, D_1, \ldots, S_0\right)$, where $\log L\left(\tilde{\mathbf{S}} \mid v_1, D_1, \ldots, S_0\right)$ is the logarithm of the likelihood, which, for a Gaussian noise model, equals

$$\log L\left(\tilde{\mathbf{S}} \mid v_1, D_1, \ldots, S_0\right) = -\frac{M}{2}\log\left(2\pi\sigma^2\right) - \frac{1}{2\sigma^2}\sum_{m=1}^{M}\left(\tilde{S}_m - S_0\sum_{n=1}^{N} v_n\, e^{-b_m D_n}\right)^2. \quad (8.28)$$

[7] Note that although Equation 8.26 contains a linear combination of terms, it is non-linear as a function of the unknown diffusion coefficient *D*. This fact justifies the term 'non-linear fitting'.

More generally, a continuous distribution of diffusivities $P(D)$, as that considered in Equation 8.20, can be estimated from the set of measurements $\tilde{\mathbf{S}}$ via inverse Laplace transformation. This can be implemented in practice using approaches such as regularised non-negative least square algorithms (Dhital *et al.*, 2016).

8.2.9 Diffusion weighted imaging

As discussed earlier, most MRI sequences can be weighted to sensitise images to diffusion along a specific direction (given by the direction along which the diffusion gradient is applied). This gives reduced signal in areas of the sample where the self-diffusion coefficient in that direction is higher.

Figure 8.5 compares axial SE echo planar imaging (EPI) (Stehling *et al.*, 1991) images of the brain, acquired with diffusion weighting along different directions. Areas of higher diffusivity, such as the ventricles, result in darker regions in the DW image. Areas with similar diffusion properties in every direction are said to be isotropic and they have the same signal characteristics on the DW images, independent of the direction of application of the diffusion gradient. On the other hand, areas where tissue structure favours water movement along a particular direction are characterised by different diffusion coefficients in different directions. In these cases the signal attenuation reflects the diffusion properties in the direction of application of the diffusion gradients. These areas are said to be *anisotropic*. Specifically, Figure 8.5 shows examples of isotropic (ventricles) and anisotropic (e.g. corpus callosum and the corticospinal tract) areas, indicated by arrows.

Diffusion-weighted imaging (DWI) offers ways of introducing different contrast in MRI, which can help the visual interpretation of clinical images. Inspection of DW MRI images constitutes merely a qualitative type of exam and is very sensitive to the choice of acquisition parameters and patient positioning in the scanner. Moreover, the MRI signal intensity also depends on the T_2 value of the tissue, as shown previously. Pathological conditions can cause an increase of the T_2 value of the affected part of the brain, resulting in an increased signal intensity of the DW images, typically acquired with an echo time of the order of 60–80 ms. This signal increase due to T_2 changes may be misinterpreted as a reduction in diffusivity or may mask an increase in diffusivity (the so-called T_2-shine through effect seen in old lacunar infarcts; Geijer *et al.*, 2001).

8.2.10 Quantitative analysis of DWI: ADC maps

As shown in Section 8.2.8, it is possible to measure the diffusion coefficient from two or more DW images, keeping all the other imaging parameters fixed. Because of the anisotropy of the living tissues, the value of the measured diffusion coefficient is very dependent on the choice of diffusion weighting, i.e. the values of diffusion gradient strength and the times, δ and Δ (i.e. ultimately on the b-factor) and also on the diffusion direction. Therefore, for complicated systems such as living tissues, it is more appropriate to talk about the *apparent diffusion coefficient* (ADC).

From a set of DW measurements and corresponding b-factors, the diffusion coefficient can be estimated voxel by voxel following one of the approaches illustrated by Equations 8.22, 8.24 or 8.27. This produces an ADC map, yielding in each voxel the average diffusion coefficient measured along the direction of the diffusion gradient in the tissue within the same voxel. Notice that the calculated ADC map is quantitative, and it has contrast that is inverted relative to a DW image. In particular, areas with high mobility of water molecules along the diffusion sensitising direction appear dark on a DW image but bright on the ADC map; this can lead to considerable confusion if DW images and ADC maps are not clearly distinguished (see Figure 8.6 in comparison with Figure 8.5).

8.2.11 Diffusion tensor imaging

We have seen (Section 8.2.2) that for anisotropic media the diffusion properties of a tracer can be described by a rank-2 tensor (i.e. the DT). Integration of the diffusion equation (Equation 8.6) based on a DT leads to a fundamental solution similar to Equation 8.3, such that the 3D spatial distribution of the tracer is Gaussian with covariance matrix proportional to the DT. Under the assumption that a similar Gaussian behaviour is adequate to describe the diffusion of water molecules in the living neural tissue, then the measured DW MRI signal can be written as a function of the DT (Basser *et al.*, 1994b), generalising Equation 8.18 to the anisotropic case to

$$S(TE, b_{ij}) = S_0(TE)\, e^{-\sum_{i,j} b_{ij} D_{ij}}, \tag{8.29}$$

where again $S_0(TE)$ is in the form of $S_0(TE) = A\, e^{-\frac{TE}{T_2}}$.

In Equation 8.29, $b_{i,j}$ is equal to

$$b_{ij} = \gamma^2\, G_i G_j \delta^2 (\Delta - \delta/3) \tag{8.30}$$

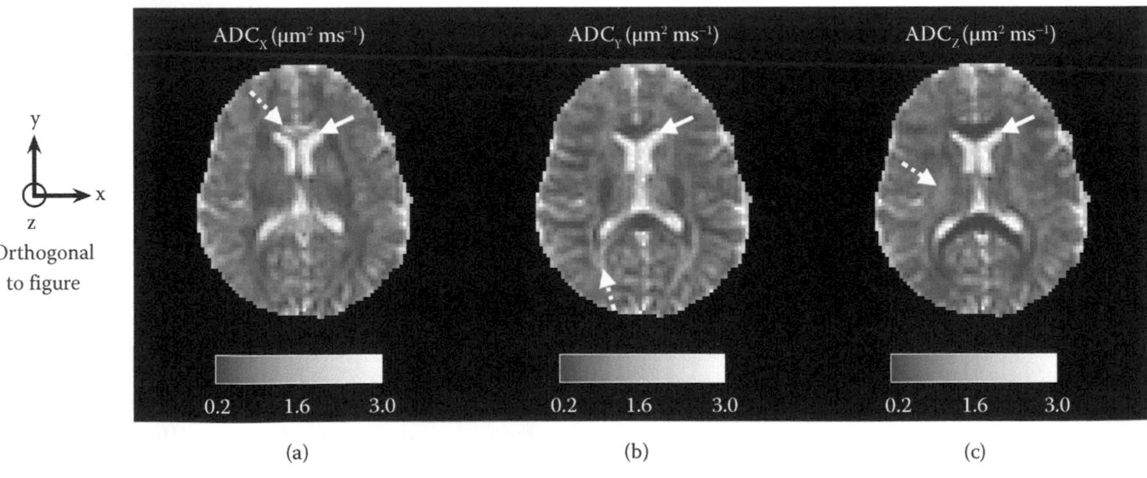

FIGURE 8.6 Apparent diffusion coefficient (ADC) maps of the same axial brain MRI slice shown in Figure 8.5. The ADC maps were obtained from the images in Figure 8.5, with diffusion gradients applied along the (a) x-axis; (b) y-axis; (c) z-axis (axes shown in the figure). The arrows are placed in the same locations as those in Figure 8.5. The solid arrows show area of isotropic diffusivity in the ventricles. The dashed arrows show areas of high neural tissue diffusivity along the applied gradient directions in each ADC map (such as corpus callosum for a gradient along x or corticospinal tract for a gradient along z). Note that the ADC is a quantitative measure, and the values shown in each voxel have the units of a diffusion coefficient. Values of ADC are of the order of 1 μm^2 ms^{-1} in the cortex (isotropic) and 3 μm^2 ms^{-1} in the ventricles (isotropic), while they are of the order of 2 μm^2 ms^{-1} in white matter parallel to axons and 0.4 μm^2 ms^{-1} perpendicular to axons.

where i and j indicate in turn x, y and z and where G_i and G_j are the gradient amplitudes along axes i and j.

The assumptions beyond the DT equation are reasonable in complex biological systems as long as the b-values are low enough for the diffusion process to look Gaussian. However, we defer to Chapter 9 for detailed discussion on the appropriateness of the DT to characterise neural tissue microstructure and for the presentation of alternative mathematical formalisms. The main advantage of using the DT rather than measuring the ADC along a single direction is that from the DT a number of underlying diffusion properties of the sample can be readily obtained, independently of the orientation of the tissue with respect to the direction of measurements, hence independent of how the subject has been oriented inside the scanner magnet and gradient coils (i.e. that are rotationally or orientationally invariant).

From a set of DW images it is possible to estimate in each voxel the six independent elements of the DT in the scanner reference frame, which depend on the chosen sampling directions x, y and z. It is always possible to transform the DT $\mathbf{D}$ into another tensor, $\mathbf{D}' = \text{diag}(\lambda_1, \lambda_2, \lambda_3)$ with off-diagonal elements equal to zero and diagonal elements reflecting the intrinsic properties of the sample, independent of the coordinate system in which they were measured:

$$\mathbf{D} = \begin{bmatrix} D_{xx} & D_{xy} & D_{xz} \\ D_{xy} & D_{yy} & D_{yz} \\ D_{xz} & D_{yz} & D_{zz} \end{bmatrix} = \begin{bmatrix} \varepsilon_1 & \varepsilon_2 & \varepsilon_3 \end{bmatrix} \begin{bmatrix} \lambda_1 & 0 & 0 \\ 0 & \lambda_2 & 0 \\ 0 & 0 & \lambda_3 \end{bmatrix} \begin{bmatrix} \varepsilon_1 & \varepsilon_2 & \varepsilon_3 \end{bmatrix}^T \quad . \tag{8.31}$$

Above, $\boldsymbol{\varepsilon}_1$, $\boldsymbol{\varepsilon}_2$ and $\boldsymbol{\varepsilon}_3$ are the three eigenvectors of $\mathbf{D}$, which are described as 3×1 column vectors. They are orthogonal and unitary (orthonormal) vectors in the scanner reference frame representing three unique directions along which molecular displacements are uncorrelated, while λ_1, λ_2 and λ_3 are the corresponding ADC values, known as the eigenvalues of the DT $\mathbf{D}$.

For each DT, the combination of eigenvectors and eigenvalues is unique and reflects the diffusion properties of the sample under investigation. The convention is to order the eigenvectors with decreasing value of their eigenvalues ($\lambda_1 \geq \lambda_2 \geq \lambda_3 \geq 0$), so that $\boldsymbol{\varepsilon}_1$ represents the principle direction of diffusivity (e.g. in Figures 8.5 and 8.6, voxels in the heart of the corpus callosum would be characterised by $\boldsymbol{\varepsilon}_1$ aligned with x). A DT can be visualised as an ellipsoid, whose axes are directed along $\boldsymbol{\varepsilon}_1$, $\boldsymbol{\varepsilon}_2$ and $\boldsymbol{\varepsilon}_3$ and whose length equals λ_1, λ_2 and λ_3 (Figure 8.7).

A DT can be broadly classified into three types, depending on the relationship among its three eigenvalues, as explained below and illustrated in Figure 8.7.

- A DT such that $\lambda_1 > \lambda_2 \approx \lambda_3$ is prolate. Prolate tensors are cylindrically symmetric if $\lambda_2 = \lambda_3$ and can be referred to as 'Zeppelins' (if $\lambda_2 = \lambda_3 \neq 0$) or 'sticks' (if $\lambda_2 = \lambda_3 = 0$) (Panagiotaki *et al.*, 2012). While the overall diffusion process in a neural tissue voxel cannot be described globally by a stick-like tensor, the stick model is often used to describe the DW signal from a single axon in more complex multicompartmental models that attempt to extend DTI (Zhang *et al.*, 2012; Kaden *et al.*, 2016).
- A DT such that $\lambda_1 \approx \lambda_2 > \lambda_3$ is oblate, and its ellipsoid has a planar shape.

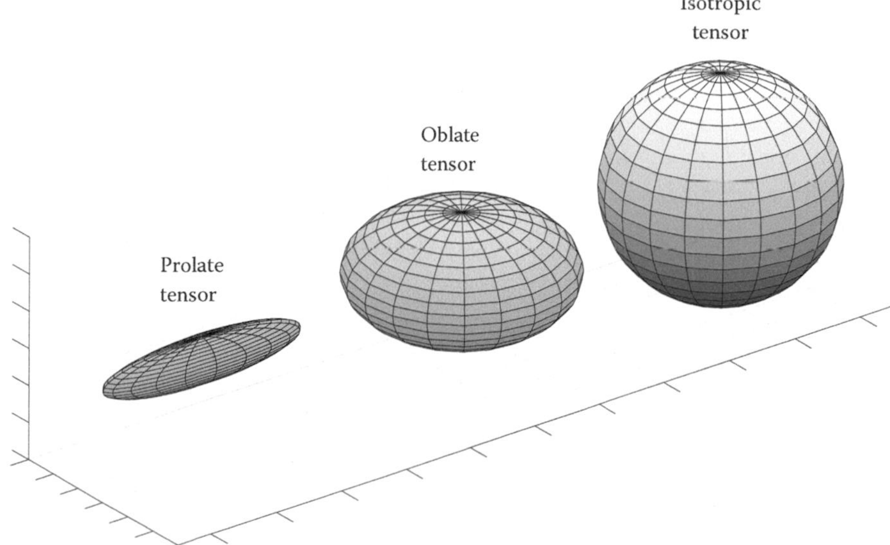

FIGURE 8.7 Examples of ellipsoids showing illustrative diffusion tensors. The figure shows three tensors: a prolate tensor ($\lambda_1 = 15$, $\lambda_2 = \lambda_3 = 3$), an oblate (planar) tensor ($\lambda_1 = \lambda_2 = 15$, $\lambda_3 = 8$) and an isotropic tensor ($\lambda_1 = \lambda_2 = \lambda_3 = 15$). The eigenvalues reported in this caption are expressed in arbitrary units for illustrative purposes.

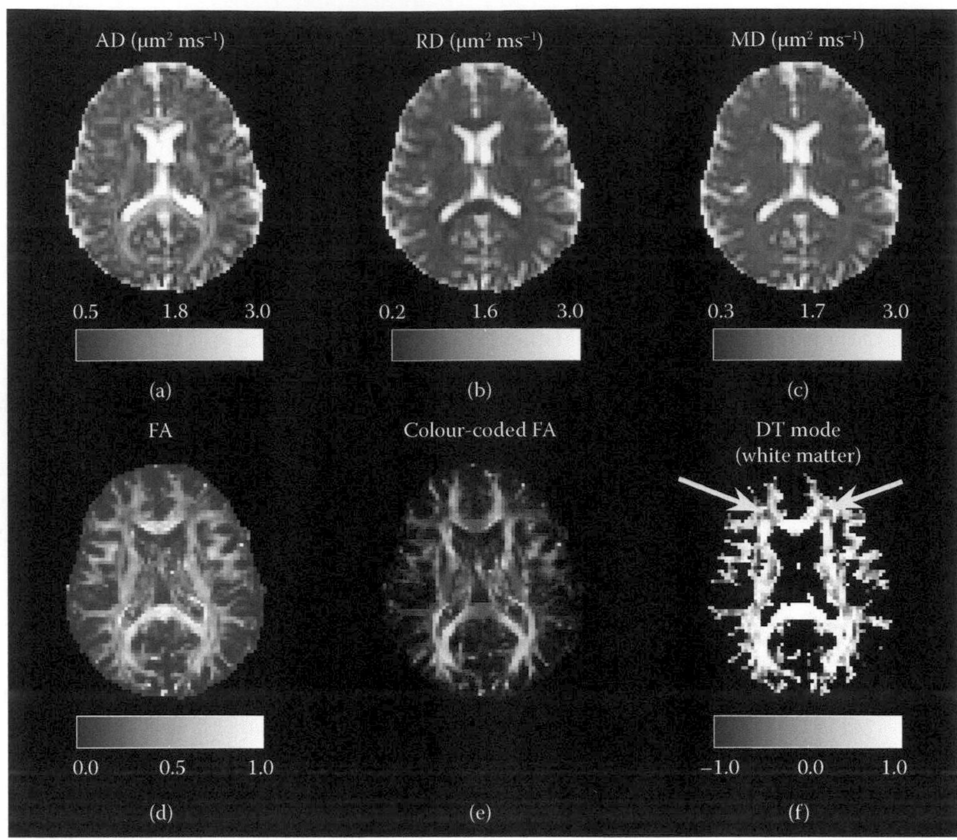

FIGURE 8.8 Examples of scalar invariants derived from the diffusion tensor (DT) and relative to the same axial brain MRI slice shown in Figure 8.6 (invariants obtained at 3T at 2.5 mm isotropic resolution). Top: (a) axial diffusivity; (b) radial diffusivity; (c) mean diffusivity. Bottom: (d) fractional anisotropy (FA); (e) colour-coded FA (the DT principal eigenvector is directed left–right for red, anterior–posterior for green, superior–inferior for blue); (f) DT mode (the mode is shown only in white matter; white arrows indicate areas of low, negative mode indicative of planar shape of the DT due to the presence of crossing fibres).

- A DT such that $\lambda_1 = \lambda_2 = \lambda_3$ is isotropic, and its ellipsoid is a sphere. An isotropic tensor can be referred to as a 'ball' (Panagiotaki *et al.*, 2012).

From a set of DW images, ADC values can be calculated that can be re-measured with a certain accuracy and precision, given the same diffusion weighting and direction. Nevertheless, each ADC value depends on the subject's orientation relative to the magnetic field gradients, which does confound comparisons across individuals and scan–rescan tests. Measuring the ADC of the corpus callosum, for example, can give slightly different results if the volunteer has been positioned with the head slightly tilted towards one side or another, as the gradient direction may or may not be perfectly parallel to the fibre direction. If the DT has been acquired and diagonalised, however, $\boldsymbol{\varepsilon}_1$ will always be aligned with the axis of the fibres and λ_1 will be the diffusion coefficient along $\boldsymbol{\varepsilon}_1$, corresponding to the maximum diffusion in each voxel (regardless of positioning). The DT thus offers the possibility of defining rotationally invariant parameters that can be compared in both cross-sectional and longitudinal studies.

8.2.12 Scalar invariants of the diffusion tensor

As we have seen, from the voxel-wise DT, three ADC maps can be produced (one for each eigenvalue λ_1, λ_2, λ_3).[8] Although it is possible to compare and analyse each component individually, a number of ways to combine the information contained in the eigenvalue maps have been devised.

8.2.12.1 Scalar invariants quantifying diffusivity

A number of invariants quantify the average diffusivity in a voxel or the rate of diffusion along specific directions. Here we report the ones that are most commonly used. Examples obtained from the healthy human brain at 3T are shown in Figure 8.8.

[8] It is important to keep in mind that the tensor also contains information on the directions (i.e. eigenvectors).

- Mean diffusivity (MD) takes the average value of the diffusion coefficients over that voxel. The most robust estimate of this is given by the average of the eigenvalues of the DT and it is normally referred to as the MD of the voxel:

$$\mathrm{MD} = \frac{\lambda_1 + \lambda_2 + \lambda_3}{3} = \frac{tr(\mathbf{D})}{3}, \tag{8.32}$$

where tr(**D**) is the trace of the *DT*[9] (see below).
- Mean ADC (< ADC >) is simply the average of ADC values obtained along three orthogonal axes, such as x, y and z:

$$< \mathrm{ADC} >= \frac{\mathrm{ADC}_x + \mathrm{ADC}_y + \mathrm{ADC}_z}{3}. \tag{8.33}$$

It is sometimes called 'average ADC'. If the influence of imaging gradients is small:

$$< \mathrm{ADC} > \approx \mathrm{MD}. \tag{8.34}$$

Since the trace is rotationally invariant, the choice of which orthogonal directions are used as x, y and z in Equation 8.33 is arbitrary.

Trace of the DT (tr(**D**)). Sometimes, rather than reporting the mean diffusivity value, the simple addition of the three eigenvalues can be reported, mathematically equivalent to the trace of the DT. From Equation 8.32 this parameter is equal to tr(**D**) = 3MD.
- Axial diffusivity (AD). The axial diffusivity is the diffusion coefficient along the principal direction of the diffusion tensor. It is equal to the DT's largest eigenvalue λ_1, i.e.

$$\mathrm{AD} = \lambda_1. \tag{8.35}$$

In coherent white matter areas such as the corpus callosum, AD is an index quantifying the rate of unrestricted diffusion along fibres. It has been reported as an indicator of neuroaxonal damage (Budde *et al.*, 2008).
- Radial diffusivity (RD). The radial diffusivity describes the rate of diffusion across the principal direction and has been reported to be sensitive to changes such as demyelination (Song *et al.*, 2005). It is calculated averaging the DT eigenvalues corresponding to the eigenvectors that are perpendicular to the principal eigenvector, that is:

$$\mathrm{RD} = \frac{\lambda_2 + \lambda_3}{2}. \tag{8.36}$$

Particular care must be taken when reporting changes of AD and RD as these are exclusively based on the properties of the DT, which ultimately is a mathematical description of the signal and does not know anatomical underpinnings. In pathological cases, e.g. when there is a severely demyelinated lesion or in the presence of axonal loss, it is possible that the principle eigenvector of the DT, $\boldsymbol{\varepsilon}_1$, is not aligned with the main direction of the underlying fibres. Comparing AD or RD between subjects would be equivalent to comparing ADC values acquired with gradient applied in different directions, hence representing a different biological substrate. For a comprehensive discussion, please see Wheeler-Kingshott and Cercignani (2009) and Wheeler-Kingshott *et al.*, (2012).

8.2.12.2 Scalar invariants quantifying shape

Other scalar invariants provide information about the shape of the diffusion tensor (Ennis and Kindlmann, 2006). The most common indices in this sense are the fractional anisotropy and the mode of the DT, whose relationship is illustrated in Figure 8.9.

- Fractional anisotropy (FA) is a normalised, dimensionless index that measures the properties of anisotropy of the DT (Basser and Pierpaoli, 1996). It conveys information about how pronounced the directional dependence of the diffusion process is. Low FA values imply similar diffusion along all directions (i.e. λ_1, λ_2 and λ_3 are similar), while high FA implies that there is a marked directional dependence such that diffusion occurs preferentially along one dominant direction (λ_1 is much bigger than λ_2 and λ_3). FA has the great advantage of being characterised by a high white–grey matter contrast and has a great visual impact, as white matter has much higher FA then grey matter (Figure 8.8). Colour-coded maps revealing the underlying, dominant orientation can be modulated in intensity by FA (Figure 8.8e). In this type of colour-coded image, blue indicates that the dominant diffusion

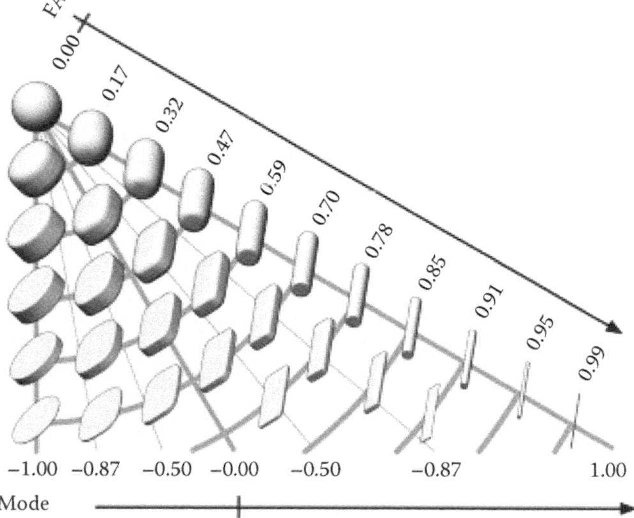

FIGURE 8.9 Shape of the diffusion tensor as FA and tensor mode vary. (Reproduced with permission from Ennis, D.B., and Kindlmann, G., *Magn. Reson. Med.*, 55(1), 136–146, 2006.)

[9] The trace of a square matrix is the sum of the elements along the diagonal. It equals the sum of the matrix eigenvalues and is rotationally invariant.

direction is along the super–inferior direction; red along the right–left direction; green along the anterior–posterior direction.

Mathematically, FA is proportional to the square root of the variance of the eigenvalues divided by the square root of the sum of the squares of the eigenvalues, giving an estimate of what proportion (fraction) of the 'magnitude' of the DT is due to anisotropic diffusion:

$$\mathrm{FA} = \sqrt{\frac{3}{2}}\,\frac{\sqrt{(\lambda_1-\mathrm{MD})^2+(\lambda_2-\mathrm{MD})^2+(\lambda_3-\mathrm{MD})^2}}{\sqrt{\lambda_1^2+\lambda_2^2+\lambda_3^2}}. \tag{8.37}$$

FA is often employed as an index of neural tissue integrity.

- The mode of the DT (mode(**D**)) measures its degree of planarity (Tricoche *et al.*, 2008) and can be used to detect areas characterised by the presence of crossing fibres (Lundell *et al.*, 2011; Grussu *et al.*, 2015) (see Figure 8.8f). It ranges between [−1;1], such that values of mode close to −1 imply that the tensor is oblate (planar), while values close to 1 imply that the tensor is prolate. Mathematically, mode(**D**) is proportional to the skewness of the DT eigenvalues λ_1, λ_2 and λ_3 and can be derived directly from the deviatoric tensor $\mathbf{D}_{\mathrm{dev}}$ of **D** (Tricoche *et al.*, 2008) as

$$\mathrm{mode}(\mathbf{D}) = \sqrt{2}\ \mathrm{skewness}(\lambda_1,\lambda_2,\lambda_3) = 3\sqrt{6}\ \det\left(\frac{\mathbf{D}_{\mathrm{dev}}}{\|\mathbf{D}_{\mathrm{dev}}\|}\right). \tag{8.38}$$

Above, the deviatoric tensor is $\mathbf{D}_{\mathrm{dev}} = \mathbf{D} - \frac{\mathrm{tr}(\mathbf{D})}{3}\mathbf{I}$, whereas the operators det(·) and $\|\cdot\|$ respectively indicate the determinant and the norm of a matrix (specifically, $\|\mathbf{A}\| = \sqrt{\mathrm{tr}(\mathbf{A}\ \mathbf{A}^{\mathrm{T}})}$ for a generic matrix **A**). Note that the mode is not defined for isotropic tensors.

8.2.13 Microscopic diffusion tensors

The orientationally invariant metrics obtained from the DT (see Section 8.2.14) have been widely used to investigate properties of grey and white matter in health and disease. However, it is well known that DT-derived metrics lack in specificity, as shown in Figures 8.10 and 8.11 for the FA dependency on fibre

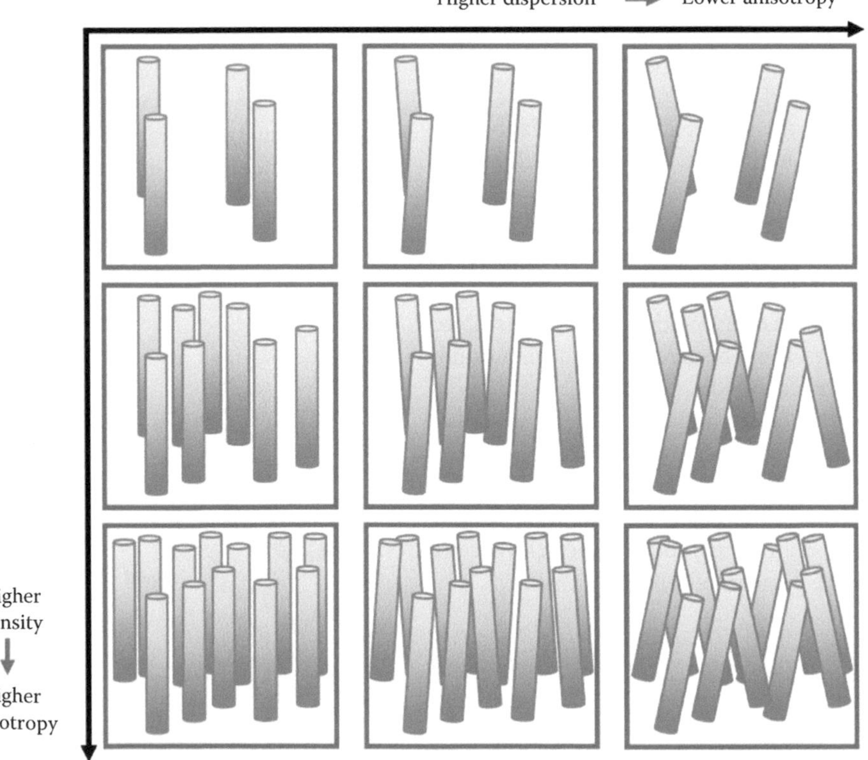

FIGURE 8.10 Schematic illustrating the opposite effects that increasing density of anisotropic structures and increasing dispersion of the orientations of the same structures have on the DT anisotropy. The cartoons are representative of white matter patches where the axons have been modelled as cylinders, which are intrinsically anisotropic. The figure shows that increasing density of the cylinders implies higher DT anisotropy, while increasing variability of the cylinder orientations implies lower DT anisotropy. It follows that a change on DT anisotropy can be caused independently by variations of the density of the underlying microscopic domains, or by a change in their configuration.

orientations. In practice, for white matter this means that axons arranged according to different orientation distributions can produce different values of DT metrics such as FA. Similarly, the same values of for example FA can be produced by fibre arrangements characterised by different combinations of neuronal density and orientation (Zhang *et al.*, 2012).

Recently, it was shown that it is possible to obtain DTs that are not confounded by the underlying orientation distribution of fibres, using advanced design of the diffusion encoding gradients (Jespersen *et al.*, 2013; Lasič *et al.*, 2014; Szczepankiewicz *et al.*, 2015) or even using conventional PGSE (Kaden *et al.*, 2015). Specifically, in the latter case, the fact that the spherical mean of the DW signal (i.e. the integral of the DW signal over all possible unit vectors in the 3D space) is independent of such an orientation distribution is exploited. This approach, known as *spherical mean technique* (Kaden *et al.*, 2015), enables the estimation of scalar invariants describing the properties of individual axonal elements, referred to as *microscopic*. As an example, Figure 8.11 shows a schematic representation of different possible white matter segments, constituted by the same set of cylindrical axonal elements arranged according to different orientation distributions, alongside values of FA and microscopic FA (or micro FA, μFA). The figure shows that as the variability of the axonal orientations (also known as axonal *orientation dispersion*; Zhang et al., 2012) increases, FA decreases, while μFA does not change. The fact that lower FA values are observed for increasing orientation dispersion is explained by the fact that increasing orientation dispersion widens the range of directions parallel to at least one axonal element (diffusion along axonal elements is relatively free). In contrast, the constant μFA reflects that μFA is a property of the individual segments, regardless of their spatial orientation.

8.2.14 Estimation of the diffusion tensor from MRI measurements

Similarly to what was previously illustrated in Section 8.2.8, the elements of the DT can be estimated in each MRI voxel from a set of DW images. Specifically, the six independent elements of the DT can be evaluated voxel by voxel, fitting the tensor model (Equation 8.29) to a set of M measurements $\tilde{\mathbf{S}}$ via linear or non-linear fitting procedures.

In the case of linear fitting, a linear system of equations in the unknowns represented by the elements of the DT can be obtained by taking the logarithm of both sides of Equation 8.29, which provides

$$\log S\left(TE,b_{ij}\right)= \log S_0\left(TE\right)- \sum_{i,j} b_{ij}\, D_{i,j}\,. \tag{8.39}$$

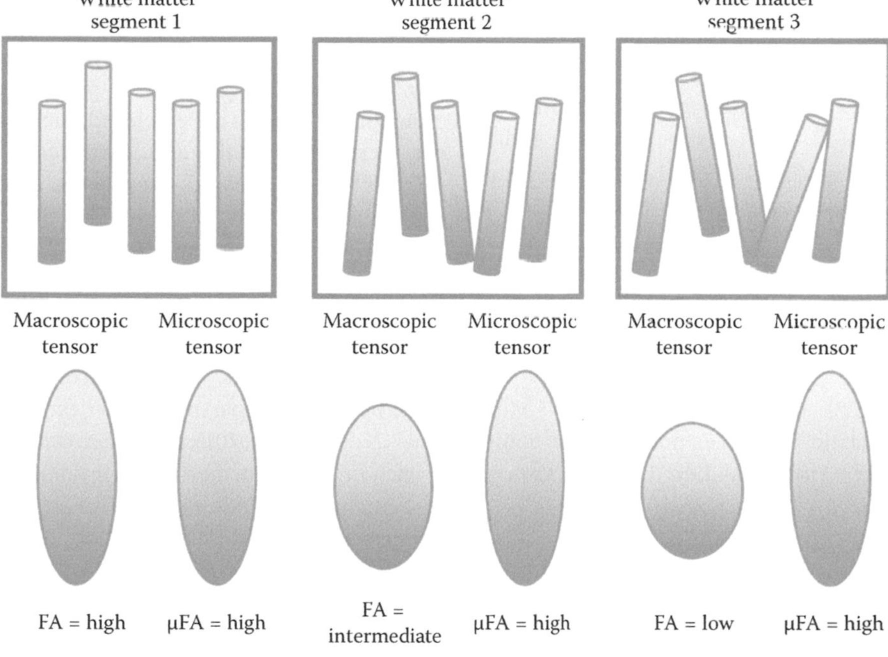

FIGURE 8.11 Illustration of the effect of orientation dispersion on the diffusion tensor (DT) and its relationship with the microscopic DT. The figure depicts three white matter segments, represented as a collection of cylinders modelling axons. From left to right, the coherence of cylinder orientations decreases, i.e. the cylinder orientation dispersion increases. The figure also shows the macroscopic and microscopic DT, alongside their FA values (indicated as μFA for the microscopic DT). The macroscopic DT describes the overall diffusion properties of the segment, while the microscopic DT describes the diffusion properties of an individual axon and of its surroundings. When all cylinders are perfectly parallel, the two tensors are the same. However, when orientation dispersion increases, the macroscopic tensor becomes less anisotropic, while the microscopic tensor is not confounded by the underlying cylinder orientation distribution. Notice that the concept of microscopic anisotropy is rather advanced and is the focus of several research groups across the world.

The term $\Sigma_{i,j} b_{i,j} D_{i,j}$ can be expanded as

$$\sum_{i,j} b_{ij}\, D_{i,j} = b_{xx}\, D_{xx} + b_{yy}\, D_{yy} + b_{zz}\, D_{zz} + \left(b_{xy} + b_{yx}\right) D_{xy} + \left(b_{xz} + b_{zx}\right) D_{x} + \left(b_{yz} + b_{zy}\right) D_{yz} = \mathbf{b} : \mathbf{D}. \tag{8.40}$$

Above, : is the Hadamard matrix product (i.e. element-wise product) and **b** is a 3×3 b-matrix, defined as

$$\mathbf{b} = \left[b_{ij}\right]_{\substack{i=x,y,z,\\ j=x,y,z}} \tag{8.41}$$

such that $b_{i,j}$ follows the expression in Equation 8.30.

Equation 8.39 can be fitted to a set of M measurements $\tilde{\mathbf{S}} = \left\{\tilde{S}_m(\mathbf{b}_m) \mid m = 1, \dots, M\right\}$ with the least squares approach using the tools of linear regression or weighted linear regression, as the latter is more robust towards the effects on homoscedasticity of the linearisation process. For instance, Basser and colleagues (Basser *et al.*, 1994a) describe the use of multivariate linear regression to estimate the DT from a non-DW image plus six or more DW measurements along non-collinear directions.

Alternatively, non-linear fitting can also be employed. Similarly to what we illustrated before for the estimation of a scalar diffusion coefficient D, under the hypothesis of additive Gaussian noise the measured signal $\tilde{S}$ can be related to the signal predicted by the DT model as

$$\tilde{S}\left(\mathbf{D}, S_0; b_{ij}\right) = S_0\, e^{-\mathbf{b}:\mathbf{D}} + \eta, \tag{8.42}$$

As before (Section 8.2.8), $S_0(TE) = A\, e^{-\frac{TE}{T_2}}$ and $\eta \sim N(0,\sigma^2)$ is a noise-related stochastic term (i.e. noise has zero mean and variance equal to σ^2). From Equation 8.42, it follows that $\tilde{S} \sim N\left(S_0\, e^{-\mathbf{b}:\mathbf{D}}, \sigma^2\right)$. The unknown elements of the DT **D** and the unknown non-DW signal S_0 can be estimated maximising the posterior probability $P\left(\mathbf{D}, S_0 \mid \tilde{\mathbf{S}}\right)$. When no prior information about the elements of **D** and about S_0 are employed, by means of Bayes' theorem this implies maximising the likelihood $L\left(\tilde{\mathbf{S}} \mid \mathbf{D}, S_0\right)$ of the measurements given the DT model, or, in practice, minimising the quantity $-\log L\left(\tilde{\mathbf{S}} \mid \mathbf{D}, S_0\right)$, with $\log L\left(\tilde{\mathbf{S}} \mid \mathbf{D}, S_0\right)$ equal to

$$\log L\left(\tilde{\mathbf{S}} \mid D, S_0\right) = -\frac{M}{2}\log\left(2\pi\sigma^2\right) - \frac{1}{2\sigma^2}\sum_{m=1}^{M}\left(\tilde{S}_m - S_0\, e^{-\mathbf{b}_m:\mathbf{D}}\right)^2. \tag{8.43}$$

Finally, we report that when the contribution of the imaging gradients is negligible, the Hadamard product in Equation 8.40 can be practically implemented as

$$\mathbf{b} : \mathbf{D} = b\, \hat{g} \cdot \left(\mathbf{D}\, \hat{g}\right) \tag{8.44}$$

where b is the b-factor (Equation 8.15), $\hat{g}$ is a 3×1 unit vector describing the direction of the diffusion encoding gradient **g**, and where · represents the dot product.

8.2.15 Other diffusion weighting approaches

The PGSE approach is currently the standard way to sensitise MRI data to water molecule diffusion, and PGSE sequences are readily available on most clinical scanners. Nonetheless, other diffusion encoding approaches have been explored in the literature and some of them constitute now an active field of research, as described in detail in Chapter 9.

In *oscillating gradient spin echo* (OGSE), oscillating, rather than pulsed, gradient waveforms are employed. OGSE sequences probe very short diffusion times, while achieving sufficient diffusion weighting to detect a signal change (Schachter *et al.*, 2000). This approach has been employed in a number of applications, as for example the study of the time dependency of the diffusion weighted signal in the healthy brain *in vivo* (Baron and Beaulieu, 2014), as well as in animal models of brain tumours (Reynaud *et al.*, 2016).

In *multiple pulsed field gradient* (mPFG) experiments, multiple diffusion directions are applied in a SE sequence with one or more refocusing pulses. Typically, two different diffusion encoding gradients are employed, such that the mPFG sequence is referred to as *double diffusion encoding* (Shemesh *et al.*, 2016). These type of sequences enable the quantification of the correlation of spin displacements due to diffusion along multiple directions (Jespersen and Buhl, 2011), which can be employed to estimate specific geometrical characteristics of the restricted, cellular space within which water diffuses (Szczepankiewicz *et al.*, 2015). Furthermore, they can be employed to disentangle the opposing effects that within-compartment water restriction and between-compartment water exchange have on the DW signal (Nilsson *et al.*, 2013; Lampinen *et al.*, 2017), potentially useful when indices of membrane permeability are of interest.

Finally, we mention innovative approaches based on *q-space trajectory imaging* (Westin *et al.*, 2016), where the diffusion encoding gradients are not limited to being linearly polarised. This approach allows diffusion sensitisation along multiple directions within a single acquisition and is useful to obtain isotropic diffusion weighting and to characterise water diffusion in terms of a distribution of diffusion tensors and of its moments.

8.3 Biological origin of diffusion changes in brain tissue

The ADC of water in biological tissues, measured with MRI, gives information on the average diffusion properties of the sample, over a volume the size of a voxel (i.e. at 3T of the order of several mm^3). Water diffusion is a microscopic effect and its quantitative value can be very different in different tissue types and in different molecular environments. Also it is important to remember that water diffusion is strongly dependent on temperature (Mills, 1973). Increasing temperature increases the thermal agitation of diffusing particles, leading to increased diffusivity. The temperature dependence of the diffusivity can be described using the Arrenhius equation (Dhital *et al.*, 2016).

In non-biological samples, anisotropy may be an intrinsic property of the sample, dependent on its molecular structure (e.g. in certain liquid crystals characterised by shaped molecules, where mobility is facilitated along one direction more than along others; Callaghan 1995). In these cases the diffusion properties of the sample in any particular direction can be measured independently of the diffusion time (Equation 8.18 holds for any diffusion time). The relationship between the observed mean square displacement, $<R^2>$, and the diffusion time, defined as the time between corresponding points of the diffusion gradients ($t_d = \Delta - \delta/3$) is linear (see Equation 8.8, where $t = t_d$). In this situation, diffusion is said to be anisotropic and unrestricted.

In biological tissues, however, the situation is more complex, and constraints other than molecular shape influence the measured diffusion coefficient. The microstructural environments are complex and difficult to model, where elements such as cell membranes create partial barriers or obstacles to the molecular motion, creating compartmentalisation, allowing selective transport of molecules depending on permeability or restricting diffusion because of boundaries. As a result of this, the measured diffusion properties are anisotropic (Moseley *et al.*, 1990b) (i.e. diffusion properties are different when assessed along different directions) and depend on the value of the diffusion time (Figure 8.12), relative to the average time, τ_c, between two successive collisions of a molecule and the boundary. If $t_d \ll \tau_c$, then most of the molecules behave as if no boundaries were present and the molecular motion appears unrestricted, with a diffusion coefficient D_{free}. If t_d is on the order of τ_c, however, any particular molecule is likely to encounter the boundary during the time of the diffusion sequence and its spatial position at time t_d will be affected. This, in turn, affects the observed mean square displacement $<R^2>$, which no longer increases linearly with the diffusion time. At very long diffusion times, $t_d \gg \tau_c$, a completely restricted diffusion regime is reached with a new effective coefficient, D_{eff}, and diffusion in this case is said to be anisotropic and restricted. In the intermediate case, an ADC can still be calculated, ignoring the non-linearities, but the estimated D will differ from both D_{free} and D_{eff}.

When averaged over all molecules, the dependency of $<R^2>$ on t_d is reflected also on the signal attenuation plot, if the b-factor is varied by changing t_d while keeping the gradient amplitude constant. The transition between free and restricted diffusion happens smoothly as the various internal structures of the sample assume a more dominant role in confining the molecule

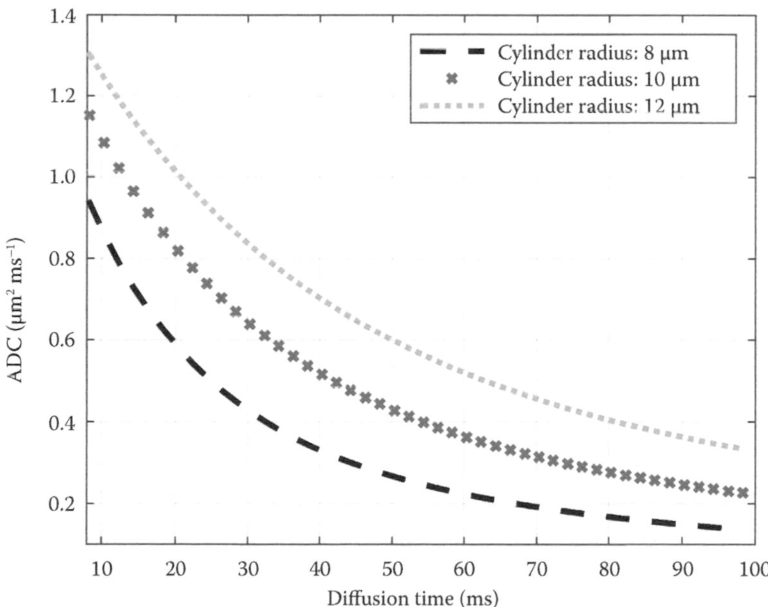

FIGURE 8.12 Example of the effect of restriction on the ADC. The figure shows the ADC of a water pool diffusing within an impermeable cylinder, with the ADC relative to a direction perpendicular to the axis of the cylinder (i.e. for a direction along which confinements restricting the random walks of spins exist). The figure was obtained from a computer simulation of a PGSE acquisition with $\delta = 5$ ms and b-values up to 3000 s mm^{-2}, adopting an intrinsic spin diffusivity of 2 μm^2 ms^{-1}. The plot shows how the ADC decreases as the diffusion time increases, due to the presence of the cylinder wall restricting the diffusion process. The plots also highlights as increasing radii imply increasing ADC, since as the radius increases a higher fraction of spins are less likely to encounter the barriers during their random walks. Note that to observe such a steep dependence of ADC within brain white matter axons, shorter diffusion times need to be used, as the axon radius is on the order of 0.5–3 μm.

movements as t_d increases (Novikov *et al.*, 2014). Because of this dependency on the diffusion time t_d, *any estimated 'apparent' diffusion coefficient depends on the measurement parameters chosen* (namely, the gradient timing).

From the above, it is clear that anything affecting the molecular environment, causing changes in molecular mobility (e.g. changes in tissue structure or in properties such as viscosity and tortuosity;[10] Latour *et al.*, 1995) or modifying tissue compartments, is likely to have a measurable effect on the DT. Pathological processes such as inflammation, oedema, cell swelling, cell necrosis, membrane damage, demyelination, cell growth, axonal loss, gliosis or axonal reorganisation are all likely to be visible on indices from DT imaging, but as all these processes in principle affect diffusion measurements, the complexity of interpreting clinical results becomes apparent.

8.4 Quantification of diffusion

8.4.1 General comments and instrumental requirements

8.4.1.1 Motion artefacts

DW techniques, sensitised to capture microscopic motion, are particularly prone to motion artefacts caused by macroscopic motion of the head or pulsation of the brain. This is can be partially overcome with motion correction techniques (e.g. navigator echo acquisition; De Crespigny *et al.*, 1995) or by implementing a fast acquisition method, able to acquire the whole image in one shot in about 80 ms, such as EPI (Stehling *et al.*, 1991). However all of these ultra-fast sequences have disadvantages such as lower SNR in the acquired images, high RF power deposition, blurring due to longer readout acquisitions and increased artefacts (due to increased sensitivity to susceptibility changes or other mechanisms). The choice of a navigated DW-SE or single-shot DW-EPI (or any other fast method) depends on the purpose of the study and the available scanning time.

It is worth knowing that rigid motion between the different DW images can be corrected via post-processing techniques now widely offered by a variety of software toolkits. However, it is important to remember that following co-registration, *the gradient directions need to be rotated* to compensate for the rotational component of the estimated rigid transformations, as failing to do so can affect the estimation of the diffusion properties (Leemans and Jones, 2009).

8.4.1.2 Gradients and eddy currents

Rapid imaging techniques such as the EPI sequence typically used for DW MRI place a number of constraints on scanner hardware, requiring rapid switching of high amplitude gradients in order to cover a large region of k-space within a time that is short relative to T_2*. In addition to these requirements, diffusion sequences involve applying high amplitude gradients for much longer periods (typically 15–30 ms, compared to the duration of a typical EPI readout gradient lobe of the order of 1 ms). These diffusion encoding gradients must have rapid rise times, to minimise unwanted T_2 decay, and must also have accurately defined, reproducible and stable amplitude. Resonant gradient technologies are not appropriate for producing the well-separated, individual gradient pulses required for diffusion encoding.

Accurate gradient calibration is essential for quantitative diffusion measurements. Image-based measurements, often used both for initial scanner calibration and ongoing quality assurance, are typically accurate to 1 pixel over a field of view (i.e. the gradient amplitude is set with an accuracy of approximately 0.5%–1%). While this is sufficient for most imaging protocols, the squared gradient term in the *b*-factor calculation makes quantitative diffusion measurements sensitive to small gradient inaccuracies, so additional calibration may be required, especially at ultra-high field and in preclinical scenarios (O'Callaghan *et al.*, 2014). Phantoms of known, isotropic diffusivity (Tofts *et al.*, 2000; Gatidis *et al.*, 2014) or ice-water phantoms (Malyarenko *et al.*, 2013) can be employed for this purpose, bearing in mind that temperature strongly affects the diffusivity (Mills, 1973) and change in temperature during experiments may confound results. Quality assurance procedure are essential (particularly in longitudinal or multicentre studies), to ensure comparability between measurements made on different dates or at different sites.

Another gradient-related problem in DW experiments is the effect of eddy currents. Current loops, induced by the rapid changes of magnetic field during gradient switching in any conducting elements close to the gradient coils (particularly in the magnet cryostat and its various thermal shields), create their own magnetic fields, which oppose the applied field (by Lenz' Law). While short-term eddy currents (with time constants of a few microseconds) are usually well compensated for by either hardware or software gradient pre-emphasis, longer-term eddy currents (with time constants of the order of tens or hundreds of microseconds) can have a significant effect on diffusion-weighted images. Firstly, eddy currents from the first DW gradient can distort the second DW gradient, leading to imperfect signal rephasing even in the absence of diffusion. Secondly, eddy currents from the second DW gradient can lead to a time-varying, background gradient throughout the EPI readout period. Depending on the orientation of the DW gradient with respect to the read and phase encode directions and its rate of variation with time, a number of characteristic image distortions are induced, most notably stretching or shearing of the image (Le Bihan *et al.*, 2006; Nunes *et al.*, 2011; Graham *et al.*, 2016). If the time constant of the eddy current is of the order of the repetition time, the effects may appear on images in subsequent *TR* periods, independently of the subsequent image DWs. Image co-registration techniques have been developed to successfully reduce distortion artefacts in DW-EPI images (for more information see works such as those by Andersson *et al.*, 2016; Andersson and Sotiropoulos, 2016;

[10] Material studies have demonstrated that the restricted behaviour of the diffusion coefficient D_{eff} reflects the tortuosity of the diffusion pathway connecting adjacent pore spaces, as shown in Latour *et al.* The tortuosity *T* is defined as the factor by which *D* is reduced from its unrestricted value: $D_{eff} = D_{free}/T$. Since the tortuosity is related to the porosity of the medium, D_{eff} gives information about the macroscopic structure of the sample in terms of the volume fraction of the interstitial spaces.

and Graham *et al.*, 2016). Moreover, gradient pre-emphasis solutions can be implemented (e.g. Papadakis *et al.*, 2000; Spees *et al.*, 2011) or further improvement can come from appropriate design of the acquisition method, as for example using twice-refocused SE sequences (Reese *et al.*, 2003; Clayden *et al.*, 2016).

8.4.1.3 Susceptibility artefacts

A further issue of all EPI scans is that of field inhomogeneities due to differences in the local magnetic susceptibility, which induce magnetic field gradients leading to geometric distortions and signal dropouts or pileups near tissue–air and tissue–bone interfaces (Le Bihan *et al.*, 2006). Susceptibility effects increase with field strength; hence the errors induced by these unknown gradients will worsen with stronger magnets such as 7T scanners. A number of approaches have been proposed to overcome this problem. In terms of image acquisition, the usage of parallel imaging techniques such as SENSE (Pruessmann *et al.*, 1999) or GRAPPA (Griswold *et al.*, 2002), the optimisation of readout (Weiskopf *et al.*, 2007) or shimming (Gu *et al.*, 2002) can all help mitigate distortions or signal dropouts and pileups. From an image processing point of view, images can be non-linearly unwarped, exploiting knowledge of the local amount of magnetic field inhomogeneity, sampled in a B_0 field map (Jezzard and Balaban, 1995; Zeng and Constable, 2002), which can be calculated in a number of ways: from gradient echo acquisitions with different *TE*, from multireference scans (Wan *et al.*, 1997) or by sampling the k-space twice but with different trajectories. Specifically, the last approach, often referred to as *blip-up/blip-down*[11] (Andersson *et al.*, 2003; Andersson and Sotiropoulos, 2016) provides the same image twice but with distortions in opposite directions. In absence of a field map or of blip-up/blip-down acquisitions, it is possible to correct the distortions using multimodal registration frameworks (Glodeck *et al.*, 2016).

8.4.1.4 Signal-to-noise ratio

Consideration of SNR issues is crucial when dealing with quantitative measurements in general, and therefore for DW measurements as well. If magnitude images are used (as opposed to complex data), the noise follows a Rician distribution (Gudbjartsson and Patz, 1995) or more complex distributions such as non-central chi (Sotiropoulos *et al.*, 2013). These approximate a Gaussian distribution at high SNR. However, at lower SNR levels, such noise characteristics lead to a minimum measurable signal intensity even in areas of pure noise (Eichner *et al.*, 2015), which can lead to errors in ADC/DT calculations unless appropriately taken into account (e.g. as shown for instance by Wheeler-Kingshott *et al.*, 2002; Koay and Basser, 2006; and Kaden *et al.*, 2015). A further issue, which is specific to diffusion measurements, is the so-called sorting bias of the tensor eigenvalues due to noise. This can lead to inaccuracy, as well as imprecision, in anisotropy measures at low signal to noise levels (Pierpaoli and Basser, 1996). Denoising techniques (Manjón *et al.*, 2013; Becker *et al.*, 2014; Veraart *et al.*, 2016) can help mitigate the issues coming from the intrinsic low SNR of DW-EPI.

8.4.1.5 Acquisition strategies

Acquisition is an important aspect of any diffusion MRI experiment.

Today, acquisitions generally rely on fast single-shot EPI readouts, which enable the acquisition of MRI slices in a fraction of second (Stehling *et al.*, 1991) and provide robustness to motion, as compared to conventional multishot SE approaches. While most commonly EPI readouts are based on Cartesian sampling, other approaches such as spiral sampling (Wilm *et al.*, 2017) have shown additional benefits (such as limited distortion), but they often rely on extra hardware (such as field cameras; Dietrich *et al.*, 2016) or acquisition of additional data to compensate for artefacts inherent to system imperfections. Single-shot sequences, such as EPI, can benefit dramatically if employed using the latest phased array coil technology and in conjunction with state-of-the-art acceleration strategies, such as *simultaneous multislice* (SMS) imaging (Setsompop *et al.*, 2013), also known as multiband (MB) imaging. SMS enables dramatic decreases of *TR* for a given slice coverage and consequently of total scan time. Nowadays, with MB factors of 3 or more, it is possible to acquire rich, multishell high resolution DW MRI data sets within 10 minutes or less, close to the limit of what could be run in a real clinical scenario.

In general, protocols for acquiring DW images in clinical settings (research or diagnosis) are always driven by a particular question and are limited by the total scanning time available. The simplest type of quantitative experiment based on DW imaging is ADC mapping. ADC mapping is useful in applications such as ischaemic stroke and may require as few as two measurements (such as one non-DW and one DW) to map ADC along a particular direction. While such a type of ADC map may be useful for inspection purposes, it is not well suited for quantitative comparison across subjects, as differences in alignment between the subject's head and the laboratory reference frame confound the value of ADC. However, with a higher number of DW measurements, as few as three, it is possible to map ADC along three orthogonal directions and, combining those, the mean ADC (<ADC>), which is a rotationally invariant index that can be considered as an estimate of MD. With even richer acquisition strategies, a full DT could be calculated, enabling the evaluation of several rotationally invariant indices such as those listed in Section 8.2.14. In the most modern systems with the latest hardware and acquisition technologies SNR levels are higher than in the past, and accurate DT indices can be obtained within a few minutes with as few as a dozen DW measurements geometrically distributed over a sphere, all corresponding to one *b*-value. However, a much higher number of DW directions potentially distributed over multiple *b*-values may be required when more advanced diffusion techniques are of interest (see Chapter 9). Nowadays, it is possible to perform DTI at resolutions of the order of 1.5 mm (isotropic) using single-shot EPI sequences that employ MB acceleration, which enable great anatomical detail even in small structures within clinically feasible times.

[11] The term *blip-up/blip-down* refers to the fact that the k-space is filled in opposite directions along the phase encode direction in two subsequent acquisitions. In practice, to achieve this, the steps of phase encoding are performed using a reversed polarity for the blips of the phase encode gradient.

While it is impossible to prescribe an acquisition protocol that suits all cases, it is advised that before the commencement of each study, appropriate repeatability experiments and SNR measurements are performed to ensure the stability of DT parameters and the sensitivity to biological variability. In each specific case, the choice of the resolution is driven by the available SNR level (which decreases as higher b-values are targeted, due to increases in TE). In general, it is suggested that when an appropriate level of SNR is achieved (on the order of 20 at $b = 0$, with SNR here defined as the ratio between the signal and the noise standard deviation), any available scan time is invested in increasing the angular resolution of the DW protocol, rather than on signal averaging.

8.4.1.6 CSF: pulsation artefacts and cardiac gating

Independently of the pulse sequence used, pulsation of the brain, caused by CSF motion synchronised to the heart beat, can introduce signal dropouts and image artefacts in DW scans. The severity of these depends on the timing of data acquisition relative to the phase of the heart cycle and is particularly prominent in the brain stem, periventricular regions (Greitz *et al.*, 1992) and spinal cord (Cohen-Adad and Wheeler-Kingshott, 2014). If a fixed repetition time is used, data collection happens independently of heart rate, and some images will be acquired during the systole (ventricle contraction), when the pulsation effect is greater (Summers *et al.*, 2006). In order to reduce this effect, cardiac triggering (also known as 'gating') can be used, so that the acquisition of the data happens during the 'flat' portion of the electrocardiogram, when the heart is relaxed (diastole). Unfortunately, this reduces the time efficiency of the sequence as the presence of a delay between acquisitions of groups of slices (Figure 8.13) implies that fewer slices can be collected within a TR period. The value of TR (which is usually expressed as a multiple or the intervals between consecutive R waves of the electrocardiogram or ECG signal, i.e. RR intervals, such that $TR = N_{RR}T_{RR}$; see Figure 8.13) becomes subject dependent and, especially if the heart cycle is irregular, TR may also vary between different diffusion gradients steps and averages. This in consequence leads to differential T_1 weighting for each b-value or gradient direction unless very large N_{RR} (≥8) are used (a larger number of N_{RR} would allow to collect enough slices for a whole brain coverage at a high resolution, guaranteeing complete relaxation of every tissue, including CSF and pathological tissue, which can be characterised by long T_1 relaxation time).

It is important to know how cardiac gating works for the scanner/sequence used because it can either trigger every QRS complex, or every *n*th QRS complex, with QRS being the sequence of Q, R and S waves in an electrocardiogram or ECG signal. Usually, cardiac gated scans trigger every *n*th QRS complex (see Figure 8.13), where an average T_{RR} interval is evaluated and proper gating happens only every *n*<*TRR*> (with <*TRR*> being an index of average heartbeat), in order to try to maintain a consistent delay between repeated acquisitions of the same slice. *Triggering every QRS complex is strongly suggested*, instead, to ensure that the data acquisition stays in phase with the cardiac cycle, even during large variations of the heartbeat.

Also, when using cardiac gating, the variable TR may make it difficult to use inversion recovery preparation for CSF suppression, which would otherwise be useful for reducing the confounding effects of very high CSF diffusion coefficient on ADC measurements in the cortex and close to the ventricles. Of course, if a fixed TR is used, CSF suppression is a viable option (Kwong *et al.*, 1991; Falconer and Narayana, 1997; Li *et al.*, 2012). However, the consequences in terms of SNR (i.e., unless a very long TR is used) and possible artefacts that can affect the data make this

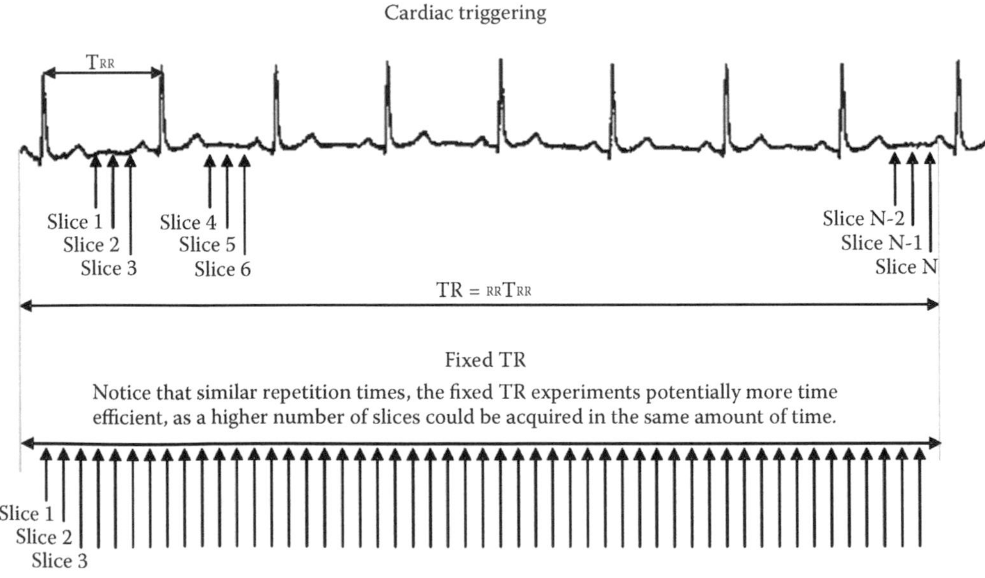

FIGURE 8.13 Comparison of the maximum number of slices that can be collected in a given TR, using either cardiac triggering or a fixed TR. The fixed TR is more efficient, but pulsation artefacts can have deleterious effects. Note that the slice numbering in the figure refers to slice acquisition time order (Slice 2 = second slice acquired, but not necessarily the second spatial position).

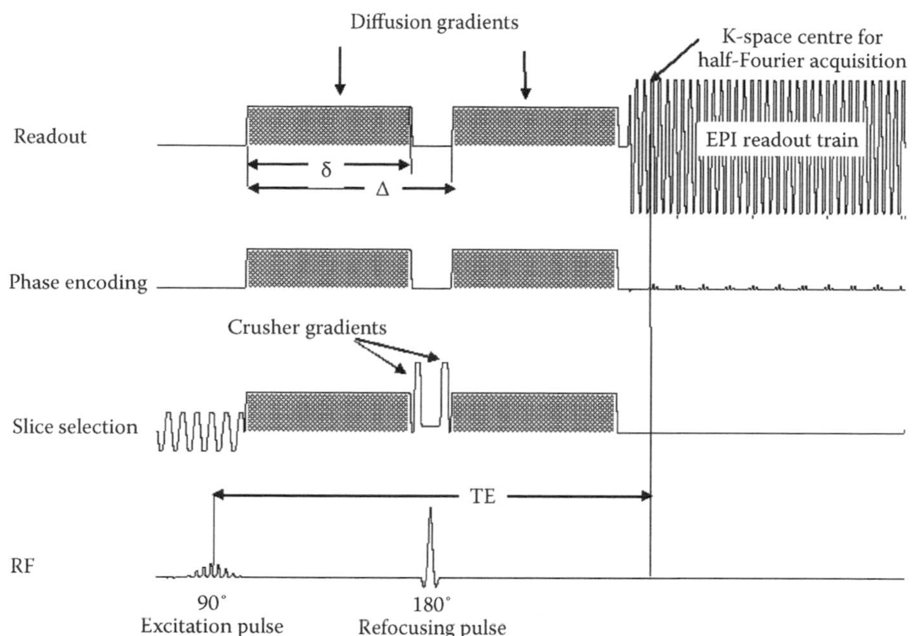

FIGURE 8.14 Typical SE DW-EPI pulse sequence. The sequence is conceptually identical to the theoretical pulse sequence shown in Figure 8.4; however, here realistic RF pulses and gradient waveforms are added as they would be used to encode the information in space and derive images. The diffusion gradients (shaded) and the main sequence parameters are indicated in the diagram. Although here the diffusion gradients are shown here as being applied along all three directions, during the acquisition they are in fact applied with amplitude and direction defined by the diffusion scheme. Note that the 90° excitation pulse is a 'spectral–spatial' pulse used for lipid suppression.

an often-unattractive option. Finally, we point out that another potential issue related to cardiac gating is that it may not be always compatible with the DW sequences supplied by major vendors.

8.4.1.7 Perfusion and diffusion

When perfusion effects[12] are present, as in grey matter tissue, at low *b*-factors the signal decay can be characterised in terms of a pseudo-diffusion coefficient, some 10 times larger than the diffusion coefficient of water in tissue (Le Bihan *et al.*, 1986; Le Bihan, 1990). This pseudo-diffusion coefficient can be explained by the *intravoxel incoherent motion* model (Le Bihan *et al.*, 1986), which suggests that at low *b*-values ($b \sim 100$ s mm^{-2}) DW images have a contrast that reflects a combination of diffusion and perfusion properties. Given the faster motion of water within the capillary bed, the contribution of perfusion to the signal attenuation can be reduced by increasing the *b*-factors used for diffusion imaging (Henkelman *et al.*, 1994). If the diffusion coefficient is calculated with a minimum *b*-value different from zero, than the isotropic perfusion effect can be reduced.

8.4.1.8 Sampling the DW signal – optimum b-values

The signal decay curve *S*(*TE*, *b*) can be sampled in many different ways, as extensively investigated over the years. It has been recognised that for a single 'apparent' diffusion component a two-point estimate is optimal in terms of SNR per unit time (Jones *et al.*, 1999). For a two-point acquisition it also becomes possible to optimise the *b* factors used for a particular tissue of interest. The optimum difference in *b*-values between the non- or minimally DW image and the highly weighted image depends on whether a single ADC or full DT measurement is to be achieved. Ignoring T_2 relaxation effects, this difference has been suggested to be optimally close to $b_{max} - b_{min} = 1.0/D$ (Jones *et al.*, 1999) in order to minimise the variance of the DT element estimates. In practice, it is common to employ values of *b* of the order of 1000 s mm^{-2} to estimate the DT in the brain.

8.4.1.9 b-Factor calculation

For the PGSE sequence the *b*-factor calculations can be based simply on the assumption that the diffusion gradients are rectangular pulses of such high amplitude that all the other imaging gradients do not contribute to the diffusion weighting. This is valid, especially if the acquisition sequence is well balanced and the imaging gradients are applied to cause minimum interaction with the diffusion gradients. For a more precise calculation of the diffusion parameters, though (e.g. see (Le Bihan, 1994)), it is preferable to include all the gradients in the *b*-factor calculations (e.g. see Mattiello *et al.*, 1994; Clark *et al.*, 1999).

8.4.1.10 Software

To date, a number of software toolkits are freely available to analyse DW MRI data and obtain ADC and DTI maps. A complete list of the available options is beyond the scope of this chapter. Briefly, we mention that the toolkits provide tools for image pre-processing (including routines for denoising, for the mitigation

[12] Microcirculation of blood in the capillary network.

of subject's motion and distortions due to eddy currents as well as off-resonance effects), for quantitative analysis (i.e. ADC and DT fitting) and for co-registration between DW MRI and anatomical scans. Software toolkits may also offer pipelines for group analysis, useful when two or more groups of controls and patients are to be compared to detect the effects that diseases have on different areas of the brain.

8.4.2 Normative values, reproducibility and multicentre studies

Several studies have investigated the reproducibility of obtaining reliable DT metrics at various field strength and have provided normative values of DT indices in the human brain. Tables 8.1 through 8.3 respectively summarise recent studies reporting

TABLE 8.1 List of studies on normative values of DTI metrics

Study	Scanner	Age	Brain area	Metric
		Range		Mean (SD)
Brander *et al.* (2010)	SiemensAvanto 1.5T	19–61 years	Basal pons (R)	FA: 0.64 (0.08)
			Basal pons (L)	FA: 0.66 (0.08)
			Corona radiata (R)	FA: 0.52 (0.04)
			Corona radiata (L)	FA: 0.55 (0.04)
			Corpus callosum (genu)	FA: 0.85 (0.04)
			Corpus callosum (splenium)	FA: 0.87 (0.05)
	Siemens Trio 3T	19–61 years	Basal pons (R)	FA: 0.67 (0.08)
			Basal pons (L)	FA: 0.67 (0.04)
			Corona radiata (R)	FA: 0.48 (0.04)
			Corona radiata (L)	FA: 0.52 (0.05)
			Corpus callosum (genu)	FA: 0.84 (0.03)
			Corpus callosum (splenium)	FA: 0.86 (0.05)
Cancelliere *et al.* (2013)	GE Signa Horizon LX 1.5T			Approximate range (from regression plot)
		0 months	Corpus callosum (genu)	FA: (0.35; 0.55); MD: (1.30; 1.70)
		12 months		FA: (0.60; 0.77); MD: (0.90; 1.30)
		24 months		FA: (0.63; 0.79); MD: (0.80; 1.20)
		50 months		FA: (0.65; 0.80); MD: (0.75; 1.10)
		100 months		FA: (0.65; 0.80); MD: (0.75; 1.10)
		Mean (SD)		Mean (95% interval)
Keller *et al.* (2013)	Siemens Avanto 1.5T	43.8 (18) years	Basal ganglia	FA: 0.17 (0.07; 0.27)
			Grey matter	FA: 0.33 (0.18; 0.48)
			Corpus callosum (genu)	FA: 0.81 (0.69; 0.94)
			Corpus callosum (splenium)	FA: 0.90 (0.81; 0.99)
		Mean (SD)		Mean (SD)
Kumar *et al.* (2013)	Siemens Magnetom Tim-Trio 3T	18.5 (4.6) years	Deep cerebellar nuclei	AD: 1.00 (0.07); RD: 0.62 (0.04)
			Superior cerebellar peduncule (R)	AD: 1.42 (0.07); RD: 0.36 (0.04)
			Superior cerebellar peduncule (L)	AD: 1.43 (0.06); RD: 0.34 (0.06)
			Anterior thalamus (R)	AD: 1.05 (0.06); RD: 0.61 (0.06)
			Anterior thalamus (L)	AD: 1.06 (0.06); RD: 0.61 (0.06)
			Occipital white matter (R)	AD: 1.25 (0.09); RD: 0.57 (0.05)
			Occipital white matter (L)	AD: 1.17 (0.17); RD: 0.57 (0.04)
			Frontal grey matter (R)	AD: 1.17 (0.05); RD: 0.67 (0.06)
			Frontal grey matter (L)	AD: 1.17 (0.06); RD: 0.67 (0.06)
			Posterior corpus callosum	AD: 1.74 (0.15); RD: 0.35 (0.16)
		Mean (Range)		Approx. 95% interval
Polders *et al.* (2011)	Philips Achieva 1.5T	27.6 (20–54) years	Corpus callosum	FA: (0.64; 0.69)
			Corticospinal tract	FA: (0.60; 0.65)
			Cingulum bundles	FA: (0.53; 0.58)
	Philips Achieva 3T	27.6 (20–54) years	Corpus callosum	FA: (0.61; 0.69)
			Corticospinal tract	FA: (0.62; 0.65)
			Cingulum bundles	FA: (0.45; 0.55)
	Philips Achieva 7T	27.6 (20–54) years	Corpus callosum	FA: (0.59; 0.68)
			Corticospinal tract	FA: (0.54; 0.63)
			Cingulum bundles	FA: (0.48; 0.58)

Note: DTI = diffusion tensor imaging; FA = fractional anisotropy; MD = mean diffusivity; AD = axial diffusivity; RD = radial diffusivity; SD = standard deviation; R = right; L = left. Metrics AD, RD and MD are provided in $\mu m^2\ ms^{-1}$. Note that values from region-of-interest (ROI) analysis are strongly dependent on the size and location of the ROIs, as well as on the operator consistency in case of manual outlining.

TABLE 8.2 List of studies on reproducibility of DTI metrics

Study	Reproducibility indices	Type of experiment
Bisdas *et al.* (2008)	ICC and Bland–Altmann score	Scan–rescan; one scanner
Bonekamp *et al.* (2007)	CV	Scan–rescan; one scanner
Cercignani *et al.* (2003)	CV	Comparison of protocols; two scanners (two vendors)
Ciccarelli *et al.* (2003)	CV	Single scan; one scanner
Duan *et al.* (2015)	ICC	Scan–rescan; one scanner
Farrell *et al.* (2007)	Accuracy and precision	Repeated scans on one subject; one scanner
Grech-Sollars *et al.* (2015)	CV	Travelling heads and phantom; eight scanners (two vendors)
Jansen *et al.* (2007)	CV	Scan–rescan; one scanner
Kamagata *et al.* (2015)	CV	Test–retest; two scanners (same model)
Landman *et al.* (2007)	Accuracy and precision	Repeated scans on one subject; one scanner
Lepomäki *et al.* (2012)	ICC and Bland–Altman score	Single scan; one scanner
Liu *et al.* (2014)	ICC, CV	Test–retest and protocol comparison; one scanner
Miao *et al.* (2015)	Angular variance, t-test statistics, CV	Protocol comparison; one scanner
Müller *et al.* (2006)	ICC, CV	Single scan; one scanner
Palacios *et al.* (2017)	CV	Travelling head and phantom; 13 scanners (3 vendors)
Papinutto *et al.* (2013)	Absolute difference	Scan–rescan; one scanner
Veenith *et al.* (2013)	CV	Scan–rescan; one scanner
Vollmar *et al.* (2010)	ICC, CV	Travelling heads, scan–rescan; two scanners (same models)

Note: ICC = intraclass correlation coefficient; CV = coefficient of variation.

TABLE 8.3 List of recent multicentre studies where DTI indices were employed as outcome measures

Study	Number of centres	Number of participants	Context	Relevant DTI metrics
Galanaud *et al.* (2012)	10	204	Traumatic brain injury	FA, AD, RD, MD
Mayo *et al.* (2017)	59	67	Alzheimer's disease	FA, MD
Mohammadi *et al.* (2012)	3	1	Design of image processing for multicentre studies	FA, MD
Müller *et al.* (2016)	8	442	Amyotrophic lateral sclerosis	FA
Nir *et al.* (2015)	59	200	Alzheimer's disease	FA, MD
Zhang *et al.* (2016)	8	1180	Schizophrenia, autism, attention deficit hyperactivity disorder	Connectivity mapping based on DTI

normative values of DT metrics (Brander *et al.*, 2010; Polders *et al.*, 2011; Kumar *et al.*, 2012; Cancelliere *et al.*, 2013; Keller *et al.*, 2013); reproducibility figures (Cercignani *et al.*, 2003; Ciccarelli *et al.*, 2003; Müller *et al.*, 2006; Bonekamp *et al.*, 2007; Farrell *et al.*, 2007; Jansen *et al.*, 2007; Landman *et al.*, 2007; Bisdas *et al.*, 2008; Vollmar *et al.*, 2010; Lepomäki *et al.*, 2012; Papinutto *et al.*, 2013; Veenith *et al.*, 2013; Liu *et al.*, 2014; Duan *et al.*, 2015; Grech-Sollars *et al.*, 2015; Kamagata *et al.*, 2015; Miao *et al.*, 2015; Palacios *et al.*, 2017); and examples of multicentre studies involving acquisition of DW images and DTI analysis (Galanaud *et al.*, 2012; Mohammadi *et al.*, 2012; Nir *et al.*, 2015; Müller *et al.*, 2016; Zhang *et al.*, 2016; Mayo *et al.*, 2017). Specifically, the studies demonstrate that DT metrics enable the reproducible characterisation of brain microstructure during development and in several neurological diseases using a fixed MRI scanner (coefficient of variation of metrics on the order of 10% or below; intraclass correlations on the order of 0.80 or above). However, even for a fixed set of acquisition parameters (i.e. *b*-value, resolution, diffusion encoding protocol), it is possible to get values of DTI metrics for the same brain regions that differ across scanners, even when the exact same individuals are scanned. For this purpose, it is advised in multisite studies to employ a travelling phantom to assess systematic intersite differences.

8.4.3 Concluding remarks

In this chapter, the molecular basis upon which diffusion MRI of the brain relies has been introduced. Moreover, a popular approach for diffusion imaging, DTI, has been described, alongside with several practical points concerned with the acquisition of DW data with modern MRI systems. While DTI offers useful metrics of microstructural changes in neurological conditions, complementary approaches extend the technique to gain a deeper insight into the cytoarchitectural properties of biological tissues, striving for more sensitive and specific diagnostic and prognostic tools that use water as a non-invasive probe of microstructure. The reader is referred to Chapter 9 for a detailed discussion on these complementary methods.

Acknowledgements

We thank Gareth J. Barker, Stefan C.A. Steens and Mark A. van Buchem for their contribution to the previous edition of the chapter, which has now been substantially reviewed and updated. We thank Torben Schneider for useful discussion.

References

Andersson JL, Graham MS, Zsoldos E, Sotiropoulos SN. Incorporating outlier detection and replacement into a non-parametric framework for movement and distortion correction of diffusion MR images. NeuroImage 2016; 141: 556–72.

Andersson JL, Skare S, Ashburner J. How to correct susceptibility distortions in spin-echo echo-planar images: application to diffusion tensor imaging. NeuroImage 2003; 20(2): 870–88.

Andersson JL, Sotiropoulos SN. An integrated approach to correction for off-resonance effects and subject movement in diffusion MR imaging. NeuroImage 2016; 125: 1063–78.

Baron CA, Beaulieu C. Oscillating gradient spin-echo (OGSE) diffusion tensor imaging of the human brain. Magn Reson Med 2014; 72(3): 726–36.

Barzykin AV. Exact solution of the Torrey-Bloch equation for a spin echo in restricted geometries. Phys Rev B 1998; 58(21): 14171.

Basser PJ, Mattiello J, LeBihan D. Estimation of the effective self-diffusion tensor from the NMR spin echo. J Magn Reson, Series B 1994a; 103(3): 247–54.

Basser PJ, Mattiello J, LeBihan D. MR diffusion tensor spectroscopy and imaging. Biophys J 1994b; 66(1): 259.

Basser PJ, Pierpaoli C. Microstructural and physiological features of tissues elucidated by quantitative-diffusion-tensor MRI. J Magn Reson, Series B 1996; 111(2): 209–19.

Bastiani M, Roebroeck A. Unraveling the multiscale structural organization and connectivity of the human brain: the role of diffusion MRI. Front Neuroanat 2015; 9: 77.

Becker S, Tabelow K, Mohammadi S, Weiskopf N, Polzehl J. Adaptive smoothing of multi-shell diffusion weighted magnetic resonance data by msPOAS. NeuroImage 2014; 95: 90–105.

Beltrachini L, Taylor ZA, Frangi AF. A parametric finite element solution of the generalised Bloch–Torrey equation for arbitrary domains. J Magn Reson 2015; 259: 126–34.

Bisdas S, Bohning D, Bešenski N, Nicholas J, Rumboldt Z. Reproducibility, interrater agreement, and age-related changes of fractional anisotropy measures at 3T in healthy subjects: effect of the applied b-value. Am J Neuroradiol 2008; 29(6): 1128–33.

Bonekamp D, Nagae LM, Degaonkar M, Matson M, Abdalla WM, Barker PB, et al. Diffusion tensor imaging in children and adolescents: reproducibility, hemispheric, and age-related differences. NeuroImage 2007; 34(2): 733–42.

Brander A, Kataja A, Saastamoinen A, Ryymin P, Huhtala H, Öhman J, et al. Diffusion tensor imaging of the brain in a healthy adult population: normative values and measurement reproducibility at 3 T and 1.5 T. Acta Radiologica 2010; 51(7): 800–7.

Budde MD, Kim JH, Liang HF, Russell JH, Cross AH, Song SK. Axonal injury detected by in vivo diffusion tensor imaging correlates with neurological disability in a mouse model of multiple sclerosis. NMR Biomed 2008; 21(6): 589–97.

Callaghan PT. *Principles of nuclear magnetic resonance microscopy.* Oxford University Press, Oxford; 1995.

Cancelliere A, Mangano F, Air E, Jones B, Altaye M, Rajagopal A, et al. DTI values in key white matter tracts from infancy through adolescence. Am J Neuroradiol 2013; 34(7): 1443–9.

Cercignani M, Bammer R, Sormani MP, Fazekas F, Filippi M. Inter-sequence and inter-imaging unit variability of diffusion tensor MR imaging histogram-derived metrics of the brain in healthy volunteers. Am J Neuroradiol 2003; 24(4): 638–43.

Ciccarelli O, Parker G, Toosy A, Wheeler-Kingshott C, Barker G, Boulby P, et al. From diffusion tractography to quantitative white matter tract measures: a reproducibility study. NeuroImage 2003; 18(2): 348–59.

Clark C, Barker G, Tofts P. An in vivo evaluation of the effects of local magnetic susceptibility-induced gradients on water diffusion measurements in human brain. J Magn Reson 1999; 141(1): 52–61.

Clark CA, Le Bihan D. Water diffusion compartmentation and anisotropy at high b values in the human brain. Magn Reson Med 2000; 44(6): 852–9.

Clayden JD, Nagy Z, Weiskopf N, Alexander DC, Clark CA. Microstructural parameter estimation in vivo using diffusion MRI and structured prior information. Magn Reson Med 2016; 75(4): 1787–96.

Cohen-Adad J, Wheeler-Kingshott C. *Quantitative MRI of the spinal cord.* Academic Press, Elsevier; Amsterdam, Netherlands; 2014.

Crank J. *The mathematics of diffusion*: Oxford University Press, Oxford; 1998.

De Crespigny AJ, Marks MP, Enzmann DR, Moseley ME. Navigated diffusion imaging of normal and ischemic human brain. Magn Reson Med 1995; 33(5): 720–8.

De Santis S, Jones DK, Roebroeck A. Including diffusion time dependence in the extra-axonal space improves in vivo estimates of axonal diameter and density in human white matter. NeuroImage 2016; 130: 91–103.

Dhital B, Labadie C, Stallmach F, Möller HE, Turner R. Temperature dependence of water diffusion pools in brain white matter. NeuroImage 2016; 127: 135–43.

Dietrich BE, Brunner DO, Wilm BJ, Barmet C, Gross S, Kasper L, et al. A field camera for MR sequence monitoring and system analysis. Magn Reson Med 2016; 75(4): 1831–40.

Duan F, Zhao T, He Y, Shu N. Test–retest reliability of diffusion measures in cerebral white matter: a multiband diffusion MRI study. J Magn Reson Imaging 2015; 42(4): 1106–16.

Dzik-Jurasz A, Domenig C, George M, Wolber J, Padhani A, Brown G, et al. Diffusion MRI for prediction of response of rectal cancer to chemoradiation. Lancet 2002; 360(9329): 307–8.

Eichner C, Cauley SF, Cohen-Adad J, Möller HE, Turner R, Setsompop K, et al. Real diffusion-weighted MRI enabling true signal averaging and increased diffusion contrast. NeuroImage 2015; 122: 373–84.

Einstein A. Über die von der molekularkinetischen Theorie der Wärme geforderte Bewegung von in ruhenden Flüssigkeiten suspendierten Teilchen. Annalen der Physik 1905; 322(8): 549–60.

Ennis DB, Kindlmann G. Orthogonal tensor invariants and the analysis of diffusion tensor magnetic resonance images. Magn Reson Med 2006; 55(1): 136–46.

Falconer JC, Narayana PA. Cerebrospinal fluid-suppressed high-resolution diffusion imaging of human brain. Magn Reson Med 1997; 37(1): 119–23.

Farrell JA, Landman BA, Jones CK, Smith SA, Prince JL, van Zijl P, et al. Effects of signal-to-noise ratio on the accuracy and reproducibility of diffusion tensor imaging–derived fractional anisotropy, mean diffusivity, and principal eigenvector measurements at 1.5 T. J Magn Reson Imaging 2007; 26(3): 756–67.

Filippi M, Cercignani M, Inglese M, Horsfield M, Comi G. Diffusion tensor magnetic resonance imaging in multiple sclerosis. Neurology 2001; 56(3): 304–11.

Frahm J, Merboldt K, Hänicke W, Haase A. Stimulated echo imaging. J Magn Resonan (1969) 1985; 64(1): 81–93.

Galanaud D, Perlbarg V, Gupta R, Stevens RD, Sanchez P, Tollard E, et al. Assessment of white matter injury and outcome in severe brain trauma: a prospective multicenter cohort. J Am Soc Anesthesiol 2012; 117(6): 1300–10.

Gatidis S, Schmidt H, Martirosian P, Schwenzer NF. Development of an MRI phantom for diffusion-weighted imaging with independent adjustment of apparent diffusion coefficient values and T2 relaxation times. Magn Reson Med 2014; 72(2): 459–63.

Geijer B, Sundgren P, Lindgren A, Brockstedt S, Ståhlberg F, Holtås S. The value of b required to avoid T2 shine-through from old lacunar infarcts in diffusion-weighted imaging. Neuroradiology 2001; 43(7): 511–7.

Glodeck D, Hesser J, Zheng L. Distortion correction of EPI data using multimodal nonrigid registration with an anisotropic regularization. Magn Resonan Imaging 2016; 34(2): 127–36.

Graham MS, Drobnjak I, Zhang H. Realistic simulation of artefacts in diffusion MRI for validating post-processing correction techniques. NeuroImage 2016; 125: 1079–94.

Grech-Sollars M, Hales PW, Miyazaki K, Raschke F, Rodriguez D, Wilson M, et al. Multi-centre reproducibility of diffusion MRI parameters for clinical sequences in the brain. NMR Biomed 2015; 28(4): 468–85.

Greitz D, Wirestam R, Franck A, Nordell B, Thomsen C, Ståhlberg F. Pulsatile brain movement and associated hydrodynamics studied by magnetic resonance phase imaging. Neuroradiology 1992; 34(5): 370–80.

Griswold MA, Jakob PM, Heidemann RM, Nittka M, Jellus V, Wang J, et al. Generalized autocalibrating partially parallel acquisitions (GRAPPA). Magn Reson Med 2002; 47(6): 1202–10.

Grussu F, Schneider T, Zhang H, Alexander DC, Wheeler-Kingshott CA. Neurite orientation dispersion and density imaging of the healthy cervical spinal cord in vivo. NeuroImage 2015; 111: 590–601.

Gu H, Feng H, Zhan W, Xu S, Silbersweig DA, Stern E, et al. Single-shot interleaved z-shim EPI with optimized compensation for signal losses due to susceptibility-induced field inhomogeneity at 3 T. NeuroImage 2002; 17(3): 1358–64.

Gudbjartsson H, Patz S. The Rician distribution of noisy MRI data. Magn Reson Med 1995; 34(6): 910–14.

Hall MG, Alexander DC. Convergence and parameter choice for Monte-Carlo simulations of diffusion MRI. IEEE Trans Med Imaging 2009; 28(9): 1354–64.

Hanyu H, Asano T, Sakurai H, Imon Y, Iwamoto T, Takasaki M, et al. Diffusion-weighted and magnetization transfer imaging of the corpus callosum in Alzheimer's disease. J Neurol Sci 1999; 167(1): 37–44.

Henkelman RM, Neil JJ, Xiang QS. A quantitative interpretation of IVIM measurements of vascular perfusion in the rat brain. Magn Reson Med 1994; 32(4): 464–9.

Jansen JF, Kooi ME, Kessels AG, Nicolay K, Backes WH. Reproducibility of quantitative cerebral T2 relaxometry, diffusion tensor imaging, and 1H magnetic resonance spectroscopy at 3.0 Tesla. Investig Radiol 2007; 42(6): 327–37.

Jelescu IO, Veraart J, Fieremans E, Novikov DS. Degeneracy in model parameter estimation for multi-compartmental diffusion in neuronal tissue. NMR Biomed 2016; 29(1): 33–47.

Jespersen SN, Buhl N. The displacement correlation tensor: microstructure, ensemble anisotropy and curving fibers. J Magn Reson 2011; 208(1): 34–43.

Jespersen SN, Lundell H, Sønderby CK, Dyrby TB. Orientationally invariant metrics of apparent compartment eccentricity from double pulsed field gradient diffusion experiments. NMR Biomed 2013; 26(12): 1647–62.

Jezzard P, Balaban RS. Correction for geometric distortion in echo planar images from B0 field variations. Magn Reson Med 1995; 34(1): 65–73.

Johansen-Berg H, Behrens TE. *Diffusion MRI: from quantitative measurement to in vivo neuroanatomy*: Academic Press, Elsevier; Amsterdam, Netherlands; 2013.

Jones D, Horsfield M, Simmons A. Optimal strategies for measuring diffusion in anisotropic systems by magnetic resonance imaging. Magn Reson Med 1999; 42(3): 515–25.

Jones DK. Diffusion MRI: *theory, methods and applications*: Oxford University Press, Oxford; 2010.

Jones DK, Knösche TR, Turner R. White matter integrity, fiber count, and other fallacies: the do's and don'ts of diffusion MRI. NeuroImage 2013; 73: 239–54.

Kaden E, Kelm ND, Carson RP, Does MD, Alexander DC. Multi-compartment microscopic diffusion imaging. NeuroImage 2016; 139: 346–59.

Kaden E, Kruggel F, Alexander DC. Quantitative mapping of the per-axon diffusion coefficients in brain white matter. Magn Reson Med 2016;75(4):1752–63.

Kamagata K, Shimoji K, Nishikori A, Tsuruta K, Yoshida M, Kamiya K, et al. Intersite reliability of diffusion tensor imaging on two 3T scanners. Magn Reson Med Sci 2015; 14(3): 227–33.

Keller J, Rulseh AM, Komárek A, Latnerová I, Rusina R, Brožová H, et al. New non-linear color look-up table for visualization of brain fractional anisotropy based on normative measurements—principals and first clinical use. PLoS One 2013; 8(8): e71431.

Kenkre V, Fukushima E, Sheltraw D. Simple solutions of the Torrey–Bloch equations in the NMR study of molecular diffusion. J Magn Reson 1997; 128(1): 62–9.

Koay CG, Basser PJ. Analytically exact correction scheme for signal extraction from noisy magnitude MR signals. J Magn Reson 2006; 179(2): 317–22.

Kumar R, Nguyen HD, Macey PM, Woo MA, Harper RM. Regional brain axial and radial diffusivity changes during development. J Neurosci Res 2012; 90(2): 346–55.

Kwong K, McKinstry R, Chien D, Crawley A, Pearlman J, Rosen B. CSF-suppressed quantitative single-shot diffusion imaging. Magn Reson Med 1991; 21(1): 157–63.

Lampinen B, Szczepankiewicz F, Westen D, Englund E, C Sundgren P, Lätt J, et al. Optimal experimental design for filter exchange imaging: apparent exchange rate measurements in the healthy brain and in intracranial tumors. Magn Reson Med 2017; 77(3): 1104–14.

Landman BA, Farrell JA, Jones CK, Smith SA, Prince JL, Mori S. Effects of diffusion weighting schemes on the reproducibility of DTI-derived fractional anisotropy, mean diffusivity, and principal eigenvector measurements at 1.5 T. NeuroImage 2007; 36(4): 1123–38.

Lasič S, Szczepankiewicz F, Eriksson S, Nilsson M, Topgaard D. Microanisotropy imaging: quantification of microscopic diffusion anisotropy and orientational order parameter by diffusion MRI with magic-angle spinning of the q-vector. Front Phys 2014; 2: 11.

Latour L, Kleinberg RL, Mitra PP, Sotak CH. Pore-size distributions and tortuosity in heterogeneous porous media. J Magn Reson, Series A 1995; 112(1): 83–91.

Le Bihan D. Magnetic resonance imaging of perfusion. Magn Reson Med 1990; 14(2): 283–92.

Le Bihan D. Diffusion, perfusion and functional magnetic resonance imaging. Journal des Maladies Vasculaires 1994; 20(3): 203–14.

Le Bihan D. Looking into the functional architecture of the brain with diffusion MRI. Nat Rev Neurosci 2003; 4(6): 469–80.

Le Bihan D, Breton E, Lallemand D, Grenier P, Cabanis E, Laval-Jeantet M. MR imaging of intravoxel incoherent motions: application to diffusion and perfusion in neurologic disorders. Radiology 1986; 161(2): 401–7.

Le Bihan D, Iima M. Diffusion magnetic resonance imaging: what water tells us about biological tissues. PLoS Biol 2015; 13(7).

Le Bihan D, Poupon C, Amadon A, Lethimonnier F. Artifacts and pitfalls in diffusion MRI. J Magn Reson Imaging 2006; 24(3): 478–88.

Leemans A, Jones DK. The B-matrix must be rotated when correcting for subject motion in DTI data. Magn Reson Med 2009; 61(6): 1336–49.

Lepomäki VK, Paavilainen TP, Hurme SA, Komu ME, Parkkola RK, Group PS. Fractional anisotropy and mean diffusivity parameters of the brain white matter tracts in preterm infants: reproducibility of region-of-interest measurements. Pediatr Radiol 2012; 42(2): 175–82.

Li L, Miller KL, Jezzard P. DANTE-prepared pulse trains: a novel approach to motion-sensitized and motion-suppressed quantitative magnetic resonance imaging. Magn Reson Med 2012; 68(5): 1423–38.

Liu X, Yang Y, Sun J, Yu G, Xu J, Niu C, et al. Reproducibility of diffusion tensor imaging in normal subjects: an evaluation of different gradient sampling schemes and registration algorithm. Neuroradiology 2014; 56(6): 497–510.

Lundell H, Nielsen JB, Ptito M, Dyrby TB. Distribution of collateral fibers in the monkey cervical spinal cord detected with diffusion-weighted magnetic resonance imaging. NeuroImage 2011; 56(3): 923–9.

Malyarenko D, Galbán CJ, Londy FJ, Meyer CR, Johnson TD, Rehemtulla A, et al. Multi-system repeatability and reproducibility of apparent diffusion coefficient measurement using an ice-water phantom. J Magn Reson Imaging 2013; 37(5): 1238–46.

Manjón JV, Coupé P, Concha L, Buades A, Collins DL, Robles M. Diffusion weighted image denoising using overcomplete local PCA. PLoS One 2013; 8(9): e73021.

Mattiello J, Basser PJ, LeBihan D. Analytical expressions for the b matrix in NMR diffusion imaging and spectroscopy. J Magn Reson, 1994; 108(2): 131–41.

Mayo CD, Mazerolle EL, Ritchie L, Fisk JD, Gawryluk JR, Initiative AsDN. Longitudinal changes in microstructural white matter metrics in Alzheimer's disease. NeuroImage: Clin 2017; 13: 330–8.

Merboldt K-D, Hanicke W, Frahm J. Self-diffusion NMR imaging using stimulated echoes. J Magn Reson 1985; 64(3): 479–86.

Miao HC, Wu MT, Kao E-f, Chiu YH, Chou MC. Comparisons of reproducibility and mean values of diffusion tensor imaging-derived indices between unipolar and bipolar diffusion pulse sequences. J Neuroimaging 2015; 25(6): 892–9.

Mills R. Self-diffusion in normal and heavy water in the range 1-45°. J Phys Chem 1973; 77(5): 685–8.

Mohammadi S, Nagy Z, Möller HE, Symms MR, Carmichael DW, Josephs O, et al. The effect of local perturbation fields on human DTI: characterisation, measurement and correction. NeuroImage 2012; 60(1): 562–70.

Moseley M, Cohen Y, Mintorovitch J, Chileuitt L, Shimizu H, Kucharczyk J, et al. Early detection of regional cerebral ischemia in cats: comparison of diffusion and T2-weighted MRI and spectroscopy. Magn Reson Med 1990a; 14(2): 330–46.

Moseley ME, Cohen Y, Kucharczyk J, Mintorovitch J, Asgari H, Wendland M, et al. Diffusion-weighted MR imaging of anisotropic water diffusion in cat central nervous system. Radiology 1990b; 176(2): 439–45.

Müller H-P, Turner MR, Grosskreutz J, Abrahams S, Bede P, Govind V, et al. A large-scale multicentre cerebral diffusion tensor imaging study in amyotrophic lateral sclerosis. J Neurol Neurosurg Psychiatry 2016; 87(6): 570–9.

Müller M, Mazanek M, Weibrich C, Dellani P, Stoeter P, Fellgiebel A. Distribution characteristics, reproducibility, and precision of region of interest-based hippocampal diffusion tensor imaging measures. Am J Neuroradiol 2006; 27(2): 440–6.

Nilsson M, Lätt J, van Westen D, Brockstedt S, Lasič S, Ståhlberg F, et al. Noninvasive mapping of water diffusional exchange in the human brain using filter-exchange imaging. Magn Reson Med 2013; 69(6): 1572–80.

Nir TM, Jahanshad N, Toga AW, Bernstein MA, Jack CR, Weiner MW, et al. Connectivity network measures predict volumetric atrophy in mild cognitive impairment. Neurobiol Aging 2015; 36: S113–S20.

Novikov DS, Jensen JH, Helpern JA, Fieremans E. Revealing mesoscopic structural universality with diffusion. Proc Natl Acad Sci 2014; 111(14): 5088–93.

Nunes RG, Drobnjak I, Clare S, Jezzard P, Jenkinson M. Performance of single spin-echo and doubly refocused diffusion-weighted sequences in the presence of eddy current fields with multiple components. Magn Reson Imag. 2011; 29(5): 659–67.

O'Callaghan J, Wells J, Richardson S, Holmes H, Yu Y, Walker-Samuel S, et al. Is your system calibrated? MRI gradient system calibration for pre-clinical, high-resolution imaging. PLoS One 2014; 9(5): e96568.

Palacios E, Martin A, Boss M, Ezekiel F, Chang Y, Yuh E, et al. Toward precision and reproducibility of diffusion tensor imaging: a multicenter diffusion phantom and traveling volunteer study. Am J Neuroradiol 2017; 38(3): 537–45.

Panagiotaki E, Schneider T, Siow B, Hall MG, Lythgoe MF, Alexander DC. Compartment models of the diffusion MR signal in brain white matter: a taxonomy and comparison. NeuroImage 2012; 59(3): 2241–54.

Papadakis NG, Martin KM, Pickard JD, Hall LD, Carpenter TA, Huang CLH. Gradient preemphasis calibration in diffusion-weighted echo-planar imaging. Magn Reson Med 2000; 44(4): 616–24.

Papinutto ND, Maule F, Jovicich J. Reproducibility and biases in high field brain diffusion MRI: an evaluation of acquisition and analysis variables. Magn Reson Imaging 2013; 31(6): 827–39.

Pierpaoli C, Basser PJ. Toward a quantitative assessment of diffusion anisotropy. Magn Reson Med 1996; 36(6): 893–906.

Polders DL, Leemans A, Hendrikse J, Donahue MJ, Luijten PR, Hoogduin JM. Signal to noise ratio and uncertainty in diffusion tensor imaging at 1.5, 3.0, and 7.0 Tesla. J Magn Reson Imaging 2011; 33(6): 1456–63.

Pruessmann KP, Weiger M, Scheidegger MB, Boesiger P. SENSE: sensitivity encoding for fast MRI. Magn Reson Med 1999; 42(5): 952–62.

Reese T, Heid O, Weisskoff R, Wedeen V. Reduction of eddy-current-induced distortion in diffusion MRI using a twice-refocused spin echo. Magn Reson Med 2003; 49(1): 177–82.

Reynaud O, Winters KV, Hoang DM, Wadghiri YZ, Novikov DS, Kim SG. Surface-to-volume ratio mapping of tumor microstructure using oscillating gradient diffusion weighted imaging. Magn Reson Med 2016; 76(1): 237–47.

Schachter M, Does M, Anderson A, Gore J. Measurements of restricted diffusion using an oscillating gradient spin-echo sequence. J Magn Reson 2000; 147(2): 232–7.

Setsompop K, Kimmlingen R, Eberlein E, Witzel T, Cohen-Adad J, McNab JA, et al. Pushing the limits of in vivo diffusion MRI for the Human Connectome Project. NeuroImage 2013; 80: 220–33.

Shemesh N, Jespersen SN, Alexander DC, Cohen Y, Drobnjak I, Dyrby TB, et al. Conventions and nomenclature for double diffusion encoding NMR and MRI. Magn Reson Med 2016; 75(1): 82–7.

Song S-K, Yoshino J, Le TQ, Lin S-J, Sun S-W, Cross AH, et al. Demyelination increases radial diffusivity in corpus callosum of mouse brain. NeuroImage 2005; 26(1): 132–40.

Sotiropoulos S, Moeller S, Jbabdi S, Xu J, Andersson J, Auerbach E, et al. Effects of image reconstruction on fiber orientation mapping from multichannel diffusion MRI: reducing the noise floor using SENSE. Magn Reson Med 2013; 70(6): 1682–9.

Spees WM, Buhl N, Sun P, Ackerman JJ, Neil JJ, Garbow JR. Quantification and compensation of eddy-current-induced magnetic-field gradients. J Magn Reson 2011; 212(1): 116–23.

Stehling MK, Turner R, Mansfield P. Echo-planar imaging: magnetic resonance imaging in a fraction of a second. Science 1991; 254(5028): 43–50.

Stejskal EO, Tanner JE. Spin diffusion measurements: spin echoes in the presence of a time-dependent field gradient. J Chem Phys 1965; 42(1): 288–92.

Summers P, Staempfli P, Jaermann T, Kwiecinski S, Kollias S. A preliminary study of the effects of trigger timing on diffusion tensor imaging of the human spinal cord. Am J Neuroradiol 2006; 27(9): 1952–61.

Szczepankiewicz F, Lasič S, van Westen D, Sundgren PC, Englund E, Westin C-F, et al. Quantification of microscopic diffusion anisotropy disentangles effects of orientation dispersion from microstructure: applications in healthy volunteers and in brain tumors. NeuroImage 2015; 104: 241–52.

Tofts PS, Lloyd D, Clark CA, Barker GJ, Parker GJ, McConville P, Baldock C, Pope JM et al. Test liquids for quantitative MRI measurements of self-diffusion coefficient in vivo. Magn Reson Med 2000; 43(3): 368–74.

Torrey HC. Bloch equations with diffusion terms. Phys Rev 1956; 104(3): 563.

Tricoche X, Kindlmann G, Westin C-F. Invariant crease lines for topological and structural analysis of tensor fields. IEEE Trans Vis Comp Graph 2008; 14(6): 1627–34.

Tsuchiya K, Osawa A, Katase S, Fujikawa A, Hachiya J, Aoki S. Diffusion-weighted MRI of subdural and epidural empyemas. Neuroradiology 2003; 45(4): 220–3.

Veenith TV, Carter E, Grossac J, Newcombe VF, Outtrim JG, Lupson V, et al. Inter subject variability and reproducibility of diffusion tensor imaging within and between different imaging sessions. PLoS One 2013; 8(6): e65941.

Veraart J, Novikov DS, Christiaens D, Ades-Aron B, Sijbers J, Fieremans E. Denoising of diffusion MRI using random matrix theory. NeuroImage 2016; 142: 394–406.

Vollmar C, O'Muircheartaigh J, Barker GJ, Symms MR, Thompson P, Kumari V, et al. Identical, but not the same: intra-site and inter-site reproducibility of fractional anisotropy measures on two 3.0 T scanners. NeuroImage 2010; 51(4): 1384–94.

Wan X, Gullberg GT, Parker DL, Zeng GL. Reduction of geometric and intensity distortions in echo-planar imaging using a multireference scan. Magn Reson Med 1997; 37(6): 932–42.

Wang Y, Sun P, Wang Q, Trinkaus K, Schmidt RE, Naismith RT, et al. Differentiation and quantification of inflammation, demyelination and axon injury or loss in multiple sclerosis. Brain 2015; 138(5): 1223–38.

Warach S, Chien D, Li W, Ronthal M, Edelman R. Fast magnetic resonance diffusion-weighted imaging of acute human stroke. Neurology 1992; 42(9): 1717–23.

Weiskopf N, Hutton C, Josephs O, Turner R, Deichmann R. Optimized EPI for fMRI studies of the orbitofrontal cortex: compensation of susceptibility-induced gradients in the readout direction. Magn Reson Mater Phys Biol Med 2007; 20(1): 39–49.

Wesbey GE, Moseley ME, Ehman RL. Translational molecular self-Diffusion in magnetic resonance imaging: II. Measurement of the self-Diffusion coefficient. Invest Radiol 1984; 19(6): 491–8.

Westin C-F, Knutsson H, Pasternak O, Szczepankiewicz F, Özarslan E, van Westen D, et al. Q-space trajectory imaging for multidimensional diffusion MRI of the human brain. NeuroImage 2016; 135: 345–62.

Wheeler-Kingshott CAM, Ciccarelli O, Schneider T, Alexander DC, Cercignani M. A new approach to structural integrity assessment based on axial and radial diffusivities. Func Neurol 2012; 27(2): 85.

Wheeler-Kingshott CA, Cercignani M. About "axial" and "radial" diffusivities. Magn Reson Med 2009; 61(5): 1255–60.

Wheeler-Kingshott CA, Parker GJ, Symms MR, Hickman SJ, Tofts PS, Miller DH, et al. ADC mapping of the human optic nerve: increased resolution, coverage, and reliability with CSF-suppressed ZOOM-EPI. Magn Reson Med 2002; 47(1): 24–31.

Wilm BJ, Barmet C, Gross S, Kasper L, Vannesjo SJ, Haeberlin M, et al. Single-shot spiral imaging enabled by an expanded encoding model: demonstration in diffusion MRI. Magn Reson Med 2017; 77(1): 83–91.

Zeng H, Constable RT. Image distortion correction in EPI: comparison of field mapping with point spread function mapping. Magn Reson Med 2002; 48(1): 137–46.

Zhang H, Schneider T, Wheeler-Kingshott CA, Alexander DC. NODDI: practical in vivo neurite orientation dispersion and density imaging of the human brain. NeuroImage 2012; 61(4): 1000–16.

Zhang J, Cheng W, Liu Z, Zhang K, Lei X, Yao Y, et al. Neural, electrophysiological and anatomical basis of brain-network variability and its characteristic changes in mental disorders. Brain 2016; 139(8): 2307–21.

9

Advanced Diffusion Models[1]

Contents

Aurobrata Ghosh[2],
Andrada Ianus[2] and
Daniel C.Alexander
University College London

9.1 Introduction

Diffusion MRI (dMRI) probes the dispersion of water-molecules typically over a few tens of milliseconds. In biological tissue, the geometry and organisation of cellular membranes determines the dispersion pattern. Thus, the measured signal is sensitive to the tissue architecture at the micron scale, orders of magnitude below the typical resolution of MRI. One of the key aims of dMRI research and applications is to extract information and obtain maps that provide insight into microscopic tissue properties at the macroscopic scale.

Diffusion tensor imaging (DTI) (Basser *et al.*, 1994a, 1994b) established dMRI's potential for investigating the microstructure *in vivo* and non-invasively. DTI indices such as fractional anisotropy, mean diffusivity, etc. (see Chapter 8), provide some quantitative insight into the fine structure of the tissue changes that arise through, for example, pathology and ageing (Abe *et al.*, 2002; Head *et al.*, 2004). Furthermore, tractography (i.e. the reconstruction of white matter pathways based on the principal directions of diffusion in the brain) emerged naturally from DTI and has today become a cornerstone of human brain connectivity mapping (Mori *et al.*, 1999; Basser *et al.*, 2000).

Nonetheless, DTI's assumption of Gaussian diffusion leads to two main limitations:

1. It does not explain departures from log-linear signal against b-value (mono-exponential signal decay) that are observed in a wide range of biological tissue; e.g. Clark and Le Bihan (2000) observed bi-exponential signal decay in both white and grey matter of the brain.
2. DTI can detect only a single fibre orientation, which is insufficient for describing complex anisotropic tissue structures containing a distribution of fibre orientations (Wiegell *et al.*, 2000; Alexander *et al.*, 2002; Frank, 2002; Tuch *et al.*, 2002), which arise frequently in white matter of the brain as well as e.g. muscle or heart tissue (Peyrat *et al.*, 2007).

However, neither limitation is fundamental to dMRI in general. Observations of signal variation more complex than what DTI assumes have prompted the development of a wide variety of alternative models to recover more sensitive and specific microscopic features of the tissue. The literature can be broadly classified as (1) signal modelling approaches, which aim to capture

[1] Edited by Mara Cercignani; reviewed by Noam Shemesh, Champalimaud Neuroscience Programme, Champalimaud Centre for the Unknown, Lisbon, Portugal

[2] These authors contributed equally and are listed alphabetically.

the variation in signal in a mathematical function with a small number of parameters and thus map properties of the dMRI signal sensitive to microstructural features and (2) biophysical modelling approaches, which aim to relate the signal variation directly to microstructural properties of the tissue and thus map estimates of intrinsic tissue properties. Both approaches fit a model in each image voxel to a set of dMRI measurements acquired with different contrasts arising from varying the acquisition parameters (gradient direction, diffusion weighting, etc.).

In this chapter, we review a variety of models in each category. We aim to explain the pros and cons of each in terms of simplicity versus realism, the amount and quality of measured data typically acquired to support each technique, as well as highlight key applications that each support. The first two sections present the most well-known signal and biophysical modelling approaches, respectively. In the next section we present several relevant software packages for analysis of dMRI data, and in the final section we discuss current and emerging research on this topic to highlight expectations for the next generation of tools.

9.2 Signal models

Signal modelling approaches strive to capture the general form of the diffusion signal, usually using function bases. Their strengths include the fact that such a generic representation is not specific to a particular kind of tissue, and the basis-formulation that often leads to efficient linear estimation. The model parameters do not directly relate to specific microstructural features but usually combine to reflect interesting features of the dispersion pattern that users can subsequently relate to microstructural properties in specific applications.

This section begins by introducing the theory of the q-space formalism, which supports recovery of a more general picture of the dispersion pattern than DTI and thus overcomes its limitations. Then it reviews signal modelling approaches that apply this formalism in practice. First it presents the estimation of the generic diffusion propagator, which recovers the general dispersion pattern, but at the expense of high acquisition requirements. Then it reviews some methods that aim to recover just the angular profile of the propagator from relatively economical acquisitions. Next it presents diffusion kurtosis imaging and finally spherical deconvolution techniques. Detailed reviews of these methods and many more can be found in work by Assemlal *et al.* (2011) and Ghosh and Deriche (2015).

Examples of the advantages of using advanced signal models over DTI are summarised in Figures 9.1 and 9.2. Figure 9.1 compares typical scalar maps derived from DTI with scalar maps derived from higher order models that show the additional contrasts that arise from avoiding the mono-exponential signal assumption in DTI. Figure 9.2 compares the DTI ellipsoids, which represent the diffusion tensors, with the angular plots from a more complex model that is able to detect fibre-crossing configurations within the voxels, thus overcoming the unimodal limitation of DTI.

9.2.1 q-Space formalism

The q-space formalism (Stejskal, 1965; Callaghan, 1991; Fieremans *et al.*, 2011) relates the signal in each voxel, from a pulsed gradient spin echo (PGSE) sequence, to the average diffusion propagator (i.e. the diffusion probability density function (PDF), P, of the displacement vector $\mathbf{r}$, over the diffusion time Δ) via the Fourier transform:

$$E(\boldsymbol{q},\Delta)= \int \mathrm{P}(\boldsymbol{r},\Delta)\, exp\,(-2\pi i \boldsymbol{q}\cdot \boldsymbol{r})\mathrm{d}\boldsymbol{r}, \tag{9.1}$$

Where $\boldsymbol{q} = \gamma\delta G\boldsymbol{g}/2\pi$, with γ the gyromagnetic ratio of the hydrogen nucleus and δ the length of the diffusion gradient pulse of magnitude G and direction $\mathbf{g}$; and $E(\boldsymbol{q}, \Delta) = S(\boldsymbol{q}, \Delta)/S(\mathbf{0})$ is the normalised dMRI signal, with $S(\mathbf{q},\Delta)$ the diffusion weighted signal and $S(\mathbf{0})$ the signal without diffusion weighting. This equation holds under the assumption that δ is negligible compared to Δ, which is rarely satisfied in practice, but even so the inverse Fourier transform of the signal retains useful structure of P. The current convention (Shemesh *et al.*, 2015) denotes PGSE as single diffusion encoding (SDE), and we use the latter terminology for the remainder of this chapter.

The q-space formalism is a powerful umbrella framework that has been exploited by numerous methods to infer microstructural properties of the tissue – either by estimating the propagator or its specific features – as will be seen in this section. Early studies that were instrumental in validating the q-space formalism in animal models and in *in vivo* clinical pathologies were proposed by de Graaf *et al.* (2001), Mori and Van Zijl (1995) and Tang *et al.* (2004). The q-space formalism also predicts diffraction-like effects in the signal from idealised pores that reflect their internal structure (Topgaard, 2013).

DTI can be thought of as a special case of this q-space formalism in which the diffusion propagator is a zero-mean trivariate Gaussian (Stejskal, 1965; Basser *et al.*, 1994b); the effect of non-negligible δ is then straightforward to model. However, the generic q-space formalism places no such constraints on the diffusion propagator, so it can recover greater details and thus reveal hidden complexity in the geometry of the underlying tissue.

9.2.2 Diffusion spectrum imaging

Diffusion spectrum imaging (DSI) aims to recover the diffusion propagator by computing the inverse Fourier transform of the signal (Wedeen *et al.*, 2000, 2005, 2008). This requires the 3D q-space to be sampled widely. Typically, in practice, more than 500 diffusion weighted images (DWIs) are acquired on a Cartesian grid within a sphere in q-space with b_{max} up to 17,000 s/mm^2 (Wedeen *et al.*, 2005). A shorter protocol was proposed by Wu and Alexander (2007), with a hybrid multi-shell encoding with a total of 100 DWIs that takes about half an hour to acquire. The signal is premultiplied to a Hann window to ensure a smooth attenuation at high $\mathbf{q}$ values. The diffusion

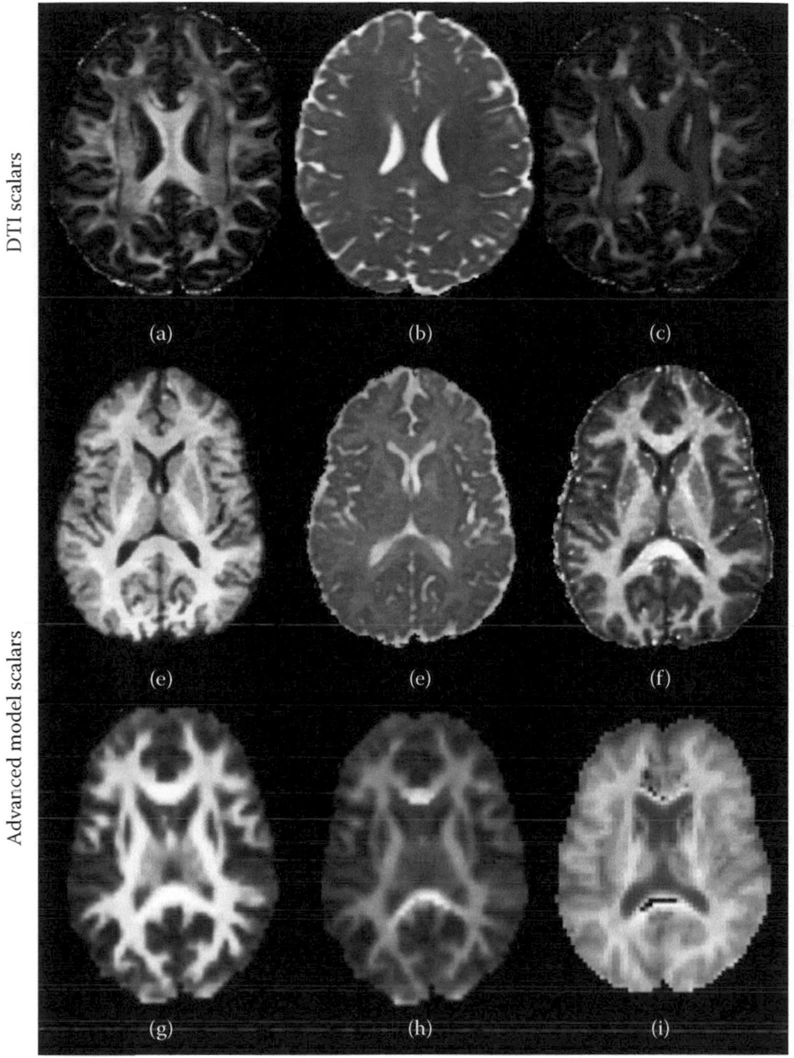

FIGURE 9.1 Scalar parameters derived from diffusion tensor imaging (DTI) and advanced signal models – diffusion spectrum imaging (DSI), mean apparent propagator MRI (MAP-MRI) and diffusion kurtosis imaging (DKI): (a) fractional anisotropy, (b) mean diffusivity, (c) Direction Encoded Colour map; (d) zero displacement probability $P0 = P(r = 0,\Delta)$, (e) mean-squared displacement (MSD), (f) root-mean-square residual between the DTI signal model and b > 2000 s/mm^2 measurements – which provides a measure of non-Gaussianity; (g) propagator anisotropy; (h) non-Gaussianity derived from MAP-MRI; (i) mean kurtosis from DKI, which also provides a measure of non-Gaussianity. Figure 9.1d through f (Reprinted from *NeuroImage*, 36, Wu, Y.-C., and Alexander, A.L., Hybrid diffusion imaging, 617–629, Copyright 2007, with permission from Elsevier); Figure 9.1g and h (Reprinted from *NeuroImage*, 127, Avram, A.V., *et al.*, Clinical feasibility of using mean apparent propagator (MAP) MRI to characterise brain tissue microstructure, 422–434, Copyright 2016, with permission from Elsevier). Figure 9.1i (Tabesh, A., *et al.*: Estimation of tensors and tensor-derived measures in diffusional kurtosis imaging. *Magn. Reson. Med.* 2011. 65. 823–836. Copyright Wiley-VCH Verlag GmbH & Co. KGaA. Reproduced with permission.)

propagator is then computed numerically by a discrete Fourier transform.

An important application of DSI is tractography, since, unlike DTI, the angular profile of the diffusion propagator can be multimodal and can therefore tease apart fibres crossing inside a voxel. This angular structure is highlighted by integrating out the radial information and is known as the *diffusion orientation distribution function* (dODF):

$$\psi(\mathbf{u}, \Delta) = \int_{\mathbb{R}^+} P(r\mathbf{u}, \Delta) r^2 \, dr, \text{ with } \|\mathbf{u}\| = 1. \tag{9.2}$$

The dODF is a probability density function, defined on the unit sphere, of a particle moving in the direction of each unit vector **u** (Wedeen *et al.*, 2000, 2005, 2008).

The radial profile of the propagator also has potential applications. It is informative about the tissue geometry through derived scalar parameters such as the zero displacement probability – also known as the *return to origin probability* – $P0 = P(\mathbf{r} = 0, \Delta)$, mean squared displacement, anisotropy, non-Gaussianity, etc. (see Figure 9.1a through f).

Assessment: DSI overcomes key limitations of DTI. The DSI dODF is able to discern multiple diffusion orientations that can

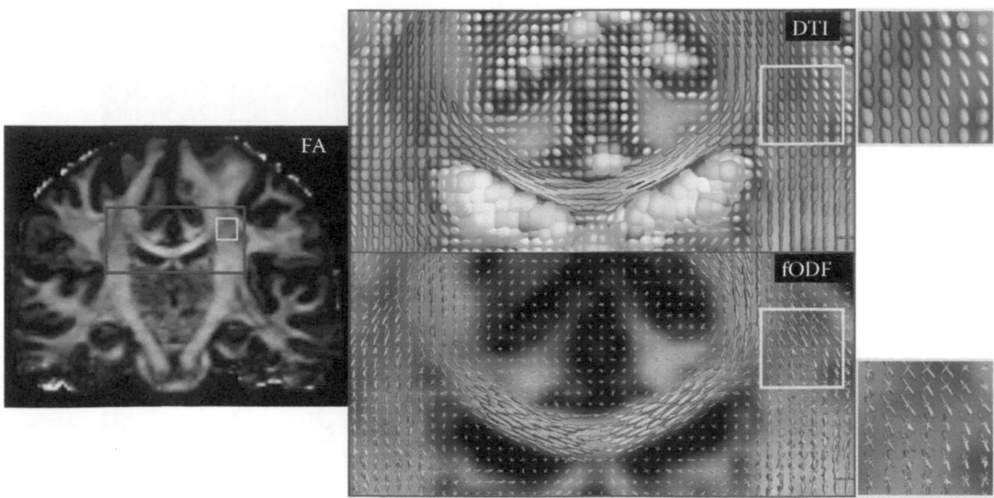

FIGURE 9.2 Comparison of DTI ellipsoids and constrained spherical deconvolution fibre orientation distribution function (fODF) angular plots on a coronal slice. The zooms, red then yellow, show progressively more details of regions that contain crossing fibres. DTI is unable to separate the different fibre populations, but the fODF detects them. Generated using MRtrix3 (From Tournier, J.D., *et al.*, *NeuroImage*, 35, 1459–1472, 2007.)

TABLE 9.1 Typical acquisitions proposed for popular signal models

Method	Total acquisitions	b-Shells (s/mm^2)	Acquisitions/Shell
DSI (Wedeen *et al.*, 2005)	515	≤ 17,000	Cubic lattice
QBI (Descoteaux *et al.*, 2007)	60–90	3000	All on single shell
CSA-QBI (Aganj *et al.*, 2010)	76	4800	All on single shell
PASMRI/MESD (Alexander, 2005)	54	1200	All on single shell
DOT (Ozarslan *et al.*, 2006)	81	1500	All on single shell
DPI (Descoteaux *et al.*, 2010)	256	1000, 2000, 4000, 6000	64, 64, 64, 64
BFOR (Hosseinbor *et al.*, 2012)	102	375, 1500, 3375, 6000, 9375	3, 12, 12, 24, 50 (HYDI (Wu and Alexander, 2007))
SPFI (Assemlal *et al.*, 2009)	64	1000, 3000	32, 32
MAP-MRI/SHORE 3D (Avram *et al.*, 2016)	98	1000, 2000, 3000, 4000, 5000, 6000	4, 7, 11, 17, 23, 31
DKI (Jensen and Helpern, 2010)	60	1000, 2000	30, 30
CSD (Tournier *et al.*, 2007)	60	3000	All on single shell
MSMT-CSD (Jeurissen *et al.*, 2014).	98	1000, 2000, 3000	17, 31, 50 + T_1 scan

Note: Note that in addition to the diffusion weighted images, these methods also require non-diffusion weighted or b0 images, which are not included in this table. DSI = diffusion spectrum imaging; QBI = q-ball imaging; CSA-QBI = constant solid angle QBI; PASMRI = persistent angular structure MRI; MESD = maximum entropy spherical deconvolution; DOT = diffusion orientation transform; DPI = diffusion propagator imaging; BFOR = Bessel Fourier orientation reconstruction; SPFI = spherical polar Fourier imaging; MAP-MRI = mean apparent propagator MRI; SHORE-3D = simple harmonic oscillator-based reconstruction and estimation; DKI = diffusion kurtosis imaging; CSD = constrained spherical deconvolution; MSMT CSD = multi-shell, multi-tissue CSD.

reveal fibre-crossings within a voxel. The study of the brain's connectivity – the human connectome (Hagmann, 2005) – greatly benefitted from this capacity. However, this is achieved at the price of very lengthy acquisition times and extremely high b-values, which make it difficult to conduct DSI clinically.

Typical acquisition (see Table 9.1) for DSI comprises 515 DWIs on a cubic lattice up to very high b-values (17,000 s/mm^2). In comparison DTI typically uses 30–60 DWIs at moderate b-values (1000 s/mm^2).

9.2.3 Angular profile of the diffusion propagator

Several methods aim to recover the angular profile of the diffusion propagator directly from a moderate number of acquisitions spread evenly on a sphere (or shell) in q-space. They attempt to overcome the penalising acquisition requirements of DSI while retaining its multimodal angular properties. Such methods support crossing-fibre tractography from clinically feasible data (Descoteaux *et al.*, 2009).

Q-ball imaging: For example, q-ball imaging (QBI) (Tuch *et al.*, 2003; Tuch, 2004) maps a single shell high angular resolution diffusion imaging (HARDI) data to a modified dODF via the Funk–Radon transform (FRT). The resulting QBI-dODF has the form $\tilde{\psi}(\mathbf{u}, \Delta) = \int_{\mathbb{R}^+} P(r\mathbf{u}, \Delta)dr$, which, in comparison to the DSI-dODF in Equation 9.2, is blurred by a zeroth-order Bessel function with width dependent on the radius of the HARDI shell. This means that higher b-values improve the QBI-dODF's ability to resolve distinct peaks. In the original QBI, 252 isotropic acquisitions on a HARDI shell with radius b = 4000 s/mm^2 are used (Tuch, 2004). The data is interpolated using spherical radial basis functions while the FRT is computed numerically. The resulting QBI-dODF is normalised numerically then rescaled by min–max normalisation for visualisation.

The use of analytical function-basis representations (e.g. spherical harmonics) improves QBI considerably (Anderson, 2005; Hess *et al.*, 2006; Descoteaux *et al.*, 2007). First, only a moderate number of spherical harmonic coefficients are necessary to represent the signal and these can be estimated linearly, implying smaller acquisition requirements. Second, the FRT can be computed analytically from the signal coefficients. And third, both the signal and the dODF have continuous and smooth representations on the sphere.

Furthermore, in constant solid angle QBI (CSA-QBI), the correctly weighted and normalised DSI-dODF (Equation 9.1) can also be computed from the spherical harmonic representation by making mild assumptions on the radial (i.e. as a function of q) signal decay (Tristan-Vega *et al.*, 2009; Aganj *et al.*, 2010). This further improves QBI's ability to discern individual fibre bundles in crossings with narrow angles.

Persistent angular structure MRI (PASMRI) aims to capture the angular profile of the diffusion propagator that tends to persist over a large range of contours (Jansons and Alexander, 2004; Alexander, 2005). Similar to QBI, PASMRI also aims to retain the flexibility of the DSI-dODF to capture multiple peaks while reducing acquisition requirements and has been used in multifibre tractography (Seunarine *et al.*, 2007). PASMRI assumes the data is acquired on a single HARDI shell. Thus, by assuming independence of the angular and radial structures of the propagator and following the principle of maximum entropy it conceives the angular profile of the ideal propagator to have the form:

$$\tilde{P}(\mathbf{u},\Delta) = exp\left(\lambda_0 + \sum_{j=1}^{N} \lambda_j \cos(\rho \mathbf{q}_j \cdot \mathbf{u})\right), \tag{9.3}$$

where ρ is radius of the contour. $\tilde{P}$, encoded by the λ_j parameters, is estimated from the signal by numerical optimisation.

Diffusion orientation transform: Other prominent single shell HARDI methods include the diffusion orientation transform (DOT) (Ozarslan *et al.*, 2006), which attempts to reconstruct an isoradius profile of the diffusion propagator. DOT is not a radial integral of the propagator (dODF); instead it is the angular profile of the generic propagator for a fixed radius **r**. DOT also uses the spherical harmonic basis representation in its computations but estimates the mono-exponential apparent diffusion coefficient (ADC) rather than the signal itself. However, DOT can be extended to multishell data and to a more general multi-exponential signal attenuation.

Assessment: QBI, PASMRI and DOT overcome the acquisition limitations of DSI and still approximate the multimodal angular profile of the propagator. They require far fewer acquisitions – on a single HARDI shell – and do not require extreme b-values, which can be unachievable for clinical scanners. Therefore, all three are estimable from clinical data (Alexander, 2005; Descoteaux *et al.*, 2007). QBI and DOT are linear and therefore fast to compute. PASMRI can be more accurate and robust for detecting fibre orientations but lacks an intuitive physical interpretation and is several orders of magnitude slower to estimate (Alexander, 2005). It can be accelerated through a generic linear function-basis representation but then loses its sensitivity and robustness (Alexander, 2005). However, all three estimate properties of the diffusion process rather than the underlying tissue geometry; hence they cannot directly infer the fibre orientations. Finally, like many of the other methods that only recover the angular profile of the diffusion propagator, QBI, PASMRI and DOT lose the radial information available in DSI that is potentially useful for inferring scalar microstructural parameters.

Typical acquisition (see Table 9.1) for QBI comprises 60–90 measurements on a single b-shell from 3000–4000 s/mm^2. For PASMRI the typical acquisition is ~60 acquisitions on a single b-shell of 1200 s/mm^2 and for DOT it is ~80 acquisitions on a single b-shell of 1500 s/mm^2.

9.2.4 Analytical diffusion propagator

These methods aim to recover the entire propagator from a moderate number of acquisitions spread over multiple HARDI shells. Hence, they overcome the acquisition constraints of DSI and recover not only its angular profile (like ODFs) but also its radial content. Therefore, like DSI, they can not only detect fibre crossings but also inform about the tissue microstructure through scalar parameters. Figure 9.1d through h presents scalar maps derived from mean apparent propagator (MAP) MRI, one such method discussed below, and DSI.

A number of methods exist that differ in their usage of basis pairs to represent the signal and propagator in order to make computation of the Fourier transform analytic and straightforward. They also differ in their coordinate representations – some use the Cartesian system while others use the spherical system. Overall, the analytical representation makes it feasible to estimate the propagator from as few as 100–200 DWIs strategically placed in a multishell HARDI acquisition, thus providing clinical applicability (Fick *et al.*, 2016).

For example, in generalised DTI (Liu *et al.*, 2003, 2004), the authors use a Cartesian representation with polynomial or higher-order tensor basis representation for the signal and Hermite tensors for the propagator along with the Gram–Charlier series approximation to estimate the propagator analytically. Simple harmonic oscillator-based reconstruction and estimation (SHORE-3D) (Ozarslan *et al.*, 2013), diffusion propagator imaging (DPI) (Descoteaux *et al.*, 2010), Bessel Fourier orientation reconstruction (BFOR) (Hosseinbor *et al.*, 2012) and spherical polar Fourier imaging (SPFI) (Assemlal *et al.*, 2009; Cheng *et al.*, 2010) all use the spherical coordinate system and the spherical Fourier transform but differ in their function-bases representations. MAP-MRI (Ozarslan *et al.*, 2013) uses the Cartesian Fourier transform and eigenfunctions of this transform to represent both the signal and the propagator.

Assessment: Analytical diffusion propagator methods make the estimation of the diffusion propagator possible from moderate multishell clinically achievable data (Avram *et al.*, 2016;

Fick *et al.*, 2016). Furthermore, these methods all enjoy the advantages of analytical representations and are all computationally viable. However, some methods perform better than others because the shapes of the chosen basis functions are more suitable for representing the dMRI data, and hence fewer basis functions are required or fewer parameters need to be estimated.

Typical acquisition (see Table 9.1) includes multiple b-shells. SHORE-3D and MAP-MRI can be estimated from 100 acquisitions on six b-shells (1000, 2000, 3000, 4000, 5000, 6000 s/mm^2). Practitioners of DPI have proposed 256 acquisitions on four b-shells (1000, 2000, 4000, 6000 s/mm^2). BFOR has been estimated from the alternate hybrid acquisition for DSI with roughly 100 acquisitions (Wu and Alexander, 2007), while SPFI has been estimated from 64 measurements on two b-shells (1000, 3000 s/mm^2).

9.2.5 Diffusion kurtosis imaging

Diffusion kurtosis imaging (DKI) (Jensen *et al.*, 2005) estimates the fourth-order cumulant of the propagator as an indicator sensitive to the microstructure of the underlying tissue. DTI estimates only the second-order cumulant (covariance) of the propagator. The higher-order cumulants quantify the deviation from Gaussianity. DKI only considers the second- and fourth-order cumulants, i.e. the diffusion tensor and the kurtosis tensor. Hence, the deviation from Gaussianity is captured by the kurtosis tensor. By extending the DTI signal model, DKI has more scalar parameters. These include the mean kurtosis (see Figure 9.1i), radial kurtosis and axial kurtosis. Numerous estimation techniques exist for estimating the DKI tensors (Tabesh *et al.*, 2011; Veraart *et al.*, 2011; Ghosh *et al.*, 2013). Recent work has tried to directly estimate DKI scalar parameters like mean kurtosis from specialised and fast acquisition schemes (Hansen *et al.*, 2013).

Assessment: DKI is clinically viable and widely used, as it is a simple way to estimate features beyond those available from DTI. Applications include stroke, traumatic brain injury and age- and disease-related neurodegeneration (Steven *et al.*, 2013). However, DKI also utilises a generic signal model and does not have any direct biophysical interpretation; hence DKI scalar parameters lack specificity. Furthermore, the DKI formulation does not explicitly apply any constraints to ensure that the tensors are consistent with a cumulant expansion of an unknown diffusion propagator. The estimated parameters are only dependent on the acquisition scheme and are not related to a propagator explicitly. Hence, they are not necessarily true cumulants and are sensitive to the precise choice of b-values.

Typical acquisition (see Table 9.1): *In vivo* applications typically use about 60 DWIs distributed over a minimum of two non-zero b-shells with a maximum of b = 3000 s/mm^2 (Jensen and Helpern, 2010).

9.2.6 Spherical deconvolution and fibre orientation distribution function

Spherical deconvolution methods strive to recover the fibre orientation distribution function (fODF). Unlike the dODF, which is a property of the diffusion process in the underlying tissue, the fODF is an intrinsic property of the tissue that indicates dominant fibre orientations.

These methods assume that the signal in each voxel is the convolution of the fODF with the signal from a single fibre population – the 'kernel' or the 'response function'. Therefore, spherical deconvolution works by deconvolving the signal with the kernel to estimate the fODF. A number of different methods can be found (Tournier *et al.*, 2004; Alexander, 2005; Anderson, 2005; Dell'Acqua *et al.*, 2007; Kaden *et al.*, 2007; Ramirez-Manzanares *et al.*, 2007). Interestingly, the mathematical formulation of PASMRI is equivalent to spherical deconvolution. That observation leads to maximum entropy spherical deconvolution (MESD) (Alexander, 2005), which uses the non-negative representation of PASMRI to estimate the fODF. A compressed sensing framework to accelerate many of these methods was given by Jian and Vemuri (2007).

One of the most popular deconvolution methods is constrained spherical deconvolution (CSD) (Tournier *et al.*, 2004, 2007, 2012), which is used extensively for tractography. CSD estimation involves two steps. First, the kernel is estimated from the dMRI data by taking an average over voxels with the most anisotropic signals, after orientational alignment, on the assumption that these give signals from a single fibre population. Then, the signal in every voxel is deconvolved with this kernel to produce the fODF. The estimation is (almost) linear and therefore fast. CSD represents the fODF in the spherical harmonic basis and benefits from estimating more spherical harmonic coefficients than the number of acquisitions by using regularisation (Tournier *et al.*, 2007). Figure 9.2 compares the angular plots of the CSD fODF with the DTI ellipsoids on a coronal slice of the brain. The advantage of CSD over DTI is highlighted in the zoomed yellow box, where sub-voxel fibre crossings are picked up by CSD while the diffusion tensor only indicates the average dominant direction.

A recent extension of CSD uses multishell HARDI data with multitissue kernels (MSMT-CSD) to distinguish between grey matter, white matter and cerebrospinal fluid (CSF) (Jeurissen *et al.*, 2014). A similar idea employing a range of tissue kernels is explored in restriction spectrum imaging (RSI) (White *et al.*, 2013), where the key difference lies in the kernel design. RSI employs a range of diffusion tensor based kernels with diffusivities chosen *a priori* to reflect different diffusion environments like restricted and hindered diffusion.

Assessment: Spherical deconvolution methods are popular for tractography and connectivity analysis since they can estimate multiple fibre orientations from single shell HARDI protocols that are widely available and require moderate acquisition times. In comparison to QBI and MAP-MRI, spherical deconvolution techniques in general are better able to resolve narrower crossings, implying a higher sensitivity and angular resolution (e.g. CSD) (Ning *et al.*, 2015). In the trade-off, however, these methods have lower specificity and tend to produce spurious peaks in the presence of signal noise. For example, CSD employs a heuristic algorithm to eliminate such false peaks. The peaks of the fODF are also intuitively simpler to associate with fibre orientations than peaks of the diffusion propagator or its angular profile. Therefore, the output of MESD is more readily interpretable than the PAS. However, MESD is computationally extremely demanding and is comparable to the PASMRI algorithm (Alexander, 2005), which is much greater than for linear methods. CSD on the other hand is more efficient and well tested (Jeurissen *et al.*, 2013). All these methods employ the same kernel for all the voxels in the brain, which may not be biophysically accurate. An alternate approach that alleviates this limitation is proposed by Kaden *et al.* (2016), where the kernel is estimated separately in each voxel. Finally, the spherical deconvolution techniques presented thus far do not account for more complex configurations such as fanning, bending, undulation, etc. It is unclear if they can accurately reflect the true structure of the fODF or produce spurious peaks in such situations. Future research is required to identify such failures and to formulate the methods better to accommodate such configurations.

Typical acquisition (see Table 9.1) for MESD comprises about 60 acquisitions on a single b-shell of 1200 s/mm^2. CSD has also been estimated from ~60 acquisitions on a single b-shell, but with a much higher value of 3000 s/mm^2. Finally, MSMT-CSD can be estimated from ~100 measurements on three b-shells (1000, 2000, 3000 s/mm^2) and additionally requires a T_1-weighted structural acquisition.

9.3 Multicompartment and biophysical models

Multicompartment and biophysical modelling approaches aim to relate various microscopic tissue properties to the dMRI signal. Thus, fitting the model to the measured data results in parameters that reflect tissue features, such as intracellular volume fraction, cell size, shape, fibre orientation, etc.

This section starts with the simplest multicompartment tensor models, which aim to disentangle the signal contribution from different tissue types. Then it presents modelling techniques that aim to estimate the axon diameter and the volume fractions of different compartments and discusses the resolution limit, i.e. the smallest diameter that can be detected for a given acquisition. It reviews more recent models that explicitly account for fibre dispersion, then it discusses model-selection approaches to seek the multicompartment models that best explain the measured signal.

9.3.1 Multicompartment tensor models

Multicompartment tensor models aim to model separately the signal contribution from two or more water pools in order to differentiate between slow and fast diffusion (e.g. intracellular–extracellular, tissue–CSF) and/or multiple fibre populations. Thus, they can overcome the main limitations of DTI to provide a better representation of the measured signal (Clark *et al.*, 2000).

A multitensor model assumes signal contributions from multiple water pools with slow or no exchange (Clark and Le Bihan, 2000; Tuch *et al.*, 2002):

$$S(\boldsymbol{g}_i) = S_0 \sum_n f_n \exp\left(-b_i \hat{\boldsymbol{g}}_i^T \boldsymbol{D}_n \hat{\boldsymbol{g}}_i\right), \text{ with} \sum_n f_n = 1 \qquad (9.4)$$

where f_n and $\boldsymbol{D}_n$ are the volume fraction and diffusion tensor characterising compartment n and $\boldsymbol{g}_i$ is the gradient of the ith measurement with corresponding b-value b_i. Various multitensor techniques have been studied in the literature to characterise the diffusion signal for different applications. In general, such models require at least a two-shell acquisition (Scherrer and Warfield, 2010).

Bi-exponential: The simplest case is a bi-exponential model, i.e. two isotropic tensors, which has been proposed to differentiate between slow and fast diffusion in the brain (Niendorf *et al.*, 1996; Clark and Le Bihan, 2000) based on data acquired at several b-values. However, the interpretation of the slow and fast diffusion is not straightforward, as restricted diffusion in a single compartment can also be well described by a bi-exponential function (Milne and Conradi, 2009), so we cannot assign the two diffusivities directly, e.g. to the intra- and extracellular space. The same model is also used as the intravoxel incoherent motion (IVIM) model (Le Bihan *et al.*, 1988), which separates a fast diffusion component arising from blood flow through capillaries that effect dMRI at low b-values ($b < 200$ s/mm^2) from a slow diffusion component arising from cellular water.

Bi and multitensor: A bi-tensor model allows anisotropy in one or both of the component diffusion tensors. Such models can be used to investigate the anisotropy of slow and fast diffusing compartments (Clark *et al.*, 2000) as well as to distinguish the contribution of different fibre populations from HARDI acquisition (Tuch *et al.*, 2002), which can be used to improve tractography (Parker and Alexander, 2003). Depending on the number of tensors, such a model can have a large number of parameters to estimate (e.g. 13 free parameters for a full bi-tensor model); thus, several model constraints have been proposed in the literature to reduce the complexity and stabilise the fit. For instance, the eigenvalues of the DTs can be *a priori* set to literature values (Tuch *et al.*, 2002) or the diffusion tensors can be assumed cylindrically symmetric (Alexander and Barker, 2005). When using two- and three-fibre models, constraining the fibre directions to be spatially consistent can also improve the robustness of tensor parameter estimates (Pasternak *et al.*, 2005, 2008; Malcolm *et al.*, 2011) and the tractography results. A bi-tensor model that constrains one of the tensors to be isotropic can also be used to mitigate the effects of CSF partial volume and to eliminate the signal contribution of free water (Pasternak *et al.*, 2009). An idealised two-tensor compartment model of white matter was proposed by Fieremans *et al.* (2011), which provides a biophysical interpretation for DKI parameters in voxels with parallel fibre configurations and can be used to compute white matter tract integrity (WMTI) metrics (Fieremans *et al.*, 2013).

Ball and stick(s): A special case of a bi- or multitensor model, ball and stick (Behrens *et al.*, 2003) aims to separate the effect of fibre orientation from the bulk diffusion. The 'stick' compartment models intra-axonal diffusion assuming mobility only in the fibre direction, i.e. zero radial diffusivity $D_\perp = 0$. The 'ball' compartment models extra-axonal diffusion as an isotropic tensor. The volume fraction parameter, f_n, in Equation 9.4 provides an estimate of fibre density and the stick orientation of fibre direction. Behrens *et al.* (2007) extended the model to multiple sticks and proposed an automatic relevance determination algorithm to detect the number of different fibre directions that are present in one voxel.

Assessment: Multitensor models have some advantage over DTI in distinguishing the signal contribution of different water pools and estimating their volume fraction, although the tensor model of each compartment does not necessarily estimate specific microstructural features. Another important application is to distinguish between fibre populations with different orientations. The multitensor model estimates a large number of parameters; thus various constraints have been proposed, which are important for practical usage. In particular, fitting the full multitensor model requires multiple b-values and careful parameter estimation. Simplifications avoid such demands. For example, the ball-and-stick model, which is a special case of a bi-tensor model, is popular due to its simplicity and microstructural relevance and only requires single shell HARDI data. The WMTI model makes use of a typical DKI acquisition with two shells (b = 1000 and 2000 s/mm^2) and 30 gradient directions each.

Typical acquisition (see Table 9.2) for bi-exponential modelling consists of measurements acquired at many b-values along one or several directions (e.g. Clark and Le Bihan [2000] use 45 b-values linearly spaced between 0 and 3800 s/mm^2 along the three principal directions). To investigate the anisotropy of slow and fast diffusion tensors, Clark *et al.* (2000) use an acquisition with three b-values up to 3500 s/mm^2 and 12 directions each. When

TABLE 9.2 Typical acquisitions proposed for popular techniques that use multicompartment and biophysical models

Method	Total acquisitions	b-Shells (s/mm^2)	Acquisitions/Shell
		Multicompartment Tensor Models	
Bi-exponential (Clark and Le Bihan, 2000)	135	≤3800	Linearly spaced b-values, three directions
Bi-tensor (fast and slow) (Clark *et al.*, 2000)	36	~1000, 2000, 3500	12, 12, 12
Bi-tensor (crossing fibres) (Tuch *et al.*, 2002; Pasternak *et al.*, 2008)	126 (Tuch *et al.*, 2002); 33 (Pasternak *et al.*, 2008)	~1000	All on single shell
Ball and stick (Behrens *et al.*, 2003, 2007)	60	1000	All on single shell
WMTI (Fieremans *et al.*, 2011)	60	1000, 2000	30, 30
		Distributions of Tensors	
TDF (Leow *et al.*, 2009)	96	~1200	All on single shell
DIAMOND (Scherrer *et al.*, 2016)	65	1000 and ≤3000	Single shell and cubic lattice
DBSI (Wang *et al.*, 2014)	99	≤3000	Cubic lattice
		Models of Restricted Diffusion	
Stanisz' model (Stanisz *et al.*, 1997)	400	≤40,000*	Four diffusion times, 50 gradient steps, ‖ and ⊥ to the fibre direction
CHARMED (Assaf and Basser, 2005) Optimised protocol (De Santis *et al.*, 2014)	16845	≤10,000 ≤8750	• Ten b-shells with 6, 6, 12, 12, 16, 16, 20, 20, 30, 30 directions • Eight b-shells with 6, 3, 4, 5, 6, 6, 7, 8 directions
AxCaliber (Assaf *et al.*, 2008)	80	–	Five diffusion times, 16 gradient steps from 0 to 300 mT/m, ⊥ to the fibre direction
ActiveAx (Alexander *et al.*, 2010) (+ dispersion) (Zhang and Alexander, 2010)	360	530, 700, 2720, 2780	90
		Explicitly Modelling Fibre Dispersion	
Crossing and Bingham dispersion (Kaden *et al.*, 2007)	60	1000	All on single shell
NDM (Jespersen *et al.*, 2006, 2009)	144	≤15,000*	Sixteen b-shells, nine different directions each
NODDI (Zhang *et al.*, 2012)	90	711, 2855	30, 60
SMT (Kaden *et al.*, 2015)	151	1000, 2500	76, 75
Multicompartment SMT (Kaden *et al.*, 2016)	270	1000, 2000, 3000	90

Note: In addition to the diffusion weighted images, these methods also require non-diffusion weighted or b0 images, which are not included in this table. WMTI = white matter tract integrity; TDF = tensor distribution function; DIAMOND = distribution of 3D anisotropic microstructural environments in diffusion-compartment imaging; DBSI = diffusion basis spectrum imaging; CHARMED = composite hindered and restricted model of diffusion; NDM = neurite density models; NODDI = neurite orientation dispersion and density imaging; SMT = spherical mean technique.

*Experiments performed *ex vivo* at room temperature, requiring much higher b-values in order to obtain similar signal attenuation.

mapping crossing fibres using a bi-tensor model, Tuch *et al.* (2002) use a single shell acquisition with $b = 1077$ s/mm^2 and 126 directions, while Pasternak *et al.* (2008) use a simpler protocol with $b = 1000$ s/mm^2 and only 33 directions. A similar acquisition with $b = 1000$ s/mm^2 and 60 directions is used for the ball-and-stick techniques (Behrens *et al.*, 2003, 2007).

9.3.2 Distributions of tensors

Various techniques aim to describe the diffusion signal using a distribution instead of a finite mixture of tensors:

$$S(g_i) = S_0 \int_{D \in \mathbb{D}} P(\mathbf{D}) \exp\left(-b_i \hat{g}_i^T \mathbf{D}_n \hat{g}_i\right) d\mathbf{D}, \text{ with } \int_{D \in \mathbb{D}} P(\mathbf{D}) d\mathbf{D} = 1, \tag{9.5}$$

where P(**D**) is the probability function and $\mathbb{D}$ represents the space of diffusion tensors. Such approaches do not require *a priori* knowledge on the number of compartments as the majority of multitensor techniques presented above.

One possible choice of P is the Wishart distribution, which is a generalisation of the gamma distribution to the space of diffusion tensors and was employed by Jian *et al.* (2007). The Wishart distribution provides a conjugate prior, so that the Laplace transform linking the signal and the distribution has a closed form, which makes computation tractable. The approach proposed by Jian *et al.* (2007) assumes the same diffusion tensor for different fibres. This limitation was overcome by Leow *et al.* (2009), who used a tensor distribution function to allow for different anisotropies, i.e. different tensor eigenvalues, along different orientations. Computational complexity can be reduced by considering only cylindrically symmetric tensors. More recently, Scherrer *et al.* (2016) proposed DIAMOND (distribution of 3D anisotropic microstructural environments in diffusion-compartment imaging), a technique that aims to capture the heterogeneity of each compartment, by modelling a distribution of tensors along each main orientation. Wang *et al.* (2014) proposed a diffusion basis spectrum imaging (DBSI) technique to separate the contributions from a sum of cylindrically symmetric anisotropic tensors and a distribution of isotropic ones.

Assessment: Various studies have proposed to use distributions of diffusion tensors to represent the measured signal. By allowing a distribution of orientations and/or diffusivities, such techniques can represent more complex fibre configurations such as dispersed, bending or fanning fibres, as well as intra-compartment variability, which cannot be represented by the simpler multitensor models. Although these approaches present a general formalism, various assumptions and constrains on the shape of the diffusion tensors as well as the number of fibre directions and/or compartments have been made in order to reduce the complexity of the problem.

Typical acquisition (see Table 9.2): While methods that estimate only the orientation distribution can be applied to single shell data (e.g. $b = 1159$ s/mm^2 with 96 directions in the paper by Leow *et al.*, 2009), other approaches require more complex acquisitions. The DIAMOND technique (Scherrer *et al.*, 2016), which aims to capture intracompartment heterogeneity, has been assessed with an acquisition consisting of a HARDI $b = 1000$ s/mm^2 shell and 30 directions together with 30 gradients with b-values between 1000 and 3000 s/mm^2 lying on the cube of constant echo time (TE), which minimises the TE, thus maximising signal-to-noise ratio (SNR). The DBSI modelling (Wang *et al.*, 2014) has been applied to a dataset with 99 diffusion measurements placed on the 3D q-space grid with maximum b-value of 3000 s/mm^2 (Wang *et al.*, 2014).

9.3.3 Models of restricted diffusion

Although the multitensor models presented above aim to describe the signal from multiple compartments, they do not capture effects such as restricted diffusion that place a hard limit on the displacement in certain directions (Milne and Conradi, 2009). Other techniques construct more specific geometric models of various tissue components and relate them to the diffusion MRI signal.

Early models: One of the first biophysical models to describe diffusion signal in nervous tissue was proposed by Stanisz *et al.* (1997). The study investigates bovine optic nerves and represents the tissue using three compartments: prolate ellipsoids (axons), spheres (glial cells) and hindered diffusion (extracellular space) with exchange between the compartments, as illustrated in Figure 9.3a through c. A different approach to understanding compartmentalisation, correlating transverse relaxation components with diffusion measurements in the frog sciatic nerve, is presented by Peled *et al.* (1999). For each main peak in the T_2 spectrum, the authors present the diffusion decay curves acquired at different b-values and diffusion times. While the decay curves for the peak with short T_2 (~78 ms) do not show clear dependence on diffusion time, the data for the peak with long T_2 (~300 ms) exhibit a strong time dependence. Thus, the authors proposed to explain the experimental data using a model of restricted diffusion inside cylinders, with a distribution of radii reflecting histological measurements. Fitting such models requires high quality data with many different measurements, and the authors use spectroscopic data from a small bore system with over 1T/m gradients. For *in vivo* experiments, simpler models that describe the key features of tissue are necessary.

CHARMED, AxCaliber and ActiveAx: Assaf's work (Assaf *et al.*, 2004; Assaf and Basser, 2005) builds on the idea of a multi-compartment model that includes restricted diffusion inside the axons and makes it feasible for clinical acquisition. Composite hindered and restricted model of diffusion (CHARMED) aims to estimate the parameters of a hindered compartment described by a diffusion tensor, as well as the volume fractions and orientations of one or two fibre populations that are assumed to have a known axon diameter distribution. The axon diameter values as well as the intracellular diffusivity perpendicular to the fibres are fixed to typical values for axons in the spinal cord. For *in vivo* brain imaging, this technique uses a rich multishell,

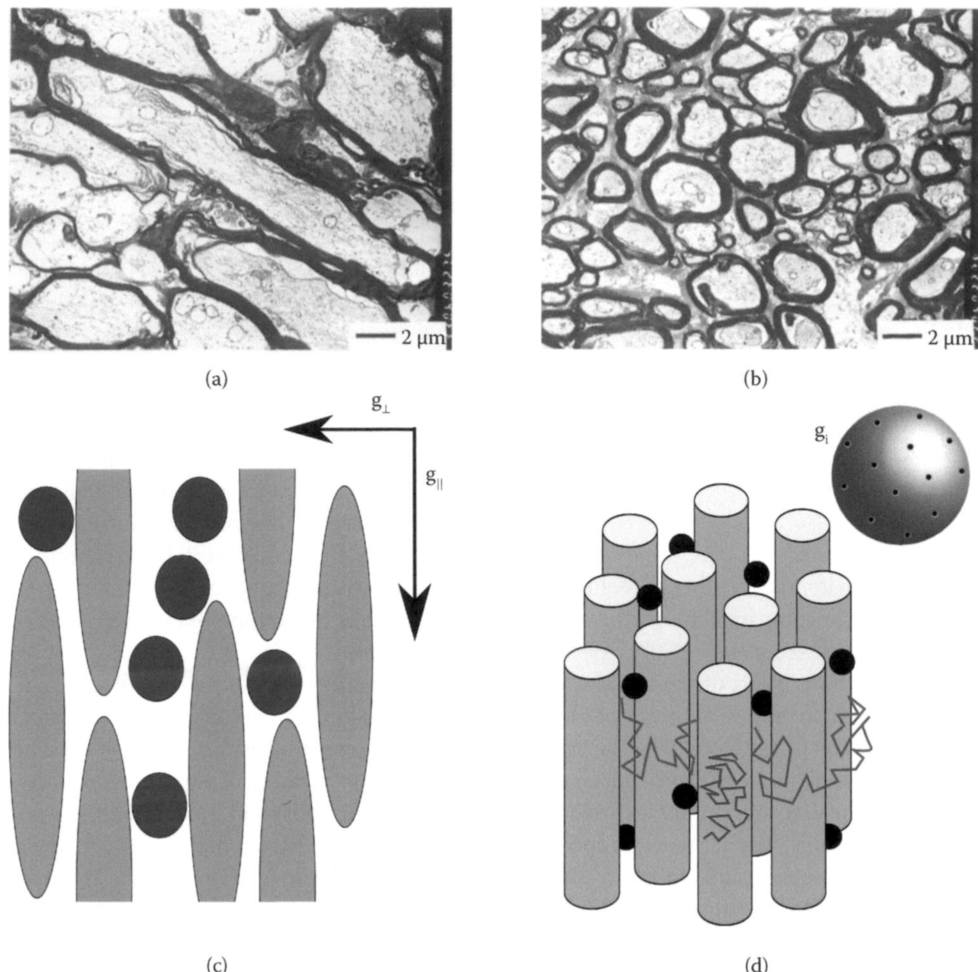

FIGURE 9.3 Electromicrograph images of bovine optic nerve in direction (a) parallel and (b) perpendicular to the axis defined by the orbit and optic chiasma. (c) Schematic representation of the three-compartment white matter model (Stanisz, G.J., *et al.*: An analytical model of restricted diffusion in bovine optic nerve. *Magn. Reson. Med.*, 1997, 37, 103–111. Copyright Wiley-VCH Verlag GmbH & Co. KGaA. Reproduced with permission). Schematic representation of a white matter model *i*th parallel cylinders used in AxCaliber and ActiveAx.

multidirectional dataset to estimate the fibre orientation(s), intracellular diffusivity parallel to the fibres, extracellular diffusivities as well as the corresponding volume fractions.

A later technique, AxCaliber (Assaf *et al.*, 2008), also uses the CHARMED model, but estimates the distribution of axon diameters, assuming a known fibre direction, from measurements acquired perpendicular to the fibres with multiple combinations of gradient strengths and diffusion times. Based on previous histological work by Aboitiz *et al.* (1992), the axon diameters are assumed to follow a gamma distribution. This technique has been further used *in vivo* to estimate the axon diameter distribution in the rat corpus callosum (Barazany *et al.*, 2009). However, this approach requires many measurements perpendicular to the nerves and assumes prior knowledge of the fibre orientation.

ActiveAx (Alexander, 2008) aims to map axon diameter and density estimates over the whole brain using an orientationally invariant acquisition protocol. Specifically, the optimised protocol requires only four HARDI *b*-shells, with the highest *b*-value close to 3000 s/mm^2. ActiveAx combines features of Stanisz's model and AxCaliber to obtain a minimal model of white matter diffusion (MMWMD), the simplest model including parameters for axon diameter and density that adequately fits the measured signal. In areas of coherent white matter fibres, the intra-axonal space is characterised using a single axon diameter, the extracellular space is described by an axially symmetric tensor and the glial cell compartment is simplified to fully restricted diffusion, as illustrated in Figure 9.3d. In this model, the intracellular diffusivity and extracellular diffusivity parallel to the axons are equal and fixed. Thus, the ActiveAx framework estimates the axon diameter index, which at least in simulation correlates well with the mean volume-weighted diameter, fibre orientation and volume fractions of the different compartments. A later study (Alexander *et al.*, 2010) explores the feasibility of estimating axon diameter index and volume fraction in the human corpus callosum *in vivo* and in the monkey brain *ex vivo* and shows a strong improvement when the available gradient strengths increase from standard clinical values (~60 mT/m) to specialist values of about 300 mT/m.

Assessment: Techniques that aim to estimate indices of compartment size, such as axon diameter, require a rich, multishell acquisition that includes high *b*-values (around 3000 s/mm^2 or above for *in vivo* imaging, where the effects of restriction dominate) and multiple diffusion times. The majority of diffusion-based axon diameter estimates reported in the literature (Assaf *et al.*, 2008; Alexander *et al.*, 2010; Dyrby *et al.*, 2012; Horowitz *et al.*, 2015; Sepehrband *et al.*, 2016) are overestimated compared with histology. This could result from a variety of factors including exchange, orientation dispersion, fibre undulation, oversimplification of the extracellular model and limited gradient strength. This effect is especially pronounced in clinical studies (Assaf *et al.*, 2008; Alexander *et al.*, 2010; Dyrby *et al.*, 2012; Horowitz *et al.*, 2015) with limited gradient strength and SNR. For this reason, the study presented by Horowitz *et al.* (2015), which aims to correlate the apparent axon diameter (AAD) in corpus callosum estimated from AxCaliber with the axonal conduction velocity (ACV) estimated from the inter-hemispheric transfer time measured by electroencephalography, has been met with scepticism by research groups working on histology (Innocenti *et al.*, 2015). The commentary by Innocenti *et al.* (2015) points out that the reported ACV values are smaller compared to previously reported numbers, an effect which might arise from the task design, while the AAD values from AxCaliber are overestimated, reflecting that diffusion-based AAD estimates are biased towards larger values. These findings raise the question of what is the smallest size we can measure with diffusion experiments, i.e. the axon diameter resolution.

To better understand this issue, Drobnjak *et al.* (2015) analysed the sensitivity of the measured signal to cylinder radius for a wide range of sequence parameters and practically achievable SNRs. The results show that for gradient strengths smaller than 60 mT/m, axon diameters less than 6 μm are indistinguishable from zero, while for stronger gradients up to 300 mT/m, which can be achieved on the Connectome scanner (McNab *et al.*, 2013), the resolution limit decreases to 2–3 μm. As the majority of axons in the brain are smaller than 3 μm (Aboitiz *et al.*, 1992), it appears that dMRI techniques are mostly sensitive to the larger diameter values from the tail of the distribution, although these axons do also contribute more signal compared to the small ones. Information regarding restriction size and packing order is also contained in the extracellular signal (Burcaw *et al.*, 2015). Accounting for these effects can also influence the estimates of axon diameter (De Santis *et al.*, 2016). Thus, the estimated parameters need to be interpreted with care when these are related to histological features such as axon diameter distribution.

Typical acquisition (see Table 9.2): The early modelling approach from Stanisz' work has been applied to high quality data acquired with four diffusion times and 50 gradient steps between 0 and 1.4 T/m, yielding *b*-values up to 40,000 s/mm^2 for *ex vivo* acquisition. The original CHARMED acquisition scheme by Assaf and Basser (2005) consists of 10 shells with *b*-values up to 10,000 s/mm^2 with the number of gradient directions increasing from 6, in the low *b*-value shells, to 30 in the high *b*-value shells. An optimised CHARMED acquisition with only 45 measurements spread over eight shells (*b* up to 8750 s/mm^2) has been proposed by De Santis *et al.* (2014). The AxCaliber imaging experiments by Assaf *et al.* (2008) were acquired with five different diffusion times and 16 gradient increments between 0 and 300 mT/m with only one direction perpendicular to the fibre orientation in spinal cord. The acquisition of ActiveAx consists of four HARDI shells with 90 directions each and with *b*-values up to 3000 s/mm^2 for the clinical set-up and up to 13,000 s/mm^2 for the preclinical one.

9.3.4 Explicit models of fibre dispersion

Various techniques proposed in the literature aim to explicitly model orientation dispersion that encompasses the effects of fibre bending, fanning or undulation, which are widespread throughout the brain (Burgel *et al.*, 2006), even in areas of coherent white matter fibres, as illustrated in Figure 9.4a. Thus, such parametric models estimate indices of orientation dispersion that cannot be straightforwardly estimated from general tensor distributions. Moreover, they aim to overcome the limitation of previously discussed models of restriction that assume parallel cylinders.

In order to combine fibre crossing and dispersion, Kaden *et al.* proposed a spherical deconvolution approach that parameterises the fibre orientation density as a finite mixture of Bingham distributions (Kaden *et al.*, 2007), which is the analogue of a Gaussian distribution for directional data. The technique uses a multicompartment model, with isotropic diffusion in the extracellular space and anisotropic diffusion (cylindrically symmetric diffusion tensor) in the intraneurite space. The model allows for up to two fibre populations that exhibit dispersion around their main direction and fixes the diffusivity values. Thus only the volume fractions and directional information is estimated from the data.

Jespersen *et al.* (2006) proposed a model that aims to estimate the volume fraction and distribution of neurites (axons and dendrites) both in white matter and grey matter. The technique uses a two-compartment model with an isotropic tensor, which represents extracellular diffusion, and a distribution of cylindrically symmetric anisotropic tensors, which describes diffusion inside axons and dendrites. The orientation distribution is written in terms of spherical harmonics; thus the model estimates the corresponding coefficients in addition to the tensor diffusivities and the volume fraction. Jespersen *et al.* (2009) expand the model to include anisotropy in the extracellular space and constrain the intraneurite model to a 'stick', in order to stabilise the fitting procedure.

Zhang and Alexander (2010) extend the cylinder restriction model proposed in ActiveAx and aim to recover the axon diameter index in the presence of dispersion. To this end, they

Histology and structure tensor analysis

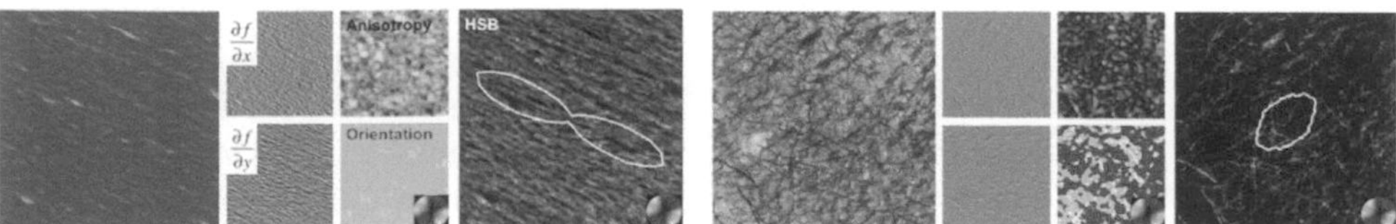

Highly anisotropic region, corpus callosum

Isotropic region, cortical grey matter

Diffusion model including dispersion, e.g. NODDI

(a)

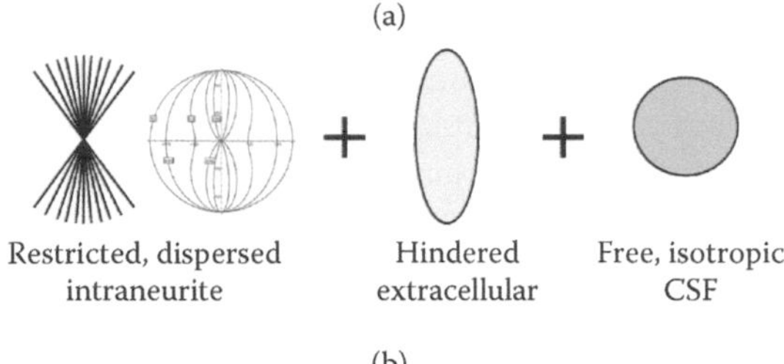

Restricted, dispersed intraneurite

Hindered extracellular

Free, isotropic CSF

(b)

In - vivo parameter maps from NODDI model

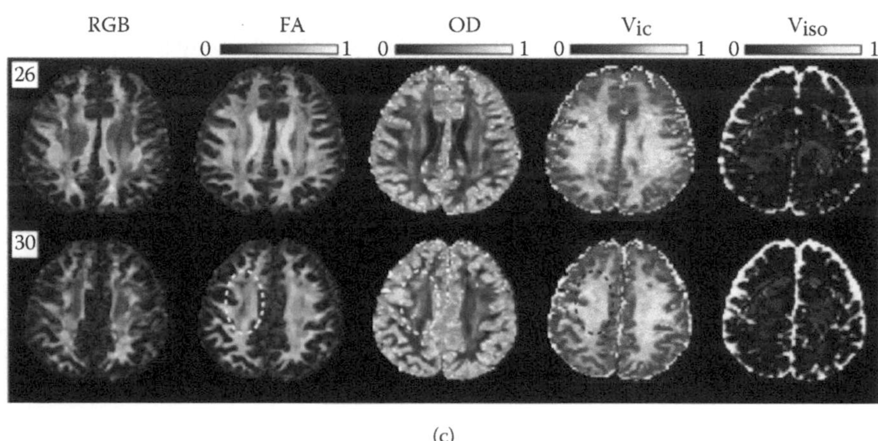

(c)

FIGURE 9.4 (a) Histological staining of neural tissue and structure tensor analysis shown for regions of highly anisotropic white matter in corpus callosum and isotropic regions of cortical grey matter. (Adapted from Budde, M.D., and Annese, J., *Front. Integr. Neurosci.*, 7, 3, 2013, under the terms of the Creative Commons Attribution License. With Permission.) (b) Schematic representation of a multicompartment model that includes orientation dispersion, such as NODDI. (c) *In vivo* parameter maps from the NODDI model that show complementary contrast to DTI fractional anisotropy in areas of bending and fanning fibres. (Reprinted from *NeuroImage*, 61, Zhang, H., *et al.*, NODDI: practical in vivo neurite orientation dispersion and density imaging of the human brain, 1000–1016, Copyright 2012, with permission from Elsevier.)

use a parametric Watson distribution, which is a cylindrically symmetric directional distribution, and estimate axon diameter index and orientation dispersion from *in vivo* human data acquired with four HARDI shells.

In later work, Zhang *et al.* (2012) developed neurite orientation dispersion and density imaging (NODDI), a multicompartment model that aims to estimate intraneurite volume fraction and orientation dispersion from a clinically feasible acquisition, as schematically illustrated in Figure 9.4b and c. NODDI models the signal contribution from three compartments: intraneurite signal represented by sticks following a Watson distribution, extraneurite signal modelled by a cylindrically symmetric tensor and CSF characterised by free isotropic diffusion. Due to its simplicity and clinically feasible data acquisition, NODDI has been used in many different studies to characterise healthy and pathological brain tissue (e.g. Eaton-Rosen *et al.*, 2014; Kunz *et al.*, 2014; Timmers *et al.*, 2014; Winston *et al.*, 2014).

The Bingham distribution, which allows for anisotropic dispersion and can represent fibre configurations such as fanning, has also been used in more recent works. Sotiropoulos *et al.* (2012) proposed the ball and rackets model, which extends the intraneurite compartment of the ball and stick model to account for fibre dispersion, while Tariq *et al.* (2016) extended the NODDI model to incorporate anisotropic orientation dispersion by replacing the Watson model of dispersion in the original NODDI with a Bingham model; this provides one extra parameter map (orientation dispersion anisotropy) beyond the original NODDI.

A different angle is considered in the spherical mean technique (SMT) (Kaden *et al.*, 2015), which factors out the effects of orientation distribution in order to estimate the per-axon parallel and radial diffusivities. SMT assumes that the signal in a given voxel can be described by a distribution of cylindrically symmetric tensors with identical eigenvalues that vary only in orientation. Thus, the diffusivity values can be calculated from the mean of the signal computed over isotropically oriented gradient directions at different b-values and requires an acquisition protocol with at least two HARDI shells (Kaden *et al.*, 2015).

Kaden *et al.* (2016) proposed a different technique to estimate intraneurite volume fraction and orientation distribution, without assuming fixed diffusivities and a parametric orientation distribution. This approach uses the spherical mean technique formulation and assumes that the signal from each fibre bundle is described by a two-compartment model, with an anisotropic tensor for the extracellular part and a stick for the intracellular part (Kaden *et al.*, 2016). This modelling approach enables a spatially variant deconvolution kernel in spherical deconvolution, which is shown to make a difference to the results.

Assessment: The models discussed above aim to provide a measure of fibre dispersion in the tissue besides other microstructural parameters such as intraneurite volume fraction or axon diameter index. The methods that estimate the volume fractions of different compartments require acquisition protocols with at least two different shells for a stable fit. In order to make the estimated parameters more robust to noise and improve the stability of fit, various modelling constraints are usually imposed, such as using cylindrically symmetric distributions, representing intraneurite compartment with a stick or using prior information to fix various parameters. A recent study investigated the effect of fixing diffusivity in NODDI, and the results show that using values that are not close to the ground truth can bias the estimated parameters (Jelescu *et al.*, 2015). Thus, the obtained quantitative maps need to be interpreted with care.

The techniques presented so far are specific cases of multicompartment models that capture various features of the tissue. In general, a wide variety of tissue models can be formed by combining different compartments that represent the main water pools in the tissue such as intra- and extra-axonal space, CSF, glial cells, etc. Ranking the models according to Bayesian information criterion, which accounts for the goodness of fit and number of parameters, shows that three compartment models with restriction along the fibre direction explain the data the best both in the rat corpus callosum, *ex vivo* (Panagiotaki *et al.*, 2012), as well as in the human corpus callosum (Ferizi *et al.*, 2014). A more recent work (Ferizi *et al.*, 2015) that includes fibre dispersion shows that these models outperform the best models with coherent fibres.

Typical acquisition (see Table 9.2): The technique proposed by Kaden *et al.* (2007), which models up to two fibre populations with dispersion around their main orientation, has been used with a dataset comprising of a single HARDI shell with $b = 1000$ s/mm^2 and 60 gradient directions. Estimating axon diameter in the presence of dispersion (Zhang and Alexander, 2010) requires a richer acquisition and has been applied to the four HARDI shell data used in the original ActiveAx study (Alexander *et al.*, 2010). The neurite density models in papers by Jespersen *et al.* (2006, 2009) have been used with a rich *ex vivo* acquisition consisting of 16 b-shells between 0 and 15,000 s/mm^2 with nine different directions each. The acquisition for NODDI (Zhang *et al.*, 2012), which is aimed at clinical studies, consists of two HARDI shells with recommended b-values around 700 and 2800 s/mm^2, with 30 and 60 gradient directions, respectively. A similar two shell acquisition can be used for Bingham NODDI (Tariq *et al.*, 2016), as well as for the SMT technique. Similarly, the multicompartment SMT technique has been used with an acquisition consisting of three HARDI shells with b-values of 1000, 2000 and 3000 s/mm^2 and 90 gradient directions each (Kaden *et al.*, 2016) but in theory will also work with a NODDI acquisition.

9.4 Software packages

Here we provide a non-exhaustive list of popular software packages useful for advanced dMRI modelling along with a list of important models they have implemented. These are available for download and use from the Internet. When possible we have included the author's description of the software.

FSL (FMRIB [Oxford Centre for Functional MRI of the Brain] Software Library) (Jenkinson *et al.*, 2012): FSL is a comprehensive library of analysis tools for functional, structural and diffusion MRI brain imaging data. FDT (FMRIB's Diffusion Toolbox) is part of FSL and includes tools for data preprocessing, local diffusion modelling and tractography. TBSS (tract-based spatial statistics) aims to improve the sensitivity, objectivity and interpretability of analysis of multisubject diffusion imaging studies, while 'eddy' is a tool for correcting eddy currents and movements in diffusion data and 'topup' is a tool for estimating and correcting susceptibility induced distortions.

Models: Ball and sticks (Behrens *et al.*, 2007), QBI-dODF (Tuch, 2004), spherical harmonic QBI-dODF (Descoteaux *et al.*, 2007), CSA-QBI (Aganj *et al.*, 2010).

Camino (Cook *et al.*, 2006): Camino is an open-source software toolkit for dMRI processing. The toolkit implements standard techniques, such as diffusion tensor fitting, mapping fractional anisotropy and mean diffusivity, deterministic and probabilistic tractography. It also contains more specialised and cutting-edge techniques, such as Monte-Carlo diffusion simulation, multifibre and HARDI reconstruction techniques, multifibre PICo, compartment models, and axon density and diameter estimation.

Models: Multitensor (Alexander and Barker, 2005), ActiveAx (Alexander, 2008), multicompartment tissue models (Panagiotaki *et al.*, 2012; Ferizi *et al.*, 2014, 2015), spherical harmonic QBI-dODF (Descoteaux *et al.*, 2007), PASMRI (Jansons and Alexander, 2004) and MESD (Alexander, 2005).

MRtrix3 (Tournier *et al.*, 2007): MRtrix3 provides a set of tools to perform various types of diffusion MRI analyses, from various forms of tractography through to next-generation group-level analyses. It is designed with consistency, performance and stability in mind and is freely available under an open-source license. It is developed and maintained by a team of experts in the field, fostering an active community of users from diverse backgrounds.

Models: CSD (Tournier *et al.*, 2007), MSMT-CSD (Jeurissen *et al.*, 2014), DKI.

DIPY (Garyfallidis *et al.*, 2014): Dipy is a free and open source software project for computational neuroanatomy, focusing mainly on dMRI analysis. It implements a broad range of algorithms for denoising, registration, reconstruction, tracking, clustering, visualisation, and statistical analysis of MRI data.

Models: CSD (Tournier *et al.*, 2007), SHORE-3D (Ozarslan *et al.*, 2013), DKI (Jensen *et al.*, 2005), CSA QBI-dODF (Aganj *et al.*, 2010), DSI (Wedeen *et al.*, 2008).

DSI Studio: DSI Studio is an open source diffusion MRI analysis tool that maps brain connections and correlates findings with neuropsychological disorders. It is a collective implementation of several methods, including DTI, QBI, DSI, generalised q-sampling imaging, q-space diffeomorphic reconstruction, diffusion MRI connectometry and generalised deterministic fibre tracking.

Models: Spherical harmonic QBI-dODF (Descoteaux *et al.*, 2007), DSI (Wedeen *et al.*, 2008).

Diffusion Kurtosis Estimator (Tabesh *et al.*, 2011): Diffusional Kurtosis Estimator (DKE) is a software tool for post-processing DKI datasets that includes a suite of command-line programs along with a graphical user interface. DKE generates a set of kurtosis (axial, mean, radial, KFA, MKT) parametric maps with a given set of DWIs acquired from a valid DKI protocol. Diffusivity (axial, mean, radial) and fractional anisotropy maps using either DKI or diffusion tensor imaging signal models are also calculated in the processing.

Models: DKI (Jensen *et al.*, 2005).

NODDI MATLAB® Toolbox (Zhang *et al.*, 2012): The NODDI MATLAB Toolbox provides MATLAB tools for estimating NODDI parameters from diffusion MRI data. NODDI is a multicompartment model that aims to estimate intraneurite volume fraction and orientation dispersion from a clinically feasible acquisition.

Models: NODDI (Zhang *et al.*, 2012).

SMT Toolbox (Kaden *et al.*, 2015): The SMT Toolbox provides tools for estimating SMT parameters from diffusion MRI data. The purpose is to map microscopic features unconfounded by the effects of fibre crossings and orientation dispersion, which are ubiquitous in the brain. This technique requires only an off-the-shelf diffusion sequence with two (or more) b-shells achievable on any standard MRI scanner. So far, SMT comes in two flavours, a microscopic tensor model and a simple multicompartment model.

Models: SMT (diffusion tensor) (Kaden *et al.*, 2015), SMT (multicompartment microscopic diffusion model) (Kaden *et al.*, 2016).

MISST (Ianus *et al.*, 2016a): MISST (Microstructure Imaging Sequence Simulation Software) is a practical diffusion MRI simulator for development, testing and optimisation of novel MR pulse sequences for microstructure imaging. MISST is based on a matrix method approach and simulates the diffusion signal for a large variety of generalised pulse sequences and tissue models.

Models: Multicompartment tissue models (Panagiotaki *et al.*, 2012) and MR pulse sequences (Ianus *et al.*, 2016a).

9.5 Current and future trends

We have presented so far a spectrum of advanced dMRI models that map greater details of the underlying tissue than DTI. As we point out above, all of these techniques have limitations and their future development remains an active area of research. In this section, we discuss key challenges and active research directions in terms of acquisition sequences, diffusion modelling and parameter estimation, hardware trends, as well as extension to multimodal acquisitions.

9.5.1 Beyond single diffusion encoding

The majority of quantitative dMRI techniques presented so far are based on standard SDE acquisitions that are widely available on clinical and preclinical scanners. Nevertheless, there is increasing interest in other diffusion encoding gradients that can provide additional contrast compared to SDE sequences.

Oscillating gradients: Oscillating diffusion encoding (ODE) sequences (Gross and Kosfeld, 1969) illustrated in Figure 9.5 replace the pulsed gradient in the standard SDE sequence with oscillating gradient waveforms, such as sine, cosine, square or trapezoidal waveforms. By increasing the oscillation frequency, ODE sequences can measure the diffusion process on shorter time scales compared to SDE sequences.

One application of ODE sequences is 'temporal diffusion spectroscopy', which requires cosine-like gradient waveforms and measures the diffusion spectrum $\boldsymbol{D}(\omega)$, i.e. the frequency dependence of the diffusion tensor (Callaghan and Stepisnik, 1995; Does *et al.*, 2003;Aggarwal *et al.*, 2012; Baron and Beaulieu, 2013; Portnoy *et al.*, 2013; Van *et al.*, 2013). A change in measured ADC or tensor parameters with frequency directly reflects the presence of restricted diffusion, without any modelling assumptions.

A quantitative estimation of restriction size can also be derived from the diffusion spectrum $\boldsymbol{D}(\omega)$, which has an explicit dependence on pore size (Stepisnik, 1993; Portnoy *et al.*, 2013; Xu *et al.*, 2014); however, this requires an acquisition that consists of cosine-like gradient waveform. A different approach to estimate microstructural parameters is to adapt models like CHARMED or MMWMD to ODE sequences and fit a multicompartment model to the acquired ODE data (Xu *et al.*, 2009; Ianus *et al.*, 2012; Siow *et al.*, 2013). Recent work (Drobnjak *et al.*, 2015) has compared the sensitivity of SDE and ODE sequences to axon diameter and has shown that low frequency oscillating gradients improve the sensitivity in the situation of dispersed fibres and/or unknown gradient orientation, while higher frequency gradients are able to better estimate the intrinsic diffusivity values (Jiang *et al.*, 2016).

Gradients with varying orientation in one measurement – what can we gain? The dMRI techniques discussed above use a

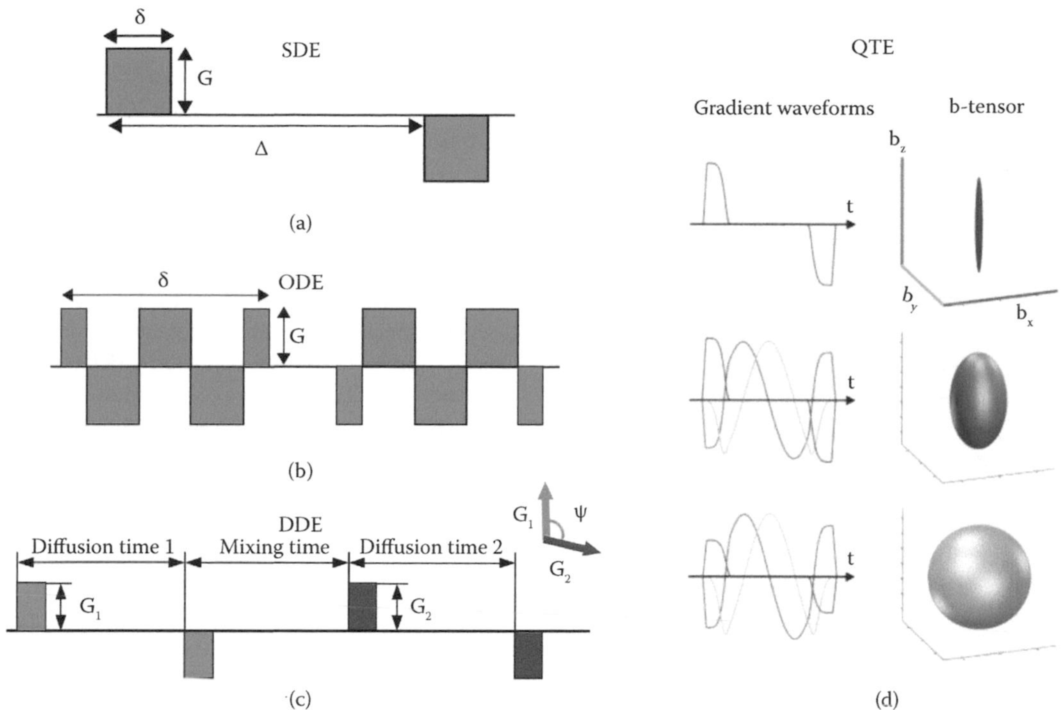

FIGURE 9.5 Schematic representation of (a) single diffusion encoding (SDE), (b) oscillating diffusion encoding (ODE) with square gradient waveforms, (c) double diffusion encoding (DDE) and (d) q-space trajectory imaging (QTI) acquisition. (Reprinted from *NeuroImage*, 135, Westin, C.F., et al., Q-space trajectory imaging for multidimensional diffusion MRI of the human brain, 345–362, Copyright 2016, with permission from Elsevier.)

collection of diffusion sequences with fixed gradient orientation during one measurement. Although such sequences have been used to estimate various microstructural parameters, they do not differentiate between more complex tissue configurations featuring microscopic anisotropy and a distribution of pore sizes without any prior information on the substrate type (Topgaard and Soderman, 2002; Ianus *et al.*, 2016b). Using sequences in which gradient orientation varies within the gradient pulse can overcome this limitation.

One way to provide sensitivity to pore anisotropy even in heterogeneous substrates with a distribution of pore shapes and sizes is to employ double diffusion encoding (DDE) sequences (Mitra, 1995; Cheng and Cory, 1999; Shemesh *et al.*, 2010). DDE sequences, illustrated in Figure 9.5c, combine two independent pairs of diffusion gradients, which can probe the correlation of molecular displacements in different directions when they have non-collinear orientations. Various approaches have been proposed to estimate microscopic anisotropy (μA) in macroscopically isotropic substrates (Cheng and Cory, 1999; Shemesh *et al.*, 2010; Shemesh and Cohen, 2011), while more recent studies have introduced rotationally invariant acquisitions that allow for a consistent estimation of μA in macroscopically anisotropic tissues, such as white matter (Lawrenz *et al.*, 2009; Jespersen *et al.*, 2013). Lawrenz *et al.* used DDE sequences on a clinical scanner to investigate microscopic anisotropy in the human brain (Lawrenz and Finsterbusch, 2015) and showed age-related differences (Lawrenz *et al.*, 2016). Combining DDE with oscillating waveforms could further improve the sensitivity to μA (Ianus *et al.*, 2016c). Such techniques provide a measure of microscopic anisotropy that is not influenced by the orientation distribution of the fibres and does not require the assumption of identical microdomains as used in SMT, which might bias the estimated parameters in areas with a distribution of pore sizes and/or be affected by partial voluming.

Another approach that has been recently proposed for quantifying μA is to combine measurements that provide isotropic and linear encoding at several diffusion weightings (Lasic *et al.*, 2014). These concepts have been generalised in the *q*-space trajectory imaging (QTI) approach (Westin *et al.*, 2016), which employs a complex acquisition (illustrated in Figure 9.5d) to disentangle variations in size from variations in shape and orientation of the underlying structures. This technique has been successfully applied to study the white matter and tumour microstructure (Szczepankiewicz *et al.*, 2014, 2016) and yields very good correlation between MRI and histology-derived tissue parameters, such as cell eccentricity and density.

All these alternative pulse sequences offer benefits to future dMRI techniques in two ways: (1) by providing insight into tissue features that can inform models we aim to estimate from simpler and more readily available data, for example providing good priors on the range of intra- and extracellular diffusivities and (2) as such sequences become more widely available on clinical scanners, by informing models that underpin future imaging techniques. For now, however, only SDE sequences are widely deployed on both clinical and research scanners.

9.5.2 Modelling considerations

A considerable effort in dMRI research is directed towards developing better modelling approaches and data analysis tools. Some of the goals are to capture more realistic tissue features, to increase the computational speed and stability of fit for real time parameter estimation or to make use of fewer data points.

Other important effects: The majority of biophysical modelling techniques that aim to estimate axon diameter assume restricted diffusion inside impermeable, parallel cylinders for intracellular space and Gaussian diffusion for the extracellular space. This is an oversimplified representation, and factors such as axonal undulation (Nilsson *et al.*, 2012), membrane permeability (Nilsson *et al.*, 2013, Nedjati-Gilani, 2014, 488), other cell types, the thickness of the myelin sheath, etc., all influence the measured signal. However, including all these features makes the models intractable. Recently, several studies have focused on improving the models for diffusion in the extracellular space, which can be characterised by analytical expressions. Thus, accounting for a time-dependent diffusivity provides a more accurate characterisation of extracellular space when measurements are acquired at different diffusion times, either with SDE or ODE sequences (Novikov *et al.*, 2014; Xu *et al.*, 2014; Burcaw *et al.*, 2015; De Santis *et al.*, 2016; Jiang *et al.*, 2016). For *in vivo* measurements of human white matter acquired with SDE sequences, a more pronounced time dependence of the extracellular diffusivity is visible for diffusion times larger than 100 ms (De Santis *et al.*, 2016). For ODE sequences, Burcaw *et al.* used *ex vivo* brain data to show that at low frequencies $\boldsymbol{D}(\omega)$ exhibits a linear dependence on ω, which is characteristic for disordered diffusion in the extracellular space, rather than a parabolic dependence on ω, which is characteristic for the intracellular diffusion (Burcaw *et al.*, 2015). Thus, accounting for time-dependent diffusivity in the extracellular space can improve the analysis for both ODE and SDE measurements and yield more accurate parameter estimates of compartment size and volume fractions.

Numerical models and machine learning: As mathematical models of dMRI aim to capture more and more effects, they become more complex and less mathematically tractable. A possible approach to improve the accuracy and tractability of such models is to circumvent analytical expressions, which cannot capture the effects of all parameters of interest, for instance membrane permeability. This can be achieved using numerical techniques to simulate the diffusion signal, e.g. Monte Carlo simulations, numerical solutions of the diffusion equation, etc. (Nilsson *et al.*, 2010). Then, the acquired signal can be related to the simulated values for different tissue configurations using various techniques such as dictionary-based approaches as well as regression techniques from machine learning (Nedjati-Gilani *et al.*, 2014). Examples of dictionary-based approaches, which also combines convex optimisation, include (Jian and Vemuri, 2007) to speed up numerous spherical deconvolution methods using appropriate dictionaries and sparsity regularisation, and the accelerated microstructure imaging via convex optimisation (AMICO) framework (Daducci *et al.*, 2014), which utilises a similar approach to speed up ActiveAx and NODDI and can be extended for other multicompartment models. Another recent dictionary-based method for axon diameter estimation was proposed by Sepehrband *et al.* (2016).

Magnetic resonance spectroscopy: MRS can measure the diffusion of various metabolites present in the intracellular space, which provides specificity towards different cellular compartments. In comparison, water-based diffusion MRI techniques, e.g. those discussed in this chapter, are confounded by the presence of water in all cellular compartments as well as in the extracellular space. Thus, developing a comprehensive model to explain the link between tissue features and the measured signal is a challenging task in water-based diffusion MRI. In MRS metabolites can be selected to target specific cell types, such as N-acetyl-aspartate and glutamate for neurons and myo-inositol and choline for glial cells. Therefore, computational models for MRS can estimate structural features such as process length, complexity and fibre diameter for different cell types with greater accuracy (Palombo *et al.*, 2016, 2017; Ligneul *et al.*, 2017). Hence, MRS can not only recover interesting microstructural features by itself but can also help to refine models for water-based diffusion MRI.

9.5.3 Experiment design

Experiment design involves deriving the best possible acquisition protocol to minimise parameter estimation error – while respecting a viable acquisition time. For example, DTI estimation is shown to improve by spreading out the gradient directions on the sphere (Jones *et al.*, 1999). The idea is generalised to multishell HARDI data in the paper by Caruyer *et al.* (2013), which is now part of the Human Connectome Project's (HCP) dMRI acquisition protocol (Van Essen *et al.*, 2012). A generic signal model-based experiment design by Caruyer and Deriche (2012) aims to optimise multishell HARDI acquisitions for model-free representation in the SPFI basis. Experiment design for DKI was considered by Poot *et al.* (2010). Interestingly, the importance of clinical viability was demonstrated by Jensen and Helpern (2010), who proposed a less optimal but shorter acquisition for DKI to reduce the total acquisition time. Alexander (2008) aimed to optimise the SDE sequence parameters for fitting a given biophysical model. This approach is very effective in reducing the acquisition time, although it requires prior information on the model parameters, which is usually obtained from a preliminary rich dataset. An extension of this procedure was applied to optimise generalised diffusion sequences (Drobnjak *et al.*, 2010; Drobnjak and Alexander, 2011) for white matter models, while other studies have used a similar approach for different models, such as bi-tensor (Farrher *et al.*, 2016), IVIM (Lemke *et al.*, 2010) or Vascular, Extracellular and Restricted Diffusion for Cytometry in Tumors (VERDICT) MRI (Panagiotaki *et al.*, 2015).

9.5.4 Hardware trends

Recent years have seen an increased effort at pushing the frontiers of the scanner hardware, which includes augmenting both the strength of the magnetic field as well as the magnetic gradients for diffusion weighting. Typical clinical scanners have magnetic fields of 1.5T and more recent ones of 3T, with gradient strengths of 40–60 mT/m. An important development for dMRI is the Connectome scanner of the HCP (Van Essen *et al.*, 2012), a 3T scanner that comes equipped with gradient strengths up to 300 mT/m, exceeding clinical scanners by almost an order of magnitude. This drastic increase in gradient strength results in shorter echo time for a given *b*-value, improving the SNR, as well as a lower limit on the measurable pore sizes, which is especially important for different biophysical modelling approaches. Moreover, the HCP data is publicly available and aims to provide the research community with high quality dMRI data, which can be used to compare, test and develop novel dMRI modelling approaches.

DW-MRI can also benefit from going to higher field, i.e. 7T instead of 3T, which provides more signal and higher spatial resolution. Although T_2 values are shorter at 7T, which is detrimental for diffusion contrast, recent work has shown the potential benefits of combining 3T and 7T data to complement higher resolution at 7T with higher diffusion weighting achieved at 3T (Heidemann *et al.*, 2010; Vu *et al.*, 2015).

9.5.5 Multimodal acquisitions and modelling

The combination of diffusion MRI measurements with complementary information from other MR contrasts or modalities potentially provides access to additional attributes of tissue microstructure not visible with any one modality individually. Estimation of the myelin g-ratio, which is closely related to development, cognition and disease (Stikov *et al.*, 2015), is one example that has gained importance recently. This involves using dMRI data that lacks information on myelin content but is sensitive to the tissue microstructure with T_1/T_2 weighted data (Dean *et al.*, 2016), myelin water imaging data (Melbourne *et al.*, 2014) or quantitative magnetisation transfer data (Stikov *et al.*, 2015) that lack microstructure information but are sensitive to myelination content. Combining these MRI modalities makes it possible to estimate the relationship between the axon size and myelin thickness, or the g-ratio, which is the ratio of the inner diameter (only axon from dMRI) to the outer diameter (axon and myelin from myelin content) of the fibre. Similarly, combinations of diffusion and susceptibility mapping, or diffusion and relaxometry, potentially provide contrasts between tissue types that neither modality can reveal alone.

9.6 Conclusions

This chapter presents a variety of dMRI techniques that aim to overcome the main limitations of DTI in order to improve the estimation of the angular and radial diffusion profiles. The methods are classified into two broad categories: signal models, which aim to capture the diffusion signal and usually represent it in terms of basis functions, and biophysical models, which aim to describe the effect of various microscopic tissue features on the measured signal. These techniques use a richer acquisition compared with DTI, both in terms of angular resolution as well as the number of different b-shells, which allows for a more accurate estimation of the underlying tissue features. This chapter also discusses the current and future trends in terms of diffusion acquisition, modelling, hardware advances and multimodal imaging.

References

Abe O, Aoki S, Hayashi N, Yamada H, Kunimatsu A, Mori H, et al. Normal aging in the central nervous system: quantitative MR diffusion-tensor analysis. Neurobiology of Aging 2002; 23: 433–41.

Aboitiz F, et al. Fiber composition of the human corpus callosum. Brain Res 1992; 598: 143–53.

Aganj I, Lenglet C, Sapiro G, Yacoub E, Ugurbil K, Harel N. Reconstruction of the orientation distribution function in single- and multiple-shell q ball imaging within constant solid angle.

Aggarwal M, et al. Probing mouse brain microstructure using oscillating gradient diffusion MRI. Magn Reson Med 2012; 67: 98–109.

Alexander DC. Maximum Entropy Spherical Deconvolution for Diffusion MRI. In: Christensen GE and Sonka M, editors. Information Processing in Medical Imaging: 19th International Conference, IPMI 2005, Glenwood Springs, CO, USA, July 10-15, 2005. Proceedings. Berlin, Heidelberg: Springer Berlin Heidelberg, 2005, pp. 76–87.

Alexander DC. A general framework for experiment design in diffusion MRI and its application in measuring direct tissue-microstructure features. Magn Reson Med 2008; 60: 439–48.

Alexander DC, Barker GJ. Optimal imaging parameters for fiber-orientation estimation in diffusion MRI. NeuroImage 2005; 27: 57–67.

Alexander DC, Barker GJ, Arridge SR. Detection and modeling of non-Gaussian apparent diffusion coefficient profiles in human brain data. Magn Reson Med 2005; 48: 331–40.

Alexander DC, et al. Orientationally invariant indices of axon diameter and density from diffusion MRI. NeuroImage 2010; 52: 1374–89.

Anderson AW. Measurement of fiber orientation distributions using high angular resolution diffusion imaging. Magn Reson Med 2005; 54: 1194–206.

Assaf Y, Basser PJ. Composite hindered and restricted model of diffusion (CHARMED) MR imaging of the human brain. NeuroImage 2005; 27: 48–58.

Assaf Y, et al. New modeling and experimental framework to characterize hindered and restricted water diffusion in brain white matter. Magn Reson Med 2004; 52: 965–78.

Assaf Y, et al. AxCaliber: a method for measuring axon diameter distribution from diffusion MRI. Magn Reson Med 2008; 59: 1347–54.

Assemlal H-E, Tschumperle D, Brun L. Efficient and robust computation of PDF features from diffusion MR signal. Med Image Anal 2009; 13: 715–729.

Assemlal H-E, et al. Recent advances in diffusion MRI modeling: angular and radial reconstruction. Med Image Anal 2011; 15: 369–96.

Avram AV, et al. Clinical feasibility of using mean apparent propagator (MAP) MRI to characterize brain tissue microstructure. NeuroImage 2016; 127: 422–34.

Barazany D, Basser PJ, Assaf Y. In vivo measurement of axon diameter distribution in the corpus callosum of rat brain. Brain 2009; 132: 1210–20.

Baron CA, Beaulieu C. Oscillating gradient spin-echo (OGSE) diffusion tensor imaging of the human brain. Magn Reson Med. 2013; 72: 726–36.

Basser PJ, Mattiello J, Lebihan D, Estimation of the effective self-diffusion tensor from the NMR Spin Echo. J Magn Reson B 1994a; 103: 247–54.

Basser PJ, Mattiello J, LeBihan D. MR diffusion tensor spectroscopy and imaging. Biophys J 1994b; 66: 259–67.

Basser PJ, et al. In vivo fiber tractography using DT-MRI data. Magn Reson Med 2000; 44: 625–32.

Behrens TE, et al. Characterization and propagation of uncertainty in diffusion-weighted MR imaging. Magn Reson Med 2003; 50: 1077–88.

Behrens TE, et al. Probabilistic diffusion tractography with multiple fibre orientations: what can we gain? NeuroImage 2007; 34: 144–55.

Budde MD, Annese J. Quantification of anisotropy and fibre orientation in human brain histological sections. Front Integr Neurosci 2013; 7: 3.

Burcaw L, Fieremans E, Novikov DS. Mesoscopic structure of neuronal tracts from time-dependent diffusion. NeuroImage 2015; 114: 18–37.

Burgel U, et al. White matter fiber tracts of the human brain: three-dimensional mapping at microscopic resolution, topography and intersubject variability. NeuroImage 2006; 29: 1092–1105.

Callaghan PT. Principles of nuclear magnetic resonancel. Oxford: Oxford University Press.

Callaghan PT, Stepisnik J. Frequency domain analysis of spin motion using modulated-gradient NMR. J Magn Reson A 1995; 117: 118–22.

Caruyer E, Deriche R. A computational framework for experimental design in diffusion MRI. CDMRI – MICCAI Workshop on Computational Diffusion MRI. Nice, France.

Caruyer E, et al. Design of multishell sampling schemes with uniform coverage in diffusion MRI. Magn Reson Med 2013; 69: 1534–40.

Cheng Y, Cory D. Multiple scattering by NMR. J Am Chem Soc 1999; 121: 7935–96.

Cheng J, et al. Model-free and analytical EAP reconstruction via spherical polar Fourier diffusion MRI. In: Medical image computing and computer-assisted intervention – MICCAI. Vol. 6361 – Part I, N. Nassir, P.W.P. Josien, A.V. Max, eds. Beijing, China: Springer; 2010, pp. 590–7.

Clark CA, Hedehus M, Moseley ME. In vivo mapping of the fast and slow diffusion tensors in human brain. Magn Reson Med 2000; 47: 623–8.

Clark CA, Le Bihan D. Water diffusion compartmentation and anisotropy at high b values in the human brain. Magn Reson Med 2000; 44: 852–9.

Cook PA, et al. Camino: diffusion MRI reconstruction and processing. 14th Scientific Meeting of the International Society for Magnetic Resonance in Medicine, Seattle, WA, p. 2759.

Daducci A, et al. Accelerated microstructure imaging via convex optimization (AMICO) from diffusion MRI data. NeuroImage 2014; 105: 32–44.

de Graaf RA, Braun KPJ, Nicolay K, Single-shot diffusion trace 1H NMR spectroscopy. Magn Reson Med 2001; 45: 741–8.

De Santis S, Jones DK, Roebroek A. Including diffusion time dependence in the extra-axonal space improves in vivo estimates of axonal diameter and density in human white matter. NeuroImage 2016; 130: 91–103.

De Santis S, et al. Improved precision in CHARMED assessment of white matter through sampling scheme optimization and model parsimony testing. Magn Reson Med 2014; 71: 661–71.

Dean DC, et al. Mapping an index of the myelin g-ratio in infants using magnetic resonance imaging. NeuroImage 2016; 132: 225–37.

Dell'Acqua F, et al. A model-based deconvolution approach to solve fiber crossing in diffusion-weighted MR imaging. IEEE Trans Biomed Eng 2007; 54: 462–72.

Descoteaux M, et al. Regularized, fast, and robust analytical Q-ball imaging. Magn Reson Med 2007; 58: 497–510.

Descoteaux M, et al. Deterministic and probabilistic tractography based on complex fibre orientation distributions. IEEE Trans Med Imaging 2009; 28: 269–86.

Descoteaux M, et al. Multiple q-shell diffusion propagator imaging. Med Image Anal 2010; 15: 603–21.

Does MD, Parsons EC, Gore JC. Oscillating gradient measurements of water diffusion in normal and globally ischemic rat brain. Magn Reson Med 2003; 49: 206–15.

Drobnjak I, Alexander DC. Optimising time-varying gradient orientation for microstructure sensitivity in diffusion-weighted MR. J Magn Reson 2011; 212: 344–354.

Drobnjak I, Siow B, Alexander DC. Optimizing gradient waveforms for microstructure sensitivity in diffusion-weighted MR. J Magn Reson 2010; 206: 41–51.

Drobnjak I, et al. PGSE, OGSE, and sensitivity to axon diameter in diffusion MRI: insight from a simulation study. Magn Reson Med 2015; 75: 688–700.

Dyrby TB, Søgaard LV, Hall MG, Ptito M, Alexander DC. Contrast and stability of the axon diameter index from microstructure imaging with diffusion MRI. Magn Reson Med 2013; 70: 711–21.

Eaton-Rosen Z, et al. Measurement of white matter maturation in the preterm brain using NODDI. ISMRM, Milan, 2014, pp. 3512.

Farrher E, et al. A new framework for the optimisation of multi-shell diffusion weighting MRI settings using a parameterised Cramér-Rao lower-bound. ISMRM, Singapore, 2016.

Ferizi U, et al. A ranking of diffusion MRI compartment models with in vivo human brain data. Magn Reson Med 2014; 72: 1785–92.

Ferizi U, et al. White matter compartment models for in vivo diffusion MRI at 300mT/m. NeuroImage 2015; 118: 468–83.

Fick RHJ, et al. MAPL: tissue microstructure estimation using Laplacian-regularized MAP-MRI and its application to HCP data. NeuroImage 2016; 134: 365–85.

Fieremans E, Jensen JH, Helpern JA. White matter characterization with diffusional kurtosis imaging. NeuroImage 2011a; 58: 177–88.

Fieremans E, et al. Novel white matter tract integrity metrics sensitive to Alzheimer Disease progression. Am J Neuroradiol 2013; 34: 2105–12.

Frank LR. Characterization of anisotropy in high angular resolution diffusion-weighted MRI. Magn Reson Med 2002; 47: 1083–99.

Garyfallidis E, Brett M, Amirbekian B, Rokem A, van der Walt S, Descoteaux M, et al. Dipy, a library for the analysis of diffusion MRI data. Frontiers in Neuroinformatics 2014; 8: 8.

Ghosh A, Deriche R. A survey of current trends in diffusion MRI for structural brain connectivity. J Neural Eng 2015; 13: 011001.

Ghosh A, Milne T, Deriche R. Constrained diffusion kurtosis imaging using ternary quartics & MLE. Magn Reson Med 2013; 71: 1581–91.

Gross B, Kosfeld R. Anwendung der spin-echo-methode der messungder selbstdiffusion. Messtechnik 1969; 77: 171–7.

Hagmann P. From diffusion MRI to brain connectomics: EPFL, PhD thesis, 2005.

Hansen B, et al. Experimentally and computationally fast method for estimation of a mean kurtosis. Magn Reson Med 2013; 69: 1754–60.

Head D, et al. Differential vulnerability of anterior white matter in nondemented aging with minimal acceleration in dementia of the Alzheimer type: evidence from diffusion tensor imaging. Cereb Cortex 2004; 14: 410–23.

Heidemann RM, et al. Diffusion imaging in humans at 7T using readout-segmented EPI and GRAPPA. Magn Reson Med 2010; 64: 9–14.

Hess CP, et al. Q-ball reconstruction of multimodal fiber orientations using the spherical harmonic basis. Magn Reson Med 2006; 56: 104–17.

Horowitz A, et al. In vivo correlation between axon diameter and conduction velocity in the human brain. Brain Struct Funct 2015; 220: 1777–88.

Hosseinbor AP, et al. Bessel Fourier Orientation Reconstruction (BFOR): an analytical diffusion propagator reconstruction for hybrid diffusion imaging and computation of q-space indices. NeuroImage 2012; 64: 650–70.

Ianuş A, Alexander DC, Drobnjak I. Microstructure Imaging Sequence Simulation Toolbox. In: Tsaftaris SA, Gooya A, Frangi AF and Prince JL (Eds.). Simulation and Synthesis in Medical Imaging: First International Workshop, SASHIMI 2016, Held in Conjunction with MICCAI 2016, Athens, Greece, October 21, 2016, Proceedings. Cham: Springer International Publishing, 2016a, 34–44.

Ianus A, Drobnjak I, Alexander DC. Model-based estimation of microscopic anisotropy using diffusion MRI: a simulation study. NMR Biomed 2016b; 29; 627–85.

Ianus A, et al. Gaussian phase distribution approximations for oscillating gradient spin echo diffusion MRI. J Magn Reson 2012; 227: 25–34.

Ianuş A, Shemesh N, Alexander DC, Drobnjak I. Double oscillating diffusion encoding and sensitivity to microscopic anisotropy. Magn Reson Med 2017; 78: 550–564.

Innocenti GM, Caminiti R, Aboitiz F. Comments on the paper by Horowitz et al. (2014). Brain Struct Funct 2015; 220: 1789–90.

Jansons KM, Alexander DC. Persistent angular structure: new insights from diffusion MRI data. Dummy version. Inf Process Med Imaging 2004; 18: 672–83.

Jelescu IO, et al. Degeneracy in model parameter estimation for multi-compartmental diffusion in neuronal tissue. NMR Biomed 2015; 29: 33–47.

Jenkinson M, et al. FSL. NeuroImage 2012; 62: 782–90.

Jensen JH, Helpern JA. MRI quantification of non-Gaussian water diffusion by kurtosis analysis. NMR Biomed 2010; 23: 698–710.

Jensen JH, et al. Diffusional kurtosis imaging: the quantification of non-gaussian water diffusion by means of magnetic resonance imaging. Magn Reson Med 2005; 53: 1432–40.

Jespersen SN, et al. Modeling dendrite density from magnetic resonance diffusion measurements. NeuroImage 2006; 34: 1473–86.

Jespersen SN, et al. Neurite density from magnetic resonance diffusion measurements at ultrahigh field: comparison with light microscopy and electron microscopy. NeuroImage 2009; 49: 205–16.

Jespersen SNHL, et al. Rotationally invariant double pulsed field gradient diffusion imaging. Proceedings of the International Society for Magnetic Resonance in Medicine, Salt Lake City, UT, p. 256.

Jeurissen B, et al. Investigating the prevalence of complex fiber configurations in white matter tissue with diffusion magnetic resonance imaging. Hum Brain Mapp 2013; 34: 2747–66.

Jeurissen B, et al. Multi-tissue constrained spherical deconvolution for improved analysis of multi-shell diffusion MRI data. NeuroImage 2014; 103: 411–26.

Jian B, Vemuri. A unified computational framework for deconvolution to reconstruct multiple fibers from diffusion weighted MRI. IEEE Trans Med Imaging 2007; 26: 1464–71.

Jian B, et al. A novel tensor distribution model for the diffusion-weighted MR signal. NeuroImage 2007; 37: 164–76.

Jiang X, et al. Quantification of cell size using temporal diffusion spectroscopy. Magn Reson Med 2016; 75: 1076–85.

Jones DK, Horsfield MA, Simmons A. Optimal strategies for measuring diffusion in anisotropic systems by magnetic resonance imaging. Magn Reson Med 1999; 42: 515–25.

Kaden E, Knosche TR, Anwander A. Parametric spherical deconvolution: inferring anatomical connectivity using diffusion MR imaging. NeuroImage 2007; 37: 474–88.

Kaden E, Kruggel F, Alexander DC. Quantitative mapping of the per-axon diffusion coefficients in brain white matter. Magn Reson Med 2015; 75: 1752–63.

Kaden E, et al. Multi-compartment microscopic diffusion imaging. NeuroImage 2016; 139: 346–59.

Kunz N, et al. Assessing white matter microstructure of the newborn with multi-shell diffusion MRI and biophysical compartment models. NeuroImage 2014; 96: 288–99.

Lasič S, Szczepankiewicz F, Eriksson S, Nilsson M, Topgaard D. Microanisotropy imaging: quantification of microscopic diffusion anisotropy and orientational order parameter by diffusion MRI with magic-angle spinning of the q-vector. Frontiers in Physics 2014; 2.

Lawrenz M, Brassen S, Finsterbusch J. Microscopic diffusion anisotropy in the human brain: age-related changes. NeuroImage 2016; 141: 313–25.

Lawrenz M, Finsterbusch J. Mapping measures of microscopic diffusion anisotropy in human brain white matter in vivo with double-wave-vector diffusion weighted imaging. Magn Reson Med 2015; 73: 773–83.

Lawrenz M, Koch MA, Finsterbusch J. A tensor model and measures of microscopic anisotropy for double-wave-vector diffusion-weighting experiments with long mixing times. J Magn Reson 2009; 202: 43–56.

Le Bihan D, Breton E, Lallemand D, Aubin ML, Vignaud J, Laval-Jeantet M. Separation of diffusion and perfusion in intravoxel incoherent motion MR imaging. Radiology 1988; 497–505.

Lemke A, et al. Towards an optimal distribution of b-values for IVIM imaging. ISMRM, Stockholm.

Leow AD, et al. The tensor distribution function. Magn Reson Med 2009; 61: 205–14.

Ligneul C, Palombo M, Valette J. Metabolite diffusion up to very high b in the mouse brain in vivo: revisiting the potential correlation between relaxation and diffusion properties. Magn Reson Imaging 2017; 77: 1390–8.

Liu C, Bammer R, Moseley ME. Generalized diffusion tensor imaging (GDTI): a method for characterizing and imaging diffusion anisotropy caused by non-Gaussian diffusion. Israel J Chem 2003; 43: 145–54.

Liu C, et al. Characterizing non-Gaussian diffusion by using generalized diffusion tensors. Magn Reson Med 2004; 51: 924–37.

Malcolm JG, Shenton ME, Rathi Y. Filtered multi-tensor tractography. IEEE Trans Med Imaging 2011; 29: 1664–75.

McNab JA, et al. The human connectome project and beyond: initial applications of 300 mT/m gradients. NeuroImage 2013; 80: 234–45.

Melbourne A, et al. Multi-modal measurement of the myelin-to-axon diameter g-ratio in preterm-born neonates and adult controls. Med Image Comput Comput Assist Interv 2014; 17: 268–75.

Milne ML, Conradi MS. Multi-exponential signal decay from diffusion in a single compartment. J Magn Reson 2009; 197: 87–90.

Mitra PP. Multiple wave-vector extensions of the NMR pulsed-field-gradient spin-echo diffusion measurement. Phys Rev B 1995; 51: 15074–8.

Mori S, Van Zijl PC. Diffusion weighting by the trace of the diffusion tensor within a single scan. Magn Reson Med 1995; 33: 41–52.

Mori S, et al. Three-dimensional tracking of axonal projections in the brain by magnetic resonance imaging. Ann Neurol 1999; 45: 265–9.

Nedjati-Gilani GL, et al. Machine learning based compartment models with permeability for white matter microstructure imaging. Med Image Comput Comput Assist Interv 2014; 17: 257–64.

Niendorf T, et al. Biexponential diffusion attenuation in various states of brain tissue: implications for diffusion-weighted imaging. Magn Reson Med 1996; 36: 847–57.

Nilsson M, et al. Evaluating the accuracy and precision of a two-compartment Karger model using Monte Carlo simulations. J Magn Reson 2010; 206: 59–67.

Nilsson M, et al. The importance of axonal undulation in diffusion MR measurements: a Monte Carlo simulation study. NMR in Biomed 2012; 25: 795–805.

Nilsson M, et al. The role of tissue microstructure and water exchange in biophysical modelling of diffusion in white matter. MAGMA 2013; 26: 345–70.

Ning L, et al. Sparse reconstruction challenge for diffusion MRI: validation on a physical phantom to determine which acquisition scheme and analysis method to use? Med Image Anal 2015; 26: 316–31.

Novikov DS, Jensen JH, Helpern JA, Fieremans E. Revealing mesoscopic structural universality with diffusion. Proceedings of the National Academy of Sciences 2014; 111: 5088–93.

Ozarslan E, et al. Resolution of complex tissue microarchitecture using the diffusion orientation transform (DOT). NeuroImage 2006; 31: 1086–103.

Ozarslan E, et al. Mean apparent propagator (MAP) MRI: a novel diffusion imaging method for mapping tissue microstructure. NeuroImage 2013; 78: 16–32.

Palombo M, et al. New paradigm to assess brain cell morphology by diffusion-weighted MR spectroscopy in vivo. PNAS 2016; 113: 6671–6.

Palombo M, Ligneul C, Valette J. Modeling diffusion of intracellular metabolites in the mouse brain up to very high diffusion-weighting: diffusion in long fibers (almost) accounts for non-monoexponential attenuation. Magn Reson Imaging 2017; 77: 343–50.

Panagiotaki E, et al. Compartment models of the diffusion MR signal in brain white matter: a taxonomy and comparison. NeuroImage 2012; 59: 2241–54.

Panagiotaki E, et al. Optimised VERDICT MRI protocol for prostate cancer characterisation. ISMRM, Toronto, 2015.

Parker GJM, Alexander DC. Probabilistic Monte Carlo Based Mapping of Cerebral Connections Utilising Whole-Brain Crossing Fibre Information. In: Taylor C and Noble JA, eds. Information Processing in Medical Imaging: 18th International Conference, IPMI 2003, Ambleside, UK, July 20–25, 2003. Proceedings. Berlin, Heidelberg: Springer Berlin Heidelberg, 2003: 684–695.

Pasternak O, Sochen N, Assaf Y. PDE based estimation and regularization of multiple diffusion tensor fields. In: Visualization and Image Processing of Tensor Fields. J. Weickert, H. Hagen, eds. Springer; 2006.

Pasternak O, et al. Variational multiple-tensor fitting of fiber-ambiguous diffusion-weighted magnetic resonance imaging voxels. Magn Reson Imaging 2008; 26: 1133–44.

Pasternak O, et al. Free water elimination and mapping from diffusion MRI. Magn Reson Med 2009; 62: 717–30.

Peled S, et al. Water diffusion, T(2), and compartmentation in frog sciatic nerve. Magn Reson Med 1999; 42: 911–18.

Peyrat JM, et al. A computational framework for the statistical analysis of cardiac diffusion tensors: application to a small database of canine hearts. IEEE Trans Med Imaging 2007; 26: 1500–14.

Poot DH, et al. Optimal experimental design for diffusion kurtosis imaging. IEEE Trans Med Imaging 2010; 29: 819–29.

Portnoy S, et al. Oscillating and pulsed gradient diffusion magnetic resonance microscopy over an extended b-value range: implications for the characterization of tissue microstructure. Magn Reson Med 2013; 69: 1131–45.

Ramirez-Manzanares A, et al. Diffusion basis functions decomposition for estimating white matter intravoxel fiber geometry. IEEE Trans Med Imaging 2007; 26: 1091–102.

Scherrer B, et al. Characterizing brain tissue by assessment of the distribution of anisotropic microstructural environments in diffusion-compartment imaging (DIAMOND). Magn Reson Med 2016; 76: 963–77.

Scherrer B, Warfield SK. Why multiple B-values are required for multi-tensor models. In: Evaluation with a Constrained Log-Euclidean Model2010: 1389–392. doi: 10.1109/ISBI.2010.5490257.

Sepehrband F, et al. Towards higher sensitivity and stability of axon diameter estimation with diffusion-weighted MRI. NMR Biomed 2016; 29: 293–308.

Seunarine KK, et al. Exploiting peak anisotropy for tracking through complex structures. 2007 IEEE 11th International Conference on Computer Vision, Rio de Janeiro, Brazil, pp. 1–8.

Shemesh N, Özarslan E, Adiri T, Basser PJ, Cohen Y. Noninvasive bipolar double-pulsed-field-gradient NMR reveals signatures for pore size and shape in polydisperse, randomly oriented, inhomogeneous porous media. The Journal of Chemical Physics 2010; 2010b; 133: 044705.

Shemesh N, Cohen Y. Microscopic and compartment shape anisotropies in gray and white matter revealed by angular bipolar double-PFG MR. Magn Reson Med 2011; 65: 1216–27.

Shemesh N, et al. Conventions and nomenclature for double diffusion encoding NMR and MRI. Magn Reson Med 2015; 75: 82–7.

Siow B, et al. Axon radius estimation with oscillating gradient spin echo (OGSE) diffusion MRI. Diffus Fundament 2013; 18: 1–6.

Sotiropoulos SN, Behrens TEJ, Jbabdi S, Ball and rackets: inferring fiber fanning from diffusion-weighted MRI. NeuroImage 2012; 60: 1412–25.

Stanisz GJ, et al. An analytical model of restricted diffusion in bovine optic nerve. Magn Reson Med 1997; 37: 103–11.

Stejskal EO. Use of spin echoes in a pulsed magnetic-field gradient to study anisotropic, restricted diffusion and flow. J Chem Phys 1965; 43: 3597–603.

Stepisnik J. Time-dependent self-diffusion by NMR spin echo. Phys B 1993; 183: 343–50.

Steven AJ, Zhuo J, Melhem ER. Diffusion Kurtosis imaging: an emerging technique for evaluating the microstructural environment of the brain. Am J Roentgenol 2013; 202: W26-33.

Stikov N, et al. In vivo histology of the myelin g-ratio with magnetic resonance imaging. NeuroImage 2015; 118: 397–405.

Szczepankiewicz F, Lasic S, van Westen D, Sundgren PC, Englund E, Westin CF, et al. Quantification of microscopic diffusion anisotropy disentangles effects of orientation dispersion from microstructure: applications in healthy volunteers and in brain tumors. Neuroimage 2015; 104: 241–52.

Szczepankiewicz F, van Westen D, Englund E, Westin CF, Stahlberg F, Latt J, et al. The link between diffusion MRI and tumor heterogeneity: Mapping cell eccentricity and density by diffusional variance decomposition (DIVIDE). Neuroimage 2016; 142: 522–32.

Tabesh A, et al. Estimation of tensors and tensor-derived measures in diffusional kurtosis imaging. Magn Reson Med 2011; 65: 823–36.

Tang XP, Sigmund EE, Song YQ. Simultaneous measurement of diffusion along multiple directions. J Am Chem Soc 2004; 126: 16336–7.

Tariq M, et al. Bingham-NODDI: mapping anisotropic orientation dispersion of neurites using diffusion MRI. NeuroImage 2016; 133: 207–23.

Timmers I, Zhang H, Bastiani M, Jansma BM, Roebroeck A, Rubio-Gozalbo ME. White matter microstructure pathology in classic galactosemia revealed by neurite orientation dispersion and density imaging. J Inherit Metab Dis 2015; 38: 295–304.

Topgaard D. Isotropic diffusion weighting in PGSE NMR: numerical optimization of the q-MAS PGSE sequence. Microporous Mesoporous Mater 2013; 178: 60–3.

Topgaard D, Soderman O. Self-diffusion in two- and three-dimensional powders of anisotropic domains: an NMR study of the diffusion of water in cellulose and starch. J Phys Chem 2002; 106: 11887–92.

Tournier JD, et al. Direct estimation of the fiber orientation density function from diffusion-weighted MRI data using spherical deconvolution. NeuroImage 2004; 23: 1176–85.

Tournier JD, Calamante F, Connelly A. Robust determination of the fibre orientation distribution in diffusion MRI: non-negativity constrained super-resolved spherical deconvolution. NeuroImage 2007; 35: 1459–72.

Tournier JD, Calamante F, Connelly A. MRtrix: diffusion tractography in crossing fiber regions. Int J Imaging Syst Technol 2012; 22: 53–66.

Tristan-Vega A, Westin CF, Aja-Fernandez S. Estimation of fiber orientation probability density functions in high angular resolution diffusion imaging. NeuroImage 2009; 47: 638–50.

Tuch DS. Q-ball imaging. Magn Reson Med 2004; 52: 1358–72.

Tuch DS, et al. High angular resolution diffusion imaging reveals intravoxel white matter fiber heterogeneity. Magn Reson Med 2002; 48: 577–82.

Tuch DS, et al. Diffusion MRI of complex neural architecture. Neuron 2003; 40: 885–95.

Van AT, Holdsworth SJ, Bammer R. In vivo investigation of restricted diffusion in the human brain with optimized oscillating diffusion gradient encoding. Magn Reson Med 2013; 71: 83–94.

Van Essen DC, et al. The human Connectome Project: a data acquisition perspective. NeuroImage 2012; 62: 2222–31.

Veraart J, Van Hecke W, Sijbers J. Constrained maximum likelihood estimation of the diffusion kurtosis tensor using a Rician noise model. Magn Reson Med 2011; 66: 678–86.

Vu AT, et al. High resolution whole brain diffusion imaging at 7T for the Human Connectome Project. NeuroImage 2015; 122: 318–31.

Wang X, et al. Diffusion basis spectrum imaging detects and distinguishes coexisting subclinical inflammation, demyelination and axonal injury in experimental autoimmune encephalomyelitis mice. NMR Biomed 2014; 27: 843–52.

Wedeen VJ, et al. Mapping fiber orientation spectra in cerebral white matter with Fourier-transform diffusion MRI. Proc Int Soc Magn Reson Med 2000; 8: 82.

Wedeen VJ, et al. Mapping complex tissue architecture with diffusion spectrum magnetic resonance imaging. Magn Reson Med 2005; 54: 1377–86.

Wedeen VJ, et al. Diffusion spectrum magnetic resonance imaging (DSI) tractography of crossing fibers. NeuroImage 2008; 41: 1267–77.

Westin CF, et al. Q-space trajectory imaging for multidimensional diffusion MRI of the human brain. NeuroImage 2016; 135: 345–62.

White NS, et al. Probing tissue microstructure with restriction spectrum imaging: histological and theoretical validation. Hum Brain Mapp 2013; 34: 327–46.

Wiegell MR, Larsson HB, Wedeen VJ. Fiber crossing in human brain depicted with diffusion tensor MR imaging. Radiology 2000; 217: 897–903.

WinstonGP, et al. Advanced diffusion imaging sequences could aid assessing patients with focal cortical dysplasia and epilepsy. Epilepsy Res 2014; 108: 336–9.

Wu Y-C, Alexander AL. Hybrid diffusion imaging. NeuroImage 2007; 36: 617–29.

Xu J, Does MD, Gore JC. Quantitative characterization of tissue microstructure with temporal diffusion spectroscopy. J Magn Reson 2009; 200: 189–97.

Xu J, et al. Mapping mean axon diameter and axonal volume fraction by MRI using temporal diffusion spectroscopy. NeuroImage 2014; 103C: 10–19.

Zhang H, Alexander DC. Axon diameter mapping in the presence of orientation dispersion with diffusion MRI. Med Image Comput Comput Assist Interv 2010; 13: 640–7.

Zhang H, et al. NODDI: practical in vivo neurite orientation dispersion and density imaging of the human brain. NeuroImage 2012; 61: 1000–16.

10

MT: Magnetisation Transfer[1]

Contents

Marco Battiston
University College London

Mara Cercignani
University of Sussex

10.1 Introduction

In 1989, Wolff and Balaban first demonstrated magnetisation transfer (*MT*) *in vivo* (Wolff and Balaban, 1989) and introduced it as a source of MR contrast alternative to T_1, T_2, and T_2*. *MT* is based on the exchange of magnetisation occurring between groups of spins characterised by different molecular environments, and ever since *MT* imaging has been widely used in clinical imaging to improve the suppression of static tissue in MR angiography and to increase lesion visibility on conventional MRI when gadolinium-based contrast agents are used (Edelman *et al.*, 1992; Finelli *et al.*, 1994). In these clinical applications *MT* is used to increase contrast in MRI images that are designed for qualitative analysis and radiological interpretation. Beyond these applications, however, *MT* has proven to be particularly powerful for characterising tissue microstructure, particularly in the brain, by applying it as a quantitative technique.

[1] Reviewed by John Sled, Department of Medical Biophysics, University of Toronto, Canada.

10.2 How to navigate this chapter

MT is a complex subject and it is impossible to provide a simple recipe for implementing data acquisition and analysis. It is important to understand the theory behind each different model before attempting to use them. This chapter therefore gives a full overview of the theoretical concepts behind *MT*. However, in order to ease its navigation, we have used labels for the following sections, as explained in the table below.

Label	Meaning
B	Basic concepts/historical perspective
T	Theory
P	Practical: this will give insights into the practical implementation
V	Validation, reproducibility, phantoms, etc.

Note that some sections may be labelled with more than one letter. We also wish to state that at the time of writing this chapter there is no established package for analysing *MT* data, and most researchers use some form of customised processing tool, although one open-source software package has been made available (Cabana *et al.*, 2015). A quick recipe for setting up an *MT* protocol is given in Box 10.1.

10.3 What is *MT*? (B)

The concept that nuclei in different chemical compounds could be coupled magnetically has been exploited in NMR spectroscopy for many years. Either the nuclei might exchange environments or they might come close enough to each other, by diffusion or molecular rotation, to exchange magnetic energy through dipole–dipole interactions. Forsen and Hoffman (1963) studied a system consisting of two sets or baths of spins, each bath being in good equilibrium with the others in the bath and being relatively weakly coupled to the other bath. They showed how saturating one peak, by applying a large amount of continuous wave (CW) power, would alter the size of the other. Modified Bloch equations were given and solved for the case of the longitudinal magnetisation growing not only from a T_1 term but also from a coupling term with saturated spins in the other (second) environment. They studied the transient behaviour of such a system, characterising the recovery times after a variety of saturation schemes. They measured exchange times for protons (i.e. the inverse of the rate at which protons exchange from one compartment to the other), in a particular two-component chemical system, of 1–4 seconds, pointing out that these are independent of the magnetic field strength.

The concept was extended to studying protons with the same resonant frequency by Edzes and Samulsky (1978a). Studies of collagen and muscle showed that cross relaxation between free water protons and restricted macromolecular protons significantly altered the T_1 of the water and caused a reduction in the water signal. The phenomenon is explained by the observation that hydrogen nuclei (protons) in water are 'mobile', while hydrogen atoms in a macromolecule experience restricted mobility, and the signal from those hydrogen atoms will decay too quickly to be seen using a clinical MRI scanner. However, the mobile protons are in constant motion and come into regular contact with the macromolecular protons, making it possible for a proton in a water molecule to exchange with a proton in a macromolecule when the water momentarily binds to the surface of the macromolecule.

This exchange of magnetisation, which can be either by direct chemical exchange of the hydrogen atom or by spin–spin interactions, forms the basis of *MT* imaging. Interacting with the coupled system is possible due to the difference in the width of the resonance lines between the two proton species: mobile protons are free to diffuse and rotate so that dipolar–dipolar interactions are averaged out, resulting in small variations of their resonant frequency and thus a very narrow resonance line. This is not the case for macromolecular protons, which have a broad resonance line (see Figure 10.1) and hence are sensitive to off-resonance irradiation. If sufficient off-resonance power is applied, then the macromolecular spins become 'saturated', a state in which the numbers of up spins and down spins are

Box 10.1 How to Set Up Your Quantitative MT Protocol

1. *Choose a suitable model.*
   The choice may be driven by your scientific hypothesis and by scan time constraints.
2. *Check whether the corresponding pulse sequence is available on your scanner.*
   Off-resonance pulses to produce MT effect are generally available in all scanners. However the option to change their amplitude and offset frequency may require some sequence programming. Similarly, balanced steady-state free precession sequences are usually available but not the option to alter the duration of the RF pulses.
3. *Check whether the model/acquisition you have chosen requires any additional acquisitions (e.g. T_1-mapping, B_1-mapping, B_0-mapping).*
4. *Set up the processing pipeline.*
   This usually requires some image alignment, the computation of any additional maps (B_1, T_1, etc.) and model-fitting (usually non-linear). There is only one quantitative MT package currently available (Cabana *et al.*, 2015), implemented in Matlab. Alternatively, you can implement your own software or contact the authors who originally described the model.
5. *Sanity check.*
   It may be useful to extract the raw data from a region of interest and plot them to verify they match the distribution expected from an MT weighted experiment.
6. *Repeat the model fitting voxel-wise to obtain the parametric maps.*

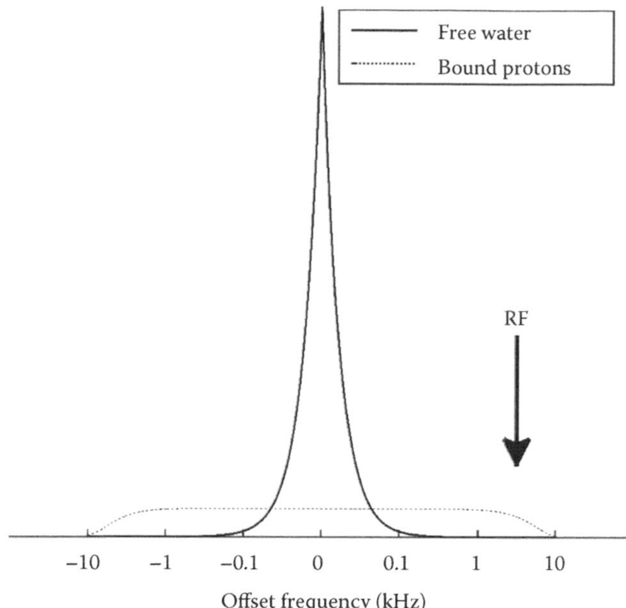

FIGURE 10.1 Symbolic representation of a spectrum of free water and bound proton pools in brain tissue. The free protons, in water, have a narrow line shape and are relatively unaffected by the RF irradiating pulses (at offsets higher than 1–2 kHz), whilst the bound macromolecular protons, having a broader line shape, are saturated (i.e. lose their magnetisation). As a result of irradiating the bound protons, magnetisation transfer between the two pools reduces the magnetisation of the free water and thus its signal intensity, producing an alternative source of contrast in the acquired image. A logarithmic scale is used for the *x*-axis to enhance the visualisation of the different line width between free and bound protons.

equalised and the net magnetisation vector is zero. If some of this saturated magnetisation is transferred to the liquid (mobile) protons by the exchange processes described above, they can become partially saturated and, consequently, the signal intensity from the observable liquid protons is reduced, producing the so-called *MT* contrast.

10.4 Clinical and neuroscientific relevance of MT imaging (B)

MT is able to probe indirectly macromolecules such as proteins and lipids, which are normally 'invisible' on MRI. This is of great interest in the brain, as myelin, the substance that wraps around the majority of the axons in both the central (CNS) and the peripheral nervous system, is a lipid–protein structure (its dry mass is approximately 70%–80% lipids and 20%–30% proteins; Laule *et al.*, 2007). In the CNS, myelin is primarily found in the white matter (WM), although it is present in smaller quantities also in the grey matter (GM). The main purpose of myelin is to act as an insulator, thus increasing the speed of action potential transmission, ensuring efficient signal travelling along axons. Thinning or disruption of myelin can result in neurological deficits due to a reduction in neuronal conductivity. Myelin damage typically occurs in demyelinating diseases (such as multiple sclerosis), but it is also believed to take place in degenerative processes secondary to neuronal or axonal damage. In addition, recent evidence suggests that myelination is partly regulated by neuronal activity (Scholz *et al.*, 2009; Wang and Young, 2014), and therefore changes to myelin can be modulated by physical and cognitive activities.

As a consequence, the ability to measure myelin *in vivo* would have tremendously important consequences, and *MT* is among the most promising MR techniques for the assessment of myelin.

10.5 MT contrast and MT ratio (B)

In routine clinical practice, *MT* is generally used to increase the contrast between fluids and other tissues with a high macromolecular content in a number of MRI applications. In musculoskeletal imaging it is used to improve contrast between cartilage and synovial fluid, while in cardiac imaging it can improve the conspicuity of the blood in white-blood imaging. In time-of-flight MR angiography it is routinely used to enhance suppression of the background brain tissue, resulting in increased small vessel visibility (Edelman *et al.*, 1992). In conjunction with gadolinium injection, *MT* can also improve the visibility of demyelinating lesions by suppressing healthy tissue (Huot *et al.*, 1997).

Early attempts to quantify the *MT* effects for tissue characterisation purposes were based on the *MT* ratio (MTR), first introduced by Dousset *et al.* (1992). They computed the MTR as the pixel intensities percentage difference between two images, one acquired with off-resonance saturation (M_s) and one without (M_0):

$$\mathrm{MTR} = \frac{M_0 - M_S}{M_0} \times 100 \tag{10.1}$$

An MTR map is obtained by applying this simple formula on a voxel-by-voxel basis, and it is typically expressed in percentage units (pu), or simply as a fraction (without the multiplicative factor in Equation 10.1). In the brain, both WM and GM have a non-zero positive MTR (with WM MTR higher than GM MTR), whose absolute value depends mostly on the amount of saturation applied, while CSF should have a value close to zero since the saturation pulses have little effect where there is no macromolecular content.

10.6 Theoretical models of *MT* in biological tissue (T)

The MTR is regarded as a 'quantitative' measurement because it provides an index that can be compared between serial scans of the same patient (providing they have been collected using the same scanner and the same pulse sequence), or between populations of subjects, and can be correlated with clinical variables. However,

TABLE 10.1 Parameters used to quantify the MT effects.

Parameter	aka	Full name	Notes	Interpretation	Method
MTR		Magnetisation Transfer Ratio	MTR = $(S_0-S_{MT})/S_0 \times 100$	Semi-quantitative parameter, whose value is proportional to $R{\cdot}F{\cdot}T_1^A$ – A change in MTR can reflect a change in any of these factors	Any method that allows collection of MT-weighted images (S_{MT})
F	PSR	Macromolecular pool ratio, pool size ratio	$F = M_0^B / M_0^A = k_f/k_rF = f/(1-f)$	Macromolecular density – often used as a myelin proxy	qMT and appropriate model fitting, which depends on acquisition method
f	BPF	Macromolecular pool fraction, bound pool fraction	$f = M_0^B/(M_0^A + M_0^B)$ $f = F/(1 + F)$	Macromolecular density – often used as a myelin proxy	
RM_0^B, RM_0^A	k_f, k_r	forward/reverse exchange rate	$RM_0^B = FRM_0^A$	Not fully known. It has been associated with metabolic function	
R		Exchange rate constant		Not fully known. It has been associated with metabolic function	
T_2^A, T_2^B	T_2^F, T_2^R	Transverse relation of the free/macromolecular pool		T_2^A is the main contributor to the observed T2. T_2^B is stable across conditions, but sensitive to WM fibre orientation	
R^A, R^B	R^F, R^R	Longitudinal relaxation rate of the free/macromolecular pool	R_B is often fixed to $1s^{-1}$ for fitting purposes	R_A is the main contributor to the observed R1	
R_{RFB}	W	Rate of absorption of external irradiation by the macromolecular pool	Function of T_2^B		
MT_{SAT}	δ	MT saturation		Similar to MTR, but adjusted for T_1 and B_1 contributions	

the MTR does not have a direct physical or biological interpretation, as it reflects a rather complex interplay between several fundamental quantities. Specifically, the MTR is dependent on T_1, as well as on the density of macromolecules. In some circumstances (e.g. demyelinating lesions) the increase in T_1 can partially mask the decrease in macromolecular density, resulting in an apparent stability of the MTR (Henkelman *et al.*, 2001). Furthermore, the MTR is highly dependent on the acquisition parameters, including the shape, the amplitude, the duration and the offset frequency of the saturating pulse. Imaging parameters such as the repetition time (TR) and the excitation flip angle can affect the MTR (Finelli and Reed, 1998). Thus there has been considerable interest in building and testing models that will predict and explain *MT* effects in biological tissues by means of quantitative biologically meaningful parameters. The *MT* literature is quite diverse, and several approaches have been proposed. Table 10.1 summarises the main *MT* parameters and their interpretation.

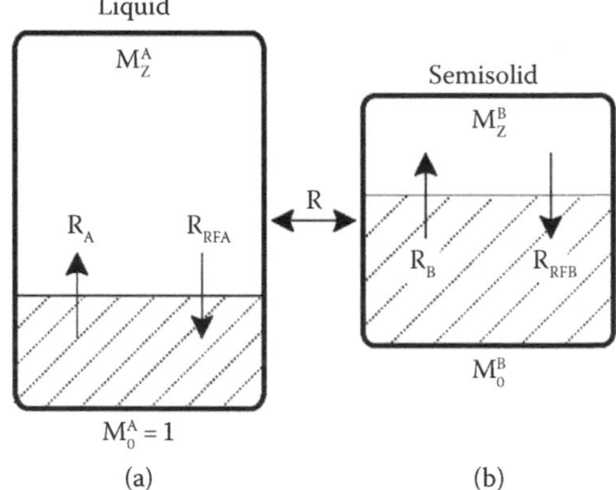

FIGURE 10.2 A model of a two-pool system, with exchange. The shaded part of each pool (depicted with dashed lines) represents magnetisation, which is not aligned longitudinally. R_A and R_B are the longitudinal relaxation rates ($R = 1/T_1$), which enable longitudinal magnetisation to recover. R_{RFA} and R_{RFB} are the saturation terms, which destroy longitudinal magnetisation. R is the exchange between the pools. (Henkelman, R.M., *et al.*: Quantitative interpretation of magnetisation transfer. *Magn. Reson. Med.* 1993. 29. 759–766. Copyright Wiley-VCH Verlag GmbH & Co. KGaA. Reproduced with permission.)

10.6.1 Two-Pool models and Henkelman's model

Early two-pool models (also called the 'two-pool Bloch model' or 'binary spin bath model') for MT, based on the Bloch equations, with extra terms for cross relaxation, were given by McConnell (1958) and Edzes (Edzes and Samulski, 1978a) and used to study a variety of biological systems. Although the Bloch model is expected to be sufficient for describing the response of the free water pool to the applied RF, there has been concern that a more complex model may be needed to accurately describe that of the restricted (macromolecular) pool (Hua and Hurst, 1995). As shown below, the Bloch formalism necessarily produces Lorentzian absorption line shapes, even though other line shapes (e.g. Gaussian or super-Lorentzian) for the macromolecular pool have shown to provide better description of the experimental data (Morrison *et al.*, 1995). Most of the quantitative *MT* (qMT) methods available to date are derived from the model proposed by Henkelman *et al.* (1993), reproduced in Figure 10.2. This model was originally developed for a CW irradiation experiment in agar gels, in which RF pulses of several seconds with constant amplitude are used to saturate the macromolecular pool. More details about this type of acquisition are provided in Section 10.7.1.

In Henkelman's model (Figure 10.2), *A* labels the liquid pool, and *B* labels the macromolecular pool. The density of spins in the two pools is M_0^A and M_0^B, respectively. M_0^A is often set equal to 1 (Henkelman *et al.*, 1993), following the normalisation of acquired data to the equilibrium reference value. The exchange terms (previously forward and backward terms) were made symmetrical to include the bound and free magnetisations explicitly. As $M_0^A >> M_0^B$, this is a pseudo-first order exchange process (meaning that M_0^A can be assumed constant and absorbed in the exchange rate constant *R*); the exchange constant from A to B can be set equal to RM_{0B}. The rate from B to A is therefore RM_{0A} to preserve compartment sizes. Assuming that the *MT* effect can be modelled using this two-pool description, the magnetisation of either pool can be described by its longitudinal component (M_z^A, M_z^B) and its transverse components ($M_x^A, M_y^A, M_x^B, M_y^B$). The exchange between pools associated with the transverse components of magnetisation can be considered negligible due to the extremely short T_2 associated with the macromolecular pool.

The coupled Bloch equations for the system can thus be written in matrix form as follows (Portnoy and Stanisz, 2007):

$$\frac{dM(t)}{dt} = A(t)M(t) + BM_0 \tag{10.2}$$

with $\mathbf{M}(t) = [M_x^A, M_y^A, M_z^A, M_z^B]^T$, and

$$A(t) = \begin{bmatrix} \frac{-1}{T_2^A} & -2\pi\Delta f & 0 & 0 \\ 2\pi\Delta f & \frac{-1}{T_2^A} & -\omega_1(t) & 0 \\ 0 & \omega_1(t) & -\left(R_A + RM_0^B\right) & RM_0^A \\ 0 & 0 & RM_0^B & -R_B + RM_0^A + R_{RFB}\left(\omega_1(t)\right) \end{bmatrix} \tag{10.3}$$

$$B = \begin{bmatrix} 0 \\ 0 \\ R_A \\ R_B \end{bmatrix}. \tag{10.4}$$

T_2^A represents the transverse relaxation time of the liquid pool, Δf represents the frequency offset of the pulse, while $\omega_1(t)$ is the time dependent amplitude of the pulse expressed in rad/s (i.e. the angular frequency of precession induced by the pulse). R_A and R_B represent the longitudinal relaxation rates of the two pools.

In the CW case considered by Henkelman *et al.*, $\omega_1(t)$ is constant and equal to ω_1, and the system admits analytical solution in the steady state, that is, when all the derivatives are equal to zero:

$$M_z^A = \frac{M_0\left(R_B RM_0^B + R_{RFB}R_A + R_B R_A + R_A RM_0^A\right)}{\left(R_{RFA} + R_A + RM_0^B\right)\left(R_{RFB} + R_B + RM_0^A\right) - RRM_0^B}, \tag{10.5}$$

where

$$R_{RFA} = \frac{\omega_1^2 T_2^A}{1 + \left(2\pi\Delta f T_2^A\right)^2} \tag{10.6}$$

$$R_{RFB} = \frac{\omega_1^2 T_2^B}{1 + \left(2\pi\Delta f T_2^B\right)^2} \tag{10.7}$$

T_2^B in Equation 10.7 represents the transverse relaxation time of the macromolecular pool. Equations 10.6 and 10.7 are derived analytically from the steady-state solution and are proportional to the Lorentzian absorption line shape implicitly assumed for each pool.

Nevertheless, the Lorentzian line shape obtained analytically from Bloch equations does not adequately describe the macromolecular component obtained experimentally. Henkelman *et al.* replaced R_{RFB} with a Gaussian line shape for agar gel (Henkelman *et al.*, 1993), while Li *et al.* (1997) show that, in CNS tissue, the spectra associated with macromolecular pool are better modelled by a super-Lorentzian:

$$R_{RFB}\left(\Delta f, \omega_1\right) = \omega_1^2 \sqrt{2\pi}\left[T_2^B \int_0^1 \frac{1}{\left|3u^2 - 1\right|} exp\left(-2\left(\frac{2\pi\Delta f T_2^B}{3u^2 - 1}\right)^2\right) du\right] \tag{10.8}$$

As typically $2\pi\Delta f T_2^A >> 1$, Equation 10.5 can be rewritten, after dividing by R_A, as

$$M_z^A = \frac{M_0^A\left(R_B\left[\frac{RM_0^B}{R_A}\right] + R_{RFB} + R_B + RM_0^A\right)}{\frac{RM_0^B}{R_A}\left(R_B + R_{RFB}\right) + \left(1 + \left[\frac{\omega_1}{2\pi\Delta f}\right]^2\left[\frac{1}{R_A T_2^A}\right]\right)\left(R_{RFB} + R_B + RM_0^A\right)} \tag{10.9}$$

Five model parameters can be extracted by fitting to the data: R_B, T_2^B, R, $(R\ M_0^B/R_A)$ and $(1/R_A\ T_2^A)$. The constraints $R_B = 1\text{s}^{-1}$ is commonly used, as steady-state signal M_z^A (Equation 10.9) shows little dependence on this parameter (Henkelman *et al.*, 1993). Solving the equations after a perturbation of the free water equilibrium magnetisation gives the observed relaxation rate ($R_A^{obs} = 1/T_1$) (in the presence of exchange, but no saturating field), in terms of R_A:

$$R_A = R_{A_{obs}} - \frac{RM_{0^B}\left(R_B - R_{Aobs}\right)}{R_B - R_{A_{obs}} + RM_{0^A}}. \tag{10.10}$$

Obtained by assuming that R_{Aobs} is the smallest eigenvalue of matrix $A(t)$ with $\omega_1(t) = 0$, when only the longitudinal components M_z^A and M_z^B are considered.

Equation 10.9 can thus be fitted, using a non-linear least squares technique, to a minimum of four measurements, obtained with variable settings of ω_1 and Δf, in a CW experiment. Independent measurement of R_{Aobs} allows fitting directly for two-pool model parameters T_2^B, R, M_0^B and T_2^A through the use of Equation 10.11.

Morrison and Henkelman (1995) also applied this two-pool model to CW measurements of MT in fresh bovine brain tissue, obtaining a good fit (Figure 10.3).

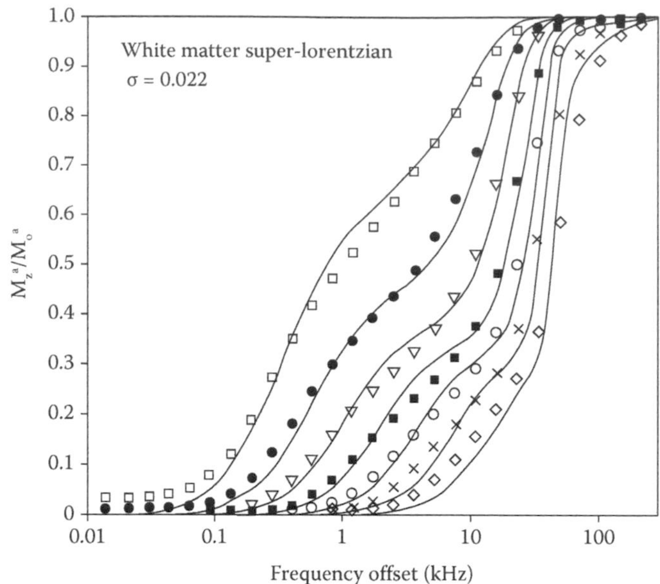

FIGURE 10.3 Continuous wave magnetisation transfer data from bovine white matter, and the fitted two-pool model of Henkelman, using a super-Lorentzian line shape The y-axis is the ratio of the A ool (i.e. the free pool) z-magnetisation to the total amount of A-pool magnetisation. Seven different saturating amplitudes and 27 different offset frequencies were used. The upper curve corresponds to $\omega 1/2\pi = 83$ Hz (i.e. $\omega 1 = 522$ rad s^{-1}, B1 = 1.95 μT), the second curve to 170 hz, and the lower to 5.34 kHz. Clinical MTR sequences have an effect equivalent to the region of the upper two curves. (Morrison, C., and Henkelman, R.M.: A model for magnetisation transfer in tissues. *Magn. Reson. Med.* 1995. 33. 475–482. Copyright Wiley-VCH Verlag GmbH & Co. KGaA. Reproduced with permission.)

10.6.2 Three-Pool model

Depending on the specific acquisition, the spatial resolution of MT-weighted datasets typically ranges from 1 to 3 mm³. As a consequence, partial volume with CSF can bias the estimation of model parameters (especially the compartment size M_0^B), due to CSF relaxation times being drastically different from parenchyma and to the fact that there is virtually no MT in CSF. In order to compensate for these effects, Mossahebi *et al.* (2015) proposed the addition of a third, non-exchanging pool (Figure 10.4). The fraction of the voxel taken up by the non-exchanging component is represented by f_{NE}, defined as

$$f_{NE} = \frac{M_0^{NE}}{M_0 + M_0^{NE}}, \tag{10.11}$$

where M_0^{NE} is the magnetisation associated with the non-exchanging compartment, and $M_0 = M_0^A + M_0^B$. A signal equation can then be derived as

$$SI = \beta M_0 \left[SI_{MT} + \frac{f_{NE}}{1 - f_{NE}} SI_{NE} \right], \tag{10.12}$$

where β is a scaling factor, and SI_{MT} and SI_{NE} are the signal equations for the MT and the non-exchanging compartments, respectively. The analytical form for SI_{MT} and SI_{NE} is determined by specific acquisition used.

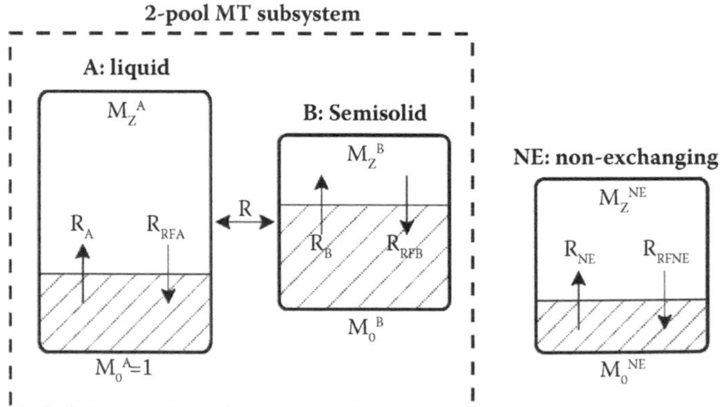

FIGURE 10.4 A three-pool model consisting of a standard two-pool subsystem (A = free water or liquid pool, B = bound protons, or semisolid pool) and a non-exchanging water pool, used to model CSF contribution. (Adapted with permission from Mossahebi, P., et al.: Removal of cerebrospinal fluid partial volume effects in quantitative magnetisation transfer imaging using a three-pool model with nonexchanging water component. Magn. Reson. Med. 2015. 74. 1317–1326. Copyright Wiley-VCH Verlag GmbH & Co. KGaA.)

10.6.3 Four-Pool models

The binary spin-bath model does not take account of water being in several different environments, as observed in multi-echo studies by MacKay and co-workers (Whittall *et al.*, 1997) and discussed in detail in Chapter 4 on proton density. According to these observations, T_2-decay is multi-exponential and explained by a component with a short T_2 (~10–20 ms), thought to be trapped within the myelin layers, and a component with longer T_2 (~100–300 ms) corresponding to intra- and extracellular water.

Attempts to provide a unified view led to the development of four-pool models (Stanisz *et al.*, 1999; Levesque and Pike, 2009). This approach assumes four communicating proton pools (myelin solids, myelin water, intra/extracellular water, and non-myelin solids), including most of the basic features of the multicompartment

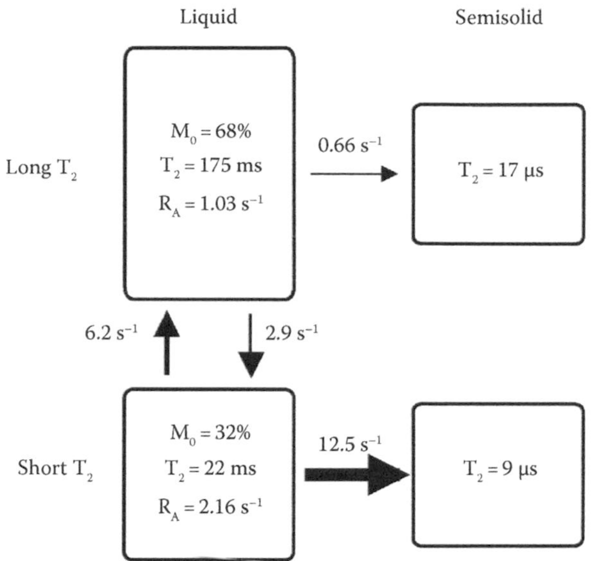

FIGURE 10.5 A four-pool model of protons in bovine optic nerve, showing free water (left-hand liquid pools) in intracellular water (long T_2 pool) and trapped in myelin (short T_2 pool). There is strong magnetisation transfer (MT) between myelin (short T_2) water and lipid protons (12.5 s–1). Thirty-two per cent of water is trapped in myelin, and this water has a T_2 = 22 ms. Exchange between the two free (water) pools makes them have the same MT ratio values about 200 ms after irradiation and is consistent with the failure to observe multi-exponential T_1 behaviour in tissue. To obtain the parameter values shown, the model was fitted to nearly half a million data points, collected from three samples of tissue, using an MT-prepared multi-echo sequence. (Stanisz, G., *et al.*: Characterizing white matter with magnetisation transfer and T(2). *Magn. Reson. Med.* 1999. 42. 1128–1136. Copyright Wiley-VCH Verlag GmbH & Co. KGaA. Reproduced with permission.)

T_2 water model and the two-pool *MT* model (see Figure 10.5). The number of parameters to be fitted requires the acquisition of a large set of data points that makes this type of model impractical for *in vivo* applications, despite attempts to use *ad hoc* acquisition strategies such as MT-prepared multi-echo sequences (Stanisz *et al.*, 1999). Even though impractical for data fitting, these complex models can be used to predict the behaviour of T_2 spectra and *MT* as a function of myelin content, thus helping the interpretation of changes observed with these quantitative MRI methods based on simplified models of WM (Stanisz *et al.*, 2005).

10.7 Measuring MT (P)

Several methods can be used to produce an MT-weighted image. They can be divided into off-resonance and on-resonance methods, as well as into steady-state and transient methods, and continuous or pulsed methods.

10.7.1 CW irradiation

In a CW experiment RF pulses of several seconds with constant amplitude are used to saturate the macromolecular pool. Typically, irradiation is applied with 0.5–10 kHz *off-resonance*. Direct saturation is minimised by the narrow bandwidth of CW irradiation. This experiment represents the 'ideal' off-resonance saturation method but is not feasible on clinical systems, since the RF transmitters are not designed for CW operation. In addition, the specific absorption rate (SAR) would breach the safety limits.

10.7.2 On-Resonance binomial pulses

The first attempts to implement MT-weighted sequences on clinical scanners used binomial *on-resonance pulses* (Hu *et al.*, 1992; Yeung and Aisen, 1992; Pike *et al.*, 1993; Schneider *et al.*, 1993) were used. These consist of 'transparent pulses', such as $1\bar{1}$ or $1\bar{2}1$, whose net effect is zero for long T_2 spins such as those in free water (the number refers to the relative tip angle, and bar means that the angle is reversed; thus $1\bar{1}$ could be 45° immediately followed by −45°). For short T_2 spins, such as those in the bound pool, the transverse magnetisation decays as soon as it is produced, and the *z*-magnetisation is not recovered by the second pulse. Thus there is no effect on mobile protons, but bound protons are saturated (their *z*-magnetisation is destroyed). Binomial pulses are easy to implement (they do not require generating a frequency offset for the transmitter) and they are considered efficient, as they produce a large signal reduction for a given RF power. However the intrinsic direct saturation (Hua and Hurst, 1995) and lack of flexibility, compared to off-resonant pulses, have contributed to a decline in their use.

10.7.3 Off-Resonance pulsed irradiation

Off-resonance pulses have almost completely replaced binomial pulses, as they allow more control over the saturation. They are shaped (typically Gaussian shape, or sinc shape with up to three lobes), with a bandwidth of a few 100 Hz, and typically play out at frequency offsets of 1–5 kHz from the mobile water peak. In 1992 the group at Philadelphia published the first clinical images of MTR using off-resonant pulses (Dousset *et al.*, 1992) and started a wave of interest in using MTR to study disease, particularly in multiple sclerosis. They used a *3D gradient echo sequence* (matrix: 256 × 128 × 28; 5 mm sections; *TR* = 100 ms; TE = 6 ms; FA = 12°); the *MT* pulse was 2 kHz off-resonance and an MTR of 43 ± 3 pu was measured in normal WM. The SAR was 0.1 W/kg. In principle, off-resonance pulses can be combined with any acquisition sequence, but due to their short scan times and low T_1 and T_2 contrast they are mainly used with (spoiled) gradient-echo (GRE) sequences. Three-dimensional gradient echoes are preferred to 2D multislice, as they prevent the incidental *MT* effects from slice selective pulses (Dixon *et al.*, 1990) and interference between adjacent slices. In order to minimise T_1-weighting while keeping the acquisition sufficiently short, typically the *TR* ranges between 25 and 50 ms, with flip angles between 5° and 15°.

Spin-echo sequences have been used for MTR, with one example of *interleaved dual echo* (Barker *et al.*, 1996) that gives intrinsically registered M_0 and M_s images (see Equation 10.1). Registered PD and T_2-weighted images are also obtained, so that regions of interest can be defined around lesions or normal-appearing structures seen in these images with standard MRI contrast. The disadvantages of this approach are that there is some T_1-weighting (since *TR* must be kept reasonably short in order to limit the overall imaging time), that a true saturated or unsaturated state for the bound pool may not be reached and that much time is spent waiting for the free pool to recover its equilibrium.

10.7.4 Steady-State free precession

In 2006, Bieri and Scheffler demonstrated that balanced steady-state free precession (bSSFP) sequences are inherently *MT* weighted (Bieri and Scheffler, 2006). This observation was prompted by the deviation often observed between the signal measured with bSSFP and theoretical predictions. The unaccounted signal loss can indeed be explained by *MT* effects, which are dependent on the choice of *TR* and flip angle. Bieri and Scheffler thus proposed to modulate the amount of *MT* weighting by varying the *TR* concurrently with the pulse duration, whereby longer pulses (and *TRs*) minimise the *MT* effect. This type of acquisition is attractive as it provides high-resolution images with significantly reduced acquisition times in brain imaging. High quality MTR maps of the brain can thus be obtained in seconds (Bieri *et al.*, 2008; Garcia *et al.*, 2012). The description of this approach for quantitative *MT* imaging is given in Section 10.7.5.

10.7.5 Selective inversion recovery

The selective inversion recovery fast spin echo (SIR-FSE) method (Gochberg *et al.*, 1997, 1999; Gochberg and Gore, 2007) aims at manipulating the liquid protons instead of the macromolecular ones. The pulse sequence is designed to ensure that at the end of each repetition, both the macromolecular and free water pools have zero *z*-magnetisation. This is achieved by rotating the magnetisation of the free water protons by the 90° pulse, followed by the series of closely spaced refocusing 180° pulses (which prevent any effective T_1 recovery). The macromolecular protons are not directly affected by the 90° pulse. However, due to the *MT* effect, they are pulled toward zero by the nulled liquid pool magnetisation on a timescale of order of magnitude of T_1. This method investigates the *MT* effect during its transient evolution and is mainly used for quantitative MT. More details are provided in Section 10.7.4.

Figure 10.6 summarises schematically the main acquisition strategies for MT-weighted MRI.

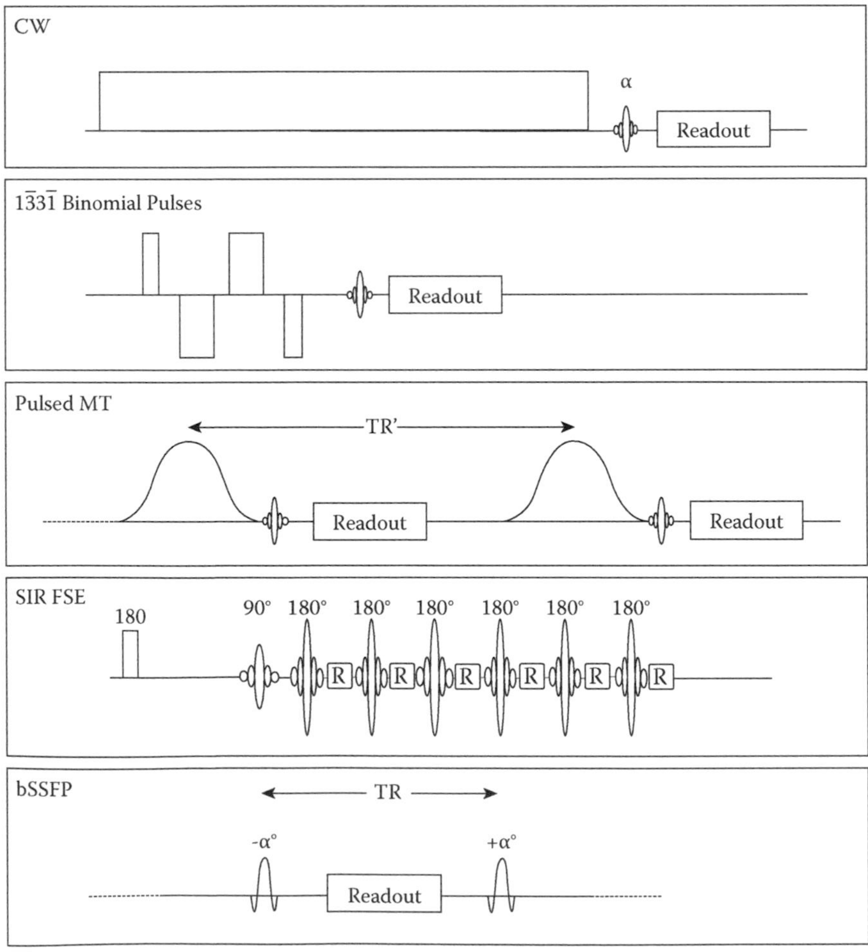

FIGURE 10.6 Schematic representation of the main acquisition strategies for MT-weighted data.

10.8 Quantitative *MT* imaging (P,T)

Due to the limitations of the MTR discussed in Section 10.5, there is great interest in directly estimating *in vivo* the *MT* parameters that describe the theoretical models of *MT* (see Section 10.5). The major design challenges are to adapt the theory (Equation 10.2) to give solutions for pulsed *MT* (or other acquisitions feasible *in vivo*) that can be computed in a reasonable time and to design imaging protocols that will collect appropriate data in an acceptable time with adequate coverage of the brain and at sufficient resolution to depict anatomical details.

10.8.1 Absorption line shape of the macromolecular pool

As mentioned in Section 10.5, the only analytical solution admitted by the Bloch equation for both pools is a Lorentzian line shape (Henkelman *et al.*, 1993). However, such a line shape is not appropriate for the macromolecular pool. A variety of line shapes for the bound pool have been used (e.g. Gaussian, super Lorentzian, Kubo-Tomita and flexible) (Morrison *et al.*, 1995; Li *et al.*, 1997; Quesson *et al.*, 1997). Lorentzian and Gaussian line shapes are easy to implement and fast to compute in a fitting procedure to determine maps of quantitative *MT* model parameters. The super-Lorentzian line shape requires numerical estimation but produces better fits in WM. For this reason, it is almost universally used for human *in vivo* applications (Sled and Pike, 2001; Yarnykh, 2002; Cercignani *et al.*, 2005). The analytical expression for this line shape is given in Equation 10.8 in the form of saturation rate term. In order to speed up the fitting procedure, it is typically computed in advance for a range of plausible values of T_2^B, which is fitted using look-up tables (Sled and Pike, 2001; Cercignani *et al.*, 2005).

In the attempt to explore the orientation dependence of *MT* parameters in WM, Pampel *et al.* (Pampel *et al.*, 2015) developed a model for a bound pool line shape that accounts for different directionality of WM fibres with respect to the static magnetic field, suggesting that regional variability observed for T_2^B arises mostly for unaccounted WM geometry more than variations in membrane composition. Pampel and co-authors have proposed an alternative line shape that explicitly considers the specific geometrical arrangement of lipid bilayers wrapped around a cylindrical axon. In order to account for multidirectionality within a voxel, a fibre orientation distribution function (fODF) (with respect to B_0) is derived using diffusion-weighted MRI and spherical deconvolution. Bingham functions are fitted to the fODF peaks to identify fibre bundles with a specific orientation. When using this novel line shape, which accounts for orientation effects in RF absorption, T_2^B shows very little variation between WM regions.

10.8.2 In Vivo implementations of Henkelman's model

As explained in Section 10.6.1, CW irradiation is impractical and generally not available for *in vivo* imaging experiments. The earliest attempts to translate this model into *in vivo* applications were based on pulsed-MT acquisitions, requiring the modification of Henkelman's model to allow for the short duration of the saturation pulses relative to T_1. Under these circumstances Equation 10.2 only admits a numerical solution (Graham and Henkelman, 1997).

Three different methods based on approximated signal equations for pulsed *MT* were presented approximately at the same time (Sled and Pike, 2001; Ramani *et al.*, 2002; Yarnykh, 2002). Here, for simplicity and in order to ease the comparison between them, we have rewritten all equations, keeping the conventions introduced by Henkelman *et al.* (1993) wherever possible. However, we remark that some authors label the A and B pools as *F* and *R*, respectively (for 'free' and 'restricted') (Sled and Pike, 2001) or *f* and *m* (for 'free' and 'macromolecular') (Yarnykh, 2002) and use the symbol *W* instead of R_{RFB} (Sled and Pike, 2001) for the rate of absorption of RF energy. The pseudo-first order exchange rates, RM_{0B} and RM_{0A}, are often referred to as k_f (or simply *k*) and k_r. See also Table 10.1 for nomenclature.

The simplest approximation was proposed by Ramani *et al.*, who used a CW power equivalent approximation (CWPE) (Ramani *et al.*, 2002). In their paper, the pulse is simply replaced by a CW irradiation with the mean square amplitude that would give the same power over the repetition period between *MT* pulses:

$$\omega_{1CWPE} = \gamma\sqrt{P_{SAT}} \tag{10.13}$$

where P_{SAT} is the mean square saturating field and γ is the gyromagnetic ratio.

P_{SAT} is equivalent to

$$P_{SAT} = p_2 B_{max}^2 \frac{\tau_{SAT}}{TR'} \tag{10.14}$$

where p_2 is the ratio of the mean square amplitude of the saturation pulses to that of a rectangular pulse of equivalent amplitude, B_{max} is the maximum amplitude of the pulse, τ_{SAT} is its duration and TR′ is the interval between successive pulses.

B_{max} can be computed with knowledge of the equivalent on-resonance flip angle (ϑ) of the pulse, being

$$\vartheta[^\circ] = \left[\frac{180}{\pi}\right]\gamma p_1 B_{max}\tau_{SAT} \tag{10.15}$$

where p_1 is the ratio of the mean amplitude of the saturation pulse to that of a rectangular pulse of the same amplitude. It is worth noting that p_1 and p_2 in Equations 10.14 and 10.15 represent correction factors for the actual pulse shape used (compared to a rectangular pulse). P_{sat} and θ can be equivalently obtained by integrating the time modulated pulse amplitude $B_1(t)$ and its square $B_1^2(t)$ over the pulse duration period τ_{SAT}.

By means of the CWPE approximation, Henkelman's steady-state model can be straightforwardly applied to the *in vivo* MRI case, neglecting the imaging elements of the pulse sequence.

Introducing the symbol $F = \frac{M_0^B}{M_0^A}$ a and substituting $RM_0^A = \frac{RM_0^B}{F}$ the equation

$$SI(\omega_1, \Delta f) = \frac{M_0\left(R_B\left[\frac{RM_0^B}{R_A}\right] + R_{RFB}(\omega_{1CWPE}, \Delta f) + R_B + \frac{RM_0^B}{F}\right)}{\left[\frac{RM_0^B}{R_A}\right]\left(R_B + R_{RFB}(\omega_{1CWPE}, \Delta f)\right) + \left(1 + \left[\frac{\omega_{1CWPE}}{2\pi\Delta f}\right]^2\left[\frac{1}{R_A T_2^A}\right]\right)\left(R_{RFB}(\omega, \Delta f) + R_B + \frac{RM_0^B}{F}\right)} \tag{10.16}$$

can be used to fit data collected using the pulsed *MT* method. We note that, in order to ease the comparison between the signal equations, we have broken with the terminology of the original papers by Ramani *et al.* (2002), where the macromolecular fraction f (with $f = F/(F+1)$) was used instead of the relative size of the macromolecular pool F. As $F \ll 1$, the two parameters tend to be similar (and related by a simple algebraic conversion) and have both been associated with myelin content (Schmierer *et al.*, 2007).

As Ramani's equation does not explicitly model the effects of the excitation pulses and *TR*, its description of the MT-weighted signal is valid only when the degree of T_1-weighting in the acquisition sequence is minimal (Cercignani and Barker, 2008).

Sled and Pike (2000) proposed a solution derived by approximating the pulse sequence as a series of periods of free precession, CW irradiation and instantaneous saturation of the free pool. During each of these periods, Equation 10.2 has either an exact or an approximate solution, and these solutions can then be concatenated by imposing the appropriate initial conditions, leading to an expression for the measured signal (under the steady-state condition), which is less expensive to compute than numerically integrating the full set of differential equations.

The effect of an *MT* pulse on the macromolecular pool is modelled as a rectangular pulse whose width is equal to the full width at half maximum (τ_{RP}) of the curve obtained by squaring the instantaneous amplitude of the *MT* pulse throughout its duration, and whose amplitude is such that the pulses have equivalent average power (rectangular pulse, or RP, approximation). The effect of the pulse on the liquid pool is modelled as an instantaneous fractional saturation of the longitudinal magnetisation. Such fractional saturation (S_{1A}) is estimated by solving (numerically) the system of Equations 10.2 when R and R_A are set to 0, meaning that exchange and relaxation are neglected during pulse application.

In matrix form, considering the longitudinal components of magnetisation only

$$M_z(t) = \begin{bmatrix} M_z^A(t) \\ M_z^B(t) \end{bmatrix} \tag{10.17}$$

Instantaneous saturation of the free pool, caused by both *MT* and excitation pulses, is described by multiplying $\mathbf{M}_z$ by the matrix **S** (where α is the excitation flip angle)

$$S = \begin{bmatrix} S_{1A}\cos\alpha & 0 \\ 0 & 1 \end{bmatrix} \tag{10.18}$$

The state of the magnetisation after a period t_1 (assuming starting time = t_0) is given by the solution to the system of Equations 10.2 through 10.4 for either free precession (FP) or CW:

$$M_z(t_0 + t_1) = exp\{A_{CW}t_1\}M_z(t_0) + [I - exp\{A_{CW}t_1\}]A_{CW^{-1}}BM_0 \tag{10.19}$$

$$M_z(t_0 + t_1) = exp\{A_{FP}t_1\}M_z(t_0) + [I - exp\{A_{FP}t_1\}]A_{FP^{-1}}BM_0, \tag{10.20}$$

with

$$A_{CW} = \begin{bmatrix} -R_A - RM_0^B & RM_0^A \\ RM_0^B & -R_B - RM_0^A - R_{RFB} \end{bmatrix}$$

$$A_{FP} = \begin{bmatrix} -R_A - RM_0^B & RM_0^A \\ RM_0^B & -R_B - RM_0^A \end{bmatrix} \tag{10.21}$$

$$B = \begin{bmatrix} -R_A & 0 \\ 0 & -R_B \end{bmatrix}$$

According to Sled and Pike's RP approximation, over the time interval TR' between application of *MT* pulses (typically the time required to excite and collect data for a single *k*-space line of a single image slice), M_z undergoes instantaneous saturation, CW irradiation for a period $\tau_{RP}/2$, FP for a period (TR' – τ_{RP}), and CW for another $\tau_{RP}/2$. After including all three steps, we can impose the equality

$$M_z(TR') = M_z(0) \tag{10.22}$$

and solve for $\mathbf{M}_z$, recalling that the signal observed at readout is

$$SI(\omega_1, \Delta f) = \beta M_z^A(TR)S_{1A}\sin\alpha \tag{10.23}$$

where β is a constant scaling factor. It is thus possible to obtain an analytical expression to model the MT-weighed signal (following numerical evaluation of factor S_{1A}).

Yarnykh (2002) proposed a technique that neglects any direct saturation of the liquid pool by assuming the offset frequency is large enough (i.e. $\Delta f > 2.5$ kHz; Portnoy and Stanisz, 2007) and approximates shaped RF pulses by an effective rectangular pulse, with the same duration as the real pulse (τ_{SAT}), and constant amplitude ω_{1eff}:

$$\omega_{1eff} = \frac{1}{\tau_{SAT}}\int \omega_1^2(t)dt \tag{10.24}$$

Following an approach similar to that of Sled and Pike (2000), Yarnykh decomposes the sequence into a period of saturation, a delay for spoiling gradients, a readout period and a relaxation period.

The solutions (in matrix form) obtained for each period are then concatenated to yield an expression for $\boldsymbol{M}_s$, the magnetisation vector immediately before a readout pulse.

The expression is then simplified using a first order approximation assuming short time intervals and low excitation flip angles. The resulting analytical solution, M_{zS}^A, for the z-component of the liquid pool magnetisation just before the readout pulse is given by

$$M_{zS}^A = \frac{\frac{1}{F+1}(A + R_A s R_{RFB})}{A + (R_A + RM_0^B) s R_{RFB} - \left(R_B + \frac{RM_0^B}{F} + s R_{RFB}\right)\frac{ln(cos\alpha)}{TR}} \quad (10.25)$$

where $s=\tau_{SAT}/TR$ is the duty cycle and

$$A = R_A R_B + R_A \frac{RM_0^B}{F} + R_B RM_0^B \quad (10.26)$$

Yarnykh expresses the signal intensity relative to the unsaturated case as the MTR:

$$MTR(R_{RFB}) = 1 - \frac{M_{zS}^A(R_{RFB})}{M_{zS}^A(R_{RFB}=0)} \approx \frac{sR_{RFB}}{P + QsR_{RFB}} \quad (10.27)$$

with

$$P = \frac{A\left(A - \left(R_B + \frac{RM_0^B}{F}\right) ln\left(\frac{cos\alpha}{TR}\right)\right)}{RM_0^B\left(A - \frac{R_B ln(cos\alpha)}{TR}\right)} \quad (10.28)$$

$$Q = \frac{A\left(R_A + RM_0^B - ln\left(\frac{cos\alpha}{TR}\right)\right)}{RM_0^B\left(A - \frac{R_B ln(cos\alpha)}{TR}\right)} \quad (10.29)$$

The parameters P, Q and T_{2B} can thus be obtained by fitting the model to a series of MTR measurements at different settings of the saturating pulse. F (f) and RM_{0B} (k) can be derived with knowledge of R_A^{obs} from

$$P \approx R_A^{obs} \frac{F+1}{F} - \frac{ln(cos\alpha)}{TR}\frac{1}{F} \quad (10.30)$$

$$Q \approx \left(R_A^{obs} - \frac{ln(cos\alpha)}{TR}\right) RM_0^{B^{-1}} + 1 \quad (10.31)$$

It has been shown (Portnoy and Stanisz, 2007; Cercignani and Barker, 2008) that these three approximations yield consistent results, providing that the experimental conditions comply with their basic assumptions. A typical set of *MT* parametric maps obtained using Ramani's approximation is shown in Figure 10.7.

10.8.3 Minimal approximation magnetisation transfer model

An alternative approach to model the free pool longitudinal magnetisation (M_z^A) during pulsed *MT* experiments consists in the numerical integration of the full set of differential equations describing the two-pool model (Graham and Henkelman, 1997). By doing so, the time evolution of the coupled system can be obtained for a given saturation scheme.

This technique, termed minimal approximation magnetisation transfer (MAMT) (Portnoy and Stanisz, 2007), has been proposed as viable strategy to model magnetisation behaviour and estimate *MT* parameters avoiding unwarranted simplifications, such as assuming steady state or neglecting imaging pulses, that are normally introduced to derive analytical expressions for M_z^A.

The magnetisation vector $\mathbf{M}(t_1)$, solution of Equation 10.2 at the time instant $t_1 = t_0 + \tau$, assumes a closed form if matrix $\mathbf{A}(t)$ is not dependent on time:

$$M(t_1) = e^{A\tau} M(t_0) + \left[e^{A\tau} - I\right] A^{-1} B M_0 \quad (10.32)$$

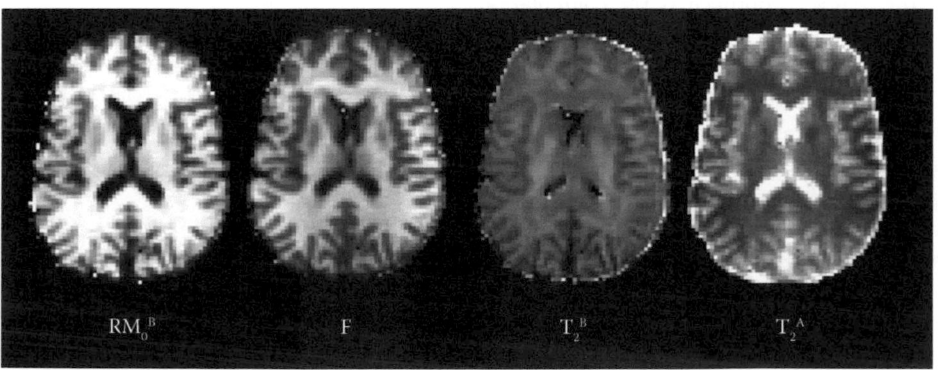

FIGURE 10.7 MT parametric maps of a healthy participant obtained *in vivo* using Ramani's approximation of Henkelman's model.

The MAMT technique integrates the system by replacing the continuous function $\omega_1(t)$ in $\boldsymbol{A}(t)$ with a discretised version:

$$\omega_1'(t)=\omega_1\left(t_{i-1}+\frac{\tau}{2}\right)=\omega_{1,i}$$
$$\forall t,(i-1)\tau=t_{i-1}\leq t<t_i=i\tau,i=1,2,3\ldots,N_d \qquad (10.33)$$

where τ represents the discretisation step and N_d the total number of steps.

Using Equation 10.33, matrix $\boldsymbol{A}$ is constant and equal to $\boldsymbol{A}_i=\boldsymbol{A}(\omega_{1,i})$ over intervals $t_{i-1}\leq t<t_i$. The magnetisation vector $\boldsymbol{M}$ (and hence M_z^F) at any time instant t can be obtained by concatenating Equation 10.32 for different intervals $[t_{i-1}, t_i]$, where $\boldsymbol{A}$ is replaced by the corresponding $\boldsymbol{A}_i$ and the initial condition $\boldsymbol{M}(t_0)$ is given by the output of the previous step, in a recursive fashion:

$$M_i=e^{A_i\tau}M_{i-1}+\left[e^{A_i\tau}-I\right]A_i^{-1}B \qquad (10.34)$$

and assuming pool magnetisations are aligned along the z-axis at the initial time t_0:

$$M_0=\begin{bmatrix}0\\0\\1\\F\end{bmatrix} \qquad (10.35)$$

The formalism described above can be adapted to experiments employing arbitrary saturation schemes and imaging pulse sequences. In addition, exact pulse shapes, timing between pulses, effects of on-resonance excitation and transient behaviour of magnetisation towards steady state can all be taken into account, within the approximations introduced by the two-pool model, when estimating *MT* parameters from a pulsed experiment (Figure 10.8).

However, care must be taken when modelling on-resonance excitation periods as the super-Lorentzian line shape usually associated with the macromolecular pool shows a singularity for $\Delta f=0$. Extrapolating absorption line shape values in the proximity of $\Delta f=0$, or neglecting the term R_{RFB} during on-resonance (equivalent of assuming no instantaneous effect of imaging pulses on macromolecular pool) are both valid solutions.

The downside of the MAMT approach is the computation time, as a matrix exponential has to be evaluated for every step used in the discretisation of the shaped RF pulse. To improve time efficiency, matrix exponentials can be computed in advance and saved in look-up tables. Enhanced performances have been obtained by replacing the full calculation of exponential matrices product with a polynomial interpolation (Müller *et al.*, 2013).

Another factor impacting the MAMT time efficiency is related to the size of the discretizing step, τ. Although computation time linearly scales with it, τ cannot be arbitrary reduced as an artefactual spike will appear in the Z-spectrum when the pulse envelope is digitised with reduced number of steps (i.e. using longer τ). However, since spikes will appear periodically spaced by $1/\tau$ in the frequency domain (with amplitude depending on the particular RF pulse used) potential errors in the fitting could be avoided by acquiring experimental data out of the spikes region (Müller *et al.*, 2013).

The MAMT model can serve as a reference against which to compare other more approximate models, for different experimental conditions and/or noise level (Portnoy and Stanisz, 2007; Cercignani and Barker, 2008). It can also be used for transient conditions as it does not require steady state. This means that qMT model parameters can be extracted by sampling the system approach to the steady state (Helms and Hagberg, 2003; Tyler and Gowland, 2005) or in general, when *MT* preparation, is combined with fast imaging technique, such as echo-planar imaging (EPI) (Battiston *et al.*, 2016; Gelderen *et al.*, 2016).

10.8.4 MT modelling of selective inversion recovery signal

A different way to quantify *MT* fundamental parameters consists in measuring the relaxation of magnetisation after selectively disturbing either the free or the bound pool for a short time (compared with the characteristic exchange time).

For the two-pool model Bloch model, longitudinal relaxation after a perturbation (either of one or both the pools) is bi-exponential (Edzes and Samulski, 1977, 1978b):

$$\frac{M_A(t)}{M_0^A}=C_1e^{-\lambda_1 t}+C_2e^{-\lambda_2 t}+1 \qquad (10.36)$$

where the relaxation rates $\lambda_{1,2}$ are

$$\lambda_{1,2}=\frac{1}{2}\Big(R_A+R_B+RM_0^B+RM_0^A$$
$$\pm\sqrt{\left(R_A-R_B+RM_0^B-RM_0^A\right)^2+4RM_0^BRM_0^A}\Big) \qquad (10.37)$$

and the coefficients $C_{1,2}$ are

$$C_{1,2}=\pm\frac{\left[\frac{M^A(0)}{M_0^A}-1\right](R_A-\lambda_{2,1})+\left[\frac{M^A(0)}{M_0^A}-\frac{M^B(0)}{M_0^B}\right]RM_0^B}{\lambda_1-\lambda_2} \qquad (10.38)$$

In principle, Equation 10.36 can be used to fit data acquired at different delays from the perturbation to estimate coefficients $C_{1,2}$ and relaxation rates $\lambda_{1,2}$, relating to the underlying two-pool model parameters through Equations 10.37 and 10.38.

While the relaxation rates λ_1 and λ_2 are intrinsic properties of the system, the coefficients C_1 and C_2 are dependent on the experimental settings (i.e. initial conditions after the perturbation, $M_z^A(0)$ and $M_z^B(0)$). Perturbation capable of selectively manipulating magnetisation of one pool will enhance the bi-exponential behaviour of the relaxation process, through the difference term $\left[\frac{M^A(0)}{M_0^A}-\frac{M^B(0)}{M_0^B}\right]$ in the numerator of Equation 10.38 (Gochberg *et al.*, 1997).

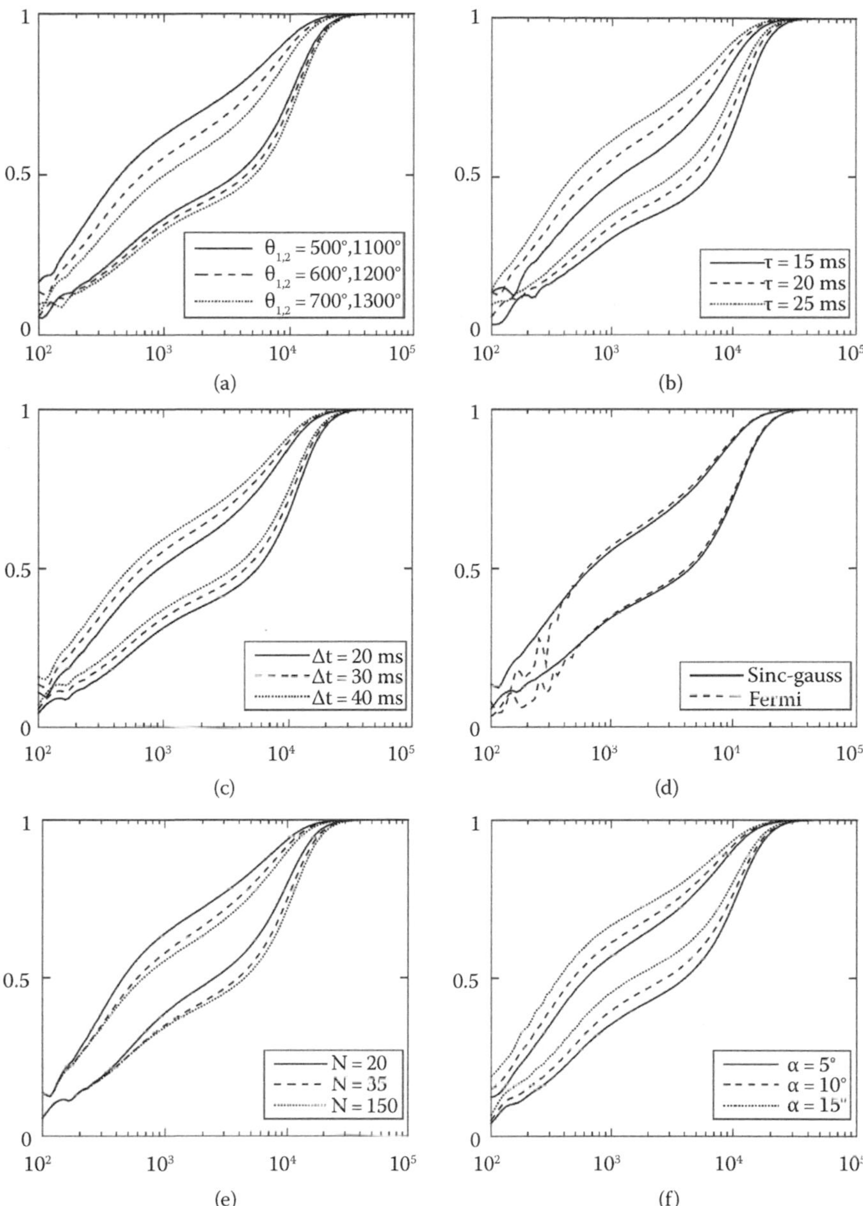

FIGURE 10.8 Minimal approximation MT (MAMT) model predictions for different pulse sequence parameters values. Plots show predicted M_z^F (normalised on a non-weighted reference value) over frequency offset (from 100 Hz to 100 kHz) at two different saturation powers (effective flip angles $\theta_1 = 600°$ and $\theta_2 = 1200°$). When not specified, other sequence parameters are as follows: pulse shape sinc-Gaussian, pulse duration 20 ms, interpulse gap 30 ms, number of pulses 150. The MAMT model allows us to account for any variation of the sequence parameters: (a) different flip angles at fixed pulse duration, (b) different pulse durations at fixed flip angles, (c) different interpulse gaps, (d) different pulse shapes, (e) different number of pulses for non-steady state acquisition and (f) presence of imaging pulses at different flip angles. A single tissue configuration is used to calculate model prediction: F = 0.13, $T2^F$ = 35 ms, T_2^B = 12 μs, RM_0^B = 2.5.

Quantitative determination of two-pool model parameters using this approach has been first demonstrated in bovine serum albumin (BSA) phantoms by Gochberg and Gore (2003), where an inversion recovery pulse capable of selectively inverting free water proton magnetisation was used in combination with an EPI readout. Selectivity to free water proton is guaranteed with a low power hard pulse whose duration is small compared to T_2^F and long compared to T_2^B (typical durations are a few milliseconds).

The approach was subsequently improved by combining free water selective inversion with a fast spin-echo readout (Gochberg and Gore, 2007), which allows shortening the *TR* below the condition $TR > 5/\lambda_2$, as longitudinal magnetisation of both pools is nearly nulled at the end of a sufficiently long (~40 ms) train of refocusing pulses. Hence, during the delay anticipating the subsequent inversion pulse (t_d), longitudinal relaxation can be approximated as a mono-exponential function with rate λ_1.

Equation 10.36 can therefore be used to fit data acquired at varying Inversion Times (TIs), accounting for the different initial conditions M_z^A (0), M_z^B (0) due to partial longitudinal recovery during t_d, as follows:

$$\left[\frac{M^{A,B}(0)}{M_0^{A,B}} = S_{A,B}\left(1-e^{-\lambda_{1,2}t_d}\right)\right] \tag{10.39}$$

where $S_{A,B}$ quantifies the effect of the inversion pulse on the free and macromolecular pool, respectively. Macromolecular fractional saturation S_B is computed numerically following assumptions on macromolecular absorption line shape and transverse relaxation time T_2^B. A reference value for S_B produced by a τ_{inv} = 1 ms inversion hard pulse is 0.83 ± 0.07 (Gochberg *et al.*, 1999), calculated assuming a Gaussian line shape, a T_2^B between 10 and 20 μs and neglecting exchange ($\lambda_1 >> 1/TR$) and relaxation during the inversion ($R_1^B << \tau_{inv}$).

The bi-exponential model in Equation 10.36 is fitted for M_0^F, C_1, C_2, λ_1 and λ_2. Two-pool model parameters can be derived using the assumption $RM_0^A >> RM_0^B$, R_A, R_B, that holds for a variety of biological tissues, leading to the following approximations:

$$\lambda_1 \approxeq RM_0^A \tag{10.40}$$

$$C_1 \approxeq \left(\frac{M^A(0)}{M_0^A} - \frac{M^B(0)}{M_0^B}\right)F \tag{10.41}$$

Once M_0^F, C_1, C_2, λ_1 and λ_2 have been determined, Equations 10.37, 10.40 and 10.41 allow the macromolecular fraction to be isolated:

$$F = \frac{C_1}{C_2 + 1 - S_B\left(1-e^{\lambda_2 t_d}\right)}. \tag{10.42}$$

A key advantage of the technique is the low power required for saturation, which makes it particularly attractive at high and ultra-high fields where high power off-resonance RF pulses are more problematic due to SAR restrictions (Dortch *et al.*, 2013). Moreover, SIR-FSE does not require any separate measurements to correct for filed inhomogeneity or to measure longitudinal relaxation time.

The main shortcoming of SIR-FSE is its lower time efficiency compared to the more traditional pulsed saturation qMT imaging techniques. Multislice imaging cannot be implemented straightforwardly, as the train of refocusing pulses employed will cause additional *MT* effect in the neighbouring slices and a recovery time in the order of ~$5/\lambda_2$ has to be added to allow contiguous slice acquisition. However, guidelines on definition of sequence parameters for optimising precision efficiency have been investigated (Li *et al.*, 2010), as well as a 3D version of the protocol (Dortch *et al.*, 2013).

10.8.5 MT Models of steady-state free precession signal

In 2008 Gloor *et al.* (2008) adapted the two-pool model to bSSFP acquisitions. The main difference in this case is that all the pulses are applied on-resonance, and therefore Δf in Equation 10.2 tends to zero.

The system of coupled equations thus reduces to

$$\begin{bmatrix} M_y^A \\ M_z^A \\ M_z^B \end{bmatrix} = \begin{bmatrix} -R_2^A & \omega_1(t) & 0 \\ -\omega_1(t) & -\left(RM_0^B + R_A\right) & RM_0^A \\ 0 & RM_0^B & -\left(RM_0^A + R_B\right) \end{bmatrix} \begin{bmatrix} M_y^A \\ M_z^A \\ M_z^B \end{bmatrix} + \begin{bmatrix} 0 \\ R_A M_0^A \\ R_B M_0^B \end{bmatrix} \tag{10.43}$$

where the component M_X^A can be excluded given the assumption that all the RF pulses are played out along the *x*-axis, so that in the transverse plane only the *y*-component contributes to the formation of the steady-state signal.

An analytical solution for M_Z^A in the steady state is obtained by following an approach similar to the one described in Section 10.8.2. Here, the on-resonance excitation is captured in the matrix $R_x(\alpha)$, containing rotation of free pool magnetisation and saturation of the bound pool, whereas phase alternation between consecutive RF pulses is modelled by the matrix $R_z(\phi = 180)$:

$$R_x(\alpha) = \begin{bmatrix} \cos\alpha & \sin\alpha & 0 \\ -\sin\alpha & \cos\alpha & 0 \\ 0 & 0 & e^{-R_{RFB}T_{RF}} \end{bmatrix} \tag{10.44}$$

$$R_z(\phi = 180) = \begin{bmatrix} -1 & 0 & 0 \\ 0 & 1 & 0 \\ 0 & 0 & 0 \end{bmatrix} \tag{10.45}$$

where α is the flip angle (played out with alternating phase between consecutive *TRs*), and $\langle R_{RFB} \rangle$ represents the mean saturation rate when Δ tends to 0, calculated over the whole duration of the RF pulse T_{RF}:

$$\left(R_{RFB}(\Delta \to 0)\right) = G(\Delta \to 0)\frac{\pi}{T_{RF}} \int_0^{T_{RF}} \omega_1^2(t)\,dt \tag{10.46}$$

Equation 10.46 requires the value of the absorption line shape *G* in the limit Δ = 0, which was estimated (through extrapolation) by Gloor and co-authors to be $G(0) = 1.4 \cdot 10^{-5} s^{-1}$, for T_2^B = 12 μs at 1.5T, for both white and GM (Gloor *et al.*, 2008).

An analytical model for two-pool bSSFP signal can be derived by assuming that during free precession intervals between RF excitations exchange and relaxation processes are decoupled and therefore can be separately described by matrices $A(t)$ and $E(t)$:

$$A(t) = \frac{1}{F+1} \begin{bmatrix} F+1 & 0 & 0 \\ 0 & 1+Fe^{-(F+1)RM_0^A} & 1-e^{-(F+1)RM_0^A} \\ 0 & F-Fe^{-(F+1)RM_0^A} & F+e^{-(F+1)RM_0^A} \end{bmatrix} \tag{10.47}$$

$$E(t) = \begin{bmatrix} e^{-R_2^A t} & 0 & 0 \\ 0 & e^{-R_1^A t} & 0 \\ 0 & 0 & e^{-R_1^B t} \end{bmatrix} \quad (10.48)$$

In the derivation it is further hypothesised that relaxation takes place before exchange; however, an almost identical solution is derived for the case where exchange takes place before relaxation (Gloor *et al.*, 2008).

Concatenating free pool excitation and bound pool saturation, relaxation and exchange, and imposing steady-state condition between subsequent on-resonance pulses, yields

$$M_\infty^+ = R_x\left(I - R_z AER_x\right)^{-1} AM_0 \quad (10.49)$$

which represents the steady-state magnetisation vector directly after the RF excitation, from which the measured signal is readily given by

$$S(\alpha, TR) = \beta M_0^A sin\alpha \frac{\left(1-e^{-R_1^A TR}\right)B + C}{A - Be^{-R_1^A TR}e^{-R_2^A TR} - \left(Be^{-R_1^A TR} - Ae^{-R_2^A TR}\right)cos\alpha} \quad (10.50)$$

Parameters A, B and C are combinations of sequence parameters (TR, T_{RF}) and two-pool model parameters, in particular F, RM_0^B, T_2^A, R_A and R_B (full definitions can be found in (Gloor *et al.*, 2008)). Bound pool transverse relaxation T_2^B instead does not contribute to the predicted signal, as macromolecular absorption line shape is sampled at a single point (Δ = 0) and therefore cannot be estimated using this method.

MT parameters are extracted exploiting the different *MT* sensitivity of the bSSFP signal to α and T_{RF}. A common protocol for qMT using bSSFP includes both sequences at fixed α and varied T_{RF} (and consequently TR), and sequences at fixed T_{RF}/TR and varied α. The greatest signal sensitivity to the *MT* effect is obtained through T_{RF}/TR variations.

Given the intrinsic dependency of the bSSFP signal (and consequently Equation 10.50) to the ratio T_1/T_2 (Scheffler and Lehnhardt, 2003), reliable estimation of R_A and R_2^A requires an independent measure of one of the two quantities. Usually R_A is estimated via an auxiliary experiment (VFA or driven equilibrium single pulse observation of T1 [DESPOT1]), whereas R_2^A is estimated together with F and RM_0^B from the non-linear fitting of Equation 10.50. Accordingly to other qMT modalities, estimated R_B shows high uncertainty and is therefore set equal to R_A (Gloor *et al.*, 2008). A common protocol for qMT using bSSFP, comprising of B_1 mapping for flip angle correction, can be kept under 10 minutes' duration.

In the first demonstration of qMT imaging using bSSFP sequence, *MT* parameters from two-pool bSSFP model were found in good correspondence with those from common quantitative MT models, with standard deviation below 3% in a repeated (five times) measurement from the same subject (Gloor *et al.*, 2008).

qMT via bSSFP provides a framework that meets many of the requirements of an optimal qMT protocol. In particular it allows reliable and reproducible quantification of *MT* model parameters exploiting the high-resolution acquisition and intrinsic high SNR of the bSSFP sequence within a clinically feasible acquisition time.

There are however several issues that can impact the overall quality of the qMT analysis. Along with the common issues regarding F estimate sensitivity to B_1 field inhomogeneity and inability to fit for R_B, others are specific to this approach.

The ambiguity in determining $G(0)$ has a strong impact on *MT* parameter estimation, especially on BPF, as it scales the mean saturation rate $\langle R_{RFB} \rangle$, resulting in linear errors on BPF (Gloor *et al.*, 2008).

Additionally, off-resonance artefacts, known as *banding artefacts*, affect bSSFP signal intensity. Off-resonance frequencies (ν) cause the magnetisation to dephase during the *TR* by an angle $\theta = 2\pi \nu TR$. For dephasing approaching $\pm\pi$, the bSSFP signal exhibits a sharp and prominent drop. This introduces a dependency in the signal amplitude on the local off-resonance effect (ν), as they translate in a constant dephasing θ over *TR*, and in the context of *MT* represents a limitation in the possibility of generating *MT* effects for varied *TR* values. As banding artefacts increase with field strength, these issues become non-negligible for fields greater than 1.5T.

10.8.5.1 Extension to Non-Balanced SSFP

The theory of MT-bSSFP can be extended to non-balanced SSFP sequences (Bieri *et al.*, 2008). This assumes particular importance when MTI protocols are performed in anatomical locations characterised by high susceptibility variations (e.g. cartilage, or musculoskeletal system), or at ultra-high magnetic fields, where image degradations due to bSSFP banding artefacts are exacerbated and may therefore hamper the interpretation of magnetisation transfer weighted bSSFP images.

Non-balanced SSFP sequences are inherently insensitive to banding artefacts and ideally could be used to robustly assess *MT* indices in such cases. However, when moving from balanced to non-balanced SSFP sequences, SNR is reduced and SSFP signal becomes dependent on the TR, a fact that could potentially lead to artefactual over- or underestimation of the underlying *MT* effect, when RF pulse elongation, and hence *TR* variations are used to generate *MT* contrast.

For a *TR* varying in the range 3–10 ms, non-balanced SSFP sequences show an increase in the signal of 4%–5% (or a decrease of 13%–16%, according to how the echo is encoded), compared to a negligible variation (<0.1%) for the balanced counterpart. Therefore to minimise signal variations unrelated to the *MT* effect to be quantified, a compromise has to be made with the size of the *MT* effect generated. By limiting the *TR* difference between MT-weighted and non-MT-weighted acquisitions, contamination in the *MT* metrics, such as the MTR, can be kept small, between 2% and 5%, at the cost of reduced MTR values (Bieri *et al.*, 2008).

This concept was experimentally validated by comparison of MTR histograms obtained using bSSFP and non-balanced SSFP sequences, which showed excellent correspondence and highly comparable tissue delineation, among different protocol variants (Bieri *et al.*, 2008).

MT-SSFP was successfully applied in the cartilage at low (1.5T) and ultra-high (7T) fields, to prove the possibility of generating high-resolution 3D MTR maps in anatomical environments or experimental settings where bSFFP-MT imaging may fail due to its sensitivity to off-resonance variations. Similarly to the theory developed for the balanced version, a quantitative description for the two-pool SSFP signal has also been formulated to allow full qMT analysis with non-balanced sequences (Gloor *et al.*, 2010).

However, this is obtained at a cost of a lower CNR in the resulting maps, due to the intrinsic lower SNR of the non-balanced protocols and to the reduced *MT* effect that can be attained in order to keep signal corruption by *TR* variations mechanism below 1% and 2% and at a cost of an increased sensitivity to flow and motion artefacts.

10.9 Interpretation and validation of *MT* parameters (T,V)

Myelin is believed to dominate the *MT* exchange process in WM, and therefore changes in the MTR have mainly been interpreted as myelin damage or loss. The interpretation of *MT* changes in the GM can be more challenging. The MTR value is proportional to the product $R \cdot F \cdot T_1^A$ (Henkelman *et al.*, 2001), and therefore can reflect changes in any of these three parameters. This limits the sensitivity of MTR, as changes to T_1 and F tend to occur in opposite directions.

Quantitative *MT* provides a number of parameters that reflect indirectly some properties of the macromolecular pool and may potentially measure differing pathological substrates. Most research has focused on demonstrating that F reflects myelin content. Animal (Ou *et al.*, 2009a, 2009b) and post-mortem (Schmierer *et al.*, 2007) studies have reached encouraging results, which support the interpretation of *F* (or *f*) as a marker of myelination. Despite these observations, *F* and measures of myelination derived from short T_2 mapping (see Chapter 4) appear largely uncorrelated (Sled *et al.*, 2004), prompting the debate on which technique better estimates myelin *in vivo*.

Among the other *MT* parameters, T_2^B has received little attention, mainly because it appears relatively stable across conditions (healthy subjects vs. patients, WM vs. GM, etc.), although its dependency on WM fibre orientation with respect to the main field is intriguing (Pampel *et al.*, 2015).

Finally, the interpretation of RM_0^B, the pseudo-first order forward exchange rate, is uncertain, as the exact mechanisms of *MT* are still unknown. Nevertheless, recent studies highlighted changes in the cortex of patients with Alzheimer's disease (Giulietti *et al.*, 2012) and in healthy participants undergoing an inflammatory challenge (Harrison *et al.*, 2015) that overlap with abnormal fluorodeoxyglucose (FDG) uptake on PET scans in analogous conditions, suggesting that RM_0^B might be sensitive to subtle inflammation or metabolic changes.

10.10 'Quick and Dirty': *MT*-Based myelin proxies from fewer than five measurements (P)

The clinical applications of quantitative models of *MT* described in the previous sessions have been limited so far due to the long acquisition times. The scan time is predominantly affected by the number of data points and/or the signal-to-noise ratio (SNR) of the underlying acquisition required to achieve a sufficient SNR in the calculated maps. Simplified approaches have been introduced to facilitate the translation of these methods to clinical work.

10.10.1 *Reduced* models

A constraint on the minimum number of images is imposed by the total number of model parameters to be estimated. The number of data points required can thus be decreased by reducing the number of model parameters to a subset of those required to fully fit the model by imposing physically meaningful constraints on the others. Yarnykh and Yuan (Yarnykh and Yuan, 2004) constrained both T_{2A} and T_{2B}, based on the following observations. First, the product $T_{2A}R_A$ is fairly constant in CNS tissues, even in the presence of pathology, such as MS lesions (Sled and Pike, 2001). Secondly, the reported T_{2B} values for the brain, both *in vivo* and *in vitro* (Morrison *et al.*, 1995; Sled and Pike, 2001), fall into the range of 9.2–12.3 μs with little difference between white and grey matter. They were thus able to reduce the number of *MT* sampling points to four (plus a T_1-mapping scan). Used in combination with a 3D high-resolution acquisition, this strategy was shown to produce high-quality maps of f and RM_{0B} (*k*).

Using further constraints, recently the same group presented a method for mapping *F* based on two measurements (an MT-weighted and a reference scan), together with the appropriate T_1 and field mapping scans (Yarnykh, 2012, 2016). The same technique has been successfully applied also to the spinal cord (Smith *et al.*, 2014).

10.10.2 *Magnetisation transfer saturation*

This method, developed by Helms *et al.* (2008b), takes one step towards the quantification of the two-pool model parameters while maintaining the simplicity characterising MTR data acquisition and analysis.

It employs a three-point acquisition scheme to quantify, together with the apparent tissue proton density (PD_{app}) and the longitudinal relaxation time (T_1), a parameter called magnetisation transfer saturation (MT_{sat}), which shows improved contrast between differently myelinated structures compared to the conventional MTR (Helms *et al.*, 2009).

The MT_{sat} parameter incorporates the effects of macromolecular pool saturation, direct (free water pool) saturation and *MT* exchange produced by a single *MT* pulse during a *TR* period. As the MTR, the MT_{sat} still represents a semi-quantitative parameter, meaning that it lacks direct biological meaning and is dependent on sequence parameters, but compared to the MTR it provides corrections for T_1 relaxation and on-resonance excitation and appears to be inherently insensitive to RF inhomogeneity (Helms *et al.*, 2008b).

A phenomenological signal equation for the pulsed *MT* spoiled GRE (SPGR) sequence can be obtained by analogy from a dual excitation sequence, as the one used in the actual flip angle imaging technique (Yarnykh, 2007), where two on-resonance excitations and relative free evolution periods occur in each TR. The signal readout by each excitation is given by

$$S_{1,2} = A\sin\alpha_{1,2}\frac{1-e^{-R_1TR}-(1-\cos\alpha_{2,1})(1-e^{-R_1TR_{1,2}})e^{-R_1TR_{2,1}}}{1-\cos\alpha_1\cos\alpha_2 e^{-R_{1,2}TR}} \quad (10.51)$$

where $\alpha_{1,2}$ denotes the two consecutive flip angles, $TR_{1,2}$ their respective time intervals, $TR = TR_1 + TR_2$ the full repetition time and $A = PD_{app}$ is a scaling factor containing the true proton density PD, the effect of coil sensitivity and the effect of T_2* relaxation during the time TE.

By assuming $R_1TR \ll 1$ and $\alpha_{1,2} \ll 1$ (in radians), Equation 10.51 can be approximated to

$$S_{1,2} \approx A\sin\alpha_{1,2}\frac{R_1TR}{\frac{\alpha_1^2}{2}+\frac{\alpha_2^2}{2}+R_1TR} \quad (10.52)$$

which resembles the rational approximation of the well-known Ernst equation for the SPGR signal (Helms *et al.*, 2008a), except for the second term at the denominator ($\alpha_2^2/2$), due to the additional excitation.

In the MT-SPGR experiment, one excitation is replaced by the *MT* pulse. Equation 10.52 can be used to describe the signal intensity in such an experiment, exploiting its analogy with the dual excitation sequence: the effect of the second excitation is replaced by the empirical parameter δ, having the meaning of a saturation term $\delta = \alpha_2^2/2 \sim 1-\cos(\alpha_2)$, describing the additional saturation imposed by the *MT* pulse on the measured signal:

$$S_{MT-SPGR} \approx A\sin\alpha\frac{R_1TR}{\frac{\alpha^2}{2}+\delta+R_1TR} \quad (10.53)$$

As the effect of on-resonance excitation and longitudinal relaxation are separately accounted for in Equation 10.53, $MT_{sat} = \delta$ is expected to be largely independent from α and T_1.

A more quantitative interpretation of $MT_{sat} = \delta$ can be obtained via an approximated analysis of the two-pool system undergoing partial progressive saturation during a train of off-resonance RF pulses and modified to account for the effect of additional small on-resonance excitation per pulse repetition time (i.e. *TR*). Details can be found in the work by Helms *et al.* (2008b).

10.11 Sources of error and factors that can alter measured *MT* parameters (V)

The measured MTR value depends on pulse sequence factors (Berry *et al.*, 1999) such as the time between *MT* pulse repeats (TR′), the pulse shape, bandwidth, duration, amplitude and offset frequency, and the degree of T_1-weighting, if any. Altering the number of slices usually alters TR′. In addition, machine imperfections such as transmitter coil non-uniformity and inaccuracy and instability in setting the flip angle will alter the amount of *MT* pulse power applied. If MTR is approximately proportional to mean pulse power, then a 10% error in the RF field B_1, arising either from poor prescan procedure or transmit field non-uniformity, would give a 20% error in MTR. Transmit non-uniformity increases the width of histograms (Tofts *et al.*, 2006) and can be improved by using body-coil excitation. The receiver gain must be the same for collection of the M_0 and M_s images. Poor or variable shimming could alter the frequency offset of the pulse from its nominal value.

Quantitative measurements of model *MT* parameters are also vulnerable to all these factors, although if B_1 variations are detected and measured, they can be used in the model fitting procedure in place of the nominal values. Sled and Pike (2000), whilst measuring and correcting for B_1 and B_0 errors, reported that a 40 Hz error in B_0 altered the estimate for k_f by 20% and the bound/free ratio F by 5%. A 10% change in B_1 produced a 20% change in F. Image noise may have a large effect on the fitted parameters, depending on what pairs of offset frequency and saturating amplitude have been chosen for the data collection procedure. Sampling scheme optimisation procedures can aid in mitigating these issues (Cercignani and Alexander, 2006; Levesque *et al.*, 2011).

Patient movement may affect the image datasets (either the M_0 and M_s images for MTR or the multiple images collected for qMTI). Registration may be able to partly correct the movement for 2D multislice data; it will probably completely correct 3D volume datasets.

10.12 Reproducibility, phantoms and quality assurance (V)

Several studies have attempted to quantify the reproducibility of MTR measurements in the brain. Early studies (Berry *et al.*, 1999; Sormani *et al.*, 2000) clearly established that, while within-site and within-sequence reproducibility of MTR is good, the use of different MR scanners is the main source of variability. Attempts to standardise the acquisition led to the proposal of a standard pulse sequence, to include a standard presaturation pulse and set of parameters, which could be implemented on scanners by different manufacturers (Barker *et al.*, 2005). Tofts *et al.* (2006) did a rigorous analysis of the potential sources of variation, concluding that transmit field non-uniformity and B_1 errors account for most of the variability, and recommended the use of the body coil transmit to achieve better uniformity. All these studies focused on MTR measured using the pulsed *MT* method. In 2011, Gloor *et al.* (2011) showed an excellent reproducibility for bSSFP MTR, also across sites.

Assessing the reproducibility of quantitative *MT* is challenging, because the results will depend on the choice of the model, on the number of sampling points, on the SNR of the raw data and other parameters such as the quality of the T_1 maps. Nevertheless, one study was published that assessed the reproducibility of quantitative magnetisation-transfer imaging parameter estimates in healthy subjects (Levesque *et al.*, 2010). The results showed that R_A and T_2^B have the lowest variability across time (for scan–rescan), while k_f has the largest.

Standard MR phantoms are not suitable for *MT* measurements, as they do not contain macromolecules likely to exhibit *MT* effects. To date, *MT* experiments have been carried out using agar gels (Henkelman *et al.*, 1993) and cross-linked BSA (Bertini *et al.*, 1998; Ou and Gochberg, 2008), a protein derived from cows. Hair conditioner makes an excellent testing material for *MT* phantoms, as it contains fatty alcohols (Xu *et al.*, 2016). All these phantoms can be used to perform regular quality assurance measurements and to compare *MT* parameters between sites. Caution, however, must be taken to ensure that all the measurements are performed at the same temperature and that the materials are not degrading over time.

10.13 Inhomogeneous *MT* (T)

In 2006 a new contrast mechanism based on MT, termed *inhomogeneous magnetisation transfer* (*ihMT*), was reported for the first time (Alsop *et al.*, 2006).

Although the origin and behaviour of the *ihMT* signal is still a matter of debate, several applications have been proposed at 1.5T, 3T and 11.7T (Girard *et al.*, 2015, 2016; Varma *et al.*, 2015a; Prevost *et al.*, 2016). *ihMT* is thought to be sensitive to biological structures showing inhomogeneously broadened spectral lines, and the first quantitative description (Varma *et al.*, 2015b) suggests an improved specificity towards myelin content compared to conventional MT.

The high specificity of *ihMT* towards myelin has been attributed to myelin's unique architecture and molecular composition among CNS tissues, which fosters the hypothesis of myelin having an inhomogeneously broadened spectral line on a timescale of several milliseconds.

In the following section, the original explanation of *ihMT* given by Varma *et al.* is briefly described. However, this is a speculative interpretation, which has been challenged by other authors (Manning *et al.*, 2016). As *ihMT* is still in its infancy, future work will better clarify its substrate.

10.13.1 Homogeneous versus inhomogeneous NMR line broadening

The broad line that characterises the semi-solid pool in conventional *MT* can be seen as the result of multiple individual lines centred at different offset frequency, where spins can move rapidly from one frequency to the other, such that different lines cannot be separated. These rapidly exchanging lines are known as *homogeneously broadened lines* and result from mechanisms such as translational and rotational motion, chemical exchange and spin diffusion. In terms of quantitative description, the hypothesis of homogeneous broadening is modelled by assuming that the applied RF power saturates the entire line equally, with strength dependent on the applied power and the absorption spectrum of the line. This effect is summarised by the parameter R_{RFB} (see Equations 10.6 through 10.8).

Non-homogeneous broadening has been suggested as a unique feature of myelin among tissues composing CNS. The densely packed sheaths of lipid bilayer composing myelin allow rotation of lipid chains around an axis perpendicular to the surface (Opella and Marassi, 2004) and result in restricted or inefficient spin diffusion and molecular motion mechanisms (Huster *et al.*, 2002), producing a significant residual dipolar coupling. This implies that spins exchange slowly between lines at a distinct offset frequency and therefore can be saturated to produce a 'hole' in the broadened line.

ihMT is able to highlight this hypothesised feature of lipid-rich membranes by isolating transfer from inhomogeneous broadened magnetisation in the presence of a larger background of homogeneous MT. *ihMT* has therefore the potential of being specific to myelinated structures.

10.13.2 Definition of *ihMT* signal and *ihMT* ratio

In order to detect *ihMT* signal, together with a normal *MT* signal, obtained at offset frequency +f (the single-side saturated signal $S(f+)$), an additional signal (the dual-side saturated signal) has to be acquired where the same total RF power is used to saturate simultaneously (or quasi simultaneously) both frequencies +f and $-f$ ($S(f\pm)$). Under the condition of symmetry for the *NMR* line, homogeneously broadened magnetisation is saturated equally by both the single and dual side experiment, and the difference between the two signals is zero. The difference signal can thus be used as indicator of *MT* from inhomogeneous line (see Figure 10.9).

The *MT* spectrum is known to be slightly asymmetric (Hua *et al.*, 2007). *MT* saturation at positive frequencies can differ from those at their respective negative values, causing *ihMT* to be non-zero also for a homogeneously broadened line. Acquiring the single side saturation both at positive and negative frequency, and the double side saturation with both the alternating orders, provides a first-order correction for this offset error in the calculation of *ihMT*. The analogue of MTR in the context of *ihMT*, the *ihMTR*, can be readily defined by normalising the difference signal on a non-*MT*-weighted signal S_0:

$$ihMT = S(f+) + S(f-) - S(f\pm) - S(f\mp) \tag{10.54}$$

$$ihMTR = \frac{ihMT}{S_0} \tag{10.55}$$

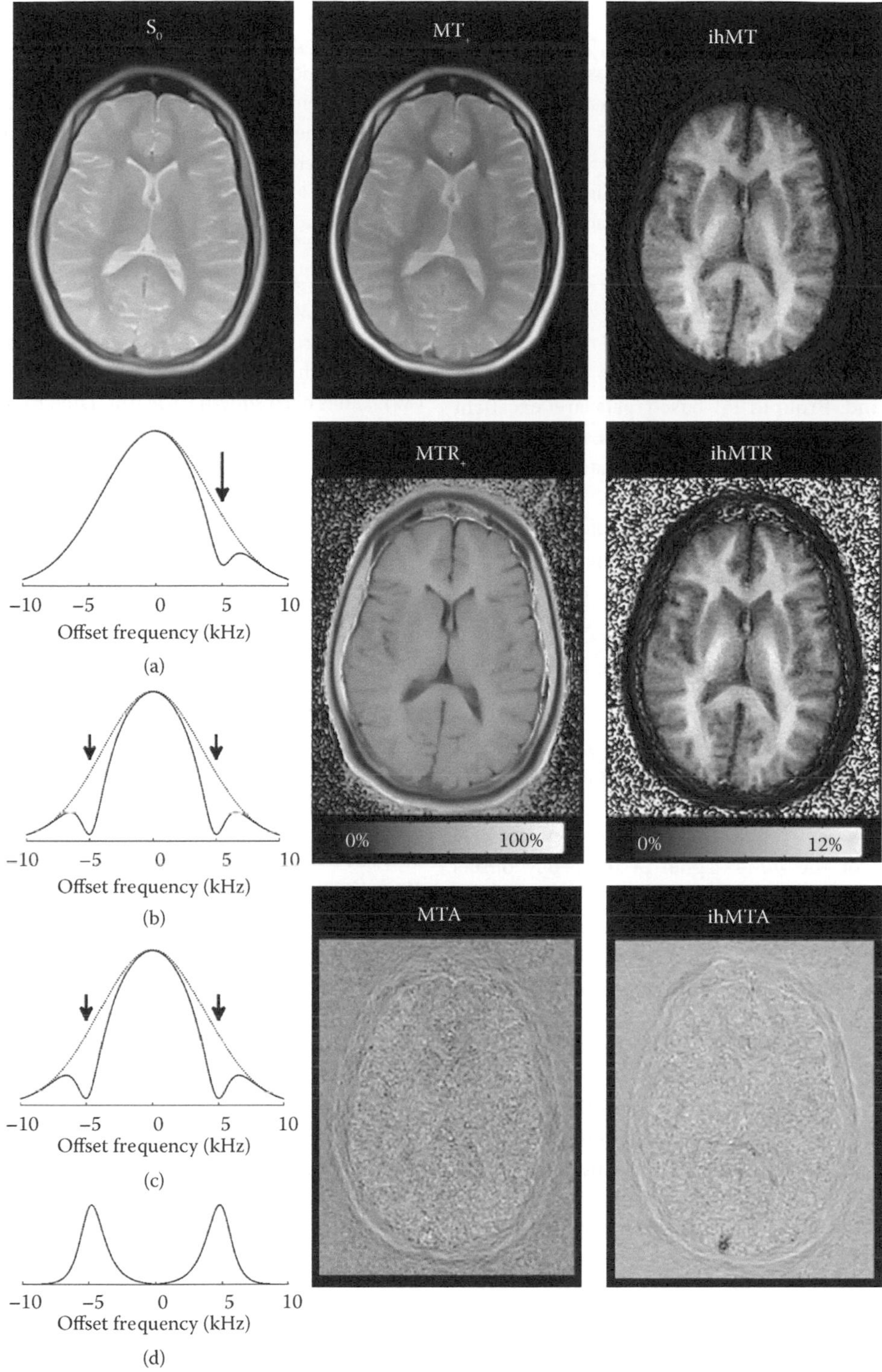

FIGURE 10.9 Example of inhomogeneous MT (ihMT) dataset obtained with the typical acquisition strategy, reproduced as a conceptual scheme in panels **a** through **d**. MT+ shows a conventional MT-weighted image acquired at single positive offset frequency (a), from which MTR is readily computed using a non-weighted images (S_0). Repetition of the same scheme, but at negative offset frequency (b), allows calculation of MT asymmetry MTA. An additional acquisition, performed with simultaneous saturation of positive and negative frequency (c), produces, by subtraction with (a+b) an ihMT weighted image and the correspondent ihMTR image. These maps arise from the enhanced dual-side saturation compared to the single-side one (d). Analogous iHMT asymmetry can be obtained from the dual saturation scheme. (Adapted with permission from Girard, O.M., *et al.*: Magnetisation transfer from inhomogeneously broadened lines (ihMT): experimental optimisation of saturation parameters for human brain imaging at 1.5 Tesla. *Magn. Reson. Med.* 2015. 73. 2111–2121. Copyright Wiley-VCH Verlag GmbH & Co. KGaA; and from Varma, G., *et al.*: Magnetisation transfer from inhomogeneously broadened lines: a potential marker for myelin. *Magn. Reson. Med.* 2015. 73. 614–622. Copyright Wiley-VCH Verlag GmbH & Co. KGaA.)

Together with the *ihMT* signal and its corresponding *ihMTR*, within the same acquisition scheme additional metrics can be derived, such as the traditional MTR, the *MT* asymmetry and asymmetry ratio, and the analogous *ihMT* asymmetry and *ihMT* asymmetry ratio (Figure 10.9).

Following the definition given in Equation 10.54, all the metrics derived from the *ihMT* signal should be divided by a factor of two when performing quantitative comparisons with their conventional *MT*-based counterparts.

10.13.3 Phantom selectivity (hair conditioner, prolipid)

Hair conditioner and Prolipid-161-based gels are excellent materials for *ihMT* phantoms, due to the presence of fatty alcohols in a lamellar structure that approximate well the *MT* properties of CNS tissues (Swanson *et al.*, 2012; Varma *et al.*, 2015a). These phantoms show a marked *ihMT* effect, while traditional *MT* phantoms (i.e. heat-denatured proteins) are silent to *ihMT* effect despite giving rise to a large *MT* signal (Varma *et al.*, 2015a).

10.13.4 *ihMT* acquisition

ihMT contrast is usually generated by means of long trains (500 ms – 1 s) of pulsed off-resonance saturation, following the scheme outlined above.

Initial investigations of sequence parameters effect have shown a maximum of the *ihMT* signal at frequency offsets around 7 kHz (Varma *et al.*, 2015a), and most studies to date have used this value as a reference. However, it has been hypothesised that the actual optimal offset frequency may depend on the orientation of WM fibres with respect to B_0, which can influence the averaging of the residual dipolar coupling believed to underpin the inhomogeneous *MT* effect.

Similarly to conventional MT, adequate RF deposition is essential for contrast generation. The pulse repetition time (the time between positive to negative frequency switching) is also believed to be a key parameter in the generation of the contrast, as it may control the mixing effect between inhomogeneous broadened lines.

Imaging is performed with single shot readout, mostly in a 2D fashion. A 3D version of the protocol, based on a GRE sequence, has been also implemented exploiting pulse amplitude modulation to simultaneously saturate both sides of the spectrum (Varma *et al.*, 2013). However, development of large coverage, time-efficient acquisition sequences, as well as optimisation of the pulsed saturation parameters, represent active and still relatively unexplored areas of investigation and research.

10.13.5 Applications of *ihMT*

ihMT imaging has been successfully applied to *in vivo* human brain and spinal cord, *ex vivo* animal tissues and at different field strengths, showing consistency in selectively highlighting tissue with elevated myelin content.

An average ratio of 2.1:1 between WM and GM *ihMTR* has been reported, consistent among different brain regions, outscoring contrast achievable with conventional MTR (Varma *et al.*, 2015a).

ihMT seems also to be reproducible among vendors and different scanner platforms and shows little, if any, dependency on the field strength (Girard *et al.*, 2014).

10.13.6 Confounding factors

A potential source of errors in the *ihMT* signal comes from the known *MT* asymmetry, due to a slight shift in the centre of the Z-spectrum and different contributions to the total saturation on both sides of the spectrum. However, *MT* asymmetry (MTA) has been shown to be unrelated to the origin of the *ihMT* contrast, as a slight correction of the carrier frequency in the pulse off-resonance experiment (+150 Hz) minimises the MTA while keeping the *ihMT* signal almost unvaried (Varma *et al.*, 2015a). Moreover, MTA shows very little contrast between WM and GM, compared to the apparent tremendous selectivity of *ihMT* to WM myelinated structures. Finally, *ihMT* and MTA are characterised by very different spectra as a function of the offset frequency: the first peaks around 6–7 kHz, while the latter decreases with frequency offset.

Similarly, potential contributions related to CEST effects are not expected to play a role in the *ihMT* signal given the relatively high offset frequency used in the acquisition of *ihMT*, compared to those usually selected for CEST studies.

Despite its relative recent discovery and yet incomplete characterisation, *ihMT* appears to be a promising mechanism to generate MR contrast specific to myelin in the CNS. Development of accurate quantitative methodologies, validation through histology studies, and comparison with existing myelin-sensitive techniques, however, are needed in order to demonstrate *ihMT* specificity to myelin and promote the development of myelin biomarkers.

10.14 Conclusions

The specificity of *MT* imaging to the macromolecular component of biological tissues makes it an attractive modality for imaging a variety of diseases in different organs of the body. The presence of myelin in the CNS makes *MT* imaging particularly suitable for investigating demyelinating disease in the brain and spinal cord. While MTR has been widely applied in clinical settings, the use of potentially more specific approaches to myelin, such as qMT, is still limited. The main factors limiting the applicability of qMT methods in clinical scenarios are the long scan times and the lack of easy-to-implement analysis packages. Optimisation of *MT* sequences and development of reduced quantitative models are therefore warranted to foster the translation of quantitative *MT* methods. New approaches, such as MT_{sat} or *ihMT*, could also provide valuable alternative ways to qMT to promote the use of *MT* in clinics. *MT* methods could also benefit from higher field strength (e.g. 7T), although some of the implementations, such as those based on off-resonance saturation, present technical challenges related to the increased SAR deposition.

References

Alsop DC, De Bazelaire C, Garcia M, Duhamel G. Inhomogenous magnetization transfer imaging: A potentially specific marker for myelin. In: 13th Annual Meeting of ISMRM, Miami, FL; 2005, p. 2224.

Barker G, Schreiber W, Gass A, Ranjeva J, Campi A, van Waesberghe J, et al. A standardised method for measuring magnetisation transfer ratio on MR imagers from different manufacturers—The EuroMT sequence. MAGMA 2005; 18(2): 76–80.

Barker GJ, Tofts PS, Gass A. An interleaved sequence for accurate and reproducible clinical measurement of magnetization transfer ratio. Magn Reson Imaging 1996; 14(4): 403–11.

Battiston M, Grussu F, Fairney J, Prados F, Ourselin S, Cercignani M, et al. In vivo quantitative Mgnetizaion Transfer in the cervical spinal cord using reduced Field-of-View imaging: A feasibility study. Proc ISMRM 2016; 2016: 306.

Berry I, Barker GJ, Barkhof F, Campi A, Dousset V, Franconi JM, et al. A multicenter measurement of magnetization transfer ratio in normal white matter. J Magn Reson Imaging 1999; 9(3): 441–6.

Bertini I, Luchinat C, Parigi G, Quacquarini G, Marzola P, Cavagna FM. Off-resonance experiments and contrast agents to improve magnetic resonance imaging. Magn Reson Med 1998; 39(1): 124–31.

Bieri O, Mamisch TC, Trattnig S, Scheffler K. Steady state free precession magnetization transfer imaging. Magn Reson Med 2008; 60(5): 1261–6.

Bieri O, Scheffler K. On the origin of apparent low tissue signals in balanced SSFP. Magn Reson Med 2006; 56(5): 1067–74.

Cabana JF, Gu Y, Boudreau M, Levesque IR, Atchia Y, Sled JG, et al. Quantitative magnetization transfer imaging made easy with qMTLab: Software for data simulation, analysis, and visualization. Conc Magn Reson A 2015; 44(5): 263–77.

Cercignani M, Alexander DC. Optimal acquisition schemes for in vivo quantitative magnetization transfer MRI. Magn Reson Med 2006; 56(4): 803–10.

Cercignani M, Barker GJ. A comparison between equations describing in vivo MT: The effects of noise and sequence parameters. J Magn Reson 2008; 191(2): 171–83.

Cercignani M, Symms MR, Schmierer K, Boulby PA, Tozer DJ, Ron M, et al. Three-dimensional quantitative magnetisation transfer imaging of the human brain. Neuroimage 2005; 27(2): 436–41.

Dixon WT, Engels H, Castillo M, Sardashti M. Incidental magnetization transfer contrast in standard multislice imaging. Magn Reson Imaging 1990; 8(4): 417–22.

Dortch RD, Moore J, Li K, Jankiewicz M, Gochberg DF, Hirtle JA, et al. Quantitative magnetization transfer imaging of human brain at 7T. NeuroImage 2013; 64: 640–9.

Dousset V, Grossman RI, Ramer KN, Schnall MD, Young LH, Gonzalez-Scarano F, et al. Experimental allergic encephalomyelitis and multiple sclerosis: Lesion characterization with magnetization transfer imaging. Radiology 1992; 182(2): 483–91.

Edelman R, Ahn S, Chien D, Li W, Goldmann A, Mantello M, et al. Improved time-of-flight MR angiography of the brain with magnetization transfer contrast. Radiology 1992; 184(2): 395–9.

Edzes HT, Samulski ET. Cross relaxation and spin diffusion in the proton NMR of hydrated collagen. 1977; 265(5594): 521–3.

Edzes HT, Samulski ET. The measurement of cross-relaxation effects in the proton NMR spin-lattice relaxation of water in biological systems: Hydrated collagen and muscle. J Magn Reson (1969) 1978; 31(2): 207–29.

Finelli DA, Hurst GC, Gullapali RP, Bellon EM. Improved contrast of enhancing brain lesions on postgadolinium, T1-weighted spin-echo images with use of magnetization transfer. Radiology 1994; 190(2): 553–9.

Finelli DA, Reed DR. Flip angle dependence of experimentally determined T1sat and apparent magnetization transfer rate constants. J Magn Reson Imaging 1998; 8(3): 548–53.

Forsén S, Hoffman RA. Study of moderately rapid chemical exchange reactions by means of nuclear magnetic double resonance. J Chem Phys 1963; 39(11): 2892–901.

Garcia M, Gloor M, Radue E-W, Stippich C, Wetzel S, Scheffler K, et al. Fast high-resolution brain imaging with balanced SSFP: Interpretation of quantitative magnetization transfer towards simple MTR. Neuroimage 2012; 59(1): 202–11.

Gelderen P, Jiang X, Duyn JH. Rapid measurement of brain macromolecular proton fraction with transient saturation transfer MRI. Magn Reson Med 2017; 77(6): 2174–85.

Girard OM, Callot V, Prevost VH, Robert B, Taso M, Ribeiro G, et al. Magnetization transfer from inhomogeneously broadened lines (ihMT): Improved imaging strategy for spinal cord applications. Magn Reson Med 2017; 77(2): 581–91.

Girard OM, Prevost VH, Varma G, Alsop DC, Duhamel G. Magnetization transfer from inhomogeneously broadened lines (ihMT): Field strength dependency. In: 22nd Annual Meeting of the ISMRM 2014, Milan; 2014, p. 4236.

Girard OM, Prevost VH, Varma G, Cozzone PJ, Alsop DC, Duhamel G. Magnetization transfer from inhomogeneously broadened lines (ihMT): Experimental optimization of saturation parameters for human brain imaging at 1.5 Tesla. Magn Reson Med 2015; 73(6): 2111–21.

Giulietti G, Bozzali M, Figura V, Spanò B, Perri R, Marra C, et al. Quantitative magnetization transfer provides information complementary to grey matter atrophy in Alzheimer's disease brains. Neuroimage 2012; 59(2): 1114–22.

Gloor M, Scheffler K, Bieri O. Quantitative magnetization transfer imaging using balanced SSFP. Magn Reson Med 2008; 60(3): 691–700.

Gloor M, Scheffler K, Bieri O. Nonbalanced SSFP-based quantitative magnetization transfer imaging. Magn Reson Med 2010; 64(1): 149–56.

Gochberg DF, Gore JC. Quantitative imaging of magnetization transfer using an inversion recovery sequence. Magn Reson Med 2003; 49(3): 501–5.

Gochberg DF, Gore JC. Quantitative magnetization transfer imaging via selective inversion recovery with short repetition times. Magn Reson Med 2007; 57(2): 437–41.

Gochberg DF, Kennan RP, Gore JC. Quantitative studies of magnetization transfer by selective excitation and T1 recovery. Magn Reson Med 1997; 38(2): 224–31.

Gochberg DF, Kennan RP, Robson MD, Gore JC. Quantitative imaging of magnetization transfer using multiple selective pulses. Magn Reson Med 1999; 41(5): 1065–72.

Graham S, Henkelman RM. Understanding pulsed magnetization transfer. J Magn Reson Imaging 1997; 7(5): 903–12.

Harrison NA, Cooper E, Dowell NG, Keramida G, Voon V, Critchley HD, et al. Quantitative magnetization transfer imaging as a biomarker for effects of systemic inflammation on the brain. Biol Psychiatry 2015; 78(1): 49–57.

Helms G, Dathe H, Dechent P. Quantitative FLASH MRI at 3T using a rational approximation of the Ernst equation. Magn Reson Med 2008a; 59(3): 667–72.

Helms G, Dathe H, Kallenberg K, Dechent P. High-resolution maps of magnetization transfer with inherent correction for RF inhomogeneity and T1 relaxation obtained from 3D FLASH MRI. Magn Reson Med 2008b; 60(6): 1396–407.

Helms G, Draganski B, Frackowiak R, Ashburner J, Weiskopf N. Improved segmentation of deep brain grey matter structures using magnetization transfer (MT) parameter maps. Neuroimage 2009; 47(1): 194–8.

Helms G, Hagberg GE. Quantification of magnetization transfer by sampling the transient signal using MT-prepared single-shot EPI. Concepts Magn Reson A 2003; 19(2): 149–52.

Henkelman R, Stanisz G, Graham S. Magnetization transfer in MRI: A review. NMR Biomed 2001; 14(2): 57–64.

Henkelman RM, Huang X, Xiang QS, Stanisz G, Swanson SD, Bronskill MJ. Quantitative interpretation of magnetization transfer. Magn Reson Med 1993; 29(6): 759–66.

Hu BS, Conolly SM, Wright GA, Nishimura DG, Macovski A. Pulsed saturation transfer contrast. Magn Reson Med 1992; 26(2): 231–40.

Hua J, Hurst GC. Analysis of on-and off-resonance magnetization transfer techniques. J Magn Reson Imaging 1995; 5(1): 113–20.

Hua J, Jones CK, Blakeley J, Smith SA, van Zijl P, Zhou J. Quantitative description of the asymmetry in magnetization transfer effects around the water resonance in the human brain. Magn Reson Med 2007; 58(4): 786–93.

Huot P, Dousset V, Hatier F, Degreze P, Carlier P, Caille J. Improvement of post-gadolinium contrast with magnetization transfer. Eur Radiol 1997; 7: 174.

Huster D, Yao X, Hong M. Membrane protein topology probed by 1H spin diffusion from lipids using solid-state NMR spectroscopy. J Am Chem Soc 2002; 124(5): 874–83.

Laule C, Vavasour IM, Kolind SH, Li DK, Traboulsee TL, Moore GW, et al. Magnetic resonance imaging of myelin. Neurotherapeutics 2007; 4(3): 460–84.

Levesque IR, Pike GB. Characterizing healthy and diseased white matter using quantitative magnetization transfer and multicomponent T2 relaxometry: A unified view via a four-pool model. Magn Reson Med 2009; 62(6): 1487–96.

Levesque IR, Sled JG, Narayanan S, Giacomini PS, Ribeiro LT, Arnold DL, et al. Reproducibility of quantitative magnetization-transfer imaging parameters from repeated measurements. Magn Reson Med 2010; 64(2): 391–400.

Levesque IR, Sled JG, Pike GB. Iterative optimization method for design of quantitative magnetization transfer imaging experiments. Magn Reson Med 2011; 66(3): 635–43.

Li JG, Graham SJ, Henkelman RM. A flexible magnetization transfer line shape derived from tissue experimental data. Magn Reson Med 1997; 37(6): 866–71.

Li K, Zu Z, Xu J, Janve VA, Gore JC, Does MD, et al. Optimized inversion recovery sequences for quantitative T1 and magnetization transfer imaging. Magn Reson Med 2010; 64(2): 491–500.

Manning AP, Chang KL, MacKay A, Michal CA. ihMT: Is it misnamed? A simple theoretical description of "inhomogeneous. In: MT 24th Annual Meeting of the ISMRM, Singapore; 2016, p. 305.

McConnell HM. Reaction rates by nuclear magnetic resonance. J Chem Phys 1958; 28(3): 430–1.

Morrison C, Mark HR. A model for magnetization transfer in tissues. Magn Reson Med 1995; 33(4): 475–82.

Morrison C, Stanisz G, Henkelman RM. Modeling magnetization transfer for biological-like systems using a semi-solid pool with a super-Lorentzian lineshape and dipolar reservoir. J Magn Reson Ser B 1995; 108(2): 103–13.

Mossahebi P, Alexander AL, Field AS, Samsonov AA. Removal of cerebrospinal fluid partial volume effects in quantitative magnetization transfer imaging using a three-pool model with nonexchanging water component. Magn Reson Med 2015; 74(5): 1317–26.

Müller DK, Pampel A, Möller HE. Matrix-algebra-based calculations of the time evolution of the binary spin-bath model for magnetization transfer. J Magn Reson 2013; 230: 88–97.

Opella SJ, Marassi FM. Structure determination of membrane proteins by NMR spectroscopy. Chem Rev 2004; 104(8): 3587–606.

Ou X, Gochberg DF. MT effects and T1 quantification in single-slice spoiled gradient echo imaging. Magn Reson Med 2008; 59(4): 835–45.

Ou X, Sun SW, Liang HF, Song SK, Gochberg DF. Quantitative magnetization transfer measured pool-size ratio reflects optic nerve myelin content in ex vivo mice. Magn Reson Med 2009b; 61(2): 364–71.

Ou X, Sun SW, Liang HF, Song SK, Gochberg DF. The MT pool size ratio and the DTI radial diffusivity may reflect the myelination in shiverer and control mice. NMR Biomed 2009a; 22(5): 480–7.

Pampel A, Müller DK, Anwander A, Marschner H, Möller HE. Orientation dependence of magnetization transfer parameters in human white matter. NeuroImage 2015; 114: 136–46.

Pike GB. Pulsed magnetization transfer contrast in gradient echo imaging: A two-pool analytic description of signal response. Magn Reson Med 1996; 36(1): 95–103.

Portnoy S, Stanisz GJ. Modeling pulsed magnetization transfer. Magn Reson Med 2007; 58(1): 144–55.

Prevost VH, Girard OM, Varma G, Alsop DC, Duhamel G. Minimizing the effects of magnetization transfer asymmetry on inhomogeneous magnetization transfer (ihMT) at ultra-high magnetic field (11.75 T). MAGMA 2016; 29(4): 699–709.

Quesson B, Thiaudière E, Delalande C, Dousset V, Chateil JF, Canioni P. Magnetization transfer imaging in vivo of the rat brain at 4.7 T: Interpretation using a binary spin-bath model with a superlorentzian lineshape. Magn Reson Med 1997; 38(6): 974–80.

Ramani A, Dalton C, Miller DH, Tofts PS, Barker GJ. Precise estimate of fundamental in-vivo MT parameters in human brain in clinically feasible times. Magn Reson Imaging 2002; 20(10): 721–31.

Scheffler K, Lehnhardt S. Principles and applications of balanced SSFP techniques. Eur Radiol 2003; 13(11): 2409–18.

Schmierer K, Tozer DJ, Scaravilli F, Altmann DR, Barker GJ, Tofts PS, et al. Quantitative magnetization transfer imaging in postmortem multiple sclerosis brain. J Magn Reson Imaging 2007; 26(1): 41–51.

Schneider E, Prost RW, Glover GH. Pulsed magnetization transfer versus continuous wave irradiation for tissue contrast enhancement. J Magn Reson Imaging 1993; 3(2): 417–23.

Scholz J, Klein MC, Behrens TE, Johansen-Berg H. Training induces changes in white-matter architecture. Nat Neurosci 2009; 12(11): 1370–1.

Sled JG, Levesque I, Santos A, Francis S, Narayanan S, Brass SD, et al. Regional variations in normal brain shown by quantitative magnetization transfer imaging. Magn Reson Med 2004; 51(2): 299–303.

Sled JG, Pike GB. Quantitative imaging of magnetization transfer exchange and relaxation properties in vivo using MRI. Magn Reson Med 2001; 46(5): 923–31.

Smith AK, Dortch RD, Dethrage LM, Smith SA. Rapid, high-resolution quantitative magnetization transfer MRI of the human spinal cord. NeuroImage 2014; 95: 106–16.

Sormani MP, Iannucci G, Rocca MA, Mastronardo G, Cercignani M, Minicucci L, et al. Reproducibilty of magnetization transfer ratio Histogram–Derived measures of the brainin healthy volunteers. Am J Neuroradiol 2000; 21(1): 133–6.

Stanisz G, Kecojevic A, Bronskill M, Henkelman R. Characterizing white matter with magnetization transfer and T2. Magn Reson Med 1999; 42(6): 1128–36.

Stanisz GJ, Odrobina EE, Pun J, Escaravage M, Graham SJ, Bronskill MJ, et al. T1, T2 relaxation and magnetization transfer in tissue at 3T. Magn Reson Med 2005; 54(3): 507–12.

Swanson S, Malyarenko D, Schmiedlin-Ren P, Adler J, Helvie K, Reingold L, et al. Lamellar liquid crystal phantoms for MT-calibration and quality control in clinical studies. In: Proceedings of the 20th Annual Meeting of ISMRM, Melbourne, Australia; 2012, p. 1378.

Tofts PS, Steens SC, Cercignani M, Admiraal-Behloul F, Hofman PA, van Osch MJ, Teeuwisse WM, Tozer DJ, van Waesberghe JH, Yeung R, Barker GJ, van Buchem MA et al. Sources of variation in multi-centre brain MTR histogram studies: Body-coil transmission eliminates inter-centre differences. MAGMA 2006; 19(4): 209–22.

Tyler DJ, Gowland PA. Rapid quantitation of magnetization transfer using pulsed off-resonance irradiation and echo planar imaging. Magn Reson Med 2005; 53(1): 103–9.

Varma G, Duhamel G, de Bazelaire C, Alsop DC. Magnetization transfer from inhomogeneously broadened lines: A potential marker for myelin. Magn Reson Med 2015a; 73(2): 614–22.

Varma G, Girard O, Prevost V, Grant A, Duhamel G, Alsop D. Interpretation of magnetization transfer from inhomogeneously broadened lines (ihMT) in tissues as a dipolar order effect within motion restricted molecules. J Magn Reson 2015b; 260: 67–76.

Varma G, Schlaug G, Alsop D. 3D acquisition of the inhomogeneous magnetization transfer effect for greater white matter contrast. In: Proceedings of the 21st Annual Meeting of ISMRM, Salt Lake City, UT; 2013.

Wang S, Young K. White matter plasticity in adulthood. Neuroscience 2014; 276: 148–60.

Whittall KP, Mackay AL, Graeb DA, Nugent RA, Li DK, Paty DW. In vivo measurement of T2 distributions and water contents in normal human brain. Magn Reson Med 1997; 37(1): 34–43.

Wolff SD, Balaban RS. Magnetization transfer contrast (MTC) and tissue water proton relaxation in vivo. Magn Reson Med 1989; 10(1): 135–44.

Xu J, Chan KW, Xu X, Yadav N, Liu G, van Zijl P. On-resonance variable delay multipulse scheme for imaging of fast-exchanging protons and semisolid macromolecules. Magn Reson Med 2017; 77(2): 730–9.

Yarnykh VL. Pulsed Z-spectroscopic imaging of cross-relaxation parameters in tissues for human MRI: Theory and clinical applications. Magn Reson Med 2002; 47(5): 929–39.

Yarnykh VL. Actual flip-angle imaging in the pulsed steady state: A method for rapid three-dimensional mapping of the transmitted radiofrequency field. Magn Reson Med 2007; 57(1): 192–200.

Yarnykh VL. Fast macromolecular proton fraction mapping from a single off-resonance magnetization transfer measurement. Magn Reson Med 2012; 68(1): 166–78.

Yarnykh VL. Time-efficient, high-resolution, whole brain three-dimensional macromolecular proton fraction mapping. Magn Reson Med 2016; 75(5): 2100–6.

Yarnykh VL, Yuan C. Cross-relaxation imaging reveals detailed anatomy of white matter fiber tracts in the human brain. Neuroimage 2004; 23(1): 409–24.

Yeung HN, Aisen AM. Magnetization transfer contrast with periodic pulsed saturation. Radiology 1992; 183(1): 209–14.

11

CEST: Chemical Exchange Saturation Transfer[1]

Mina Kim
University College London

Moritz Zaiss
Max-Planck-Institute of biological cybernetics, Tübingen

Stefanie Thust
University College London

Xavier Golay
University College London

Contents

11.1 Principle of chemical exchange saturation transfer imaging

11.1.1 Chemical exchange saturation transfer mechanism

Chemical exchange saturation transfer (CEST) has recently emerged as an alternative contrast mechanism for MRI (Aime *et al.*, 2002; Goffeney *et al.*, 2001; Ward *et al.*, 2000; Zhang *et al.*, 2003). In conventional CEST-MRI, the proton magnetisation of molecules is saturated by a frequency selective radiofrequency (RF) irradiation pulse. The saturation is transferred to the water proton pool via labile protons of the solute, such as amide (NH) and hydroxyl (OH) (Ling *et al.*, 2008; van Zijl *et al.*, 2003; Zhou *et al.*, 2003a, 2003b), as first demonstrated by Wolff and Balaban (1990). Through chemical exchange or dipolar interactions, saturated labile protons are repeatedly replaced by non-saturated water protons, leading to an accumulation of saturated protons in the water pool (see Figure 11.1). After a few seconds of RF irradiation, this gives rise to an observable signal reduction in the water pool. Highest sensitivity to proton transfer is achieved if the solute has a high concentration of exchangeable protons, and the exchange rate from solute to water k_{sw} is relatively large while allowing time for sufficient saturation. The relationship between irradiation power and exchange rate affecting saturation efficiency is discussed later in this chapter.

[1] Edited by Mara Cercignani.

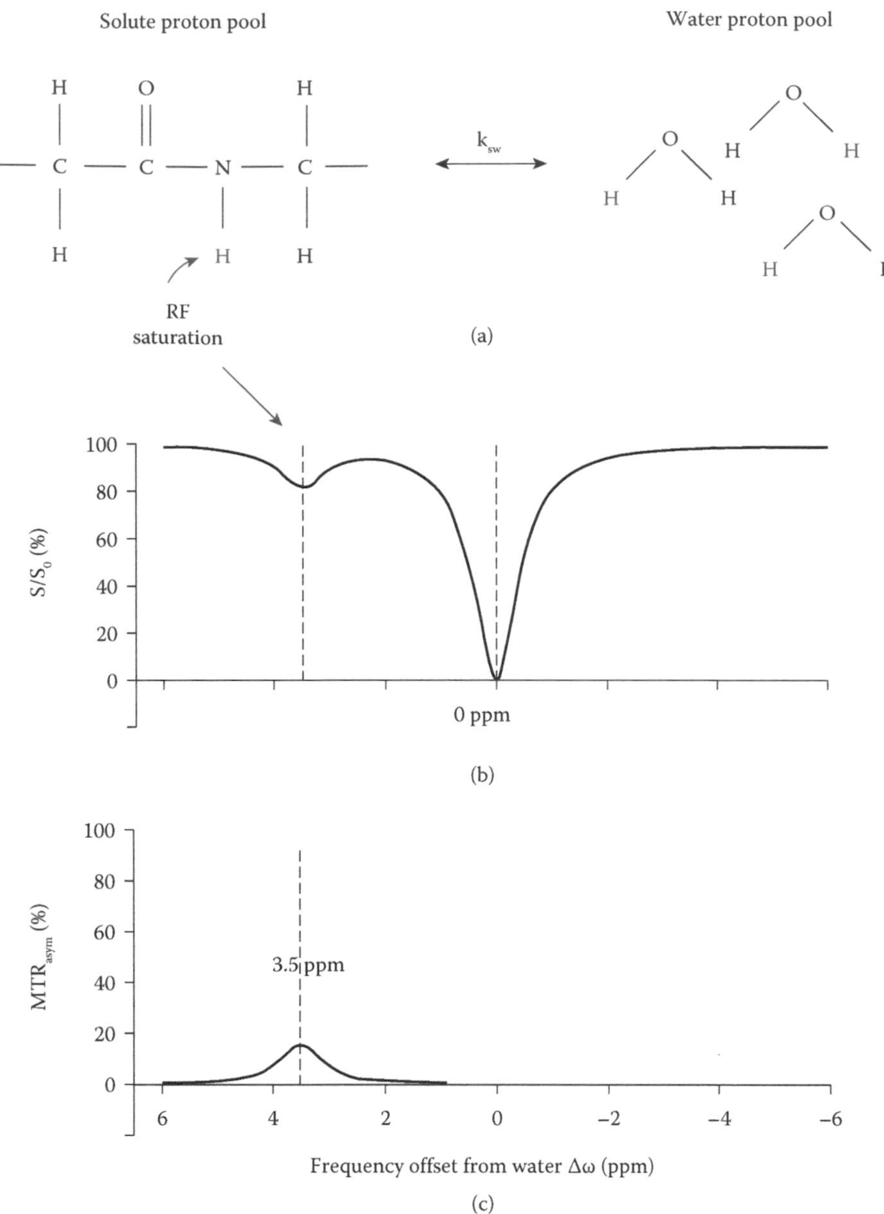

FIGURE 11.1 CEST principles and measurement approach. **(a)** Magnetisation of solute protons is saturated at their resonance frequency and the saturated protons (red) are transferred to water at exchange rate k_{sw}. Meanwhile, non-saturated protons (black) in the water pool return to the solute pool. After a sufficient time, this effect can be observed as reduction of the water signal. **(b)** In order to measure this effect, saturation pulses are applied as a function of irradiation frequency, generating a so-called *Z*-spectrum (or chemical exchange saturation transfer spectrum). When irradiating the water protons at 4.75 ppm (ω_0), the signal disappears due to direct water saturation (DS) and the water frequency is assigned to 0 ppm in *Z*-spectra. Conventionally, magnetisation transfer ratio asymmetry (MTR_{asym}) analysis of the *Z*-spectrum has been used to remove the effect of direct water saturation. S_{sat} and S_0 are the saturated and non-saturated intensities.

11.1.2 CEST versus magnetisation transfer

CEST-MRI experiments are performed in a similar manner to magnetisation transfer (MT) MRI (see Chapter 10) but with important differences (see Figure 11.2). While MT contrast is based on magnetisation exchange between water protons and protons in solid or semi-solid environment, CEST contrast originates from chemical exchange between labile protons and water protons. In MT-MRI, the broad spectral line shape of the semi-solid components can be saturated by a variety of pulses over a large frequency range extending up to ±100 kHz. In contrast, selective CEST peaks, e.g. resonances of interest such as NH and OH groups, only appear if much narrower bandwidth of the RF irradiation is used. In general, CEST is observed in a small chemical shift range of less than 10 ppm from water, but may also be registered outside this range when using an exogenous CEST agent.

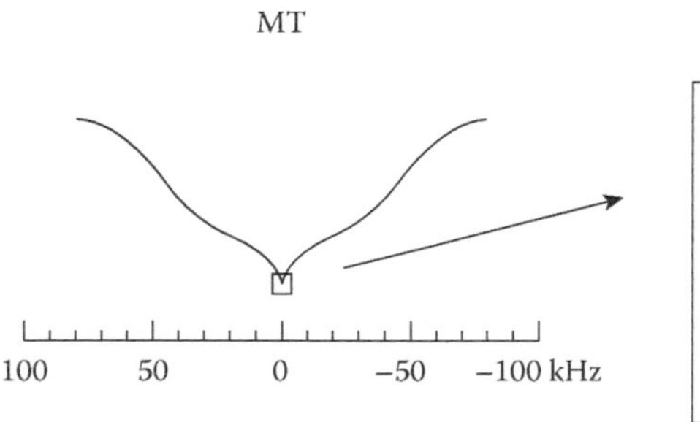

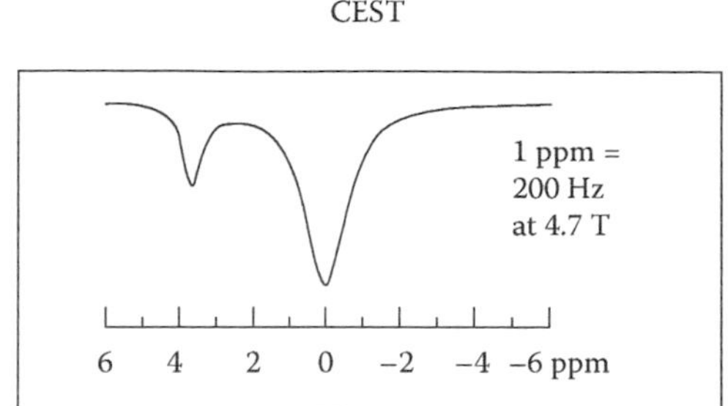

- Range: ±100 kHz
- Approximately symmetric
- Solid-like macromolecules
- Chemical exchange, etc.

- Range: ± 6–7 ppm
- Individual frequency (Asymmetric)
- Mobile molecules
- Chemical exchange

FIGURE 11.2 Comparison between magnetisation transfer (MT) and chemical exchange saturation transfer (CEST). In MT-MRI, the broad spectral line shape of the semi-solid components can be saturated by a variety of pulses over a large frequency range of about ±100 kHz. In contrast, selective CEST peaks, e.g. resonances of interest such as NH and OH groups, only appear if a much narrower bandwidth of the RF irradiation is used. (Reproduced from the *Progr. Nuclear Magn. Reson. Spect.*, 48, Zhou, J., and van Zijl, P.C., Chemical exchange saturation transfer imaging and spectroscopy, 109–136, Copyright 2006, with permission from Elsevier.)

11.2 Classification of CEST contrast

Endogenous and exogenous CEST agents are usually classified in two main groups: diamagnetic (diaCEST) and paramagnetic (paraCEST) CEST agents. The members of each class can be further divided into subgroups depending on other criteria such as molecular size, endogenous occurrence and type of molecular construct. In particular, there are a few groups involving specific labile protons, molecules or mechanisms, and therefore those groups are categorised into subgroups, such as amide proton transfer CEST (APT-CEST) for peptides/proteins (Zhou *et al.*, 2003a, 2003b), gagCEST for glycosaminoglycans (Ling *et al.*, 2008), glucoCEST for glucose (Chan *et al.*, 2012; Walker-Samuel *et al.*, 2013), gluCEST for glutamate (Cai *et al.*, 2012), glycoCEST for glycogen (van Zijl *et al.*, 2007), lipoCEST for liposomes (Terreno *et al.*, 2007) and miCEST for myo-inositol (Haris *et al.*, 2011). However, it should be noted that overlap exists between the classes of CEST agents, due in part to the overlap in resonance frequencies between certain molecules. In addition, for exogenous CEST agents, both macromolecular and liposome applications can be used as e.g. diaCEST and paraCEST agents.

11.2.1 Diamagnetic CEST agents

CEST using diamagnetic molecules with low-molecular-weight was first proposed by Balaban and co-workers as exogenous CEST agents (Ward *et al.*, 2000). For diaCEST compounds, the range is generally 10 ppm from water (hydroxyl, amide, amine and imino groups); however, it can be further extended up to 19 ppm for hydrogen-bonded groups such as in enzymes and salicylic acids. The CEST experiments may be restricted for two reasons. First, application of a presaturation pulse within 10 ppm away from the bulk water signal results in a certain degree of water saturation, limiting the amount of power used for presaturation without severe reduction in signal-to-noise ratio (SNR). Secondly, detectable exchange rates are limited by the labelling efficiency, which is dependent on the feasible RF power B_{1max}: the CEST signal vanishes if the exchange rate is much larger than $\gamma \cdot B_1$. Additionally, selective CEST peak detection is restricted by the line width of the CEST peak and the chemical shift: if $k < \Delta\omega$, selective peaks can be detected, and if $k > \Delta\omega$ (intermediate and fast regime), CEST becomes similar to $T_{1\rho}$ or T_{2ex} experiments. Selectivity can be improved by a larger chemical shift separation between the two exchanging pools, which can be realised by higher magnetic field strength but also by using an agent with larger chemical shift from water (see below paraCEST agents).

11.2.2 Paramagnetic CEST agents

Lanthanide(III) complexes have been suggested as paraCEST agents based on two different kinds of mobile protons: (a) protons of water molecules coordinated to the Ln(III) ion and (b) mobile protons belonging to the ligand structure. A representative example of a paraCEST agent, a tricationic Eu(III) macrocyclic complex, was first reported by Sherry's group (Zhang *et al.*, 2001). Its metal-bound water protons are in slow exchange, which is reported as k_{sw} of about 2600 s^{-1} and $\Delta\omega$ = 50 ppm. To date, much faster exchanging paraCEST agents have been developed as reviewed in the literature (Woods *et al.*, 2006).

11.3 CEST acquisition protocols

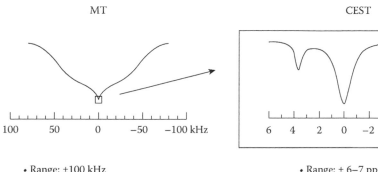

In CEST imaging, a prepulse sequence is applied to indirectly measure the presence of the small exchangeable proton pool via the large solvent proton pool by saturating the small pool and allowing magnetisation to be transferred between the pools. The examples of CEST imaging parameters used for human brain are shown in Table 11.1.

11.3.1 Continuous wave irradiation

In the classical CEST imaging, a low-bandwidth continuous-wave (CW, Figure 11.3a) RF pulse is used prior to fast image readout. CEST data are acquired by sweeping this frequency-selective saturation pulse through a range of offsets, thereby collecting a full *z*-spectrum and, differently from MT, on both sides of the water frequency. For a clinically applicable exam, alternative methods to full *Z*-spectrum acquisition have been proposed in order to reduce imaging time. One of the methods is to obtain high-resolution images following saturation at selected offsets around the resonance of interest in combination with a reference *z*-spectrum (Mougin *et al.*, 2010; Zhou *et al.*, 2008). Another method is to decrease the acquisition time of CEST imaging by reducing *k*-space sampling, so-called keyhole CEST, which has been applied for dynamic contrast imaging (Varma *et al.*, 2012).

11.3.2 Pulsed irradiation

On clinical scanners, CW RF irradiation may not be applicable due to limited maximal RF pulse duration and hardware limits on the duty cycle due to the solid-state amplifiers used. A train of pulsed RF irradiations (Figure 11.3b), therefore, has been suggested to resolve those issues, albeit with a lower saturation efficiency (Sun *et al.*, 2011; Zhou *et al.*, 2003a). Some studies have shown that pulsed-CEST MRI might reach comparable contrast as CW-CEST MRI for imaging of diamagnetic CEST agents undergoing slow/intermediate chemical exchange (exchange rate <50 s^{-1}) (Aime *et al.*, 2005; Dixon *et al.*, 2010a). Furthermore, it has been reported that block RF pulses are less adequate than Gaussian-shaped RF pulses. Pulsed-CEST MRI can be suitable for clinical translation of the so-called pH-weighted APT MRI, where the endogenous amide proton exchange rate has been estimated to be about 30 s^{-1} (Sun *et al.*, 2011).

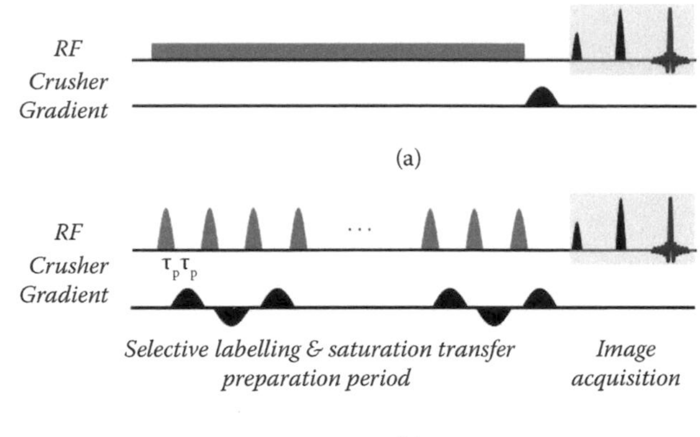

FIGURE 11.3 The CEST imaging sequence is composed of a frequency-selective irradiation and saturation module followed by fast image acquisition sequence such as echo-planar imaging (EPI). Two RF irradiation schemes are shown: (a) conventional long continuous-wave (CW) RF irradiation versus (b) repetitive RF irradiation pulse train. Bipolar crusher gradient is proposed to suppress free induction decay (FID) and minimise unwanted spin echoes. (Sun, P.Z., *et al.*, Investigation of optimising and translating pH-sensitive pulsed-chemical exchange saturation transfer (CEST) imaging to a 3T clinical scanner. *Magn. Reson. Med.* 2008. 60. 834–841. Copyright Wiley-VCH Verlag GmbH & Co. KGaA. Reproduced with permission.)

TABLE 11.1 CEST Imaging parameters used for human brain

B_0 (T)	Prepulse				Acquisition sequence	Scan time (matrix)	Reference
	Shape	Power (uT)	Duration (ms)	Offset frequency			
7.0	Pulsed; 150 Gaussian pulses	0.6, 0.9	3750	−4 to 4 ppm with various intervals[a]	2D centric-reordered gradient echo	4.07 min for each B_1 amplitude (128 × 112)	Zaiss *et al.* (2015a, 2016)
7.0	Pulsed; 16 sinc-Gaussian pulses	3.4	20	−20 to 20 ppm with various intervals[b]	Single-shot turbo gradient echo	14.5 min (77 × 77)	Xu *et al.* (2016)
7.0	CW; block pulse	3.5	1000	−7 to 7 ppm with 0.5 ppm intervals	Single-shot 3D turbo field echo	13.55 min for two scans (101 × 101)	Dula *et al.* (2011)
3.0	CW; half blackman window-shaped ramp-up B_1-down	Various[c]	Various[c]	32 offsets, pairs up to ±5 kHz	Single shot spin-echo EPI	2 min per *z*-spectrum (96 × 96)	Scheidegger *et al.* (2014)
3.0	Pulsed; 50 Gaussian pulses	0.55[d]	2000	−4.5 to 4.5 ppm with 0.3 ppm intervals	Spin-echo EPI	2 min (64 × 64)	Tee *et al.* (2014)
3.0	Pulsed; four block pulses	2	200 each	−6 to 6 ppm with 0.5 ppm intervals	3D GRASE, multislice spin-echo	8.40 min[e] (96 × 96)	Zhu *et al.* (2010)

Note: CEST = chemical exchange saturation transfer; CW = continuous wave.
[a] ±4 ppm to ±3 ppm in steps of 0.1 ppm, from ±2.75 ppm to ±2 ppm in steps of 0.25 ppm, ±1.8 ppm to ±1.2 ppm in steps of 0.1 ppm, ±0.5 ppm, ±0.25 ppm.
[b] ±20 ppm in steps of 1 ppm, ±4 ppm in steps of 0.5 ppm.
[c] B_1 = 0.5 μT (NEX = 4) and 1.5 μT (NEX = 1) and saturation time T_{sat} = 200 ms, B_1 = 3 μT (NEX = 1) and 6 μT (NEX = 1) and T_{sat} = 100 ms. These power levels are more optimal for study of rapidly exchanging lines from amine and hydroxyl protons.
[d] Equivalent continuous saturation B_1 value.
[e] For a whole brain.

11.4 CEST data analysis methods

11.4.1 Asymmetry analysis

In conventional CEST experiments, the water signal in each voxel is obtained by acquiring a full *Z*-spectrum (Bryant, 1996), also known as the CEST spectrum (Ward and Balaban, 2000). The *Z*-spectrum displays both exchange-mediated and direct RF saturation effects on the water peak as a function of saturation frequency offset relative to water, which is assigned to be at 0 ppm. Once the centre of the *Z*-spectrum representing the water resonance frequency is determined, the magnitude of the CEST effect can be quantified as MT asymmetry ratio (MTR_{asym}), defined as

$$MTR_{asym}(\Delta\omega) = \frac{S(-\Delta\omega) - S(\Delta\omega)}{S_0} \tag{11.1}$$

where $\Delta\omega$ is the shift difference between the irradiation frequency and the water frequency. S and S_0 are the saturated and non-saturated intensities.

This classical and simple procedure suffers however from many drawbacks, linked to (1) B_0 field inhomogeneities; (2) presence of direct lipid signal; (3) nuclear Overhauser enhancement (NOE) effect' and (4) water relaxation influences.

1. Due to the steep slope of the direct saturation curve, even a small B_0 field difference and a concomitant shift in the *Z*-spectrum may cause a large change in MTR_{asym}. *In vivo*, this effect will lead to the occurrence of artefactual signal spikes in CEST images. A few approaches have been suggested to overcome B_0 inhomogeneity issues (Zhou *et al.*, 2003; Kim *et al.*, 2009; Tagao *et al.*, 2016; Schuenke *et al.*, 2017).
2. If strong lipid signals are apparent in a voxel, the acquired *Z*-spectrum will be a superposition of water and lipids *Z*-spectra. This compromises asymmetry analysis, as lipids resonate on the opposite side of the water peak than the amide protons. The use of proper lipid suppression can help reduce these effects (Dixon *et al.*, 2010a; Mougin *et al.*, 2010; Sun *et al.*, 2005; Zhu *et al.*, 2017, Zhang *et al.*, 2017).
3. NOE-mediated exchange signals, originating from lipids or proteins, will further bias the CEST signal, as these effects are present upfield from water and in the same range as the standard metabolites of interest.
4. As CEST effects are mediated by the water pool, water relaxation and semi-solid MT can affect the CEST signal. This is the so-called spillover effect, which dilutes the actual MTR_{asym} signal; see details in section 11.5.5.

Although CEST observation of solutes and particles in the millimolar to nanomolar range has been shown based on asymmetry spectrum analysis, full-spectral sampling may not be optimal for clinical studies due to limited scan time and SNR. A few reports have suggested to obtain CEST data for only a few number of necessary frequencies around $\pm\Delta\omega$ in multiple scans, while the water frequency is determined from an additional full *Z*-spectrum (Kim *et al.*, 2009; Zhou *et al.*, 2008), yet these are prone to potential validity problems in cases of large B_0 inhomogeneities.

11.4.2 Model-based analysis

In order to quantify the CEST effect, a model-based approach has been proposed that allows separation of the contributions from various metabolites in a more systematic way than the conventional MTR_{asym} method (Jones *et al.*, 2012). In order to quantify the individual parameters making up the CEST signal, such as the metabolite concentration or the exchange rate, an exchange model can be fitted to the measured data in principle (Liu *et al.*, 2013). However, *in vivo*, this approach is complicated by the presence of many metabolites, leading to a model with a large unknown number of highly correlated parameters, made more difficult by the presence of noise. Recently, a probabilistic approach based on a Bayesian analysis has been proposed to resolve those issues, by determining the parameters that best explain the observed data under the assumption of a generative forward model (Chappell *et al.*, 2009, 2013). Aside from these exhaustive model-based analyses, hybrid methodologies, based on multi-Lorentzian fitting of *Z*-spectra, while not permitting full quantification, still allow separate estimation of the different CEST peak amplitudes *in vivo* (Desmond *et al.*, 2014; Jones *et al.*, 2012; Windschuh *et al.*, 2015; Zaiss *et al.*, 2011, 2015a).

11.5 Quantitative CEST

11.5.1 Definition of quantification

In MRI, there are two definitions of quantitative imaging. The first is what we call *relative quantification definition*, which means to have a reproducible number in each voxel for a certain imaging method, for example, 2% MTR_{asym} in CEST MRI or 20% enhancement in T_1-weighted imaging. A relative quantification parameter normally depends on the used measurement technique. It is easier to achieve but weaker than the second definition, which can be defined as *absolute quantification*. The latter definition aims at mapping physically meaningful parameters such as a relaxation time, an absolute concentration, a pH value or an apparent diffusion constant. An absolute quantification parameter, by definition, shall not or only minimally depend on the measurement settings. In the following, we will discuss both kinds of quantification in the field of CEST.

Relative quantification of CEST signal is often defined by MTR_{asym} (see Equation 11.1), which was first described by Zhou *et al.* (2004) for steady-state irradiation and neglecting of spillover effects:

$$MTR_{asym} = \frac{f_s k_{sw} \alpha}{R_{1w} + f_s k_{sw}} \tag{11.2}$$

where R_{1w} is the water relaxation rate, f_s is the relative proton fraction, k_{sw} is the exchange rate and α is the labelling efficiency. Employing $R_{1\rho}$ theory, this first estimation could be improved and also direct saturation could be included (Jin *et al.*, 2011; Zaiss and Bachert, 2013):

$$MTR_{asym} = \cos^2\theta \frac{R'_{ex} R_{1obs}}{R_{1\rho w}(R_{1\rho w} + R'_{ex})} \overset{\text{no spillover}}{=} \frac{f_s k_{sw} \alpha}{R_{1w} + \alpha f_s k_{sw}} \quad (11.3)$$

which contains the parameters important for direct saturation, namely, the observed longitudinal relaxation rate R_{1obs}, the longitudinal relaxation rate in the rotating frame of water $R_{1\rho w}$ and the tilt angle of the effective field $\theta = \tan^{-1}(\omega^1/\Delta\omega)$, but also the exchange-dependent relaxation rate R'_{ex} which is defined on resonance of the CEST pool as

$$R'_{ex} = \sin^2\theta \cdot R_{ex} = f_s k_{sw} \underbrace{\frac{\omega_1^2}{\omega_1^2 + k_{sw}(k_{sw} + R_{2s})}}_{\alpha} \quad (11.4)$$

It should be noted that the $\sin^2\theta$ term is included in R'_{ex}; only then the labelling efficiency α (Equation 11.3) can be derived. R_{ex} is the key for *absolute quantification*: it contains the most important parameters of the CEST pool, which are (1) exchange rate from solute to water k_{sw}; (2) relative proton fraction of the solute pool f_s; (3) transverse relaxation rate R_{2s}. It also contains the saturation power $\omega_1 = \gamma \cdot B_1$, which determines the labelling efficiency α. Trott and Palmer were first to express exchange processes outside of the fast-exchange limit by R_{ex} (Trott and Palmer, 2002); Jin *et al.* and others translated this $R_{1\rho}$-based approach to the field of CEST (Jin *et al.*, 2011, 2012; Wu *et al.*, 2015; Yuan *et al.*, 2012), where the benefits of the approach were shown by several groups (Sun *et al.*, 2016; Zaiss and Bachert, 2013). It is worth noting that Equation 11.4 is a simplification for off-resonant pools (chemical shift > exchange rate); a more complete expression also valid for pools closer to water at lower field strength is given in Equation 11.16. Previous studies have shown that most quantification methods make implicit or explicit use of R_{ex} (Jin *et al.*, 2011, 2012; Sun *et al.*, 2016; Trott and Palmer, 2002; Wu *et al.*, 2015; Yuan *et al.*, 2012; Zaiss and Bachert, 2013). R_{ex} has a physical meaning, being the exchange-induced part of the longitudinal relaxation rate in the rotating frame, and depends mostly on CEST pool properties, such as exchange rate and relative concentration.

11.5.2 The CEST model: Bloch–McConnell equations

Going from the CEST phenomenon to a quantitative understanding of the underlying exchange processes we first need to understand the principle dependencies. These are given by the dynamic of the magnetisation vectors of the water pool $\overrightarrow{M}_w$ and a coupled solute pool $\overrightarrow{M}_s$ described by the Bloch–McConnell differential equations (McConnell, 1958), which can be written as

$$\frac{d}{dt}\overrightarrow{M} = A \cdot \overrightarrow{M} + \overrightarrow{C} \quad (11.5)$$

With the coupled magnetisation vectors of water and solute

$$\overrightarrow{M} = (M_{xw}, M_{yw}, M_{zw}, M_{xs}, M_{ys}, M_{zs})^T \quad (11.6)$$

the Bloch–McConnell matrix A

$$A = \begin{bmatrix} -R_{2w} - k_{ws} & -\Delta\omega_w & 0 & +k_{sw} & 0 & 0 \\ +\Delta\omega_w & -R_{2w} - k_{ws} & +\omega_1 & 0 & +k_{sw} & 0 \\ 0 & -\omega_1 & -R_{1w} - k_{ws} & 0 & 0 & +k_{sw} \\ +k_{ws} & 0 & 0 & -R_{2s} - k_{sw} & -\Delta\omega_s & 0 \\ 0 & +k_{ws} & 0 & +\Delta\omega_s & -R_{2s} - k_{sw} & +\omega_1 \\ 0 & 0 & +k_{ws} & 0 & -\omega_1 & -R_{1s} - k_{sw} \end{bmatrix} \quad (11.7)$$

and the constant vector

$$\overrightarrow{C} = (0, 0, R_{1w}M_{0w}, 0, 0, R_{1s}M_{0s})^T \quad (11.8)$$

The first step for quantification is correct normalisation: if signals are normalised by the signal of fully relaxed magnetisation M_0, a so-called *Z*-value can be defined. It is proportional to the *z*-magnetisation after irradiation relative to the M_0 magnetisation and thus well defined

$$Z(\Delta\omega) = \frac{S(\Delta\omega)}{S_0} = \frac{M_z(\Delta\omega)}{M_0} \quad (11.9)$$

Solving the underlying Bloch–McConnell equations (Equation 11.5) (McConnell, 1958) for two exchanging pools upon RF irradiation yields the following formula for the normalised *z*-magnetisation after irradiation at a certain frequency offset Δω, with a certain RF saturation power B_1 and saturation duration t_p (Jin *et al.*, 2011; Zaiss and Bachert, 2013):

$$Z(\Delta\omega, t_p) = Z^{ss}(\Delta\omega) + (\cos^2\theta \cdot Z_i - Z^{ss}(\Delta\omega))e^{-R_{1\rho}(\Delta\omega)t_p} \quad (11.10)$$

which describes a mono-exponential decay from an initial magnetisation $Z_i = M_i/M_0$ towards a saturated steady-state magnetisation

$$Z^{ss}(\Delta\omega) = \cos^2\theta \frac{R_{1obs}}{R_{1\rho}(\Delta\omega)} \quad (11.11)$$

Thus the whole CEST experiment can be understood quantitatively if the effective off-resonant relaxation rate in the rotating frame $R_{1\rho}$ and the observed longitudinal relaxation rate R_{1obs} is known (Zaiss *et al.*, 2015b). Further, it has been shown that direct saturation effects as well as semi-solid magnetisation transfer effects can be superimposed with CEST effects to yield the full relaxation rate (Zaiss *et al.*, 2015b):

$$R_{1\rho} = R_{1\rho,water} + R_{1\rho,MT} + R'_{ex} \quad (11.12)$$

The rate $R_{1\rho,water}$ (also called R_{eff}) for the case of direct saturation exclusively (Trott and Palmer, 2002, 2004; Zaiss and Bachert, 2013) is determined by the water relaxation, whereas $R_{1\rho,MT}$ contains the MT pool parameters. As shown in Equation 11.4, the term R_{ex} contains all the dependencies on the saturation parameters of the pure CEST effect. Before discussion on the dependencies of R_{ex}, the influence of direct saturation is investigated in the next two sections.

11.5.3 Effect of saturation duration and power

All mentioned effects and their shape in the Z-spectrum strongly depend on the parameters of RF irradiation (Figure 11.4a and b): with increasing RF irradiation time t_p (Figure 11.4a) the CEST effect increases, during what is called here the *transient state*, and reaches a steady state after a time equal to several times T_1. The same is observed for magnetisation transfer and direct saturation effects. Thus T_1 gives us a measure for the required saturation time t_p of the system under investigation: The steady-state signal is reached after long saturation periods ($t_p >> T_1$). With increasing RF amplitude B_1 (Figure 11.4b) the CEST peak also increases in amplitude and width, which is due to more effective labelling of the CEST pool, as described below (Figure 11.5). However, after reaching a maximal CEST effect, direct water saturation and semi-solid MT, which are also scaled by B_1, become dominant and dilute the CEST effect again (Figure 11.4b and c). This effect is called *spillover dilution* and is explained in Section 11.5.4. Therefore, generally, to measure a specific CEST effect from a defined tissue will require proper optimisation of all parameters at hand. Table 11.1 shows the different optimal B_1 values that were reported, depending on the metabolite and tissue of interest and the exchange rate; typical B_1 values for effective saturation in clinical MRI systems are on the order of a few μT.

11.5.4 Direct water saturation, spillover dilution and T_1 scaling

The indirect measurement of small exchangeable pools via the water pool signal allows amplification, but at the same time the CEST signal is modulated by all relaxation and concomitant effects affecting the mediator. The first issue comes from the direct water saturation: when approaching the water resonance frequency, the water is saturated directly by the RF irradiation pulse, and because of this saturation the CEST effects become less apparent (Figure 11.4b and c) – this phenomenon is called *spillover dilution*. It means that less water magnetisation is left for preparation by saturation transfer and can be understood by the water relaxation terms in the denominator of Equation 11.3. Spillover dilution is larger for short T_2 and high B_1 amplitudes. The second issue is the overlay with the strong concomitant semi-solid MT (see Section 11.1.2) that can induce similar dilution effects as the spillover (solid line in Figure 11.4c). The third issue is the T_1 scaling of CEST effects. The longitudinal relaxation time of water T_{1w}

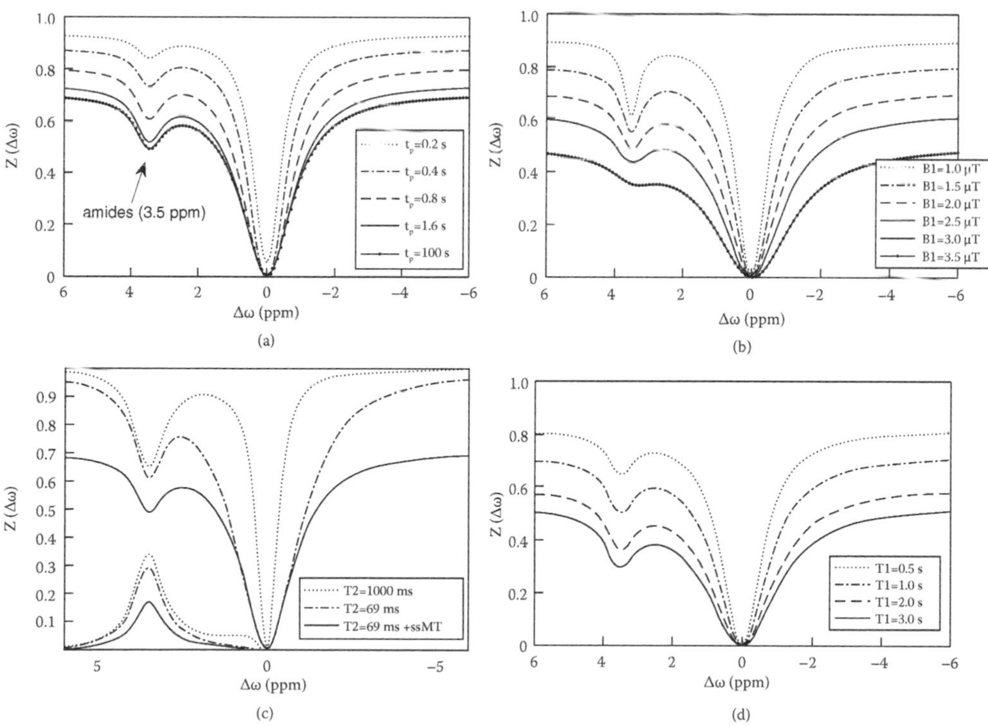

FIGURE 11.4 Simulated Z-spectra of a three-pool system, an amide CEST pool, MT pool and a water pool at B_0 = 7T as a function of irradiation parameter t_p and B_1 (a,b) and relaxation parameters T_2, MT, T_1 (c,d). Simulation parameters: B_1 = 2 μT, t_p = 5 s, R_{1w} = 0.9225 Hz, R_{2w} = 14.49 Hz, f_s: 0.01, k_{sw} = 50 Hz, R_{2s} = 66.66 Hz, δ_s = 3.5 ppm, f_{mt} = 0.1390, k_{mt} = 23 Hz, R_{1mt} = 0.9225 Hz, R_{2mt} = 100,000 Hz; simulation methods available on www.cest-sources.org

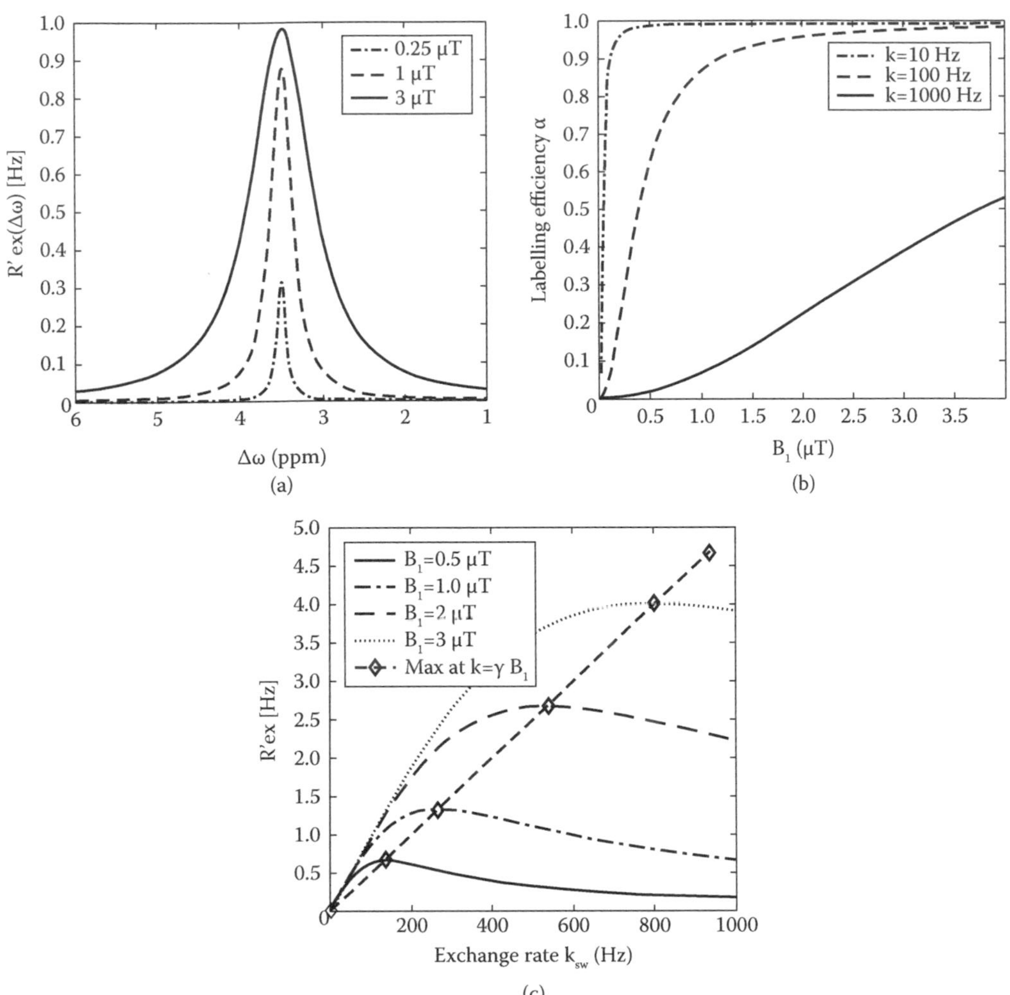

FIGURE 11.5 (a) The exchange dependent relaxation rate $R_{ex}(\Delta\omega)$ is a Lorentzian function with amplitude and line width depending on both k and B_1 (Equations 11.17 and 11.19). (b) Labelling efficiency is a measure of how well a species is labelled: higher exchange rates require higher power. (c) R_{ex} as a function of exchange rate show to peak at $k_{sw} = \gamma B_1$. For constant B_1 CEST effects can decrease despite k increases.

limits the accumulation of saturation in the water pool and by this determines the size of all saturation transfer effects (Figure 11.4d). In steady state, longer T_{1w} leads to larger apparent CEST effects, since $R_{1w} = 1/T_{1w}$ appears in the denominator of Equation 11.3.

11.5.5 Relaxation compensation techniques

Following the theory different evaluation metrics can partially remove these unwanted influences using a reference value. More generally an evaluation metric can be defined using a label scan M_{zlab} and a reference scan M_{zref}. There are several approaches for getting a good reference value, such as baseline estimation or Lorentzian fitting (Jin *et al.*, 2012; Zaiss *et al.*, 2015a), or an extrapolated semisolid magnetisation transfer reference (Heo *et al.* 2016); for the special case of asymmetry $Z_{lab} = Z(+\Delta\omega)$ and $Z_{ref} = Z(-\Delta\omega)$.

With this we can generally define

$$MTR_{lind} = Z_{ref} - Z_{lab} \tag{11.13}$$

The linear difference method effectively removes the baseline signal of direct water saturation. MTR_{asym} is a linear difference method where Z_{ref} is the signal at the opposite frequency. However, CEST effects are indirectly diluted by the background, which is why a linear difference is still affected by water relaxation (see Equation 11.3). This is because the water signal is already partially saturated by the background effects; thus depending on the amount of this background saturation, the CEST saturation is less effective than without background saturation. This can be understood and solved by the inverse difference metric AREX, which, with additional T_1 normalisation, becomes a relaxation-compensated CEST metric (compare Equations 11.11 and 11.12):

$$AREX = R_{1obs} \cdot \left(\frac{1}{Z_{lab}} - \frac{1}{Z_{ref}} \right) \tag{11.14}$$

Applying this to Equation 11.11 explains its name, as AREX yields in steady-state the apparent R_{ex} (see Equation 11.4):

$$AREX \approx R'_{ex}(\Delta\omega) \tag{11.15}$$

This was shown to be able to remove relaxation influences in simulations and *in vitro* (Rerich *et al.*, 2015; Sun *et al.*, 2016; Wu *et al.*, 2015; Zaiss *et al.*, 2015a, 2015b, 2014); *in vivo* Khlebnikov *et al.* reported removal of the linear correlation of CEST effects with T_1 in different tissues (Khlebnikov *et al.*, 2017). However, it is important to note that at lower fields and larger saturation, the simple approximation of Equation 11.3 is not valid anymore and large AREX signals can appear close to water according to Equation 11.16; see also (Heo *et al.*, 2016b). In addition if very small signals are inversely subtracted this is numerically instable; thus interpretation of the inverse difference relies on decent SNR and the correct limit of R_{ex}. In terms of SNR, it has been reported that the inverse metric and linear metric are equal (Jiang *et al.*, 2016). This can be understood by error propagation: for lower and lower *Z*-values, the linear difference yields smaller effects with similar statistical errors; the inverse difference yields larger effects with larger statistical errors. When relaxation changes are small or compensated for, a CEST metric fulfils the requirements of the relative CEST quantification definition for the following conditions: B_0, B_1, saturation time and normalisation are not altered from measurement to measurement or subject to subject.

We only showed here the steady-state approach to estimate R_{ex}, while it can also be determined during transient state according to Equation 11.10. These are quantification techniques using saturation time, originally shown by McMahon *et al.* (2006) and refined for more general application (Roeloffs *et al.*, 2015; Sun, 2012; Vinogradov *et al.*, 2012; Zaiss and Bachert, 2013).

11.5.6 R_{ex}, labelling efficiency and optimal CEST effect

Having access to R_{ex} as the key parameter of CEST, the behaviour of R_{ex} can now be investigated. Including the actual frequency offset and introducing the labelling efficiency α, R'_{ex} can be written as described elsewhere (Zaiss and Bachert, 2013):

$$R'_{ex}(\Delta\omega) = \sin^2\theta \cdot R_{ex}(\Delta\omega) =$$

$$= f_s k_{sw} \frac{\delta\omega_s^2}{\omega_1^2 + \Delta\omega^2} \frac{\omega_1^2}{\Gamma^2/4 + \Delta\omega_s^2} + f_s R_{2s} \frac{\omega_1^2}{\Gamma^2/4 + \Delta\omega_s^2} \tag{11.16}$$

$$+ f_s k_{sw} \sin^2\theta \frac{R_{2s}(R_{2s} + k_{sw})}{\Gamma^2/4 + \Delta\omega_s^2}$$

Only for larger chemical shifts > k_{sw} this can be further simplified to

$$R'_{ex}(\Delta\omega) = k_{sw} f_s \cdot \alpha \frac{\Gamma^2/4}{\Gamma^2/4 + \Delta\omega_s^2} \tag{11.17}$$

where $\Delta\omega_s$ is the frequency offset with respect to the CEST pool **s**. The labelling efficiency can be approximated by the following (Zaiss and Bachert, 2013):

$$\alpha = \frac{\omega_1^2}{\omega_1^2 + k_{sw}(k_{sw} + R_{2s})} \tag{11.18}$$

Only if the $\sin^2\theta$ term is included in R'_{ex} in Equation 11.16 can the known labelling efficiency α be derived.

In the large shift limit, R'_{ex} is a Lorentzian function (Equation 11.17) of $\Delta\omega_s$ with its maximum at $\Delta\omega_s = 0$ and line width

$$\Gamma = 2\sqrt{\frac{k_{sw} + R_{2s}}{k_{sw}} \omega_1^2 + (k_{sw} + R_{2s})^2}. \tag{11.19}$$

R'_{ex} as a function of the saturation offset shows a Lorentzian behaviour, where the peak width increases with saturation power (Equation 11.19, Figure 11.5a). $R'_{ex}(\Delta\omega)$ for a given exchange rate k_{sw} increases with B_1 amplitude and its amplitude approaches a plateau (Figure 11.5b); this behaviour is determined by the labelling efficiency (Equation 11.18). R'_{ex} as a function of k_{sw} for a specific B_1 shows a maximum at $k_{sw} \approx \gamma \cdot B_1 = \omega_1$ (Figure 11.5c).

R_{ex} which is free of dependencies on water relaxation, describes a physical parameter in Hz: the exchange-dependent relaxation rate. However, R_{ex} is still a relative quantification parameter as it depends on the used saturation scheme. In the following, we will show how the dependency on saturation parameters can be used to get access to fundamental parameters: the exchange rate and the concentration of the CEST pool.

11.6 Absolute quantification techniques full Bloch–McConnell fitting

As shown below, quantification of exchange rate and concentration from CEST data requires *Z*-spectrum data acquired with several irradiation powers B_1. Then simultaneous fitting of such multi-B_1 *Z*-spectra using the full numerical Bloch–McConnell solution allows in principle for full quantification. Although such approaches have been performed (Chappell *et al.*, 2013; Liu *et al.*, 2013), they always depend on how the pool model is established in terms of number of exchangeable pools, dispersion of chemical shifts and other *a priori* information.

11.6.1 Quantification of exchange by multiple B_1

As stated above, labelling and thus R_{ex} depend on the saturation amplitude B_1. This is also known as *B_1 dispersion of CEST effects*. The exact behaviour of this dispersion depends on the actual exchange rate (see Equation 11.18); thus measurements of the dispersion or labelling curves provide insight to the actual exchange rate. This was first shown by McMahon *et al.* as quantification of exchange by saturation power (McMahon *et al.*, 2006). By determining two or more values of R_{ex} for different

B_1, the exchange rate k_{sw} and the concentration of exchangeable labile protons f_s can be estimated by fitting the R_{ex} data using Equation 11.3 directly. As this equation requires general knowledge of spillover terms, again the inverse metric can be used to get a general applicable approach valid in steady state by fitting AREX data directly to Equation 11.17. Such a multi-B_1 approach allowing full quantification was elegantly derived by Dixon *et al.* (2010b) by showing that the problem is transformable to a linear function in $(1/\omega_1)^2$. This reveals the relationship very clearly; using the reference and label scan this can be written as follows (Meissner *et al.*, 2015):

$$y\left(\frac{1}{\omega_1^2}\right)=\frac{1}{\frac{1}{Z_{lab}}-\frac{1}{Z_{ref}}}=\frac{R_{1obs}}{f_s k_{sw}}+\frac{R_{1obs}k_{sw}}{f_s}\cdot\frac{1}{\omega_1^2} \quad (11.20)$$

Determination of this linear function, the so-called Ω-plot, allows then calculating both k_{sw} and f_s. R_{2s} was neglected here but can and must be taken into account especially for slow exchange (Meissner *et al.*, 2015; Wu *et al.*, 2015). In principle, the line width of a CEST peak in the *Z*-spectrum can be also used to determine the exchange rate through the width of R_{ex}, a function of exchange rate and B_1, as given by Equation 11.19 (Figure 11.2a). This has not yet been shown directly *in vivo* but is often used implicitly if full *Z*-spectra fits are performed.

11.6.2 Quantification in pulsed CEST

As mentioned above, on clinical scanners RF irradiation is often limited to pulse trains consisting of multiple short pulses. This has of course a profound effect on the optimisation of the detectable CEST signal, but efficient saturation can still be achieved (Sun *et al.*, 2011), even with quasi-continuous RF saturation using two separate transmission channels (Togao *et al.*, 2015). The question is how to interpret resulting pulsed CEST effects quantitatively.

Pulsed saturation brings at least four additional effects to CEST spectra:

1. The shape of the pulse affects the labelling dynamic and efficiency.
2. Exchange occurs during the pulse but also in the delay between saturation pulses.
3. Rotation transfer effects can occur (due to the relatively long T_2 values of the labile protons, their magnetisation can be rotated by arbitrary flip angles).
4. The bandwidth of the pulse can influence the width of the saturation profile.

The first point leads to the crucial question of whether the CW theory can be used to interpret pulsed CEST data. The answer, however, is not straightforward and depends on many parameters. For slow exchange rates (<50 Hz) and a train of inversion pulses, it was shown that the CW theory can be applied by using B_1 power equivalents (Tee *et al.*, 2012; Zu *et al.*, 2011):

$$\omega_{1,cwpe}=\sqrt{\frac{\int_0^{t_p}\omega_1^2(t)dt}{t_p+t_d}} \quad (11.21)$$

This power equivalent is generally helpful as especially the background effects, direct saturation and semi-solid MT effects, follow this equivalent power definition and thus it makes comparison of Z-spectra of cw and different pulsed irradiation schemes possible.

For faster exchange rates it was shown by Meissner *et al.* that a modified analytical theory can be used that incorporates the pulse shape and duty-cycle explicitly (Meissner *et al.*, 2015):

$$\begin{aligned}R_{ex,shaped-pulses}&=DC\cdot\frac{1}{t_p}\int_0^{t_p}R_{ex}(\omega_1(t))dt\\&\approx DC\cdot f_s k_{sw}\cdot c_1\cdot\frac{\omega_1^2}{\omega_1^2+k_{sw}(k_{sw}+R_{2s})\cdot c_2^2}\end{aligned} \quad (11.22)$$

with the form factors given by the Gaussian pulse shape $c_1=\frac{\sigma\sqrt{2\pi}}{t_p}$ and $c_2=c_1\cdot\sqrt[4]{2}$, while $B_1=\omega_1/\gamma$ is the mean amplitude of a single pulse, thus equal to the flip angle divided by t_p. Note that it can also be transformed to the Ω-plot formalism of Equation 11.21 (Meissner *et al.*, 2015).

Regarding Issue 2, it was shown experimentally (Friedman *et al.*, 2010; Jones *et al.*, 2013; Roeloffs *et al.*, 2015) that in pulsed CEST the dynamic in between the saturation pulses also gives insight to the actual exchange rate. Thus, by variation of the delay time t_d, the exchange rate can be determined using a variable delay multipulse experiment as described previously (Jones *et al.*, 2013; Xu *et al.*, 2014). There is no complete analytical description of this process; thus data in that regime can only be interpreted by the full Bloch-McConnell equations. With this technique, Jones *et al.* (Jones *et al.*, 2013) were able to determine the amide exchange rate *in vivo* to be approximately 30 Hz and the NOE exchange rate to be about 17 Hz; the fits of the delay time behaviour are shown in Figure 11.6.

For general description of the exchange dynamics in the case of a train of short pulses, the full Bloch-McConnell equations must be used. This is especially true if the so-called rotation transfer regime is reached (Issue 3). In particular, it was shown that the actual flip angle of the saturation pulse has a direct effect on the pulsed CEST labelling efficiency, with optimum close to 180° pulses (Sun *et al.*, 2011; Zu *et al.*, 2011). Generally, different flip angles can be used in chemical exchange rotation transfer experiments (Zu *et al.*, 2012) where 180° and 360° pulses are used to isolate the pure CEST effect (Zu *et al.*, 2013). Related to that is the previously shown frequency-labelled exchange transfer

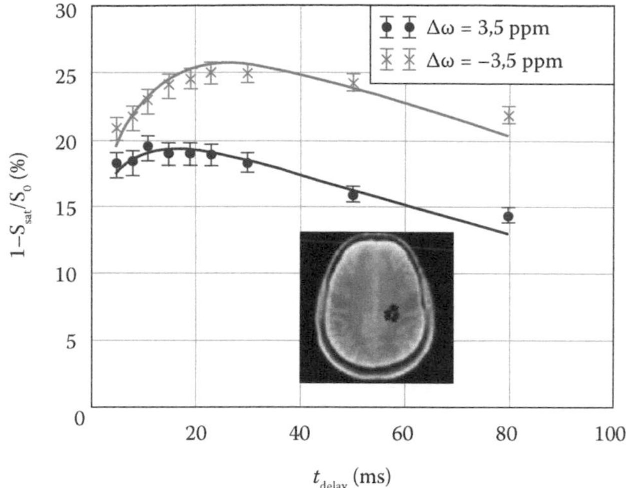

FIGURE 11.6 Saturation transfer data as a function of interpulse delay time in the variable-delay multipulse CEST sequence for a region of interest in white matter in the human brain. The data points are fitted to a two-pool Bloch model. The dependencies for the amide proton transfer (APT) CEST spectral range (3.3 to 3.7 ppm) and corresponding upfield NOE range (−3.3 to −3.7 ppm) resemble those found in protein solution, with the NOE range building up slower but both APT-CEST and NOE decaying at the same rate, determined by the T_1 of water protons. (Reproduced from *NeuroImage*, 77, Jones, C.K., *et al.*, Nuclear overhauser enhancement (NOE) imaging in the human brain at 7T, 114–124, Copyright 2013, with permission from Elsevier.)

technique, which uses different rotations of the CEST pool magnetisation in the transverse plain to isolate different CEST effects (Friedman *et al.*, 2010; Yadav *et al.*, 2013).

Finally, Issue 4 depends on the field strength and the pulse duration, but even at clinical field strengths of 3T pulses of 100 ms can be shaped so as to reduce the bandwidth far below the CEST peak width (Equation 11.19). Yet labelling can be more efficient as potentially more than one pool can be labelled using a single pulse, depending on its bandwidth.

11.6.3 Calibration approaches and ratiometric approaches

As discussed above, uncorrected CEST effects can be affected by many unwanted influences. An elegant way to avoid this are the so-called ratiometric approaches. Here the ratio of two CEST peaks is used as a robust signal that is independent of concentration and can be e.g. pH calibrated (Ward and Balaban, 2000). This was first shown *in vivo* for exogenous CEST agents (Longo *et al.*, 2011) but was also demonstrated for endogenous amine and amide peaks (McVicar *et al.*, 2014). Moreover, a single peak and two different powers can be used to obtain a pH calibration (Longo *et al.*, 2014; Rerich *et al.*, 2015; Sun *et al.*, 2014).

Under the assumption that the concentration of the CEST pool is constant or even known, CEST effects can also be directly interpreted using exchange rate–pH dependency. This then also allows pH imaging, with applications such as in stroke (Sun *et al.*, 2011; Zaiss *et al.*, 2014; Zhou *et al.*, 2003b) (see Section 11.7).

11.6.4 Dependence of the exchange rate constant on pH, buffer condition, and temperature

The exchange rate shown as a quantitative parameter in the Bloch–McConnell is not independent from its milieu and depends on its direct chemical environment. The (first-order) exchange rate constant k_{sw} in aqueous solution can be divided into an acid- and base-catalysed proton exchange k_a and k_b, respectively, and a constant k_0 depending on the environment (pH buffer and other solutes) (Englander *et al.*, 1972):

$$k_{sw} = k_a \cdot \left[H_3O^+\right] + k_b \cdot \left[OH^-\right] + k_0 \tag{11.23}$$

Focusing now on a typical base-catalysed exchange process, everything except k_b is neglected in the following. Combination of Equation 11.24

$$k_{sw}\left(pH,T\right) = k_b(T) \cdot \left[OH^-\right]\left(pH,T\right) = k_b(T) \cdot \frac{mol}{l} \cdot 10^{pH-pK_w(T)} \tag{11.24}$$

with the Arrhenius equation (Bai *et al.*, 1993) and the Van't Hoff equation (Atkins and de Paula, 2006) allows to determine the exchange rate constant of a base-catalysed exchange process as a function of pH and T:

$$k_{sw}\left(pH,T\right) = k_c(298.15\ ^\circ K) \cdot \frac{mol}{l} \cdot 10^{pH-14+\frac{E_{A,b}+\Delta H_R^0}{R \cdot ln10}\left(\frac{1}{298.15\ ^\circ K}-\frac{1}{T}\right)} \tag{11.25}$$

with the constant $k_c(298.15\ ^\circ K)$ being the collision frequency factor at 298.15°K, E_A the activation energy, $\Delta H_R^0 = 55.84 \frac{kJ}{mol}$ the standard reaction enthalpy (Atkins and de Paula, 2006) for the self-dissociation of water and $R = 8.314 \frac{J}{mol^\circ K}$ the perfect gas constant. The same equations can be derived for acid- and water-catalysed proton exchange.

11.7 Applications of CEST

11.7.1 Stroke and hypoxia: pH reduction

With such a strong dependency on pH, CEST imaging found an ideal application in the identification of tissue ischaemia. Numerous studies have shown that pH changes can be visualised through CEST contrast (Huang *et al.*, 2015; Li *et al.*, 2015; Sun *et al.*, 2012; Zaiss *et al.*, 2014; Zhou *et al.*, 2003b) and furthermore that CEST imaging has the potential to spatially map pH (McVicar *et al.*, 2014; Zhou *et al.*, 2003b), based on the assumptions present in the previous section.

With a mostly base-catalysed saturation transfer *in vivo* (see Section 11.6.4), reduced saturation transfer as a sign of acidosis can be observed in areas of brain ischaemia within 2 hours (Dai *et al.*, 2014). In animal models, a consistent correlation of APT signal with apparent diffusion coefficient (ADC) maps has been demonstrated in hyperacute stroke (Huang *et al.*, 2015), but pH changes can occur earlier without ADC evidence of cellular depolarisation (Zhou and van Zijl, 2011). A localised pH reduction may precede ADC changes and extend beyond the boundaries of final infarction, indicating that APT-CEST can depict viable ischaemic tissue (Sun *et al.*, 2007).

Translation onto clinical MRI systems, despite their limitations on RF pulse duration and duty cycle, appears feasible to achieve an overall contrast similar to that of continuous wave saturation used in preclinical systems (Sun *et al.*, 2011). In human studies, some success has been achieved in generating quantitative pH maps for hyperacute stroke studies within a clinically justifiable time frame, with a demonstration of sufficient overlap with the final infarct core in patients (Tee *et al.*, 2014; Tietze *et al.*, 2014) (see Figure 11.7).

11.7.2 Tumour cellularity based on APT

The majority (80%) of brain tumours are gliomas, whereby glioblastoma (GBM) constitutes the commonest and most malignant type with a mean survival in the region of 15 months (Schwartzbaum *et al.*, 2006). Conventional MRI imaging suffers from limited sensitivity and specificity, and although advanced techniques such as diffusion, perfusion and MRS add information they still lack some diagnostic accuracy (Lacerda and Law, 2009; Wang *et al.*, 2016).

The heightened protein and peptide content of brain tumours compared to normal tissue can be visualised through their increased chemical exchange between solute protein and peptide groups and bulk water even using simple metrics such as the MTR_{asym} in the region of 3.5 ppm at clinical field strength (3T), the so-called APT contrast (van Zijl *et al.*, 2003; Zhou *et al.*, 2003a). Furthermore, a distinction of gliomas grades has been achieved in patients (see Figure 11.8); with GBM and WHO III–graded glioma showing significantly higher contrast compared with WHO II glioma (Wen *et al.*, 2010; Zhou *et al.*, 2008, 2013). A further potential application of endogenous CEST imaging is the distinction of recurrent tumour from chemoradiation effects, which remains a common challenge in post-treatment glioblastoma assessment, as both conditions may demonstrate contrast enhancement as a sign of blood–brain barrier breakdown (Ma *et al.*, 2016; Zhou *et al.*, 2011).

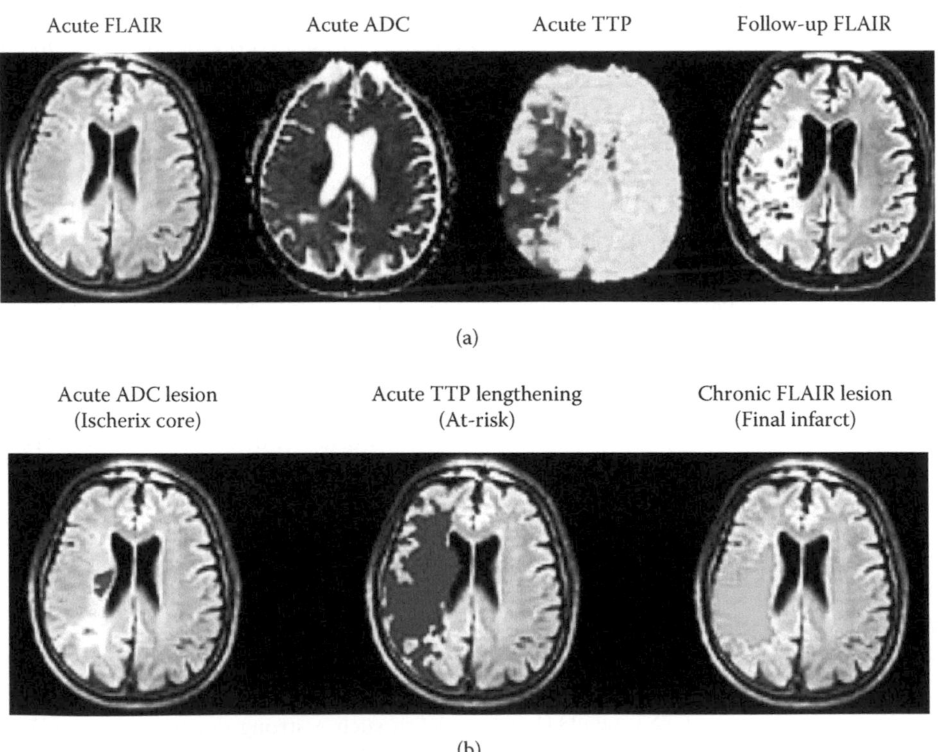

FIGURE 11.7 Analysis pipeline: (a) Example of acute T_2 fluid attenuated inversion recovery (T2 FLAIR), apparent diffusion coefficient (ADC), time-to-peak (TTP) and follow-up (>1 month) T2 FLAIR image. (b) Semi-automated segmentation is used to define the ischaemic core, which outlines hypointensity in the acute ADC map, the at-risk tissue, defined as lengthening on the TTP map, and the final infarct volume, defined as hyperintensity on the follow-up T2 FLAIR. (Tietze, A., *et al.*, Assessment of ischemic penumbra in patients with hyperacute stroke using amide proton transfer (APT) chemical exchange saturation transfer (CEST) MRI. *NMR Biomed.* 2014. 27. 163–174. Copyright Wiley-VCH Verlag GmbH & Co. KGaA. Reproduced with permission.)

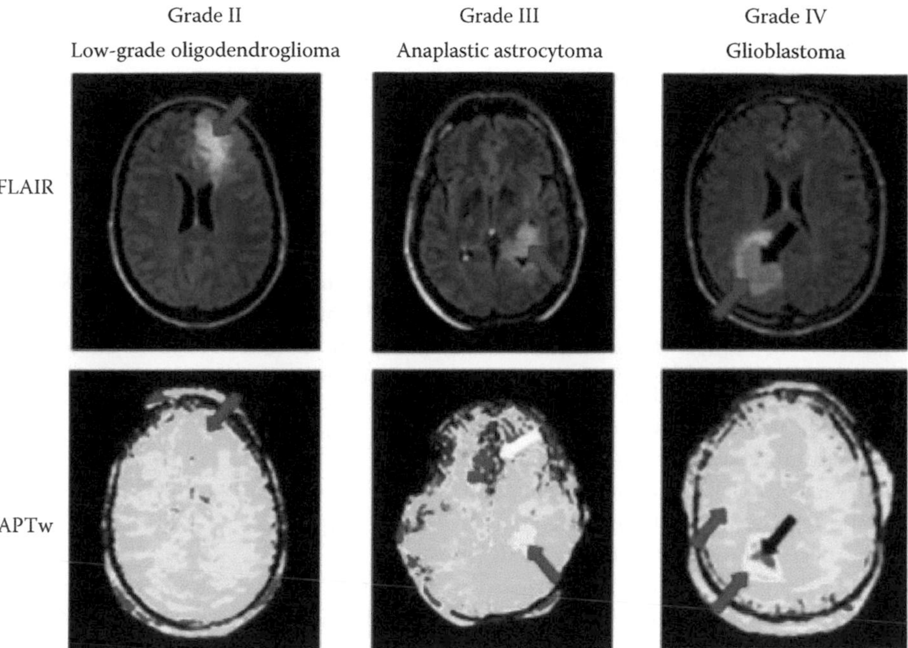

FIGURE 11.8 APT-weighted and conventional MR images for patients with a low-grade oligodendroglioma (WHO Grade II), anaplastic astrocytoma (WHO Grade III) and glioblastoma (WHO Grade IV). (Zhou, J., *et al.*, Three-dimensional amide proton transfer MR imaging of gliomas: initial experience and comparison with gadolinium enhancement. *J. Magn. Reson. Imaging.* 2013. 38. 1119–1128. Copyright Wiley-VCH Verlag GmbH & Co. KGaA. Reproduced with permission.)

It should be noted that NOE is a significant contributor to APT-weighted contrast in addition to asymmetry in the semi-solid MTC effect. Therefore, there is a high demand on developing methods to remove confounding MT asymmetry and NOE to understand the full potential of APT. Nonetheless, some have postulated that NOE may be associated with tumour cellularity via altered protein synthesis in proliferating cells as NOE-based signal represents a better contrast in glioma as compared to APT-based signal (Heo *et al.*, 2016a; Paech *et al.*, 2015; Zaiss *et al.*, 2016).

11.7.3 Neurodegeneration: Protein folding, protein turnover and accumulation, myo-inositol, glutamate

In addition to protein content, a potentially more interesting determinant of saturation transfer for CEST imaging is the folding of brain and body proteins. Protein abnormalities are characteristic associations of nearly all neurodegenerative diseases (Yerbury *et al.*, 2016), in which structural MRI often lacks sensitivity in early disease. The mapping of protein conformations and aggregates could therefore be extremely valuable in this context. A plausible relationship exists between the signal strength of NOE peaks in the *Z*-spectra and protein structure (Aguirre *et al.*, 2015; Braun, 1987). In support of this, a correlation of NOE-mediated saturation transfer with the structural state of bovine serum albumin (BSA) was observed recently (Zaiss *et al.*, 2014; Goerke *et al.*, 2015). Myo-inositol, a marker of glial cell proliferation that has been shown to increase in early Alzheimer's disease, has been recently mapped using CEST in a mouse model (Haris *et al.*, 2013). Finally, glutamate may also be considered as a future imaging biomarker, as it has been implicated as a neurotoxic metabolite in many neurodegenerative conditions including Alzheimer's and frontotemporal dementia (Marjanska *et al.*, 2005). As such, CEST imaging may become an important imaging technique to visualise protein denaturation, brain stress and ageing processes through visualising the structural integrity of proteins in cells or accumulation of specific metabolites in future.

11.7.4 GlucoCEST studies in tumours

As seen above, the native CEST contrast can reflect on a number of metabolites specifically, in addition to general proteasome. In this context, glucose may be an interesting diaCEST agent. Tumour cells preferentially use anaerobic glucose breakdown, even in the presence of abundant oxygen (the 'Warburg effect'). As such, CEST imaging may be used to depict the saturation exchange between glucose hydroxyl protons (GlucoCEST) and water along the *Z*-spectrum in the region of 1.2–3.0 ppm. In animal experiments, the difference in GlucoCEST signal before and after injection of glucose has shown remarkable spatial overlap with fluorodeoxyglucose maps, supporting its validity (Walker-Samuel *et al.*, 2013).

In patients, dynamic injection of glucose has been shown to provide similar early-phase contrast to dynamic contrast enhanced T_1-weighted MRI, therefore most likely reflecting local blood flow, vascular permeability and volume of the extracellular space. Whether or not dynamic GlucoCEST can also reflect on glucose metabolism, and especially in brain tumours, remains a point of controversy. The first translation of dynamic glucose enhanced MRI into human glioma observed variations in GlucoCEST contrast across glioma components over time (Xu *et al.*, 2015).

11.7.5 Additional hydroxyl-CEST applications: Glycosaminoglycans

Finally, apart from brain applications, hydroxyl-based CEST has been used in other parts of the body. One particular class of sugars detectable through saturation transfer is polysaccharides, known as glycosaminoglycans (GAGs). In particular, the gradual loss through degradation of GAGs in osteoarthritis (OA), a degenerative joint disease affecting nearly 80% of individuals by the time they reach age 65 (Bradley *et al.*, 1991), may serve as an early indication of disease progression (Singh *et al.*, 2012). As GAGs contain a large number of exchangeable protons, and in particular hydroxyl groups, imaging through these exchangeable hydroxyl groups, known as *gagCEST*, has been suggested as a potential image-based biomarker in OA (Ling *et al.*, 2008). However, due to the very close proximity of the GAG hydroxyl groups with the water peak, typically spanning 0.9–1.9 ppm (Ling *et al.*, 2008), most of the applications so far have been reported at high field strength (>3T).

11.7.6 Reliability and reproducibility of CEST

While numerous studies have demonstrated the potential of CEST imaging, it is crucial to investigate its reproducibility as a quantitative imaging method in clinical practice. To date, a few test–retest reproducibility studies have reported good inter- and intra-session reliability of the CEST signal for whole tumour region of interest (ROI) (Togao *et al.*, 2015), *in vitro* BSA phantom and brain regions (grey matter, white matter and ventricle ROIs) of healthy volunteers (Schmidt *et al.*, 2016), and intervertebral lumbar discs of healthy volunteers and disc degenerative disease patients (Deng *et al.*, 2016). However, Schmidt *et al.* showed that intra-ROI covariances of MTR_{asym} (3.5 ppm) were sensitive to the choice of fitting algorithms used (Schmidt *et al.*, 2016). Therefore, further investigation is desirable on the reproducibility of CEST imaging with optimal data processing methods.

11.8 Conclusion

Following from this chapter, it becomes clear that CEST is a very versatile and potentially very powerful new quantitative imaging method to add to the existing arsenal available to radiologists. However, it is also still an expanding field, with new developments, both on the theoretical and practical side, being developed at a very high rate by the CEST community. As such, in contrast to MTR (Chapter XX and YY) or quantitative MT (Chapter YY), CEST has not found a definitive place yet in clinical practice, and the imaging readouts still require a serious amount of work for it to become really practical. This chapter therefore concentrated on the theoretical aspects of this new contrast, while offering a glimpse of its new potential applications, spanning the whole range of organs, from brain to musculoskeletal through general oncological applications.

Grant Support: Max Planck Society, German Research Foundation (DFG, grant no. ZA 814/2-1, support to M.Z.), European Union's Horizon 2020 Research and Innovation Programme (grant no. 667510, support to M.Z., M-K., S.T., X.G.), and Department of Health's NIHR - funded Biomedical Research Centre at University College London (S.T., X.G.).

References

Aguirre C, Cala O, Krimm I. Overview of probing protein-ligand interactions using NMR. Curr Protoc Protein Sci 2015; 81: 17 (18): 1–24.

Aime S, Barge A, Delli Castelli D, Fedeli F, Mortillaro A, Nielsen FU, et al. Paramagnetic lanthanide(III) complexes as pH-sensitive chemical exchange saturation transfer (CEST) contrast agents for MRI applications. Magn Reson Med 2002; 47: 639–48.

Aime S, Carrera C, Delli Castelli D, Geninatti Crich S, Terreno E. Tunable imaging of cells labeled with MRI-PARACEST agents. Angew Chem Int Ed Engl 2005; 44: 1813–15.

Atkins PW, de Paula J. Atkins' physical chemistry. Freeman & Company, New York, 2006.

Bai Y, Milne JS, Mayne L, Englander SW. Primary structure effects on peptide group hydrogen exchange. Proteins 1993; 17: 75–86.

Bradley JD, Brandt KD, Katz BP, Kalasinski LA, Ryan SI. Comparison of an antiinflammatory dose of ibuprofen, an analgesic dose of ibuprofen, and acetaminophen in the treatment of patients with osteoarthritis of the knee. N Engl J Med 1991; 325: 87–91.

Braun W. Distance geometry and related methods for protein structure determination from NMR data. Q Rev Biophys 1987; 19: 115–57.

Bryant RG. The dynamics of water-protein interactions. Annu Rev Biophys Biomol Struct 1996; 25: 29–53.

Cai K, Haris M, Singh A, Kogan F, Greenberg JH, Hariharan H, et al. Magnetic resonance imaging of glutamate. Nat Med 2012; 18: 302–6.

Chan KW, McMahon MT, Kato Y, Liu G, Bulte JW, Bhujwalla ZM, et al. Natural D-glucose as a biodegradable MRI contrast agent for detecting cancer. Magn Reson Med 2012; 68: 1764–73.

Chappell MA, Donahue MJ, Tee YK, Khrapitchev AA, Sibson NR, Jezzard P, et al. Quantitative Bayesian model-based analysis of amide proton transfer MRI. Magn Reson Med 2013; 70: 556–67.

Chappell MA, Groves AR, Whitcher B, Woolrich MW. Variational Bayesian inference for a nonlinear forward model. IEEE Trans Signal Process 2009; 57: 223–36.

Dai Z, Ji J, Xiao G, Yan G, Li S, Zhang G, et al. Magnetization transfer prepared gradient echo MRI for CEST imaging. PLoS One 2014; 9: e112219.

Deng M, Yuan J, Chen WT, Chan Q, Griffith JF, Wang YX. Evaluation of glycosaminoglycan in the lumbar disc using chemical exchange saturation transfer MR at 3.0 Tesla: reproducibility and correlation with disc degeneration. Biomed Environ Sci 2016; 29: 47–55.

Desmond KL, Moosvi F, Stanisz GJ. Mapping of amide, amine, and aliphatic peaks in the CEST spectra of murine xenografts at 7 T. Magn Reson Med 2014; 71: 1841–53.

Dixon WT, Hancu I, Ratnakar SJ, Sherry AD, Lenkinski RE, Alsop DC. A multislice gradient echo pulse sequence for CEST imaging. Magn Reson Med 2010a; 63: 253–6.

Dixon WT, Ren J, Lubag AJ, Ratnakar J, Vinogradov E, Hancu I, et al. A concentration-independent method to measure exchange rates in PARACEST agents. Magn Reson Med 2010b; 63: 625–32.

Dula AN, Asche EM, Landman BA, Welch EB, Pawate S, Sriram S, et al. Development of chemical exchange saturation transfer at 7 T. Magn Reson Med 2011; 66: 831–8.

Englander SW, Downer NW, Teitelbaum H. Hydrogen exchange. Annu Rev Biochem 1972; 41: 903–24.

Friedman JI, McMahon MT, Stivers JT, Van Zijl PC. Indirect detection of labile solute proton spectra via the water signal using frequency-labeled exchange (FLEX) transfer. J Am Chem Soc 2010; 132: 1813–15.

Goerke S, Zaiss M, Kunz P, et al. Signature of protein unfolding in chemical exchange saturation transfer imaging. NMR Biomed 2015; 28(7): 906–13.

Goffeney N, Bulte JW, Duyn J, Bryant LH Jr., van Zijl PC. Sensitive NMR detection of cationic-polymer based gene delivery systems using saturation transfer via proton exchange. J Am Chem Soc 2001; 123: 8628–9.

Haris M, Cai K, Singh A, Hariharan H, Reddy R. In vivo mapping of brain myo-inositol. NeuroImage 2011; 54: 2079–85.

Haris M, Singh A, Cai K, Nath K, Crescenzi R, Kogan F, et al. MICEST: a potential tool for non-invasive detection of molecular changes in Alzheimer's disease. J Neurosci Methods 2013; 212: 87–93.

Heo HY, Jones CK, Hua J, Yadav N, Agarwal S, Zhou J, et al. Whole-brain amide proton transfer (APT) and nuclear overhauser enhancement (NOE) imaging in glioma patients using low-power steady-state pulsed chemical exchange saturation transfer (CEST) imaging at 7T. J Magn Reson Imaging 2016a; 44: 41–50.

Heo H-Y, Zhang Y, Jiang S, Lee D-H, Zhou J. Quantitative assessment of amide proton transfer (APT) and nuclear overhauser enhancement (NOE) imaging with extrapolated semisolid magnetization transfer reference (EMR) signals: II. Comparison of three EMR models and application to human brain glioma at 3 Tesla. Magn Reson Med 2016; 75(4): 1630–39.

Heo HY, Lee DH, Zhang Y, Zhao X, Jiang S, Chen M, et al. Insight into the quantitative metrics of chemical exchange saturation transfer (CEST) imaging. Magn Reson Med 2017;77(5):1853–65.

Huang D, Li S, Dai Z, Shen Z, Yan G, Wu R. Novel gradient echo sequence-based amide proton transfer magnetic resonance imaging in hyperacute cerebral infarction. Mol Med Rep 2015; 11: 3279–84.

Jiang W, Zhou IY, Wen L, Zhou X, Sun PZ. A theoretical analysis of chemical exchange saturation transfer echo planar imaging (CEST-EPI) steady state solution and the CEST sensitivity efficiency-based optimization approach. Contrast Media Mol Imaging 2016; 11: 415–23.

Jin T, Autio J, Obata T, Kim SG. Spin-locking versus chemical exchange saturation transfer MRI for investigating chemical exchange process between water and labile metabolite protons. Magn Reson Med 2011; 65: 1448–60.

Jin T, Wang P, Zong X, Kim SG. Magnetic resonance imaging of the Amine-Proton EXchange (APEX) dependent contrast. NeuroImage 2012; 59: 1218–27.

Jones CK, Huang A, Xu J, Edden RA, Schar M, Hua J, et al. Nuclear overhauser enhancement (NOE) imaging in the human brain at 7T. NeuroImage 2013; 77: 114–24.

Jones CK, Polders D, Hua J, Zhu H, Hoogduin HJ, Zhou J, et al. In vivo three-dimensional whole-brain pulsed steady-state chemical exchange saturation transfer at 7 T. Magn Reson Med 2012; 67: 1579–89.

Khlebnikov V, Polders D, Hendrikse J, Robe PA, Voormolen EH, Luijten PR, et al. Amide proton transfer (APT) imaging of brain tumors at 7 T: the role of tissue water T1 -Relaxation properties. Magn Reson Med 2017; 77: 1525–32.

Kim M, Gillen J, Landman BA, Zhou J, van Zijl PC. Water saturation shift referencing (WASSR) for chemical exchange saturation transfer (CEST) experiments. Magn Reson Med 2009; 61: 1441–50.

Lacerda S, Law M. Magnetic resonance perfusion and permeability imaging in brain tumors. Neuroimaging Clin N Am 2009; 19: 527–57.

Li H, Zu Z, Zaiss M, Khan IS, Singer RJ, Gochberg DF, et al. Imaging of amide proton transfer and nuclear overhauser enhancement in ischemic stroke with corrections for competing effects. NMR Biomed 2015; 28: 200–9.

Ling W, Regatte RR, Navon G, Jerschow A. Assessment of glycosaminoglycan concentration in vivo by chemical exchange-dependent saturation transfer (gagCEST). Proc Natl Acad Sci U S A 2008; 105: 2266–70.

Liu D, Zhou J, Xue R, Zuo Z, An J, Wang DJ. Quantitative characterization of nuclear Overhauser enhancement and amide proton transfer effects in the human brain at 7 tesla. Magn Reson Med 2013; 70: 1070–81.

Longo DL, Dastru W, Digilio G, Keupp J, Langereis S, Lanzardo S, et al. Iopamidol as a responsive MRI-chemical exchange saturation transfer contrast agent for pH mapping of kidneys: in vivo studies in mice at 7 T. Magn Reson Med 2011; 65: 202–11.

Longo DL, Sun PZ, Consolino L, Michelotti FC, Uggeri F, Aime S. A general MRI-CEST ratiometric approach for pH imaging: demonstration of in vivo pH mapping with iobitridol. J Am Chem Soc 2014; 136: 14333–6.

Ma B, Blakeley JO, Hong X, Zhang H, Jiang S, Blair L, et al. Applying amide proton transfer-weighted MRI to distinguish pseudoprogression from true progression in malignant gliomas. J Magn Reson Imaging 2016; 44: 456–62.

Marjanska M, Curran GL, Wengenack TM, Henry PG, Bliss RL, Poduslo JF, et al. Monitoring disease progression in transgenic mouse models of Alzheimer's disease with proton magnetic resonance spectroscopy. Proc Natl Acad Sci U S A 2005; 102: 11906–10.

McConnell HM. Reaction rates by nuclear magnetic resonance. J Chem. Phys. 1958; 28: 430–1.

McMahon MT, Gilad AA, Zhou J, Sun PZ, Bulte JW, van Zijl PC. Quantifying exchange rates in chemical exchange saturation transfer agents using the saturation time and saturation power dependencies of the magnetization transfer effect on the magnetic resonance imaging signal (QUEST and QUESP): Ph calibration for poly-L-lysine and a starburst dendrimer. Magn Reson Med 2006; 55: 836–47.

McVicar N, Li AX, Goncalves DF, Bellyou M, Meakin SO, Prado MA, et al. Quantitative tissue pH measurement during cerebral ischemia using amine and amide concentration-independent detection (AACID) with MRI. J Cereb Blood Flow Metab 2014; 34: 690–8.

Meissner JE, Goerke S, Rerich E, Klika KD, Radbruch A, Ladd ME, et al. Quantitative pulsed CEST-MRI using Omega-plots. NMR Biomed 2015; 28: 1196–208.

Mougin OE, Coxon RC, Pitiot A, Gowland PA. Magnetization transfer phenomenon in the human brain at 7 T. NeuroImage 2010; 49: 272–81.

Paech D, Burth S, Windschuh J, Meissner JE, Zaiss M, Eidel O, et al. Nuclear overhauser enhancement imaging of glioblastoma at 7 Tesla: region specific correlation with apparent diffusion coefficient and histology. PLoS One 2015; 10: e0121220.

Rerich E, Zaiss M, Korzowski A, Ladd ME, Bachert P. Relaxation-compensated CEST-MRI at 7 T for mapping of creatine content and pH – preliminary application in human muscle tissue in vivo. NMR Biomed 2015; 28: 1402–12.

Roeloffs V, Meyer C, Bachert P, Zaiss M. Towards quantification of pulsed spinlock and CEST at clinical MR scanners: an analytical interleaved saturation-relaxation (ISAR) approach. NMR Biomed 2015; 28: 40–53.

Scheidegger R, Wong ET, Alsop DC. Contributors to contrast between glioma and brain tissue in chemical exchange saturation transfer sensitive imaging at 3 Tesla. NeuroImage 2014; 99: 256–68.

Schmidt H, Schwenzer NF, Gatidis S, Kustner T, Nikolaou K, Schick F, et al. Systematic evaluation of amide proton chemical exchange saturation transfer at 3 T: effects of protein concentration, pH, and acquisition parameters. Invest Radiol 2016; 51: 635–46.

Schuenke P, Windschuh J, Roeloffs V, Ladd ME, Bachert P, Zaiss M. Simultaneous mapping of water shift and B1 (WASABI)-Application to field-Inhomogeneity correction of CEST MRI data. Magn Reson Med. 2017; 77(2): 571–80.

Schwartzbaum JA, Fisher JL, Aldape KD, Wrensch M. Epidemiology and molecular pathology of glioma. Nat Clin Pract Neurol 2006; 2: 494–503; quiz 1 p following 516.

Singh A, Haris M, Cai K, Kassey VB, Kogan F, Reddy D, et al. Chemical exchange saturation transfer magnetic resonance imaging of human knee cartilage at 3 T and 7 T. Magn Reson Med 2012; 68: 588–94.

Sun PZ. Simplified quantification of labile proton concentration-weighted chemical exchange rate (k(ws)) with RF saturation time dependent ratiometric analysis (QUESTRA): normalization of relaxation and RF irradiation spillover effects for improved quantitative chemical exchange saturation transfer (CEST) MRI. Magn Reson Med 2012; 67: 936–42.

Sun PZ, Benner T, Kumar A, Sorensen AG. Investigation of optimizing and translating pH-sensitive pulsed-chemical exchange saturation transfer (CEST) imaging to a 3T clinical scanner. Magn Reson Med 2008; 60: 834–41.

Sun PZ, Longo DL, Hu W, Xiao G, Wu R. Quantification of iopamidol multi-site chemical exchange properties for ratiometric chemical exchange saturation transfer (CEST) imaging of pH. Phys Med Biol 2014; 59: 4493–504.

Sun PZ, Wang E, Cheung JS. Imaging acute ischemic tissue acidosis with pH-sensitive endogenous amide proton transfer (APT) MRI – correction of tissue relaxation and concomitant RF irradiation effects toward mapping quantitative cerebral tissue pH. NeuroImage 2012; 60: 1–6.

Sun PZ, Wang E, Cheung JS, Zhang X, Benner T, Sorensen AG. Simulation and optimization of pulsed radio frequency irradiation scheme for chemical exchange saturation transfer (CEST) MRI-demonstration of pH-weighted pulsed-amide proton CEST MRI in an animal model of acute cerebral ischemia. Magn Reson Med 2011; 66: 1042–8.

Sun PZ, Xiao G, Zhou IY, Guo Y, Wu R. A method for accurate pH mapping with chemical exchange saturation transfer (CEST) MRI. Contrast Media Mol Imaging 2016; 11: 195–202.

Sun PZ, Zhou J, Sun W, Huang J, van Zijl PC. Suppression of lipid artifacts in amide proton transfer imaging. Magn Reson Med 2005; 54: 222–5.

Sun PZ, Zhou J, Sun W, Huang J, van Zijl PC. Detection of the ischemic penumbra using pH-weighted MRI. J Cereb Blood Flow Metab 2007; 27: 1129–36.

Tee YK, Harston GW, Blockley N, Okell TW, Levman J, Sheerin F, et al. Comparing different analysis methods for quantifying the MRI amide proton transfer (APT) effect in hyperacute stroke patients. NMR Biomed 2014; 27: 1019–29.

Tee YK, Khrapitchev AA, Sibson NR, Payne SJ, Chappell MA. Evaluating the use of a continuous approximation for model-based quantification of pulsed chemical exchange saturation transfer (CEST). J Magn Reson 2012; 222: 88–95.

Terreno E, Cabella C, Carrera C, Delli Castelli D, Mazzon R, Rollet S, et al. From spherical to osmotically shrunken paramagnetic liposomes: an improved generation of LIPOCEST MRI agents with highly shifted water protons. Angew Chem Int Ed Engl 2007; 46: 966–8.

Tietze A, Blicher J, Mikkelsen IK, Ostergaard L, Strother MK, Smith SA, et al. Assessment of ischemic penumbra in patients with hyperacute stroke using amide proton transfer (APT) chemical exchange saturation transfer (CEST) MRI. NMR Biomed 2014; 27: 163–74.

Togao O, Hiwatashi A, Keupp J, Yamashita K, Kikuchi K, Yoshiura T, et al. Scan-rescan reproducibility of parallel transmission based amide proton transfer imaging of brain tumors. J Magn Reson Imaging 2015; 42: 1346–53.

Togao O, Keupp J, Hiwatashi A, et al. Amide proton transfer imaging of brain tumors using a self-corrected 3D fast spin-echo dixon method: Comparison With separate B0 correction. Magn Reson Med. 2016.

Trott O, Palmer AG 3rd. R1rho relaxation outside of the fast-exchange limit. J Magn Reson 2002; 154: 157–60.

Trott O, Palmer AG 3rd. Theoretical study of R(1rho) rotating-frame and R2 free-precession relaxation in the presence of n-site chemical exchange. J Magn Reson 2004; 170: 104–12.

van Zijl PC, Jones CK, Ren J, Malloy CR, Sherry AD. MRI detection of glycogen in vivo by using chemical exchange saturation transfer imaging (glycoCEST). Proc Natl Acad Sci U S A 2007; 104: 4359–64.

van Zijl PC, Zhou J, Mori N, Payen JF, Wilson D, Mori S. Mechanism of magnetization transfer during on-resonance water saturation. A new approach to detect mobile proteins, peptides, and lipids. Magn Reson Med 2003; 49: 440–9.

Varma G, Lenkinski RE, Vinogradov E. Keyhole chemical exchange saturation transfer. Magn Reson Med 2012; 68: 1228–33.

Vinogradov E, Soesbe TC, Balschi JA, Sherry AD, Lenkinski RE. pCEST: positive contrast using chemical exchange saturation transfer. J Magn Reson 2012; 215: 64–73.

Walker-Samuel S, Ramasawmy R, Torrealdea F, Rega M, Rajkumar V, Johnson SP, et al. In vivo imaging of glucose uptake and metabolism in tumors. Nat Med 2013; 19: 1067–72.

Wang Q, Zhang H, Zhang J, Wu C, Zhu W, Li F, et al. The diagnostic performance of magnetic resonance spectroscopy in differentiating high-from low-grade gliomas: a systematic review and meta-analysis. Eur Radiol 2016; 26: 2670–84.

Ward KM, Aletras AH, Balaban RS. A new class of contrast agents for MRI based on proton chemical exchange dependent saturation transfer (CEST). J Magn Reson 2000; 143: 79–87.

Ward KM, Balaban RS. Determination of pH using water protons and chemical exchange dependent saturation transfer (CEST). Magn Reson Med 2000; 44: 799–802.

Wen Z, Hu S, Huang F, Wang X, Guo L, Quan X, et al. MR imaging of high-grade brain tumors using endogenous protein and peptide-based contrast. NeuroImage 2010; 51: 616–22.

Windschuh J, Zaiss M, Meissner JE, Paech D, Radbruch A, Ladd ME, et al. Correction of B1-inhomogeneities for relaxation-compensated CEST imaging at 7 T. NMR Biomed 2015; 28: 529–37.

Wolff SD, Balaban RS. NMR imaging of labile proton exchange. J Magn Reson 1990; 86: 164–9.

Woods M, Woessner DE, Sherry AD. Paramagnetic lanthanide complexes as PARACEST agents for medical imaging. Chem Soc Rev 2006; 35: 500–11.

Wu R, Xiao G, Zhou IY, Ran C, Sun PZ. Quantitative chemical exchange saturation transfer (qCEST) MRI – omega plot analysis of RF-spillover-corrected inverse CEST ratio asymmetry for simultaneous determination of labile proton ratio and exchange rate. NMR Biomed 2015; 28: 376–83.

Xu J, Yadav NN, Bar-Shir A, Jones CK, Chan KW, Zhang J, et al. Variable delay multi-pulse train for fast chemical exchange saturation transfer and relayed-nuclear overhauser enhancement MRI. Magn Reson Med 2014; 71: 1798–812.

Xu X, Yadav NN, Knutsson L, Hua J, Kalyani R, Hall E, et al. Dynamic glucose-enhanced (DGE) MRI: translation to human scanning and first results in glioma patients. Tomography 2015; 1: 105–14.

Xu X, Yadav NN, Zeng H, Jones CK, Zhou J, van Zijl PC, et al. Magnetization transfer contrast-suppressed imaging of amide proton transfer and relayed nuclear overhauser enhancement chemical exchange saturation transfer effects in the human brain at 7T. Magn Reson Med 2016; 75: 88–96.

Yadav NN, Jones CK, Hua J, Xu J, van Zijl PC. Imaging of endogenous exchangeable proton signals in the human brain using frequency labeled exchange transfer imaging. Magn Reson Med 2013; 69: 966–73.

Yerbury JJ, Ooi L, Dillin A, Saunders DN, Hatters DM, Beart PM, et al. Walking the tightrope: proteostasis and neurodegenerative disease. J Neurochem 2016; 137: 489–505.

Yuan J, Zhou J, Ahuja AT, Wang YX. MR chemical exchange imaging with spin-lock technique (CESL): a theoretical analysis of the Z-spectrum using a two-pool R(1rho) relaxation model beyond the fast-exchange limit. Phys Med Biol 2012; 57: 8185–200.

Zaiss M, Angelovski G, Demetriou E, McMahon MT, Golay X, Scheffler K. QUESP and QUEST revisited - fast and accurate quantitative CEST experiments. Magn Reson Med 2017.

Zaiss M, Bachert P. Exchange-dependent relaxation in the rotating frame for slow and intermediate exchange – modeling off-resonant spin-lock and chemical exchange saturation transfer. NMR Biomed 2013; 26: 507–18.

Zaiss M, Schmitt B, Bachert P. Quantitative separation of CEST effect from magnetization transfer and spillover effects by Lorentzian-line-fit analysis of z-spectra. J Magn Reson 2011; 211: 149–55.

Zaiss M, Windschuh J, Goerke S, Paech D, Meissner JE, Burth S, et al. Downfield-NOE-suppressed amide-CEST-MRI at 7 Tesla provides a unique contrast in human glioblastoma. Magn Reson Med 2017; 77(1): 196–208.

Zaiss M, Windschuh J, Paech D, Meissner JE, Burth S, Schmitt B, et al. Relaxation-compensated CEST-MRI of the human brain at 7T: unbiased insight into NOE and amide signal changes in human glioblastoma. NeuroImage 2015a; 112: 180–8.

Zaiss M, Xu J, Goerke S, Khan IS, Singer RJ, Gore JC, et al. Inverse Z-spectrum analysis for spillover-, MT-, and T1 -corrected steady-state pulsed CEST-MRI – application to pH-weighted MRI of acute stroke. NMR Biomed 2014; 27: 240–52.

Zaiss M, Zu Z, Xu J, Schuenke P, Gochberg DF, Gore JC, et al. A combined analytical solution for chemical exchange saturation transfer and semi-solid magnetization transfer. NMR Biomed 2015b; 28: 217–30.

Zhang S, Merritt M, Woessner DE, Lenkinski RE, Sherry AD. PARACEST agents: modulating MRI contrast via water proton exchange. Acc Chem Res 2003; 36: 783–90.

Zhang S, Keupp J, Wang X, et al. Z-spectrum appearance and interpretation in the presence of fat: Influence of acquisition parameters. Magn Reson Med. September 2017. doi:10.1002/mrm.26900.

Zhang S, Winter P, Wu K, Sherry AD. A novel europium(III)-based MRI contrast agent. J Am Chem Soc 2001; 123: 1517–18.

Zhou J, Blakeley JO, Hua J, Kim M, Laterra J, Pomper MG, et al. Practical data acquisition method for human brain tumor amide proton transfer (APT) imaging. Magn Reson Med 2008; 60: 842–9.

Zhou J, Lal B, Wilson DA, Laterra J, van Zijl PC. Amide proton transfer (APT) contrast for imaging of brain tumors. Magn Reson Med 2003a; 50: 1120–6.

Zhou J, Payen JF, Wilson DA, Traystman RJ, van Zijl PC. Using the amide proton signals of intracellular proteins and peptides to detect pH effects in MRI. Nat Med 2003b; 9: 1085–90.

Zhou J, Tryggestad E, Wen Z, Lal B, Zhou T, Grossman R, et al. Differentiation between glioma and radiation necrosis using molecular magnetic resonance imaging of endogenous proteins and peptides. Nat Med 2011; 17: 130–4.

Zhou J, van Zijl PC. Chemical exchange saturation transfer imaging and spectroscopy. Progr NMR Spectr 2006; 48: 109–36.

Zhou J, van Zijl PC. Defining an acidosis-based ischemic penumbra from pH-weighted MRI. Transl Stroke Res 2011; 3: 76–83.

Zhou J, Wilson DA, Sun PZ, Klaus JA, Van Zijl PC. Quantitative description of proton exchange processes between water and endogenous and exogenous agents for WEX, CEST, and APT experiments. Magn Reson Med 2004; 51: 945–52.

Zhou J, Zhu H, Lim M, Blair L, Quinones-Hinojosa A, Messina SA, et al. Three-dimensional amide proton transfer MR imaging of gliomas: initial experience and comparison with gadolinium enhancement. J Magn Reson Imaging 2013; 38: 1119–28.

Zhu H, Jones CK, van Zijl PC, Barker PB, Zhou J. Fast 3D chemical exchange saturation transfer (CEST) imaging of the human brain. Magn Reson Med 2010; 64: 638–44.

Zu Z, Janve VA, Li K, Does MD, Gore JC, Gochberg DF. Multi-angle ratiometric approach to measure chemical exchange in amide proton transfer imaging. Magn Reson Med 2012; 68: 711–19.

Zu Z, Janve VA, Xu J, Does MD, Gore JC, Gochberg DF. A new method for detecting exchanging amide protons using chemical exchange rotation transfer. Magn Reson Med 2013; 69: 637–47.

Zu Z, Li K, Janve VA, Does MD, Gochberg DF. Optimizing pulsed-chemical exchange saturation transfer imaging sequences. Magn Reson Med 2011; 66: 1100–8.

12

MRS: ^{1}H Spectroscopy[1]

Contents

Yan Li and
Sarah J. Nelson
University of California San Francisco

12.1 Introduction

Conventional MRI can delineate structural abnormalities and changes in vascular parameters but fails to characterise features at the molecular and cellular levels. Magnetic resonance spectroscopy (MRS) is another application of the nuclear magnetic resonance phenomena that is able to differentiate the chemical composition of small tissue metabolites. The molecules in the brain are mobile and they mostly have narrow linewidths, which makes it easy to differentiate from each other (see Figure 12.1). Compared with tissue water that is the target for conventional MRI and is approximately 41 molar in the brain, these metabolites have much lower concentrations, which are in the range of 0.5–15 mM. This means that MRS techniques have relatively low sensitivity and have required significant technical development in order to adequately detect changes in brain metabolism that are associated with normal and pathological conditions.

The first *in vivo* MRS in the brain was reported in 1983 and used ^{31}P nuclei to assess phosphorus metabolism (Bottomley *et al.*, 1983; Cady *et al.*, 1983). Although measurements of ^{31}P metabolites can provide valuable information about cellular energetics, the higher sensitivity of protons and their abundant biological presence have made ^{1}H MRS the method of choice for most applications in the brain. With the increased availability of MR systems with higher field strength and more sensitive radiofrequency (RF) coils, MRS has become more commonly applied in clinical and preclinical research settings. Point resolved spectroscopy (PRESS) (Bottomley, 1987) and stimulated echo acquisition mode (STEAM) (Frahm *et al.*, 1989) are commonly used to limit the signals obtained to subregions of the brain but require optimised methods for water and lipid suppression in order to provide good quality data. Phase encoding or other *k*-space sampling methods may also be applied in conjunction with whole slice or volume selection to provide 1D, 2D or 3D arrays of spectra.

This chapter begins by reviewing the physical principles of MRS and the biological significance of brain metabolites. This is

[1] Reviewed by Franklyn Howe.

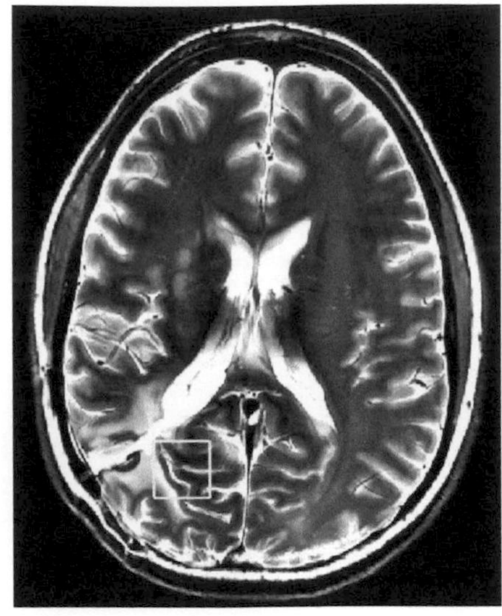

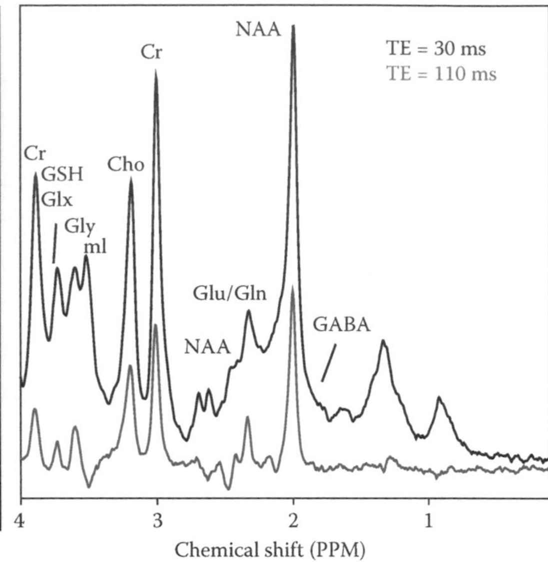

FIGURE 12.1 Single-voxel short (black) and long (grey) echo time ^{1}H magnetic resonance spectroscopy (MRS) from a patient with brain tumour at 7T. Each spectra dataset was acquired with semi-LASER (sLASER, localised adiabatic spin-echo refocusing) localisation, VAPOR water suppression, voxel size = 2 × 2 × 2 cm^{3}, TR = 3 s and 64 averages.

followed by a summary of the methodologies used for acquiring, post-processing and quantifying ^{1}H MR spectra, together with practical considerations for designing protocols for obtaining such data. The final section presents a number of clinical applications and provides examples of how the results can be valuable for diagnosing different pathologies, assessing prognosis and monitoring response to treatment.

12.2 Physical principles

When experiencing a magnetic field, protons precess at the resonance (Larmor) frequency. The negative-charged electrons surrounding the nucleus can shield or oppose the external magnetic field, resulting in different resonance frequencies based on chemical functional groups. The amount of shift in resonance is called *chemical shift*. Nuclei within a molecule may have a range of different chemical shifts, with their resonance frequency being dependent on their molecular environment and the external field strength. Chemical shift is expressed in units of parts per million (ppm), which are independent of the strength of the main magnetic field. Effects that influence chemical shift include pH and temperature.

To obtain the MRS signal, a RF pulse is applied perpendicularly to the main magnetic field in order to perturb thermal equilibrium. This causes the coherent precession of the nuclear magnetic moments of the protons in the *x–y* plane and induces a voltage in the receiver coil. The MRS signal, which is typically referred to as a *free-induction decay* (FID), is acquired in the time domain and then Fourier transformed into the frequency domain to give the spectrum. The spectrum shows peaks corresponding to the resonance frequencies of the magnetically distinct proton environments with the sample. The area underneath a spectral peak is proportional to the total number of protons in the sample. Metabolite levels can thus be measured by quantifying the intensities of peaks in the spectrum, which will be described in more detail in Section 12.6.

Some brain metabolites, such as glutamate, have resonances split into several small lines, making them appear as multiplets in the spectrum. This phenomenon is referred to as J-coupling (spin–spin coupling) and is caused by the interaction between electrons and the adjacent nucleus through a small number of chemical bonds. Unlike chemical shift, J-coupling is independent of the applied magnetic field strength and may make the quantification of ^{1}H spectra more complex. The presence of multiplets may be valuable for identifying the signatures of particular chemical species and can be used in conjunction with more advanced spectral editing strategies to separate metabolites with overlapping peaks. This is discussed in Section 12.4.4.

12.3 Biological significance of brain metabolites

The majority of the studies that have been performed to date have used long echo time (TE) methods (>100 ms) in order to provide robust estimate of the major metabolites that have high signal-to-noise ratio (SNR). These include choline-containing compounds (Cho), creatine (Cr) and N-acetyl aspartate (NAA), lactate (Lac) and lipid (Lip). Shorter TE sequences (<40 ms) yield additional information from other metabolites in the spectrum, such as glutamate (Glu), glutamine (Gln), myo-inositol (mI) and macromolecules. An example of ^{1}H MRS from human brain that shows the major metabolites is given in Figure 12.1. A full list of brain metabolic compounds, their chemical shifts, J-coupling and solution spectra were presented in a previous publication (Govindaraju *et al.*, 2000). The biological and clinical significance of these metabolites is discussed below.

NAA is an amino acid that is only present in brain and spinal cord. It is synthesised in the mitochondria of the neuron and hydrolysed in oligodendrocytes. It has also been suggested to function as an osmolyte (Baslow, 2003). Since NAA is considered to be a neuronal marker, any pathology with a loss of neurons or function results in reduced levels of NAA. The main peak resonates at 2.01 ppm in ^{1}H spectra and originates from the methyl protons of NAA. The unresolved N-acetylaspartylglutamate (NAAG), which is located at 2.04 ppm, also provides a contribution to the observed peak (Frahm *et al.*, 1991; Pouwels and Frahm, 1997). NAAG is a neurotransmitter that plays a key role in cell signalling (Baslow, 2000).

The *in vivo* Cho peak resonates at 3.22 ppm and represents the total Cho brain stores, including Cho, phosphocholine and glycerophosphocholine. These compounds play important roles in phospholipid metabolism, with the elevation of Cho indicating increased membrane turnover. Cho has been observed to increase in brain malignancy (Fulham *et al.*, 1992; Ott *et al.*, 1993), in demyelinating processes (Matthews *et al.*, 1991; Richards, 1991), inflammation (Brenner *et al.*, 1993) and several other pathologies. Reduced Cho has been reported in hepatic encephalopathy (Kreis *et al.*, 1992).

The Cr peaks that resonate as singlets at 3.0 (CH_2 protons) and 3.9 ($N(CH_3)$ protons) ppm include contributions from creatine and phosphocreatine and are associated with energy metabolism. The level of Cr is typically considered as a marker of cellular bioenergetics. The singlet of Cr at 3.0 ppm is frequently used as a baseline reference level for other metabolites, since it was originally thought to be relatively unaffected in pathological conditions. More recently, a number of studies have shown regional differences in normal brain (Jacobs *et al.*, 2001), as well as both increased and reduced Cr in tumour (Hattingen *et al.*, 2008; Howe *et al.*, 2003) and the absence of Cr in Cr deficiency syndromes (Cecil *et al.*, 2001; Stockler *et al.*, 1994).

Glu (excitatory) and gamma-aminobutyric acid (GABA, inhibitory) are the main neurotransmitters in the brain, while Gln is an important energy source in mitochondria. The levels of Glu and GABA in normal brain are maintained by the Glu–Gln and GABA–Gln cycles (Bak *et al.*, 2006; Schousboe *et al.*, 2013). Although these metabolites can be detected in spectra acquired with short TE, peak overlap can make it difficult to isolate individual components. To accommodate this limitation, current analyses often use an index known as Glx that represents a combination of Glu and Gln (Kreis *et al.*, 1992), while GABA is usually detected using spectral editing methods (Mullins *et al.*, 2014) (see Section 12.4.4).

mI is predominantly located within glial cells, is involved in cellular signalling (Fisher *et al.*, 1992) and acts as an osmoregulator (Fisher *et al.*, 2002). It has been suggested as a glial marker for *in vivo* MRS studies (Brand *et al.*, 1993). The mI peaks often overlap with the peak of glycine (Gly), which is an inhibitory neurotransmitter and resonates as a singlet at 3.56 ppm. Because of the difficulty in separating these components, mIG (the sum of mI and Gly) is often referred to in the literature.

Lactate is the end product of anaerobic glucose metabolism and is hardly visible in normal brain spectra because of its relatively low concentration (~0.5 mM). Increased lactate indicates hypoxia or poor perfusion in pathological conditions. The lipids in the cell membrane do not significantly contribute to the lipid resonance in proton spectra due to their very short T_2 relaxation times. Lipid peaks that are detected at 1.3 ppm come from methylene groups in long alkyl chains that result from the formation of necrosis or as peaks that are aliased from subcutaneous lipid. These peaks overlap with the methyl group resonances of lactate and have meant that J-difference spectral editing methods are typically applied to separate the signals from Lac from Lip peaks (Star-Lack *et al.*, 1998).

12.4 Acquisition of ¹H MRS data

The methods used to obtain *in vivo* ^{1}H MRS data combine volume selection with saturation of unwanted signals and *k*-space localisation. Because an extra time/frequency dimension of data is acquired and the concentration of the metabolites is much lower than that of water, the acquisition time is typically longer than for other types of MR data. The following addresses spatial localisation, strategies for suppressing water and lipid signals, methods for obtaining single and multivoxel data, and advanced techniques that provide improved spatial coverage and rapid acquisition times.

12.4.1 Spatial localisation

The most widely clinically used methods for localising ^{1}H signals to a specific region of the body are PRESS (Bottomley, 1987) and STEAM (Frahm *et al.*, 1989). The former applies a double spin echo sequence and has the advantage of providing twice the SNR, while the latter utilises three 90° pulses and allows shorter TE (Moonen *et al.*, 1989). The accuracy of localisation in these methods depends upon the efficiency of the RF pulses used for volume selection. Adiabatic pulses are often preferred because they are relatively insensitive to B_1 inhomogeneity (Tannus and Garwood, 1997). Excitation and refocusing RF pulses that satisfy the adiabatic condition have been implemented in the LASER pulse sequence (Garwood and DelaBarre, 2001; Slotboom *et al.*, 1991) and its simplified form, 'sLASER' (Scheenen *et al.*, 2008a, 2008b). They produce a more uniform excitation profile but at the cost of relatively longer repetition time (TR) and minimum TE. Another acquisition that has been designed for short TE MRS at ultra-high field (7T or greater) is SPECIAL (spin-echo full-intensity acquired localised spectroscopy) (Mlynarik *et al.*, 2006). This combines 1D image-selected *in vivo* (Ordidge *et al.*, 1986) and 2D spin echo localisation.

12.4.2 Water and lipid suppression

The water signal provides an extremely large peak that dominates *in vivo* proton spectra and is typically suppressed in order to reveal changes in the much smaller metabolite peaks (Figure 12.2). This is most commonly achieved prior to spatial localisation by using chemical shift selective (CHESS) pulse (Haase *et al.*, 1985). In practice, multiple repetitions of CHESS pulses are frequently required in conjunction with optimised flip angles and delays in order to achieve an adequate suppression (Ernst and Hennig, 1995). The recently developed VAPOR (variable power RF pulses with optimised relaxation delays) uses a combination of seven pulses to minimise sensitivity to B_1 inhomogeneity and T_1 variations (Tkac *et al.*, 1999). Alternatives are to only excite the frequency of interest, such as spectral–spatial pulses with the metabolite of interest in the pass band and water/lipid in the stop band (Star-Lack *et al.*, 1997b), or to incorporate frequency selective RF inversion pulses inside PRESS, such as band-selective inversion with gradient dephasing (Star-Lack *et al.*, 1997a) and Mescher-Garwood (Mescher *et al.*, 1998).

Outer volume saturation (OVS) pulses (Duyn *et al.*, 1993) are commonly applied to eliminate unwanted lipid signals from outside the volume of interest. This is particularly important for covering regions such as the scalp that contain high lipid volume. A number of different types of pulses may be used in several different spatial configurations (Osorio *et al.*, 2009; Tran *et al.*, 2000). Depending on the location and size of the selected volume, optimisation of the location and orientation of such pulses may be quite complicated and automated methods for prescribing their configuration are beneficial for clinical applications (see Section 12.5.7). Other options that have been implemented for lipid suppression are the application of an inversion recovery pulse for lipid nulling (Bydder and Young, 1985) or limiting the excitation pulses to act upon a small range of frequencies that correspond to the tissue metabolites of interest.

12.4.3 Single-Voxel MRS and multivoxel MRS imaging

The basic elements required to obtain single-voxel ^{1}H MRS data comprise water suppression, lipid suppression and spatial localisation. Typical voxel sizes are 4–27 cm^3. This is most commonly available methodology that is applied in the clinical setting with a total acquisition time of 5–10 min. It has the advantage of being simple to implement but covering multiple anatomical regions and consistent placement for specific anatomic locations are challenging. Magnetic resonance spectroscopic imaging (MRSI) is used to evaluate the spatial distribution of different metabolites. The

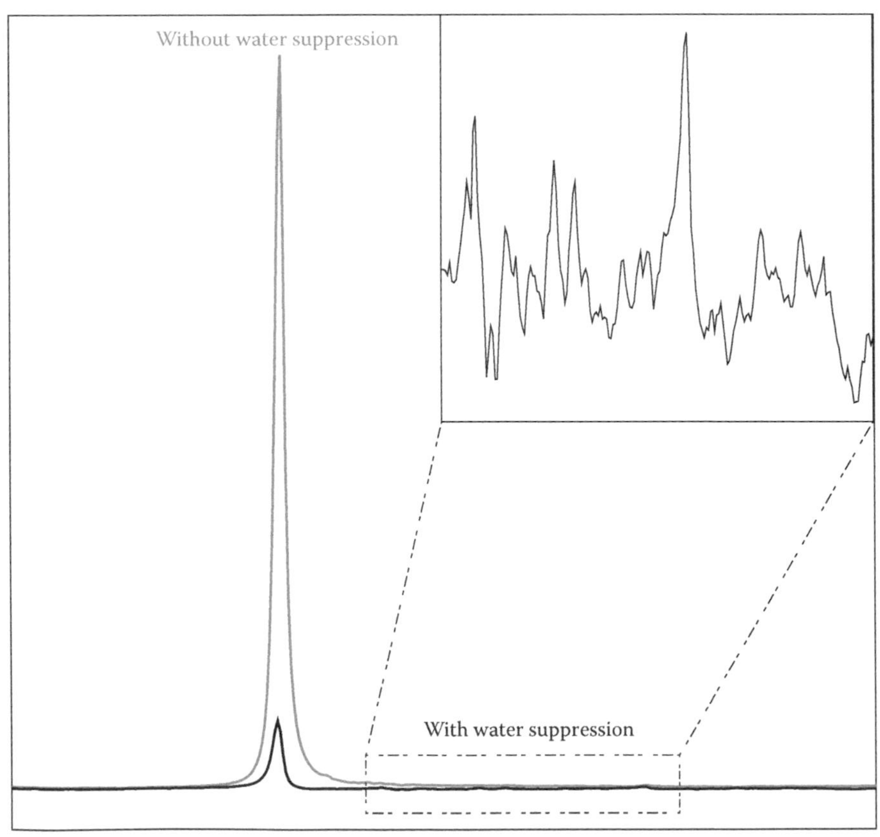

FIGURE 12.2 Single-voxel ^{1}H MRS from a healthy volunteer without and with CHESS water suppression at 7T (PRESS localisation, TE/TR = 35/2000 ms, 1 average, voxel size = 2 × 2 × 2 cm^3; total acquisition time 2 s, respectively). Compared to water, brain metabolites have much lower concentrations and typically 64–128 averages are required for a voxel of 2 × 2 × 2 cm^3 to obtain ^{1}H MRS with good signal-to-noise ratio (SNR). TE = echo time; TR = repetition time.

initial studies used standard phase-encoding procedures to generate the spatial information (Maudsley *et al.*, 1983), but the time required to cover large regions of the brain (20–30 min) is a limiting factor for clinical applications. Most MRSI acquisitions use PRESS to preselect a rectangular volume, which reduces the field of view required and eliminates signals from unwanted regions of the brain. An alternative MRSI spectral localisation method that uses a spin-echo sequence for a single-slice excitation is able to offer shorter TE but places a higher demand on the methods used for water and lipid suppression (Duyn *et al.*, 1993).

12.4.4 Techniques for J-Coupled metabolites

The spin echo sequence can refocus chemical shift and inhomogeneity, but the relative phase difference for the coupled resonances will depend on TE and J, the coupling constant. By acquiring a series of equally spaced TEs, 2D J-resolved spectroscopy can separate J-coupling information from chemical shift by encoding in two different dimensions (Bruch and Bruch, 1982). The addition of the second frequency dimension also allows the separation of lactate from lipid (Li *et al.*, 2007; Thomas *et al.*, 1996). Averaging across spectra with different TE values that are obtained using this acquisition scheme also offers unobstructed detection of Glu at 3T. This approach has been termed *TE-averaged PRESS* (Hurd *et al.*, 2004). Since data are acquired at multiple echo times, estimation of the T_2 relaxation times of uncoupled resonances can be performed and used in the absolute quantification of metabolite levels.

The separation of partially overlapping resonances in MR spectra may be achieved by spectral editing methods that take advantage of the coupling patterns of its protons. The BASING (Star-Lack *et al.*, 1997a) and MEGA (Mescher *et al.*, 1998) methods are the two most common acquisition schemes that are applied in conjunction with different editing pulses to obtain unobstructed GABA (Mullins *et al.*, 2014), lactate (Park *et al.*, 2011; Star-Lack *et al.*, 1998) and glutathione (Srinivasan *et al.*, 2010b; Terpstra *et al.*, 2003). Examples of detecting these metabolites using spectral editing methods are shown in Figure 12.3.

12.4.5 Speeding up the acquisition of MRSI data

The additional spatial coverage provided by 3D MRSI allows metabolic parameters from regions of pathology to be compared with those from contralateral normal brain. It is also possible to use the anatomic images to register the data from serial scans and analyse changes in metabolism that are associated with disease progression and response to therapy. In order for 3D MRSI data to be acquired in a clinically feasible scan time it is necessary to take advantage of fast MRSI techniques that use a combination of phase encoding and alternative *k*-space sampling methods to provide adequate coverage. Reviews and comparisons on these fast techniques can be found in a number of different publications (Pohmann *et al.*, 1997; Posse *et al.*, 2013; Zhu and Barker, 2011; Zierhut *et al.*, 2009). The most commonly used techniques are shown in Box 12.1

12.5 Key factors in determining data quality

The specialised methods used to obtain ^1H MRS data and the need to distinguish between signals coming from multiple metabolites place a major emphasis on optimising the hardware and software capabilities of the scanner. The first step is to select the field strength and RF coils that can provide the best balance between sensitivity and uniformity of the B_0 and B_1 fields. Other critical factors are the implementation of selection pulses that reduce chemical shift artefacts, of methods that minimise lipid contamination and of strategies for reducing the impact of patient motion. The values of TR and TE that are chosen for a specific application must be tailored to provide optimal detection of the metabolites of interest within the time available. Once the data acquisition parameters have been selected, the success of the scan is dependent upon the operator being able to define a prescription that covers the region of interest and places the OVS bands in the appropriate location. While this can be done reproducibly by an experienced technologist, the reliability is much improved by using an automated procedure.

12.5.1 Field strength and use of phased array coils

The majority of scanners being used to perform brain studies at the current time have a field strength of 3T and have been shown to provide increases in SNR and spectral resolution over the prior clinical standard of 1.5T for both single voxel ^1H MRS and 3D MRSI (Barker *et al.*, 2001; Gonen *et al.*, 2001; Li *et al.*, 2006; Srinivasan *et al.*, 2004). The combined effect of using a 3T MR scanner with an eight-channel phased array coil was seen to provide an increase in the SNR of metabolite peaks of more than 2.33 times the corresponding acquisition performed at 1.5T with a quadrature volume head coil (Li *et al.*, 2006). More recently, there has been an increase in the number of scanners in research imaging centres that have ultra-high field strength (7T). Studies that have been performed using ^1H MRS over field strengths ranging from 1.5T to 7T have shown a linear gain in SNR and decrease in spectral line width in ppm (Mekle *et al.*, 2009; Otazo *et al.*, 2006). This improvement in sensitivity is critical for applications that require metabolic data to be obtained with shorter acquisition times or improved spatial resolution. The use of phased array coils (Moyher *et al.*, 1995) with high numbers of channels not only increases the SNR of ^1H MRS but also facilitates the implementation of parallel imaging and provides increased coverage without increasing the total acquisition time.

12.5.2 B_0 and B_1 inhomogeneity

^1H MRS is highly susceptible to magnetic field inhomogeneity. This results in broadening of peak linewidths, which are inversely proportional to T_2*, and a shift in the frequency of the peaks, which can affect water suppression and spectral editing. Regions near boundaries with large susceptibility differences, such as tissue–bone or tissue–air interfaces, are challenging to study using ^1H MRS. Concerns on variations in susceptibility and B_1 are increased

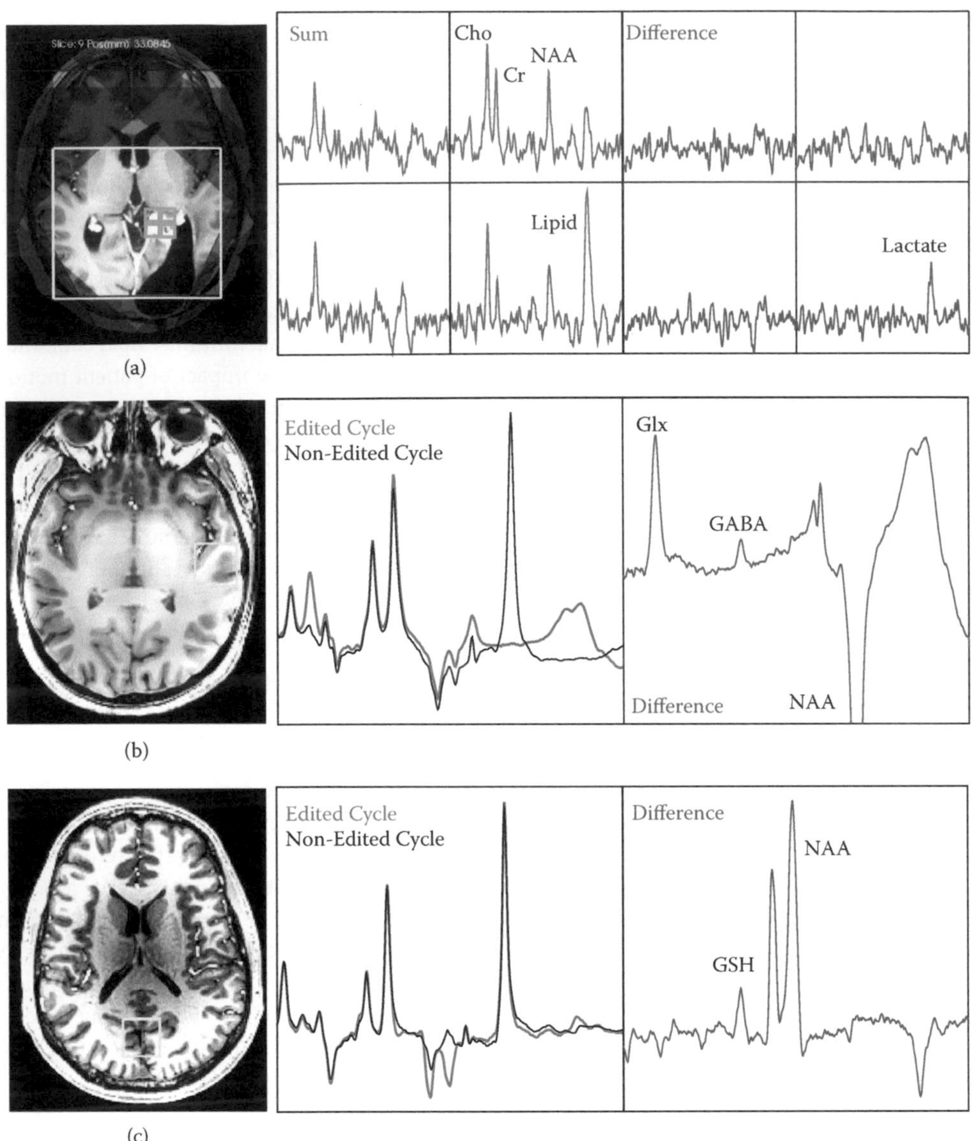

FIGURE 12.3 Spectral Editing ^{1}H MRS: (a) 3D Lactate-edited MRSI from a patient with glioma at 3T (PRESS localisation, CHESS water suppression, TE/TR = 144/1250 ms, matrix size = 18 × 18 × 16, spatial resolution = 1 × 1 × 1 cm^3, flyback trajectory in S/I, total acquisition time ~10 min); (b) GABA-edited MRS from a healthy volunteer at 7T (PRESS localisation, CHESS water suppression, TE/TR = 68/2000 ms, spatial resolution = 2.5 × 2.5 × 2 cm^3, matrix size = 8 × 8 × 1, total acquisition time ~4.5 min); **(c)** Glutathione (GSH)-edited MRS from a healthy volunteer at 7T (sLASER localisation, VAPOR water suppression, TE/TR = 68/3000 ms, spatial resolution = 2 × 2 × 2 cm^3, 64 steps editing on/64 steps editing off; total acquisition time ~6.5 min). GABA = gamma-aminobutyric acid.

when using a scanner with high field strength. To compensate for local variations in the magnetic field most MR systems are equipped with shim coils and software routines that are capable of providing automatic first- and second-order shimming (Gruetter, 1993; Gruetter and Tkac, 2000; Kim *et al.*, 2002). Higher order (third and fourth) shims (Pan *et al.*, 2012a) or dynamic shimming (Juchem *et al.*, 2010) have been recommended for 7T scanners but are not yet widely available. Factors contributing to intensity non-uniformity in MRI (Belaroussi *et al.*, 2006), such as RF coil inhomogeneity, pulse sequence and the nature of tissue itself, also cause problems for 3D MRSI data. The non-uniform RF excitation can be reduced by using adiabatic excitation and refocusing pulses, which are able to provide a more uniform flip angle over a range of B_1 values. Parallel transmission, or multitransmit technology, is recommended to improve the spatial homogeneity of RF pulses at 3T and above (Katscher and Bornert, 2006; Ugurbil, 2014).

12.5.3 Volume selection and chemical shift artefacts

It is well known that PRESS and other spatial selection techniques result in spatial variations in metabolite ratios at the edges of the

Box 12.1 Common techniques for accelerating MRSI Acquisition

1. Ellipsoidal *k*-space sampling reduces the number of rectangular phase encoding steps by cutting the corners of *k*-space (Maudsley *et al.*, 1994). It is easy to implement but increases the effective voxel size and hence leads to more severe partial volume effects.
2. Echo-planar spectroscopic imaging (EPSI) (Cunningham *et al.*, 2005; Posse *et al.*, 1995) and spiral *k*-space sampling are able to localise signals more rapidly than standard rectilinear phase encoding (Adalsteinsson *et al.*, 1998). These significantly reduce the acquisition time by using time-varying gradients to map spatial and spectral dimensions during a single readout. Of the two methods, EPSI is relatively simple to implement and has become the most prevalent fast 3D MRSI for evaluating brain pathologies. The issues that need to be considered when designing echo planar or spiral *k*-space trajectories are the gradient performance, the trade-off between spectral bandwidth and spatial resolution, and the SNR efficiency.
3. Parallel imaging techniques that were originally developed for speeding up MRI acquisitions. The basic premise of these techniques is that it is possible to use the sensitivity profiles from multiple receive coils to encode spatial information. The most common strategies are sensitivity encoding (SENSE) (Pruessmann *et al.*, 1999) and generalised auto-calibrating partially parallel acquisitions (GRAPPA) (Griswold *et al.*, 2002). The same principles have been applied in a research setting for speeding up MRSI in conjunction with rectangular phase encoding (Dydak *et al.*, 2001; Ozturk-Isik *et al.*, 2006), with elliptical phase-encoding MRSI (Banerjee *et al.*, 2009; Ozturk-Isik *et al.*, 2009) and with EPSI (Lin *et al.*, 2007; Sabati *et al.*, 2014). Problems that are encountered in using these methods are similar to those observed with phase-sensitive imaging applications and include an increase in artefacts due to large geometry (g)-factors, inaccurate sensitivity profiles and lipid contamination.
4. Other advanced sampling methods, such as compressed sensing have been applied to obtain MRSI data in cases where there is a relatively high signal. Examples are in accelerating ^{13}C MRSI for hyperpolarised metabolic imaging (Hu *et al.*, 2008) and in using high-resolution SPICE (spectroscopic imaging by exploiting spatiospectral correlation) MRSI (Lam *et al.*, 2016; Ma *et al.*, 2016). These methods are still under development and have not yet been used in clinical studies.

excited volume. This phenomenon is referred to as a *chemical shift artefact* or *chemical shift misregistration* (Figure 12.4). The factors that determine the magnitude of the effect are the size of the selected region, the field strength, the bandwidth of the RF pulse being used and the frequency offset. One way of reducing the impact of the artefact is to increase the size of the prescribed PRESS box by a factor so that it is larger than the region of interest (overPRESS) (Li *et al.*, 2006) and then to use very selective suppression (VSS) pulses (Tran *et al.*, 2000) to suppress signal from the outer regions where the artefact occurs. This has the added advantage of sharpening up the edges of the selected region and can also be used to further shape the region of interest as shown schematically in Figure 12.5. Alternative strategies that are able to compensate for this effect are to correct the differences in metabolite levels in the edge voxels during post-processing (Li *et al.*, 2006) and to use high bandwidth adiabatic (Tannus and Garwood, 1997) or spectral/spatial pulses (Star-Lack *et al.*, 1997b) to more precisely select the region of interest.

12.5.4 Lipid contamination

Lipids are not present in the healthy brain, but inappropriate or inaccurate prescription of the selected volume can result in the spectrum being contaminated with signals from subcutaneous tissue. While this can be minimised for single-voxel ¹H MRS data by careful choice of the selected volume and judicious placement of OVS bands, the effect becomes more significant for 3D MRSI because phase-encoding procedures are used to generate the spatial information (see Section 12.4.3) and the shape of the point spread function (PSF) for typical acquisitions is such that contributions from incompletely suppressed lipid peaks can bleed into voxels within the selected volume. While OVS bands can be used to reduce this effect, manual prescription of multiple oblique planes can be time-consuming and prone to error. Given that voxels from regions of necrosis have been shown to have lipid peaks, it may be difficult to distinguish between signals from regions of pathology and signals that correspond to artefacts. Shimming variations can also cause frequency and phase shifts of the lipid signals originating from outside the region of interest. This results in lipid peaks that appear in unexpected locations in the spectra, leading to additional phasing and measurement problems. Methods for suppressing lipid are discussed in more detail in Section 12.4.2.

12.5.5 Motion artefacts

Unlike MRI, where the impact of motion can be readily observed by visual inspection, it may be challenging to distinguish artefacts that are associated with head movement during the acquisition of ¹H MRS data from other factors that influence data quality. Spectra suffering from this effect have signals that come from parts of the brain that were not intended to be excited. This causes a reduction in the SNR of metabolite peaks, an increase in linewidth, poor water suppression and imprecise localisation (Kreis, 2004). To minimise these

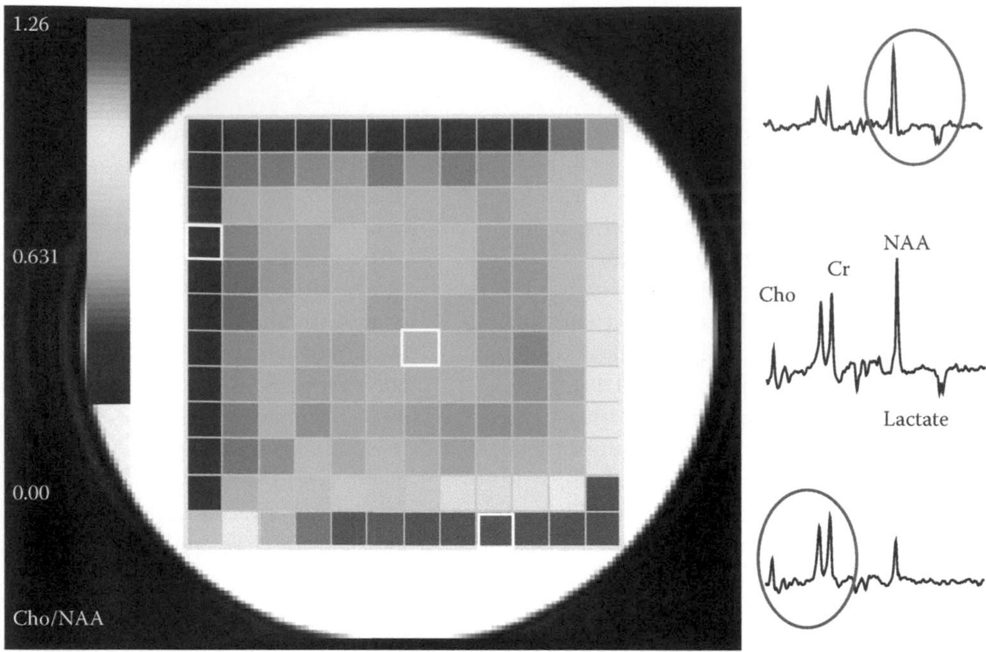

FIGURE 12.4 Example of chemical shift artefact: 3D MRS imaging data acquired from a standard GE MRS phantom that contains 12.5 mM NAA, 10 mM Cr, 3 mM Cho, 12.5 mM Glu, 7.5 mM mI, and 5 mM lactate with TE/TR = 144/1250 ms, matrix = 18 × 18 × 12 and an overPRESS factor of 1.0 (i.e. without scanning a volume larger than interest) at 3T. The spectra from the top right corner have low Cho/NAA while those for the opposite have higher levels. NAA = N-acetyl aspartate; Cr = creatine; Cho = choline; Glu = glutamate; mI = myo-inositol.

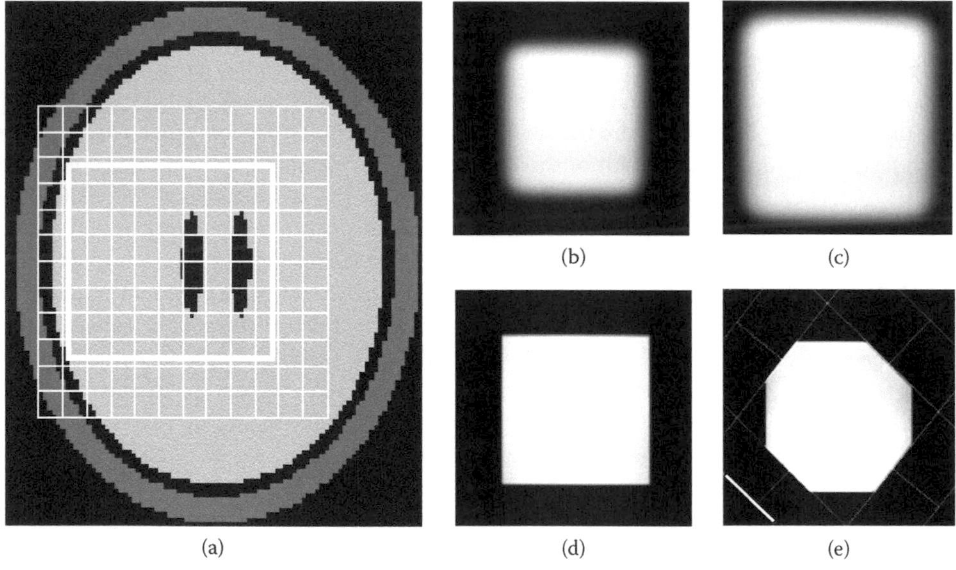

FIGURE 12.5 Improving PRESS volume selection using overPRESS and VSS bands. The required volume and grid size are shown in (a), with an image of what would be actual excited excited volume in (b). If the size of the PRESS volume is increased (c) the VSS bands can be used to sharpen up the edges of the selected volume and to eliminate regions that would contain chemical shift artefacts (d). Additional VSS bands can be used to further modify the shape of the region of interest (e), which may be important in reducing lipid contamination. VSS = very selective suppression.

effects, subjects should be placed in a comfortable position with padding between the holder and their head. The scanner operator should also check carefully and frequently, with the imaging sequences obtained before and after the MRS data being examined, to see if there has been movement during the scan. A variety of prospective and retrospective methods for motion correction have been reported, such as optical tracking (Zaitsev *et al.*, 2010), integrated image-based adaptive motion correction module (PROMO) with PRESS sequence (Keating *et al.*, 2010), EPI-based navigators (Hess *et al.*, 2012) and strategies for phase/frequency correction (Bhattacharyya *et al.*, 2007; Gabr *et al.*, 2006).

12.5.6 Choice of TE and TR

Although there are more peaks present and higher SNR in short TE spectra, peak overlap and the increase in the magnitude of signals from lipid and macromolecules may make it difficult to identify individual components. It is for this reason that longer TEs (80–144 ms) are still most commonly utilised in clinical studies to evaluate the major brain metabolites. Another factor that influences the choice of TE for applications to patients with brain tumours is that there are differences in T_2 relaxation times that result in the ratio of Cho/NAA at longer TE being able to better differentiate the lesion from normal brain (Li *et al.*, 2008). For metabolites with complicated coupling patterns, the choice of TE is typically driven by the values that gives the unobstructed peak or peaks with optimal SNR. Examples are using a TE of 97 ms to detect 2-hydroxyglutarate (Choi *et al.*, 2012), and implementing J-coupling dependent acquisitions with a TE of 144 ms for obtaining lactate-edited spectra (Star-Lack *et al.*, 1998). T_1 relaxation times of brain metabolites range from 1 to 2 s (Traber *et al.*, 2004). For single voxel ^1H MRS, multiple signal averages are typically acquired to obtain adequate SNR; TR values are chosen to be 2 s or longer. For ^1H MRSI data, multiple excitations are required to perform the phase encoding and often a single signal average is acquired; TR values chosen for patient studies are typically shorter (1–2 s) so that the overall acquisition time is clinically feasible.

12.5.7 Methods for automatic prescription

When a specific anatomic region is of interest, differences in head shape, brain anatomy and positioning between examinations may make it challenging to prescribe the same or similar locations for acquiring single-voxel ^1H MRS. This adds to the uncertainty in comparing the results obtained between patients and even from the same patient in serial scans. One approach to overcome this effect in longitudinal studies is to first register the MR images to the baseline scan and then obtain a localiser at the appropriate oblique angle that can be used for prescribing single-voxel ^1H MRS from a similar location (Hancu *et al.*, 2005). Although retrospective voxel shifting can be applied for 3D ^1H MRSI, choosing exactly the same slice or selected volume, as well as placing multiple OVS bands to cover regions with high lipid signal while maintaining the signals that are interested, requires a skilled scanner operator. In a clinical setting it is not always possible to have the same technologist perform serial examinations and so, even with the most experienced individuals, there may be interoperator variability in defining regions of anatomy. Methods for automatic prescription of the volume of interest and/or of the OVS bands have been developed (Bian *et al.*, 2015; Ozhinsky *et al.*, 2011, 2013 Yung *et al.*, 2011) and have shown their ability to facilitate the acquisition process and offer consistent data quality for comparative analysis.

12.6 Post-Processing and estimation of metabolite levels

One of the challenges for using ^1H MRS in clinical studies has been that specialised tools are needed to reconstruct and quantify the resulting data. These comprise a combination of methods used for analysis of *ex vivo* NMR spectra and spatial reconstruction strategies that were developed for state-of-the-art MRI techniques. A number of recent reviews of *in vivo* MRS have included discussions of the key issues that need to be considered in order to produce quantitative results (Alger, 2010; Buonocore and Maddock, 2015; Graveron-Demilly, 2014; Posse *et al.*, 2013; Zhu and Barker, 2011). While several open-source software packages for analysing ^1H MRS data from the brain have been implemented by research groups, few of them are general enough to cover the entire range of possible acquisition methods and applications. The following focuses first on the processing steps required for quantifying individual ^1H spectra and then on the more challenging additional aspects associated with multivoxel datasets.

12.6.1 Factors to consider for individual spectra

Key steps in reconstructing and processing the raw data to generate spectra and quantify the peaks are summarised in Figure 12.6. The first task is to convert from the manufacturer specific format into a standardised format that can be read by the software package being used. The complex signals observed from a single ^1H resonance from a uniform sample in a homogenous magnetic field corresponds to a decaying exponential with phase that is described by the resonance frequency of the corresponding chemical species. After Fourier transform this corresponds to a peak with a shape described by a Lorentzian function.

$$y(t) = c\ exp\ (-kt)\ exp\ (2\pi i\omega_o t)) \tag{12.1}$$

$$Y(\omega) = (ck/(k^2 + 4\pi^2(\omega - \omega_o)^2), 2\pi(\omega - \omega_o) / c\ (k^2 + 4\pi^2(\omega - \omega_o)^2)) \tag{12.2}$$

where c is the intensity, k is the decay constant and ω_o is the resonance frequency. Molecules with multiple ^1H resonances and/or mixed samples can be represented as a weighted sum of individual functions that produce a spectrum with peaks at distinct frequency locations. Figure 12.7 shows an example of simulated data with two distinct peaks. For *in vivo* spectra, where there is often considerable peak overlap, tissue heterogeneity or spatial variations in the underlying B_0 magnetic field, the observed peak shapes are often distorted (de Graaf and Bovee, 1990) and alternative functions are used to represent them. Options that have been proposed are the Gaussian and more general Voigt distribution (Gillies *et al.*, 2006; Marshall *et al.*, 1997), or a unique line shape function in LCModel (Provencher, 1993).

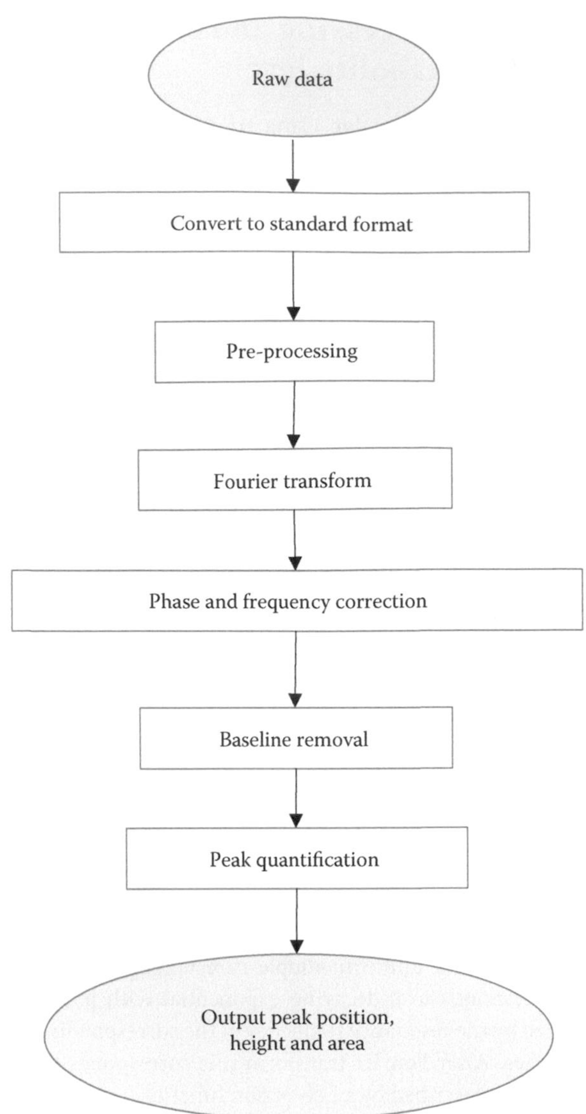

FIGURE 12.6 Schematic of steps that may be applied to process the raw MRS data obtained from the scanner and to generate estimates of peak parameters. Preprocessing may entail removal of residual water, apodization and zero filling in the time domain.

The metabolite peaks are described empirically by their frequency position height and width at half height. Variations in the RF pulses used for excitation or delays in sampling may cause frequency dependent variations in the phase of individual peaks, which also needs to be corrected in order to provide upright peaks that can be clearly identified upon visual inspection. The real part of the spectrum is preferred for visualising multiple peaks because, as is seen in Figure 12.7, they are narrower than in the corresponding magnitude spectrum. Once they have been phased, the ability to detect the resonances of interest depends upon the accuracy of the shim and hence width of individual peaks.

Another component that needs to be considered and may be a limiting factor in detecting and quantifying signals from individual peaks is the contribution from random noise. This is assumed to have a Gaussian distribution with constant variance at each point. The SNR of a peak is used to provide a sense of the reliability of quantification and is described by its height divided by the standard deviation of the noise in the spectrum. As is shown in Figure 12.7, apodization of the time domain signal with a Lorentzian, Gaussian or other low-pass filter significantly reduces the contribution from random noise but at the expense of broadening individual peaks, hence increasing the likelihood of there being overlap between them.

In addition to the metabolites of interest, the *in vivo* signal includes contributions from residual (partially suppressed) water and from macromolecules, which correspond to relatively broad peaks in the spectrum. Depending on the reliability of the volume selection methods used, there may also be contamination from subcutaneous lipids (Ebel *et al.*, 2001). The general signal equation for the acquired FID or echo and the spectrum can thus be represented as follows:

$$y(t) = rw(t) + m(t) + l(t) + \sum_{j=1}^{n} c_j \, exp\,(2\pi\phi_j)\; p_j(k_j, \omega_j, t) + e(t) \quad (12.3)$$

where rw(t) is the residual water, m(t) is the component from macromolecules, l(t) is the contribution from lipids, c_j represents the intensity, ϕ_j the phase, k_j the width and ω_j the location of peak j (frequency); $e(t)$ represents the random noise and t varies from zero to the end of the acquisition window. Direct Current (DC) offsets in the hardware also can contribute to the baseline and can be reduced by subtracting the DC offset from the FID or using phase cycling (Drost *et al.*, 2002). This effect is not included in Equation 12.3. Given the linear nature of the Fourier transform these components can be represented in either domain. The general equation for the frequency domain is thus as follows.

$$Y(\omega) = RW(\omega) + M(\omega) + L(\omega) + \phi\; c_j \, exp\,(2\pi\phi_j)\; P_j(k_j, \omega_j, \omega) + E(\omega) \quad (12.4)$$

When the data are acquired with phased array coils, an additional step is required to combine the signals from each of the elements (see Figure 12.8). To obtain the optimal SNR, this must be performed in a phase-sensitive manner with amplitudes weighted based upon the SNR of individual components. For single-voxel data, these parameters are typically obtained by acquiring a small number of excitations without water suppression and using the large water peak as a reference. For multivoxel data, limitations on acquisition time may make it difficult to acquire an additional dataset and so other strategies need to be considered (see Section 12.6.7). A relatively simple approach that works for single- and multivoxel data is to use the phase and amplitude of the first point of the time domain signal in each voxel as the weighting factors (Brown, 2004). This may fail if lipid contamination in the spectra is similar or greater in magnitude to the signal from residual water. If the SNR of a reference peak such as NAA or residual water is high enough for all elements, it can also be used to estimate the phase and amplitude parameters (Maril and Lenkinski, 2005). Other more complex strategies that have been shown to perform well in single-voxel data include whitened singular value decomposition (WSVD) (Rodgers and Robson, 2010) and generalised least squares (An *et al.*, 2013).

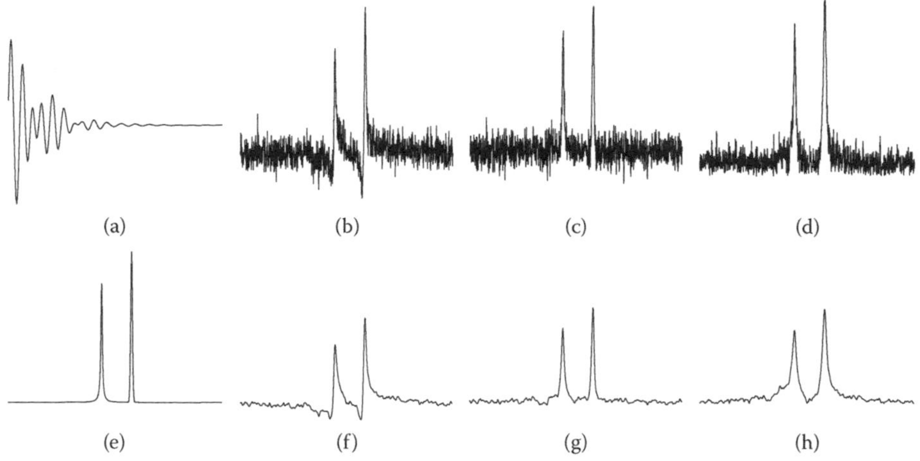

FIGURE 12.7 Simulations of data with two Lorentzian peaks. (a) The free-induction decay (FID)/echo; (e) the real part of its Fourier transform. Spectra (b,c,d) represent the raw data with random noise added prior to phasing, after phasing and in magnitude mode. Corresponding spectra (f,g,h) represent the same data but apodized by a Lorentzian function prior to the transform. Note the reduced intensity and broader peaks in the apodized data.

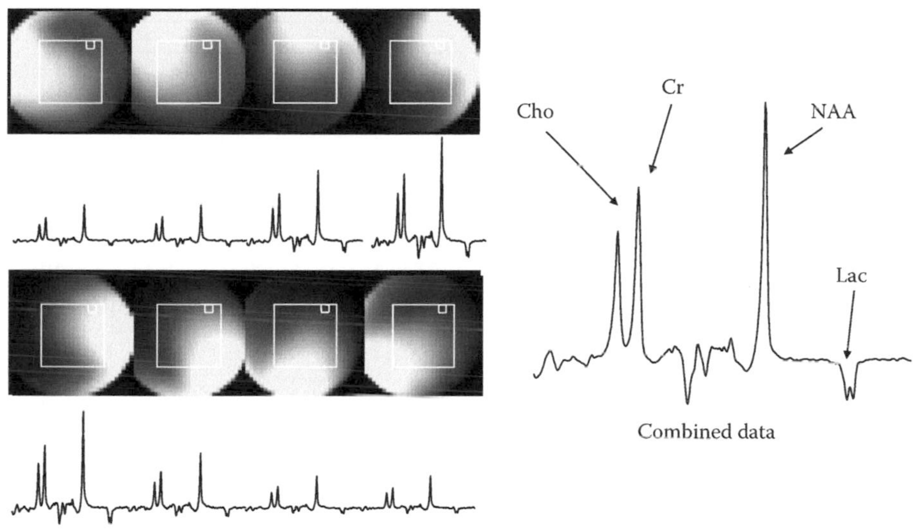

FIGURE 12.8 Data from an MRS phantom that were obtained with the body transmit and 8-channel receive coils. The individual coil calibration images show the location where the voxel comes from and shows the varying intensities in each. These spectra have been prephased and then combined in the panel on the right side to produce one spectrum with higher SNR.

12.6.2 Analysis of baseline components

The baseline components associated with residual water, macromolecular peaks and lipid contamination need to be removed in order to obtain accurate estimates of the intensities of the metabolite peaks of interest. Two approaches are commonly applied. The first of these is based upon eliminating specific frequency components using the Hankel singular value decomposition (HSVD) or related algorithms (Coron *et al.*, 2001; Vanhamme *et al.*, 1998). The second approach assumes that the baseline varies slowly underneath the peaks of interest and applies local polynomial fitting routines to extrapolate beneath them (Nelson, 2001; Provencher, 2001). The first method is typically implemented in the time domain and the second method in the frequency domain. In both cases it is important that the reference frequency be set correctly. This may be done by using an unsuppressed water acquisition obtained at the same time, the residual water peak from the spectrum itself or by matching the location of one or more of the major metabolite peaks to a predefined template defined from prior knowledge of their locations (Nelson, 2001). Approaches specific to removal of the macromolecules include the use of specific models of the components that are generated from simulated or experimental data

(Gottschalk *et al.*, 2008; Hofmann *et al.*, 2001; Kreis *et al.*, 2005; Ratiney *et al.*, 2004; Seeger *et al.*, 2003; Seeger *et al.*, 2001).

12.6.3 Estimating peak parameters

A large number of peak fitting procedures have been proposed and applied to ^{1}H spectra from the brain (de Graaf and Bovee, 1990; Elster *et al.*, 2005; Kanowski *et al.*, 2004; Mierisova and Ala-Korpela, 2001; Reynolds *et al.*, 2006; Slotboom *et al.*, 1998; Soher and Maudsley, 2004; Vanhamme *et al.*, 2001; Vanhamme *et al.*, 1999; Wilson *et al.*, 2011). Some of these methods are applied in the time domain and others in the frequency domain. As indicated previously, the echo time used, field strength and quality of the shimming are key factors in defining the complexity of the problem. The most commonly used software package for analysing single-voxel ^{1}H spectra from the brain is LCModel (Provencher, 1993, 2001), which does frequency domain analysis with an integrated baseline estimation and a sum of predefined model functions (basis set), which may be obtained by theoretical simulations, empirical measurements or a mixture of both (Cudalbu *et al.*, 2008; Hofmann *et al.*, 2002). The input to the LCModel package assumes that there is no spectral apodization in the time domain. The goodness of fit at specific frequencies is typically evaluated by examining the local differences between the actual and fitted signal or by using the Cramer-Rao bounds (Cavassila *et al.*, 2001) that are produced by the algorithm. Figure 12.9 shows an example of the fitted spectra and output parameters that are obtained. Other packages that are available as open-source software are QUEST (Ratiney *et al.*, 2004), jMRUI (Naressi *et al.*, 2001), AQSES (Poullet *et al.*, 2007) and TARQUIN (Wilson *et al.*, 2011).

12.6.4 Normalisation of signals

Once estimates of peak heights, areas and linewidths have been obtained for individual metabolites, the next step is to compare the results obtained with reference values. Depending on the application that is being considered and the hypothesis being tested, comparisons may be made with metabolite levels obtained (1) from voxels in contralateral brain in the same individual; (2) from results obtained in prior scans; or (3) from similar anatomical locations in a population of control subjects. One of the challenges in evaluating the data is how to scale the results in order to make

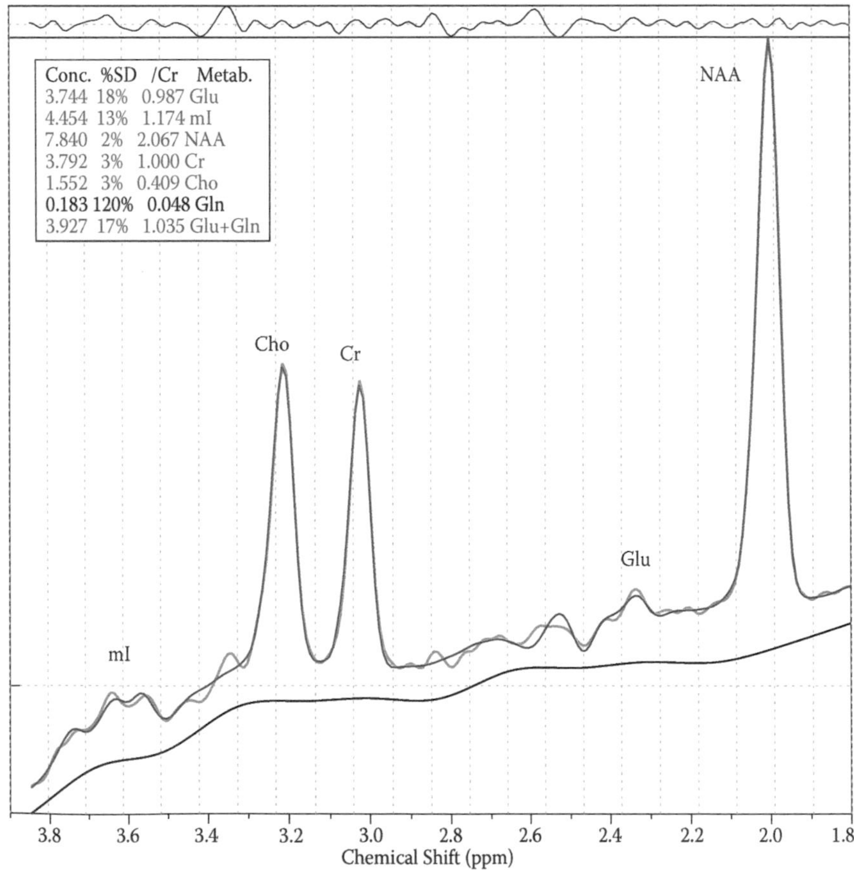

FIGURE 12.9 Single-voxel TE-averaged PRESS MRS at 3T quantified by LCModel from a healthy volunteer. The volume of interest is located in the parietal white matter with a size of 2 × 2 × 2 cm^3. The TE-averaged PRESS was acquired with 64 steps with a time increment of 2.5 ms starting at TE = 35 ms and TR of 2 s (Hurd *et al.*, 2004). The estimates of metabolite concentrations, Cramer-Rao lower bounds (CRLB) and concentrations (mM) relative to Cr from LCModel (without corrections of T_1/T_2 relaxation times) using an *in vitro* basis set of individual metabolites are included. The line in black is the fitted baseline.

them comparable. The simplest way of approaching this is to select one of the peaks in the same spectrum and to express the intensities of other peaks as ratios relative to the reference peak. As was indicated previously, the Cr peak at 3.0 ppm is often used for this purpose. Interpretation of the values obtained must therefore consider that there may be variations in the intensity of the reference peak, as well as in levels of the other metabolites (Li *et al.*, 2003). Another option that has been proposed is to normalise the metabolite intensities using an unsuppressed water spectrum from the same spatial location (Barker *et al.*, 1993; Ernst *et al.*, 1993). Both of these strategies depend upon the signals from the reference peak (metabolite or unsuppressed water) coming from the same selected volume, but the alteration in concentration and relaxation times of the reference peak can cause the variation on normalised signals between tissues. Although the size of the chemical shift artefact is relatively small for most single-voxel data and can be minimised by using pulses with large bandwidth or implementing the over-PRESS strategy (see Section 12.5.3), its effect should be considered for analysing multivoxel data (see Section 12.6.9).

Another factor that may complicate interpretation of the results is that the relatively coarse spatial resolution of the ¹H MRS data means that most voxels will contain a mixture of white matter, grey matter and cerebrospinal fluid (CSF) for healthy controls. Segmentation of anatomic images and resampling of component masks may be used to predict the fractions of each in a voxel (McLean *et al.*, 2000). While this can correct for the reduction in signal intensity that is caused by the CSF component in a single voxel, separating contributions from white and grey matter require the acquisition of data from multiple voxels. Having a reference atlas of normal metabolite levels (see Section 12.7) would be beneficial for testing hypotheses about changes in metabolite levels that are due to pathology.

12.6.5 Absolute quantification

While analysis of relative peak intensities can provide useful information about prognosis and for evaluating treatment-induced changes in metabolite levels, the biological interpretation of the data may be enhanced by estimating their absolute concentration. There is a long history of studies that have focused on the design of acquisition protocols and making measurements necessary for correcting peak intensities to account for variation in experimental conditions (Alger *et al.*, 1993; Barker *et al.*, 1993; Danielsen and Henriksen, 1994; Jansen *et al.*, 2006; Jost *et al.*, 2005; Kreis *et al.*, 1993; Michaelis *et al.*, 1993; Tofts and Wray, 1988). As is the case with the signal from standard MR images, the signals determined from spectral fitting are dependent upon the T_1 and T_2 relaxation times of the metabolites ($j = 1, \ldots, M$) being considered and corrections need to be applied to compare with results that are obtained with different TR and TE.

$$C_j \propto (1 - exp\,(-\,TR/T1_j))exp\,(-\,TE/T2_j) \qquad (12.5)$$

For *in vivo* applications it is rare that there is time available within a single examination to perform the multiple acquisitions that would be necessary to estimate T_1 and T_2 values, especially for multivoxel datasets. To compensate for this, most studies that provide absolute concentrations use literature values that have been estimated from normal brain (Table 12.1) (Ethofer *et al.*, 2003; Mlynarik *et al.*, 2001; Rutgers *et al.*, 2003; Traber *et al.*, 2004; Zaaraoui *et al.*, 2007). While this is a step closer to absolute concentration there are known to be differences in relaxation times in regions of pathology (Li *et al.*, 2008), and so in some cases the corrections may actually reduce the contrast between normal and abnormal metabolite signals.

In addition to having prior knowledge of relaxation times, the estimation of absolute concentrations requires a means of correcting for experimental factors such as coil loading and variations in the RF pulses that are used. Strategies that have been proposed for calibrating the signal for these effects include using an internal calibration such as unsuppressed water, an external standard that is placed near the head and a phantom with electrical conductivity matched to the head and is placed in a similar position so that equivalent reference data can be obtained immediately after the patient/volunteer scan (Jansen *et al.*, 2006). Many of the early studies that examined the reliability of methods for absolute concentration were performed at 1.5T and

TABLE 12.1 T_1/T_2 Relaxation Times (in seconds) of Metabolites at 3T

Population	Location	Cho	Cr	NAA
Healthy volunteers	Occipital GM[a]	1.30/0.207	1.46/0.152	1.47/0.247
	Occipital WM[a]	1.08/0.187	1.24/0.156	1.35/0.295
	Parietal WM[b]	1.06/0.169	1.38/0.139	1.38/0.249
	Motor cortex GM[c]	1.14*/0.222	1.11*/0.121	1.34*/0.247
	Caudate[d]	/0.219	/0.163	/0.254
	Thalamus[d]	/0.198	/0.135	/0.229
Glioblastoma	T_2 hyperintensity[b]	1.00**/0.209	1.37**/0.157	1.38**/0.235

Note: GM = grey matter; WM = white matter; Cho = choline; Cr = creatine; NAA = N-acetyl aspartate.
*Volume of interest includes occipital white matter, motor cortex and frontolateral cortex.
**Calculated from patient with high-grade glioma.
[a] Mlynarik *et al.* (2001).
[b] Li *et al.* (2008).
[c] Traber *et al.* (2004).
[d] Zaaraoui *et al.* (2007).

at a time when volume transmit–receive coils were being used for brain studies. The transition to the routine use of 3T scanners, in conjunction with body coil transmit and phased array multichannel receivers, has meant that the reception profile of the receive coils should also be considered in comparing spectra from different regions of the brain (Natt *et al.*, 2005), see also the discussion in Chapter 2, section 2.1.10.

12.6.6 Analysis of multivoxel datasets

In this case the raw data acquired have both time and *k*-space dimensions and the reconstruction process needs to cope with the complexities of translating these into frequency and spatial domains. Figure 12.10 describes the key steps that are involved. With standard rectilinear phase encoding one FID or echo is obtained at each *k*-space point ($\underline{k}$). Apodization and Fourier transform in time gives an intermediate array of *k*-space spectra, and a spatial array of spectra is produced by a further multidimensional Fourier transform. Note that the field of view and PSF of the resulting spectral array are fixed by the gradient step size and matrix size, but the location of the grid may be changed based upon the well-known Fourier shift theorem and the size of the spectral array may be increased by zero filling in the *k*-space dimensions. Low-pass filtering in *k*-space is often applied to increase the SNR of the spectra but at the cost of decreasing the true spatial resolution.

When the raw data are acquired with echo planar (Cunningham *et al.*, 2005; Ebel *et al.*, 2001), spiral (Adalsteinsson *et al.*, 1998; Sarkar *et al.*, 2002) or other types of non-uniform *k*-space sampling (e.g. Schirda *et al.*, 2016), a preprocessing step that interpolates the data onto a standard rectangular grid is typically required. The resulting 4D array can then be reconstructed in a similar manner as for standard phase-encoded data. When parallel imaging or compressed sensing strategies are applied, reconstruction strategies that are similar to those developed for standard MRI are used. Added complications are the requirement for producing a phase-sensitive output and the modest size of the acquisition matrices used for ^{1}H MRSI (Dydak *et al.*, 2001, 2003; Sabati *et al.*, 2014; Tsai *et al.*, 2008). In either case, having an accompanying dataset that is obtained without water suppression is helpful for providing a high SNR reference performing the analysis but adds an extra burden in terms of scan time. The other choice is to obtain water spectra with low spatial resolution, which is frequently used on Philips scanners.

12.6.7 Coil combination

As was indicated previously, a critical step for reconstructing multivoxel data obtained with phased array coils is to combine

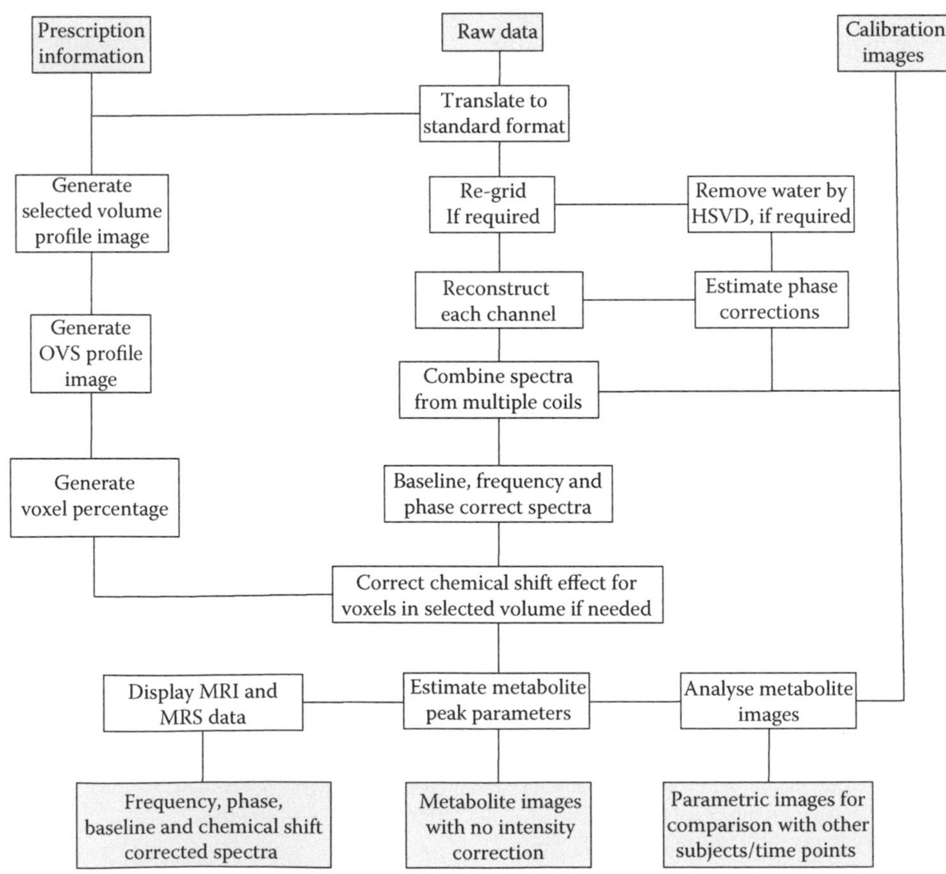

FIGURE 12.10 Schematic of the steps required to process and quantify MRSI data. In addition to the raw data, the prescription information and calibration images are a critical part of the process.

the signals from individual channels in a manner that optimises the SNR of the resulting spectral array. This requires making sure that the arrays of spectra from different coils are phased in a similar manner and then summing them together with weights that enhance the components with high signal and de-emphasise those with low signal. The combined spectral array may be represented as follows:

$$S(\underline{X},\omega)=\sum_{i=1}^{M} a_i(\underline{X})\cdot f_i(\underline{X})\cdot s_i(\underline{X},\omega) \quad (12.6)$$

where $s_i(\underline{x}, \omega)$ is the spectrum from coil i at spatial location $\underline{x}$, $a_i(\underline{x})$ is the amplitude weighting factor, $\phi_i(\underline{x})$ is the corresponding phase offset and M is the number of coils. Note that the phase offsets vary as a function of the coil geometry and should be determined on a voxel-by-voxel basis.

The voxel size and large coverage associated with ¹H MRSI means that there are always regions of the array where the SNR is very low in some channels and it is usually not possible to use metabolite peaks to estimate the phase offsets. Alternatives that have been shown to provide robust results are to use residual water peaks from the same dataset if they have sufficiently high SNR (Dong and Peterson, 2007) or to use a separate water reference scan (Sabati *et al.*, 2014). The amplitude weighting factors ($a_i(\underline{x})$) can be obtained from proton density weighted calibration images, which are typically acquired for use in parallel reconstructions for the imaging portion of the examination.

12.6.8 Correcting for chemical shift and voxel intensities

One of the advantages of using multivoxel and in particular volumetric ¹H MRSI is the ability to compare metabolite levels between regions in the same individual. As with single-voxel spectra, this may be done with ratios of metabolite levels, of metabolite levels relative to a water reference or with metabolite levels corrected for variations in coil reception profiles. Note that if any of the voxels on the edge of the selected volume are impacted by the chemical shift artefact, the metabolite levels must also be corrected for this effect. The shift in the spatial location of the selected volume in dimension j (j = 1, 2, 3) for an offset of $\delta\omega$ from the reference frequency is given by

$$\Delta B_j = \delta\omega \, W_j \,/BW_j \quad (12.7)$$

where W_j is the width of the selected volume and BW_j is the bandwidth of the corresponding selection pulse. The correction that is required for each point in the spectrum can be estimated from its frequency offset, the bandwidths of the selection pulses and the size of the selected volume. Examples of chemical shift corrected simulated and phantom data are seen in Figure 12.11. Coil intensity corrections are determined from the calibration images and the percentage of each voxel that is excited is determined by the size of the selected volume, the field of view and the encoding matrix, together with the location of the OVS bands, their orientation and their widths. An example of anatomic images with the outlines of these regions superimposed upon them is seen in Figure 12.12. Clearly it is important that the parameters defining

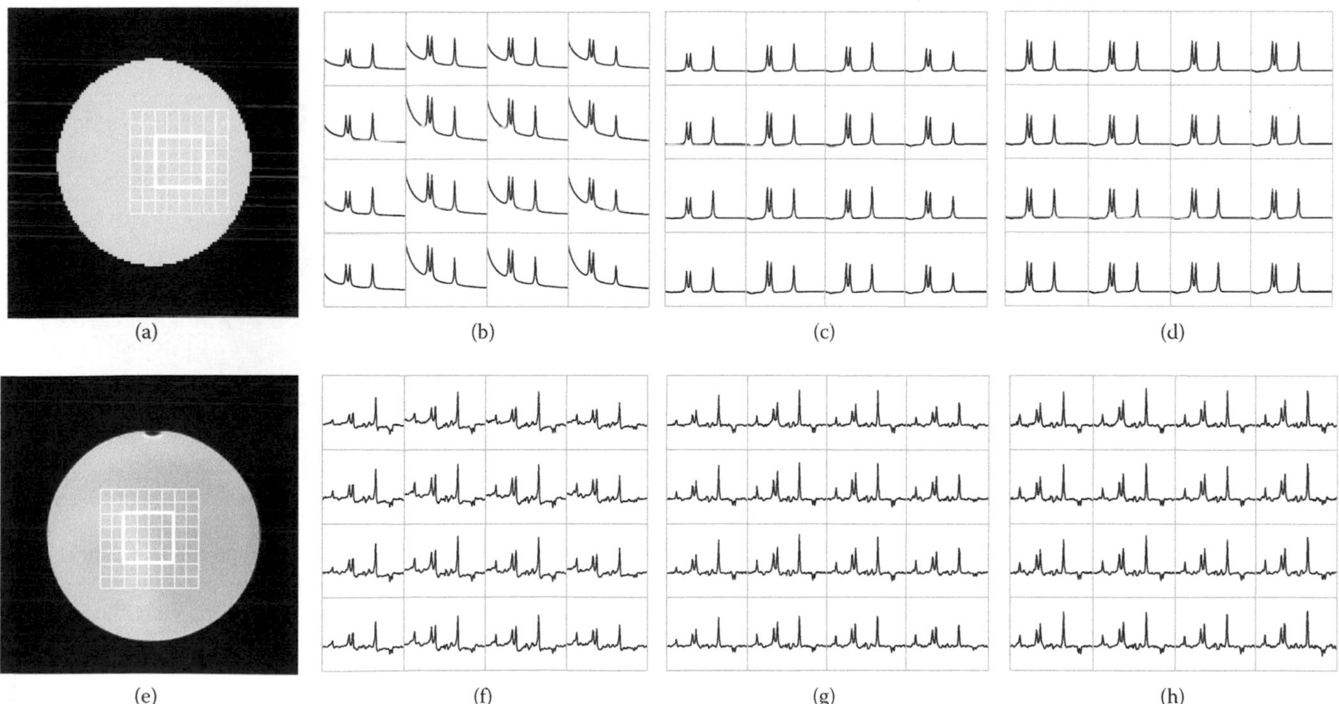

FIGURE 12.11 Simulated (a) and phantom data (e) prior to baseline removal and phasing (b, f), after baseline removal and phasing (c, g) and after additional correction at each frequency for the chemical shift artefact (d, h). Note the difference in the relative peak intensities on the edge voxels prior to correction.

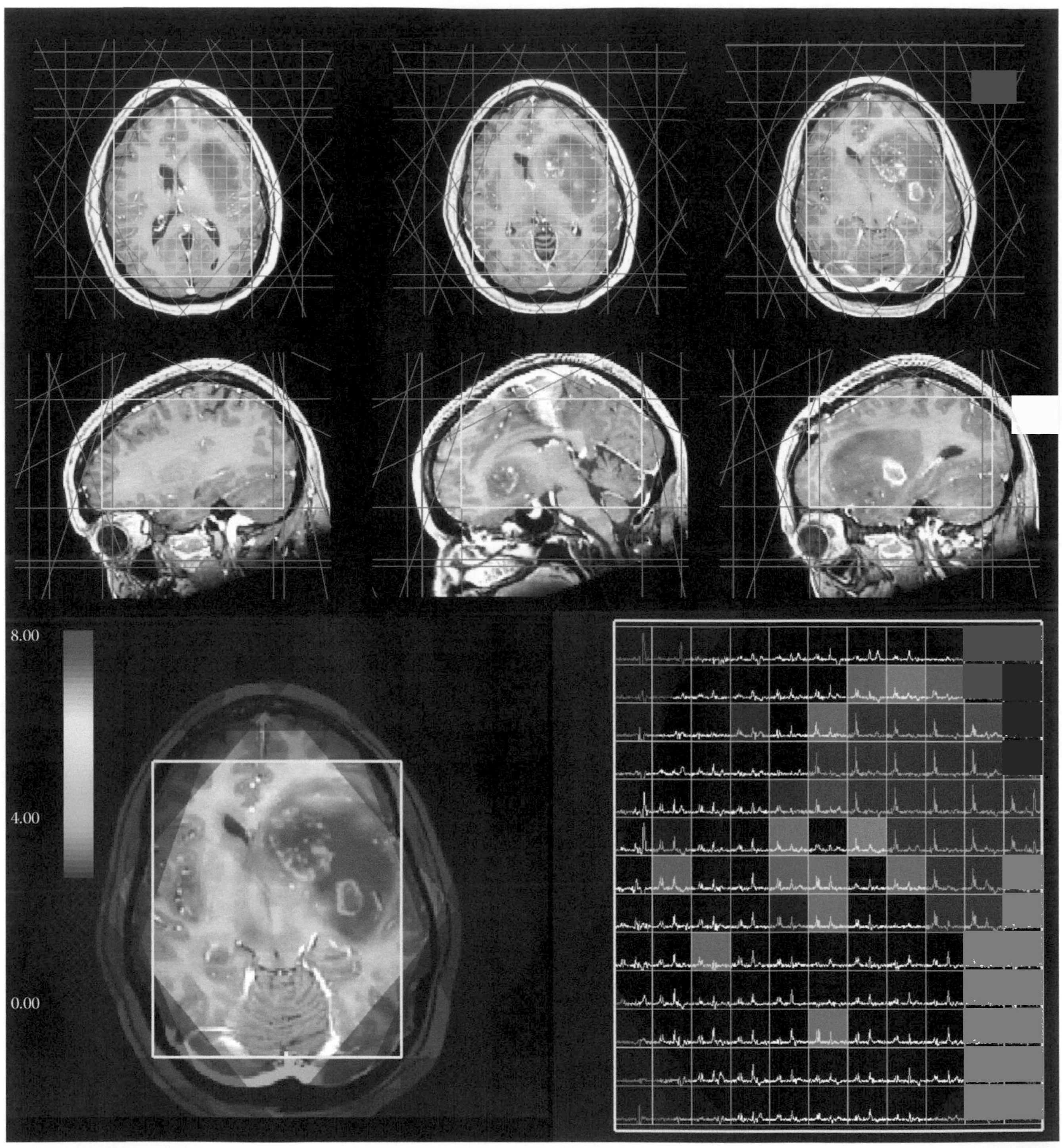

FIGURE 12.12 Prescription for large PRESS selected volume, 3D EPSI scan at 3T (TE/TR = 144/1250 ms, matrix = 18×18×16, nominal spatial resolution = 1 cm^3, two cycles, total acquisition time ~13 min) and output from the SIVIC analysis package. The yellow box shows the PRESS volume desired, the white grid shows individual voxels and the brown lines represent the edges of the multiple outer volume saturation (OVS) bands that were used to eliminate unwanted lipid signals. One slice of the lactate edited 3D echo-planar spectroscopic imaging spectral array from within the selected volume is shown together with the corresponding anatomic image and colour overlays of the choline to NAA index (CNI) map that was generated automatically for this subject. The colour look-up table shows the CNI scale that is represented in the overlay, with the red regions of the map representing voxels for which the CNI was highly abnormal (around 8 standard deviations away from normal), the amber colour where the CNI was around 6 and the green region where it was around 4. The regions of the image and spectral array for which the OVS bands suppressed signal are seen as dark overlays.

the prescription are stored in conjunction with the raw data and are able to be accessed by the processing routines.

12.6.9 Generation of metabolite maps

Once a combined array of spectra has been produced, the next step in the analysis is to remove any residual baseline components and estimate peak parameters. This can be done on a voxel-by-voxel basis by using any of the strategies mentioned in Section 12.6.3, but bearing in mind that there are always some of the spectra with metabolites that have relatively low SNR and so the performance of the fitting routines will be dependent upon this (Soher *et al.*, 2000). Another issue to consider is that, unless parallel processing is applied in a batch mode, the computational time for the non-linear fitting routines used in these packages may produce a delay in providing the results.

For situations where the peak overlap is minimal, such as long TE or edited spectra, it is possible to take a simpler approach that can be fully automated (Nelson, 2001). This first does phase and frequency correction and baseline subtraction. For voxels having low SNR, the phase and frequencies are interpolated from surrounding voxels. This is followed by finding maxima within specific predefined frequency ranges and using them to estimate peak locations, heights and areas by integration. The results obtained are more robust than the non-linear fitting routines and allow for stable estimates of parameters in regions where the more complex algorithms would have failed.

For interpretation, the spectral arrays and estimated metabolite parameters can be viewed as greyscale images or as colour overlays that are superimposed upon the spectral grid and corresponding anatomic images. Note that careful inspection of spectra quality and fitting accuracy is required before generating metabolic maps. Flexible displays such as those developed for the SIVIC package (Crane *et al.*, 2013) allow for the direct correlation of anatomic and metabolic data, the rapid detection of abnormal regions (see Figure 12.12) and the ability to generate DICOM screen captures that can be used to summarise the results obtained. This is essential for radiological interpretation and for being able to incorporate the metabolite maps into the clinical data pathway for use in making treatment decisions.

12.6.10 Metrics for tissue characterisation and evaluation of serial studies

As with the single voxel case, segmentation of the anatomic MR images allows for the identification of the fractions of grey matter, white matter, CSF and any lesions that can be identified (Hetherington *et al.*, 1996; Laudadio *et al.*, 2005; Malucelli *et al.*, 2009; Maudsley *et al.*, 2006; Tal *et al.*, 2012). A number of different approaches can be taken in order to generate parametric maps that highlight abnormal voxels. One possibility is to assume that metabolite levels are constant in grey matter and white matter and to use all of the voxels estimate high SNR model spectra for them ($S_{wm}(\underline{x})$ and $S_{gm}(\underline{x})$). If $f_{wm}(\underline{x})$ and $f_{gm}(\underline{x})$ are the fractions of white and grey matter, a difference spectrum can be generated for each voxel using the following equation to highlight regions with significant abnormality.

$$S_{diff}(\underline{x}) = S(\underline{x}) - f_{wm}(\underline{x}) \cdot S_{wm}(\underline{x}) - f_{gm}(\underline{x}) \cdot S_{gm}(\underline{x}) \quad (12.8)$$

Similar expressions could be applied to difference maps for individual metabolites in order to provide a greyscale or colour image that can be directly compared with the anatomic data.

Another fairly simple approach that combines information from two (or more) different metabolite maps is to assume their ratio r_{12} is constant in normal brain and to estimate its value using linear regression over voxels that are in normal brain as follows:

$$M_2(\underline{x}) = r_{12}.M_1(\underline{x}) + \varepsilon(\underline{x}) \quad (12.9)$$

where $\varepsilon(\underline{x})$ is random noise. The deviation in relative metabolite levels from normal in all voxels can then be measured by determining their perpendicular distance away from the line with slope r_{12}. Dividing this metric by the standard deviation of the distances from the line for normal voxels provides a *z*-score map that can be interpreted as an indicator of the probability of the voxel being abnormal. A modified version of this strategy has been applied to estimates of Cho and NAA from patients with brain tumours (McKnight *et al.*, 2002). High positive values of what has been termed the *Cho to NAA index* or CNI indicate that a voxel has a high likelihood of corresponding to tumour (see Figure 12.12) and have been shown to be extremely valuable for tracking response to therapy and predicting survival (see Section 12.8.1). A more general approach that uses both MRI and MRSI data for classifying normal and abnormal tissues is represented by nosologic imaging (Luts *et al.*, 2009).

For analysing serial scans from the same individual it is beneficial to be able to make direct comparisons between signals obtained at different time points on a voxel-by-voxel basis. For volumetric ¹H MRSI data this can be achieved retrospectively (see Figure 12.13) by using the anatomic images to perform a rigid alignment between time points and applying the transformation matrix that is obtained to the metabolite maps and spectral data (Nelson, 2001; Nelson *et al.*, 2016b). Limitations of this approach are that one can only compare voxels that were inside the selected volumes from both scans. Figure 12.14 shows an example of baseline and follow-up scans from a patient with a brain tumour that have been aligned in this manner. A similar approach can be used to compare metabolite levels between subjects, but in this case the data need to be aligned to a common space such as an atlas using non-rigid registration (Maudsley *et al.*, 2006). A more robust solution is to use the automatic prescription techniques described in Section 12.5.7 to prospectively acquire ¹H MRSI data from the same region at each time point (Bian *et al.*, 2015; Ozhinsky *et al.*, 2011, 2013).

12.7 ¹H MRS data from normal brain

Metabolite levels/concentrations that have been obtained from ¹H MRS/MRSI may vary with excitation profiles (e.g. chemical shift

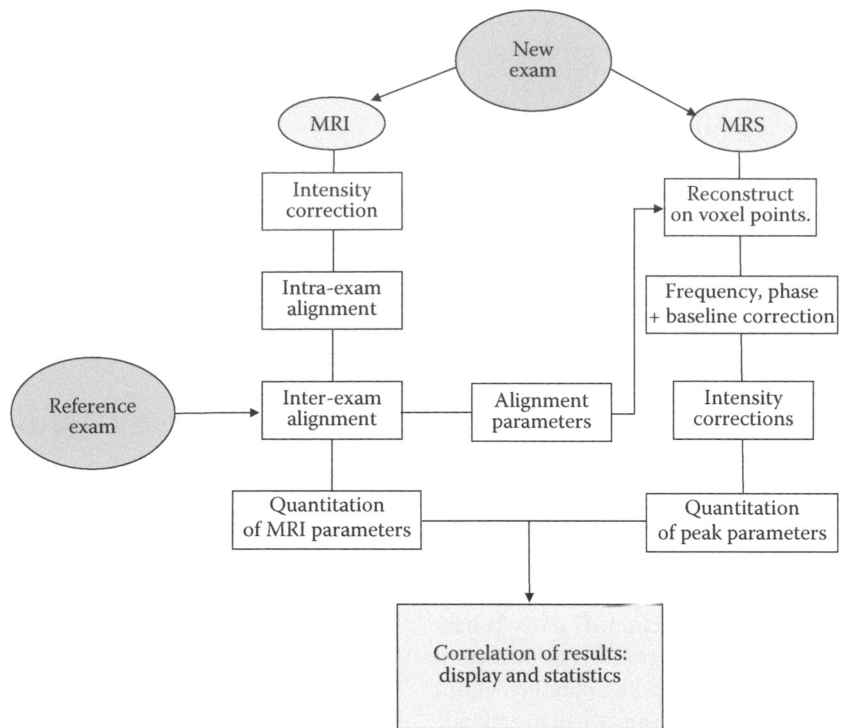

FIGURE 12.13 Schematic of the procedure used to retrospectively register images and spectra from serial examinations.

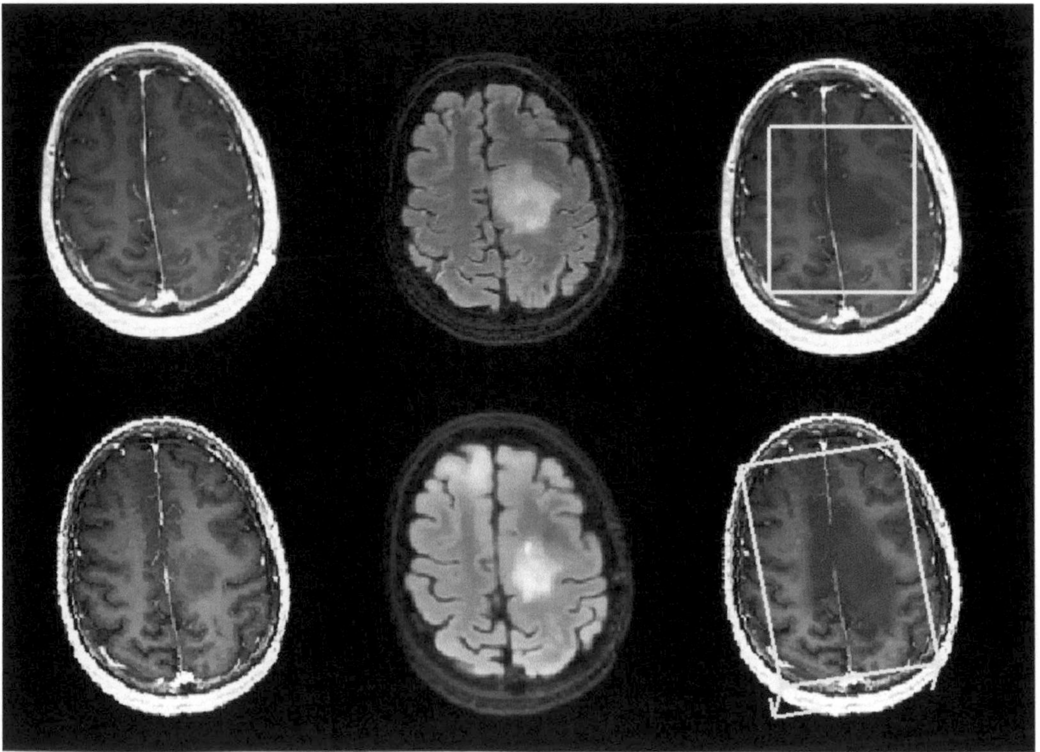

FIGURE 12.14 Results obtained using retrospective registration depicted in Figure 12.13 for a follow-up relative to the baseline examination on the same subjects. The post-gadolinium T_1-weighted images and the T_2-weighted FLAIR images from this patient with a non-enhancing brain tumour show very little change, but the metabolite image overlays demonstrate an increase in the size of the metabolic region that has high CNI shown in red. Subsequent follow-up showed that patient progressed rapidly on both anatomic and metabolic images. The yellow outline shows the spatial extent of the selected volumes.

artefact), acquisition parameters (e.g. TE/TR), processing methods (e.g. the use of spatial filtering), quantification methods (e.g. correction of T_1/T_2 relaxation times), as well as the age of the subject (Brooks *et al.*, 2001; Kreis *et al.*, 2002) and the chosen region of interest (Goryawala *et al.*, 2016; Schuff *et al.*, 2001). These factors can make it difficult to compare the results between different studies. Although there is lack of widely available normative data, several studies have focused on whole brain MRSI and described metabolic profiles for short and long TE at 3T (Ding *et al.*, 2015; Goryawala *et al.*, 2016; Maudsley *et al.*, 2009).

Variations in hardware and physiologic differences between subjects can both cause variations in estimates of metabolite levels. Tests of reproducibility are critical for providing confidence about the significance of the differences within or between subjects and to interpret changes associated with specific pathologies. Studies of single-voxel ^{1}H MRS reproducibility have reported intrasubject coefficients of variation (CVs) in the range of 3%–8% for Cho, Cr and NAA and below 15% for Glu and mI. Intersubject CVs were found to be relatively larger (<10% and <20%) (Bednarik *et al.*, 2015; Brooks *et al.*, 1999; de Matos *et al.*, 2016; Terpstra *et al.*, 2016). CVs for GABA or GABA+macromolecules obtained from spectral editing were between 8% and 15% (Geramita *et al.*, 2011; Mikkelsen *et al.*, 2016; Near *et al.*, 2014; Prinsen *et al.*, 2017). Reproducibility for 3D MRSI showed similar results and confirmed that there are variations in the different regions of the brain (Ding *et al.*, 2015; Langer *et al.*, 2007; Maudsley *et al.*, 2010; Veenith *et al.*, 2014).

Nowadays, more and more clinical studies are performed at multicentres on single or multivendor platforms. The variations for both single-voxel ^{1}H MRS and 3D MRSI between sites and vendors have been evaluated at 1.5T (Traber *et al.*, 2006), at 3T (Sabati *et al.*, 2015) and at 7T (van de Bank *et al.*, 2015). These studies showed comparable manner across sites and vendors. While there is not yet a widely available atlas of metabolite values in normal brain, several groups have been working on their own solutions and future studies should be directed at making this happen.

12.8 Clinical applications

^{1}H MRS provides important biochemical information and can be applied to any diseases that cause a change in the metabolic function of brain tissue. A recent publication from the international MRS consensus group has reviewed the methods and practical implementation of ^{1}H MRS to the central nervous system (Oz *et al.*, 2014). In the following, we focus on several of the most common applications: brain tumours, multiple sclerosis, neurodegenerative diseases, epilepsy, psychiatric disorders and intracranial infections.

12.8.1 Brain tumours

One of the first applications of *in vivo* ^{1}H MRS demonstrated large increases in Cho and reductions of NAA between spectra from tumour and that from normal brain (Bruhn *et al.*, 1989). Comparisons of *ex vivo* spectra from tumour specimens with *in vivo* data from the same patient confirmed these observations (Gill *et al.*, 1990). Another early finding was that, in addition to changes in Cho and NAA, spectra from high-grade lesions often had regions with peaks corresponding to lipid and lactate. A large number of research groups have subsequently confirmed these results (Li *et al.*, 2015) and have applied single- or multivoxel MRS to (1) compare metabolite levels in brain tumours with those in other focal lesions; (2) characterise the metabolic signatures of different types of tumour; (3) evaluate response to therapy; and (4) predict outcome.

Studies that have been directed at using metabolic parameters to distinguish between different types of brain lesions have typically applied pattern recognition and multivariate analysis techniques to identify unique characteristics for each subgroup. These have successful in differentiating brain tumours from regions of pathology that correspond to abscesses and demyelination (Hourani *et al.*, 2008; Majos *et al.*, 2009; Moller-Hartmann *et al.*, 2002; Poptani *et al.*, 1999). The accuracies of classifiers used in a group of 84 patients for differentiating tumour and non-tumour using mI/NAA (from short TE MRS) and Cho/NAA (from long TE MRS) were found to be 82% and 79%, respectively (Majos *et al.*, 2009).

^{1}H MRS has also been applied to evaluate the metabolic signatures of different types of brain tumours. Spectra from meningioma were found to have a marked reduction of NAA and/or Cr, increased Cho and the appearance of alanine (Cho *et al.*, 2003). Spectra from metastases were reported to exhibit very low levels of Cr in the lesion itself (Ishimaru *et al.*, 2001; Majos *et al.*, 2004) and relatively normal metabolite levels within the peritumoral regions of T_2 hyperintensity (Wijnen *et al.*, 2012). Studies using short TE spectra to estimate metabolite levels were able to non-invasively distinguish between these and other types of brain tumour (Garcia-Gomez *et al.*, 2009; Guzman-De-Villoria *et al.*, 2014; Julia-Sape *et al.*, 2012; Opstad *et al.*, 2007; Tate *et al.*, 2006; Vicente *et al.*, 2013).

For studies of glioma, which are the most common primary tumours in adults, levels of Cho, Cr, NAA, Lac and Lip from long TE MRS were found to provide added value relative to advanced MRI methods for differentiating between low-grade and high-grade lesions (de Fatima Vasco Aragao *et al.*, 2014; Guzman-De-Villoria *et al.*, 2014). *Ex vivo* proton high-resolution magic angle spinning spectroscopy from image-guided tissue samples have shown that there are significant differences between the levels of other metabolites between gliomas of Grades 2, 3 and 4, as well as between primary and secondary lesions (Elkhaled *et al.*, 2014). Metabolites that were highlighted in this study included mI, Glu, phosphocholine and glycerophosphocholine. Analysis of *in vivo* data obtained using short TE acquisitions showed that the level of mI/Cr was increased in low-grade astrocytoma but decreased in glioblastoma relative to levels in healthy controls (Castillo *et al.*, 2000). The presence of increased mI in contralateral normal brain was also observed in patients with untreated glioblastoma (Kallenberg *et al.*, 2009). A significant increase in Glx was found in patients with oligodendroglioma relative to those with low-grade astrocytoma (Rijpkema *et al.*, 2003).

A recent finding that has led to increased interest in using ^{1}H MRS for diagnostic purposes has been that the oncometabolite 2-hydroxyglutarate is associated with the presence of isocitrate dehydrogenase enzymes (Dang *et al.*, 2009) and can be detected *in vivo* using specialised acquisition methods (Andronesi *et al.*, 2013; Choi *et al.*, 2012; Emir *et al.*, 2016) (Figure 12.15a). The recent integration of molecular characteristics such as mutations in the isocitrate dehydrogenase gene into the WHO classification of glioma and the corresponding differences in prognosis that have been observed provides increased motivation for integrating metrics that describe the levels of *in vivo* metabolic markers into the management of such patients (Louis *et al.*, 2016).

Another critical application for ^{1}H MRS is the ability to predict survival. The majority of studies that have been performed to date have focused on patients with glioblastoma, which is the most common and most malignant type of glioma with a median overall survival of 15 months (Stupp *et al.*, 2005). A series of studies have examined the predictive value of metabolic parameters from 3D long TE MRSI (Figure 12.15b) in patients with newly diagnosed glioblastoma at different time points (such as before and after receiving surgical resection or radiation therapy) and for different types of chemotherapy (such as temozolomide and anti-angiogenic agents) (Crawford *et al.*, 2009; Li *et al.*, 2013; Nelson *et al.*, 2016a, 2016b; Saraswathy *et al.*, 2009). Patients with high levels of lactate and lipid in the region with abnormal CNI (McKnight *et al.*, 2001) had worse overall survival in these studies.

The differentiation between progression and treatment-induced effects is one of the most challenging questions for the clinical evaluation of patients with brain tumours. Several studies have attempted to evaluate MRS in conjunction with advanced MR imaging techniques, but the results have been varied (Chernov *et al.*, 2013; Matsusue *et al.*, 2010; Schlemmer *et al.*, 2001; Seeger *et al.*, 2013). A recent *ex vivo* study illustrated the ability of mI/Cho to differentiate tumour from treatment effects, but this has not yet been validated *in vivo* (Srinivasan *et al.*, 2010a). More recently, the positron emission tomography (PET) tracer 18F-fluoroethyltyrosine has identified as a promising agent for assessing response to treatment for patients with glioma (Albert *et al.*, 2016). These studies have suggested using a multimodality metabolic imaging examination to evaluate these patients.

12.8.2 Multiple sclerosis

Multiple sclerosis is a common, chronic, inflammatory disorder of the central nervous system with the pathophysiological processes of demyelination, remyelination, gliosis and axonal loss. ^{1}H MRS is able to add biochemical information for glial markers (mI and Cho), the neuronal marker NAA and the excitatory neurotransmitter Glu. An example of 3D MRSI obtained from a patient with relapsing multiple sclerosis is shown in Figure 12.16. Compared to the values in the normal-appearing white matter, the levels of Cho/NAA and mI/Cr are higher in the lesions, while Glu/Cr is decreased. Several recent publications have reviewed the application of ^{1}H MRS and its related metabolic patterns in different types of multiple sclerosis (Chang *et al.*, 2013; Miller *et al.*, 2014;

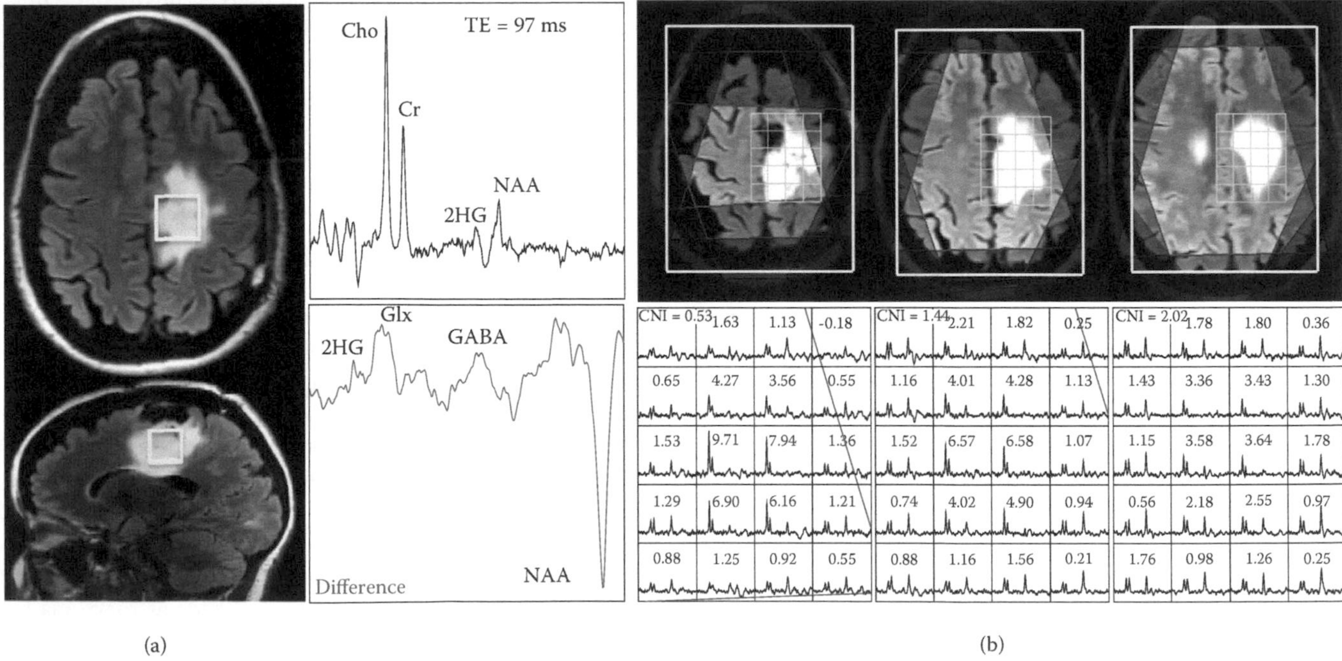

FIGURE 12.15 Single-voxel ^{1}H MRS (a) and 3D MRSI (b) at 3T from a patient with recurrent grade 2 IDH-mutant astrocytoma. The 2HG peak was resolved in PRESS (TE1/TE2/TR = 32/65/2000 ms, voxel size = 2 × 2 × 2 cm^3, 128 averaging) (a, top) and difference spectra of 2HG-edited BASING PRESS (TE/TR = 68/2000 ms, voxel size = 2 × 2 × 2 cm^3, 64 cycles editing-on and 64 cycles editing-off) (a, bottom). 3D MRSI (1 cm^3) provides the spatial extent of metabolite information.

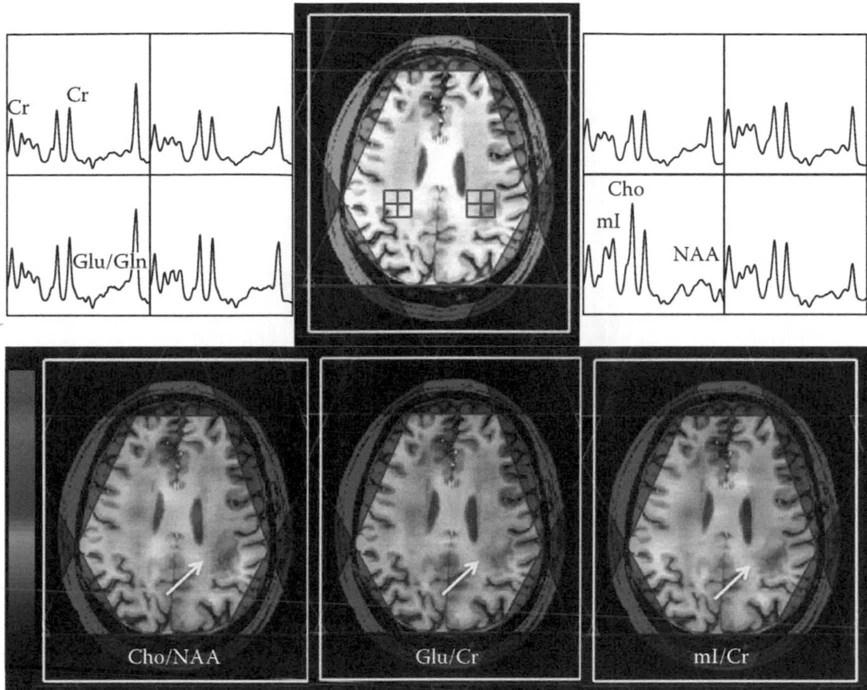

FIGURE 12.16 T_1-weighed images, MRSI quantified using LCModel and maps of Cho/NAA [0–0.4], Glu/Cr [0–1.50] and mI/Cr [0–2.0] overlaid on 7T T_1-weighted images from a patient with relapsing multiple sclerosis. Higher Cho/NAA, lower Glu/Cr and higher mI/Cr can be found in the lesions relative to the values in the contralateral normal-appearing white matter.

Rovira and Alonso, 2013). With the availability of ultra-high field strength, it is also possible to evaluate GABA and GSH in patients with multiple sclerosis within a clinical relevant acquisition time (Prinsen *et al.*, 2017; Srinivasan *et al.*, 2010b).

12.8.3 Neurodegenerative diseases

Neurodegenerative diseases are not currently curable and can cause increasing levels of debilitation, including mental (such as Alzheimer's disease) and/or movement disorders (such as Huntington's disease). A decrease in NAA is a consistent finding of ^{1}H MRS in neurodegenerative disease (Federico *et al.*, 1997; Graff-Radford and Kantarci, 2013; Sturrock *et al.*, 2010). The level observed is consistent with the severity of neuropathologic findings for Alzheimer's disease (Kantarci *et al.*, 2008) and is able to predict progression from mild cognitive impairment to Alzheimer's disease (Chao *et al.*, 2005; Metastasio *et al.*, 2006). These and other findings concerning the relationship of brain metabolites to Alzheimer's disease have been discussed in two recent reviews (Gao and Barker, 2014; Graff-Radford and Kantarci, 2013).

12.8.4 Epilepsy

Epilepsy is one of the most common neurologic disorders and affects patients of all ages. Since the early 1990s (Connelly *et al.*, 1994; Hugg *et al.*, 1993), ^{1}H MRS has been utilised to investigate patients with temporal lobe epilepsy. Most studies were performed using relatively long TE and reported reduced NAA or NAA relative to other metabolites (Cr, Cho+Cr) compared to healthy controls (Hugg *et al.*, 1993; Riederer *et al.*, 2006; Zubler *et al.*, 2003). Abnormal metabolism is not only localised in the affected and/or contralateral temporal lobe (Cendes *et al.*, 1997; Connelly *et al.*, 1994; Kuzniecky *et al.*, 1998; Zubler *et al.*, 2003) but is also widespread in the limbic and subcortical regions (Mueller *et al.*, 2011; Pan *et al.*, 2012b).

Within the patients who have temporal lobe epilepsy, a substantial portion (>30%) of patients are drug-resistant. In these cases, surgical resection of the epileptogenic zone that is responsible for seizure onset is the only possibility for obtaining a cure. Pre-operative ^{1}H MRS was shown to provide additional value for lateralisation and location (Cendes *et al.*, 1997; Kuzniecky *et al.*, 1998; Willmann *et al.*, 2006). Metabolic changes observed in these studies were found to be associated with histopathological abnormalities (Fountas *et al.*, 2012; Hammen *et al.*, 2008). Following surgical resection, the level of NAA/Cr was normalised relative to the contralateral temporal lobe (Simister *et al.*, 2009). A recent ^{1}H MRSI study performed using a 7T MR scanner showed that a positive outcome after surgical resection was associated with the extent of resection of the region of abnormal NAA/Cr (Pan *et al.*, 2013). For cases of MRI-negative temporal lobe epilepsy, the level of NAA/Cr was less reduced than that for temporal lobe epilepsy (Woermann *et al.*, 1999). Figure 12.17 illustrates an example of 7T 3D MRSI from a patient with malformation of cortical development and epilepsy.

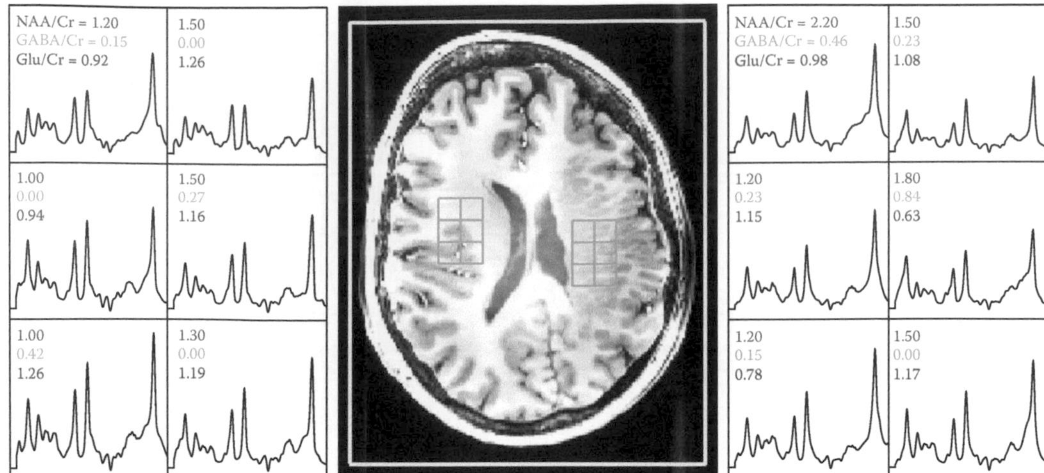

FIGURE 12.17 T_1-weighted images and 3D MRSI at 7T (TE/TR = 20/2000 ms, matrix size = 18 × 22 × 8, spatial resolution = 1 cm^3) from a patient with malformation of cortical development and epilepsy. The fitted spectra and metabolite ratios from LCModel were plotted in the selected regions. Since there is tissue difference between grey matter and white matter on metabolite levels, tissue correction is important to compare the lesion and normal brain tissue.

Neurotransmitters are also believed to play important roles in epilepsy. Short TE or spectral editing sequences have been applied to detect these metabolites, but the number of studies is limited and there have been inconsistencies in the results that have been obtained. Decreased thalamic and frontal GABA was found in patients with juvenile myoclonic epilepsy (Hattingen *et al.*, 2014; Petroff *et al.*, 2001), and low occipital GABA was associated with poor seizure control for patients with complex partial seizures (Petroff *et al.*, 1996). It was also reported that there was no difference in GABA for patients with idiopathic generalised epilepsy but with an elevation of Glx and a reduction of NAA in the frontal lobe (Simister *et al.*, 2003). For patients with temporal lobe epilepsy, there were no significant differences in Glx/Cr and GABA/Cr detected in the temporal lobe (Simister *et al.*, 2009). Because of the challenges associate with acquiring good quality ^{1}H MRS data from the temporal lobes, improved acquisition methods, high field strength and the combination with other imaging modalities (such as PET/MR) are likely to be necessary for providing additional information.

12.8.5 Psychiatric disorders

The applications of ^{1}H MRS to psychiatric disorders, such as schizophrenia, major depressive disorder and anxiety disorder, have grown rapidly over recent years. Multiple studies have used single- or multivoxel, short/long TE or spectral editing spectroscopic acquisitions for these patients. These have reported differences in metabolite parameters in a number of brain regions, as well as changes that are correlated with response to the treatment. The published literature on the use of ^{1}H MRS for evaluating different types of psychiatric diseases was summarised in two recent reviews (Dager *et al.*, 2008; Maddock and Buonocore, 2012). A recent study that applied ultra-high field (7T) ^{1}H 3D MRSI to patients with major depressive disorder has reported that metabolite levels normalised after patients receive mindfulness-based cognitive therapy (Li *et al.*, 2016). Another 7T pilot study (unpublished) showed the difference on GABA and NAA between seven male healthy controls and six male patients who had recent-onset schizophrenia or schizoaffective disorder, based on DSM-IV criteria, and were on stable doses of an antipsychotic medication (Figure 12.18). These studies have suggested that ^{1}H MRS has an important role in understanding the underlying psychopathological and therapeutic mechanisms of psychiatric diseases.

12.8.6 Intracranial infections

Intracranial infections can progress rapidly and cause life-threatening symptoms. Accurate diagnosis is crucial for treating patients. ^{1}H MRS has been shown to be valuable for studying intracranial infections (Foerster *et al.*, 2007). Brain abscesses are focal brain infections and, as indicated previously, metabolite levels are helpful in differentiating abscesses from other cystic lesions (Chang *et al.*, 1998). Pyogenic brain abscesses have spectra with abnormal levels of succinate, acetate, alanine, amino acids and lactate (Burtscher and Holtas, 1999; Chang *et al.*, 1998), while tuberculous abscesses have lipid and lactate (Gupta *et al.*, 2001). The results from single-voxel MRS were compared with data obtained using MRSI and showed that there were similar metabolite patterns in the pyogenic brain abscess but that the latter provided more spatial information (Hsu *et al.*, 2013).

A number of studies have demonstrated marked abnormalities in brain metabolites between patients infected with human immunodeficiency virus (HIV) and healthy controls (Tate

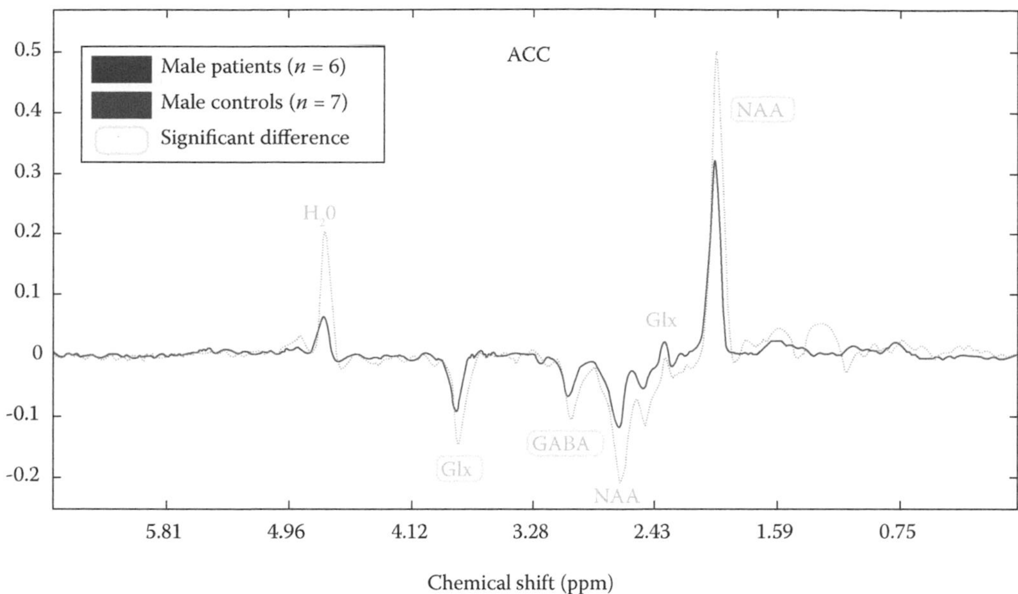

FIGURE 12.18 Averaged difference spectra from GABA-edited MRS (TE/TR = 68/2000 ms, 2 × 2 × 2 cm^3, 64 cycles editing-on and 64 cycles editing-off) at 7T in the regions of anterior cingulate cortex (ACC) from seven male healthy controls and six male patients who had recent-onset schizophrenia or schizoaffective disorder based on DSM-IV criteria and were on stable doses of an antipsychotic medication. The levels of NAA, GABA and Glx in the patients were relatively decreased compared to those in healthy controls. (Courtesy of Dr. Adam Elkhaled, University of California, San Francisco.)

et al., 2011). Patients with asymptomatic HIV infection had significantly higher Cho and low NAA/Cho (Meyerhoff *et al.*, 1999; Tarasow *et al.*, 2003). The levels of NAA or NAA/Cr were also reduced in patients with impaired cognition (Meyerhoff *et al.*, 1993; Moller *et al.*, 1999; Paley *et al.*, 1995), and low Glu or Glx was associated with cognitive deficiency (Ernst *et al.*, 2010; Mohamed *et al.*, 2010).

12.9 Future directions

Limitations of the current methodologies for obtaining ^1H MRS data are due to complicated acquisition procedures, low SNR, inferior post-processing tools available on the scanner and the lack of normative data on levels of brain metabolites. Many of these difficulties have been resolved and solutions implemented in a research setting. The translational and dissemination of convenient fast acquisition, post-processing and quantification processes that use multichannel coils and higher field strength (Nelson *et al.*, 2013) are critical for facilitating the broader use of ^1H MRS in a clinical setting. Implementing standardised acquisition procedures across different MR vendors and the availability of a database of nominal data are especially important for facilitating multicentre clinical trials. A multimodality approach that combines quantitative information from brain ^1H MRS with results obtained using other MR imaging and PET techniques is also likely to be helpful for understanding the underlying disease mechanisms and directing patient care.

References

Adalsteinsson E, Irarrazabal P, Topp S, Meyer C, Macovski A, Spielman DM. Volumetric spectroscopic imaging with spiral-based k-space trajectories. Magn Reson Med 1998; 39: 889–98.

Albert NL, Weller M, Suchorska B, Galldiks N, Soffietti R, Kim MM, et al. Response Assessment in Neuro-Oncology working group and European Association for Neuro-Oncology recommendations for the clinical use of PET imaging in gliomas. Neuro Oncol 2016; 18: 1199–208.

Alger JR. Quantitative proton magnetic resonance spectroscopy and spectroscopic imaging of the brain: a didactic review. Top Magn Reson Imaging 2010; 21: 115–28.

Alger JR, Symko SC, Bizzi A, Posse S, DesPres DJ, Armstrong MR. Absolute quantitation of short TE brain 1H-MR spectra and spectroscopic imaging data. J Comput Assist Tomogr 1993; 17: 191–9.

An L, Willem van der Veen J, Li S, Thomasson DM, Shen J. Combination of multichannel single-voxel MRS signals using generalized least squares. J Magn Reson Imaging 2013; 37: 1445–50.

Andronesi OC, Rapalino O, Gerstner E, Chi A, Batchelor TT, Cahill DP, et al. Detection of oncogenic IDH1 mutations using magnetic resonance spectroscopy of 2-hydroxyglutarate. J Clin Invest 2013; 123: 3659–63.

Bak LK, Schousboe A, Waagepetersen HS. The glutamate/GABA-glutamine cycle: aspects of transport, neurotransmitter homeostasis and ammonia transfer. J Neurochem 2006; 98: 641–53.

Banerjee S, Ozturk-Isik E, Nelson SJ, Majumdar S. Elliptical magnetic resonance spectroscopic imaging with GRAPPA for imaging brain tumors at 3 T. Magn Reson Imaging 2009; 27: 1319–25.

Barker PB, Hearshen DO, Boska MD. Single-voxel proton MRS of the human brain at 1.5T and 3.0T. Magn Reson Med 2001; 45: 765–9.

Barker PB, Soher BJ, Blackband SJ, Chatham JC, Mathews VP, Bryan RN. Quantitation of proton NMR spectra of the human brain using tissue water as an internal concentration reference. NMR Biomed 1993; 6: 89–94.

Baslow MH. Functions of N-acetyl-L-aspartate and N-acetyl-L-aspartylglutamate in the vertebrate brain: role in glial cell-specific signaling. J Neurochem 2000; 75: 453–9.

Baslow MH. N-acetylaspartate in the vertebrate brain: metabolism and function. Neurochem Res 2003; 28: 941–53.

Bednarik P, Moheet A, Deelchand DK, Emir UE, Eberly LE, Bares M, et al. Feasibility and reproducibility of neurochemical profile quantification in the human hippocampus at 3 T. NMR Biomed 2015; 28: 685–93.

Belaroussi B, Milles J, Carme S, Zhu YM, Benoit-Cattin H. Intensity non-uniformity correction in MRI: existing methods and their validation. Med Image Anal 2006; 10: 234–46.

Bhattacharyya PK, Lowe MJ, Phillips MD. Spectral quality control in motion-corrupted single-voxel J-difference editing scans: an interleaved navigator approach. Magn Reson Med 2007; 58: 808–12.

Bian W, Li Y, Crane JC, Nelson SJ. Towards Robust Reproducibility Study for MRSI via Fully Automated Reproducible Imaging Positioning. Proceedings of the 23rd Annual Meeting of ISMRM. Toronto, Canada, 2015.

Bottomley PA. Spatial localization in NMR spectroscopy in vivo. Ann N Y Acad Sci 1987; 508: 333–48.

Bottomley PA, Hart HR, Edelstein WA, Schenck JF, Smith LS, Leue WM, et al. NMR imaging/spectroscopy system to study both anatomy and metabolism. Lancet 1983; 2: 273–4.

Brand A, Richter-Landsberg C, Leibfritz D. Multinuclear NMR studies on the energy metabolism of glial and neuronal cells. Dev Neurosci 1993; 15: 289–98.

Brenner RE, Munro PM, Williams SC, Bell JD, Barker GJ, Hawkins CP, et al. The proton NMR spectrum in acute EAE: the significance of the change in the Cho:Cr ratio. Magn Reson Med 1993; 29: 737–45.

Brooks JC, Roberts N, Kemp GJ, Gosney MA, Lye M, Whitehouse GH. A proton magnetic resonance spectroscopy study of age-related changes in frontal lobe metabolite concentrations. Cereb Cortex 2001; 11: 598–605.

Brooks WM, Friedman SD, Stidley CA. Reproducibility of 1H-MRS in vivo. Magn Reson Med 1999; 41: 193–7.

Brown MA. Time-domain combination of MR spectroscopy data acquired using phased-array coils. Magn Reson Med 2004; 52: 1207–13.

Bruch RC, Bruch MD. Two-dimensional J-resolved proton NMR spectroscopy of oligomannosidic glycopeptides. J Biol Chem 1982; 257: 3409–13.

Bruhn H, Frahm J, Gyngell ML, Merboldt KD, Hanicke W, Sauter R, et al. Noninvasive differentiation of tumors with use of localized H-1 MR spectroscopy in vivo: initial experience in patients with cerebral tumors. Radiology 1989; 172: 541–8.

Buonocore MH, Maddock RJ. Magnetic resonance spectroscopy of the brain: a review of physical principles and technical methods. Rev Neurosci 2015; 26: 609–32.

Burtscher IM, Holtas S. In vivo proton MR spectroscopy of untreated and treated brain abscesses. AJNR Am J Neuroradiol 1999; 20: 1049–53.

Bydder GM, Young IR. MR imaging: clinical use of the inversion recovery sequence. J Comput Assist Tomogr 1985; 9: 659–75.

Cady EB, Costello AM, Dawson MJ, Delpy DT, Hope PL, Reynolds EO, et al. Non-invasive investigation of cerebral metabolism in newborn infants by phosphorus nuclear magnetic resonance spectroscopy. Lancet 1983; 1: 1059–62.

Castillo M, Smith JK, Kwock L. Correlation of myo-inositol levels and grading of cerebral astrocytomas. AJNR Am J Neuroradiol 2000; 21: 1645–9.

Cavassila S, Deval S, Huegen C, van Ormondt D, Graveron-Demilly D. Cramer-Rao bounds: an evaluation tool for quantitation. NMR Biomed 2001; 14: 278–83.

Cecil KM, Salomons GS, Ball WS, Jr., Wong B, Chuck G, Verhoeven NM, et al. Irreversible brain creatine deficiency with elevated serum and urine creatine: a creatine transporter defect? Ann Neurol 2001; 49: 401–4.

Cendes F, Caramanos Z, Andermann F, Dubeau F, Arnold DL. Proton magnetic resonance spectroscopic imaging and magnetic resonance imaging volumetry in the lateralization of temporal lobe epilepsy: a series of 100 patients. Ann Neurol 1997; 42: 737–46.

Chang KH, Song IC, Kim SH, Han MH, Kim HD, Seong SO, et al. In vivo single-voxel proton MR spectroscopy in intracranial cystic masses. AJNR Am J Neuroradiol 1998; 19: 401–5.

Chang L, Munsaka SM, Kraft-Terry S, Ernst T. Magnetic resonance spectroscopy to assess neuroinflammation and neuropathic pain. J Neuroimmune Pharmacol 2013; 8: 576–93.

Chao LL, Schuff N, Kramer JH, Du AT, Capizzano AA, O'Neill J, et al. Reduced medial temporal lobe N-acetylaspartate in cognitively impaired but nondemented patients. Neurology 2005; 64: 282–9.

Chernov MF, Ono Y, Abe K, Usukura M, Hayashi M, Izawa M, et al. Differentiation of tumor progression and radiation-induced effects after intracranial radiosurgery. Acta Neurochir Suppl 2013; 116: 193–210.

Cho YD, Choi GH, Lee SP, Kim JK. (1)H-MRS metabolic patterns for distinguishing between meningiomas and other brain tumors. Magn Reson Imaging 2003; 21: 663–72.

Choi C, Ganji SK, DeBerardinis RJ, Hatanpaa KJ, Rakheja D, Kovacs Z, et al. 2-hydroxyglutarate detection by magnetic resonance spectroscopy in IDH-mutated patients with gliomas. Nat Med 2012; 18: 624–9.

Connelly A, Jackson GD, Duncan JS, King MD, Gadian DG. Magnetic resonance spectroscopy in temporal lobe epilepsy. Neurology 1994; 44: 1411–7.

Coron A, Vanhamme L, Antoine JP, Van Hecke P, Van Huffel S. The filtering approach to solvent peak suppression in MRS: a critical review. J Magn Reson 2001; 152: 26–40.

Crane JC, Olson MP, Nelson SJ. SIVIC: open-Source, Standards-Based Software for DICOM MR Spectroscopy Workflows. Int J Biomed Imaging 2013; 2013: 169526.

Crawford FW, Khayal IS, McGue C, Saraswathy S, Pirzkall A, Cha S, et al. Relationship of pre-surgery metabolic and physiological MR imaging parameters to survival for patients with untreated GBM. J Neurooncol 2009; 91: 337–51.

Cudalbu C, Cavassila S, Rabeson H, van Ormondt D, Graveron-Demilly D. Influence of measured and simulated basis sets on metabolite concentration estimates. NMR Biomed 2008; 21: 627–36.

Cunningham CH, Vigneron DB, Chen AP, Xu D, Nelson SJ, Hurd RE, et al. Design of flyback echo-planar readout gradients for magnetic resonance spectroscopic imaging. Magn Reson Med 2005; 54: 1286–9.

Dager SR, Corrigan NM, Richards TL, Posse S. Research applications of magnetic resonance spectroscopy to investigate psychiatric disorders. Top Magn Reson Imaging 2008; 19: 81–96.

Dang L, White DW, Gross S, Bennett BD, Bittinger MA, Driggers EM, et al. Cancer-associated IDH1 mutations produce 2-hydroxyglutarate. Nature 2009; 462: 739–44.

Danielsen ER, Henriksen O. Absolute quantitative proton NMR spectroscopy based on the amplitude of the local water suppression pulse. Quantification of brain water and metabolites. NMR Biomed 1994; 7: 311–8.

de Fatima Vasco Aragao M, Law M, Batista de Almeida D, Fatterpekar G, Delman B, Bader AS, et al. Comparison of perfusion, diffusion, and MR spectroscopy between low-grade enhancing pilocytic astrocytomas and high-grade astrocytomas. AJNR Am J Neuroradiol 2014; 35: 1495–502.

de Graaf AA, Bovee WM. Improved quantification of in vivo 1H NMR spectra by optimization of signal acquisition and processing and by incorporation of prior knowledge into the spectral fitting. Magn Reson Med 1990; 15: 305–19.

de Matos NM, Meier L, Wyss M, Meier D, Gutzeit A, Ettlin DA, et al. Reproducibility of neurochemical profile quantification in pregenual cingulate, anterior midcingulate, and bilateral posterior insular subdivisions measured at 3 Tesla. Front Hum Neurosci 2016; 10: 300.

Ding XQ, Maudsley AA, Sabati M, Sheriff S, Dellani PR, Lanfermann H. Reproducibility and reliability of short-TE whole-brain MR spectroscopic imaging of human brain at 3T. Magn Reson Med 2015; 73: 921–8.

Dong Z, Peterson B. The rapid and automatic combination of proton MRSI data using multi-channel coils without water suppression. Magn Reson Imaging 2007; 25: 1148–54.

Drost DJ, Riddle WR, Clarke GD. Proton magnetic resonance spectroscopy in the brain: report of AAPM MR task group #9. Med Phys 2002; 29: 2177–97.

Duyn JH, Gillen J, Sobering G, van Zijl PC, Moonen CT. Multisection proton MR spectroscopic imaging of the brain. Radiology 1993; 188: 277–82.

Dydak U, Pruessmann KP, Weiger M, Tsao J, Meier D, Boesiger P. Parallel spectroscopic imaging with spin-echo trains. Magn Reson Med 2003; 50: 196–200.

Dydak U, Weiger M, Pruessmann KP, Meier D, Boesiger P. Sensitivity-encoded spectroscopic imaging. Magn Reson Med 2001; 46: 713–22.

Ebel A, Soher BJ, Maudsley AA. Assessment of 3D proton MR echo-planar spectroscopic imaging using automated spectral analysis. Magn Reson Med 2001; 46: 1072–8.

Elkhaled A, Jalbert L, Constantin A, Yoshihara HA, Phillips JJ, Molinaro AM, et al. Characterization of metabolites in infiltrating gliomas using ex vivo (1)H high-resolution magic angle spinning spectroscopy. NMR Biomed 2014; 27: 578–93.

Elster C, Schubert F, Link A, Walzel M, Seifert F, Rinneberg H. Quantitative magnetic resonance spectroscopy: semi-parametric modeling and determination of uncertainties. Magn Reson Med 2005; 53: 1288–96.

Emir UE, Larkin SJ, de Pennington N, Voets N, Plaha P, Stacey R, et al. Noninvasive quantification of 2-Hydroxyglutarate in human gliomas with IDH1 and IDH2 mutations. Cancer Res 2016; 76: 43–9.

Ernst T, Hennig J. Improved water suppression for localized in vivo 1H spectroscopy. J Magn Reson B 1995; 106: 181–6.

Ernst T, Jiang CS, Nakama H, Buchthal S, Chang L. Lower brain glutamate is associated with cognitive deficits in HIV patients: a new mechanism for HIV-associated neurocognitive disorder. J Magn Reson Imaging 2010; 32: 1045–53.

Ernst T, Kreis R, Ross BD. Absolute quantitation of water and metabolites in the human brain. I. Compartments and water. J Magn Reson B 1993; 102: 1–8.

Ethofer T, Mader I, Seeger U, Helms G, Erb M, Grodd W, et al. Comparison of longitudinal metabolite relaxation times in different regions of the human brain at 1.5 and 3 Tesla. Magn Reson Med 2003; 50: 1296–301.

Federico F, Simone IL, Lucivero V, Iliceto G, De Mari M, Giannini P, et al. Proton magnetic resonance spectroscopy in Parkinson's disease and atypical parkinsonian disorders. Mov Disord 1997; 12: 903–9.

Fisher SK, Heacock AM, Agranoff BW. Inositol lipids and signal transduction in the nervous system: an update. J Neurochem 1992; 58: 18–38.

Fisher SK, Novak JE, Agranoff BW. Inositol and higher inositol phosphates in neural tissues: homeostasis, metabolism and functional significance. J Neurochem 2002; 82: 736–54.

Foerster BR, Thurnher MM, Malani PN, Petrou M, Carets-Zumelzu F, Sundgren PC. Intracranial infections: clinical and imaging characteristics. Acta Radiol 2007; 48: 875–93.

Fountas KN, Tsougos I, Gotsis ED, Giannakodimos S, Smith JR, Kapsalaki EZ. Temporal pole proton preoperative magnetic resonance spectroscopy in patients undergoing surgery for mesial temporal sclerosis. Neurosurg Focus 2012; 32: E3.

Frahm J, Bruhn H, Gyngell ML, Merboldt KD, Hanicke W, Sauter R. Localized high-resolution proton NMR spectroscopy using stimulated echoes: initial applications to human brain in vivo. Magn Reson Med 1989; 9: 79–93.

Frahm J, Michaelis T, Merboldt KD, Hanicke W, Gyngell ML, Bruhn H. On the N-acetyl methyl resonance in localized 1H NMR spectra of human brain in vivo. NMR Biomed 1991; 4: 201–4.

Fulham MJ, Bizzi A, Dietz MJ, Shih HH, Raman R, Sobering GS, et al. Mapping of brain tumor metabolites with proton MR spectroscopic imaging: clinical relevance. Radiology 1992; 185: 675–86.

Gabr RE, Sathyanarayana S, Schar M, Weiss RG, Bottomley PA. On restoring motion-induced signal loss in single-voxel magnetic resonance spectra. Magn Reson Med 2006; 56: 754–60.

Gao F, Barker PB. Various MRS application tools for Alzheimer disease and mild cognitive impairment. AJNR Am J Neuroradiol 2014; 35: S4–11.

Garcia-Gomez JM, Luts J, Julia-Sape M, Krooshof P, Tortajada S, Robledo JV, et al. Multiproject-multicenter evaluation of automatic brain tumor classification by magnetic resonance spectroscopy. MAGMA 2009; 22: 5–18.

Garwood M, DelaBarre L. The return of the frequency sweep: designing adiabatic pulses for contemporary NMR. J Magn Reson 2001; 153: 155–77.

Geramita M, van der Veen JW, Barnett AS, Savostyanova AA, Shen J, Weinberger DR, et al. Reproducibility of prefrontal gamma-aminobutyric acid measurements with J-edited spectroscopy. NMR Biomed 2011; 24: 1089–98.

Gill SS, Thomas DG, Van Bruggen N, Gadian DG, Peden CJ, Bell JD, et al. Proton MR spectroscopy of intracranial tumours: in vivo and in vitro studies. J Comput Assist Tomogr 1990; 14: 497–504.

Gillies P, Marshall I, Asplund M, Winkler P, Higinbotham J. Quantification of MRS data in the frequency domain using a wavelet filter, an approximated Voigt lineshape model and prior knowledge. NMR Biomed 2006; 19: 617–26.

Gonen O, Gruber S, Li BS, Mlynarik V, Moser E. Multivoxel 3D proton spectroscopy in the brain at 1.5 versus 3.0 T: signal-to-noise ratio and resolution comparison. AJNR Am J Neuroradiol 2001; 22: 1727–31.

Goryawala MZ, Sheriff S, Maudsley AA. Regional distributions of brain glutamate and glutamine in normal subjects. NMR Biomed 2016; 29: 1108–16.

Gottschalk M, Lamalle L, Segebarth C. Short-TE localised 1H MRS of the human brain at 3 T: quantification of the metabolite signals using two approaches to account for macromolecular signal contributions. NMR Biomed 2008; 21: 507–17.

Govindaraju V, Young K, Maudsley AA. Proton NMR chemical shifts and coupling constants for brain metabolites. NMR Biomed 2000; 13: 129–53.

Graff-Radford J, Kantarci K. Magnetic resonance spectroscopy in Alzheimer's disease. Neuropsychiatr Dis Treat 2013; 9: 687–96.

Graveron-Demilly D. Quantification in magnetic resonance spectroscopy based on semi-parametric approaches. MAGMA 2014; 27: 113–30.

Griswold MA, Jakob PM, Heidemann RM, Nittka M, Jellus V, Wang J, et al. Generalized autocalibrating partially parallel acquisitions (GRAPPA). Magn Reson Med 2002; 47: 1202–10.

Gruetter R. Automatic, localized in vivo adjustment of all first- and second-order shim coils. Magn Reson Med 1993; 29: 804–11.

Gruetter R, Tkac I. Field mapping without reference scan using asymmetric echo-planar techniques. Magn Reson Med 2000; 43: 319–23.

Gupta RK, Vatsal DK, Husain N, Chawla S, Prasad KN, Roy R, et al. Differentiation of tuberculous from pyogenic brain abscesses with in vivo proton MR spectroscopy and magnetization transfer MR imaging. AJNR Am J Neuroradiol 2001; 22: 1503–9.

Guzman-De-Villoria JA, Mateos-Perez JM, Fernandez-Garcia P, Castro E, Desco M. Added value of advanced over conventional magnetic resonance imaging in grading gliomas and other primary brain tumors. Cancer Imaging 2014; 14: 35.

Haase A, Frahm J, Hanicke W, Matthaei D. 1H NMR chemical shift selective (CHESS) imaging. Phys Med Biol 1985; 30: 341–4.

Hammen T, Hildebrandt M, Stadlbauer A, Doelken M, Engelhorn T, Kerling F, et al. Non-invasive detection of hippocampal sclerosis: correlation between metabolite alterations detected by (1)H-MRS and neuropathology. NMR Biomed 2008; 21: 545–52.

Hancu I, Blezek DJ, Dumoulin MC. Automatic repositioning of single voxels in longitudinal 1H MRS studies. NMR Biomed 2005; 18: 352–61.

Hattingen E, Luckerath C, Pellikan S, Vronski D, Roth C, Knake S, et al. Frontal and thalamic changes of GABA concentration indicate dysfunction of thalamofrontal networks in juvenile myoclonic epilepsy. Epilepsia 2014; 55: 1030–7.

Hattingen E, Raab P, Franz K, Lanfermann H, Setzer M, Gerlach R, et al. Prognostic value of choline and creatine in WHO grade II gliomas. Neuroradiology 2008; 50: 759–67.

Hess AT, Andronesi OC, Tisdall MD, Sorensen AG, van der Kouwe AJ, Meintjes EM. Real-time motion and B0 correction for localized adiabatic selective refocusing (LASER) MRSI using echo planar imaging volumetric navigators. NMR Biomed 2012; 25: 347–58.

Hetherington HP, Pan JW, Mason GF, Adams D, Vaughn MJ, Twieg DB, et al. Quantitative 1H spectroscopic imaging of human brain at 4.1 T using image segmentation. Magn Reson Med 1996; 36: 21–9.

Hofmann L, Slotboom J, Boesch C, Kreis R. Characterization of the macromolecule baseline in localized (1)H-MR spectra of human brain. Magn Reson Med 2001; 46: 855–63.

Hofmann L, Slotboom J, Jung B, Maloca P, Boesch C, Kreis R. Quantitative 1H-magnetic resonance spectroscopy of human brain: influence of composition and parameterization of the basis set in linear combination model-fitting. Magn Reson Med 2002; 48: 440–53.

Hourani R, Brant LJ, Rizk T, Weingart JD, Barker PB, Horska A. Can proton MR spectroscopic and perfusion imaging differentiate between neoplastic and nonneoplastic brain lesions in adults? AJNR Am J Neuroradiol 2008; 29: 366–72.

Howe FA, Barton SJ, Cudlip SA, Stubbs M, Saunders DE, Murphy M, et al. Metabolic profiles of human brain tumors using quantitative in vivo 1H magnetic resonance spectroscopy. Magn Reson Med 2003; 49: 223–32.

Hsu SH, Chou MC, Ko CW, Hsu SS, Lin HS, Fu JH, et al. Proton MR spectroscopy in patients with pyogenic brain abscess: MR spectroscopic imaging versus single-voxel spectroscopy. Eur J Radiol 2013; 82: 1299–307.

Hu S, Lustig M, Chen AP, Crane J, Kerr A, Kelley DA, et al. Compressed sensing for resolution enhancement of hyperpolarized 13C flyback 3D-MRSI. J Magn Reson 2008; 192: 258–64.

Hugg JW, Laxer KD, Matson GB, Maudsley AA, Weiner MW. Neuron loss localizes human temporal lobe epilepsy by in vivo proton magnetic resonance spectroscopic imaging. Ann Neurol 1993; 34: 788–94.

Hurd R, Sailasuta N, Srinivasan R, Vigneron DB, Pelletier D, Nelson SJ. Measurement of brain glutamate using TE-averaged PRESS at 3T. Magn Reson Med 2004; 51: 435–40.

Ishimaru H, Morikawa M, Iwanaga S, Kaminogo M, Ochi M, Hayashi K. Differentiation between high-grade glioma and metastatic brain tumor using single-voxel proton MR spectroscopy. Eur Radiol 2001; 11: 1784–91.

Jacobs MA, Horska A, van Zijl PC, Barker PB. Quantitative proton MR spectroscopic imaging of normal human cerebellum and brain stem. Magn Reson Med 2001; 46: 699–705.

Jansen JF, Backes WH, Nicolay K, Kooi ME. 1H MR spectroscopy of the brain: absolute quantification of metabolites. Radiology 2006; 240: 318–32.

Jost G, Harting I, Heiland S. Quantitative single-voxel spectroscopy: the reciprocity principle for receive-only head coils. J Magn Reson Imaging 2005; 21: 66–71.

Juchem C, Nixon TW, Diduch P, Rothman DL, Starewicz P, de Graaf RA. Dynamic shimming of the human brain at 7 Tesla. Concepts Magn Reson Part B Magn Reson Eng 2010; 37B: 116–28.

Julia-Sape M, Coronel I, Majos C, Candiota AP, Serrallonga M, Cos M, et al. Prospective diagnostic performance evaluation of single-voxel 1H MRS for typing and grading of brain tumours. NMR Biomed 2012; 25: 661–73.

Kallenberg K, Bock HC, Helms G, Jung K, Wrede A, Buhk JH, et al. Untreated glioblastoma multiforme: increased myo-inositol and glutamine levels in the contralateral cerebral hemisphere at proton MR spectroscopy. Radiology 2009; 253: 805–12.

Kanowski M, Kaufmann J, Braun J, Bernarding J, Tempelmann C. Quantitation of simulated short echo time 1H human brain spectra by LCModel and AMARES. Magn Reson Med 2004; 51: 904–12.

Kantarci K, Knopman DS, Dickson DW, Parisi JE, Whitwell JL, Weigand SD, et al. Alzheimer disease: postmortem neuropathologic correlates of antemortem 1H MR spectroscopy metabolite measurements. Radiology 2008; 248: 210–20.

Katscher U, Bornert P. Parallel RF transmission in MRI. NMR Biomed 2006; 19: 393–400.

Keating B, Deng W, Roddey JC, White N, Dale A, Stenger VA, et al. Prospective motion correction for single-voxel 1H MR spectroscopy. Magn Reson Med 2010; 64: 672–9.

Kim DH, Adalsteinsson E, Glover GH, Spielman DM. Regularized higher-order in vivo shimming. Magn Reson Med 2002; 48: 715–22.

Kreis R. Issues of spectral quality in clinical 1H-magnetic resonance spectroscopy and a gallery of artifacts. NMR Biomed 2004; 17: 361–81.

Kreis R, Ernst T, Ross BD. Absolute quantitation of water and metabolites in the human brain. II. Metabolite concentrations. J Magn Reson B 1993; 102: 9–19.

Kreis R, Hofmann L, Kuhlmann B, Boesch C, Bossi E, Huppi PS. Brain metabolite composition during early human brain development as measured by quantitative in vivo 1H magnetic resonance spectroscopy. Magn Reson Med 2002; 48: 949–58.

Kreis R, Ross BD, Farrow NA, Ackerman Z. Metabolic disorders of the brain in chronic hepatic encephalopathy detected with H-1 MR spectroscopy. Radiology 1992; 182: 19–27.

Kreis R, Slotboom J, Hofmann L, Boesch C. Integrated data acquisition and processing to determine metabolite contents, relaxation times, and macromolecule baseline in single examinations of individual subjects. Magn Reson Med 2005; 54: 761–8.

Kuzniecky R, Hugg JW, Hetherington H, Butterworth E, Bilir E, Faught E, et al. Relative utility of 1H spectroscopic imaging and hippocampal volumetry in the lateralization of mesial temporal lobe epilepsy. Neurology 1998; 51: 66–71.

Lam F, Ma C, Clifford B, Johnson CL, Liang ZP. High-resolution (1) H-MRSI of the brain using SPICE: data acquisition and image reconstruction. Magn Reson Med 2016; 76: 1059–70.

Langer DL, Rakaric P, Kirilova A, Jaffray DA, Damyanovich AZ. Assessment of metabolite quantitation reproducibility in serial 3D-(1)H-MR spectroscopic imaging of human brain using stereotactic repositioning. Magn Reson Med 2007; 58: 666–73.

Laudadio T, Pels P, De Lathauwer L, Van Hecke P, Van Huffel S. Tissue segmentation and classification of MRSI data using canonical correlation analysis. Magn Reson Med 2005; 54: 1519–29.

Li BS, Wang H, Gonen O. Metabolite ratios to assumed stable creatine level may confound the quantification of proton brain MR spectroscopy. Magn Reson Imaging 2003; 21: 923–8.

Li Y, Chen AP, Crane JC, Chang SM, Vigneron DB, Nelson SJ. Three-dimensional J-resolved H-1 magnetic resonance spectroscopic imaging of volunteers and patients with brain tumors at 3T. Magn Reson Med 2007; 58: 886–92.

Li Y, Jakary A, Gillung E, Eisendrath S, Nelson SJ, Mukherjee P, et al. Evaluating metabolites in patients with major depressive disorder who received mindfulness-based cognitive therapy and healthy controls using short echo MRSI at 7 Tesla. MAGMA 2016; 29: 523–33.

Li Y, Lupo JM, Parvataneni R, Lamborn KR, Cha S, Chang SM, et al. Survival analysis in patients with newly diagnosed glioblastoma using pre- and postradiotherapy MR spectroscopic imaging. Neuro Oncol 2013; 15: 607–17.

Li Y, Osorio JA, Ozturk-Isik E, Chen AP, Xu D, Crane JC, et al. Considerations in applying 3D PRESS H-1 brain MRSI with an eight-channel phased-array coil at 3 T. Magn Reson Imaging 2006; 24: 1295–302.

Li Y, Park I, Nelson SJ. Imaging tumor metabolism using in vivo magnetic resonance spectroscopy. Cancer J 2015; 21: 123–8.

Li Y, Srinivasan R, Ratiney H, Lu Y, Chang SM, Nelson SJ. Comparison of T(1) and T(2) metabolite relaxation times in glioma and normal brain at 3T. J Magn Reson Imaging 2008; 28: 342–50.

Lin FH, Tsai SY, Otazo R, Caprihan A, Wald LL, Belliveau JW, et al. Sensitivity-encoded (SENSE) proton echo-planar spectroscopic imaging (PEPSI) in the human brain. Magn Reson Med 2007; 57: 249–57.

Louis DN, Perry A, Reifenberger G, von Deimling A, Figarella-Branger D, Cavenee WK, et al. The 2016 World Health Organization classification of tumors of the central nervous system: a summary. Acta Neuropathol 2016; 131: 803–20.

Luts J, Laudadio T, Idema AJ, Simonetti AW, Heerschap A, Vandermeulen D, et al. Nosologic imaging of the brain: segmentation and classification using MRI and MRSI. NMR Biomed 2009; 22: 374–90.

Ma C, Lam F, Ning Q, Johnson CL, Liang ZP. High-resolution 1 H-MRSI of the brain using short-TE SPICE. Magn Reson Med 2017; 77: 467–79.

Maddock RJ, Buonocore MH. MR spectroscopic studies of the brain in psychiatric disorders. Curr Top Behav Neurosci 2012; 11: 199–251.

Majos C, Aguilera C, Alonso J, Julia-Sape M, Castaner S, Sanchez JJ, et al. Proton MR spectroscopy improves discrimination between tumor and pseudotumoral lesion in solid brain masses. AJNR Am J Neuroradiol 2009; 30: 544–51.

Majos C, Julia-Sape M, Alonso J, Serrallonga M, Aguilera C, Acebes JJ, et al. Brain tumor classification by proton MR spectroscopy: comparison of diagnostic accuracy at short and long TE. AJNR Am J Neuroradiol 2004; 25: 1696–704.

Malucelli E, Manners DN, Testa C, Tonon C, Lodi R, Barbiroli B, et al. Pitfalls and advantages of different strategies for the absolute quantification of N-acetyl aspartate, creatine and choline in white and grey matter by 1H-MRS. NMR Biomed 2009; 22: 1003–13.

Maril N, Lenkinski RE. An automated algorithm for combining multivoxel MRS data acquired with phased-array coils. J Magn Reson Imaging 2005; 21: 317–22.

Marshall I, Higinbotham J, Bruce S, Freise A. Use of Voigt lineshape for quantification of in vivo 1H spectra. Magn Reson Med 1997; 37: 651–7.

Matsusue E, Fink JR, Rockhill JK, Ogawa T, Maravilla KR. Distinction between glioma progression and post-radiation change by combined physiologic MR imaging. Neuroradiology 2010; 52: 297–306.

Matthews PM, Francis G, Antel J, Arnold DL. Proton magnetic resonance spectroscopy for metabolic characterization of plaques in multiple sclerosis. Neurology 1991; 41: 1251–6.

Maudsley AA, Darkazanli A, Alger JR, Hall LO, Schuff N, Studholme C, et al. Comprehensive processing, display and analysis for in vivo MR spectroscopic imaging. NMR Biomed 2006; 19: 492–503.

Maudsley AA, Domenig C, Govind V, Darkazanli A, Studholme C, Arheart K, et al. Mapping of brain metabolite distributions by volumetric proton MR spectroscopic imaging (MRSI). Magn Reson Med 2009; 61: 548–59.

Maudsley AA, Domenig C, Sheriff S. Reproducibility of serial whole-brain MR spectroscopic imaging. NMR Biomed 2010; 23: 251–6.

Maudsley AA, Hilal SK, Perman WH, Simon HE. Spatially resolved high-Resolution spectroscopy by 4-Dimensional Nmr. J Magn Reson 1983; 51: 147–52.

Maudsley AA, Matson GB, Hugg JW, Weiner MW. Reduced phase encoding in spectroscopic imaging. Magn Reson Med 1994; 31: 645–51.

McKnight TR, Noworolski SM, Vigneron DB, Nelson SJ. An automated technique for the quantitative assessment of 3D-MRSI data from patients with glioma. J Magn Reson Imaging 2001; 13: 167–77.

McKnight TR, von dem Bussche MH, Vigneron DB, Lu Y, Berger MS, McDermott MW, et al. Histopathological validation of a three-dimensional magnetic resonance spectroscopy index as a predictor of tumor presence. J Neurosurg 2002; 97: 794–802.

McLean MA, Woermann FG, Barker GJ, Duncan JS. Quantitative analysis of short echo time (1)H-MRSI of cerebral gray and white matter. Magn Reson Med 2000; 44: 401–11.

Mekle R, Mlynarik V, Gambarota G, Hergt M, Krueger G, Gruetter R. MR spectroscopy of the human brain with enhanced signal intensity at ultrashort echo times on a clinical platform at 3T and 7T. Magn Reson Med 2009; 61: 1279–85.

Mescher M, Merkle H, Kirsch J, Garwood M, Gruetter R. Simultaneous in vivo spectral editing and water suppression. NMR Biomed 1998; 11: 266–72.

Metastasio A, Rinaldi P, Tarducci R, Mariani E, Feliziani FT, Cherubini A, et al. Conversion of MCI to dementia: role of proton magnetic resonance spectroscopy. Neurobiol Aging 2006; 27: 926–32.

Meyerhoff DJ, Bloomer C, Cardenas V, Norman D, Weiner MW, Fein G. Elevated subcortical choline metabolites in cognitively and clinically asymptomatic HIV+ patients. Neurology 1999; 52: 995–1003.

Meyerhoff DJ, MacKay S, Bachman L, Poole N, Dillon WP, Weiner MW, et al. Reduced brain N-acetylaspartate suggests neuronal loss in cognitively impaired human immunodeficiency virus-seropositive individuals: in vivo 1H magnetic resonance spectroscopic imaging. Neurology 1993; 43: 509–15.

Michaelis T, Merboldt KD, Bruhn H, Hanicke W, Frahm J. Absolute concentrations of metabolites in the adult human brain in vivo: quantification of localized proton MR spectra. Radiology 1993; 187: 219–27.

Mierisova S, Ala-Korpela M. MR spectroscopy quantitation: a review of frequency domain methods. NMR Biomed 2001; 14: 247–59.

Mikkelsen M, Singh KD, Sumner P, Evans CJ. Comparison of the repeatability of GABA-edited magnetic resonance spectroscopy with and without macromolecule suppression. Magn Reson Med 2016; 75: 946–53.

Miller TR, Mohan S, Choudhri AF, Gandhi D, Jindal G. Advances in multiple sclerosis and its variants: conventional and newer imaging techniques. Radiol Clin North Am 2014; 52: 321–36.

Mlynarik V, Gambarota G, Frenkel H, Gruetter R. Localized short-echo-time proton MR spectroscopy with full signal-intensity acquisition. Magn Reson Med 2006; 56: 965–70.

Mlynarik V, Gruber S, Moser E. Proton T (1) and T (2) relaxation times of human brain metabolites at 3 Tesla. NMR Biomed 2001; 14: 325–31.

Mohamed MA, Barker PB, Skolasky RL, Selnes OA, Moxley RT, Pomper MG, et al. Brain metabolism and cognitive impairment in HIV infection: a 3-T magnetic resonance spectroscopy study. Magn Reson Imaging 2010; 28: 1251–7.

Moller HE, Vermathen P, Lentschig MG, Schuierer G, Schwarz S, Wiedermann D, et al. Metabolic characterization of AIDS dementia complex by spectroscopic imaging. J Magn Reson Imaging 1999; 9: 10–18.

Moller-Hartmann W, Herminghaus S, Krings T, Marquardt G, Lanfermann H, Pilatus U, et al. Clinical application of proton magnetic resonance spectroscopy in the diagnosis of intracranial mass lesions. Neuroradiology 2002; 44: 371–81.

Moonen CT, von Kienlin M, van Zijl PC, Cohen J, Gillen J, Daly P, et al. Comparison of single-shot localization methods (STEAM and PRESS) for in vivo proton NMR spectroscopy. NMR Biomed 1989; 2: 201–8.

Moyher SE, Vigneron DB, Nelson SJ. Surface coil MR imaging of the human brain with an analytic reception profile correction. J Magn Reson Imaging 1995; 5: 139–44.

Mueller SG, Ebel A, Barakos J, Scanlon C, Cheong I, Finlay D, et al. Widespread extrahippocampal NAA/(Cr+Cho) abnormalities in TLE with and without mesial temporal sclerosis. J Neurol 2011; 258: 603–12.

Mullins PG, McGonigle DJ, O'Gorman RL, Puts NA, Vidyasagar R, Evans CJ, et al. Current practice in the use of MEGA-PRESS spectroscopy for the detection of GABA. NeuroImage 2014; 86: 43–52.

Naressi A, Couturier C, Devos JM, Janssen M, Mangeat C, de Beer R, et al. Java-based graphical user interface for the MRUI quantitation package. MAGMA 2001; 12: 141–52.

Natt O, Bezkorovaynyy V, Michaelis T, Frahm J. Use of phased array coils for a determination of absolute metabolite concentrations. Magn Reson Med 2005; 53: 3–8.

Near J, Ho YC, Sandberg K, Kumaragamage C, Blicher JU. Long-term reproducibility of GABA magnetic resonance spectroscopy. NeuroImage 2014; 99: 191–6.

Nelson SJ. Analysis of volume MRI and MR spectroscopic imaging data for the evaluation of patients with brain tumors. Magn Reson Med 2001; 46: 228–39.

Nelson SJ, Kadambi AK, Park I, Li Y, Crane J, Olson M, et al. Association of early changes in 1H MRSI parameters with survival for patients with newly diagnosed glioblastoma receiving a multimodality treatment regimen. Neuro Oncol 2017;19: 430–39.

Nelson SJ, Li Y, Lupo JM, Olson M, Crane JC, Molinaro A, et al. Serial analysis of 3D H-1 MRSI for patients with newly diagnosed GBM treated with combination therapy that includes bevacizumab. J Neurooncol 2016; 130: 171–179.

Nelson SJ, Ozhinsky E, Li Y, Park I, Crane J. Strategies for rapid in vivo 1H and hyperpolarized 13C MR spectroscopic imaging. J Magn Reson 2013; 229: 187–97.

Opstad KS, Ladroue C, Bell BA, Griffiths JR, Howe FA. Linear discriminant analysis of brain tumour (1)H MR spectra: a comparison of classification using whole spectra versus metabolite quantification. NMR Biomed 2007; 20: 763–70.

Ordidge RJ, Connelly A, Lohman JAB. Image-Selected Invivo Spectroscopy (Isis) - a New technique for spatially selective Nmr-Spectroscopy. J Magn Reson 1986; 66: 283–94.

Osorio JA, Xu D, Cunningham CH, Chen A, Kerr AB, Pauly JM, et al. Design of cosine modulated very selective suppression pulses for MR spectroscopic imaging at 3T. Magn Reson Med 2009; 61: 533–40.

Otazo R, Mueller B, Ugurbil K, Wald L, Posse S. Signal-to-noise ratio and spectral line width improvements between 1.5 and 7 Tesla in proton echo-planar spectroscopic imaging. Magn Reson Med 2006; 56: 1200–10.

Ott D, Hennig J, Ernst T. Human brain tumors: assessment with in vivo proton MR spectroscopy. Radiology 1993; 186: 745–52.

Oz G, Alger JR, Barker PB, Bartha R, Bizzi A, Boesch C, et al. Clinical proton MR spectroscopy in central nervous system disorders. Radiology 2014; 270: 658–79.

Ozhinsky E, Vigneron DB, Chang SM, Nelson SJ. Automated prescription of oblique brain 3D magnetic resonance spectroscopic imaging. Magn Reson Med 2013; 69: 920–30.

Ozhinsky E, Vigneron DB, Nelson SJ. Improved spatial coverage for brain 3D PRESS MRSI by automatic placement of outer-volume suppression saturation bands. J Magn Reson Imaging 2011; 33: 792–802.

Ozturk-Isik E, Chen AP, Crane JC, Bian W, Xu D, Han ET, et al. 3D sensitivity encoded ellipsoidal MR spectroscopic imaging of gliomas at 3T. Magn Reson Imaging 2009; 27: 1249–57.

Ozturk-Isik E, Crane JC, Cha S, Chang SM, Berger MS, Nelson SJ. Unaliasing lipid contamination for MR spectroscopic imaging of gliomas at 3T using sensitivity encoding (SENSE). Magn Reson Med 2006; 55: 1164–9.

Paley M, Wilkinson ID, Hall-Craggs MA, Chong WK, Chinn RJ, Harrison MJ. Short echo time proton spectroscopy of the brain in HIV infection/AIDS. Magn Reson Imaging 1995; 13: 871–5.

Pan JW, Duckrow RB, Gerrard J, Ong C, Hirsch LJ, Resor SR, Jr., et al. 7T MR spectroscopic imaging in the localization of surgical epilepsy. Epilepsia 2013; 54: 1668–78.

Pan JW, Lo KM, Hetherington HP. Role of very high order and degree B0 shimming for spectroscopic imaging of the human brain at 7 tesla. Magn Reson Med 2012a; 68: 1007–17.

Pan JW, Spencer DD, Kuzniecky R, Duckrow RB, Hetherington H, Spencer SS. Metabolic networks in epilepsy by MR spectroscopic imaging. Acta Neurol Scand 2012b; 126: 411–20.

Park I, Chen AP, Zierhut ML, Ozturk-Isik E, Vigneron DB, Nelson SJ. Implementation of 3 T lactate-edited 3D 1H MR spectroscopic imaging with flyback echo-planar readout for gliomas patients. Ann Biomed Eng 2011; 39: 193–204.

Petroff OA, Hyder F, Rothman DL, Mattson RH. Homocarnosine and seizure control in juvenile myoclonic epilepsy and complex partial seizures. Neurology 2001; 56: 709–15.

Petroff OA, Rothman DL, Behar KL, Mattson RH. Low brain GABA level is associated with poor seizure control. Ann Neurol 1996; 40: 908–11.

Pohmann R, von Kienlin M, Haase A. Theoretical evaluation and comparison of fast chemical shift imaging methods. J Magn Reson 1997; 129: 145–60.

Poptani H, Kaartinen J, Gupta RK, Niemitz M, Hiltunen Y, Kauppinen RA. Diagnostic assessment of brain tumours and non-neoplastic brain disorders in vivo using proton nuclear magnetic resonance spectroscopy and artificial neural networks. J Cancer Res Clin Oncol 1999; 125: 343–9.

Posse S, Otazo R, Dager SR, Alger J. MR spectroscopic imaging: principles and recent advances. J Magn Reson Imaging 2013; 37: 1301–25.

Posse S, Tedeschi G, Risinger R, Ogg R, Le Bihan D. High speed 1H spectroscopic imaging in human brain by echo planar spatial-spectral encoding. Magn Reson Med 1995; 33: 34–40.

Poullet JB, Sima DM, Simonetti AW, De Neuter B, Vanhamme L, Lemmerling P, et al. An automated quantitation of short echo time MRS spectra in an open source software environment: AQSES. NMR Biomed 2007; 20: 493–504.

Pouwels PJ, Frahm J. Differential distribution of NAA and NAAG in human brain as determined by quantitative localized proton MRS. NMR Biomed 1997; 10: 73–8.

Prinsen H, de Graaf RA, Mason GF, Pelletier D, Juchem C. Reproducibility measurement of glutathione, GABA, and glutamate: towards in vivo neurochemical profiling of multiple sclerosis with MR spectroscopy at 7T. J Magn Reson Imaging 2017; 45: 187–198.

Provencher SW. Estimation of metabolite concentrations from localized in vivo proton NMR spectra. Magn Reson Med 1993; 30: 672–9.

Provencher SW. Automatic quantitation of localized in vivo 1H spectra with LCModel. NMR Biomed 2001; 14: 260–4.

Pruessmann KP, Weiger M, Scheidegger MB, Boesiger P. SENSE: sensitivity encoding for fast MRI. Magn Reson Med 1999; 42: 952–62.

Ratiney H, Coenradie Y, Cavassila S, van Ormondt D, Graveron-Demilly D. Time-domain quantitation of 1H short echo-time signals: background accommodation. MAGMA 2004; 16: 284–96.

Reynolds G, Wilson M, Peet A, Arvanitis TN. An algorithm for the automated quantitation of metabolites in in vitro NMR signals. Magn Reson Med 2006; 56: 1211–9.

Richards TL. Proton MR spectroscopy in multiple sclerosis: value in establishing diagnosis, monitoring progression, and evaluating therapy. AJR Am J Roentgenol 1991; 157: 1073–8.

Riederer F, Bittsansky M, Schmidt C, Mlynarik V, Baumgartner C, Moser E, et al. 1H magnetic resonance spectroscopy at 3 T in cryptogenic and mesial temporal lobe epilepsy. NMR Biomed 2006; 19: 544–53.

Rijpkema M, Schuuring J, van der Meulen Y, van der Graaf M, Bernsen H, Boerman R, et al. Characterization of oligodendrogliomas using short echo time 1H MR spectroscopic imaging. NMR Biomed 2003; 16: 12–18.

Rodgers CT, Robson MD. Receive array magnetic resonance spectroscopy: whitened singular value decomposition (WSVD) gives optimal Bayesian solution. Magn Reson Med 2010; 63: 881–91.

Rovira A, Alonso J. 1H magnetic resonance spectroscopy in multiple sclerosis and related disorders. Neuroimaging Clin N Am 2013; 23: 459–74.

Rutgers DR, Kingsley PB, van der Grond J. Saturation-corrected T 1 and T 2 relaxation times of choline, creatine and N-acetyl aspartate in human cerebral white matter at 1.5 T. NMR Biomed 2003; 16: 286–8.

Sabati M, Sheriff S, Gu M, Wei J, Zhu H, Barker PB, et al. Multivendor implementation and comparison of volumetric whole-brain echo-planar MR spectroscopic imaging. Magn Reson Med 2015; 74: 1209–20.

Sabati M, Zhan J, Govind V, Arheart KL, Maudsley AA. Impact of reduced k-space acquisition on pathologic detectability for volumetric MR spectroscopic imaging. J Magn Reson Imaging 2014; 39: 224–34.

Saraswathy S, Crawford FW, Lamborn KR, Pirzkall A, Chang S, Cha S, et al. Evaluation of MR markers that predict survival in patients with newly diagnosed GBM prior to adjuvant therapy. J Neurooncol 2009; 91: 69–81.

Sarkar S, Heberlein K, Hu X. Truncation artifact reduction in spectroscopic imaging using a dual-density spiral k-space trajectory. Magn Reson Imaging 2002; 20: 743–57.

Scheenen TW, Heerschap A, Klomp DW. Towards 1H-MRSI of the human brain at 7T with slice-selective adiabatic refocusing pulses. MAGMA 2008a; 21: 95–101.

Scheenen TW, Klomp DW, Wijnen JP, Heerschap A. Short echo time 1H-MRSI of the human brain at 3T with minimal chemical shift displacement errors using adiabatic refocusing pulses. Magn Reson Med 2008b; 59: 1–6.

Schirda CV, Zhao T, Andronesi OC, Lee Y, Pan JW, Mountz JM, et al. In vivo brain rosette spectroscopic imaging (RSI) with LASER excitation, constant gradient strength readout, and automated LCModel quantification for all voxels. Magn Reson Med 2016; 76: 380–90.

Schlemmer HP, Bachert P, Herfarth KK, Zuna I, Debus J, van Kaick G. Proton MR spectroscopic evaluation of suspicious brain lesions after stereotactic radiotherapy. AJNR Am J Neuroradiol 2001; 22: 1316–24.

Schousboe A, Bak LK, Waagepetersen HS. Astrocytic control of biosynthesis and turnover of the neurotransmitters glutamate and GABA. Front Endocrinol (Lausanne) 2013; 4: 102.

Schuff N, Ezekiel F, Gamst AC, Amend DL, Capizzano AA, Maudsley AA, et al. Region and tissue differences of metabolites in normally aged brain using multislice 1H magnetic resonance spectroscopic imaging. Magn Reson Med 2001; 45: 899–907.

Seeger A, Braun C, Skardelly M, Paulsen F, Schittenhelm J, Ernemann U, et al. Comparison of three different MR perfusion techniques and MR spectroscopy for multiparametric assessment in distinguishing recurrent high-grade gliomas from stable disease. Acad Radiol 2013; 20: 1557–65.

Seeger U, Klose U, Mader I, Grodd W, Nagele T. Parameterized evaluation of macromolecules and lipids in proton MR spectroscopy of brain diseases. Magn Reson Med 2003; 49: 19–28.

Seeger U, Mader I, Nagele T, Grodd W, Lutz O, Klose U. Reliable detection of macromolecules in single-volume 1H NMR spectra of the human brain. Magn Reson Med 2001; 45: 948–54.

Simister RJ, McLean MA, Barker GJ, Duncan JS. Proton MRS reveals frontal lobe metabolite abnormalities in idiopathic generalized epilepsy. Neurology 2003; 61: 897–902.

Simister RJ, McLean MA, Barker GJ, Duncan JS. Proton MR spectroscopy of metabolite concentrations in temporal lobe epilepsy and effect of temporal lobe resection. Epilepsy Res 2009; 83: 168–76.

Slotboom J, Boesch C, Kreis R. Versatile frequency domain fitting using time domain models and prior knowledge. Magn Reson Med 1998; 39: 899–911.

Slotboom J, Mehlkopf AF, Bovee WMMJ. A single-Shot localization pulse sequence suited for coils with inhomogeneous Rf Fields using adiabatic slice-Selective Rf pulses. J Magn Reson 1991; 95: 396–404.

Soher BJ, Maudsley AA. Evaluation of variable line-shape models and prior information in automated 1H spectroscopic imaging analysis. Magn Reson Med 2004; 52: 1246–54.

Soher BJ, Vermathen P, Schuff N, Wiedermann D, Meyerhoff DJ, Weiner MW, et al. Short TE in vivo (1)H MR spectroscopic imaging at 1.5 T: acquisition and automated spectral analysis. Magn Reson Imaging 2000; 18: 1159–65.

Srinivasan R, Phillips JJ, Vandenberg SR, Polley MY, Bourne G, Au A, et al. Ex vivo MR spectroscopic measure differentiates tumor from treatment effects in GBM. Neuro Oncol 2010a; 12: 1152–61.

Srinivasan R, Ratiney H, Hammond-Rosenbluth KE, Pelletier D, Nelson SJ. MR spectroscopic imaging of glutathione in the white and gray matter at 7 T with an application to multiple sclerosis. Magn Reson Imaging 2010b; 28: 163–70.

Srinivasan R, Vigneron D, Sailasuta N, Hurd R, Nelson S. A comparative study of myo-inositol quantification using LCmodel at 1.5 T and 3.0 T with 3 D 1H proton spectroscopic imaging of the human brain. Magn Reson Imaging 2004; 22: 523–8.

Star-Lack J, Nelson SJ, Kurhanewicz J, Huang LR, Vigneron DB. Improved water and lipid suppression for 3D PRESS CSI using RF band selective inversion with gradient dephasing (BASING). Magn Reson Med 1997a; 38: 311–21.

Star-Lack J, Spielman D, Adalsteinsson E, Kurhanewicz J, Terris DJ, Vigneron DB. In vivo lactate editing with simultaneous detection of choline, creatine, NAA, and lipid singlets at 1.5 T using PRESS excitation with applications to the study of brain and head and neck tumors. J Magn Reson 1998; 133: 243–54.

Star-Lack J, Vigneron DB, Pauly J, Kurhanewicz J, Nelson SJ. Improved solvent suppression and increased spatial excitation bandwidths for three-dimensional PRESS CSI using phase-compensating spectral/spatial spin-echo pulses. J Magn Reson Imaging 1997b; 7: 745–57.

Stockler S, Holzbach U, Hanefeld F, Marquardt I, Helms G, Requart M, et al. Creatine deficiency in the brain: a new, treatable inborn error of metabolism. Pediatr Res 1994; 36: 409–13.

Stupp R, Mason WP, van den Bent MJ, Weller M, Fisher B, Taphoorn MJ, et al. Radiotherapy plus concomitant and adjuvant temozolomide for glioblastoma. N Engl J Med 2005; 352: 987–96.

Sturrock A, Laule C, Decolongon J, Dar Santos R, Coleman AJ, Creighton S, et al. Magnetic resonance spectroscopy biomarkers in premanifest and early Huntington disease. Neurology 2010; 75: 1702–10.

Tal A, Kirov, II, Grossman RI, Gonen O. The role of gray and white matter segmentation in quantitative proton MR spectroscopic imaging. NMR Biomed 2012; 25: 1392–400.

Tannus A, Garwood M. Adiabatic pulses. NMR Biomed 1997; 10: 423–34.

Tarasow E, Wiercinska-Drapalo A, Kubas B, Dzienis W, Orzechowska-Bobkiewicz A, Prokopowicz D, et al. Cerebral MR spectroscopy in neurologically asymptomatic HIV-infected patients. Acta Radiol 2003; 44: 206–12.

Tate AR, Underwood J, Acosta DM, Julia-Sape M, Majos C, Moreno-Torres A, et al. Development of a decision support system for diagnosis and grading of brain tumours using in vivo magnetic resonance single voxel spectra. NMR Biomed 2006; 19: 411–34.

Tate DF, Khedraki R, McCaffrey D, Branson D, Dewey J. The role of medical imaging in defining CNS abnormalities associated with HIV-infection and opportunistic infections. Neurotherapeutics 2011; 8: 103–16.

Terpstra M, Cheong I, Lyu T, Deelchand DK, Emir UE, Bednarik P, et al. Test-retest reproducibility of neurochemical profiles with short-echo, single-voxel MR spectroscopy at 3T and 7T. Magn Reson Med 2016; 76: 1083–91.

Terpstra M, Henry PG, Gruetter R. Measurement of reduced glutathione (GSH) in human brain using LCModel analysis of difference-edited spectra. Magn Reson Med 2003; 50: 19–23.

Thomas MA, Ryner LN, Mehta MP, Turski PA, Sorenson JA. Localized 2D J-resolved 1H MR spectroscopy of human brain tumors in vivo. J Magn Reson Imaging 1996; 6: 453–9.

Tkac I, Starcuk Z, Choi IY, Gruetter R. In vivo 1H NMR spectroscopy of rat brain at 1 ms echo time. Magn Reson Med 1999; 41: 649–56.

Tofts PS, Wray S. A critical assessment of methods of measuring metabolite concentrations by NMR spectroscopy. NMR Biomed 1988; 1: 1–10.

Traber F, Block W, Freymann N, Gur O, Kucinski T, Hammen T, et al. A multicenter reproducibility study of single-voxel 1H-MRS of the medial temporal lobe. Eur Radiol 2006; 16: 1096–103.

Traber F, Block W, Lamerichs R, Gieseke J, Schild HH. 1H metabolite relaxation times at 3.0 tesla: measurements of T1 and T2 values in normal brain and determination of regional differences in transverse relaxation. J Magn Reson Imaging 2004; 19: 537–45.

Tran TK, Vigneron DB, Sailasuta N, Tropp J, Le Roux P, Kurhanewicz J, et al. Very selective suppression pulses for clinical MRSI studies of brain and prostate cancer. Magn Reson Med 2000; 43: 23–33.

Tsai SY, Otazo R, Posse S, Lin YR, Chung HW, Wald LL, et al. Accelerated proton echo planar spectroscopic imaging (PEPSI) using GRAPPA with a 32-channel phased-array coil. Magn Reson Med 2008; 59: 989–98.

Ugurbil K. Magnetic resonance imaging at Ultrahigh fields. IEEE Trans Biomed Eng 2014; 61: 1364–79.

van de Bank BL, Emir UE, Boer VO, van Asten JJ, Maas MC, Wijnen JP, et al. Multi-center reproducibility of neurochemical profiles in the human brain at 7 T. NMR Biomed 2015; 28: 306–16.

Vanhamme L, Fierro RD, Van Huffel S, de Beer R. Fast removal of residual water in proton spectra. J Magn Reson 1998; 132: 197–203.

Vanhamme L, Sundin T, Hecke PV, Huffel SV. MR spectroscopy quantitation: a review of time-domain methods. NMR Biomed 2001; 14: 233–46.

Vanhamme L, Van Huffel S, Van Hecke P, van Ormondt D. Time-domain quantification of series of biomedical magnetic resonance spectroscopy signals. J Magn Reson 1999; 140: 120–30.

Veenith TV, Mada M, Carter E, Grossac J, Newcombe V, Outtrim J, et al. Comparison of inter subject variability and reproducibility of whole brain proton spectroscopy. PLoS One 2014; 9: e115304.

Vicente J, Fuster-Garcia E, Tortajada S, Garcia-Gomez JM, Davies N, Natarajan K, et al. Accurate classification of childhood brain tumours by in vivo (1)H MRS - a multicentre study. Eur J Cancer 2013; 49: 658–67.

Wijnen JP, Idema AJ, Stawicki M, Lagemaat MW, Wesseling P, Wright AJ, et al. Quantitative short echo time 1H MRSI of the peripheral edematous region of human brain tumors in the differentiation between glioblastoma, metastasis, and meningioma. J Magn Reson Imaging 2012; 36: 1072–82.

Willmann O, Wennberg R, May T, Woermann FG, Pohlmann-Eden B. The role of 1H magnetic resonance spectroscopy in pre-operative evaluation for epilepsy surgery. A meta-analysis. Epilepsy Res 2006; 71: 149–58.

Wilson M, Reynolds G, Kauppinen RA, Arvanitis TN, Peet AC. A constrained least-squares approach to the automated quantitation of in vivo (1)H magnetic resonance spectroscopy data. Magn Reson Med 2011; 65: 1–12.

Woermann FG, McLean MA, Bartlett PA, Parker GJ, Barker GJ, Duncan JS. Short echo time single-voxel 1H magnetic resonance spectroscopy in magnetic resonance imaging-negative temporal lobe epilepsy: different biochemical profile compared with hippocampal sclerosis. Ann Neurol 1999; 45: 369–76.

Yung KT, Zheng W, Zhao C, Martinez-Ramon M, van der Kouwe A, Posse S. Atlas-based automated positioning of outer volume suppression slices in short-echo time 3D MR spectroscopic imaging of the human brain. Magn Reson Med 2011; 66: 911–22.

Zaaraoui W, Fleysher L, Fleysher R, Liu S, Soher BJ, Gonen O. Human brain-structure resolved T(2) relaxation times of proton metabolites at 3 Tesla. Magn Reson Med 2007; 57: 983–9.

Zaitsev M, Speck O, Hennig J, Buchert M. Single-voxel MRS with prospective motion correction and retrospective frequency correction. NMR Biomed 2010; 23: 325–32.

Zhu H, Barker PB. MR spectroscopy and spectroscopic imaging of the brain. Meth Mol Biol 2011; 711: 203–26.

Zierhut ML, Ozturk-Isik E, Chen AP, Park I, Vigneron DB, Nelson SJ. (1)H spectroscopic imaging of human brain at 3 Tesla: comparison of fast three-dimensional magnetic resonance spectroscopic imaging techniques. J Magn Reson Imaging 2009; 30: 473–80.

Zubler F, Seeck M, Landis T, Henry F, Lazeyras F. Contralateral medial temporal lobe damage in right but not left temporal lobe epilepsy: a (1)H magnetic resonance spectroscopy study. J Neurol Neurosurg Psychiatry 2003; 74: 1240–4.

13

Multinuclear MR Imaging and Spectroscopy[1]

Wieland A. Worthoff, Aliaksandra Shymanskaya, Chang-Hoon Choi, Jörg Felder, Ana-Maria Oros-Peusquens and N. Jon Shah
Institute of Neuroscience and Medicine (INM-4) Forschungszentrum Jülich

Contents

13.1 Introduction

Due to its naturally high abundance in the human body, magnetic resonance imaging (MRI) or spectroscopy (MRS) is, in general, based on the hydrogen (^{1}H) nucleus. However, with the increased availability of ultra-high field MR systems (≥7T), the use of non-proton X-nuclei such as ^{23}Na or ^{31}P, is now more feasible. The use of these X-nuclei is particularly interesting for clinicians due to their seemingly central role in many cellular processes.

Although metabolic information can also be obtained through the use of positron emission tomography (PET), this method is invasive and detects radiopharmaceutical uptake and possibly the first metabolite conversion step. In contrast, the X-nuclei possess non-zero nuclear spin and are sufficiently abundant in the human body or may be exogenously supplied to generate measurable MR signals. X-nuclei MR(S)I methods excel, in particular, in the analysis of entire cascades of chemical processes and are consequently an extremely versatile and proficient instrument with which to reveal these mechanisms, reactions and interactions, which is metabolism. Common X-nuclei to be investigated are ^{2}H, ^{7}Li, ^{13}C, ^{17}O, ^{19}F, ^{23}Na, ^{31}P, ^{35}Cl and ^{39}K (see also Table 13.1). It is worth noting that it is possible to combine MRI with other imaging modalities, such as hybrid PET-MR imaging, to allow simultaneous, twofold investigation of metabolism in one measurement session (Shah *et al.*, 2013). However, X-nuclei are challenging targets since their inherent natural abundances in the human body and gyromagnetic ratios are lower than those of ^{1}H, resulting in reduced MR sensitivity. Additionally, their relaxation rates also tend to differ significantly compared to those of protons (in particular quadrupolar nuclei tend to exhibit significantly shorter relaxation times). These facts necessitate the use of technically advanced quantification methods to gain useful information. Table 13.1 summarises properties of biologically common X-nuclei adapted from de Graaf (2007). This chapter introduces the use of common X-nuclei with MRI/S and highlights methods for detection and quantification followed by examples of biomedical applications.

13.2 Phosphorus

Phosphorus-31 (^{31}P) is one of the key nuclei in the human brain and plays an important role in tissue energy metabolism and membrane integrity. Exploring ^{31}P using non-invasive MRS and/or MRI, offers unique insight into both static and dynamic aspects of metabolites.

[1] Editor: Nicholas G. Dowell Reviewer: Jannie P Winjen, University Medical Center Utrecht, The Netherlands

TABLE 13.1 Magnetic properties of most frequent nuclei in biological tissue with non-zero spin

Nuclide	Spin	Gyromagnetic Ratio $\gamma/2\pi$ [MHz/T]	Magnetic Dipole Moment μ_z/μ_N	Natural Abundance [%]	Relative Sensitivity [%]
^{1}H	1/2	42.58	2.7928	99.9885	100
^{2}H	1	6.54	0.8574	0.0115	0.97
^{7}Li	3/2	16.55	3.2565	92.41	29.4
^{13}C	1/2	10.71	0.7024	1.11	1.59
^{17}O	5/2	−5.77	−1.8938	0.038	2.91
^{19}F	1/2	40.05	2.6269	100	83.2
^{23}Na	3/2	11.27	2.2177	100	9.27
^{31}P	1/2	17.25	1.1316	100	6.65
^{35}Cl	3/2	4.18	0.8219	75.76	0.47
^{39}K	3/2	1.83	0.3915	93.258	0.05

Source: Adapted from de Graaf, R.A., *In Vivo NMR Spectroscopy: Principles and Techniques,* 2nd ed, John Wiley & Sons, Chichester, 2007; Stone, N.J., *Atom Data Nucl. Data,* 90, 75–176, 2005.

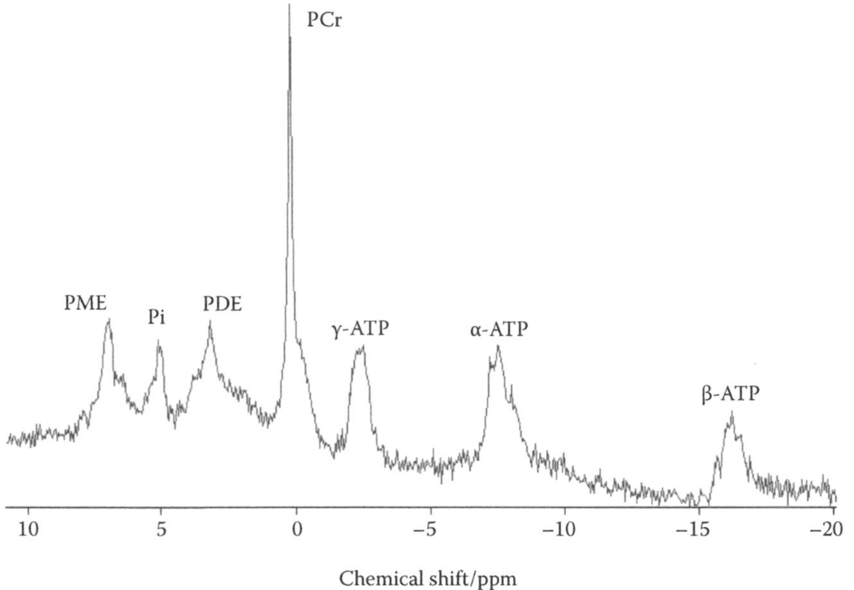

FIGURE 13.1 *In vivo* brain ^{31}P spectrum of a healthy volunteer.

13.2.1 ^{31}P spectrum and challenges

^{31}P MRS provides a spectrum (Figure 13.1) consisting of seven distinct resonance peaks with a relatively large chemical shift range (~30 ppm). These peaks correspond to: three variants of adenosine triphosphate (ATP), phosphocreatine (PCr), phosphodiester (PDE), inorganic phosphate (Pi) and phosphomonoester (PME) containing phosphocholine and phosphoethanomine. The quality of the spectrum depends on a number of factors, e.g. B_0 and B_1 homogeneity and spectral separation of the peaks. Abnormalities in metabolic concentrations are closely linked to various pathological and neurodegenerative conditions.

Conducting *in vivo* ^{31}P MR for brain applications is a challenging task compared to ^{1}H measurements due to several reasons: intrinsically low MR sensitivity (~6.6% that of ^{1}H), low natural abundance (~few millimoles), approximately 2.4 times lower gyromagnetic ratio, relatively long T_1 (order of a few seconds) resulting in long acquisition time, and short T_2 (order of tens of milliseconds) resulting in rapidly decaying signals (Lei *et al.*, 2003; Ren *et al.*, 2015). General improvements are achieved by increasing the B_0 field strength (Qiao *et al.*, 2006) or by applying advanced techniques, such as multivoxel interleaved acquisition schemes (Niess *et al.*, 2016) or nuclear Overhauser enhancement (NOE) (Lagemaat *et al.*, 2016; Lei *et al.*, 2003).

13.2.2 Acquisition techniques for ^{31}P

In order to conduct ^{31}P MRS experiments, a well-designed RF system is essential. Double-resonant RF coils are typically used where the ^{1}H coil is utilised for shimming to provide sufficient B_0 field homogeneity (~few Hz in the full width at half maximum) and for anatomical/scout imaging to guide localisation. The ^{1}H decoupling, NOE or polarisation transfer technique may also be employed to improve spectral quality of ^{31}P (Klomp *et al.*, 2008). Concepts using LCC-traps (van de Bank *et al.*, 2015) and multilayer array coils (Mirkes *et al.*, 2016) have recently been introduced with significantly improved double-resonant RF coil performance in terms of SNR or B_1 field homogeneity.

For ^{31}P brain applications in particular, pulse-acquire FID-based sequences such as image selective *in vivo* spectroscopy (ISIS) or 2D/3D chemical shift imaging (CSI) are preferable due to the short T_2 of *in vivo* ^{31}P metabolites. Numerous techniques have been developed and added onto these original sequences to improve the quality of ^{31}P spectra. For example, Meyerspeer *et al.* used a conventional slice selective excitation, combined with localisation by B_1 insensitive adiabatic selective refocusing sequence, in order to achieve localised dynamic measurements in exercising muscle (Meyerspeer *et al.*, 2012). In addition, Chmelik *et al.* combined a 1D ISIS adiabatic slice selection with conventional phase encoding as a 2D pulse-acquired CSI sequence to minimise chemical shift displacement error (Chmelik *et al.*, 2008). Furthermore, a proton-observed phosphorus editing method and an adiabatic multi-echo technique were introduced to enhance detection sensitivity of *in vivo* phospholipid metabolites (van der Kemp *et al.*, 2013; Wijnen *et al.*, 2016).

Using spatially tailored frequency selective irradiation pulses, for example, either the ATP or Pi peak is saturated and energy is exchanged among ATP, Pi and PCr, leading to a decreasing signal intensity of the PCr peak (Befroy *et al.*, 2012; Du *et al.*, 2007). A similar technique using band inversion transfer has recently been introduced with benefits of lower specific absorption rate and without the need for prolonged presaturation pulse (Ren *et al.*, 2015).

13.2.3 Quantification and applications

Understanding the static and dynamic properties of energy metabolites from the ^{31}P spectrum requires accurate spectral quantification. ^{31}P resonance peaks can simply be integrated after zero- and first-order phase and baseline corrections and be spectrally fitted by application of a jMRUI AMARES algorithm (Vanhamme *et al.*, 1997) or the LCModel (Provencher, 1993) software package.

The most common method used to interpret acquired ^{31}P data quantitatively is to compare ratios between metabolites. The PCr peak tends to be set as a reference and its ratio to others is measured. ^{31}P peak ratios of various brain tumours were determined for differentiation and significant changes were observed in PME/PDE, PDE/Pi, PME/PCr and PDE/PCr (Ha *et al.*, 2013) according to tumour type and grade. Yuksel *et al.* monitored the ratio of PCr/ATP of patients with bipolar disorder during visual stimulation and found that the PCr level was maintained while the ATP value decreased, resulting in an increase of the PCr/ATP ratio. This result was shown to be opposite to that of a healthy control group (Yuksel *et al.*, 2015).

Intracellular pH and free cellular magnesium concentration in the brain can be calculated by measuring the chemical shift differences between the pH-sensitive Pi peak relative to the pH-insensitive PCr peak and between α- and β-ATP, respectively (Ren *et al.*, 2015). The cellular redox state using the intracellular NAD^+/NADH ratio can also be implied (Lu *et al.*, 2014).

Du *et al.* examined the ATP metabolism in the human brain by measuring the kinetic network and metabolic fluxes of ATP, PCr and Pi (Du *et al.*, 2007). In the brain energy metabolism diagram shown in Kemp and Due *et al.*, the ATP metabolism cycle consisting of ATP production and consumption is modulated essentially by ATP synthase and creatine kinase reactions (Du *et al.*, 2007; Kemp, 2000). ATP kinetics reflecting unidirectional metabolic exchange reaction rates of the PCr↔ATP↔Pi network can indirectly be measured by a magnetisation transfer technique (Befroy *et al.*, 2012). This information is used to determine the concentrations of PCr and Pi that are further processed to calculate ATP metabolic fluxes. This method has also been applied for cerebral abnormalities evaluation in schizophrenia and it was found that the flux rate constant and the intracellular pH were considerably decreased while no significant changes were found in metabolite concentration ratios (Du *et al.*, 2014). Quantitative ^{31}P MR is, therefore, a useful tool to detect brain activity and biochemical alternations as well as for early diagnosis of neurodegenerative diseases.

13.3 Carbon

Carbon nuclei form the basis of organic compounds and are excellent markers for metabolic investigations. Carbon-13 (^{13}C) MRS is an outstanding tool with which to monitor a large number of metabolites, metabolic fluxes and metabolic conversion of the entire tricarboxylic acid cycle. The most important peaks in a ^{13}C spectrum are the carbons connected to N-acetyl aspartate (NAA), γ-aminobutyric acid (GABA), glutamate, glutamine, inositol and glucose (Rodrigues and Cerdán, 2005). In addition, ^{13}C MRS allows one to access information relating to the structural and chemical composition of many relevant organic compounds.

^{13}C resonances are distributed over a large chemical shift range, providing an excellent spectral resolution and a high specificity. However, like other X-nuclei, due to their low natural abundance and gyromagnetic ratio, the detection of ^{13}C is challenging (Table 13.1). Furthermore, carbon spectral peaks are subject to coupling effects induced by being bound to ^1H, causing peak splitting and resulting in a loss in SNR. Nevertheless, the natural abundance of ^{13}C resonances at 1.5T provide more than adequate resolution to quantify glutamate and glutamine (Blüml, 1999) without the technical difficulties (peak splitting, overlap of peaks, contamination through macromolecules) associated with ^1H MRS.

13.3.1 Acquisition techniques for ^{13}C

^{13}C spectra tend to be acquired in combination with ^{13}C-enriched substrates administered orally or by infusion (Moreno *et al.*, 2001) in order to improve the MR sensitivity of ^{13}C. Administration of ^{13}C-labelled tracer molecules increases the available MR signal and thus overcomes their low natural abundance, which allows processes of tracer uptake to be monitored in dynamic studies.

Several MR sequence techniques similar to ^{31}P are also available to increase the sensitivity of ^{13}C *in vivo* and for localisation (Gruetter *et al.*, 2003). Broadband ^{1}H decoupling or polarisation transfer is essentially employed in order to suppress the J-coupling of the ^{1}H spin and the ^{13}C nuclei (Klomp *et al.*, 2006). Additionally, use of the NOE technique has been found to enhance the SNR and distortionless enhanced polarisation transfer has also been introduced to minimise phase distortions in the spectra (Doddrell *et al.*, 1982).

With reference to hardware, a number of $^{1}H/^{13}C$ double-resonant RF coils have been developed to enable the decoupling technique while maintaining the RF coil quality (Klomp *et al.*, 2006; Roig *et al.*, 2015).

More recently, a hyperpolarised ^{13}C technique has emerged, which is dominantly combined with the dynamic nuclear polarisation (DNP) principle (Golman *et al.*, 2003). This combination by means of DNP can significantly boost the sensitivity of hyperpolarised carbon-13 (Ardenkjaer-Larsen *et al.*, 2003) and thus augment the clinical applicability (Brindle *et al.*, 2011).

13.3.2 Applications

^{13}C MRS was used to investigate Canavan's disease and a considerable amount of rate reduction of NAA synthesis was found in patients, although NAA concentration is, in general, extremely stable (Moreno *et al.*, 2001).

Lin *et al.* measured natural abundance and post-[1-^{13}C] glucose infusion ^{13}C MRS to study patients with Alzheimer's disease. It was found that compared to the healthy control, glutamate level and NAA were reduced, which is linked with the glutamate–glutamine cycle and the glutamate neurotransmitter rate (Lin *et al.*, 2003).

Blüml *et al.* examined paediatric patients with leukodystrophies and mitochondrial disorders and demonstrated changes in the ^{13}C enhancement pattern with and without administration of ^{13}C glucose (Blüml *et al.*, 2001). Additional reviews have described ^{13}C MRI and MRS applications (Golman and Petersson, 2006; Kurhanewicz *et al.*, 2008; Mason and Krystal, 2006; Ross *et al.*, 2003).

13.4 Fluorine

Fluorine (^{19}F) has the second-highest NMR sensitivity of all naturally occurring nuclides. However, detection of fluorine in healthy, living organisms is not carried out due to its absence (except for a slight presence in blood thought to arise from fluorine in drinking water and fluorinated toothpaste) in soft tissue, its fast decay and its broad signal spectrum in hard structures (bone, teeth, nail and hair). Despite this, the absence of an endogenous signal gives rise to an excellent contrast-to-noise ratio when targeting fluorinated molecules supplied exogenously.

A major application of ^{19}F MR is the investigation of fluorinated drugs or anaesthesia to reveal the dynamics of uptake, biodistribution and metabolism *in vivo* (pharmacokinetics and pharmacodynamics). Direct quantitative results can be obtained for oxygen tension (pO_2), e.g. to detect cellular hypoxia, extracellular pH using exogenous ^{19}F probes and intracellular metal ions (e.g. analyte Ca^{2+}). Spectroscopy using fludeoxyglucose (FDG) has been used to measure brain metabolism and has shown to be sufficient to detect three groups of metabolites (FDG, α-group and β-group of fluorodeoxymannose) and compares reliably with ^{18}F-labelled FDG-PET (Brix *et al.*, 1996). Examples from oncology, where the conversion of 5-fluorouracil (5-FU) into the major MR-visible FU catabolites α-fluoro-β-ureidopropionic acid and fluoro-β-alanine during chemotherapy can be followed by 19F-MRS/MRSI were given by van Gorp *et al.* (2015) and van Laarhoven *et al.* (2007).

Technical challenges arise from the requirement of dedicated hardware to detect the ^{19}F signal as well as from the fact that the chemical shift range of fluorinated organic compounds is large and displays complex coupling mechanisms. To facilitate successful detection, ^{19}F probe heads should ideally be double resonant so that the proton signal can be used for scouting and shimming. In order to maintain a good SNR they must be equipped with both dedicated preamplifiers as well as T/R switches. It should be pointed out that most commercial spectrometers create proton and ^{19}F signals with a single high band electronic, and consequently experiments employing decoupling or making use of the NOE, with these two nuclei, require extra attention towards system configuration and also require careful filtering to remove interferences. The ^{19}F nucleus exhibits a large chemical shift range (~300 ppm), which is exquisitely sensitive to the electronic microenvironment. Although this is beneficial in terms of spectral resolution, there are several difficulties in employing ^{19}F spectroscopy. Primarily problems arise due to the complex coupling mechanism displayed in fluorine. That is to say, ^{19}F–^{19}F coupling does not necessarily attenuate with the number of bonds and large J-coupling values are assumed to arise from through-space coupling mechanisms. Covering the complete spectral bandwidth of ^{19}F with high resolution imposes several technological challenges. Major obstacles are (a) high digitisation bandwidth, (b) high transmit power and bandwidth, e.g. for adiabatic inversion pulses, and (c) difficulties of creating a spin lock over the wide range of ^{19}F frequencies. Due to these technological challenges and the fact that only a finite quantity of imaging moieties can be assessed based on the registered MR signal intensity, many ^{19}F experiments employ imaging methods. The application of spectroscopic techniques was reviewed by Battiste and Newmark (2006), Heerschap (2016) and Lindon and Wilson (2007); imaging methods were described by Doi *et al.* (2009) and Prior *et al.* (1992) and quantitative methods were reviewed by Chen *et al.* (2010).

13.5 Sodium

^{23}Na imaging is a MR imaging technique, used primarily for the analysis of metabolic processes in tissue. ^{23}Na can be found in intra- and extracellular space, with additional information about pathologies decoded in its compartmental distribution. The main challenges of ^{23}Na imaging are low sensitivity and fast relaxation times. These factors make the use of high magnetic fields and sophisticated imaging sequences necessary. Although this review is focused on brain imaging, ^{23}Na imaging is also performed to investigate heart (Ouwerkerk *et al.*, 2008; Yushmanov *et al.*, 2009), kidney (Haneder *et al.*, 2011; Maril *et al.*, 2006) and prostate (Hausmann *et al.*, 2012; Near and Bartha, 2010) disease, as well as cartilage imaging (Reddy *et al.*, 1998).

As ^{23}Na is the second most abundant element in living organisms and is a part of the sodium–potassium exchange across cell membranes; it requires energy in the form of ATP for the maintenance of a constant concentration gradient across the membrane. The intracellular concentration of ^{23}Na is low (10–15 mmol/L), while the extracellular concentration is about 140 mmol/L (Modo and Bulte, 2011). A resting potential, responsible for signal transmission in nerves and muscles, is generated by ions in the intra- and extracellular space. As intracellular sodium concentration is much lower than the extracellular concentration, changes in the intracellular concentration can be masked by the changes in extracellular concentration, even though intracellular volume is larger than extracellular volume.

Intracellular sodium concentration change can be a sign of disease. For example, an increase in intracellular sodium concentration can be observed in tumours or cytotoxic oedema (Winkler, 1990), and extracellular sodium concentration increases sharply in pathologies such as vasogenic oedema (Sharma, 2005). Paramagnetic shift reagents such as $Dy(PPP_i)_2^{7-}$, $Dy(TTHA)^{3-}$ and $Tm(DOTP)_5^-$ (Bansal *et al.*, 1993; Gupta and Gupta, 1982; Naritomi *et al.*, 1987) can be used as contrast agents for the selective detection of intracellular sodium, but they are highly toxic (Allis *et al.*, 1991).

As shown in Table 13.1, ^{23}Na imaging is challenging due to both low sensitivity, originating from the low gyromagnetic ratio, as well as low natural abundance and fast relaxation times caused by the quadrupole moment of the ^{23}Na nucleus. In the human brain, the highest ^{23}Na density-weighted signal usually originates from the cerebrospinal fluid (CSF) in ventricles, central canal, vitreous humour (VH), and subarachnoid space, whereas a high multiple quantum-weighted signal is created in tissue, i.e. grey (GM) and white matter (WM).

13.5.1 Image acquisition techniques

Unlike ^1H, ^{23}Na imaging sequences must compensate for fast relaxation with signal acquisition at short echo times, which makes spin-echo sequences inept for sodium imaging (Konstandin and Nagel, 2014). Furthermore, the separation between signals of intra- and extracellular sodium is also desirable for accurate analysis of pathology. Two basic MRI techniques used for the signal separation *in vivo* are inversion recovery and triple-quantum filtering (TQF). Inversion recovery allows differences in T_1 to be enhanced and enables the separation of signals from different compartments with different T_1 constants. Comparatively, TQF sodium MRI allows signals from different physiological structures to be separated on grounds of different degenerations in the energy level structure of the nuclei. TQF makes use of the fact that multiple-quantum coherences evolve mainly in the spatially restricted intracellular space, due to a nuclear spin 3/2 and non-spherical charge distribution in the nucleus (Fiege *et al.*, 2013b). The created electric quadrupole moment interacts with electric field gradients in the biological environment, which are characterised through the distribution of the local electric field gradients created by water molecules or complex polysaccharides.

The energy level structures of ^{23}Na depend strongly on the environment of the nuclei; several cases are presented in Figure 13.2. In an isotropic environment with non-restricted $^{23}Na^+$ the energy levels are equally spaced and the resultant signal decays mono-exponentially; this behaviour can be observed in CSF and unrestricted extracellular sodium. If the quadrupole frequency ω_Q is non-zero in time as found in highly anisotropic

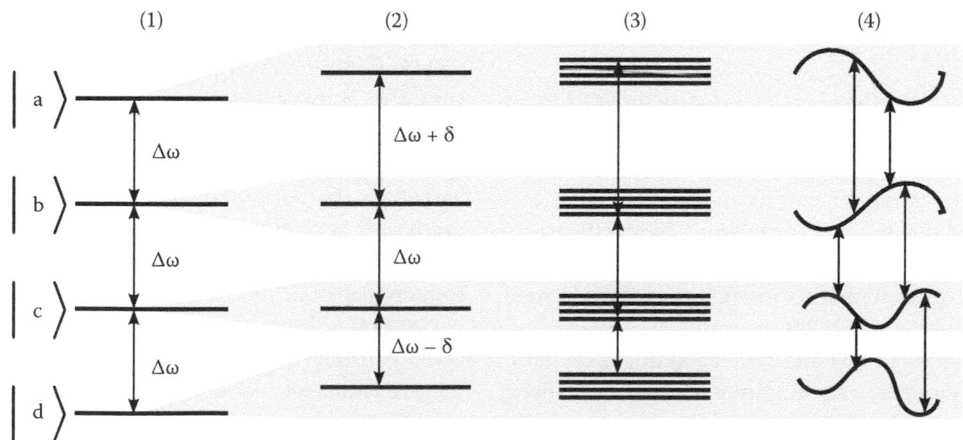

FIGURE 13.2 ^{23}Na NMR spectra depend on the effects of the environment on the energy level structure: (1) is for nuclei in a homogeneous environment, which exhibits mono-exponential relaxation, (2) is present in samples with high inhomogeneity, (3) occurs in anisotropic environments with non-zero rest average quadrupole interaction frequency ω_Q and (4) exists in isotropic environment with a zero average ω_Q. (Reproduced from Shah, N. J., *et al.*, *NMR Biomed.*, 29, 162–174, 2016. With permission.)

samples, the separation between energy levels is altered. The restricted environment represents a so-called slow motion limit with two possibilities: in an anisotropic environment, ordered tissue is characterised by the non-zero rest average ω_Q (Van der Maarel, 2003). This is the case for the restricted extracellular sodium, which experiences additional influence in structured domains and decays bi-exponentially. In an isotropic environment, such as intracellular space with a zero average ω_Q, which is non-zero for a given moment of time, ^{23}Na nuclei exhibit bi-exponential relaxation, where transitions from 3/2 to 1/2 and −1/2 to −3/2 spin states decay faster with T_{2f} than the central −1/2 to 1/2 transition with relaxation time T_{2s} (Rooney and Springer, 1991). The concentration of sodium in tissue (TSC) provides substantial information on a number of pathologies.

TSC sequences often employ gradient echo techniques; however, modern non-Cartesian acquisition schemes allow one to acquire signal without generation of an echo, which leads to high signal-to-noise ratio (SNR) and short acquisition times (Shah *et al.*, 2016). Different *k*-space sampling schemes using twisted projection imaging (TPI) and density-adapted 3D-projection reconstruction imaging methods also improve SNR (Konstandin and Nagel, 2014). However, TPI was proven to be a more robust technique at high fields for ^{23}Na concentration mapping compared to (radial) gradient echo (GRE) or density adapted radial imaging (Romanzetti *et al.*, 2006). TPI-SENSE is a combination of TPI trajectories with sensitivity encoding techniques (Qian *et al.*, 2009). Madelin *et al.* (2012) demonstrated that compressed sensing can lead to a reduction in acquisition time, without significant image quality reduction. For imaging of the intra- and extracellular sodium signals, multiple quantum filtering can be used (Keltner *et al.*, 1994), which usually consist of three hard pulses (Chung and Wimperis, 1990) and adheres to a scheme that isolates multi–quantum filtered signal contributions, such as multiplex phase cycling (Ivchenko *et al.*, 2003). The SISTINA sequence allows simultaneous acquisition of single and triple quantum coherences (Fiege *et al.*, 2013a).

13.5.2 ^{23}Na relaxometry

As mentioned previously, the energy level structure of sodium nuclei is impacted by their surroundings; thus relaxation, which mainly occurs via quadrupolar coupling effects, is altered accordingly. Quantification of the ^{23}Na signal at non-zero echo times can be biased due to signal relaxation. To investigate the relaxation behaviour several models, mainly representing mono-exponential or bi-exponential decay in brain tissue, can be used. For example, Fleysher *et al.* (2009) measured two echoes with long echo times (TE_1 = 12 ms) and corresponding SNR. From the signal intensity values, assuming mono-exponential decay it was possible to calculate the long relaxation component T_{2s}^* in CSF and tissue. Winkler *et al.* (1989) utilised four short echo times (TE = 3, 6, 9, 12 ms) for mono-exponential function fitting to determine fast relaxation time T_{2f}^*. By identifying the voxels with similar decay, constants are considered to consist of the same tissue composition. Bartha and Menon (2004) determined the long component of T_{2s}^* with a gradient-echo sequence with 10 echoes in the range from 3.8 to 68.7 ms. Perman *et al.* (1989) obtained T_{2f}^* and T_{2s}^* from bi-exponential, non-linear, least-squares function fit to eight echo times.

In addition to signal relaxation, it is also necessary to consider the impact of longitudinal relaxation time. Ouwerkerk *et al.* (2003) determined T_1 under consideration of mono-exponential decay using images acquired with twisted-projection MR imaging repeated with different TR values and a non-linear square fit routine, while the method suggested by Reetz *et al.* (2012) and Stobbe and Beaulieu (2006) uses variation of inversion-recovery time and of inversion pulse lengths to determine T_1 relaxation. However, the mono-exponential decay model has been proven to be insufficient in brain tissue, and a bi-exponential model is required to overcome this (Stobbe and Beaulieu, 2006).

13.5.3 Quantification of ^{23}Na concentration

Quantification of ^{23}Na content in tissue generally requires appropriate techniques, depending on whether total tissue concentration or compartmental concentrations are to be extracted. Methods differ in how contributions are segmented and what is used as reference. For example, Ouwerkerk *et al.* measured total tissue ^{23}Na concentration using high resolution ^{1}H images for tissue segmentation, where the concentration was determined in a post-processing step through signal amplitude normalisation using two saline phantoms containing a known NaCl concentration and placed in the field of view (FOV) (Ouwerkerk *et al.*, 2003). Fleysher *et al.* determined tissue sodium concentration, intracellular sodium concentration, intracellular sodium volume fraction and intracellular sodium molar fraction through the acquisition of single quantum and triple quantum weighted images (Fleysher *et al.*, 2013). Signal decay was simulated with a two-compartment tissue model and corrections for flip angle distribution and B_0 inhomogeneity were applied.

13.5.4 Applications

^{23}Na concentration is tightly regulated in the human body, both with respect to total sodium concentration as well as compartmental sodium concentrations. Thus, changes in these concentrations are important in the diagnosis of existing or emerging pathology. In particular metabolic diseases instigating cell death such as stroke or disorders altering membrane resistivity, e.g. dementia or epilepsy, are promising targets for quantitative sodium metabolic imaging and spectroscopy.

Cerebral ischaemia, for example, in stroke, causes a reduced activity or interruption of the sodium-potassium pump due to the disruption of blood circulation. This leads to a depletion of the oxygen and glucose supply in the affected regions of the brain and results in an increase in TSC and eventually in cell death. TSC can therefore act as a marker for tissue viability in the ramification of such an event (Boada *et al.*, 2012; Hussain *et al.*, 2009). From studies on stroke subjects, including a cross correlation with histochemical analysis of tissue from the infarcted region, it can be established that regions containing viable tissue can be discriminated from critically damaged tissue considering a TSC threshold. For example, in patients with clinically confirmed stroke, TSC in the infarcted regions was found to exceed 70 mmol/L after 6 hours (Thulborn *et al.*, 1999). Furthermore, by measuring the temporal change of TSC it is, in principle, possible to estimate the stroke onset time and thus qualify patients for thrombolytic therapy (Jones *et al.*, 2006).

Multiple sclerosis (MS) is an autoimmune disease affecting myelinated axons in the central nervous system. A main contributor to the symptoms of the disease is the disruption of conductivity of neuronal signals along the myelinated axons due to inflammation and consequent demyelisation. Myelinated axons act as lines of communication based on action potentials and signals are propagated, in the main part, by sodium ions. Consequently, impaired or malfunctioning sodium channels disrupt the normal flow of information (Smith, 2007; Waxman, 2006). In patients suffering with MS, GM, WM and WM lesions have been shown to exhibit elevated sodium concentrations, with the sodium concentration in GM increasing with the severity of the symptoms of disease (Inglese *et al.*, 2010; Paling *et al.*, 2013; Zaaraoui *et al.*, 2012). Due to the differences in the way MS occurs in individuals, four distinct courses have been identified: clinically isolated syndrome, relapsing–remitting MS, primary progressive MS and secondary progressive MS. Depending on the classification of MS, increased sodium concentrations are observed in specific areas of the brain. For example, secondary progressive MS patients are known to experience increased sodium concentrations in waste regions of the brain including, but not limited to, motor and visual cortices. In contrast, primary progressive MS patients, however, experience raised sodium levels in the region of the motor cortex only (Maarouf *et al.*, 2014).

Huntington's disease (HD) is a hereditary disease causing the progressive destruction of a specific subgroup of nerve cells in the brain, such as caudate nucleus and putamen (Vonsattel *et al.*, 1985). The breakdown of the nerve cells results in elevated concentrations of sodium. This can be seen both in areas with known pathological changes as well as in regions that are otherwise seemingly unaffected. A study conducted by Reetz *et al.* (2012) compared sodium tissue concentrations across the brains of patients with HD and a healthy control group. All HD subjects exhibited significantly elevated sodium concentrations in the bilateral striatum, putamen, pallidum, thalamus, hippocampus, insula, precuneus and occipital cortex and, most dominantly, in the bilateral caudate. Nevertheless, a few brain areas were found to have remained unaffected. These included the amygdala, pre- or post-central gyrus, frontal or temporal cortices or cerebellum (as seen in Figure 13.3). Intriguingly, ^{23}Na MR imaging was able to detect increased tissue sodium concentration in affected areas of the brain, as well as in structurally non-affected regions, which makes this method appealing for early clinical diagnosis.

Intracranial neoplasms, i.e. brain tumours, are characterised by pathological cell division and undesired tissue growth, with gliomas being the most common malignant example. Prior to cell division, in patients suffering with an intracranial neoplasm, the cell membrane potential undergoes a change due to depolarisation. An influx of sodium causes the net charge of the cell to be less negative than usual and this influx of sodium manifests itself in raised intracellular sodium concentration and TSC, respectively (Ouwerkerk *et al.*, 2003). Thus, the measurement of sodium concentrations is a viable biomarker for emerging and active tumour growth. Nevertheless, it should be noted that features in the vicinity of the tumour, such as oedema and necrotic regions, might also contribute to detected changes in sodium concentration. These changes are not directly accredited to tumour growth;thus efforts must be taken to correctly associate measured concentration changes in patients. Fiege *et al.* (2013a) compared ^{23}Na MRI to ^{18}F-FET-PET and proton MRI for three tumour patients and demonstrated that in each case pathology manifests itself in the sodium images through increased TSC in the region of the tumour, seen in the UTE part of the SISTINA data set, as well as in the TQF image contrast (see Figure 13.4). In TQF images regions of hypointensity are expected in case of necrosis and regions of hyperintensity in the case of developing tumour tissue (Madelin and Regatte, 2013).

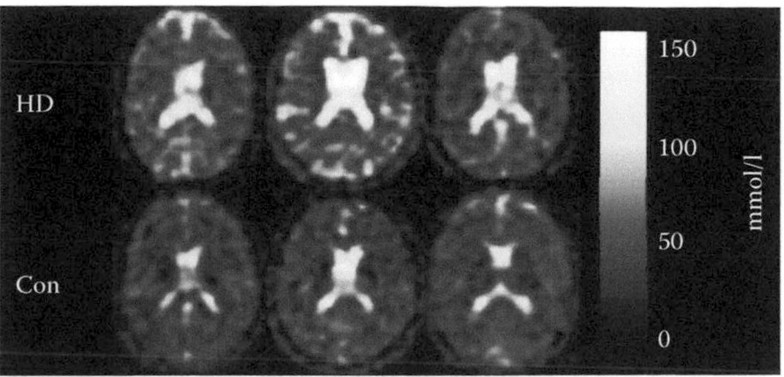

FIGURE 13.3 Patients suffering from Huntington's disease (HD) show increased sodium levels in various regions of the brain compared to healthy controls. (Reproduced from Reetz, K., *et al.*, *NeuroImage*, 63, 517–24, 2012. With permission.)

13.6 Lithium

^{7}Li treatments are carried out in acute mania and bipolar disorders, and lithium MRS is the only technique available to measure ^{7}Li concentration in the brain *in vivo*. Because of the low signal strength, applications focus on localised MRS using double spin echoes. In a simplified model, bi-exponential decay is assumed and tentatively assigned to intra- and extracellular lithium. In the rat brain at 7T the values for T_2 relaxation were found to be 14.8 ± 4.3 ms (intracellular) and 295 ± 61 ms (extracellular) with a mean intracellular fraction of 54.5 ± 6.7% (Komoroski *et al.*, 2013). Developments are under way to map the spatial distribution of ^{7}Li in the brain (Ramaprasad, 2005). A review of biomedical applications was given by Komoroski (2005).

Among other MRS applications, ^{7}Li investigations in humans were summarised by Lindquist and Komoroski (2005) and by

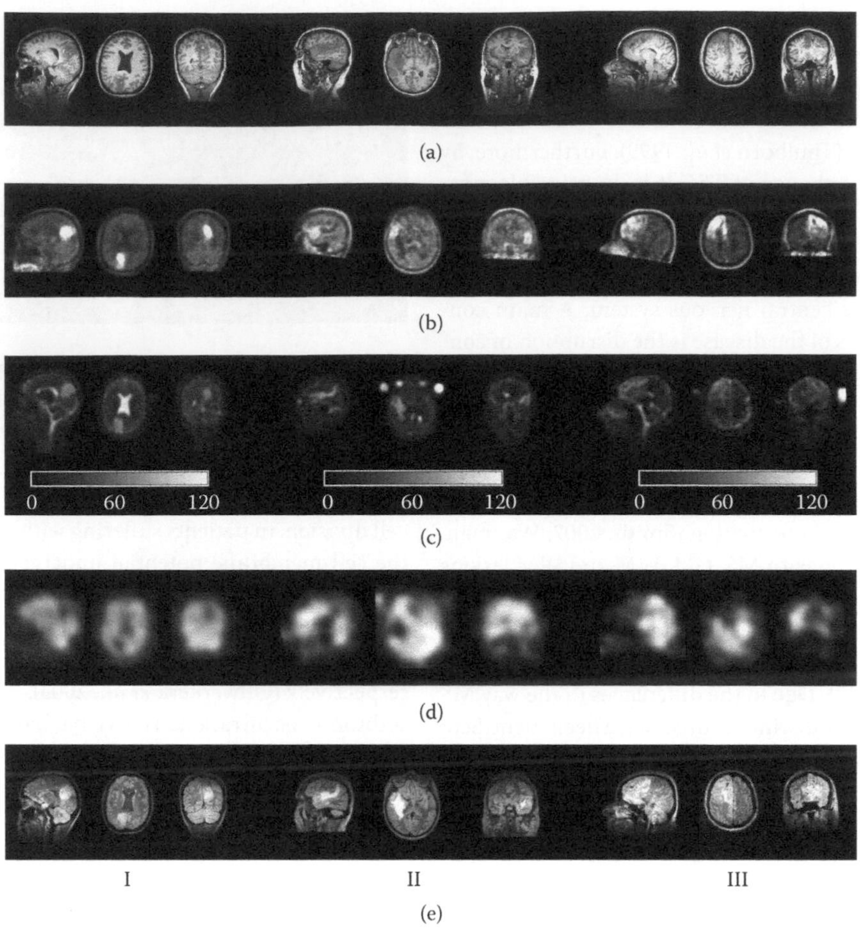

FIGURE 13.4 Comparison of MR and positron emission tomography (PET) measurements of three brain tumour patients: (a) T_1-weighted MP-RAGE: dark shadows; (b) FET PET: increased uptake; (c) SISTINA UTE: hyperintense (colour scale of sodium concentrations in mmol/L); (d) SISTINA TQF: hypointense; (e) FLAIR: hyperintense in the region of the tumour. Reproduced from Fiege D. P., *et al.*, *Magn. Reson. Med.*, 69, 1691–1696, 2013a. With permission.)

Mason and Krystal (2006), who provides more specific focus on psychiatric diseases. More recently Machado-Vieira *et al.* (2016) conducted a study investigating brain lithium levels in patients at 3T and an earlier report based on investigations at 1.5T was published by Lyoo (2003).

13.7 Potassium and chlorine

In vivo transport and homeostasis of K^+ and Cl^- are relevant to the study of pathophysiological processes in several diseases. The maintenance of transmembrane ionic gradients is critical to preserve synaptic efficacy, and the concentrations and fluxes of K and Cl play a major role therein (Chamma *et al.*, 2012). K^+ is predominantly present in the intracellular space, whereas Cl^- is found mainly in the extracellular space. Imaging of these nuclei is a challenging task due to their low tissue concentration and rapid relaxation: for ^{39}K at 7T $T_2^* = 2.4$ ms (fast) and 12.9 ms (slow component) and $T_1 = 14$ ms (Augath *et al.*, 2009; Stevens *et al.*, 1992); for ^{35}Cl in the brain at 7T $T_2^* = 1.2$ ms (fast) $T_2^* = 7$ ms (slow), $T_1 = 9.2$ ms (Nagel *et al.*, 2014). The fact that not all K in biological tissue is NMR visible makes quantification of this nucleus even more difficult. In the brain, only about 60% of the present potassium was estimated to be NMR visible (Adam *et al.*, 1987). Even at very high fields only proof of principle imaging of these two elements has been recently performed in humans (Atkinson *et al.*, 2014; Nagel *et al.*, 2014; Umathum *et al.*, 2013; Weber *et al.*, 2016), adding to previous reports of spectroscopy (Adam *et al.*, 1987; Stevens *et al.*, 1992; Wellard *et al.*, 1993) and imaging of small animals (Augath *et al.*, 2009).

13.8 Oxygen

13.8.1 Imaging and spectroscopy

Oxygen-17 (^{17}O) is the only MRI-visible, stable isotope of oxygen (for NMR-relevant properties see Table 13.1). Despite its very low natural abundance (0.037%) and low sensitivity, in vivo ^{17}O MRI/S is a developing alternative to well-established methodologies to study metabolism (Gordji-Nejad *et al.*, 2014). The only molecule that can be well detected by *in vivo* ^{17}O MRI techniques at natural abundance is ^{17}O-water, given the high percentage of water in tissue. Fortunately, this is exactly the molecule of major interest. Production of metabolic water is the final step of oxidative phosphorylation and the cerebral metabolic rate of oxygen consumption ($CMRO_2$) is an excellent marker to reflect the metabolic health of the brain. The possibility to spatially resolve and quantify metabolic ^{17}O-water produced after respiration of ^{17}O gas is a unique advantage of *in vivo* ^{17}O imaging.

13.8.2 Direct detection

Interest in the use of ^{17}O MRI/S for *in vivo* measurements of $CMRO_2$ began in the late 1980s (Arai *et al.*, 1990, 1991; Mateescu, 2003; Pekar *et al.*, 1991), paralleling the ^{15}O-based PET technique (Mintun *et al.*, 1984; Ter-Pogossian *et al.*, 1970). Since then, several examples of *in vivo* ^{17}O MRS/MRI have been demonstrated and are reviewed by e.g. Zhu (2005a), Zhu and Chen (2011) Gordji-Nejad *et al.* (2014) and Zhu and Chen (2016). In contrast to ^{15}O-PET imaging, which uses a very short-lived isotope of oxygen (half-life of about 2 minutes for ^{15}O) requiring on-site production, MRI investigates the stable isotope ^{17}O. $CMRO_2$ values obtained by direct detection of respiration-delivered ^{17}O agree well with those of quantitative PET measurements (Atkinson and Thulborn, 2010; Hoffmann *et al.*, 2011), for example $CMRO_2$ values of 1.42 ± 0.05 and 0.75 ± 0.11 μmol/g/min for GM and WM, respectively, obtained with ^{17}O (Atkinson and Thulborn, 2010) and $CMRO_2 = 1.44 \pm 0.21$ μmol/g/min (GM), $CMRO_2 = 0.56 \pm 0.09$ μmol/g/min (WM) from PET (Leenders *et al.*, 1990).

High-resolution $CMRO_2$ quantification by ^{17}O-MRI remains challenging due to the low signal: even if increases in ^{17}O-water concentration of around 20% over natural abundance can be achieved in brain tissue by inhaling ^{17}O-enriched gas for 20–30 minutes (Hoffmann *et al.*, 2011), the final concentration of ^{17}O-water is still well below 0.1%. The relaxation times of ^{17}O were found to be magnetic field independent ($T_2^* \sim 1.8$ ms at fields from 3T to 16.4T; $T_2 \sim 3$ ms, $T_1 = 4.8$–6.8 ms for fields up to 16.4T) (Borowiak *et al.*, 2014; Wiesner *et al.*, 2016; Zhu *et al.*, 2001), which is consistent with the concept that the dominant relaxation mechanism is the quadrupolar interaction for this nucleus.

The ^{17}O NMR sensitivity at 9.4T is approximately four times higher than what is observed at 4.7T (Zhu *et al.*, 2001). This is largely due to the fact that sensitivity is not diminished by unfavourable changes in relaxation times, as can be the case for protons. Quantifying $CMRO_2$ by direct detection is, however, difficult due to partial volume effects and to low resolution of the baseline and metabolic information. In order to quantify 'metabolically active' voxels and thus to determine the accurate concentration changes, partial volume effects should be corrected. Figure 13.5 shows a simple method for this (Gordji-Nejad *et al.*, 2014): a high-resolution quantitative map of $H_2{}^{17}O$ distribution at natural abundance is derived from a quantitative map of $H_2{}^{16}O$ distribution acquired at 3T (Oros-Peusquens *et al.*, 2014). The conversion from $H_2{}^{16}O$ in percentage units to $H_2{}^{17}O$ in mg/g tissue is performed

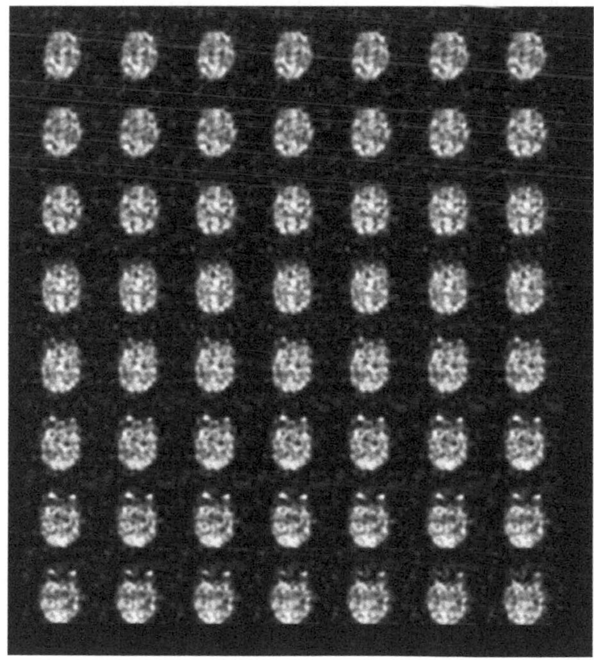

(a)

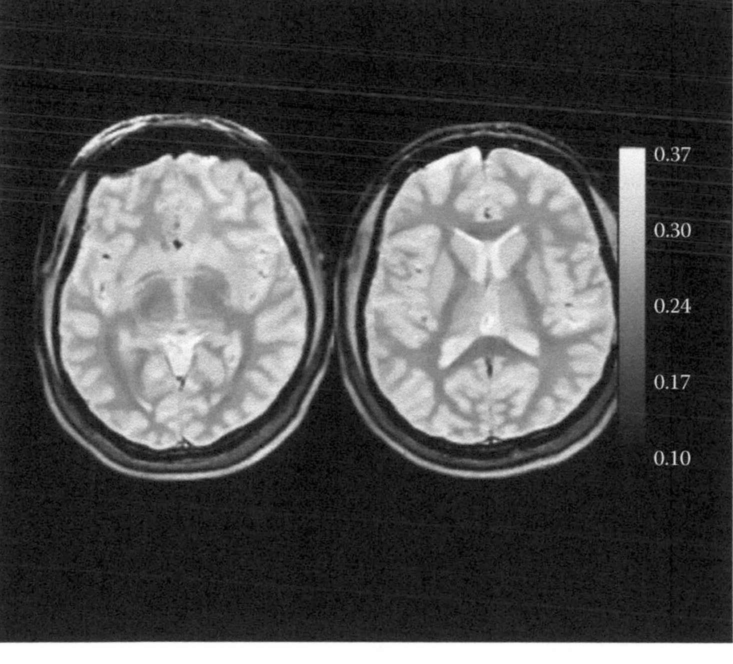

(b)

FIGURE 13.5 (a) Twisted projection imaging images of natural abundance ^{17}O-water at 9.4T with $(5\ \text{mm})^3$ nominal resolution acquired in 10 min. (b) Map of ^{17}O-water, in units of mg/g tissue, derived from a $1 \times 1 \times 1.5\ \text{mm}^3$ map of ^{16}O-water acquired at 3T with ^1H MRI. Reproduced from Oros-Peusquens, A. M., *et al.*, *Nucl. Instr. Methods Phys. Res. A.*, 734, 185–190, 2014. With permission.)

using 1.04 g/cm^3 mass density of tissue and 0.037% natural abundance for ^{17}O. The molecular mass of $H_2{}^{17}O$ water is approximately 19 g/mol, resulting in a conversion factor to μmol ^{17}O water/g tissue of 52.6. The mean water content values are 70% (WM) and 83% (GM) with a physiological variation of around 1.5% in the healthy population (Oros-Peusquens *et al.*, 2014). Similar constancy can be translated to the $H_2{}^{17}O$ baseline (with mean values of 0.24 mg/g tissue in WM and 0.29 mg/g tissue in GM).

The main progress in direct imaging of ^{17}O is expected from improved RF technology, such as multichannel receive array coils and/or fast acquisition scheme, e.g. compressed sensing (Maguire *et al.*, 2015). Given the fact that imaging of ^{17}O by direct detection has been reported at 1.5T in the pioneering work of Fiat *et al.* (1993, 2004) and seems feasible to some extent already at 3T (Borowiak *et al.*, 2014), measurements at ultra-high fields (Budinger and Bird, 2017) hold much promise.

13.8.3 Indirect detection

Scalar coupling between ^1H and ^{17}O on the water molecule, characterised by the coupling constant J = 92.5 Hz (Burnett and Zeltmann, 1974), affects the T_2 and the rotating frame spin-lattice relaxation time ($T_1\rho$) of protons, which were found to be pH dependent (Meiboom, 1961). Based on this, ^{17}O concentration changes can be indirectly measured from the changes in proton relaxation rates T_2 and/or $T_1\rho$.

In order to increase sensitivity to changes driven by ^{17}O water in the proton T_2 or $T_1\rho$, these are measured after irradiation at ^{17}O frequency or after spin-lock irradiation slightly off-resonance from the proton frequency, respectively. Both methods effectively reduce or remove the effect of J-coupling between proton and ^{17}O (de Crespigny *et al.*, 2000; Mellon *et al.*, 2009; Reddy *et al.*, 1996; Ronen *et al.*, 1997; Stolpen *et al.*, 1997). Oxygen delivery, cerebral blood flow and local oxygen consumption were thus determined through indirect ^{17}O measurements (de Crespigny *et al.*, 2000; Kwong *et al.*, 1991; Mellon *et al.*, 2009; Reddy *et al.*, 1996; Ronen and Navon, 1994; Ronen *et al.*, 1997, 1998; Stolpen *et al.*, 1997). The average $CMRO_2$ values obtained by indirect detection methods are also consistent with the ^{15}O PET literature (Leenders *et al.*, 1990).

From a practical point of view, the $T_1\rho$-based method (Mellon *et al.*, 2009) is the only one that does not require additional hardware (coil or RF amplifier) and can be performed easily in a clinical setup. However, sequences for mapping $T_1\rho$ are not currently available from manufacturers as a standard. Furthermore, contributions to $T_1\rho$ contrast in biological specimens are not fully understood and can compete with those caused by ^{17}O.

13.9 Deuterium

Deuterium is a stable, although rare (natural abundance of 0.0156%), isotope with spin-1 nucleus and small quadrupole moment (Table 13.1). Its main effect on water is to lengthen T_1 of ^1H by suppressing dipolar interaction in the HDO water molecule. This mechanism can potentially allow for indirect detection and imaging of deuterium. The T_1 of ^2H in heavy water (D_2O, $T_1 \sim 400$ ms) is an order of magnitude shorter than that of ^1H in H_2O (Koenig *et al.*, 1975; Mantsch *et al.*, 1977), thus allowing for efficient signal averaging. Multicompartment T_2 relaxation has been observed in cat brain tissue (9, 43 and 369 ms) (Ewy *et al.*, 1986). The major advantages of D_2O as a tracer are that it is non-radioactive and is capable of quickly accessing all body compartments, tissues and cell types. Currently, direct detection ^2H imaging is mainly used for ophthalmological applications, where large concentrations of heavy water can be safely reached *in vivo* and its power as a tracer has been long recognised (Kinsey *et al.*, 1942).

There is an exciting range of biological effects of deuterium, reviewed by Kushner *et al.* (1999), which include, among others, effects on the nervous system, mitosis and membrane function. D_2O provides highly unique avenues for exploring basic mechanisms of biological phenomena (Sunde *et al.*, 2009) and can provide us with a more complete picture of the microenvironment of water in biological tissue (Koenig *et al.*, 1975). Thus, using deuterium NMR and MRI it has been found that it is possible to distinguish water populations in cortical bone (pore and collagen-bound water) (Ong *et al.*, 2012) as well as in excised, *in vivo* and *in situ* rat brain tissue (Assaf and Cohen, 1996; Assaf *et al.*, 1997) (three diffusion pools in single quantum and triple-quantum filtered ^2H NMR were identified). In addition, iron overload in mice was found to influence the T_1 relaxation values of ^2H *in vivo* (Irving *et al.*, 1987). Given the power of ^2H as a label, administered either as heavy water or e.g. labelled glucose, its applications are many. To mention only a few, deuterium can be used to monitor cell proliferation (but detected by methods other than MRS, for example mass spectrometry; Busch *et al.*, 2007), as well as metabolism and fat turnover in obesity (Brereton *et al.*, 1986).

References

Adam WR, Koretsky AP, Weiner MW. Measurement of tissue potassium in vivo using ^{39}K nuclear magnetic resonance. Biophys J 1987; 51: 265–71.

Allis JL, Seymour AML, Radda GK. Absolute quantification of intracellular Na+ using triple-quantum-filtered ^{23}Na NMR. *J Magn Reson* 1991; 93: 71–6.

Arai T, Mori K, Nakao S, Watanabe K, Kito K, Aoki M, et al. In vivo oxygen-17 nuclear magnetic resonance for the estimation of cerebral blood flow and oxygen consumption. Biochem Biophys Res Commun 1991; 179: 954–61.

Arai T, Nakao S, Mori K, Ishimori K, Morishima I, Miyazawa T, et al. Cerebral oxygen utilization analyzed by the use of oxygen-17 and its nuclear magnetic resonance. Biochem Biophys Res Commun 1990; 169: 153–8.

Ardenkjaer-Larsen JH, Fridlund B, Gram A, Hansson G, Hansson L, Lerche MH, et al. Increase in signal-to-noise ratio of >10,000 times in liquid-state NMR. Proc Natl Acad Sci 2003; 100: 10158–63.

Assaf Y, Cohen Y. Detection of different water populations in brain tissue using ^{2}H single- and double-quantum-filtered diffusion NMR spectroscopy. J Magn Reson Imaging B 1996; 112: 151–9.

Assaf Y, Navon G, Cohen Y. In vivo observation of anisotropic motion of brain water using ^{2}H double quantum filtered NMR spectroscopy. Magn Reson Med 1997; 37: 197–203.

Atkinson IC, Claiborne TC, Thulborn KR. Feasibility of 39-potassium MR imaging of a human brain at 9.4 Tesla. Magn Reson Med 2014; 71: 1819–25.

Atkinson IC, Thulborn KR. Feasibility of mapping the tissue mass corrected bioscale of cerebral metabolic rate of oxygen consumption using 17-oxygen and 23-sodium MR imaging in a human brain at 9.4 T. Neuroimage 2010; 51: 723–33.

Augath M, Heiler P, Kirsch S, Schad LR. In vivo ^{39}K, ^{23}Na and ^{1}H MR imaging using a triple resonant RF coil setup. J Magn Reson 2009; 200: 134–6.

Bansal N, Germann MJ, Seshan V, Shires GT 3rd, Malloy CR, Sherry AD. Thulium 1,4,7,10-tetraazacyclododecane-1,4,7,10-tetrakis(methylene phosphonate) as a ^{23}Na shift reagent for the in vivo rat liver. Biochemistry 1993; 32: 5638–43.

Bartha R, Menon RS. Long component time constant of ^{23}Na T_2* relaxation in healthy human brain. Magn Reson Med 2004; 52: 407–410.

Battiste J, Newmark RA. Applications of F-19 multidimensional NMR. Prog Nucl Magn Reson Spectrosc 2006; 48: 1–23.

Befroy DE, Rothman DL, Petersen KF, Shulman GI. 31P-magnetization transfer magnetic resonance spectroscopy measurements of in vivo metabolism. Diabetes 2012; 61: 2669–78.

Blüml S. In vivo quantitation of cerebral metabolite concentrations using natural abundance ^{13}C MRS at 1.5T. J Magn Reson 1999; 136: 219–225.

Blüml S, Moreno A, Hwang JH, Ross BD. 1-^{13}C glucose magnetic resonance spectroscopy of pediatric and adult brain disorders. NMR Biomed 2001; 14: 19–32.

Boada FE, Qian YX, Nemoto E, Jovin T, Jungreis C, Jones SC, et al. Sodium MRI and the assessment of irreversible tissue damage during hyper-acute stroke. Transl Stroke Res 2012; 3: 236–45.

Borowiak R, Groebner J, Haas M, Hennig J, Bock M. Direct cerebral and cardiac ^{17}O-MRI at 3 Tesla: Initial results at natural abundance. MAGMA 2014; 27: 95–9.

Brereton IM, Irving MG, Field J, Doddrell DM. Preliminary studies on the potential of in vivo deuterium NMR spectroscopy. Biochem Bioph Res Co 1986; 137: 579–84.

Brindle KM, Bohndiek SE, Gallagher FA, Kettunen MI. Tumor imaging using hyperpolarized ^{13}C magnetic resonance spectroscopy. Magn Reson Med 2011; 66: 505–19.

Brix G, Bellemann ME, Haberkorn U, Gerlach L, Lorenz WJ. Assessment of the biodistribution and metabolism of 5-fluorouracil as monitored by ^{18}F PET and ^{19}F MRI: A comparative animal study. Nucl Med Biol 1996; 23: 897–906.

Budinger TF, Bird MD. MRI and MRS of the human brain at magnetic fields of 14T to 20T: Technical feasibility, safety, and neuroscience horizons. Neuroimage 2017. doi: 10.1016/j.neuroimage.2017.01.067.

Burnett LJ, Zeltmann AH. 1H–^{17}O spin-spin coupling constant in liquid water. J Chem Phys 1974; 60: 4636–7.

Busch R, Neese RA, Awada M, Hayes GM, Hellerstein MK. Measurement of cell proliferation by heavy water labeling. Nat Protoc 2007; 2: 3045–57.

Chamma I, Chevy Q, Poncer JC, Lévi S. Role of the neuronal K-Cl co-transporter KCC2 in inhibitory and excitatory neurotransmission. Front Cell Neurosci 2012; 6: 5.

Chen J, Lanza GM, Wickline SA. Quantitative magnetic resonance fluorine imaging: Today and tomorrow. Wires Nanomed Nanobi 2010; 2: 431–40.

Chmelik M, Schmid AI, Gruber S, Szendroedi J, Krssak M, Trattnig S, et al. Three-dimensional high-resolution magnetic resonance spectroscopic imaging for absolute quantification of ^{31}P metabolites in human liver. Magn Reson Med 2008; 60: 796–802.

Chung CW, Wimperis S. Optimum detection of Spin-3/2 biexponential relaxation using multiple-quantum filtration techniques. J Magn Reson 1990; 88: 440–7.

de Crespigny AJ, D'Arceuil HE, Engelhorn T, Moseley ME. MRI of focal cerebral ischemia using ^{17}O-labeled water. Magn Reson Med 2000; 43: 876–83.

de Graaf RA. In Vivo NMR Spectroscopy: Principles and Techniques. 2nd Ed: John Wiley & Sons, Chichester, 2007.

Doddrell DM, Pegg DT, Bendall MR. Distortionless enhancement of NMR signals by polarization transfer. J Magn Reson 1982; 48: 323–7.

Doi Y, Shimmura T, Kuribayashi H, Tanaka Y, Kanazawa Y. Quantitative ^{19}F imaging of nmol-level F-nucleotides/-sides from 5-FU with T_2 mapping in mice at 9.4T. Magn Reson Med 2009; 62: 1129–39.

Du F, Cooper AJ, Thida T, Sehovic S, Lukas SE, Cohen BM, et al. In vivo evidence for cerebral bioenergetic abnormalities in Schizophrenia measured using ^{31}P magnetization transfer spectroscopy. JAMA Psychiat 2014; 71: 19–27.

Du F, Zhu XH, Qiao H, Zhang X, Chen W. Efficient in vivo ^{31}P magnetization transfer approach for noninvasively determining multiple kinetic parameters and metabolic fluxes of ATP metabolism in the human brain. Magn Reson Med 2007; 57: 103–14.

Ewy CS, Babcock EE, Ackerman JJH. Deuterium nuclear magnetic resonance spin-imaging of D_2O: A potential exogenous MRI label. Magn Reson Imaging 1986; 4: 407–11.

Fiat D, Dolinsek J, Hankiewicz J, Dujovny M, Ausman J. Determination of regional cerebral oxygen consumption in the human: ^{17}O natural abundance cerebral magnetic resonance imaging and spectroscopy in a whole body system. Neurol Res 1993; 15: 237–48.

Fiat D, Hankiewicz J, Liu S, Trbovic S, Brint S. 17O magnetic resonance imaging of the human brain. Neurol Res 2004; 26: 803–8.

Fiege DP, Romanzetti S, Mirkes CC, Brenner D, Shah NJ. Simultaneous single-quantum and triple-quantum-filtered MRI of ^{23}Na (SISTINA). Magn Reson Med 2013a; 69: 1691–6.

Fiege DP, Romanzetti S, Tse DHY, Brenner D, Celik A, Felder J, et al. B_0 insensitive multiple-quantum resolved sodium imaging using a phase-rotation scheme. J Magn Reson 2013b; 228: 32–6.

Fleysher L, Oesingmann N, Brown R, Sodickson DK, Wiggins GC, Inglese M. Noninvasive quantification of intracellular sodium in human brain using ultrahigh-field MRI. NMR Biomed 2013; 26: 9–19.

Fleysher L, Oesingmann N, Stoeckel B, Grossman RI, Inglese M. Sodium long-Component T_2* mapping in human brain at 7 tesla. Magn Reson Med 2009; 62: 1338–41.

Golman K, Olsson LE, Axelsson O, Mansson S, Karlsson M, Petersson JS. Molecular imaging using hyperpolarized ^{13}C. Br J Radiol 2003; 76: S118–27.

Golman KP, Petersson JS. Metabolic imaging and other applications of hyperpolarized ^{13}C. Acad Radiol 2006; 13: 932–42.

Gordji-Nejad A, Mollenhoff K, Oros-Peusquens AM, Pillai DR, Shah NJ. Characterizing cerebral oxygen metabolism employing oxygen-17 MRI/MRS at high fields. MAGMA 2014; 27: 81–93.

Gruetter R, Adriany G, Choi IY, Henry P, Lei H, Oz G. Localized in vivo ^{13}C NMR spectroscopy of the brain. NMR Biomed 2003; 16: 313–338.

Gupta RK, Gupta P. Direct observation of resolved resonances from intracellular and extracellular ^{23}Na Ions in NMR-studies of intact-Cells and tissues using dysprosium(Iii) tripolyphosphate as paramagnetic shift-reagent. Biophys J 1982; 37: A76.

Ha DH, Choi S, Oh JY, Yoon SK, Kang MJ, Kim KU. Application of ^{31}P MR spectroscopy to the brain tumors. Korean J Radiol 2013; 14: 477–86.

Haneder S, Konstandin S, Morelli JN, Nagel AM, Zoellner FG, Schad LR, et al. Quantitative and qualitative ^{23}Na MR imaging of the human kidneys at 3 T: Before and after a water load. Radiology 2011; 260: 857–65.

Hausmann D, Konstandin S, Wetterling F, Haneder S, Nagel AM, Dinter DJ, et al. Apparent diffusion coefficient and sodium concentration measurements in human prostate tissue via hydrogen-1 and sodium-23 magnetic resonance imaging in a clinical setting at 3T. Invest Radiol 2012; 47: 677–82.

Heerschap A. In Vivo F-19 magnetic resonance spectroscopy. eMagRes 2016; 5: 1283–9.

Hoffmann SH, Begovatz P, Nagel AM, Umathum R, Schommer K, Bachert P, et al. A measurement setup for direct ^{17}O MRI at 7 T. Magn Reson Med 2011; 66: 1109–15.

Hussain MS, Stobbe RW, Bhagat YA, Emery D, Butcher KS, Manawadu D, et al. Sodium imaging intensity increases with time after human ischemic stroke. Ann Neurol 2009; 66: 55–62.

Inglese M, Madelin G, Oesingmann N, Babb JS, Wu W, Stoeckel B, et al. Brain tissue sodium concentration in multiple sclerosis: A sodium imaging study at 3 Tesla. Brain 2010; 133: 847–57.

Irving MG, Brereton IM, Field J, Doddrell DM. In vivo determination of body iron stores by natural-abundance deuterium magnetic resonance spectroscopy. Magn Reson Med 1987; 4: 88–92.

Ivchenko N, Hughes CE, Levitt MH. Multiplex phase cycling. J Magn Reson 2003; 160: 52–8.

Jones SC, Kharlamov A, Yanovski B, Kim DK, Easley KA, Yushmanov VE, et al. Stroke onset time using sodium MRI in rat focal cerebral ischemia. Stroke 2006; 37: 883–8.

Keltner JR, Wong ST, Roos MS. Three-dimensional triple-quantum-filtered imaging of 0.012 and 0.024 M sodium-23 using short repetition times. J Magn Reson B 1994; 104: 219–29.

Kemp GJ. Non-invasive methods for studying brain energy metabolism: What they show and what it means. Dev Neurosci 2000; 22: 418–28.

Kinsey VE, Grant M, Cogan DG. Water movement and the eye. Arch Ophthalmol 1942; 27: 242–52.

Klomp DWJ, Renema WK, van der Graaf M, de Galan BE, Kentgens AP, Heerschap A. Sensitivity-enhanced ^{13}C MR spectroscopy of the human brain at 3 Tesla. Magn Reson Med 2006; 55: 271–8.

Klomp DWJ, Wijnen JP, Scheenen TWJ, Heerschap A. Efficient ^{1}H to ^{31}P polarization transfer on a clinical 3T MR system. Magn Reson Med 2008; 60: 1298–305.

Koenig SH, Hallenga K, Shporer M. Protein-water interaction studied by solvent ^{1}H, ^{2}H, and ^{17}O magnetic relaxation. Proc Natl Acad Sci U S A 1975; 72: 2667–71.

Komoroski RA. Biomedical applications of ^{7}Li NMR. NMR Biomed 2005; 18: 67–73.

Komoroski RA, Lindquist DM, Pearce JM. Lithium compartmentation in brain by ^{7}Li MRS: Effect of total lithium concentration. NMR Biomed 2013; 26: 1152–7.

Konstandin S, Nagel AM. Measurement techniques for magnetic resonance imaging of fast relaxing nuclei. Magn Reson Mater Phy 2014; 27: 5–19.

Kurhanewicz J, Bok R, Nelson SJ, Vigneron DB. Current and potential applications of clinical ^{13}C MR spectroscopy. J Nucl Med 2008; 49: 341–4.

Kushner DJ, Baker A, Dunstall TG. Pharmacological uses and perspectives of heavy water and deuterated compounds. Can J Physiol Pharmacol 1999; 77: 79–88.

Kwong KK, Hopkins AL, Belliveau JW, Chesler DA, Porkka LM, McKinstry RC, et al. Proton NMR imaging of cerebral blood flow using $H_2{}^{17}O$. Magn Reson Med 1991; 22: 154–8.

Lagemaat MW, van de Bank BL, Sati P, Li S, Maas MC, Scheenen TW. Repeatability of ^{31}P MRSI in the human brain at 7 T with and without the nuclear Overhauser effect. NMR Biomed 2016; 29: 256–63.

Leenders KL, Perani D, Lammertsma AA, Heather JD, Buckingham P, Healy MJ, et al. Cerebral blood flow, blood volume and oxygen utilization. Normal values and effect of age. Brain 1990; 113 (Pt 1): 27–47.

Lei H, Zhu XH, Zhang XL, Ugurbil K, Chen W. In vivo ^{31}P magnetic resonance spectroscopy of human brain at 7 T: An initial experience. Magn Reson Med 2003; 49: 199–205.

Lin AP, Shic F, Enriquez C, Ross BD. Reduced glutamate neurotransmission in patients with Alzheimer's disease-an in vivo ^{13}C magnetic resonance spectroscopy study. MAGMA 2003; 16: 29–42.

Lindon JC, Wilson ID. 19F NMR Spectroscopy: Applications in Pharmaceutical Studies. In: eMagRes. Chichester, UK: Wiley; 2007.

Lindquist DM, Komoroski RA. MRI and MRS of nuclei other than 1H. In: Jagannathan N, editor. Biomedical Magnetic Resonance: Proceedings of the International Workshop. New Delhi, India: Jaypee Brothers Medical Publishers (P), 2005: 217–32.

Lu M, Zhu XH, Zhang Y, Chen W. Intracellular redox state revealed by in vivo ^{31}P MRS measurement of NAD+ and NADH contents in brains. Magn Reson Med 2014; 71: 1959–72.

Maarouf A, Audoin B, Konstandin S, Rico A, Soulier E, Reuter F, et al. Topography of brain sodium accumulation in progressive multiple sclerosis. Magn Reson Mater Phy 2014; 27: 53–62.

Machado-Vieira R, Otaduy MC, Zanetti MV, De Sousa RT, Dias VV, Leite CC, et al. A selective association between central and peripheral lithium levels in remitters in bipolar depression: A 3T-^{7}Li magnetic resonance spectroscopy study. Acta Psychiatr Scand 2016; 133: 214–20.

Madelin G, Regatte RR. Biomedical applications of sodium MRI in vivo. J Magn Reson Imaging 2013; 38: 511–29.

Maguire ML, Geethanath S, Lygate CA, Kodibagkar VD, Schneider JE. Compressed sensing to accelerate magnetic resonance spectroscopic imaging: Evaluation and application to ^{23}Na-imaging of mouse hearts. J Cardiovasc Magn Reson 2015; 17: 45.

Mantsch HH, Saito H, Smith ICP. Deuterium magnetic-resonance, applications in chemistry, physics and biology. Prog Nucl Magn Reson Spectrosc 1977; 11: 211–71.

Maril N, Rosen Y, Reynolds GH, Ivanishev A, Ngo L, Lenkinski RE. Sodium MRI of the human kidney at 3 Tesla. Magn Reson Med 2006; 56: 1229–34.

Mason GFK, J.H. MR spectroscopy: Its potential role for drug development for the treatment of psychiatric diseases. NMR Biomed 2006; 19: 690–701.

Mateescu GD. Functional Oxygen-17 Magnetic Resonance Imaging and Localized Spectroscopy. In: Wilson DF, Evans SM, Biaglow J and Pastuszko A, editors. Adv Exp Med Biol. Boston, MA: Springer, 2003: 213–18.

Meiboom S. Nuclear magnetic resonance study of the proton transfer in water. J Chem Phys 1961; 34: 375–388.

Mellon EA, Beesam RS, Baumgardner JE, Borthakur A, Witschey WR, 2nd, Reddy R. Estimation of the regional cerebral metabolic rate of oxygen consumption with proton detected ^{17}O MRI during precision $^{17}O_2$ inhalation in swine. J Neurosci Meth 2009; 179: 29–39.

Meyerspeer M, Robinson S, Nabuurs CI, Scheenen T, Schoisengeier A, Unger E, et al. Comparing localized and nonlocalized dynamic ^{31}P magnetic resonance spectroscopy in exercising muscle at 7 T. Magn Reson Med 2012; 68: 1713–23.

Mintun MA, Raichle ME, Martin WR, Herscovitch P. Brain oxygen utilization measured with ^{15}O radiotracers and positron emission tomography. J Nucl Med 1984; 25: 177–87.

Mirkes C, Shajan G, Chadzynski G, Buckenmaier K, Bender B, Scheffler K. ^{31}P CSI of the human brain in healthy subjects and tumor patients at 9.4 T with a three-layered multinuclear coil: Initial results. MAGMA 2016; 29: 579–89.

Modo MMJJ, Bulte JWM. Magnetic Resonance Neuroimaging: Methods and Protocols. New York: Humana Press, 2011.

Moreno A, Ross BD, Blüml S. Direct determination of the N-acetyl-L-aspartate synthesis rate in the human brain by ^{13}C MRS and [1-^{13}C]glucose infusion. J Neurochem 2001; 77: 347–50.

Nagel AM, Lehmann-Horn F, Weber MA, Jurkat-Rott K, Wolf MB, Radbruch A, et al. In vivo ^{35}Cl MR imaging in humans: A feasibility study. Radiology 2014; 271: 585–95.

Naritomi H, Kanashiro M, Sasaki M, Kuribayashi Y, Sawada T. In vivo measurements of intra- and extracellular Na+ and water in the brain and muscle by nuclear magnetic resonance spectroscopy with shift reagent. Biophys J 1987; 52: 611–6.

Near J, Bartha R. Quantitative sodium MRI of the mouse prostate. Magn Reson Med 2010; 63: 822–7.

Niess F, Fiedler GB, Schmid AI, Goluch S, Kriegl R, Wolzt M, et al. Interleaved multivoxel ^{31} P MR spectroscopy. Magn Reson Med 2016; 77: 921–7.

Ong HH, Wright AC, Wehrli FW. Deuterium nuclear magnetic resonance unambiguously quantifies pore and collagen-bound water in cortical bone. J Bone Miner Res 2012; 27: 2573–81.

Oros-Peusquens AM, Keil F, Langen KJ, Herzog H, Stoffels G, Weiss C, et al. Fast and accurate water content and T_2* mapping in brain tumours localised with FET-PET. Nucl Instrum Meth A 2014; 734: 185–90.

Ouwerkerk R, Bleich KB, Gillen JS, Pomper MG, Bottomley PA. Tissue sodium concentration in human brain tumors as measured with ^{23}Na MR imaging. Radiology 2003; 227: 529–37.

Ouwerkerk R, Bottomley PA, Solaiyappan M, Spooner AE, Tomaselli GF, Wu KC, et al. Tissue sodium concentration in myocardial infarction in humans: A quantitative ^{23}Na MR imaging study. Radiology 2008; 248: 88–96.

Paling D, Solanky BS, Riemer F, Tozer DJ, Wheeler-Kingshott CA, Kapoor R, et al. Sodium accumulation is associated with disability and a progressive course in multiple sclerosis. Brain 2013; 136: 2305–17.

Pekar J, Ligeti L, Ruttner Z, Lyon RC, Sinnwell TM, van Gelderen P, et al. In vivo measurement of cerebral oxygen consumption and blood flow using ^{17}O magnetic resonance imaging. Magn Reson Med 1991; 21: 313–9.

Perman WH, Thomasson DM, Bernstein MA, Turski PA. Multiple short-Echo (2.5-Ms) quantitation of invivo sodium T_2 relaxation. Magn Reson Med 1989; 9: 153–160.

Prior MJW, Maxwell RJ, Griffiths JR. Fluorine-19F NMR Spectroscopy and Imaging In-Vivo. In: Rudin M, editor. In-Vivo Magnetic Resonance Spectroscopy III: In-Vivo MR Spectroscopy: Potential and Limitations. Berlin, Heidelberg: Springer, 1992: 101–130.

Provencher SW. Estimation of metabolite concentrations from in vivo Proton NMR Spectra. Magn Reson Med 1993; 30: 672–9.

Qian YX, Stenger VA, Boada FE. Parallel imaging with 3D TPI trajectory: SNR and acceleration benefits. Magn Reson Imaging 2009; 27: 656–63.

Qiao H, Zhang X, Zhu XH, Du F, Chen W. In vivo ^{31}P MRS of human brain at high/ultrahigh fields: A quantitative comparison of NMR detection sensitivity and spectral resolution between 4 T and 7 T. Magn Reson Imaging 2006; 24: 1281–6.

Reddy R, Insko EK, Noyszewski EA, Dandora R, Kneeland JB, Leigh JS. Sodium MRI of human articular cartilage in vivo. Magn Reson Med 1998; 39: 697–701.

Reddy R, Stolpen AH, Charagundla SR, Insko EK, Leigh JS. 17O-decoupled ^{1}H detection using a double-tuned coil. Magn Reson Imaging 1996; 14: 1073–8.

Reetz K, Romanzetti S, Dogan I, Sass C, Werner CJ, Schiefer J, et al. Increased brain tissue sodium concentration in Huntington's Disease—a sodium imaging study at 4 T. Neuroimage 2012; 63: 517–24.

Ren J, Sherry AD, Malloy CR. (31)P-MRS of healthy human brain: ATP synthesis, metabolite concentrations, pH, and T_1 relaxation times. NMR Biomed 2015; 28: 1455–62.

Rodrigues TBC, S. 13C MRS: An outstanding tool for metabolic studies. Concept Magnetic Res A 2005; 27: 1–16.

Roig ES, Magill AW, Donati G, Meyerspeer M, Xin L, Ipek O, et al. A double-quadrature radiofrequency coil design for proton-decoupled carbon-13 magnetic resonance spectroscopy in humans at 7T. Magn Reson Med 2015; 73: 894–900.

Romanzetti S, Halse M, Kaffanke J, Zilles K, Balcom BJ, Shah NJ. A comparison of three SPRITE techniques for the quantitative 3D imaging of the ^{23}Na spin density on a 4T whole-body machine. J Magn Reson 2006; 179: 64–72.

Ronen I, Lee J-H, Merkle H, Ugurbil K, Navon G. Imaging $H_2^{17}O$ distribution in a phantom and measurement of metabolically produced $H_2^{17}O$ in live mice by proton NMR. NMR Biomed 1997; 10: 333–40.

Ronen I, Merkle H, Ugurbil K, Navon G. Imaging of $H_2^{17}O$ distribution in the brain of a live rat by using proton-detected ^{17}O MRI. Proc Natl Acad Sci U S A 1998; 95: 12934–9.

Ronen I, Navon G. A new method for proton detection of $H_2^{17}O$ with potential applications for functional MRI. Magn Reson Med 1994; 32: 789–93.

Rooney WD, Springer CS. A comprehensive approach to the analysis and interpretation of the resonances of spins 3/2 from living systems. NMR Biomed 1991; 4: 209–26.

Ross BD, Lin A, Harris K, Bhattacharya P, Schweinsburg B. Clinical experience with ^{13}C MRS in vivo. NMR Biomed 2003; 16: 358–69.

Shah NJ, Oros-Peusquens AM, Arrubla J, Zhang K, Warbrick T, Mauler J, et al. Advances in multimodal neuroimaging: Hybrid MR-PET and MR-PET-EEG at 3 T and 9.4 T. J Magn Reson 2013; 229: 101–15.

Shah NJ, Worthoff WA, Langen KJ. Imaging of sodium in the brain: A brief review. NMR Biomed 2016; 29: 162–74.

Sharma R. Sodium weighted clinical brain magnetic resonance imaging at 4.23 tesla and inversion recovery pulse sequence. Informatica Medica Slovenica 2005; 10: 56–72.

Smith K. Sodium channels and multiple sclerosis: Roles in symptom production, damage and therapy. Brain Pathol 2007; 17: 345–345.

Stevens A, Paschalis P, Schleich T. Sodium-23 and potassium-39 nuclear magnetic resonance relaxation in eye lens. Examples of quadrupole ion magnetic relaxation in a crowded protein environment. Biophys J 1992; 61: 1061–75.

Stobbe R, Beaulieu C. Sodium relaxometry (part 2): Towards the characterisation of the sodium NMR environment in the human brain using a novel relaxometry technique. Proc Intl Soc Mag Reson Med 2006; 436.

Stolpen AH, Reddy R, Leigh JS. ^{17}O-decoupled proton MR spectroscopy and imaging in a tissue model. J Magn Reson 1997; 125: 1–7.

Stone NJ. Table of nuclear magnetic dipole and electric quadrupole moments. Atom Data Nucl Data 2005; 90: 75–176.

Sunde EP, Setlow P, Hederstedt L, Halle B. The physical state of water in bacterial spores. Proc Natl Acad Sci U S A 2009; 106: 19334–9.

Ter-Pogossian MM, Eichling JO, Davis DO, Welch MJ. The measure in vivo of regional cerebral oxygen utilization by means of oxyhemoglobin labeled with radioactive oxygen-15. J Clin Invest 1970; 49: 381–91.

Thulborn KR, Gindin TS, Davis D, Erb P. Comprehensive MR imaging protocol for stroke management: Tissue sodium concentration as a measure of tissue viability in nonhuman primate studies and in clinical studies. Radiology 1999; 213: 156–66.

Umathum R, Rosler MB, Nagel AM. In vivo ^{39}K MR imaging of human muscle and brain. Radiology 2013; 269: 569–76.

van de Bank BL, Orzada S, Smits F, Lagemaat MW, Rodgers CT, Bitz AK, et al. Optimized ^{31}P MRS in the human brain at 7 T with a dedicated RF coil setup. NMR Biomed 2015; 28: 1570–8.

van der Kemp WJM, Boer VO, Luijten PR, Stehouwer BL, Veldhuis WB, Klomp DWJ. Adiabatic multi-echo ^{31}P spectroscopic imaging (AMESING) at 7T for the measurement of transverse relaxation times and regaining of sensitivity in tissues with short T_2* values. NMR Biomed 2013; 26: 1299–307.

Van der Maarel JRC. Thermal relaxation and coherence dynamics of spin 3/2. I. Static and fluctuating quadrupolar interactions in the multipole basis. Concept Magnetic Res A 2003; 19a: 97–116.

van Gorp JS, Seevinck PR, Andreychenko A, Raaijmakers AJ, Luijten PR, Viergever MA, et al. 19F MRSI of capecitabine in the liver at 7 T using broadband transmit-receive antennas and dual-band RF pulses. NMR Biomed 2015; 28: 1433–42.

van Laarhoven HW, Klomp DW, Rijpkema M, Kamm YL, Wagener DJ, Barentsz JO, et al. Prediction of chemotherapeutic response of colorectal liver metastases with dynamic gadolinium-DTPA-enhanced MRI and localized ^{19}F MRS pharmacokinetic studies of 5-fluorouracil. NMR Biomed 2007; 20: 128–40.

Vanhamme L, van den Boogaart A, Van Huffel S. Improved method for accurate and efficient quantification of MRS data with use of prior knowledge. J Magn Reson 1997; 129: 35–43.

Vonsattel JP, Myers RH, Stevens TJ, Ferrante RJ, Bird ED, Richardson EP, Jr. Neuropathological classification of Huntington's disease. J Neuropathol Exp Neurol 1985; 44: 559–77.

Waxman SG. Axonal conduction and injury in multiple sclerosis: The role of sodium channels. Nat Rev Neurosci 2006; 7: 932–41.

Weber MA, Nagel AM, Marschar AM, Glemser P, Jurkat-Rott K, Wolf MB, et al. 7-T ^{35}Cl and ^{23}Na MR imaging for detection of mutation-dependent alterations in muscular edema and fat fraction with sodium and chloride concentrations in muscular periodic paralyses. Radiology 2016; 280: 848–59.

Wellard RM, Shehan BP, Adam WR, Craik DJ. NMR measurement of ^{39}K detectability and relaxation constants in rat tissue. Magn Reson Med 1993; 29: 68–76.

Wiesner HM, Balla DZ, Shajan G, Scheffler K, Ugurbil K, Chen W, et al. 17O relaxation times in the rat brain at 16.4 Tesla. Magn Reson Med 2016; 75: 1886–93.

Wijnen JP, Klomp DW, Nabuurs CIHC, de Graaf RA, van Kalleveen IML, van der Kemp WJM, et al. Proton observed phosphorus editing (POPE) for in vivo detection of phospholipid metabolites. NMR Biomed 2016; 29: 1222–30.

Winkler SS. 23Na magnetic-resonance brain imaging. Neuroradiology 1990; 32: 416–20.

Winkler SS, Thomasson DM, Sherwood K, Perman WH. Regional T_2 and sodium concentration estimates in the normal human-brain by ^{23}Na MR imaging at 1.5 T. JCAT 1989; 13: 561–6.

Zhu XH, Zhang N, Zhang Y, Zhang X, Ugurbil K, Chen W. In vivo ^{17}O NMR approaches for brain study at high field. NMR Biomed 2005; 18: 83–103.

Yuksel C, Du F, Ravichandran C, Goldbach JR, Thida T, Lin P, et al. Abnormal high-energy phosphate molecule metabolism during regional brain activation in patients with bipolar disorder. Mol Psychiatr 2015; 20: 1079–84.

Yushmanov VE, Kharlamov A, Yanovski B, LaVerde G, Boada FE, Jones SC. Inhomogeneous sodium accumulation in the ischemic core in rat focal cerebral ischemia by ^{23}Na MRI. J Magn Reson Imaging 2009; 30: 18–24.

Zaaraoui W, Konstandin S, Audoin B, Nagel AM, Rico A, Malikova I, et al. Distribution of brain sodium accumulation correlates with disability in multiple sclerosis: A cross-sectional ^{23}Na MR imaging study. Radiology 2012; 264: 859–67.

Zhu XH, Chen W. In vivo oxygen-17 NMR for imaging brain oxygen metabolism at high field. Prog Nucl Magn Reson Spectrosc 2011; 59: 319–35.

Zhu XH, Chen W. In vivo ^{17}O MRS imaging-Quantitative assessment of regional oxygen consumption and perfusion rates in living brain. Anal Biochem 2016; 529: 171–8.

Zhu XH, Merkle H, Kwag JH, Ugurbil K, Chen W. 17O relaxation time and NMR sensitivity of cerebral water and their field dependence. Magn Reson Med 2001; 45: 543–9.

Zhu XH, Zhang N, Zhang Y, Zhang X, Ugurbil K, Chen W. In vivo ^{17}O NMR approaches for brain study at high field. *NMR Biomed* 2005; 18: 83–103.

14

T_1-Weighted DCE MRI[1]

Contents

Leonidas Georgiou and
David L. Buckley
University of Leeds

14.1 Introduction

Dynamic contrast-enhanced (DCE) MRI is designed to measure tissue haemodynamics by monitoring the distribution of an intravenously administered injection of contrast agent *in vivo*, through its effect on the tissue's longitudinal relaxation time (T_1). Quantitative analysis of signal intensity–time series can provide measurements of tissue perfusion, such as cerebral blood flow (CBF) and cerebral blood volume (CBV), that characterise the brain microvasculature, but it can also characterise blood–brain barrier (BBB) integrity by providing measurements of permeability and information about the distribution volume of the contrast agent in the extracellular extravascular space (EES). This chapter provides an overview of DCE-MRI techniques and focuses on their application to assess BBB breakdown in brain pathologies.

14.1.1 Blood–Brain barrier

The BBB is a highly selective multicellular semipermeable membrane structure that separates the circulating blood from the extracellular fluid in the central nervous system. Similar to most anatomical barriers, the BBB is formed by endothelial cells, which are however unique compared to those identified in other tissues. The endothelial cells of the BBB owe their exclusiveness to the continuous intercellular tight junctions, the lack of fenestrations (small pores), low expression of leucocyte adhesion molecules and the extremely low rates of both paracellular and transcellular transcytosis (Abbott *et al.*, 2006; Obermeier *et al.*, 2013). These properties allow the BBB to actively regulate the influx and efflux of exogenous and endogenous molecules and thus maintain a microenvironment that allows neuronal circuits to function properly.

[1] Edited and reviewed by Paul S. Tofts.

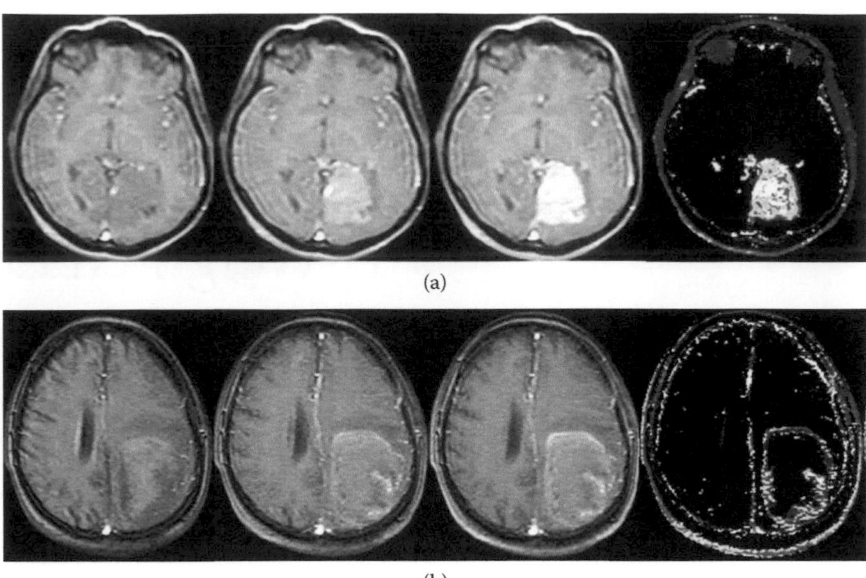

FIGURE 14.1 DCE images from a patient with meningioma (a) collected at t = 0, 8 and 260 s following administration of gadopentetate dimeglumine (Gd-DTPA). Note how easily blood–brain barrier disruption may be identified in this case, where the contrast agent rapidly leaks through the barrier and appears to distribute relatively homogeneously within the lesion. However, the corresponding K^{trans} map on the right demonstrates the intrinsic heterogeneity of the lesion. (b) Dynamic contrast-enhanced images from a patient with a Grade 4 glioma at t = 0, 16 and 260 s post-injection of Gd-DTPA, with the respective K^{trans} map estimated using quantitative analysis. The different characteristics of the two brain lesions, (a) and (b), are visible from both the enhancement profiles and the parametric maps.

14.1.2 BBB disruption

The development and maintenance of the BBB is governed by cellular and non-cellular elements (e.g. astrocytes, pericytes, microglial cells, neurons and extracellular matrix components) that interact with the endothelial cells, forming an interactive cellular complex. BBB breakdown leads to unregulated exchange of molecules, ions and/or cells and therefore agitates the normal function of neuronal processes. The mechanism of BBB disruption and the consequences are manifold and differ depending on the pathology.

Identifying BBB breakdown and its relation to disease is challenging. However, this disruption enables the brain interstitium to become locally accessible to low-molecular weight MRI contrast agents.[2] On T_1-weighted images the accumulation of contrast agents causes an increase in tissue signal intensity (Figure 14.1) and therefore provides an alternative tool to probe brain pathologies *in vivo*, such as intracranial neoplasm, ischaemia, pneumococcal meningitis, multiple system atrophy, type II diabetes, multiple sclerosis (MS), cognitive impairment, small vessel disease, etc. (Heye *et al.*, 2014).

14.1.3 Basis of contrast enhancement following bolus injection

The types of low-molecular-weight contrast agents used in most clinical applications in the brain are paramagnetic gadolinium-based contrast agents with either linear chemical structures, Gd-DTPA (gadopentetate dimeglumine), Gd-DTPA-BMA (gadodiamide), Gd-DTPA-BMEA (gadoversetamide) and Gd-BOPTA (gadobenate dimeglumine) or cyclic structures, Gd-DOTA (gadoterate meglumine), Gd-BT-DO3A (gadobutrol) and Gd-HP-DO3A (gadoteridol). Until recently, these agents were all considered to be very safe for human use. However, since 2010 the European Medicines Agency (EMA) have recommended against the use, or at least minimisation of the use, of these agents in patients with severe kidney problems or patients undergoing liver transplants. Such patients are at risk of a condition known as nephrogenic systemic fibrosis believed to be caused by gadolinium contained in these agents. Moreover, in March 2017 the EMA recommended suspension of the marketing authorisation for the four linear agents because of increasing evidence that small amounts of gadolinium may accumulate in the brain following multiple contrast agent administrations. As a result it seems likely that future studies will employ cyclic agents only.

When the bolus[3] of contrast agent passes into the blood circulation, it is temporarily confined within the vascular space and remains there for many cardiac cycles, a phase known as the *first pass* (Padhani, 2002). In the absence of an intact BBB and due to the contrast agent concentration difference created between the vascular space and the brain interstitium, the contrast agent dif-

[2] In healthy brain, contrast agents do not pass to the brain interstitium, except in a few regions such as the choroid plexus (Gibby, 2000).

[3] A bolus is a short injection, typically completed in a few seconds using a power injector (see Section 14.3.5).

fuses from the blood pool into the EES of the brain. The rate and extent of accumulation depend on the underlying pathophysiology, such as the tissue perfusion, the capillary permeability and surface area of the leaking vessels. The contrast agent does not enter the cells constituting the brain tissue and therefore the distribution volume is the EES. Over a period of time lasting many minutes to hours, the contrast agent diffuses back into the vasculature from where it is excreted (usually via the kidneys).

On T_1-weighted images the T_1 shortening, as a result of the accumulation of the contrast agent predominantly in the brain EES,[4] causes an increase in the MRI signal ('signal enhancement'). In a DCE-MRI experiment, the accumulation and washout of the contrast agent is therefore observed as a change in MR signal intensity. Since the kinetics (e.g. rate of enhancement, peak enhancement and signal decay) of the distribution of the agent depend on the pathology, DCE-MRI enables the study of brain pathophysiology *in vivo*, through the analysis of the temporal characteristics of the signal intensity–time series (Gribbestad *et al.*, 2005).

14.2 Analysis of DCE-MRI data

Numerous techniques have been applied to the analysis of signal–time curves in order to measure BBB disruption. These range from simple visual assessment of spatiotemporal enhancement patterns, to semi-quantitative and quantitative techniques through to the use of complex tracer kinetic models. Furthermore, for semi-quantitative and quantitative techniques there are usually two approaches for making measurements: a predefined region of interest (ROI) and a voxel-by-voxel analysis that generates parametric maps. In the following sections the most commonly used quantification techniques will be discussed with an emphasis on compartmental tracer kinetic model analysis.

14.2.1 Visual assessment of enhancement patterns

This technique is based upon the subjective evaluation of the signal–time curve, in which a curve is classified according to an evaluation system or by describing the enhancement characteristics in post-contrast T_1-weighted images at predefined time points. For signal–time series there are four general curve patterns: (1) no enhancement; (2) slow enhancement, where the signal rises slowly within the region of interest for the duration of the scan; (3) fast enhancement followed by a plateau; and (4) rapid enhancement followed by a washout phase. In a study investigating newly formed MS lesions, the initial enhancement patterns in lesions were determined from the first post-contrast dynamic T_1-weighted volume and were distinguished as nodular (homogeneous hyperintensity [i.e. brightness] throughout the lesion), closed ring (complete peripheral hyperintense rim enclosing a hypointense [i.e. dark] centre), and open ring (incomplete hyperintense rim with semilunar [half-moon] configuration). Furthermore, the dynamic enhancement pattern was defined as either centrifugal (enhancing from the centre to the periphery) or centripetal (enhancing from the periphery to the centre) (Gaitán *et al.*, 2011). This information alone does not, however, provide detailed insights into the pathophysiology studied.

14.2.2 Semi-Quantitative analysis of enhancement series

Perhaps the simplest approach to provide reliable and sensitive information about the kinetics of contrast accumulation is through empirical description of the enhancement curves using a series of metrics. Such features include the peak of the enhancement curve, initial wash-in slope, time to maximum signal, washout slope and initial area under the enhancement curve (IAUC) over a predefined period of time (e.g. for the first 90 s) (Parker and Buckley, 2005). This simplified approach, however, cannot differentiate the physiological from the physical factors affecting signal enhancement. During MRI planning, features including imaging sequence used, scanner parameters, tuning and scaling factors of the machine, native T_1 of the tissue of interest, dose and mode of administration (e.g. bolus injection, infusion) of contrast agent used, patient's systemic status (e.g. cardiac output) as well as the non-linear relationship between the contrast agent's concentration to signal intensity changes all play a key role in shaping the signal intensity–time curves measured. These emphasise the need to normalise this technique in order to take into account the differences in the amplitude or the slope of the enhancement because of the choice of physical factors used either in different sites or multiple MRI sessions (Gribbestad *et al.*, 2005; Padhani, 2002).

Attempts to overcome these limitations might include normalisation with respect to a reference tissue, which is assumed to be healthy and not affected by the procedures followed (e.g. muscle), or using concentration–time series instead of signal intensity. Even then, the interpretation of these parameters remains vague. For example IAUC or initial wash-in slope reflect kinetics that are governed by a combination of blood flow, endothelial permeability, blood volume and EES volume (Evelhoch, 1999; Walker-Samuel *et al.*, 2006). Furthermore, signal enhancement curves even within a single voxel may represent signal from a blood vessel, the EES or a combination of these two. It is therefore difficult to differentiate the signal coming from each compartment and thus investigate the interaction between those compartments. This is where tracer kinetic modelling may be beneficial and provide novel insights into pathology.

14.2.3 Quantitative analysis using tracer kinetic modelling

Quantitative analysis can involve the use of tracer kinetic models that describe the passage of a bolus of contrast agent through the tissue of interest. These models aim to portray

[4] In some cases of brain pathology (e.g. tumour) the contribution from contrast agent in the blood vessels might also be significant.

the temporal features observed in a concentration–time curve, through parameters that reflect physiological processes in vivo such as CBF, BBB leakiness and the volume of the EES and the vascular spaces (Table 14.1).

Most of the tracer kinetic models used in DCE-MRI consider the tissue (e.g. brain) to be made up of a number of compartments (Figure 14.2). The number, as well as any assumption about the compartments, depend on the properties of the contrast agent used as well as the tissue physiology or pathology studied. In the case of a healthy intact BBB, low molecular weight contrast agents are mostly confined within the vascular compartment (v_b). However, BBB disruption enables the contrast agent to diffuse from the vascular space into the EES (v_e). The rate of diffusion depends on the local blood flow (F_b), the BBB permeability (*P*) and surface area (*S*) (Tofts, 1997; Tofts *et al.*, 1999). The exchange of contrast agent between the vascular and EES compartments is often described by a transfer constant, K^{trans}, whose interpretation depends on the assumptions made in the model; it is usually a combination of BBB permeability surface–area product (*PS*) and CBF. The contrast agent diffuses back into the vascular space and is eventually eliminated from the blood circulation via renal excretion. At any given time, the overall concentration of contrast agent within a voxel or a region of interest is given by the weighted sum of the concentrations within each individual compartment, which may be expressed as

TABLE 14.1 Quantities and symbols used in tracer kinetic modelling

Quantity	Definitions	Unit
C_a	Tracer concentration in arterial whole blood[a]	mM[b]
C_e	Tracer concentration in EES[c]	mM
C_p	Tracer concentration in arterial blood plasma[a]	mM
C_t	Tracer concentration in tissue	mM
C_v	Tracer concentration in venous whole blood	mM
E	Initial extraction fraction	None
Hct	Haematocrit	None
F_b	Perfusion (or flow) of whole blood per unit volume of tissue, CBF[d]	ml min^{-1} ml
F_p	Perfusion (or flow) of blood plasma per unit volume of tissue	ml min^{-1} ml^{-1}
K^{trans}	Volume transfer constant between blood plasma and EES	min^{-1}
P	Total permeability of capillary wall	cm min^{-1}
PS	Permeability surface area product per unit volume of tissue	ml min^{-1} ml^{-1}
S	Surface area per unit volume of tissue	cm^{-1}
v_b	Whole blood volume per unit volume of tissue, CBV[e]	None
v_p	Plasma blood volume per unit volume of tissue	None
v_e	EES volume per unit volume of tissue	None

[a] $C_a = (1-Hct)\ C_p$.
[b] mM = 1 mmol/l.
[c] EES = extravascular extracellular space.
[d] CBF = cerebral blood flow.
[e] CBV = cerebral blood volume.

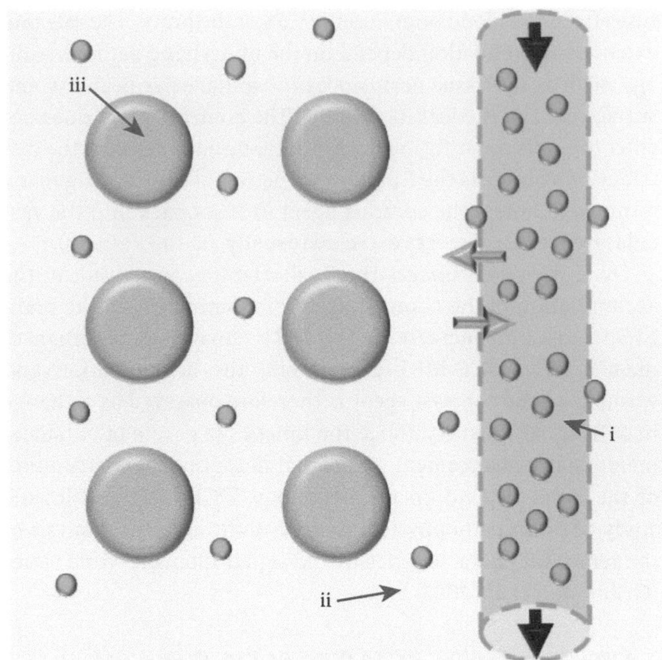

FIGURE 14.2 A general model of tissue compartmentalisation. The bold black arrows illustrate the direction of blood flow through the vessel (i), with fractional volume, v_p. Grey arrows represent the bi-directional leakage of the contrast agent (green circles) from the vascular to the extracellular extravascular space (EES (ii) volume v_e) and vice versa. Most contrast agents do not enter the intracellular space (iii gold discs).

$$C_t(t) = v_p C_p(t) + v_e C_e(t) \tag{14.1}$$

where C_t is the concentration in the entire tissue, and C_p and C_e are the concentrations in the vascular plasma space and EES, respectively. The functional parameters described above provide a simplified attempt to model the mechanisms that influence contrast agent distribution in the region of interest. There are however several factors that are not considered, e.g. plug flow in vessels and contrast agent mixing within compartments, and more complex modelling approaches attempt to take these factors into account.

Most tracer kinetic models describe the exchange of contrast agent between compartments with a simple rate equation, i.e. the diffusive transport of a substance across a semipermeable membrane. This equation describes the flux of the contrast agent to be driven by the concentration gradient between the two compartments, with a rate constant that depends on the properties (e.g. restrictiveness) of the membrane (Kety, 1951). Therefore, the flux, *J*, of the contrast agent between two compartments may be expressed as

$$J = PS\ (C_1 - C_2) \tag{14.2}$$

where $(C_l - C_2)$ is the concentration difference between the compartments. In the case of BBB breakdown, the restrictiveness of the barrier is disrupted, allowing the contrast agent to diffuse from the vascular to the EES compartment. It is important to note here that P takes into account any form of passive or active transport (e.g. paracellular and transcellular) across the semi-permeable barrier.

The next step is to describe how the proposed compartments interact with one another. This is achieved by characterising the concentration of contrast agent in each compartment as the sum of inlets and outlets, which transport the contrast agent in and out of the compartment. For example, an outlet from one compartment can form the inlet of another. This is the foundation of compartmental modelling and assumes that the response of a compartment to an influx is linear and stationary, i.e. proportional to the dose administered and independent of the time of arrival. Furthermore, for these to hold true, it is assumed that no contrast agent is created or destroyed inside the compartment. The kinetics of the contrast agent in each compartment are then expressed in the form of differential equations that describe the flux at the inlets and outlets. When these are solved, Equation 14.1 is used to describe the concentration profile in the combined system. This forms the basis of compartmental tracer kinetic modelling. The complexity of compartmental models and the number of parameters used to describe them depend on several factors (quality of data acquired, the sampling frequency, the length of acquisition, the pathophysiology of the disease, etc.). These limitations have to be considered in order to generate a realistic and reliable model. In the following paragraphs we will introduce the most common tracer kinetic models used to assess BBB integrity.

14.2.3.1 The Tofts model

One of the most widely used models is based on the approach proposed by Kety, who described the distribution of an inert gas in the lungs (Kety, 1951). Using Equation 14.2 and substituting the terminology commonly used in DCE-MRI (Table 14.1), the rate of accumulation and washout of a contrast agent in the EES, under the assumption of instantaneous and well-mixed tracer, can be described using the general form of the rate equation

$$v_e \frac{dC_e(t)}{dt} = K^{trans}(C_p(t) - C_e(t)) \tag{14.3}$$

The physiological interpretation of the volume transfer constant, K^{trans}, depends on the balance between the BBB permeability and blood flow in the tissue. In the general case of mixed perfusion and permeability regime, K^{trans} is given by

$$K^{trans} = EF_p \tag{14.4}$$

where F_p is the plasma flow (volume of blood plasma per minute per unit volume of tissue: ml min^{-1} ml^{-1}), and E is the extraction fraction of the contrast agent (i.e. the fraction of tracer extracted to v_e in a single transit through the capillary bed) given by

$$E = \frac{PS}{PS + F_p} \tag{14.5}$$

PS is measured in volume per minute per volume of tissue (ml min^{-1} ml^{-1}). If the delivery of the contrast agent to the EES is limited by flow (i.e. $PS >> F_p$) then

$$K^{trans} = F_p \tag{14.6}$$

Conversely, if the delivery of contrast agent to the EES is limited by BBB permeability ($F_p >> PS$) then

$$K^{trans} = PS \tag{14.7}$$

Tracer kinetic modelling using the approach suggested above forms the basis of models most widely used by researchers to describe the leakage of contrast agent through the BBB. Quantitative analysis in the brain was first introduced in the early 1990s (Brix *et al.*, 1991; Larsson *et al.*, 1990; Tofts and Kermode, 1991). A consensus paper aiming to standardise the quantities and symbols used was published almost a decade later (Tofts *et al.*, 1999).

The conventional Tofts model (Tofts and Kermode, 1991), first implemented to describe the kinetics of the tracer in patients with MS, assumed a negligible contribution to C_t from the vascular space (v_p~0). Solving Equation 14.3 and substituting to Equation 14.1, a one-compartment model for weakly vascularised lesions is derived

$$C_t(t) = K^{trans} \int_0^t C_p(\tau)\, exp\left(-\frac{K^{trans}(t-\tau)}{v_e}\right) d\tau \tag{14.8}$$

An example of the application of Tofts model is shown in Figure 14.3, for two enhancing lesions in the brain that exhibit different enhancement profiles.

14.2.3.2 The extended Tofts model

Whilst vp~0 assumption may hold true in pathologies where the contribution of the vascular space to the overall concentration remains small, in diseases such as tumours the contribution of v_p has to be considered. In the highly perfused regime, $F_p \sim \infty$[5] (Sourbron and Buckley, 2011), the extended Tofts model simply includes the concentration of contrast agent in the blood plasma

$$C_t(t) = v_p C_p(t) + K^{trans} \int_0^t C_p(\tau)\, exp\left(-\frac{K^{trans}(t-\tau)}{v_e}\right) d\tau \tag{14.9}$$

The effect of the vascular component of the extended Tofts model is shown in Figure 14.3.

[5] For example from a measurement perspective when F_p is large enough so that any bolus broadening in the tissue is too small to be measurable, either because of a slow injection, low temporal resolution or noise in the data.

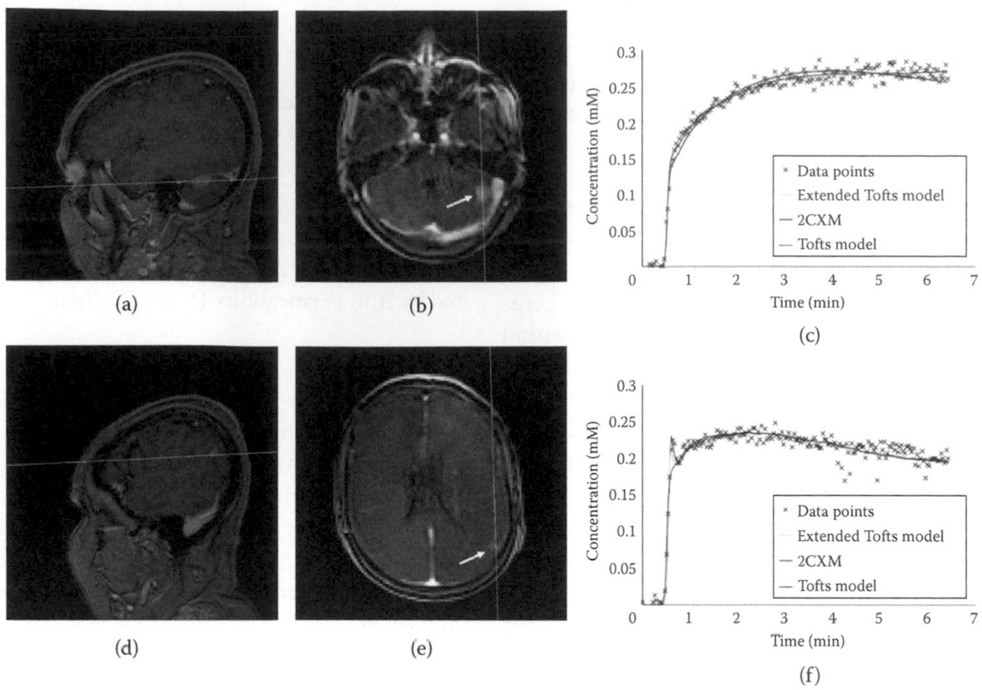

FIGURE 14.3 Examples of quantitative analysis using Tofts model, extended Tofts model (ETM) and two-compartment exchange model (2CXM) on dynamic data from two enhancing lesions (top and bottom row) in the brain: (a, d) high spatial resolution gradient echo images acquired in the sagittal plane and used as a reference; (b, e) one dynamic post-contrast image for each enhanced lesion following a bolus-injection of Gadovist; (c, f) illustrate the application of different models in the enhancing lesions. For the slowly enhancing lesion (c) 2CXM: F_b = 0.24 ml min^{-1} ml^{-1}, PS = 0.02 ml min^{-1} ml^{-1}, v_e = 0.09, v_p = 0.18; ETM: K^{trans} = 0.21 min^{-1}, v_e = 0.38, v_p = 0.00; Tofts: K^{trans} = 0.21 min^{-1}, v_e = 0.38. For the fast enhancing lesion (f) 2CXM: F_b = 0.67 ml min^{-1} ml^{-1}, PS = 0.22 ml min^{-1} ml^{-1}, v_e = 0.11, v_p = 0.06; ETM: K^{trans} = 0.27 min^{-1}, v_e = 0.28, v_p = 0.02; Tofts: K^{trans} = 0.28 min^{-1}, v_e = 0.29. The data were acquired on a 3T Philips MRI system.

14.2.3.3 The Patlak model

The two models described above assume a bi-directional transport of contrast agent between the vascular space and the EES (Figure 14.2). There are however cases where the interaction between the two compartments can be further simplified by assuming a unidirectional transport of tracer, from blood plasma into the EES. This assumption can be acceptable if the duration of the dynamic scan is short and therefore the flux of the contrast agent into v_e is not sufficient to fill that space. The Patlak model (Patlak *et al.*, 1983) describes this scenario and can be expressed as

$$C_t(t) = v_p C_p(t) + K^{trans} \int_0^t C_p(\tau)\, d\tau \qquad (14.10)$$

Equation 14.10 can also be expressed in linear form, by dividing each term with $C_p(t)$, and has been used by several researchers, as it allows simple and fast data post-processing.

14.2.3.4 The Two-Compartment exchange model

The models presented so far represent boundary regimes of the more general two-compartment exchange model (2CXM): (1) weakly vascularised (Equation 14.8) and (2) highly perfused (Equations 14.9 and 14.10) lesions. The 2CXM can be applied to the general case of a mixed perfusion and permeability regime (Brix *et al.*, 2004; Sourbron and Buckley, 2011, 2013). Where there are sufficient data to support it, 2CXM allows separate estimates of PS and F_p to be extracted from the data. The solution can be expressed as the convolution (⊗) of an impulse response function, $H(t)$, with the arterial plasma concentration, the so-called arterial input function (AIF). For a 2CXM, $H(t)$ is bi-exponential with terms (α, β, A, F_p) that relate to physiological parameters

$$\begin{aligned} C_t(t) &= F_p C_p(t) \otimes \left(Ae^{-\alpha t} - (1-A)e^{-\beta t}\right) \\ v_p &= F_p / (A \cdot (\alpha - \beta) + \beta) \\ v_e &= PS \cdot F_p / (\alpha\beta v_p) \\ PS &= [v_p(\alpha + \beta - \alpha\beta v_p / F_p) - F_p] \end{aligned} \qquad (14.11)$$

Implementation of the 2CXM in Figure 14.3 demonstrates how the more complex model can fit lesions that follow different kinetic profiles.

These are the most commonly used tracer kinetic models in quantitative analysis of BBB functionality. The choice of a model that best describes the acquired data depends on several factors. In the following paragraphs, we summarise the acquisition requirements and sources of errors that often lead to inappropriate implementation of these models and may generate inaccurate and inconsistent results.

14.3 Data acquisition requirements

DCE-MRI uses T_1-weighted images to detect the short-range effect of the contrast agent, as it transits through the tissue, on the T_1 relaxation properties of water in that tissue, e.g. a lesion in the brain. The acquisition protocol plays a major role in determining the kinetic profile of the contrast agent and hence the quantitative technique used for analysis. In this section we will address some of the requirements for appropriate data acquisition.

14.3.1 From signal to concentration

To monitor the behaviour of a contrast agent using quantitative analysis, we have to provide a link between changes in signal (relative to baseline) and the concentration of the tracer. A simplified approach is to assume that signal (or relative signal) changes are directly proportional to contrast agent concentration; however this linearity breaks down at high concentrations. The degree of non-linearity depends upon acquisition parameters (e.g. flip angle, repetition time, native T_1 of the lesion). The usual approach to this issue is to measure the baseline T_1 relaxation of the lesion or tissue and monitor the change in T_1 during the dynamic acquisition. The contrast agent concentration $C(t)$, at a time, t, may be related to change in T_1 relaxation rate as follows

$$C(t)=\frac{\frac{1}{T_1(t)}-\frac{1}{T_{10}}}{r_1} \tag{14.12}$$

where r_1 is the longitudinal relaxivity[6] of the contrast agent and T_{10} is the longitudinal relaxation time in the absence of contrast agent.

Although this approach seems simple in theory, there are practical difficulties in the measurement of T_1 that led to the adoption of different approaches. For reasons that will become clear in the following section, the T_1 measurement must be fast and accurate, in order to sample the concentration of the contrast agent at a high temporal rate. Furthermore, the measurement techniques need to be reliable over a wide range of T_1 values, since the tissues initially have a long relaxation time before contrast administration (e.g. of the order of seconds at 1.5T in blood), which can reduce enormously (e.g. down to 100 ms) as the bolus of contrast agent enters the blood. To fulfil these requirements the most common approach is to measure the baseline T_1 using an established T_1 quantification technique and then acquire image data of the tissue during bolus administration using a fast T_1-weighted sequence. Using the precontrast T_{10} and the equation that describes the MRI signal in terms of the sequence parameters used, the dynamic T_1 may be estimated indirectly.

Numerous techniques are available for T_1 quantification including inversion recovery prepared imaging sequences or multiple acquisitions with different repetition times (TR) or flip angles (θ). More details about these techniques can be found in Chapter 5 of this book. The most common data acquisition techniques in DCE-MRI are based on 3D-gradient echo sequences, due to their short acquisition times. The signal intensity obtained from a commonly used gradient echo sequence with spoiling of the transverse magnetisation is

$$S=\Omega PD\frac{\sin(\theta)\left(1-\exp(-TR/T_1)\right)}{\left(1-\cos(\theta)\exp(-TR/T_1)\right)}\exp\left(-TE/T_2{}^*\right) \tag{14.13}$$

where Ω depends on the scanner settings, PD is the proton density and TE is the echo time. When a short TE is used, the effect of $T_2{}^*$ on the signal is negligible. Hence, using the dynamic signal acquired before the contrast administration (S_0) and the precontrast T_{10} of the tissue, we may estimate the dynamic T_1 using Equation 14.13. Assuming that the contrast agent has no effect on PD, we may use the change in longitudinal relaxation rate to estimate the concentration of the contrast agent (Equation 14.12).

14.3.2 Spatio-Temporal requirements

The temporal frequency at which the signal–time series should be sampled is crucial for tracer kinetic modelling and defines the choice of model as well as the accuracy of quantitative analysis. This is of particular importance when sampling the passage of the bolus of contrast agent through a supplying artery (known as the AIF), especially during the early phase, following bolus administration, where the signal exhibits rapid temporal changes. Temporal undersampling can introduce uncertainties into parameter estimates, and previous studies have suggested that to contain these errors within 10% of the true value, the AIF should be sampled every 1 s and the tissue at least every 4 s (Cramer and Larsson, 2014; Henderson *et al.*, 1998). These suggestions are not universal but instead may depend on the tissue of interest and model applied, as well as the duration of the bolus administration.

In general, it is recommended that the sampling interval be shorter than the timescale of the processes to be measured (Henderson *et al.*, 1998; Lopata *et al.*, 2007). For example the mean transit time[7] of Gd-DTPA through the capillary bed in grey and white matter is of the order of 1–2 seconds, whereas in brain tumours (e.g. metastases, meningioma and lymphoma) it is of the order of tens of seconds (Sourbron *et al.*, 2009). To obtain a measurement of CBF a temporal resolution of the order

[6] Relaxivity is a measure of the contrast agent's ability to catalyse proton relaxation. Its value depends on field strength, the chemical properties of the contrast agent, temperature and tissue composition.

[7] Mean transit time is the time taken for the average contrast agent molecule to enter a compartment through an inlet and leave through an outlet (Sourbron and Buckley, 2013).

of seconds is required in these cases.[8] Conversely, the mean transit time of the same contrast agent in the EES of tumours is of the order of minutes; therefore if the focus is BBB permeability the temporal resolution may be relaxed (Sourbron, 2010; Sourbron *et al.*, 2009). The total acquisition time also needs careful consideration since it determines the range of information one can extract from quantitative analysis. For the measurement of CBF a short acquisition (one that includes the early recirculation passes of the bolus of contrast agent in blood) is sufficient; however for BBB permeability and v_e measurements longer acquisitions are required (Sourbron and Buckley, 2013), particularly in slowly enhancing lesions such as MS (Kermode *et al.*, 1990; Tofts, 1996).

The need for high temporal resolution has led the majority of researchers to adopt 2D or 3D gradient echo techniques. Technical developments such as parallel imaging and other acceleration schemes have made these requirements more feasible. Despite this, high temporal resolution DCE data are usually acquired at the expense of spatial resolution. For example Larsson et al., with a saturation recovery gradient recalled sequence (on a 3T MRI scanner) obtained four slices with a spatial resolution of $2.5 \times 3.1 \times 8$ mm^3 and temporal resolution of 1 s (Larsson *et al.*, 2009), when the American College of Radiology (ACR) guidelines recommend the acquisition of high spatial data with at least $1 \times 1.2 \times 5$ mm^3 resolution (see ACR guide for clinical image quality). Although temporal sampling is important, spatial resolution is required if measuring the AIF from an intracerebral artery or when the heterogeneity of the pathology or small scale lesions needs to be probed, since partial volume effects will lead to inaccurate parameter estimates. Closely linked to both temporal and spatial resolution is total coverage of the imaging volume. In pathologies where the entire brain needs to be monitored (e.g. patients with MS) a compromise between temporal and spatial resolution is inevitable. Furthermore, the imaging volume needs to include a relatively large feeding artery or vein from where the blood signal can be sampled (see Section 14.3.4).

14.3.3 Image contrast-to-noise ratio

The contrast-to-noise ratio (CNR) is key for DCE-MRI data. CNR is a measure of change in signal intensity following contrast administration referenced to baseline noise and depends on the tissue type (native T_1), sequence parameters (e.g. flip angle, TR), the administered dose of contrast agent and the field strength of the MRI scanner. All these should be considered during sequence optimisation, since the quality of the parameter estimates depends upon the quality of the signal–time series.

[8] Strictly this is true if the AIF is a delta function. In reality, the AIF's finite width and tissue broadening may slightly relax these temporal requirements. It is important, however, to understand that sampling requirements are introduced to address the need to sample the AIF itself.

14.3.4 Arterial input function

A core requirement, and often the most challenging part, of tracer kinetic modelling is to provide an accurate representation of the AIF, the delivered contrast agent dose. As described in Section 14.3.2 the temporal sampling of the AIF has a major impact on the accuracy of the parameter estimates as well as the choice of model. Furthermore, spatial resolution limitations can introduce partial volume effects in the generated AIF that will propagate as errors to the parameter estimates. This is particularly problematic in DCE-MRI of the brain because the feeding arteries are relatively small. Hence, although the AIF should be sampled directly at the inlet to the pathology, in practice it is performed further upstream in the arterial tree where the diameter of the vasculature is larger. In the majority of the studies, an AIF is sampled from the internal carotid artery (Larsson *et al.*, 2009) and in some cases signal from a venous outflow, usually the sagittal sinus, is used to correct for partial volume artefacts (Hansen *et al.*, 2009; Sourbron *et al.*, 2009). An example of AIF sampling and rescaling using a venous outflow is shown in Figure 14.4. In cases when the acquired images do not enable arterial measures, the sagittal sinus is sampled and used as a surrogate AIF, following the assumption that there is little dispersion between arteries and veins in the brain (Jelescu *et al.*, 2011; Li *et al.*, 2000). Slice positioning is crucial in obtaining an accurate AIF, and the MRI planning should aim to acquire the signal from a feeding artery simultaneously with lesion sampling. Positioning the most caudal slices orthogonally to the internal carotid artery or acquiring separate slices at the level of the artery may be used to reduce partial volume effects (Cramer and Larsson, 2014).

There are however cases where limitation in the acquired data prohibit the measurement of an individual AIF (e.g. partial volume and inflow effects; see Section 14.4). The simplest way to account for that is to assume an AIF derived from the time course of the mean blood plasma concentration from a number of volunteers. Numerous studies have reported functional forms of AIFs that can be used as a substitute (Horsfield *et al.*, 2009; Orton *et al.*, 2009; Parker *et al.*, 2006; Tofts and Kermode, 1991) but do not take into account individual variability, for example due to differences in cardiac cycle, bolus duration and contrast agent dose. Another proposed solution compares the tissue of interest curve to that of a reference region, thereby eliminating the need for direct AIF measurement (Yankeelov *et al.*, 2005), and has been applied in rat gliomas (Quarles *et al.*, 2012). Despite the alternatives an accurately measured AIF is superior with respect to the quality of the parameter estimates (McGrath *et al.*, 2009; Port *et al.*, 2001; Yankeelov *et al.*, 2007).

14.3.5 Contrast agent administration

An important decision in DCE-MRI is the choice of contrast agent, the administered dose and the injection protocol. The pharmacokinetic properties of the tracer used may have an impact on the kinetic profile of the pathology investigated.

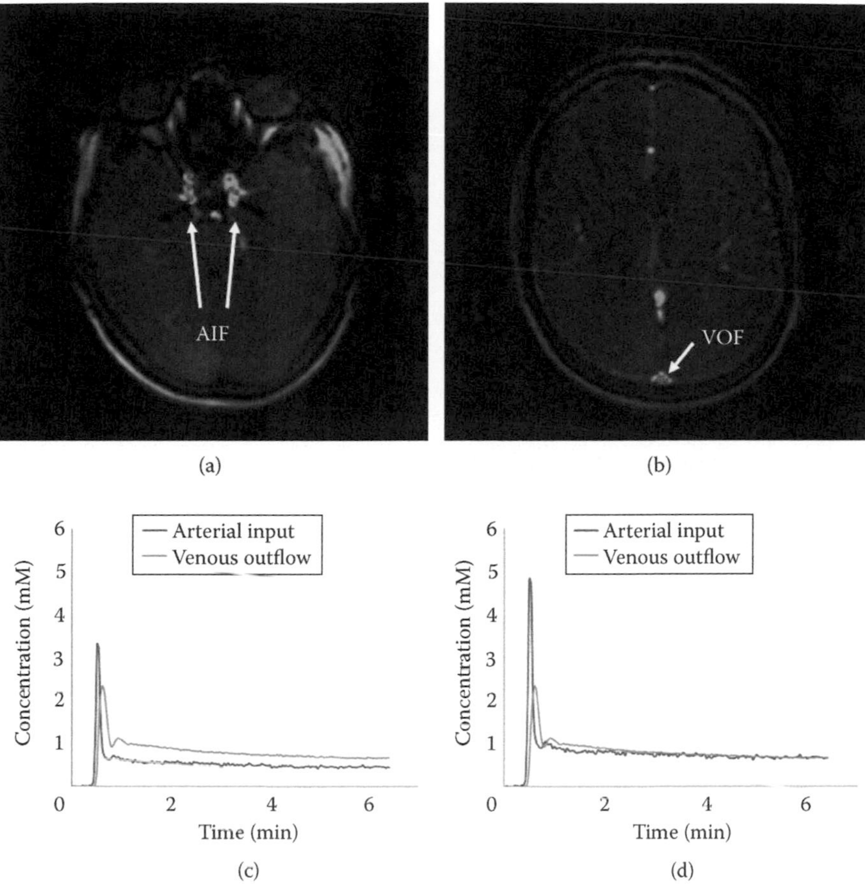

FIGURE 14.4 Examples of a typical arterial input function (AIF) and venous outflow function (VOF) in DCE data. (a, b) correspond to dynamic images following a bolus injection of Gadovist (0.1 mmol/kg at 5 ml/s), at t = 30 s (a) and t = 34 s (b). MR images are presented in colour to make AIF (sampled from the carotid arteries) and VOF (sampled from the sagittal sinus) more visible. Depending on the temporal resolution (~2 s was used for these data) of the acquisition, the contrast agent may be first seen to reach the arteries and then the veins. The concentration in the arteries will typically be underestimated due to partial volume effects (c). The VOF is often used to rescale the AIF (d), by matching the areas under the AIF and VOF curve. The data were acquired on a 3T Philips MRI system.

For example, contrast agents that partially bind to albumin (e.g. MS-325, gadofosveset) should be avoided when BBB permeability is the subject of investigation since estimates of permeability will reflect leakage of both albumin-bound and -unbound contrast agent (Richardson *et al.*, 2015). Moreover, the dose of contrast agent used is of particular importance in brain DCE-MRI because of the restricted leakage through the BBB, which makes detection and quantification of subtle changes in BBB permeability difficult. A higher dose (or imaging at a higher field strength) might help in these cases; however the dose should be carefully chosen and should also depend on the dynamic range of the MR sequence, since at high concentrations the signal reaches a saturation regime. The duration of the injection is also relevant; contrast agents are administered either as a bolus or as an infusion, although the former is the most common practice for quantitative analysis and shows higher sensitivity (Tofts and Berkowitz, 1994). For bolus administration, the desirable duration is less than 10 s to generate a sharp AIF and reduce error in the parameter estimates, but this comes at the expense of a requirement for high temporal resolution sampling to capture the AIF (Henderson *et al.*, 1998).

14.4 Sources of error

The conclusion derived from the previous section is that although quantitative analysis has the potential to enhance our understanding of BBB disruption and its relation to disease, there is a series of practical requirements that need to be met in order to achieve a realistic description of the underlying physiology. In this section we aim to summarise the most common sources of error undermining the accuracy of analysis and therefore hinder widespread adoption in clinical practice.

14.4.1 Signal non-linearity

One of the most significant sources of error is the relation between signal intensity changes and contrast agent concentration. As mentioned in Section 14.3.1, a common assumption is that of a linear relationship, but this is not true at high concentrations, particularly in the arterial blood during the early bolus passes. Despite this, many studies adopt this approach, either due to acquisition limitations (e.g. absence of estimates of native tissue T_1 relaxation) or the ease of using a direct relation between signal changes and concentration, and often introduce systematic errors into the kinetic parameters estimated. A more accurate approach is to perform T_1 measurements and relate changes in T_1 to concentration, but this does not come without errors. Temporal sampling requirements (discussed in Section 14.3.2) imply that the T_1 measurement is performed in a short time. The most common approach is to obtain a T_1 measurement before contrast administration, using a robust quantification technique, such as an inversion recovery approach, or faster techniques based on gradient echo acquisitions with variable flip angles, and subsequently estimate the dynamic T_1 as described in Section 14.3.1. However, B_1 inhomogeneity can cause the actual flip angle (during both the T_1 measurement and the dynamic acquisition) in tissue to be different from the prescribed one. The non-uniformity of the transmit radiofrequency depends on several factors (e.g. location in the anatomy, field strength, non-ideal slab profile) and can introduce errors both in the T_1 measurement and in signal conversion to dynamic T_1 (Equation 14.13). These errors will propagate into the analysis and create a bias in the pharmacokinetic parameters. The issue may be addressed by mapping the spatially varying flip angle and correcting for these non-uniformities or by using the bookend T_1 approach, which involves T_1 measurements at multiple time points during the acquisition. The extra information acquired can be used to calibrate Equation 14.13 (Cron *et al.*, 1999). Another possible error may come from T_2* effects, which may have a significant impact at high contrast agent concentrations, even when a very short echo time is used. The response of the signal may be different if a wide range of concentrations is encountered during the dynamic acquisition and a dual-echo sequence can be used to correct for these errors (Kleppestø *et al.*, 2014).

14.4.2 Water exchange

Closely related to errors in the estimated concentration and T_1 measurement are confounds related to water exchange between the compartments being studied. Equation 14.12 assumes that the tissue contains a single water population that undergoes homogeneous T_1 changes following contrast agent accumulation. However, water in tissues is found in different environments (i.e. intracellular, interstitial and intravascular) and each may have different MR properties, such as multiple T_1 relaxation times (Hazlewood *et al.*, 1974). Water is not stationary but instead moves continuously from one environment to another, and hence the T_1 measured is an average weighted by the time water resides in each subspace. A single T_1 value is a valid representation of the system if the ratio of the exchange rate of water between the compartments to the difference in the intrinsic T_1 relaxation rates in each subspace is high (Donahue *et al.*, 1994). However, if the relaxation rate difference increases (e.g. due to contrast agent accumulation), the relaxation properties of each subspace may become different and should be taken into account (Donahue *et al.*, 1997). This is particularly important in the brain, for example during the first pass of the bolus when the contrast agent remains intravascular with little or no permeation into the interstitium, but may be less relevant if the lesion under investigation is a tumour where the degree of first pass extraction is higher (Buckley, 2002; Larsson *et al.*, 2001). If the effect of water exchange is significant then this will lead to underestimated CBF and CBV. Methods to account for vascular–interstitial water exchange (Larsson *et al.*, 2001; Schwarzbauer *et al.*, 1997) and cellular–interstitial water exchange (Buckley *et al.*, 2008; Yankeelov *et al.*, 2003) have been proposed either by incorporating these into the analysis or by minimising the dependence of the measurement on water exchange (e.g. short TR, high flip angle, lower contrast agent dose).

14.4.3 AIF Measurement

AIF measurement is also prone to errors, and a common issue is inflow artefact, which causes the signal in the arterial blood to appear enhanced in precontrast images and the generated enhancement curve to therefore be underestimated (Roberts *et al.*, 2011). Inflow effects also propagate into the blood T_1 measurement and can lead to errors in CBF and blood-volume estimates. This bias can be minimised by optimising the acquisition protocol (e.g. choose a 3D acquisition over a 2D, use a non-selective saturation prepulse) or by appropriate positioning of the excitation slab such that the sampled artery runs for a long distance through it (Buckley *et al.*, 2004). Another requirement for the AIF measurement is to sample the signal from a location close to the tissue in order to minimise dispersion errors that are difficult to correct (Calamante *et al.*, 2003). However, arteries in the brain are very small and this can lead to partial-volume errors, particularly when a high temporal sampling strategy is required, at the expense of spatial resolution.

14.4.4 Haematocrit

It is also important to note that concentration estimates in the blood reflect the blood plasma compartment, rather than whole blood, since the contrast agent does not enter blood cells (e.g. red blood cells).[9] The AIF measurement therefore needs to be corrected to describe concentration in plasma (c_p) rather than whole blood (c_b), using the blood haematocrit in $c_p = c_b/(1-Hct)$. In many studies *Hct* measurement is ignored and a standard value is used, but this can lead to errors in kinetic parameters since individual patient variability or even *Hct* changes over time within the same patients due to the effects of treatment are not taken into account. Furthermore, the measured *Hct* usually reflects the large vessel *Hct*, but small vessel *Hct* is also needed since it is

[9] Water exchange between red blood cells and plasma is rapid and a single T_1 value is used to characterise the relaxation properties of blood (Herbst and Goldstein, 1989).

a more accurate representation of plasma volume in capillaries (Tofts *et al.*, 2012). Large and small vessel *Hct* values are probably different; therefore the use of the former instead of the latter will introduce a small systematic error in the pharmacokinetic parameters.

14.4.5 Temporal resolution and scan duration

The temporal resolution requirements for quantitative analysis have been discussed in Section 14.3.2 and they are closely related to spatial resolution, volume coverage and CNR. The compromise between these interrelated properties depends on the modelling approach as well as the pathology studied. For example, to model CBF and CBV, a high temporal resolution is essential and the duration of the acquisition can be reduced. In this scenario the enhancement curve is possibly an accurate representation of the underlying kinetics; however the precision of the estimates will be compromised due to increased noise. On the contrary, if permeability and EES are the subjects of interest, longer acquisitions are required but the sampling rate can be relaxed. In terms of pathology, if the lesion is highly vascularised (e.g. tumour), a simple one-compartmental model will describe the kinetics of the contrast agent with less accuracy compared to a more complex model that takes into account the contribution from the vascular space. Since the simple model uses fewer parameters to describe the underlying physiology, the parameters will be contaminated from processes that are not incorporated into the model, resulting in systematic errors in the pharmacokinetic estimates. In the case of slow enhancing lesions (e.g. MS), a long scanning duration is needed for the one compartmental model to be applied accurately and provide accurate EES estimates.

Temporal resolution requirements are particularly problematic in brain DCE-MRI because of the smaller anatomical scales of the pathologies, where a higher spatial resolution is needed. A possible solution to this conflicting requirement is to incorporate both high spatial and high temporal resolution acquisitions in one single dynamic protocol. Studies have showed that the sampling rate is more important during the early bolus passes of the contrast agent when blood and tissue signals exhibit rapid temporal changes but can be relaxed later, allowing the acquisition of high spatial resolution data (Jelescu *et al.*, 2011; Georgiou *et al.*, 2017). These studies demonstrate the errors encountered when the initial part of the AIF is undersampled or partly sampled, as well as the effect of the tissue curve type (e.g. slow or fast enhancing pathology) on the parameter estimates.

14.4.6 Other sources of error

Other sources of error include general imaging artefacts such as point-spread functions due to *k*-space undersampling or aliasing effects and are dictated by the imaging parameters, motion-induced artefacts that can be corrected by image registration and signal intensity drift, which becomes more problematic in long acquisitions (Armitage *et al.*, 2011). Moreover, the ROI selection process can introduce bias in the parameter estimates if it is not performed objectively and a similar approach is used for all patients. In addition, ROI analysis does not take into account the spatial heterogeneity of lesions; instead it averages over such effects, which might hide additional information about the progression of the pathology. Figure 14.1 demonstrates examples of parametric maps for permeability, which shows the heterogeneous nature of lesions. Parametric maps on the other hand allow regional variabilities to be investigated but are more prone to noise, lack the precision and ease of ROI analysis, and are more difficult to quantify.

14.4.7 Reproducibility

Another important consideration is that of reproducibility. Most of the discussion so far has focused on measurement errors that can cause variability in the parameter estimates; however inherent physiological changes between measurements may also have an effect. It is therefore crucial to incorporate reproducibility studies of normal and pathological tissues into clinical research, in order to assess the expected physiological variation, prior to or in parallel with attempts to monitor treatment effects or the progression of a pathology (Galbraith *et al.*, 2002; Padhani *et al.*, 2002; Wong *et al.*, 2017).

It is clear that there are numerous practical issues that need careful consideration and design prior to a DCE-MRI study. The difficulty comes from the fact that most of these issues have conflicting requirements and a compromise is inevitable. From the researcher's perspective, however, it is necessary to recognise these limitations and take all sensible actions to minimise errors or, more importantly, avoid introducing additional errors when these can be avoided. It is worth mentioning that recent studies investigating the reproducibility of quantitative analysis using various commercially available analysis software for DCE-MRI have found considerable variability among both quantitative and semi-quantitative pharmacokinetic parameters. This is due to differences in the analytical approaches used by each vendor (Beuzit *et al.*, 2016; Heye *et al.*, 2013a, 2013b). This highlights the necessity for standardisation, before establishing DCE-MRI parameters as widely incorporated imaging biomarkers.

14.5 Clinical applications

DCE-MRI in the brain has been applied in a wide range of pathologies using different sampling schemes, scanning protocols and analytical approaches. In this section we provide a summary of some of the studies conducted, with the aim to reveal the advantages and limitations of the techniques used, provide a range of parameter estimates and how these relate to the pathology studied.

14.5.1 Enhancement patterns in MS

The simplest approach for analysing DCE-MRI data in the brain is visual analysis of enhancement patterns seen on T_1-weighted images. As described in Section 14.2.1 enhancement patterns may be grouped according to their appearance on the first dynamic

T_1-weighted image or after examining the dynamic pattern of how the enhancement develops. Since the temporal resolution can be relaxed in these approaches, images may be acquired at a high spatial resolution.

Gaitan et al. investigated the development and expansion of newly formed MS lesions in patients, through the evolution of BBB permeability (Gaitán *et al.*, 2011, 2013). They reported that nodular lesions (as they appeared on the first post-contrast T_1-weighted image) were small in size and enhanced centrifugally, whereas ring-like lesions were larger and enhanced centripetally, with the majority becoming nodular following the 60-minute acquisition. Furthermore, on follow-up MRI scans over several days they identified that lesions enhancing centrifugally became larger and evolved to centripetal. These findings suggest that nodular or ring-like lesions were confounded by both size and time of scanning and that the different enhancement patterns observed actually represent different stages of MS, rather than distinct lesion types.

Moreover, the different stages suggest that BBB disruption initially occurs in the central vein, whereas the peripheral vessels within the affected area maintain an intact BBB. Demyelination and tissue damage then spreads radially as mirrored in the centrifugal dynamic enhancement patterns. As the lesion expands due to disease progression, the BBB around peripheral veins is disrupted and the enhancement pattern changes to centripetal. It was speculated that the reduction in central enhancement may be due to closing or partial closing of the BBB in the central vein and a reduction in blood perfusion of the central core because of the redirection of blood flow to the peripheral capillaries where BBB remains disrupted. This pattern of MS lesion progression has been demonstrated using immunohistochemistry in patients with MS (Henderson *et al.*, 2009). Despite the information provided by visual inspection of enhancement patterns, the absence of quantitative information, for example the role of perfusion in newly formed MS lesions, imposes limitations on the conclusions that could be extracted.

14.5.2 Area under the curve in stroke and gliomas

Semi-quantitative analysis of signal intensity–time series has been employed to test for subtle BBB leakiness, for example in patients with lacunar stroke and control patients with cortical ischaemic stroke (Wardlaw *et al.*, 2009). Maximum signal enhancement in white matter and cerebrospinal fluid was significantly higher in lacunar compared to cortical stroke patients; however, the cause of increased leakiness in lacunar stroke patients could not be identified. In another study the area under the signal intensity–time curve in lacunar stroke and leucoaraiosis patients was found to be higher compared to controls. Furthermore, the area under the curve was found to be dependent on leucoaraiosis grade. Although these metrics are sensitive and model independent, interpretation is not without ambiguity. For example simulations have shown that the area under the curve increases with increased permeability, increased EES or increased blood flow (Walker-Samuel *et al.*, 2006).

Quantitative techniques could provide novel insights into whether this is caused due to CBF changes arising from small-vessel wall thickening or as a consequence of white matter damage and increased BBB permeability. In patients with gliomas, concentration–time series were used to differentiate tumour grade using the IAUC and other semi-quantitative parameters such as tumour enhancing fraction (EnF), initial gradient of the EnF/IAUC curve, the rate at which pixels are eliminated from the enhancing tumour as the threshold value of IAUC is increased (∂EnF) and Pronin's and Tofts' measures (Mills *et al.*, 2009). All metrics proved useful in differentiating low and high grade gliomas; however their relationship and hence their interpretation with respect to physiological processes within the tumour was not identifiable.

14.5.3 Tracer kinetic modelling of subtle BBB changes

Quantitative analysis through tracer kinetic modelling has been widely used in brain DCE-MRI. In many cases, studies report the normal range of BBB leakage by analysing subtle changes in healthy brain or normal-appearing tissue in patients. The range of K^{trans} values reported for normal brain varies depending on the acquisition techniques and the model used. In the majority of the cases where the Patlak model was used, K^{trans} ranged between 0.1 and 1.0 × 10^{-3} min^{-1} (Cramer and Larsson, 2014; Heye *et al.*, 2016; Taheri *et al.*, 2011; Zhang *et al.*, 2012). DCE-MRI protocols for measuring subtle changes in BBB using the Patlak model use a wide range of temporal resolutions for the dymanic acquisition (e.g. 1.25–73 s), but the majority of these studies adopted a prolonged acquisition time, e.g. >15 min. As mentioned in Section 14.2.3, the Patlak model assumes unidirectional transfer of contrast agent from blood to brain, i.e. the flux of the contrast agent into v_e is not sufficient, for v_e to be quantifiable. This assumption is often true for short acquisitions but can also be true for longer acquisitions if the permeability is small.

Cramer *et al.* demonstrated that for low BBB permeability a prolonged acquisition generates more accurate and precise results than shorter acquisitions, whereas the temporal resolution becomes more significant as the permeability is increased. This could explain the higher permeability estimated for normal brain (K^{trans} = 5.2 × 10^{-3} min^{-1}) with a 2CXM by Larsson *et al.*, who used a total acquisition time of only 180 s (Larsson *et al.*, 2009). The extended Tofts model also systematically overestimates low BBB permeability values when the duration of the acquisition is short, particularly for highly perfused lesions, even when the temporal resolution is 1.2 s (Cramer and Larsson, 2014; Zhang *et al.*, 2012). The BBB permeability estimates in normal brain are in close agreement with those reported in PET (Iannotti *et al.*, 1987) and demonstrate the potential of DCE-MRI to detect subtle changes in BBB leakage even at the early stages of diseases such as vascular cognitive impairment and Alzheimer's dementia.

14.5.4 Tracer kinetic modelling in stroke

Another useful application of quantitative DCE-MRI is in ischaemic stroke patients. Merali *et al.* studied the evolution of BBB permeability in acute ischaemic stroke by examining patients in the hours (range: 1.3–90.7 hrs) following the onset of stroke symptoms. Using a short acquisition (<5 min) and Patlak model they showed that hyperacute stroke (<6 hours) had a significantly lower K^{trans} ($7.2 \pm 3.7 \times 10^{-3}$ min^{-1}) compared to the acute phase (6–48 hours; $K^{trans} = 9.4 \pm 6.9 \times 10^{-3}$ min^{-1}), with continuous BBB leakage for up to 90.7 hours (Merali *et al.*, 2017). These findings contradict suggestions of a biphasic behaviour of BBB disruption, i.e. a period of increased permeability followed by a return to baseline and then a second period of increase, found in rats (Huang *et al.*, 1999; Rosenberg *et al.*, 1998) but agree with a more recent study in rat models for stroke (Abo-Ramadan *et al.*, 2009).

One major issue in patients with acute ischaemic stroke (AIS) is the risk of haemorrhagic transformation (HT), which limits the general use of thrombolytic therapies, despite the fact that it is an effective treatment. In a series of publications it was demonstrated that patients with HT had significantly higher K^{trans} values compared to those that did not experience haemorrhagic complications (Kassner *et al.*, 2005, 2009; Thornhill *et al.*, 2010). These findings suggest that DCE-MRI techniques assessing BBB permeability could be potentially used as a surrogate for treatment planning at early stages of AIS. The range of K^{trans} estimates reported for AIS in these studies vary between 3 and 8.4×10^{-3} min^{-1} and all have used similar protocols and the Patlak model, whereas for mild ischaemic stroke permeability is lower, 0.6×10^{-3} min^{-1} (Heye *et al.*, 2016). Vidarsson *et al.* showed that when using the linear expression of the Patlak model in AIS patients, it is not only important to acquire for long enough to be able to measure BBB permeability but in addition, because of the delay in some infarcts, it is important to delay the starting point for quantitative analysis in order to decouple the permeability estimates from wash-in kinetics that occur during the first pass of the bolus (Vidarsson *et al.*, 2009), which could lead to overestimations of K^{trans}.

In addition to its usefulness for estimating permeability there are cases where estimating CBF is also important, for example to examine reperfusion after a stroke. Larsson *et al.* used model free deconvolution to estimate CBF (temporal resolution 1 s and duration of acquisition 180 s) in two patients with stroke. The two patients demonstrated vastly different CBF values, 0.09 and 1.4 ml min^{-1} ml^{-1} compared to an average of 0.59 ml min^{-1} ml^{-1} for insular grey matter (this matches the range reported in PET, 0.40–0.60 ml min^{-1} ml^{-1}). This might illustrate different courses of reperfusion after a stroke (the patient with low CBF value suffered language impairment, with signs of severe arteriosclerotic disease, whereas the other patient recovered faster) (Larsson *et al.*, 2008). In another study using a similar approach to Larsson, in three patients 4–5 days following the stroke event the mean CBF was estimated to be 0.56 ml min^{-1} ml^{-1} (Nadav *et al.*, 2017). The authors also demonstrated that a high temporal resolution was required to avoid errors in the estimation of CBF, whereas long scan durations were required to improve BBB permeability estimates.

14.5.5 Tracer kinetic modelling in multiple sclerosis

One of the first applications of DCE-MRI in the brain was in patients with MS lesions, by two separate groups, Larsson and Tofts. Using a one-compartment tracer kinetic model (both assumed that the contribution from the vascular space was negligible) they reported BBB permeability estimates between 2.4 and 24×10^{-3} min^{-1} (Larsson and Tofts, 1992). This large variation was attributed to the heterogeneity in MS lesions studied, which included both slow and rapidly enhancing, very leaky acute MS plaques. In addition, the two studies differed in their assumptions about the AIF (Tofts used a population average input derived from 10 healthy volunteers, whereas Larsson used a measured AIF after direct arterial catheterisation), the relaxivity of Gd-DTPA and different MR acquisition techniques.

In a more recent application of the conventional Tofts model, Jelescu *et al.* used an uninterrupted two-part acquisition protocol that consisted of an initial phase at high temporal resolution (5 s) and a second phase at low temporal resolution (32 s), a technique that preserved both high spatial resolution data needed for small lesions and the high temporal sampling rate required to sample the rapid temporal changes in the AIF during the early bolus passes (Jelescu *et al.*, 2011). The mean K^{trans} estimated using the dual acquisition was $9 \pm 2 \times 10^{-3}$ min^{-1} and, using simulations, they demonstrated that when the early passes of the AIF are undersampled K^{trans} is overestimated by up to 33%. Model selection also affects the accuracy of K^{trans} estimates. Cramer *et al.*, using both simulations and data from patients with MS, showed that as BBB permeability increased, the Patlak and extended Tofts model progressively underestimated permeability with respect to a 2CXM (Cramer and Larsson, 2014). For example, using the Patlak model and a T_1-sampling method with temporal resolution of 3.5 min, Taheri *et al.* estimated a mean K^{trans} for MS lesions of 2.3×10^{-3} min^{-1} (Taheri *et al.*, 2011).

Quantitative assessment of BBB leakage in MS patients may prove very important in monitoring relapse activity or treatment effect. For example, a study demonstrated that BBB permeability estimates were higher in periventricular normal-appearing white matter (NAWM), an area prone to development of new MS lesions, in MS patients with one or more clinical relapses in a period of 3 months compared to patients with no relapses within the same period. Furthermore, the permeability in non-enhancing lesions was lower in MS patients that showed no relapse and underwent immunomodulatory treatment (Cramer *et al.*, 2014).

CBF measurement may also play a crucial role in MS lesion detection. It is hypothesised that BBB disruption is initiated by inflammation, a consequence of which is local change in CBF. Ingrisch *et al.*, using a high temporal resolution acquisition and fitting the DCE data using an uptake model,[10] showed that in contrast-enhanced lesions in MS patients CBF was higher compared to NAWM (0.23 vs. 0.16 ml min^{-1} ml^{-1}) (Ingrisch *et al.*, 2012). Furthermore, in previous dynamic susceptibility-contrast MRI studies it was suggested that CBF is significantly reduced in the NAWM of patients with MS compared to healthy controls (Inglese *et al.*, 2008), although this was not observed in a more recent publication using quantitative DCE-MRI (Ingrisch *et al.*, 2016).

14.5.6 Tracer kinetic modelling in neoplasms[11]

Perhaps the majority of clinical applications of quantitative analysis in DCE-MRI of the brain are found in studies that involve intracranial neoplasm (Heye *et al.*, 2014). Haris *et al.*, using K^{trans} estimates derived from the conventional Tofts model, was able to differentiate infective (2.10 ± 0.46 min^{-1}) from neoplastic brain lesions. Furthermore, their estimates allowed them to distinguish high-grade from low-grade gliomas (1.24 ± 0.16 vs. 0.75 ± 0.19 min^{-1}, respectively). The grading of gliomas was confirmed after histopathology, where high-grade gliomas had a higher vascular endothelial growth factor and microvessel density (Haris *et al.*, 2008a). Differentiation between low-grade (I, II) and high-grade (III, IV) gliomas has also been demonstrated in other studies using the extended Tofts model, but the K^{trans} estimates reported differ vastly from Haris *et al.*, for example in the range of 0.002–0.21 min^{-1} (from low- to high-grade gliomas) (Zhang *et al.*, 2012). These large differences might be due to model choice, since the conventional Tofts model does not take into account the contribution to the lesion from the vascular space, which might hold true for patients with MS but not in the case of tumours where angiogenesis is a characteristic. Furthermore, Haris *et al.* used a population average AIF based on the data of Weinmann *et al.* and this may have introduced additional errors into the parameter estimates (McGrath *et al.*, 2009; Singh *et al.*, 2007; Weinmann *et al.*, 1984).

Quantitative DCE-MRI has also been used to monitor treatment effect in glioblastomas (Ferl *et al.*, 2010), where K^{trans} exhibited a significant decrease following treatment with the antiangiogenic drug bevacizumab, the differentiation of histopathologic grade in astrocytoma (Jia *et al.*, 2013), the characterisation of BBB leakage in patients with neurofibromatosis type-2 related vestibular schwannomas (Li *et al.*, 2012), and the correlation of K^{trans} estimates with the expression of matrix metalloproteinase 9 in brain tuberculomas (Haris *et al.*, 2008b). The more complex 2CXM has also been used to derive both CBF and BBB permeability estimates in patients with meningioma, lymphoma, astrocytoma and glioblastoma (F_b: 0.42–0.56, *PS*: 0.02–0.03 ml min^{-1} ml^{-1}) (Larsson *et al.*, 2009; Sourbron *et al.*, 2009). CBF and BBB permeability estimates can be very important in decision-making for treatment using drugs, since they can provide information about the expected bioavailability of the administered drug by examining how this is delivered to the pathology in the brain.

14.6 Challenges and new directions

It is clear that T_1-weighted DCE-MRI has seen a wide range of application in the brain. In a recent systematic review that focused on the assessment of the BBB using these techniques, the authors report on 70 studies with a total of 417 animals and 1564 humans (Heye *et al.*, 2014). However, the adoption of these techniques evident during the last decade was not achieved in a structured and organised manner. Studies showed a considerable heterogeneity with respect to both image acquisition methods and post-processing strategies. This heterogeneity is partly expected because of the mixed nature of pathologies, which may exhibit vastly different enhancement profiles over similar timescales, but in many cases the variability was introduced by the choice of analytical technique. To allow interstudy comparison and widespread clinical adoption, investigators need to achieve an agreed consensus.

Steps in that direction have already been made through well-documented recommendations for various technical aspects of quantitative analysis, as well as the proposal of a unified system of tracer kinetic models, with standardised names and symbols (Leach *et al.*, 2012; Sourbron and Buckley, 2011, 2013; Tofts *et al.*, 1999). Hence, with respect to method development and other theoretical considerations, DCE-MRI techniques have matured. Furthermore, technical developments, such as faster image acquisition with improved image quality, will likely promote the use of quantitative techniques even in clinical practice. Improvements in data quality will also allow the implementation of more sophisticated models that account for more of the processes known to occur *in vivo*. If DCE-MRI techniques are standardised and the quality of both the acquisition and analysis are raised, it will be possible to promote more robust and reproducible imaging biomarkers with numerous clinical applications in the brain; from assessing tumour grade and patient prognosis, monitoring treatment effect and response to therapy, to quantifying subtle changes in BBB disruption related to ageing, dementia or brain microvascular disease (Jain, 2013).

References

Abbott NJ, Rönnbäck L, Hansson E. Astrocyte–endothelial interactions at the blood–brain barrier. Nature Rev Neurosci 2006; 7: 41–53.

Abo-Ramadan U, Durukan A, Pitkonen M, Marinkovic I, Tatlisumak E, Pedrono E, et al. Post-ischemic leakiness of the blood-brain barrier: a quantitative and systematic assessment by Patlak plots. Exp Neurol 2009; 219: 328–33.

[10] The model assumes that the concentration of the contrast agent in the interstitial space remains small due to a low permeability, large interstitial space, short acquisition time or a combination of these. This is similar to Patlak except the tissue is not highly perfused (i.e. $F_p \neq \infty$).

[11] A neoplasm is a new, often uncontrolled growth of abnormal tissue, often a tumour.

Armitage PA, Farrall AJ, Carpenter TK, Doubal FN, Wardlaw JM. Use of dynamic contrast-enhanced MRI to measure subtle blood-brain barrier abnormalities. Magn Reson Imaging 2011; 29: 305–14.

Beuzit L, Eliat P-A, Brun V, Ferré J-C, Gandon Y, Bannier E, et al. Dynamic contrast-enhanced MRI: study of inter-software accuracy and reproducibility using simulated and clinical data. J Magn Reson Imaging 2016; 43: 1288–300.

Brix G, Kiessling F, Lucht R, Darai S, Wasser K, Delorme S, et al. Microcirculation and microvasculature in breast tumors: pharmacokinetic analysis of dynamic MR image series. Magn Reson Med 2004; 52: 420–9.

Brix G, Semmler W, Port R, Schad LR, Layer G, Lorenz WJ. Pharmacokinetic parameters in CNS Gd-DTPA enhanced MR imaging. J Comput Assist Tomogr 1991; 15: 621–8.

Buckley DL. Transcytolemmal water exchange and its affect on the determination of contrast agent concentration in vivo. Magn Reson Med 2002; 47: 420–4.

Buckley DL, Kershaw LE, Stanisz GJ. Cellular-interstitial water exchange and its effect on the determination of contrast agent concentration in vivo: dynamic contrast-enhanced MRI of human internal obturator muscle. Magn Reson Med 2008; 60: 1011–9.

Buckley DL, Roberts C, Parker GJM, Logue JP, Hutchinson CE. Prostate cancer: evaluation of vascular characteristics with dynamic contrast-enhanced T1-weighted MR Imaging—Initial experience. Radiology 2004; 233: 709–15.

Calamante F, Yim PJ, Cebral JR. Estimation of bolus dispersion effects in perfusion MRI using image-based computational fluid dynamics. NeuroImage 2003; 19: 341–53.

Cramer SP, Larsson HB. Accurate determination of blood–Brain barrier permeability using dynamic contrast-Enhanced T1-Weighted MRI: a simulation and in vivo study on healthy subjects and multiple sclerosis patients. J Cereb Blood Flow Metab 2014; 34: 1655–65.

Cramer SP, Simonsen H, Frederiksen JL, Rostrup E, Larsson HBW. Abnormal blood-brain barrier permeability in normal appearing white matter in multiple sclerosis investigated by MRI. NeuroImage: Clinical 2014; 4: 182–9.

Cron GO, Santyr G, Kelcz F. Accurate and rapid quantitative dynamic contrast-enhanced breast MR imaging using spoiled gradient-recalled echoes and bookend T1 measurements. Magn Reson Med 1999; 42: 746–53.

Donahue KM, Burstein D, Manning WJ, Gray ML. Studies of Gd-DTPA relaxity and proton exchange rates in tissues. Magn Res Med 1994; 32: 66–76.

Donahue KM, Weisskoff RM, Burstein D. Water diffusion and exchange as they influence contrast enhancement. J Magn Reson Imaging 1997; 7: 102–10.

Evelhoch JL. Key factors in the acquisition of contrast kinetic data for oncology. J Magn Reson Imaging 1999; 10: 254–9.

Ferl GZ, Xu L, Friesenhahn M, Bernstein LJ, Barboriak DP, Port RE. An automated method for nonparametric kinetic analysis of clinical DCE-MRI data: application to glioblastoma treated with bevacizumab. Magn Reson Med 2010; 63: 1366–75.

Gaitán MI, Shea CD, Evangelou IE, Stone RD, Fenton KM, Bielekova B, et al. Evolution of the blood-brain barrier in newly forming multiple sclerosis lesions. Ann Neurol 2011; 70: 22–29.

Gaitán MI, Sati P, Inati SJ, Reich DS. Initial investigation of the blood-brain barrier in MS lesions at 7 tesla. Mult Scler J 2013; 19: 1068–73.

Galbraith SM, Lodge MA, Taylor NJ, Rustin GJS, Bentzen SS, Stirling JJ, et al. Reproducibility of dynamic contrast-enhanced MRI in human muscle and tumours: comparison of quantitative and semi-quantitative analysis. NMR Biomed 2002; 15: 132–42.

Georgiou L, Sharma N, Broadbent DA, Wilson DJ, Dall BJ, Gangi A, Buckley DL. Estimating breast tumor blood flow during neoadjuvant chemotherapy using interleaved high temporal and high spatial resolution MRI. Magn Reson Med 2017; In Press.

Gibby WA. MRI Contrast Agents. In: Zimmerman RA, Gibby WA, Carmody RF, editors. Neuroimaging: Clinical and Physical Principles. New York, NY: Springer New York; 2000. p 313-364.

Gribbestad IS, Gjesdal KI, Nilsen G, Lundgren S, Hjelstuen MHB, Jackson A. An Introduction to Dynamic Contrast-Enhanced MRI in Oncology. In: Jackson A, Buckley DL, Parker GJM, editors. Dynamic Contrast-Enhanced Magnetic Resonance Imaging in Oncology. Berlin, Heidelberg: Springer Berlin Heidelberg; 2005. p 1–22.

Hansen AE, Pedersen H, Rostrup E, Larsson HBW. Partial volume effect (PVE) on the arterial input function (AIF) in T1-weighted perfusion imaging and limitations of the multiplicative rescaling approach. Magn Reson Med 2009; 62: 1055–9.

Haris M, Gupta RK, Singh A, Husain N, Husain M, Pandey CM, et al. Differentiation of infective from neoplastic brain lesions by dynamic contrast-enhanced MRI. Neuroradiology 2008a; 50: 531–40.

Haris M, Husain N, Singh A, Awasthi R, Rathore RKS, Husain M, et al. Dynamic contrast-enhanced (DCE) derived transfer coefficient (k trans) is a surrogate marker of matrix metalloproteinase 9 (MMP-9) expression in brain tuberculomas. J Magn Reson Imaging 2008b; 28: 588–97.

Hazlewood CF, Chang DC, Nichols BL, Woessner DE. Nuclear magnetic resonance transverse relaxation times of water protons in skeletal muscle. Biophys J 1974; 14: 583–606.

Henderson APD, Barnett MH, Parratt JDE, Prineas JW. Multiple sclerosis: distribution of inflammatory cells in newly forming lesions. Ann Neurol 2009; 66: 739–53.

Henderson E, Rutt BK, Lee TY. Temporal sampling requirements for the tracer kinetics modeling of breast disease. Magn Reson Imaging 1998; 16: 1057–73.

Herbst MD, Goldstein JH. A review of water diffusion measurement by NMR in human red blood cells. Am J Physiol 1989; 256: C1097–104.

Heye AK, Culling RD, Valdés Hernández MDC, Thrippleton MJ, Wardlaw JM. Assessment of blood-brain barrier disruption using dynamic contrast-enhanced MRI. A systematic review. NeuroImage Clin 2014; 6: 262–74.

Heye AK, Thrippleton MJ, Armitage PA, Valdés Hernández MdC, Makin SD, Glatz A, et al. Tracer kinetic modelling for DCE-MRI quantification of subtle blood-brain barrier permeability. NeuroImage 2016; 125: 446–55.

Heye T, Davenport MS, Horvath JJ, Feuerlein S, Breault SR, Bashir MR, et al. Reproducibility of dynamic contrast-enhanced MR Imaging. Part I. Perfusion characteristics in the female pelvis by using Multiple computer-aided diagnosis perfusion analysis solutions. Radiology 2013a; 266: 801–11.

Heye T, Merkle EM, Reiner CS, Davenport MS, Horvath JJ, Feuerlein S, et al. Reproducibility of dynamic contrast-enhanced MR imaging. Part II. Comparison of intra- and interobserver variability with manual region of interest placement versus semiautomatic lesion segmentation and histogram analysis. Radiology 2013b; 266: 812–21.

Horsfield MA, Thornton JS, Gill A, Jager HR, Priest AN, Morgan B. A functional form for injected MRI Gd-chelate contrast agent concentration incorporating recirculation, extravasation and excretion. Phys Med Biol 2009; 54: 2933–49.

Huang ZG, Xue D, Preston E, Karbalai H, Buchan AM. Biphasic opening of the blood-brain barrier following transient focal ischemia: effects of hypothermia. Can J Neurol Sci 1999; 26: 298–304.

Iannotti F, Fieschi C, Alfano B, Picozzi P, Mansi L, Pozzilli C, et al. Simplified, noninvasive PET measurement of blood-brain barrier permeability. J Comput Assist Tomogr. Vol 11, 1987: 390–7.

Inglese M, Adhya S, Johnson G, Babb JS, Miles L, Jaggi H, et al. Perfusion magnetic resonance imaging correlates of neuropsychological impairment in multiple sclerosis. J Cereb Blood Flow Metab 2008; 28: 164–71.

Ingrisch M, Sourbron S, Herberich S, Schneider MJ, Kümpfel T, Hohlfeld R, et al. Dynamic contrast-enhanced magnetic resonance imaging suggests normal perfusion in normal-appearing white matter in multiple sclerosis. Invest Radiol 2016; 52: 135–41.

Ingrisch M, Sourbron S, Morhard D, Ertl-Wagner B, Kümpfel T, Hohlfeld R, et al. Quantification of perfusion and permeability in multiple sclerosis. Invest Radiol 2012; 47: 252–8.

Jain R. Measurements of tumor vascular leakiness using DCE in brain tumors: clinical applications. NMR Biomed 2013; 26: 1042–9.

Jelescu IO, Leppert IR, Narayanan S, Araújo D, Arnold DL, Pike GB. Dual-temporal resolution dynamic contrast-enhanced MRI protocol for blood-brain barrier permeability measurement in enhancing multiple sclerosis lesions. J Magn Reson Imaging 2011; 33: 1291–300.

Jia Z, Geng D, Liu Y, Chen X, Zhang J. Microvascular permeability of brain astrocytoma with contrast-enhanced magnetic resonance imaging: correlation analysis with histopathologic grade. Chin Med J 2013; 126: 1953–6.

Kassner A, Roberts T, Taylor K, Silver F, Mikulis D. Prediction of hemorrhage in acute ischemic stroke using permeability MR imaging. AJNR Am J Neuroradiol 2005; 26: 2213–7.

Kassner A, Roberts TPL, Moran B, Silver FL, Mikulis DJ. Recombinant tissue plasminogen activator increases blood-brain barrier disruption in acute ischemic stroke: an MR imaging permeability study. AJNR Am J Neuroradiol 2009; 30: 1864–69.

Kermode AG, Tofts PS, Thompson AJ, MacManus DG, Rudge P, Kendall BE, et al. Heterogeneity of blood-brain barrier changes in multiple sclerosis: an MRI study with gadolinium-DTPA enhancement. Neurology 1990; 40: 229–35.

Kety SS. The theory and applications of the exchange of inert gas at the lungs and tissues. Pharmacol Rev 1951; 3: 1–41.

Kleppestø M, Larsson C, Groote I, Salo R, Vardal J, Courivaud F, et al. T2*-correction in dynamic contrast-enhanced MRI from double-echo acquisitions. J Magn Reson Imaging 2014; 39: 1314–9.

Larsson HB, Tofts PS. Measurement of blood-brain barrier permeability using dynamic Gd-DTPA scanning-a comparison of methods. Magn Reson Med 1992; 24: 174–6.

Larsson HBW, Courivaud F, Rostrup E, Hansen AE. Measurement of brain perfusion, blood volume, and blood-brain barrier permeability, using dynamic contrast-enhanced T1-weighted MRI at 3 tesla. Magn Reson Med 2009; 62: 1270–81.

Larsson HBW, Hansen AE, Berg HK, Rostrup E, Haraldseth O. Dynamic contrast-enhanced quantitative perfusion measurement of the brain using T1-weighted MRI at 3T. J Magn Reson Imaging 2008; 27: 754–62.

Larsson HBW, Rosenbaum S, Fritz-Hansen T. Quantification of the effect of water exchange in dynamic contrast MRI perfusion measurements in the brain and heart. Magn Reson Med 2001; 46: 272–81.

Larsson HBW, Stubgaard M, Frederiksen JL, Jensen M, Henriksen O, Paulson OB. Quantitation of blood-brain barrier defect by magnetic resonance imaging and gadolinium-DTPA in patients with multiple sclerosis and brain tumors. Magn Reson Med 1990; 16: 117–31.

Leach MO, Morgan B, Tofts PS, Buckley DL, Huang W, Horsfield Ma, et al. Imaging vascular function for early stage clinical trials using dynamic contrast-enhanced magnetic resonance imaging. Eur Radiol 2012; 22: 1451–64.

Li KL, Buonaccorsi G, Thompson G, Cain JR, Watkins A, Russell D, et al. An improved coverage and spatial resolutiona-using dual injection dynamic contrast-enhanced (ICE-DICE) MRI: a novel dynamic contrast-enhanced technique for cerebral tumors. Magn Reson Med 2012; 68: 452–62.

Li KL, Zhu XP, Waterton J, Jackson A. Improved 3D quantitative mapping of blood volume and endothelial permeability in brain tumors. J Magn Reson Imaging 2000; 12: 347–57.

Lopata RGP, Backes WH, van den Bosch PPJ, van Riel NAW. On the identifiability of pharmacokinetic parameters in dynamic contrast-enhanced imaging. Magn Reson Med 2007; 58: 425–9.

McGrath DM, Bradley DP, Tessier JL, Lacey T, Taylor CJ, Parker GJM. Comparison of model-based arterial input functions for dynamic contrast-enhanced MRI in tumor bearing rats. Magn Reson Med 2009; 61: 1173–84.

Merali Z, Huang K, Mikulis D, Silver F, Kassner A. Evolution of blood-brain-barrier permeability after acute ischemic stroke. PLoS One 2017; 12: e0171558.

Mills SJ, Soh C, O'Connor JPB, Rose CJ, Buonaccorsi GA, Cheung S, et al. Tumour enhancing fraction (EnF) in glioma: relationship to tumour grade. Eur Radiol 2009; 19: 1489–98.

Nadav G, Liberman G, Artzi M, Kiryati N, Bashat DB. Optimization of two-compartment-exchange-model analysis for dynamic contrast-enhanced mri incorporating bolus arrival time. J Magn Reson Imaging 2017; 45: 237–49.

Obermeier B, Daneman R, Ransohoff RM. Development, maintenance and disruption of the blood-brain barrier. Nat Med 2013; 19: 1584–96.

Orton MR, Miyazaki K, Koh D-M, Collins DJ, Hawkes DJ, Atkinson D, et al. Optimizing functional parameter accuracy for breath-hold DCE-MRI of liver tumours. Phys Med Biol 2009; 54: 2197–215.

Padhani AR. Dynamic contrast-enhanced MRI in clinical oncology: current status and future directions. J Magn Reson Imaging 2002; 16: 407–22.

Padhani AR, Hayes C, Landau S, Leach MO. Reproducibility of quantitative dynamic MRI of normal human tissues. NMR Biomed 2002; 15: 143–53.

Parker GJM, Buckley DL. Tracer Kinetic Modelling for T1-Weighted DCE-MRI. In: Jackson A, Buckley DL and Parker GJM, editors. *Dynamic Contrast-Enhanced MRi in oncology*. Berlin: Springer, 2005: 81–92.

Parker GJM, Roberts C, Macdonald A, Buonaccorsi GA, Cheung S, Buckley DL, et al. Experimentally-derived functional form for a population-averaged high-temporal-resolution arterial input function for dynamic contrast-enhanced MRI. Magn Reson Med 2006; 56: 993–1000.

Patlak CS, Blasberg RG, Fenstermacher JD. Graphical evaluation of Blood-to-Brain transfer constants from multiple-Time uptake data. J Cereb Blood Flow Metab 1983; 3: 1–7.

Port RE, Knopp MV, Brix G. Dynamic contrast-enhanced MRI using Gd-DTPA: interindividual variability of the arterial input function and consequences for the assessment of kinetics in tumors. Magn Reson Med 2001; 45: 1030–8.

Quarles CC, Gore JC, Xu L, Yankeelov TE. Comparison of dual-echo DSC-MRI- and DCE-MRI-derived contrast agent kinetic parameters. Magn Reson Imaging 2012; 30: 944–53.

Richardson OC, Bane O, Scott MLJ, Tanner SF, Waterton JC, Sourbron SP, et al. Gadofosveset-based biomarker of tissue albumin concentration: technical validation in vitro and feasibility in vivo. Magn Reson Med 2015; 73: 244–53.

Roberts C, Little R, Watson Y, Zhao S, Buckley DL, Parker GJM. The effect of blood inflow and B(1)-field inhomogeneity on measurement of the arterial input function in axial 3D spoiled gradient echo dynamic contrast-enhanced MRI. Magn Reson Med 2011; 65: 108–19.

Rosenberg GA, Estrada EY, Dencoff JE. Matrix metalloproteinases and TIMPs are associated with blood-brain barrier opening after reperfusion in rat brain. Stroke 1998; 29: 2189–95.

Schwarzbauer C, Morrissey SP, Deichmann R, Hillenbrand C, Syha J, Adolf H, et al. Quantitative magnetic resonance imaging of capillary water permeability and regional blood volume with an intravascular {MR} contrast agent. Magn Reson Med 1997; 37: 769–77.

Singh A, Haris M, Rathore D, Purwar A, Sarma M, Bayu G, et al. Quantification of physiological and hemodynamic indices using T1 dynamic contrast-enhanced MRI in intracranial mass lesions. J Magn Reson Imaging 2007; 26: 871–80.

Sourbron S. Technical aspects of MR perfusion. Eur J Radiol 2010; 76: 304–13.

Sourbron S, Ingrisch M, Siefert A, Reiser M, Herrmann K. Quantification of cerebral blood flow, cerebral blood volume, and blood-brain-barrier leakage with DCE-MRI. Magn Reson Med 2009; 62: 205–17.

Sourbron SP, Buckley DL. Tracer kinetic modelling in MRI: estimating perfusion and capillary permeability. Phys Med Biol 2012; 57: R1–R33.

Sourbron SP, Buckley DL. Classic models for dynamic contrast-enhanced MRI. NMR Biomed 2013; 26: 1004–27.

Taheri S, Gasparovic C, Shah NJ, Rosenberg GA. Quantitative measurement of blood-brain barrier permeability in human using dynamic contrast-enhanced MRI with fast T1 mapping. Magn Reson Med 2011; 65: 1036–42.

Thornhill RE, Chen S, Rammo W, Mikulis DJ, Kassner A. Contrast-enhanced MR imaging in acute ischemic stroke: T2* measures of blood-brain barrier permeability and their relationship to T1 estimates and hemorrhagic transformation. AJNR Am J Neuroradiol 2010; 31: 1015–22.

Tofts PS. Optimal detection of blood-brain barrier defects with Gd-DTPA MRI—The influences of delayed imaging and optimised repetition time. Magn Reson Imaging 1996; 14: 373–80.

Tofts PS. Modeling tracer kinetics in dynamic Gd-DTPA MR imaging. J Magn Reson Imaging 1997; 7: 91–101.

Tofts PS, Berkowitz BA. Measurement of capillary permeability from the Gd enhancement curve: a comparison of bolus and constant infusion injection methods. Magn Reson Imaging 1994; 12: 81–91.

Tofts PS, Brix G, Buckley DL, Evelhoch JL, Henderson E, Knopp MV, et al. Estimating kinetic parameters from dynamic contrast-enhanced T(1)-weighted MRI of a diffusable tracer: standardized quantities and symbols. J Magn Reson Imaging 1999; 10: 223–32.

Tofts PS, Cutajar M, Mendichovszky IA, Peters AM, Gordon I. Precise measurement of renal filtration and vascular parameters using a two-compartment model for dynamic contrast-enhanced MRI of the kidney gives realistic normal values. Eur Radiol 2012; 22: 1320–30.

Tofts PS, Kermode AG. Measurement of the blood-brain barrier permeability and leakage space using dynamic MR imaging. 1. Fundamental concepts. Magn Reson Med 1991; 17: 357–67.

Vidarsson L, Thornhill RE, Liu F, Mikulis DJ, Kassner A. Quantitative permeability magnetic resonance imaging in acute ischemic stroke: how long do we need to scan? Magn Reson Imaging 2009; 27: 1216–22.

Walker-Samuel S, Leach M, Collins D. Evaluation of response to treatment using DCE-MRI: the relationship between initial area under the gadolinium curve (IAUGC) and quantitative pharmacokinetic analysis. Phys Med Biol 2006; 51: 3593–602.

Wardlaw JM, Doubal F, Armitage P, Chappell F, Carpenter T, Muñoz Maniega S, et al. Lacunar stroke is associated with diffuse Blood-Brain barrier dysfunction. Ann Neurol 2009; 65: 194–202.

Weinmann HJ, Laniado M, Mützel W. Pharmacokinetics of GdDTPA/dimeglumine after intravenous injection into healthy volunteers. Physiol Chem Phys Med NMR 1984; 16: 167–72.

Wong SM, Jansen JFA, Zhang CE, Staals J, Hofman PAM, van Oostenbrugge RJ, et al. Measuring Subtle leakage of the blood-Brain barrier in cerebrovascular disease with DCE-MRI: test-retest reproducibility and its influencing factors. J Magn Reson Imaging 2017; 46(1):159–166.

Yankeelov TE, Cron GO, Addison CL, Wallace JC, Wilkins RC, Pappas BA, et al. Comparison of a reference region model with direct measurement of an AIF in the analysis of DCE-MRI data. Magn Reson Med 2007; 57: 353–61.

Yankeelov TE, Luci JJ, Lepage M, Li R, Debusk L, Lin PC, et al. Quantitative pharmacokinetic analysis of DCE-MRI data without an arterial input function: a reference region model. Magn Reson Imaging 2005; 23: 519–29.

Yankeelov TE, Rooney WD, Li X, Springer CS. Variation of the Relaxographic "Shutter-Speed" for transcytolemmal water exchange affects the CR Bolus-Tracking curve shape. Magn Reson Med 2003; 50: 1151–69.

Zhang N, Zhang L, Qiu B, Meng L, Wang X, Hou BL. Correlation of volume transfer coefficient Ktrans with histopathologic grades of gliomas. J Magn Reson Imaging 2012; 36: 355–63.

15

Functional and Metabolic MRI[1]

Claudine J. Gauthier
Concordia University

Audrey P. Fan
Stanford University

Contents

15.1 Blood oxygenation level dependent (BOLD) contrast and physiology

The BOLD contrast phenomenon arises from the fact that oxygenated and deoxygenated haemoglobin have different magnetic properties: oxyhaemoglobin is diamagnetic while deoxyhaemoglobin (dHb) is paramagnetic. Due to the paramagnetic nature of dHb, water spins in blood that is not fully oxygenated experience increased dephasing, which attenuates the T_2*-weighted signal from venous blood and tissue containing dHb (Ogawa *et al.*, 1990a, 1990b). During performance of a functional task, active brain regions consume additional oxygen to function, causing a local increase in dHb concentration. Concomitantly, nearby blood vessels experience vasodilation and an increase in local blood flow. Because the inflowing blood is fully oxygenated, the increased blood flow effectively dilutes the local dHb concentration and creates an increase in the BOLD signal.

The BOLD signal therefore reflects a combination of changes in oxidative metabolism, blood flow and blood volume in response to neural activity. While this haemodynamic response is expected to be tightly regulated in healthy brains, changes in neurovascular coupling in ageing and disease cause problems for the direct comparison of BOLD signals across groups (Ances *et al.*, 2008, 2009; Gauthier *et al.*, 2013; Hutchison *et al.*, 2013; Liu *et al.*, 2013; Peng *et al.*, 2014). This difficulty stems from the physiological ambiguity of the BOLD signal and the fact that BOLD measures a relative signal change from an unknown baseline. Therefore, any change in coupling or in the baseline cerebral blood flow (CBF), cerebral blood volume (CBV) or oxygen usage is expected to result in a different BOLD signal change. Accurate interpretation of neural activity in task-based functional MRI (fMRI) experiments and resting state connectivity approaches thus requires consideration of contributions from vascular components and baseline oxidative metabolism on the BOLD signal.

One of the advantages of MRI, however, is its versatility. While the BOLD signal is sensitive and valid in many contexts, alternative quantitative techniques can be used in cases where BOLD comparisons are problematic. In fact, every physiological subcomponent of the BOLD signal can be measured using MRI. This chapter describes the main techniques and recent advances to measure these subcomponents.

15.2 Cerebral blood flow

The most established quantitative, functional MRI alternative to BOLD is measurement of cerebral blood flow (CBF) using a class of methods called *arterial spin labelling* (ASL) (Detre *et al.*, 2012). In most ASL techniques, inflowing blood is magnetically tagged at the level of the carotid arteries with radiofrequency (RF) inversion. Once the spins are inverted, for the 'tag' image, there is a prescribed post-label delay (PLD) time to allow the tagged inflowing blood to reach the imaging volume. An image of the brain is acquired, in which the tagged (inverted) blood has a lower signal intensity than the blood that was already in the imaging volume (the static signal). For the subsequent

[1] Edited by Mara Cercignani; reviewed by Richard Wise, Brain Research Imaging Centre, School of Psychology, Cardiff University, Cardiff, UK

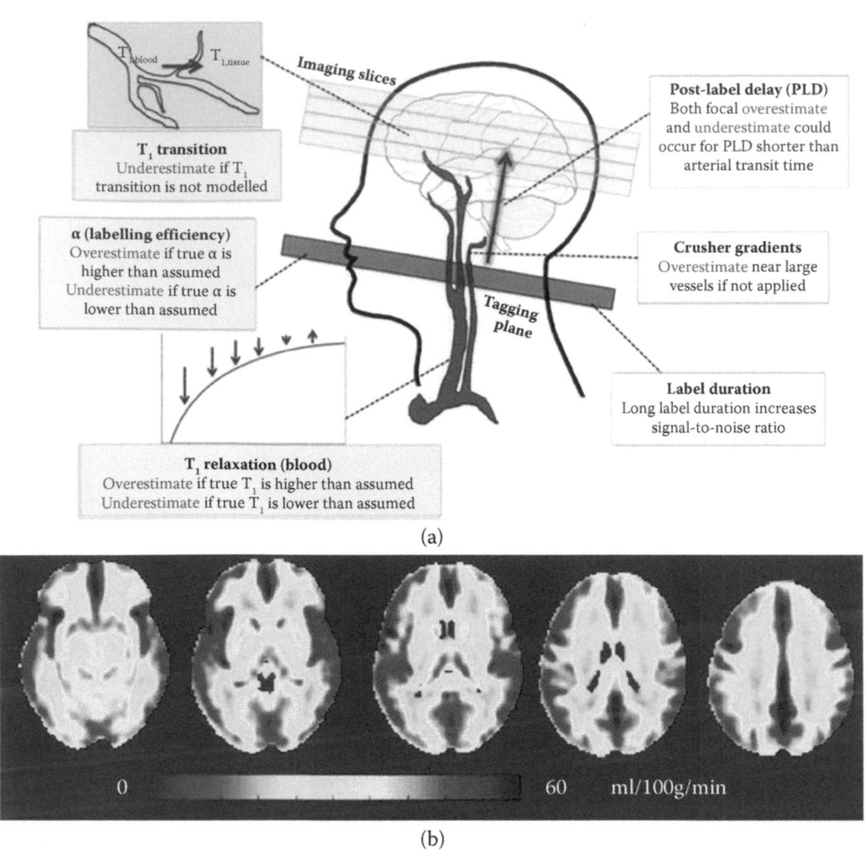

FIGURE 15.1 (a) Schematic of arterial spin labelling (ASL) technique for cerebral blood flow (CBF) measurements and considerations for accurate quantification. (Reproduced with permission under the STM Guidelines from Fan, A.P., *et al., J. Cereb. Blood Flow Metab.*, 36, 842–861, 2016a.) (b) Mean CBF measured by ASL with consensus parameters including post-label delay of 2025 sec, averaged in 20 healthy young volunteers.

('control') image, the procedure is repeated without the tag, so that all the blood has the same signal intensity. By subtracting the control and tag images, we can isolate the flow-dependent component of the signal from inflowing spins tagged at the level of the neck (Detre *et al.*, 2012; Williams *et al.*, 1992). Rapid acquisition of a separate image with fully relaxed longitudinal magnetisation allows calibration of perfusion images into CBF with units of ml/100 g/min (Figure 15.1).

The two main families of ASL sequences in current use are pulsed and continuous techniques (Edelman *et al.*, 1994; Williams *et al.*, 1992; Wong *et al.*, 1998). In pulsed ASL sequences (PASL), a large slab of blood is inverted by a single RF pulse at the level of the neck (Edelman *et al.*, 1994; Wong *et al.*, 1998). Continuous sequences label a thinner slab for a longer period of time (Williams *et al.*, 1992). True continuous sequences are technically challenging, which has led to recent development of pseudo-continuous ASL (pCASL) techniques that use a train of short RF pulses to achieve near-continuous labelling. The pCASL labelling provides superior signal-to-noise ratio (SNR) and implicit control over label duration and is compatible with the RF hardware of standard machines (Dai *et al.*, 2008; Detre *et al.*, 2012; Wu *et al.*, 2007). Efforts to standardise techniques across the field have yielded a consensus that pCASL implementations are preferred due to SNR advantages over PASL and suggested a set of parameters that are readily implementable for clinical applications (Alsop *et al.*, 2015).

In many clinical situations, CBF quantification with ASL is challenging and requires advanced methodology. Patients with cerebrovascular disorders such as steno-occlusive disease or moyamoya disease often have abnormal collateral pathways with long arterial transit times. If the long transit delays are not accounted for, the PLD of the ASL scan may be too short to capture the labelled arterial blood, resulting in CBF underestimation (if the tagged spins have not arrived at the time of imaging) or overestimation (if spins remain in large vessels and create hyperintense signal). To address these challenges, multidelay ASL acquires perfusion signals at several PLD times (Dai *et al.*, 2012). Multidelay approaches allow direct mapping of arterial transit time to correct CBF values for elongated arterial transit times but comes at the cost of longer acquisition time or reduced spatial resolution. Velocity-selective ASL, which tags spins based on velocity encoding and its flow properties instead of its location, has also shown robust CBF measurements in challenging, slow flow conditions (Wong *et al.*, 2006). Advanced ASL techniques and considerations for clinical applications were reviewed by Detre *et al.* (2009) and Petersen *et al.* (2006).

TABLE 15.1 MRI Techniques for quantitative cerebral physiology

Physiology	Method	Normal ranges	Advantages/Disadvantages
Cerebral blood flow (CBF)	Single-delay arterial spin labelling	40–100 ml/100 g/min in grey matter; 10–15 ml/100 g/min in white matter	Quantitative and has established parameters for clinical use but prone to error in pathologies with long arterial transit times
	Multidelay arterial spin labelling	40–100 ml/100 g/min in grey matter; 10–15 ml/100 g/min in white matter	Maps arterial transit time and corrects CBF for prolonged delays but requires longer scan time or lower spatial resolution
Cerebrovascular reactivity (CVR)	BOLD CVR	0.2%–0.3% ΔBOLD/mmHg $\Delta ETCO_2$	Sensitive to haemodynamic changes due to a gas or drug challenge but represents a combination of blood volume, oxygenation, flow and metabolic effects
	ASL CVR	3%–7% ΔCBF/mmHg $\Delta ETCO_2$	Specific to cerebral blood flow; lower SNR but similar contrast to noise
Cerebral blood volume (CBV)	VASO	20%–30% change 5–7 ml blood/100 g parenchyma baseline	Functional imaging tool that is sensitive to total blood volume changes but requires contrast agent for baseline CBV measurement and is confounded by CBF, BOLD, CSF and inflow effects
	iVASO	0.7–2 ml blood/100 mg parenchyma (arterial CBV)	Quantitative assessment of arterial CBV but has low SNR and is sensitive to physiological confounders and whether crushers are in the sequence
	VERVE	16% change	Specific to venous blood volume, which is more related to BOLD signal, but has only been used to study changes in blood volume
Cerebral metabolic rate of oxygen ($CMRO_2$): evoked change	Hypercapnia method	13%–25% $CMRO_2$ increase (motor and visual tasks)	Quantitative change in $CMRO_2$ but requires gas calibration and assumption that metabolism stays constant during hypercapnia
	Hyperoxia method	5%–20% $CMRO_2$ increase	Quantitative change in $CMRO_2$ and hyperoxia is better tolerated by patient but underestimates M, requires gas calibration and assumption of baseline oxygen extraction fraction
	Generalised calibration model	~30% $CMRO_2$ increase	Quantitative change in $CMRO_2$ with arbitrary combination of hypercapnia and hyperoxia but requires gas calibration
$CMRO_2$: baseline	Respiratory calibration	145–185 μmol O_2/100 g/min	Baseline $CMRO_2$ mapping; requires multiple gas calibrations and is sensitive to low-SNR ASL measurements
	T_2' and quantitative BOLD (extravascular techniques)	160 μmol O_2/100 g/min	Baseline $CMRO_2$ mapping without contrast or gas challenge, but quantitative BOLD model is complex and difficult to separate oxygenation from CBV and field inhomogeneity effects
	Intravascular T_2	125–130 μmol O_2/100 g/min	Baseline $CMRO_2$ mapping but mostly restricted to larger draining vessels and has low SNR in tissue voxels
	Magnetic susceptibility	153–160 μmol O_2/100 g/min	Baseline $CMRO_2$ mapping, but high spatial resolution necessary for smaller vessels and sensitive to residual orientation-dependent effects from susceptibility mapping

Note: BOLD = Blood oxygenation level dependent; SNR = signal to noise ratio; $ETCO_2$ = end-tidal CO_2 concentration; ASL = arterial spin labelling; VASO = vascular space occupancy; iVASO = inflow VASO; CSF = cerebrospinal fluid; VERVE = venous refocussing for volume estimation.

ASL measurements have the advantage over BOLD of yielding an absolute and unambiguous measure of a physiological quantity. However, widespread adoption of ASL as a functional contrast mechanism has been slowed because ASL is an intrinsically low-SNR technique. In all of its implementations, ASL relies on image subtractions with small effect sizes (on the order of a fraction of a percent) to provide functional contrast. This low-SNR regime is prone to additive errors in measurements or requires multiple averages, which comes at the cost of longer acquisition time or reduced spatial resolution of CBF maps. Background suppression methods may enhance SNR by decreasing the signal intensity of static tissue (through spin inversion) without a proportional penalty to the ASL difference signal (Garcia *et al.*, 2005).

15.3 Cerebral blood volume

Several MRI techniques exist to measure CBV changes. While some methods require the injection of a contrast agent such as monocrystalline iron oxide nanoparticle, other non-invasive techniques use endogenous contrast mechanisms. The most widely used of these techniques is vascular space occupancy (VASO), which yields a measure of total blood volume (Lu *et al.*, 2003). The VASO technique uses a non-specific inversion recovery sequence to null the blood signal and subtract it from non-blood signal. During brain activation, blood volume increases, creating a functional increase in VASO signal as more spins move into the nulled blood compartment. VASO can thus be used as a physiologically specific, quantitative marker of CBV changes in functional activation (Donahue *et al.*, 2006; Lu and van Zijl, 2012; Lu *et al.*, 2003).

Recent work indicates that VASO may be particularly relevant for laminar fMRI (Guidi *et al.*, 2016; Huber *et al.*, 2015, 2016). Laminar fMRI with BOLD is intrinsically limited because the dominant BOLD signal changes occur in large pial veins near the cortical surface. VASO, on the other hand, has been shown to yield greater spatial specificity to neural activity, especially at higher field strengths, e.g. 7T (Huber *et al.*, 2015, 2016). At higher fields, however, standard VASO yields reduced contrast due to the similarity between the T_1 of blood and tissue, field and excitation inhomogeneities, and larger inflow effects due to longer T_1 times (Huber *et al.*, 2014). A modification of the original technique, called slice-saturation slab-inversion vascular space occupancy (SS-SI-VASO), has been developed to address some of these high-field limitations (Huber *et al.*, 2014).

Standard VASO techniques, while specific to CBV and more quantitative than BOLD, still represent a relative change from an unknown baseline. An extension of VASO, called *inflow VASO* (iVASO) was developed to measure baseline arterial blood volume (Donahue *et al.*, 2010; Hua *et al.*, 2011). This measurement is done by inverting the blood signal outside the imaging slice or slab and acquiring the image at the blood nulling time of the inflowing blood. So far, iVASO remains a nascent technique used in a handful of application studies in populations with schizophrenia and cancer (Hua *et al.*, 2016; Talati *et al.*, 2016; Wu *et al.*, 2016).

To more specifically target the CBV changes that contribute to BOLD, a sequence called *venous refocussing for volume estimation* (VERVE) can be used to measure venous blood volume changes (Chen and Pike, 2009, 2010; Stefanovic and Pike, 2005). This sequence uses the dependence of venous blood on T_2 and on the refocussing interval in a Carr–Purcell–Meiboom–Gill (CPMG) sequence to isolate CBV changes specific to the venous compartment. VERVE has been used to better characterise the power law relationship thought to exist between flow and (venous) blood volume (Grubb *et al.*, 1974), for task-based and hypercapnia-based flow changes (Chen and Pike, 2009, 2010).

15.4 Cerebrovascular reactivity

Neurovascular coupling depends on the compliance of blood vessels and their responsiveness to vasodilatory stimuli (Girouard and Iadecola, 2006). Cerebrovascular reactivity is the vasodilatory response, typically measured using a haemodynamic imaging method, to a vascular challenge such as CO_2 inhalation. Cerebrovascular reactivity is thought to be affected in ageing (Gauthier *et al.*, 2013) and in a variety of diseases, including carotid occlusion (De Vis *et al.*, 2015a), cerebrovascular disease (Mandell *et al.*, 2008; Rodan *et al.*, 2015), coronary disease (Anazodo *et al.*, 2015), moyamoya disease (Donahue *et al.*, 2013) and dementia (Cantin *et al.*, 2011).

Cerebrovascular reactivity (CVR) is typically measured as the BOLD signal change in response to controlled mild hypercapnia (mild increases in inhaled CO_2 concentrations) (Anazodo *et al.*, 2015; Ances *et al.*, 2009; De Vis *et al.*, 2015b; Donahue *et al.*, 2012; Driver *et al.*, 2010; Gauthier *et al.*, 2013; Liu *et al.*, 2013; Smeeing *et al.*, 2016; Tancredi and Hoge, 2013). CO_2 is a potent vasodilator and inspiration of even moderate amounts of CO_2 result in global changes in CBF that alter the BOLD signal. In CVR challenges, hypercapnia is administered typically either by using a fixed inspired CO_2 concentration (Anazodo *et al.*, 2015; Ances *et al.*, 2009; Hare *et al.*, 2013; Liu *et al.*, 2013; Warnert *et al.*, 2014), most often around 5% CO_2, or using a device which controls end-tidal concentrations using a feed-forward (Bhogal *et al.*, 2014; De Vis *et al.*, 2015a, 2015b; Driver *et al.*, 2010; Gauthier *et al.*, 2013) or a feedback (Wise *et al.*, 2007) approach. It is also possible to induce hypercapnia with breath holds, which are easier to implement but more challenging to control in patients (Murphy *et al.*, 2011; Tancredi and Hoge, 2013). Alternatively, intravenous or oral administration of acetazolamide can be used instead of respiratory interventions to induce vasodilation (Bokkers *et al.*, 2011; Smeeing *et al.*, 2016). Recently, CVR measurements based on natural fluctuations in CO_2 in resting-state BOLD were proposed by Liu *et al.* (2016c), which would remove the need for an external challenge.

While BOLD is sensitive and relatively easy to implement, ASL enables a more quantitative and physiologically specific approach to measure CVR. However, global flow changes such as those caused by blood gas modulations provide unique challenges to ASL techniques. During vasodilation, blood velocity increases, which can impact the transit time to the imaging voxel relative to the prescribed labelling scheme (Donahue *et al.*, 2012). Tancredi *et al.* have shown that pCASL sequences may provide improved accuracy and sensitivity to CBF changes as compared to PASL (Tancredi *et al.*, 2012). This is likely due to the fact that blood flow velocity increases in the carotid arteries results in problematic tag compression effects in PASL sequences. While it is possible that this increased velocity could reduce labelling efficiency in pCASL acquisitions, pCASL has been shown to yield comparable CVR results to PET (Heijtel *et al.*, 2014).

15.5 Oxygen metabolism in the brain

15.5.1 Calibrated fMRI

Calibrated fMRI refers to a family of techniques to estimate oxidative metabolism in the brain using a BOLD signal model and respiratory calibration. The evolution of calibrated fMRI has followed two waves of development. Originally, techniques were developed to estimate the change in the cerebral rate of oxygen metabolism ($CMRO_2$) evoked by a task. More recently, new techniques have been proposed to measure oxidative metabolism at baseline. Task-evoked measurements provide a quantitative fractional estimate of functional $CMRO_2$ changes, which can be compared across groups such as older compared to younger adults (Ances *et al.*, 2009; Gauthier *et al.*, 2013; Mohtasib *et al.*, 2012) or in certain diseases such as dementia and HIV (Ances *et al.*, 2011). Baseline oxidative metabolism mapping also has a wide range of clinical applications, from studying the stroke penumbra where oxygen extraction fraction (OEF) is expected to be elevated, to carotid occlusion, ageing, dementia and other cerebrovascular diseases (Aanerud *et al.*, 2012; De Vis *et al.*, 2015a, 2015b; Ishii *et al.*, 1996a; Lu *et al.*, 2011; Yamauchi *et al.*, 2009).

The calibrated fMRI framework was first proposed by Davis *et al.* (1998) and the BOLD signal model was described in greater depth by Hoge *et al.* (1999). In this technique, a hypercapnia manipulation is used to evoke a purely vascular BOLD signal change. One of the main underlying assumptions in this technique is that breathing low concentrations of CO_2 around 5% evokes vasodilation and increases CBF without affecting oxidative metabolism. All calibrated fMRI techniques using hypercapnia assume isometabolism throughout the breathing task, although this assumption is still debated (Yablonskiy, 2011). During the hypercapnia challenge, a dual echo sequence is often used to simultaneously measures CBF with ASL (at a short echo) and BOLD at a longer echo (~30 ms at 3T) that optimises the BOLD signal. The CBF and BOLD measurements are used within the biophysical model (Davis *et al.*, 1998; Hoge *et al.*, 1999) to estimate a calibration parameter called *M*, which represents the maximum possible BOLD signal change as a percentage. The M parameter is the BOLD signal that would occur if all the dHb present in a voxel at baseline were completely diluted in the volunteer, yielding fully oxygen-saturated venous blood. The Davis model relates these quantities as follows:

$$\frac{CMRO_2}{CMRO_2|_0} = \left(1 - \frac{\Delta BOLD / BOLD_0}{M}\right)^{1/\beta} \cdot \left(\frac{CBF}{CBF_0}\right)^{1-\alpha/\beta}$$

Three parameters are assumed in this model. The α parameter represents the coupling exponent between CBF and CBV, as CBV changes during neuronal activity are strongly correlated with changes in CBF. From initial hypercapnia experiments in anesthetised primate models, CBV and CBF were found to have a power law relationship (Grubb *et al.*, 1974). Recent studies have remeasured the exponent relating CBV and CBF, e.g. with VERVE, to focus on venous instead of total blood volume (Chen and Pike, 2010). The β parameter represents the field strength-dependent influence of dHb on transverse relaxation (Boxerman *et al.*, 1995) and typically is assumed to be 1.3–1.5 at 3T (Bulte *et al.*, 2009; Gauthier *et al.*, 2013; Hare *et al.*, 2015). Finally, the baseline O_2 saturation of arterial blood is typically assumed to be 1. Advanced modelling approaches now treat α and β as heuristic constants without a specific physiological interpretations (Gagnon *et al.*, 2016; Griffeth and Buxton, 2011).

Given the potential discomfort to patients of breathing CO_2, several groups have explored use of hyperoxia rather than hypercapnia for calibration. The hyperoxia method (Chiarelli *et al.*, 2007) uses a model similar to the original hypercapnia model and also assumes that the breathing manipulation does not affect the rate of oxidative metabolism. Because increased inspired O_2 concentration does not have a large effect on CBF, the CBF change is usually assumed to be zero in this technique. Most groups use a 50%–60% O_2 concentration to avoid significant vasoconstriction (Chiarelli *et al.*, 2007; Mark *et al.*, 2011) and reduced CBF that may occur in fixed-inspired hyperoxia manipulations due to concomitant hypocapnia (Fan *et al.*, 2016b; Gauthier and Hoge, 2012a). In the hyperoxia model, M is calculated using measured end-tidal O_2 concentrations, the BOLD signal and assumed constants including the baseline OEF, which is used to calculate the venous dHb concentration. These assumptions represent an important weakness of the method and recent work has focused on reducing reliance on these assumed parameters. Once the M value has been calculated from the hyperoxia calibration, the evoked $CMRO_2$ change can also be determined from the Davis model (Davis *et al.*, 1998). The hyperoxia technique has been used to study cognition (Goodwin *et al.*, 2009) and cognitive changes with ageing (Mohtasib *et al.*, 2012). The hyperoxia method was also shown to have reduced measurement variance compared to the hypercapnia technique (Gauthier and Hoge, 2012a; Mark *et al.*, 2011). The lower variability likely comes because flow changes are assumed rather than measured, but this assumption also leads to systematic underestimation of the M parameter that should be addressed (Gauthier and Hoge, 2012a).

Gauthier and Hoge subsequently proposed the generalised calibration model (GCM) (Gauthier and Hoge, 2012a). The GCM model extends the hyperoxia calibration model, to account for the impact of changes in blood flow in relating the arterial and venous concentrations of O_2. This modification allows the GCM model to be used for hypercapnia, hyperoxia and any arbitrary combination of these gases. Because both flow and end-tidal values are used to calculate M, the GCM results have the advantage of being more stable (Gauthier and Hoge, 2012a) and more similar to directly measured M values (Gauthier *et al.*, 2011) than previous models.

15.5.2 Baseline $CMRO_2$ mapping with respiratory calibration and BOLD

Baseline metabolism is thought to be affected in a variety of diseases, but there is no gold standard to non-invasively measure $CMRO_2$ with MRI. The main established technique is triple ^{15}O-PET, which uses ^{15}O-labelled H_2O, O_2 and CO to measure OEF, CBF and $CMRO_2$ (Ishii *et al.*, 1996a, 1996b; Ito *et al.*, 2004; Nagata *et al.*, 2000; Yamaguchi *et al.*, 1986). PET studies using this technique have shown metabolic changes in stroke (Yamaguchi *et al.*, 1986), dementia (Nagata *et al.*, 2000) and Alzheimer's disease (Ishii *et al.*, 1996a). Unfortunately, due to the short half-life (2 min) of ^{15}O-radiotracers, the use of ^{15}O-PET requires specialised equipment, complex setup and invasive blood sampling for quantification and is challenging in clinical settings. In recent years, advanced MRI techniques based on calibrated fMRI models and other physical contrasts have been developed to quantify baseline oxygen metabolism.

In the hyperoxia and GCM biophysical models for BOLD, the value of the resting OEF is explicitly assumed. Instead of assuming baseline OEF, several groups have modified the BOLD model to measure OEF, rather than assume it, through use of multiple respiratory calibrations. This allows calibrated fMRI to be used not only for mapping task-evoked metabolism but also baseline metabolism (Bulte *et al.*, 2012; Gauthier and Hoge, 2012b; Wise *et al.*, 2013). Two methods were simultaneously proposed to measure OEF using calibrated fMRI models with slightly different underlying assumptions.

The dual-calibration method was proposed by Bulte *et al.* (2012). For this technique, two breathing manipulations (hypercapnia and hyperoxia) are required with simultaneous measurement of BOLD, ASL and end-tidal O_2 values. The BOLD and ASL signals during the hypercapnia manipulation are first used to measure M using the Davis model (Davis *et al.*, 1998). This M value is then combined with the BOLD and end-tidal O_2 measurements from a hyperoxia manipulation to estimate baseline OEF, instead of assuming OEF as in the original hyperoxia model. OEF can then be combined with resting CBF from a baseline ASL measurement to estimate resting $CMRO_2$ through the Fick principle (Kety and Schmidt, 1948). This approach relies on two sets of assumptions (from the hypercapnia and hyperoxia models) to measure M and OEF and has been used to study ageing and carotid occlusion (De Vis *et al.*, 2015a, 2015b). The technical considerations of the dual-calibration approach and their impact on estimates of metabolism have been explored by Blockley *et al.* (2015).

Gauthier and Hoge proposed an alternate approach to map baseline $CMRO_2$ based on the GCM, which measures OEF and M simultaneously by using two or more breathing manipulations (Gauthier and Hoge, 2012b). The GCM can be expressed as the relationship between measured quantities (BOLD, ASL and end-tidal O_2), with both M and OEF, yielding an equation with two unknowns. This equation can be constructed for any breathing manipulation with changes in CO_2, O_2 or both (Figure 15.2). Therefore, by performing two or more breathing challenges, both OEF and M can be simultaneously estimated. Because the measurements rely on a non-linear combination of low-SNR measurements such as ASL, three gases are sometimes used to increase the reliability of estimates (Fan *et al.*, 2016b; Gauthier *et al.*, 2012; Gauthier and Hoge, 2012b). The reproducibility and reliability of this technique has recently been explored by Lajoie *et al.* (2016) and Fan *et al.* (2016b).

A similar generalised, calibrated fMRI approach was subsequently proposed by Wise *et al.* (2013) to simultaneously measure M, OEF and $CMRO_2$ using more complex breathing manipulations. These breathing manipulations combine several levels of O_2 and CO_2 inhalations to provide a more complete picture of the relationship between BOLD, CBF, end-tidal O_2 and the metabolic parameters to be quantified. It furthermore allows

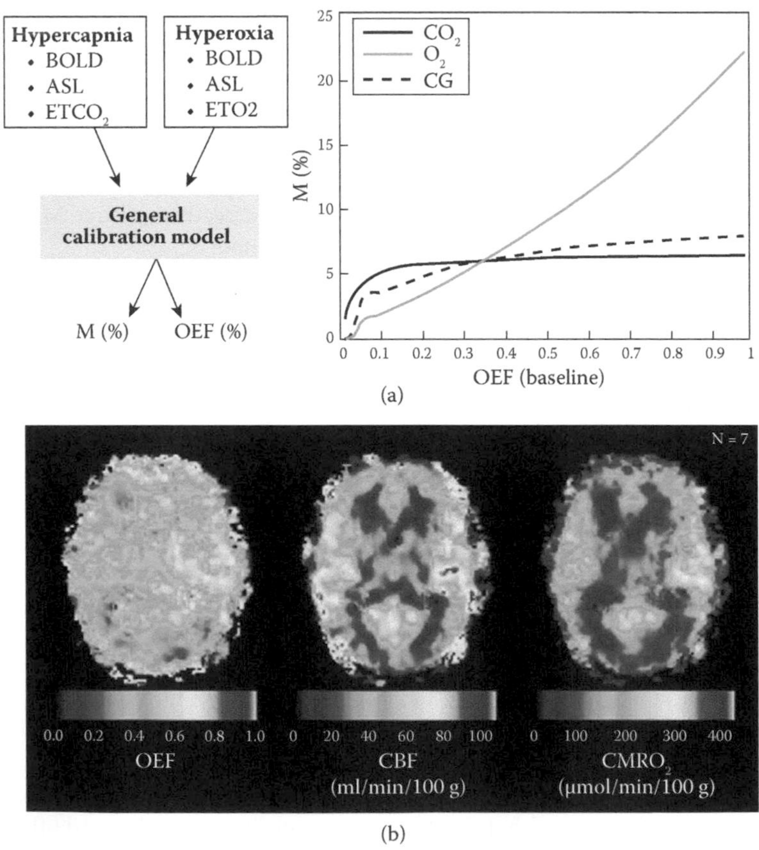

FIGURE 15.2 Baseline cerebral metabolic rate of oxygen ($CMRO_2$) by respiratory calibration of BOLD. (a) The generalised calibration model uses BOLD, cerebral blood flow and end-tidal gas measurements to relate the M (%) parameter to underline baseline oxygen extraction fraction (OEF) (%) (CG: carbogen). (b) The estimate of OEF in seven healthy volunteers is combined with baseline CBF by the Fick principle to achieve $CMRO_2$ at rest. (Reproduced from *NeuroImage*, 60, Gauthier, C.J., and Hoge, R.D., Magnetic resonance imaging of resting OEF and $CMRO_2$ using a generalised calibration model for hypercapnia and hyperoxia, 1212–1225, Copyright 2012, with permission from Elsevier.)

estimation of the α and β parameters or a heuristic combination of the two (Merola *et al.*, 2016). This generalised model has also been implemented in a Bayesian framework with promising results for physiologically accurate and stable estimates of M, OEF and $CMRO_2$ (Germuska *et al.*, 2016).

15.5.3 Baseline $CMRO_2$ mapping – non-respiratory techniques

15.5.3.1 Quantitative BOLD and R2′ maps of tissue oxygenation

Baseline oxygenation mapping can also be achieved through direct measurement of magnitude signal and relaxation times (T_2^*, T_2, T_2') in tissue, which are modulated by local field inhomogeneities created by dHb molecules in microvasculature. While several MRI relaxation times are sensitive to oxygenation level, studies suggest that the T_2' parameter is the most directly related to oxygenation (Yablonskiy and Haacke, 1994). T_2' is the reversible component of transverse relaxation, and is defined as $1/T_2' = 1/T_2^* - 1/T_2$. New hybrid sequences that combine gradient and spin echo (multi-echo) acquisitions allow estimation of T_2^*, T_2, and T_2' from the same scan (Ni *et al.*, 2015; Yablonskiy and Haacke, 1997). These hybrid sequences enable mapping of relaxation parameters that are sensitive to the underlying oxygenation state of the brain.

A major challenge of extravascular BOLD methods is that relaxation parameters are not specific to brain oxygenation. For instance, even T_2' is the product of blood volume and dHb-induced frequency shifts. As a result, complex biophysical models are often required to interpret the BOLD signal in terms of baseline oxygenation. Early quantitative BOLD (qBOLD) approaches have focused on the T_2' signal from gradient- and spin-echo acquisitions. These methods model capillary vessels in brain parenchyma as a network of randomly oriented cylinders to describe MRI signal dephasing in the presence of dHb. By fitting the signal model at each voxel, qBOLD techniques create parametric maps of venous oxygen saturation (SvO_2) and $CMRO_2$. Some qBOLD implementations assume a single extravascular tissue compartment (An and Lin, 2000; An *et al.*, 2001), while others also consider blood and cerebrospinal fluid (CSF) compartments in the model fit to each voxel (Christen *et al.*, 2012; He and Yablonskiy, 2007; He *et al.*, 2008).

In the future, BOLD methods for oxygenation imaging may synergise well with novel fingerprinting approaches. Christen *et al.* proposed a vascular fingerprint that creates a dictionary of signal curves (from gradient- and spin-echo hybrid scans) for various CBV, mean vessel radius and SvO_2 (Christen *et al.*, 2014). The measured signal curve for each voxel is matched to a dictionary curve, which reveals a specific quantitative SvO_2 (%) for tissue in the voxel. The accuracy of vascular fingerprinting depends on whether the biophysical model that governs these signal curves accurately represents cerebral physiology.

15.5.3.2 Intravascular R2-Based measurements of oxygenation

Instead of investigating the signal in extravascular brain tissues, intravascular MRI approaches seek to quantify T_2 relaxation directly in venous blood. If more oxygen is extracted, more dHb molecules are present in the venous blood, leading to lower T_2 values. Once the blood T_2 relaxation is measured, a biophysical model allows conversion of venous blood T_2 to quantitative SvO_2 (%), if haematocrit is also known (van Zijl *et al.*, 1998).

The main challenge is to isolate pure venous blood signal for T_2 measurement, because most brain voxels represent a mixture of CSF, tissue and blood signal. Many of the first intravascular oxygenation studies chose to focus on large veins with voxels that contained only pure venous blood (Golay *et al.*, 2001; Oja *et al.*, 1999).

The most widely tested intravascular approach is T_2-relaxation under spin tagging (TRUST), which measures T_2 in the sagittal sinus to assess global SvO_2 (Lu and Ge, 2008). TRUST MRI applies inversion pulses to collect images with and without labelling of venous blood at different echo times. This approach is similar to that of ASL, except with a labelling plane superior to the imaging volume to target outflowing venous blood signal. In this manner, signal contributions from CSF and static tissue are subtracted out, and the T_2 measurement is made only for venous blood in the sagittal sinus. TRUST is fast and gives absolute, global SvO_2 (%) values in minutes that have been calibrated in different physiological conditions (Lu *et al.*, 2012) and tested across multiple sites (Liu *et al.*, 2016b). Recent efforts have used velocity-encoding gradients to target T_2-based oxygenation measurements in smaller veins that are more representative of local brain function (Krishnamurthy *et al.*, 2014).

T_2 methods have also been extended to map oxygenation in brain tissues, i.e. from the microvasculature in each voxel. Quantitative imaging of extraction of oxygen and tissue consumption (QUIXOTIC) MRI uses velocity-selective RF pulses to select for venular blood (Bolar *et al.*, 2011). These pulses use known cut-off velocities and acceleration properties of blood as it passes through the microvasculature to create maps of only venular blood for T_2 and SvO_2 measurement. The main limitation of QUIXOTIC is low SNR, because typical tissue voxels only have 5% CBV. Improvements to this oxygenation mapping technique have been proposed to remove contamination from diffusion and increase the SNR of the oxygenation measurements (Guo and Wong, 2012).

15.5.3.3 Susceptibility MRI of Blood Oxygenation

A different MR contrast mechanism for oxygenation derives magnetic susceptibility shifts within veins, due to the presence of dHb, compared to the surrounding brain tissue. This susceptibility shift creates magnetic field perturbations ($\Delta B_{vein\text{-}tissue}$) that manifest on MRI phase images. MRI phase images thus provide information about susceptibility changes that enable quantification of the underlying SvO_2 in individual vessels (Haacke *et al.*, 1997; Weisskoff and Kiihne, 1992).

Although magnetic susceptibility is linearly related to OEF, there is no direct way to image susceptibility by MRI. The relationship between magnetic field and MRI phase with the underlying susceptibility depends on the vessel orientation and geometry in a complex and non-local manner. For this reason, susceptibility measurement is non-trivial and requires solution of a difficult mathematical inversion problem. Furthermore, sufficient spatial resolution must be achieved to resolve smaller cerebral veins, which lengthens the MRI scan time.

Susceptibility-based studies of oxygenation have been reviewed by Wehrli *et al.* (2016). The first phase MRI oxygenation studies estimated susceptibility by approximating cerebral veins as long cylinders parallel to the main magnetic field. For the parallel vessel geometry, there is a simple relationship between measured phase in the vein and its susceptibility. This approach, MR susceptometry, has been used to study oxygenation in large draining veins such as the internal jugular vein (Fernandez-Seara *et al.*, 2006) and sagittal sinus of the brain (Jain *et al.*, 2010). Similar to TRUST MRI, global SvO_2 measurements from MRI phase are fast and reproducible. These fast susceptometry methods can be combined with whole-brain flow in the same sequence to quantify functional $CMRO_2$ changes with high temporal resolution (Barhoum *et al.*, 2015; Rodgers *et al.*, 2013).

Recent studies have also sought to assess phase-based oxygenation in smaller veins (Fan *et al.*, 2011), which is expected to be more reflective of local brain physiology. For these smaller vessels, it will be particularly important to correct for partial volume effects (Hsieh *et al.*, 2015; McDaniel *et al.*, 2016) and potential orientation effects if the vessel is tilted relative to the main field (Langham *et al.*, 2009).

To extend susceptibility-based SvO_2 assessment to veins of arbitrary curvature and orientation, quantitative susceptibility mapping (QSM) can directly reconstruct the 3D susceptibility distribution from measured field maps. Once the QSM map is created, susceptibility differences can be converted to absolute SvO_2 along all resolved cerebral vessels, i.e. a brain oxygenation venogram (Fan *et al.*, 2014, 2015a). To ensure accurate and robust measurements, more work needs to be done to understand potential OEF underestimation from the QSM reconstruction process and from second order effects of flowing spins in the vessels (Xu *et al.*, 2014). Nascent methods that use QSM to assess tissue-based (instead of vessel-based) baseline $CMRO_2$ are also under development (Zhang *et al.*, 2015, 2016) but require multiple respiratory challenges in their current implementation.

15.5.3.4 Applications and future directions

For baseline oxygenation MRI to be clinically useful, its acquisition must be easily implementable with relatively short scan time. BOLD acquisitions have gained increasing usage in the clinic, in part because of the popularity of resting-state functional scans. However, because BOLD maps are a complex combination of cerebral physiology, BOLD studies in patients with ischaemia (Dani *et al.*, 2010; Mandell *et al.*, 2008) and tumour (Taylor *et al.*, 2001) have required scans in different gas states to isolate the oxygenation information. Quantitative respiratory-calibration methods also require multiple gas inhalations and would benefit from shorter protocols.

Due to its short acquisition time (~30 sec), global T_2 measurements for quantitative OEF assessment in the sagittal sinus by TRUST has been applied in many patient populations. These cohorts include neonates (Liu *et al.*, 2014), volunteers of different ages (Peng *et al.*, 2014) and patients with neurodegenerative diseases such as multiple sclerosis (Ge *et al.*, 2012). Dynamic whole-brain OEF measurements by MRI susceptometry have been studied in obstructive sleep apnoea during a breath-hold task (Rodgers *et al.*, 2016). Phase susceptometry also showed that neonates with congenital heart disease have impaired brain physiology, similar to premature infants, that can predict eventual white matter damage (Jain *et al.*, 2014).

While global measurements are fast and reliable, ultimately regional OEF information is necessary to assess many brain disorders. Susceptibility contrast is straightforward to obtain from a gradient echo MRI scan and can provide local oxygenation information from individual veins. Susceptibility weighted imaging has gained popularity to image oxygen disturbance in the affected versus healthy hemispheres of stroke patients (Jensen-Kondering and Bohm, 2013; Xia *et al.*, 2014), in traumatic brain injury (Liu *et al.*, 2016a) and in multiple sclerosis (Fan *et al.*, 2015b). Whole-brain mapping of oxygenation must consider challenging image reconstructions due to nearby susceptibility artefacts (e.g. from haemorrhagic blood products). At the same time, obtaining local susceptibility-based OEF information may benefit from development of complementary extravascular (tissue-based) techniques. Nonetheless, early patient studies with baseline oxygenation MRI show promising physiological findings and point to future technical development to translate new methods to clinical practice.

References

Aanerud J, Borghammer P, Chakravarty MM, Vang K, Rodell AB, Jonsdottir KY, *et al.* Brain energy metabolism and blood flow differences in healthy aging. J Cereb Blood Flow Metab 2012; 32: 1177–87.

Alsop DC, Detre JA, Golay X, Gunther M, Hendrikse J, Hernandez-Garcia L, et al. Recommended implementation of arterial spin-labeled perfusion MRI for clinical applications: a consensus of the ISMRM perfusion study group and the European consortium for ASL in dementia. Magn Reson Med 2015; 73: 102–16.

An H, Lin W. Quantitative measurements of cerebral blood oxygen saturation using magnetic resonance imaging. J Cereb Blood Flow Metab 2000; 20: 1225–36.

An H, Lin W, Celik A, Lee YZ. Quantitative measurements of cerebral metabolic rate of oxygen utilization using MRI: a volunteer study. NMR Biomed 2001; 14: 441–7.

Anazodo UC, Shoemaker JK, Suskin N, Ssali T, Wang DJ, St Lawrence KS. Impaired cerebrovascular function in coronary artery disease patients and recovery following cardiac rehabilitation. Front Aging Neurosci 2015; 7: 224.

Ances B, Vaida F, Ellis R, Buxton R. Test-retest stability of calibrated BOLD-fMRI in HIV- and HIV+ subjects. NeuroImage 2011; 54: 2156–62.

Ances BM, Leontiev O, Perthen JE, Liang C, Lansing AE, Buxton RB. Regional differences in the coupling of cerebral blood flow and oxygen metabolism changes in response to activation: implications for BOLD-fMRI. NeuroImage 2008; 39: 1510–21.

Ances BM, Liang CL, Leontiev O, Perthen JE, Fleisher AS, Lansing AE, et al. Effects of aging on cerebral blood flow, oxygen metabolism, and blood oxygenation level dependent responses to visual stimulation. Hum Brain Mapp 2009; 30: 1120–32.

Barhoum S, Langham MC, Magland JF, Rodgers ZB, Li C, Rajapakse CS, et al. Method for rapid MRI quantification of global cerebral metabolic rate of oxygen. J Cereb Blood Flow Metab 2015; 35: 1616–22.

Bhogal AA, Siero JC, Fisher JA, Froeling M, Luijten P, Philippens M, et al. Investigating the non-linearity of the BOLD cerebrovascular reactivity response to targeted hypo/hypercapnia at 7T. NeuroImage 2014; 98: 296–305.

Blockley NP, Griffeth VE, Stone AJ, Hare HV, Bulte DP. Sources of systematic error in calibrated BOLD based mapping of baseline oxygen extraction fraction. NeuroImage 2015; 122: 105–13.

Bokkers RP, van Osch MJ, Klijn CJ, Kappelle LJ, Hendrikse J. Cerebrovascular reactivity within perfusion territories in patients with an internal carotid artery occlusion. J Neurol Neurosurg Psychiatry 2011; 82: 1011–16.

Bolar DS, Rosen BR, Sorensen AG, Adalsteinsson E. QUantitative Imaging of eXtraction of oxygen and TIssue consumption (QUIXOTIC) using venular-targeted velocity-selective spin labeling. Magn Reson Med 2011; 66: 1550–62.

Boxerman JL, Hamberg LM, Rosen BR, Weisskoff RM. MR contrast due to intravascular magnetic susceptibility perturbations. Magn Reson Med 1995; 34: 555–66.

Bulte DP, Drescher K, Jezzard P. Comparison of hypercapnia-based calibration techniques for measurement of cerebral oxygen metabolism with MRI. Magn Reson Med 2009; 61: 391–8.

Bulte DP, Kelly M, Germuska M, Xie J, Chappell MA, Okell TW, et al. Quantitative measurement of cerebral physiology using respiratory-calibrated MRI. NeuroImage 2012; 60: 582–91.

Cantin S, Villien M, Moreaud O, Tropres I, Keignart S, Chipon E, et al. Impaired cerebral vasoreactivity to CO2 in Alzheimer's disease using BOLD fMRI. NeuroImage 2011; 58: 579–87.

Chen JJ, Pike GB. BOLD-specific cerebral blood volume and blood flow changes during neuronal activation in humans. NMR Biomed 2009; 22: 1054–62.

Chen JJ, Pike GB. MRI measurement of the BOLD-specific flow-volume relationship during hypercapnia and hypocapnia in humans. NeuroImage 2010; 53: 383–91.

Chiarelli PA, Bulte DP, Wise R, Gallichan D, Jezzard P. A calibration method for quantitative BOLD fMRI based on hyperoxia. NeuroImage 2007; 37: 808–20.

Christen T, Pannetier NA, Ni WW, Qiu D, Moseley ME, Schuff N, et al. MR vascular fingerprinting: a new approach to compute cerebral blood volume, mean vessel radius, and oxygenation maps in the human brain. NeuroImage 2014; 89: 262–70.

Christen T, Schmiedeskamp H, Straka M, Bammer R, Zaharchuk G. Measuring brain oxygenation in humans using a multiparametric quantitative blood oxygenation level dependent MRI approach. Magn Reson Med 2012; 68: 905–11.

Dai W, Garcia D, de Bazelaire C, Alsop DC. Continuous flow-driven inversion for arterial spin labeling using pulsed radio frequency and gradient fields. Magn Reson Med 2008; 60: 1488–97.

Dai W, Robson PM, Shankaranarayanan A, Alsop DC. Reduced resolution transit delay prescan for quantitative continuous arterial spin labeling perfusion imaging. Magn Reson Med 2012; 67: 1252–65.

Dani KA, Santosh C, Brennan D, McCabe C, Holmes WM, Condon B, et al. T2*-weighted magnetic resonance imaging with hyperoxia in acute ischemic stroke. Ann Neurol 2010; 68: 37–47.

Davis TL, Kwong KK, Weisskoff RM, Rosen BR. Calibrated functional MRI: mapping the dynamics of oxidative metabolism. Proc Natl Acad Sci U S A 1998; 95: 1834–9.

De Vis JB, Hendrikse J, Bhogal A, Adams A, Kappelle LJ, Petersen ET. Age-related changes in brain hemodynamics; a calibrated MRI study. Hum Brain Mapp 2015a; 36: 3973–87.

De Vis JB, Petersen ET, Bhogal A, Hartkamp NS, Klijn CJ, Kappelle LJ, et al. Calibrated MRI to evaluate cerebral hemodynamics in patients with an internal carotid artery occlusion. J Cereb Blood Flow Metab 2015b; 35: 1015–23.

Detre JA, Rao H, Wang DJ, Chen YF, Wang Z. Applications of arterial spin labeled MRI in the brain. J Magn Reson Imaging 2012; 35: 1026–37.

Detre JA, Wang J, Wang Z, Rao H. Arterial spin-labeled perfusion MRI in basic and clinical neuroscience. Curr Opin Neurol 2009; 22: 348–55.

Donahue MJ, Ayad M, Moore R, van Osch M, Singer R, Clemmons P, et al. Relationships between hypercarbic reactivity, cerebral blood flow, and arterial circulation times in patients with moyamoya disease. J Magn Reson Imaging 2013; 38: 1129–39.

Donahue MJ, Lu H, Jones CK, Edden RA, Pekar JJ, van Zijl PC. Theoretical and experimental investigation of the VASO contrast mechanism. Magn Reson Med 2006; 56: 1261–73.

Donahue MJ, Sideso E, MacIntosh BJ, Kennedy J, Handa A, Jezzard P. Absolute arterial cerebral blood volume quantification using inflow vascular-space-occupancy with dynamic subtraction magnetic resonance imaging. J Cereb Blood Flow Metab 2010; 30: 1329–42.

Donahue MJ, Strother MK, Hendrikse J. Novel MRI approaches for assessing cerebral hemodynamics in ischemic cerebrovascular disease. Stroke 2012; 43: 903–15.

Driver I, Blockley N, Fisher J, Francis S, Gowland P. The change in cerebrovascular reactivity between 3 T and 7 T measured using graded hypercapnia. NeuroImage 2010; 51: 274–9.

Edelman RR, Siewert B, Darby DG, Thangaraj V, Nobre AC, Mesulam MM, et al. Qualitative mapping of cerebral blood flow and functional localization with echo-planar MR imaging and signal targeting with alternating radio frequency. Radiology 1994; 192: 513–20.

Fan AP, Benner T, Bolar DS, Rosen BR, Adalsteinsson E. Phase-based regional oxygen metabolism (PROM) using MRI. Magn Reson Med. 2011; 67(3): 669–78..

Fan AP, Bilgic B, Gagnon L, Witzel T, Bhat H, Rosen BR, et al. Quantitative oxygenation venography from MRI phase. Magn Reson Med 2014; 72: 149–59.

Fan AP, Evans KC, Stout JN, Rosen BR, Adalsteinsson E. Regional quantification of cerebral venous oxygenation from MRI susceptibility during hypercapnia. NeuroImage 2015a; 104: 146–55.

Fan AP, Govindarajan ST, Kinkel RP, Madigan NK, Nielsen AS, Benner T, et al. Quantitative oxygen extraction fraction from 7-Tesla MRI phase: reproducibility and application in multiple sclerosis. J Cereb Blood Flow Metab 2015b; 35: 131–9.

Fan AP, Jahanian H, Holdsworth SJ, Zaharchuk G. Comparison of cerebral blood flow measurement with [15O]-water positron emission tomography and arterial spin labeling magnetic resonance imaging: a systematic review. J Cereb Blood Flow Metab 2016a; 36: 842–61.

Fan AP, Schafer A, Huber L, Lampe L, von Smuda S, Moller HE, et al. Baseline oxygenation in the brain: correlation between respiratory-calibration and susceptibility methods. NeuroImage 2016b; 125: 920–31.

Fernandez-Seara MA, Techawiboonwong A, Detre JA, Wehrli FW. MR susceptometry for measuring global brain oxygen extraction. Magn Reson Med 2006; 55: 967–73.

Gagnon L, Sakadzic S, Lesage F, Pouliot P, Dale AM, Devor A, et al. Validation and optimization of hypercapnic-calibrated fMRI from oxygen-sensitive two-photon microscopy. Philos Trans R Soc Lond B Biol Sci 2016; 371.

Garcia DM, Duhamel G, Alsop DC. Efficiency of inversion pulses for background suppressed arterial spin labeling. Magn Reson Med 2005; 54: 366–72.

Gauthier CJ, Desjardins-Crépeau L, Madjar C, Bherer L, Hoge RD. Absolute quantification of resting oxygen metabolism and metabolic reactivity during functional activation using QUO2 MRI. NeuroImage 2012; 63: 1353–63.

Gauthier CJ, Hoge RD. A generalized procedure for calibrated MRI incorporating hyperoxia and hypercapnia. Hum Brain Mapp 2012a; 34: 1053–69.

Gauthier CJ, Hoge RD. Magnetic resonance imaging of resting OEF and CMRO2 using a generalized calibration model for hypercapnia and hyperoxia. NeuroImage 2012b; 60: 1212–25.

Gauthier CJ, Madjar C, Desjardins-Crepeau L, Bellec P, Bherer L, Hoge RD. Age dependence of hemodynamic response characteristics in human functional magnetic resonance imaging. Neurobiol Aging 2013; 34: 1469–85.

Gauthier CJ, Madjar C, Tancredi FB, Stefanovic B, Hoge RD. Elimination of visually evoked BOLD responses during carbogen inhalation: implications for calibrated MRI. NeuroImage 2011; 54: 1001–11.

Ge Y, Zhang Z, Lu H, Tang L, Jaggi H, Herbert J, et al. Characterizing brain oxygen metabolism in patients with multiple sclerosis with T2-relaxation-under-spin-tagging MRI. J Cereb Blood Flow Metab 2012; 32: 403–12.

Germuska M, Merola A, Murphy K, Babic A, Richmond L, Khot S, et al. A forward modelling approach for the estimation of oxygen extraction fraction by calibrated fMRI. NeuroImage 2016; 139: 313–23.

Girouard H, Iadecola C. Neurovascular coupling in the normal brain and in hypertension, stroke, and Alzheimer disease. J Appl Physiol 2006; 100: 328–35.

Golay X, Silvennoinen MJ, Zhou J, Clingman CS, Kauppinen RA, Pekar JJ, et al. Measurement of tissue oxygen extraction ratios from venous blood T(2): increased precision and validation of principle. Magn Reson Med 2001; 46: 282–91.

Goodwin JA, Vidyasagar R, Balanos GM, Bulte D, Parkes LM. Quantitative fMRI using hyperoxia calibration: reproducibility during a cognitive Stroop task. NeuroImage 2009; 47: 573–80.

Griffeth VE, Buxton RB. A theoretical framework for estimating cerebral oxygen metabolism changes using the calibrated-BOLD method: modeling the effects of blood volume distribution, hematocrit, oxygen extraction fraction, and tissue signal properties on the BOLD signal. NeuroImage 2011; 58: 198–212.

Grubb RL Jr., Raichle ME, Eichling JO, Ter-Pogossian MM. The effects of changes in PaCO2 on cerebral blood volume, blood flow, and vascular mean transit time. Stroke 1974; 5: 630–9.

Guidi M, Huber L, Lampe L, Gauthier CJ, Moller HE. Lamina-dependent calibrated BOLD response in human primary motor cortex. NeuroImage 2016.

Guo J, Wong EC. Venous oxygenation mapping using velocity-selective excitation and arterial nulling. Magn Reson Med 2012; 68: 1458–71.

Haacke EM, Lai S, Reichenbach JR, Kuppusamy K, Hoogenraad FG, Takeichi H, et al. In vivo measurement of blood oxygen saturation using magnetic resonance imaging: a direct validation of the blood oxygen level-dependent concept in functional brain imaging. Hum Brain Mapp 1997; 5: 341–6.

Hare HV, Blockley NP, Gardener AG, Clare S, Bulte DP. Investigating the field-dependence of the Davis model: calibrated fMRI at 1.5, 3 and 7T. NeuroImage 2015; 112: 189–96.

Hare HV, Germuska M, Kelly ME, Bulte DP. Comparison of CO2 in air versus carbogen for the measurement of cerebrovascular reactivity with magnetic resonance imaging. J Cereb Blood Flow Metab 2013; 33: 1799–805.

He X, Yablonskiy DA. Quantitative BOLD: mapping of human cerebral deoxygenated blood volume and oxygen extraction fraction: default state. Magn Reson Med 2007; 57: 115–26.

He X, Zhu M, Yablonskiy DA. Validation of oxygen extraction fraction measurement by qBOLD technique. Magn Reson Med 2008; 60: 882–8.

Heijtel DF, Mutsaerts HJ, Bakker E, Schober P, Stevens MF, Petersen ET, et al. Accuracy and precision of pseudo-continuous arterial spin labeling perfusion during baseline and hypercapnia: a head-to-head comparison with (1)(5)O H(2)O positron emission tomography. NeuroImage 2014; 92: 182–92.

Hoge RD, Atkinson J, Gill B, Crelier GR, Marrett S, Pike GB. Investigation of BOLD signal dependence on cerebral blood flow and oxygen consumption: the deoxyhemoglobin dilution model. Magn Reson Med 1999; 42: 849–63.

Hsieh CY, Cheng YC, Xie H, Haacke EM, Neelavalli J. Susceptibility and size quantification of small human veins from an MRI method. Magn Reson Imaging 2015; 33: 1191–204.

Hua J, Brandt AS, Lee S, Blair NI, Wu Y, Lui S, et al. Abnormal grey matter arteriolar cerebral blood volume in schizophrenia measured with 3D inflow-based vascular-space-occupancy MRI at 7T. Schizophr Bull 2016.

Hua J, Qin Q, Donahue MJ, Zhou J, Pekar JJ, van Zijl PC. Inflow-based vascular-space-occupancy (iVASO) MRI. Magn Reson Med 2011; 66: 40–56.

Huber L, Goense J, Kennerley AJ, Trampel R, Guidi M, Reimer E, et al. Cortical lamina-dependent blood volume changes in human brain at 7 T. NeuroImage 2015; 107: 23–33.

Huber L, Ivanov D, Handwerker DA, Marrett S, Guidi M, Uludag K, et al. Techniques for blood volume fMRI with VASO: from low-resolution mapping towards sub-millimeter layer-dependent applications. NeuroImage 2016.

Huber L, Ivanov D, Krieger SN, Streicher MN, Mildner T, Poser BA, et al. Slab-selective, BOLD-corrected VASO at 7 Tesla provides measures of cerebral blood volume reactivity with high signal-to-noise ratio. Magn Reson Med 2014; 72: 137–48.

Hutchison JL, Shokri-Kojori E, Lu H, Rypma B. A BOLD perspective on age-related neurometabolic-flow coupling and neural efficiency changes in human visual cortex. Front Psychol 2013; 4: 244.

Ishii K, Kitagaki H, Kono M, Mori E. Decreased medial temporal oxygen metabolism in Alzheimer's disease shown by PET. J Nucl Med 1996a; 37: 1159–65.

Ishii K, Sasaki M, Kitagaki H, Sakamoto S, Yamaji S, Maeda K. Regional difference in cerebral blood flow and oxidative metabolism in human cortex. J Nucl Med 1996b; 37: 1086–8.

Ito H, Kanno I, Kato C, Sasaki T, Ishii K, Ouchi Y, et al. Database of normal human cerebral blood flow, cerebral blood volume, cerebral oxygen extraction fraction and cerebral metabolic rate of oxygen measured by positron emission tomography with 15O-labelled carbon dioxide or water, carbon monoxide and oxygen: a multicentre study in Japan. Eur J Nucl Med Mol Imaging 2004; 31: 635–43.

Jain V, Buckley EM, Licht DJ, Lynch JM, Schwab PJ, Naim MY, et al. Cerebral oxygen metabolism in neonates with congenital heart disease quantified by MRI and optics. J Cereb Blood Flow Metab 2014; 34: 380–8.

Jain V, Langham MC, Wehrli FW. MRI estimation of global brain oxygen consumption rate. J Cereb Blood Flow Metab 2010; 30: 1598–607.

Jensen-Kondering U, Bohm R. Asymmetrically hypointense veins on T2*w imaging and susceptibility-weighted imaging in ischemic stroke. World J Radiol 2013; 5: 156–65.

Kety SS, Schmidt CF. The effects of altered arterial tensions of carbon dioxide and oxygen on cerebral blood flow and cerebral oxygen consumption of normal young men. J Clin Invest 1948; 27: 484–92.

Krishnamurthy LC, Liu P, Ge Y, Lu H. Vessel-specific quantification of blood oxygenation with T2-relaxation-under-phase-contrast MRI. Magn Reson Med 2014; 71: 978–89.

Lajoie I, Tancredi FB, Hoge RD. Regional reproducibility of BOLD calibration parameter M, OEF and resting-state CMRO2 measurements with QUO2 MRI. PLoS One 2016; 11: e0163071.

Langham MC, Magland JF, Floyd TF, Wehrli FW. Retrospective correction for induced magnetic field inhomogeneity in measurements of large-vessel hemoglobin oxygen saturation by MR susceptometry. Magn Reson Med 2009; 61: 626–33.

Liu J, Xia S, Hanks R, Wiseman N, Peng C, Zhou S, et al. Susceptibility weighted imaging and Mapping of Micro-Hemorrhages and Major Deep Veins after Traumatic Brain injury. J Neurotrauma 2016a; 33: 10–21.

Liu P, Dimitrov I, Andrews T, Crane DE, Dariotis JK, Desmond J, et al. Multisite evaluations of a T2 -relaxation-under-spin-tagging (TRUST) MRI technique to measure brain oxygenation. Magn Reson Med 2016b; 75: 680–7.

Liu P, Hebrank AC, Rodrigue KM, Kennedy KM, Section J, Park DC, et al. Age-related differences in memory-encoding fMRI responses after accounting for decline in vascular reactivity. NeuroImage 2013; 78: 415–25.

Liu P, Huang H, Rollins N, Chalak LF, Jeon T, Halovanic C, et al. Quantitative assessment of global cerebral metabolic rate of oxygen (CMRO2) in neonates using MRI. NMR Biomed 2014; 27: 332–40.

Liu P, Li Y, Pinho M, Park DC, Welch BG, Lu H. Cerebrovascular reactivity mapping without gas challenges. NeuroImage 2016c; 146: 320–6.

Lu H, Ge Y. Quantitative evaluation of oxygenation in venous vessels using T2-Relaxation-Under-Spin-Tagging MRI. Magn Reson Med 2008; 60: 357–63.

Lu H, Golay X, Pekar JJ, Van Zijl PC. Functional magnetic resonance imaging based on changes in vascular space occupancy. Magn Reson Med 2003; 50: 263–74.

Lu H, van Zijl PC. A review of the development of Vascular-Space-Occupancy (VASO) fMRI. NeuroImage 2012.

Lu H, Xu F, Grgac K, Liu P, Qin Q, van Zijl P. Calibration and validation of TRUST MRI for the estimation of cerebral blood oxygenation. Magn Reson Med 2012; 67: 42–9.

Lu H, Xu F, Rodrigue KM, Kennedy KM, Cheng Y, Flicker B, et al. Alterations in cerebral metabolic rate and blood supply across the adult lifespan. Cereb Cortex 2011; 21: 1426–34.

Mandell DM, Han JS, Poublanc J, Crawley AP, Stainsby JA, Fisher JA, et al. Mapping cerebrovascular reactivity using blood oxygen level-dependent MRI in Patients with arterial steno-occlusive disease: comparison with arterial spin labeling MRI. Stroke 2008; 39: 2021–8.

Mark CI, Fisher JA, Pike GB. Improved fMRI calibration: precisely controlled hyperoxic versus hypercapnic stimuli. NeuroImage 2011; 54: 1102–11.

McDaniel P, Bilgic B, Fan AP, Stout JN, Adalsteinsson E. Mitigation of partial volume effects in susceptibility-based oxygenation measurements by joint utilization of magnitude and phase (JUMP). Magn Reson Med 2016.

Merola A, Murphy K, Stone AJ, Germuska MA, Griffeth VE, Blockley NP, et al. Measurement of oxygen extraction fraction (OEF): an optimized BOLD signal model for use with hypercapnic and hyperoxic calibration. NeuroImage 2016; 129: 159–74.

Mohtasib RS, Lumley G, Goodwin JA, Emsley HC, Sluming V, Parkes LM. Calibrated fMRI during a cognitive Stroop task reveals reduced metabolic response with increasing age. NeuroImage 2012; 59: 1143–51.

Murphy K, Harris AD, Wise RG. Robustly measuring vascular reactivity differences with breath-hold: normalising stimulus-evoked and resting state BOLD fMRI data. NeuroImage 2011; 54: 369–79.

Nagata K, Maruya H, Yuya H, Terashi H, Mito Y, Kato H, et al. Can PET data differentiate Alzheimer's disease from vascular dementia? Ann N Y Acad Sci 2000; 903: 252–61.

Ni W, Christen T, Zun Z, Zaharchuk G. Comparison of R2' measurement methods in the normal brain at 3 Tesla. Magn Reson Med 2015; 73: 1228–36.

Ogawa S, Lee TM, Kay AR, Tank DW. Brain magnetic resonance imaging with contrast dependent on blood oxygenation. Proc Natl Acad Sci U S A 1990a; 87: 9868–72.

Ogawa S, Lee TM, Nayak AS, Glynn P. Oxygenation-sensitive contrast in magnetic resonance image of rodent brain at high magnetic fields. Magn Reson Med 1990b; 14: 68–78.

Oja JM, Gillen JS, Kauppinen RA, Kraut M, van Zijl PC. Determination of oxygen extraction ratios by magnetic resonance imaging. J Cereb Blood Flow Metab 1999; 19: 1289–95.

Peng SL, Dumas JA, Park DC, Liu P, Filbey FM, McAdams CJ, et al. Age-related increase of resting metabolic rate in the human brain. NeuroImage 2014; 98: 176–83.

Petersen ET, Zimine I, Ho YC, Golay X. Non-invasive measurement of perfusion: a critical review of arterial spin labelling techniques. Br J Radiol 2006; 79: 688–701.

Rodan LH, Poublanc J, Fisher JA, Sobczyk O, Wong T, Hlasny E, et al. Cerebral hyperperfusion and decreased cerebrovascular reactivity correlate with neurologic disease severity in MELAS. Mitochondrion 2015; 22: 66–74.

Rodgers ZB, Jain V, Englund EK, Langham MC, Wehrli FW. High temporal resolution MRI quantification of global cerebral metabolic rate of oxygen consumption in response to apneic challenge. J Cereb Blood Flow Metab 2013; 33: 1514–22.

Rodgers ZB, Leinwand SE, Keenan BT, Kini LG, Schwab RJ, Wehrli FW. Cerebral metabolic rate of oxygen in obstructive sleep apnea at rest and in response to breath-hold challenge. J Cereb Blood Flow Metab 2016; 36: 755–67.

Smeeing DP, Hendrikse J, Petersen ET, Donahue MJ, de Vis JB. Arterial spin labeling and blood oxygen level-dependent MRI cerebrovascular reactivity in cerebrovascular disease: a systematic review and meta-analysis. Cerebrovasc Dis 2016; 42: 288–307.

Stefanovic B, Pike GB. Venous refocusing for volume estimation: VERVE functional magnetic resonance imaging. Magn Reson Med 2005; 53: 339–47.

Talati P, Rane S, Donahue MJ, Heckers S. Hippocampal arterial cerebral blood volume in early psychosis. Psychiatry Res 2016; 256: 21–25.

Tancredi FB, Gauthier CJ, Madjar C, Bolar DS, Fisher JA, Wang DJ, et al. Comparison of pulsed and pseudocontinuous arterial spin-labeling for measuring CO(2) -induced cerebrovascular reactivity. J Magn Reson Imaging 2012.

Tancredi FB, Hoge RD. Comparison of cerebral vascular reactivity measures obtained using breath-holding and CO2 inhalation. J Cereb Blood Flow Metab 2013; 33: 1066–74.

Taylor NJ, Baddeley H, Goodchild KA, Powell ME, Thoumine M, Culver LA, et al. BOLD MRI of human tumor oxygenation during carbogen breathing. J Magn Reson Imaging 2001; 14: 156–63.

van Zijl PC, Ulug AM, Eleff SM, Ulatowski JA, Traystman RJ, Oja JM, et al. [Quantitative assessment of blood flow, blood volume and blood oxygenation effects in functional magnetic resonance imaging]. Duodecim 1998; 114: 808–9.

Warnert EA, Harris AD, Murphy K, Saxena N, Tailor N, Jenkins NS, et al. In vivo assessment of human brainstem cerebrovascular function: a multi-inversion time pulsed arterial spin labelling study. J Cereb Blood Flow Metab 2014; 34: 956–63.

Wehrli FW, Fan AP, Rodgers ZB, Englund EK, Langham MC. Susceptibility-based time-resolved whole-organ and regional tissue oximetry. NMR Biomed 2016.

Weisskoff RM, Kiihne S. MRI susceptometry: image-based measurement of absolute susceptibility of MR contrast agents and human blood. Magn Reson Med 1992; 24: 375–83.

Williams DS, Detre JA, Leigh JS, Koretsky AP. Magnetic resonance imaging of perfusion using spin inversion of arterial water. Proc Natl Acad Sci U S A 1992; 89: 212–16.

Wise RG, Harris AD, Stone AJ, Murphy K. Measurement of OEF and absolute CMRO: MRI-based methods using interleaved and combined hypercapnia and hyperoxia. NeuroImage 2013; 83C: 135–47.

Wise RG, Pattinson KT, Bulte DP, Chiarelli PA, Mayhew SD, Balanos GM, et al. Dynamic forcing of end-tidal carbon dioxide and oxygen applied to functional magnetic resonance imaging. J Cereb Blood Flow Metab 2007; 27: 1521–32.

Wong EC, Buxton RB, Frank LR. Quantitative imaging of perfusion using a single subtraction (QUIPSS and QUIPSS II). Magn Reson Med 1998; 39: 702–8.

Wong EC, Cronin M, Wu WC, Inglis B, Frank LR, Liu TT. Velocity-selective arterial spin labeling. Magn Reson Med 2006; 55: 1334–41.

Wu WC, Fernandez-Seara M, Detre JA, Wehrli FW, Wang J. A theoretical and experimental investigation of the tagging efficiency of pseudocontinuous arterial spin labeling. Magn Reson Med 2007; 58: 1020–7.

Wu Y, Agarwal S, Jones CK, Webb AG, van Zijl PC, Hua J, et al. Measurement of arteriolar blood volume in brain tumors using MRI without exogenous contrast agent administration at 7T. J Magn Reson Imaging 2016; 44: 1244–55.

Xia S, Utriainen D, Tang J, Kou Z, Zheng G, Wang X, et al. Decreased oxygen saturation in asymmetrically prominent cortical veins in patients with cerebral ischemic stroke. Magn Reson Imaging 2014; 32: 1272–6.

Xu B, Liu T, Spincemaille P, Prince M, Wang Y. Flow compensated quantitative susceptibility mapping for venous oxygenation imaging. Magn Reson Med 2014; 72: 438–45.

Yablonskiy DA. Cerebral metabolic rate in hypercapnia: controversy continues. J Cereb Blood Flow Metab 2011; 31: 1502–3.

Yablonskiy DA, Haacke EM. Theory of NMR signal behavior in magnetically inhomogeneous tissues: the static dephasing regime. Magn Reson Med 1994; 32: 749–63.

Yablonskiy DA, Haacke EM. An MRI method for measuring T2 in the presence of static and RF magnetic field inhomogeneities. Magn Reson Med 1997; 37: 872–6.

Yamaguchi T, Kanno I, Uemura K, Shishido F, Inugami A, Ogawa T, et al. Reduction in regional cerebral metabolic rate of oxygen during human aging. Stroke 1986; 17: 1220–8.

Yamauchi H, Nishii R, Higashi T, Kagawa S, Fukuyama H. Hemodynamic compromise as a cause of internal border-zone infarction and cortical neuronal damage in atherosclerotic middle cerebral artery disease. Stroke 2009; 40: 3730–5.

Zhang J, Liu T, Gupta A, Spincemaille P, Nguyen TD, Wang Y. Quantitative mapping of cerebral metabolic rate of oxygen (CMRO2) using quantitative susceptibility mapping (QSM). Magn Reson Med 2015; 74: 945–52.

Zhang J, Zhou D, Nguyen TD, Spincemaille P, Gupta A, Wang Y. Cerebral metabolic rate of oxygen (CMRO2) mapping with hyperventilation challenge using quantitative susceptibility mapping (QSM). Magn Reson Med 2016.

16

ASL: Blood Perfusion Measurement Using Arterial Spin Labelling[1]

Contents

Lisa A. van der Kleij
University Medical Center Utrecht

Esben Thade Petersen
Danish Research Centre for Magnetic Resonance

16.1 Introduction

The vascular system ensures a sufficient supply of oxygen and nutrients to the tissue, while at the same time it removes waste, transports heat, hormones and other messengers throughout the body. This complex transport system thereby ensures the homeostasis of the body and its organs. The steady-state nutritive delivery of blood to the tissue's capillary bed is termed *perfusion*. Imaging modalities that can accurately measure perfusion for diagnostic purposes, e.g. to assess whether an ischaemic organ is viable, are therefore of vital importance. Perfusion imaging modalities includes positron emission tomography (PET), single-photon emission computerised tomography, computed tomography and magnetic resonance imaging (MRI). These methods are typically all in conjunction with the injection of an intravenous tracer. Clinically, MRI is a popular tool for perfusion imaging, due to the additional ability of obtaining structural images with superior soft tissue contrast that is often needed for evaluation of the patient. Two MRI perfusion approaches exist, the first being bolus tracking after the injection of an exogenous endovascular tracer, whereas the second technique called *arterial spin labelling* (ASL) relies on magnetically labelled water protons as an endogenous tracer. ASL is particularly appealing because the injection of a contrast agent can be avoided, which is important in patients with kidney failure or in the paediatric population. In addition, MRI does not involve radiation and this combined with the complete non-invasiveness of ASL makes it a forceful tool for perfusion studies of healthy volunteers and in patient groups requiring repetitive follow-ups. ASL was invented in 1992 (Williams *et al.*, 1992). Yet for many years, it remained a research tool for the few, partly due to hardware and software constraints but mainly due to a rather low signal-to-noise ratio (SNR) and a consequently poor robustness of the technique. The interplay in recent years between refined hardware solutions and optimised sequences have elevated ASL into widespread use, particularly within the neuroscience field, while clinical use is also rapidly increasing.

This chapter will focus on the different implementations of ASL and how to quantify perfusion from the acquired data, including the problems and pitfalls potentially experienced

[1] Edited by Nicholas G. Dowell and reviewed by Dr. Laura Parkes, Centre for Imaging Sciences, University of Manchester, UK.

during the process. Then we revisit research and clinical applications of ASL. Finally, we will look into the future directions for ASL.

16.2 Perfusion imaging

16.2.1 Terminology and units

Blood flows via the arteries and through the smaller arterioles into the capillary bed. It is the blood delivery at this level, which enables the exchange of oxygen and nutrients to the tissue, that we call *perfusion*. The study of the fluid dynamics of the blood is termed *haemodynamics*, while the study of the actual flow properties of blood and its elements is termed haemorrheology. Both naturally influence the actual perfusion, but in perfusion imaging we aim to be independent of the underlying haemodynamic system and the underlying blood composition in our perfusion estimate. Cerebral blood flow (*CBF*) is the primary parameter we use to describe perfusion and it has the unit of *ml of blood per 100 g of tissue per min*, or [ml/100g/min]. This measure is a rate [s^{-1}] rather than a volume-flux [m^3/s], which is convenient because our techniques measure *CBF* in voxels of arbitrary volume. In addition to *CBF*, one often finds other parameters mentioned in the perfusion imaging literature. These includes cerebral blood volume (*CBV*), which is a dimensionless parameter [%] or [ml/100g], describing the fraction of blood within a voxel and the mean transit time (MTT), which is the average time [s] it takes the blood to traverse the vasculature of the voxel. These are also important clinical parameters, probing the status of the underlying hemodynamic system. For instance, an increased *CBV* in a patient with large-vessel disease could suggest that autoregulation has dilated the vessels in order to maintain sufficient *CBF* to the affected tissue.

16.2.2 Quantification principles

Two types of tracer exist for tracking blood perfusion of the tissue: freely diffusible tracers and intravascular tracers. Freely diffusible tracers, which exit the intravascular space unhindered upon arrival to the tissue's vascular bed, diffuse freely through the brain's otherwise very selective blood–brain barrier. Freely diffusible tracers such as nitrous oxide (N_2O) were used by the early pioneers (Kety and Schmidt, 1948). Their distribution throughout the full tissue volume causes the MTT to be on the order of minutes at normal physiological flow rates of 40–60 [ml/100g/min] for the brain, thereby allowing ample time for the dynamic arterial and venous blood sampling typically needed for quantification using Fick's principle:

$$\frac{dC_t(t)}{dt} = CBF \cdot \left(C_a(t) - C_v(t)\right) \tag{16.1}$$

where $C_t(t)$, $C_a(t)$ and $C_v(t)$ are the concentrations of tracer in the tissue, arterial and venous blood, respectively.

Intravascular tracers on the other hand, remain in the vasculature due to their relatively large molecular structure, which limits their distribution volume to the vascular volume. Contrary to the freely diffusible tracer, the distribution of intravascular tracers offers an opportunity of measuring *CBV* in addition to *CBF*. The central volume theorem states that the ratio between distribution volume and flow is equal to the mean transit time (hence, for intravascular tracers, *MTT* = *CBV*/*CBF*). Accordingly, if any two parameters are known, we can calculate the remaining one. The gadolinium-based bolus tracking experiment (Ostergaard *et al.*, 1996) exploits this by estimating the tracer concentration by means of MRI during its first passage of both the arterial and the tissue space. Knowing both the arterial input function $C_a(t)$ and the tissue concentration curve $C_t(t)$, R(t) · *CBF* is calculated using deconvolution:

$$C_t(t) = CBF \cdot \int_0^t C_a(\tau) R(t-\tau) d\tau \tag{16.2}$$

where R(t) is the residue function describing the fraction of contrast remaining in the system at time t after arrival, and it is therefore a decaying function with R(0) = 1. The result of the deconvolution; R(t) · *CBF*, equals *CBF* at t = 0. In addition, the area under the tissue concentration curves allows the assessment of *CBV*:

$$CBV = \frac{\int_{-\infty}^{\infty} C_t(t)dt}{\int_{-\infty}^{\infty} C_a(t)dt} \tag{16.3}$$

The two tracer techniques described above form the basis of the applied perfusion techniques of today. Hybrid solutions exist in the case where restricted leakage or permeability to the tracer exists, which can be quantified by means of the appropriate modelling of R(t) (Tofts and Kermode, 1991). Which scheme ASL in fact belongs to is somewhat ambivalent, as we will come back to later.

16.3 ASL: Acquisition schemes

The basic ASL perfusion measurement consists of two parts, the 'label' and 'control'. By subtracting the label from the control, the stationary tissue signal is eliminated and the resulting difference map ΔM becomes proportional to the blood inflow from the label region and thereby it reflects local perfusion (Figures 16.1 and 16.2). The first is the label part, where the magnetisation of the inflowing blood has been manipulated, typically by inverting or saturating it prior to the subsequent post-labelling delay (PLD) and final image readout. The second part is the control part, where essentially nothing is done to the inflowing blood; the images are 'just' acquired with similar conditions as for the label experiment. The difference signal *ΔM(t)* depends on the underlying flow kinetics (Figure 16.3) and at which time (PLD) it was sampled. Often a single PLD is used, chosen such that the labelled blood spins had enough time to

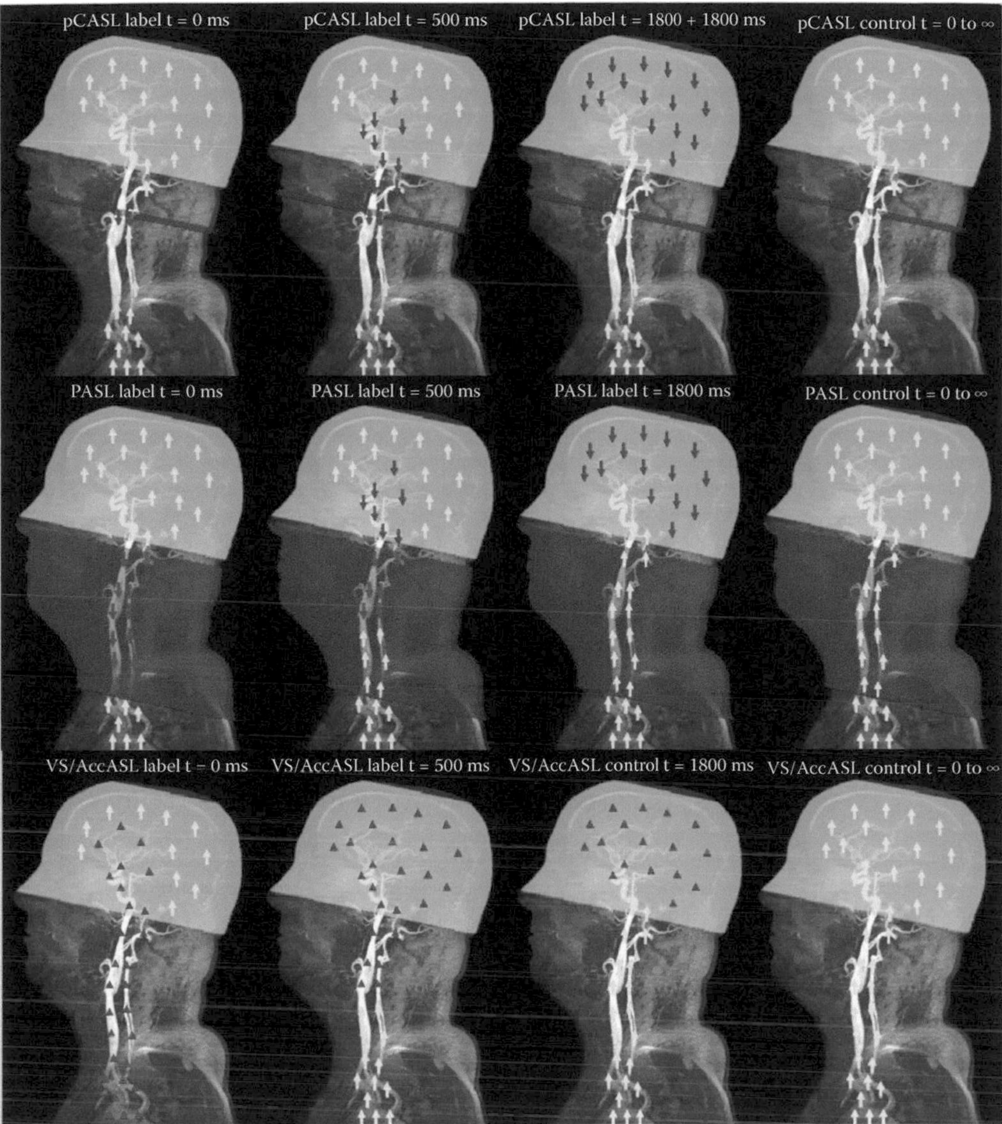

FIGURE 16.1 **Upper row**: Pseudo-continuous arterial spin labelling (ASL) labelling in a plane located perpendicular to the feeding vessels in the neck. The label plane is kept on for typically 1800 ms, during which time the blood passing through the label plane is inverted. Subsequently, a post-labelling delay (PLD) of 1800 ms is applied to allow the labelled blood to clear from the vasculature and reach the tissue before imaging. During the control experiment on the right, the blood magnetisation is kept undisturbed and otherwise the experiment is similar to the label experiment. Subtraction of control-label cancels the static tissue signal and leaves us with the magnetisation difference ΔM caused by perfusion. **Middle row**: Pulsed ASL (PASL) labels a slab of blood in the neck region at a single instance of time (10–15 ms), after which the PLD is inserted before acquisition. To get a well-defined bolus duration in a PASL experiment, a QUIPSSII saturation pulse can be applied within the label region. It has to be applied before the trailing edge of the bolus leaves the label region, e.g. at 500 ms depicted in the figure. **Bottom row**: Velocity selective ASL or acceleration selective ASL labels blood above a certain velocity or acceleration threshold, respectively. Thereby, the labelling also occurs within the imaging region itself, resulting in a very short arterial transit time. In these techniques, the label module causes a saturation, rather than an inversion. Throughout the figure: blue arrows represent non-inverted blood, while red arrows or triangles represents inverted and saturated blood. Red areas are the spatial selective regions, while the green area is the imaging region.

reach the capillaries within the voxels of interest, before image acquisition. Other sequences acquire data at multiple PLDs to allow sampling the entire kinetic curve and through this a more quantitative perfusion estimate. As ΔM is approximately 1% of the full signal, it requires averaging of multiple label-control pairs to gain a sufficient SNR.

A large range of ASL techniques are described in the literature, but they can essentially be categorised into four main types based on how they spatially label the blood. The four types are (1) continuous ASL, a plane-selective method; (2) pulsed ASL, a slab-selective method; (3) velocity and acceleration selective ASL, which are vascular-selective; and finally (4) territorial ASL,

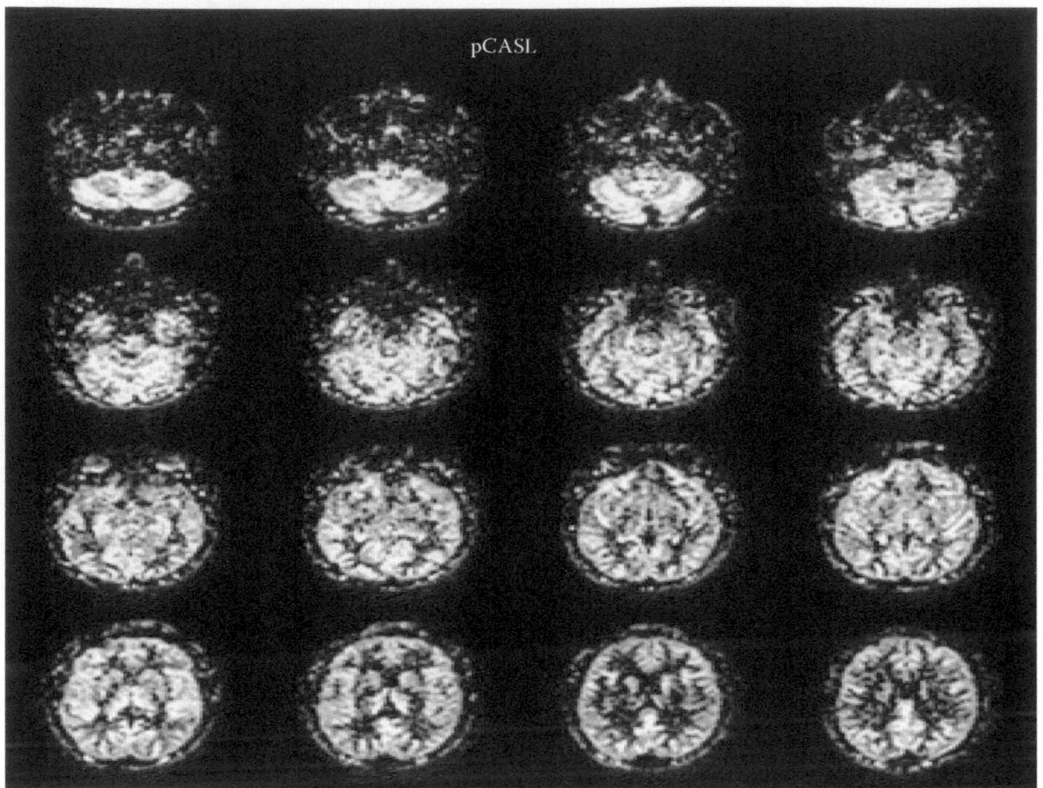

FIGURE 16.2 Example of the resulting ΔM perfusion maps obtained using the pseudo-continuous ASL (pCASL) sequence in a healthy volunteer.

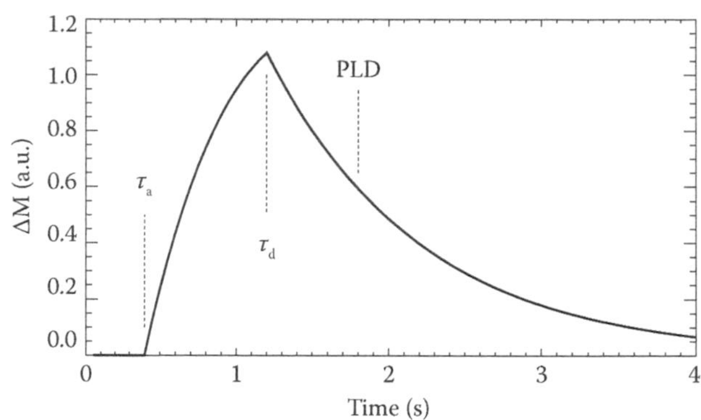

FIGURE 16.3 A kinetic model for PASL assuming fast exchange and therefore a single well-mixed compartment. Having acquired data from multiple PLDs, the cerebral blood flow (*CBF*), arterial transit time (ATT, marked as τ_a in the figure) as well as the bolus duration (τ_d–τ_a) can be fitted using Equation 16.7. If only a single PLD is acquired, one aims at placing it late enough to ensure that the entire bolus has arrived and Equation 16.9 is subsequently used for quantification.

which is vessel-selective. They will be discussed in detail next, but regardless of the chosen labelling approach, the readout scheme used and whether a single or multiple PLDs are sampled is not different between the methods.

16.3.1 Continuous ASL (plane-selective)

The first labelling category is called *continuous ASL* or CASL. As the name suggests, the inflowing blood is continuously labelled in a label plane below the imaging region (Figure 16.1). It is based on a continuous flow-driven adiabatic inversion of the blood water while it passes the label plane, a method that was previously used for angiography (Dixon *et al.*, 1986). Subsequently it was applied as a method for non-invasive perfusion imaging (Williams *et al.*, 1992), the first of its kind. The typical CASL sequence applies a continuous low amplitude radiofrequency (RF) pulse while applying a magnetic field gradient in the flow direction. By optimising the RF amplitude and gradient strength such that they match the average flow velocity of the feeding vessel, a labelling efficiency α, in the range of 70%–90% (Zhang *et al.*, 1993; Trampel *et al.*, 2004) can be obtained. A typical label duration is 1.5–2 seconds, after which a PLD of a similar duration is applied, to ensure that the trailing edge of the generated bolus will arrive to the tissue before the readout. However, with respect to the image slice, such a long off-resonance labelling pulse causes saturation of the macromolecular pool within the image region, similarly to magnetisation transfer (MT) imaging (Henkelman *et al.*, 1993). If this effect is not entirely replicated in the control experiment, the artificial signal intensity difference caused by MT will cause an overestimation of the perfusion. Various approaches have been proposed to overcome this, ranging from the use of localised labelling coils to the application of two closely spaced inversion planes during the control experiment. Small external labelling coils, with a limited RF field only covering the feeding vessels of the neck, avoid MT effects within the brain region (Zhang *et al.*, 1995). On the other hand, the double adiabatic inversion approach (Alsop and Detre, 1998) aims at keeping the inflowing blood untouched by inverting and reinverting the blood while at the same time keeping the RF power and frequency offset similar to that of the label experiment. Still, CASL provokes issues with the RF amplifiers and the application of long-lasting pulses at a 100% duty cycle. Recently an elegant solution to this was proposed (Dai *et al.*, 2008). It is called *pseudo-continuous ASL* or pCASL and it relies on an RF pulse train in combination with varying gradients that together ensure the same flow-driven adiabatic inversion as in the CASL experiment. A great benefit, besides the more amplifier-friendly duty cycle, is the fact that the control experiment is achieved simply by switching the polarity of every other pulse in the pulse label train (Figure 16.4a). This ensures similar MT effects between the label and control experiment while leaving the inflowing blood undisturbed. For these reasons, this technique has attracted great interest and it is without doubt the preferred ASL technique today.

16.3.2 Pulsed ASL (slab-selective)

The second labelling category is called *pulsed ASL* or PASL. Here a slab of typically 10–15 cm of blood is inverted in the neck region at a single instance in time (Figure 16.1), often using an adiabatic slab-selective inversion pulse (Edelman *et al.*, 1994). Depending on the velocity of the blood in the feeding vessels, the generated bolus has a typical temporal duration of 0.5–0.8 s, but contrary to CASL, the exact duration is not known. Subsequently, a PLD of a similar duration as used in CASL is applied, to ensure that the trailing edge of the generated bolus will arrive to the tissue before the readout (Figure 16.4b). Again, off-resonance labelling pulses cause saturation of the signal in the imaging region through MT; therefore, the control experiment has to be

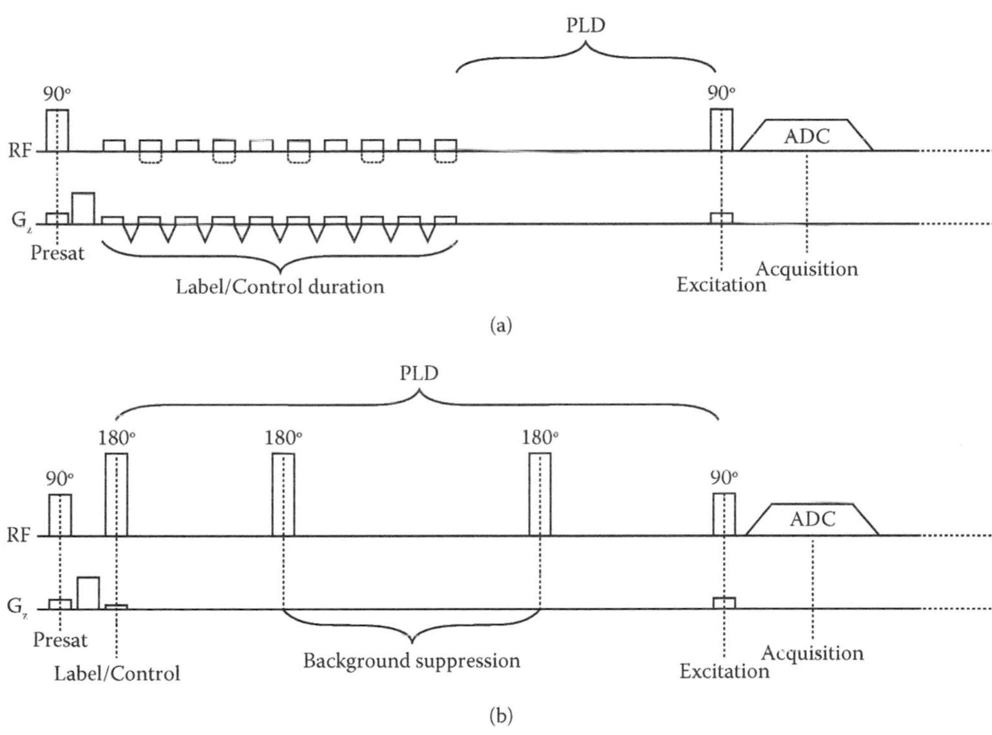

FIGURE 16.4 (a) Schematic of the pCASL sequence, which starts with a presaturation of the image region to avoid any spin history. Subsequently a train of pulses and gradients that ensure a flow-driven adiabatic inversion of the blood passing the label plane performs the pseudo-continuous labelling. For the control experiment, every other pulse phase is toggled by 180° (dotted lines in the pulse train) to cancel the labelling, while keeping magnetisation transfer effects similar to the label experiment. Then the PLD is added before the final excitation and readout (ADC = analogue to digital conversion). (b) Schematic of the PASL sequence. Again, a presaturation is applied to the image region prior to labelling, which is done using an approximately 10-cm-wide slab selective inversion in the neck region. The control experiment does not invert the blood, but it aims to keep the magnetisation transfer effects similar in the two experiments. In the PASL sequence, two global selective background suppression pulses are displayed that aim at nulling the static tissue signal of e.g. grey and white matter at the time of the excitation pulse, while keeping the ΔM signal undisturbed. This can also be applied in the pCASL sequence, and it is an efficient technique to make the sequence more robust to motion and physiological noise. Using background suppression, it is ideally only the perfusion-weighted signal, ΔM, which is measured and therefore subtraction issues originating from the static tissue are avoided.

matched in this regard, similar to the CASL sequence. In fact, the wide variety of PASL sequences described in the literature (Petersen *et al.*, 2006) mainly differ in how they deal with this compensation. Two main PASL labelling approaches exist. One uses the original slab selection proximal to the image slices proposed by Edelman *et al.* (1994), which since has been refined in many versions to circumvent MT issues (Wong *et al.*, 1997; Golay *et al.*, 1999). The second approach, proposed independently by two groups (Kim, 1995; Kwong *et al.*, 1995), centres both the label and inversion slab directly on the image region. By only varying the slab thickness between the label and control experiment, the MT effects cancel out. In today's PASL sequences, the MT effect is however negligible due to the application of a presaturation pulse scheme that saturates the static tissue in the imaging region prior to the labelling pulse.

Compared to the CASL sequences, PASL typically obtains a higher inversion efficiency of above 95%. However, the bolus duration is limited by the inversion slab thickness and the blood velocity in the arteries. A slab thickness above 15 cm is typically not feasible due to the physical extent of the body coil used for labelling. In addition, unlike in CASL where the bolus is generated in a narrow plane, in PASL the trailing edge of the bolus travels longer than the leading edge, resulting in additional T_1 relaxation of the inverted blood. Altogether, this results in CASL being superior to PASL in terms of SNR. PASL did have some advantage over CASL in terms of ease of implementation, but the popularity of PASL has waned since the introduction of pCASL, which is just as easy to implement on the existing scanner hardware as PASL. However, as we will see in later sections, there are some benefits of PASL when it comes to sampling at multiple time points.

16.3.3 Velocity and acceleration selective asl (vascular-selective)

The third labelling category is not spatially selective in the traditional sense, but rather, the labelling depends on the motion of the blood within the vasculature (Figure 16.1). Two such approaches exist, the first one being velocity selective ASL or VSASL, where the blood moving faster than a predefined velocity threshold – typically set in the range of 1–4 cm/s – is saturated by a velocity selective labelling pulse (Wong *et al.*, 2006; Wu and Wong, 2007). During the control experiment, a similar velocity selective labelling pulse module is applied but with a very high or infinite velocity threshold. This causes all the blood to remain undisturbed while still compensating for direct saturation effects caused by the pulses. An appropriate post-labelling delay is applied before the readout as in CASL and PASL. The generated label has approximately half the amplitude as compared to CASL and PASL due to a saturation effect rather than an inversion, which would result in a lower SNR in the perfusion estimates. However, VSASL has the great advantage of labelling inside the image region itself, which means that the bolus arrives to the tissue roughly instantaneously after labelling. The very short arterial transit time (ATT) minimises the T_1 relaxation of the blood bolus prior to arrival to the tissue. As a result, SNR is gained as compared to CASL and PASL where ATTs up to 1 s are commonly observed. In healthy volunteers, VSASL has a performance just below pCASL, whereas in patients with long ATT, VSASL outperforms pCASL (Qiu *et al.*, 2012; Schmid *et al.*, 2015).

The second approach works in a similar manner; however, instead of targeting the velocity, it is the acceleration and/or the deceleration of the moving blood that determines the saturation effect. Acceleration selective ASL or AccASL thereby targets both the accelerating and decelerating blood caused by the cardiac cycle but also velocity changes caused by anatomical changes of the vasculature (Schmid *et al.*, 2014, 2016). Such deceleration exists for instance at the transition from arterioles down to the capillaries, where the blood flow slows down to very low velocities. This technique seems to label the blood even closer to the tissue than VSASL while still labelling the larger vasculature (Schmid *et al.*, 2016), resulting in an increased SNR as compared to VSASL. While the initial AccASL studies looked promising (Schmid *et al.*, 2014, 2015, 2016), with regard to SNR, further studies are needed to establish *CBF* estimation accuracy.

The MRI vendors does currently not provide VSASL or AccASL sequences, but processing can be done using existing processing pipelines described below.

16.3.4 Territorial ASL (vessel-selective)

The fourth and last labelling category is aimed at labelling the individual feeding blood vessels to obtain information about the extent of the perfusion territories they feed (Hartkamp *et al.*, 2013). As such, territorial ASL or TASL provides direct clinical information about cerebral blood flow as well as about changes in the perfusion territories. These changes can for example occur as a result of arteriosclerosis obstructing a vessel. The original approach was based on PASL, with a dedicated pencil beam (Davies and Jezzard, 2003) or selection slab (Hendrikse *et al.*, 2004) covering the vessel of interest. More recently, the pCASL labelling scheme was adapted to allow a sinusoidal spatial variation of the labelling efficiency within the pCASL labelling plane (Wong, 2007). Flow territory maps of the brain's main feeding vessels can be generated by scanning a series of different orientations and spatial frequencies (Figure 16.5). This method has also been adapted for selective labelling of a single vessel at a time, even inside the brain (Helle *et al.*, 2010), based on an earlier CASL approach using similar principles (Werner *et al.*, 2005).

The MRI vendors do not currently provide TASL sequences, and processing of territorial data is done using own tools or tools made available by the provider of the sequence.

16.3.5 Controlling the acquired ASL signal

All the above-mentioned labelling categories have the freedom to choose any readout method available, to scan with or without vascular signal crushers, or to scan with or without background suppression. An overview of the recommended acquisition parameters for a basic ASL study in both healthy and patients

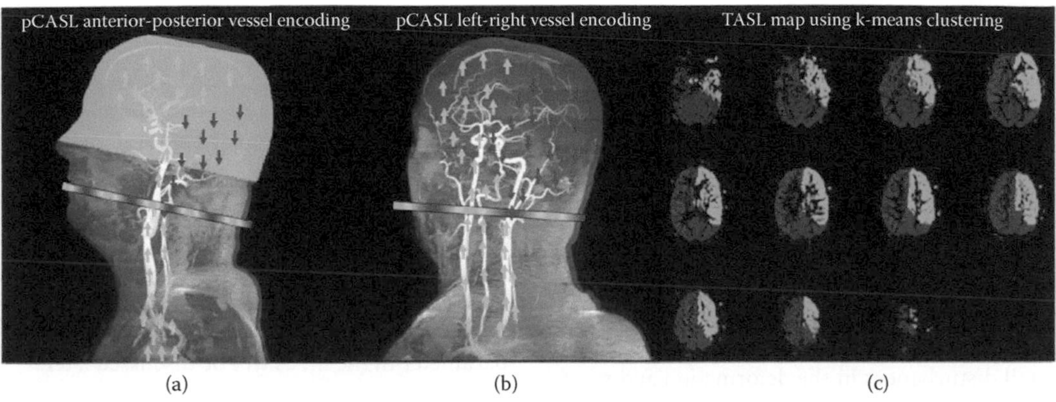

FIGURE 16.5 By applying time varying gradients within the labelling plane during the pCASL labelling train, spatially varying (sinusoidal) label efficiency is achieved. (a) Encoding in an anterior–posterior direction allows labelling of the vertebral arteries while keeping the carotids unaffected. (b) Encoding in a left–right direction allows labelling of the right carotid and vertebral artery, while the left side is kept undisturbed. (c) Territorial ASL (TASL) map based on *k*-means clustering of three different encoding directions as well as the standard label and control. This combination allows robust planning-free TASL.

TABLE 16.1 Suggested acquisition parameters

Parameter	What and why!
Field strength	*3 T* if available because it provides higher SNR than 1.5 *T*, and in addition the T_1 of blood is longer at higher field strength, which also improves the SNR.
Hardware	*Body coil* for RF transmission. *Parallel imaging* using eight or more receive coils for a fast read-out with a typical speed-up factor of 2–3.
Sequence	*pCASL* because it is made available by all vendors and it is the current workhorse for ASL due to a very good SNR. If possible to adjust, one should aim at an average labelling gradient of 1 mT/m, a slice-selective labelling gradient of 10 mT/m and an average B_1 of 1.5 μT. For quantification, one needs a proton density weighted or M_0 map in addition to the actual ASL scan. This is typically a scan without labelling, with a long TR (>5 s) and otherwise using the same matrix, TE, etc., as in the pCASL scan, to keep the distortions the same in the two images.
Label duration	A label duration of *1800 ms* is recommended.
Post-labelling delay (PLD)	The *PLD* should be long enough for the labelled blood water to reach the tissue, yet short enough to avoid an unacceptable signal loss due to T_1 decay of the blood. For clinical scanning, a PLD of *2000 ms* is recommended, while it can be reduced to *1800 ms* in healthy adult subjects below 70 years of age, if needed. Multiple PLDs enable determination of ATT.
Background suppression	Background suppression is *strongly recommended*, because it reduces noise from, for example, motion. However, always remember to acquire an M_0 scan without background suppression for the subsequent quantification.
Spatial resolution	*3–4 mm* in plane resolution with a slice thickness of *4–8 mm for sufficient SNR.*
Readout	*3D* gradient echo and spin echo, *3D* stack of spirals or *multislice 2D EPI.* 3D multishot (four or more shots) sequences are often preferred for standard clinical assessment. Multislice EPI is recommended for functional ASL studies because the readout matches the often-accompanied BOLD acquisition and they can be done as single shot, that is within each TR. In fact, the control image can provide simultaneous BOLD contrast if background suppression is switched off. Current single-shot 3D readout options have an unacceptable point-spread function for any clinical or functional evaluation. Both the shortest possible TR and TE should be chosen for the given sequence.
Planning of the labelling plane	Often the label plane is fixed at a predefined distance below the image stack. If possible, the labelling plane should be placed perpendicular to the feeding arteries (carotid and vertebral arteries) with the aid of a fast phase contrast angiogram.

Note: SNR = signal-to-noise ratio; RF = radiofrequency; ASL = arterial spin labelling; pCASL = pseudo-continuous ASL; ATT = arterial transit time; EPI = echo planar imaging; BOLD = blood oxygen level dependent.

are provided in Table 16.1. Following these guidelines will ensure data acquisition in line with the recommendations from the current ASL consensus paper (Alsop *et al.*, 2015). Several software packages are available for ASL processing, including FSL-BASIL (Chappell *et al.*, 2009), the MATLAB-based ASLtbx (Wang *et al.*, 2008) and the ASAP toolbox (Mato Abad *et al.*, 2016). These can directly deal with quantification of pCASL and PASL data and can be adjusted to process VSASL and AccASL data.

16.3.5.1 Background suppression

ASL considers difference images ΔM, where the signal is in the order of 1% of the equilibrium signal of the tissue. The small ΔM introduces a challenge to avoid artefacts in the form of severe subtraction errors should the subject move even slightly between subsequent label and control acquisitions. These typically manifest as very high intensity signals at the edge of the brain but also between interfaces of different tissues. Four solutions exist:

1. *Prospective motion correction*, which updates the image slab in the event of motion, is available as a navigator-based approach or with external tracking, e.g. using optics.
2. *Retrospective motion correction*, where the volumes are realigned before the subtraction, is more common, and it is included in most processing pipelines. A limitation is that small disturbances in the deformation and steady-state signal still exist between the corrected label-control pairs, in particular where 2D multislice readout is used. In addition, for 3D sequences where multiple shots typically are used to fill the entire *k*-space, the impact of motion occurs within the volume acquisition, which makes retrospective correction impossible.
3. *Background suppression* offers a robust solution for this issue. By nulling the static tissue signal at the time of the readout (Dixon *et al.*, 1991), only the ΔM signal is sampled, making the technique much more robust to motion and physiological noise (Ye, Frank, *et al.*, 2000). Motion will in this case blur the resulting ΔM map, rather than destroying it with severe motion artefacts. The principle relies on the application of up to several global inversion pulses applied appropriately during the post-labelling delay as depicted in Figure 16.4b. Each inversion pulse allows the suppression of one tissue type or T_1 value. This technique turns out to be crucial for segmented 3D acquisitions in ASL. While prospective motion correction can be combined with background suppression, retrospective motion correction has limited success due to the very low signal intensity in suppressed images.
4. *Outlier detection* methods are typically part of the ASL processing pipeline, where the control-label pairs that are considered outliers are discarded. This could be based on temporal signal variation, but it can also be based on the magnitude of rotation and translation obtained from the retrospective motion correction. This enables rescue of datasets where severe artefacts contaminated a few subtraction pairs and that otherwise would render the overall perfusion image useless if they were included in the average.

16.3.5.2 Bolus saturation

For the PASL, VSASL and AccASL labelling schemes, we do not have control over the temporal duration of the bolus, as is the case in CASL where it is predefined. This poses a problem when we subsequently want to quantify *CBF*, because an unknown bolus duration will introduce an error in the estimate. Therefore, these techniques should employ bolus saturation schemes that, after a predefined time from labelling, cut off the trailing edge of the bolus. In PASL a well-defined bolus duration is assured by applying a saturation slab within the label region approximately 5–800 ms after labelling (or control), assuming that all the feeding vessels still contained label inside the saturation region at the time it was applied. In Figure 16.1 (middle row), it would correspond to the application of a saturation of the red label region at around 500 ms after labelling. This is often referred to as *QUIPSSII* (Wong *et al.*, 1998). A similar approach can be applied for VSASL and AccASL. However, here it is the actual label module that is applied to saturate the blood in both the label and the control experiment. If multiple PLDs is acquired, an alternative to the bolus saturation schemes is to fit the bolus duration as a parameter in the fit, as will be discussed later.

16.3.5.3 Vascular crushers

In addition to motion and physiological noise, another artefact needs consideration at the time of acquisition. It is the potential signal contamination from the feeding arterial vessels. The definition of perfusion is the blood delivery rate to the tissue's capillary bed and, as such, it does not include the signal from the larger arterial vessels, which may or may not feed other regions outside the observed voxel. Arterial signal contamination can be limited by choosing a post-labelling delay long enough to ensure that the blood has cleared from the arterial compartment by the time of acquisition. However, if the *CBF* quantification relies on data from multiple PLDs it would often be necessary to spoil the arterial signal, which is especially prominent at short PLDs. This is done by applying bipolar crusher gradients, which dephase the signal in faster-moving vessels based on their velocity (Ye *et al.*, 1997). They are typically applied right after the excitation pulse and prior to the readout. Alternatively, it can be a dedicated crusher module similarly to the label module of VSASL, which is applied prior to the excitation pulse for both the label and the control experiment. An alternative is to keep the vascular signal and include it into the kinetic model (Chappell *et al.*, 2009).

16.3.5.4 The readout

The chosen readout is independent of the labelling scheme and the choice therefore relates to questions such as what resolution is needed and what underlying point-spread function is allowed. This again needs to be considered with respect to overall SNR, whether vascular crushers are needed and whether multiple readouts are needed within each repetition time.

The traditional readouts were multislice 2D echo planar imaging (EPI) and spiral, which both have the capacity for whole brain coverage in a very short time, 400–600 ms. These techniques have the benefit that vascular crushers are easily implemented between the excitation and the readout. As well, they allow repeated low flip-angle readouts, also called a *Look-Locker* readout. The Look-Locker readout enables the acquisition of multiple PLDs within each repetition time (TR) (Günther *et al.*, 2001), which facilitates fitting of the entire kinetic curve (Figure 16.3). This approach is most feasible for PASL because here the labelling takes place in the order of few milliseconds and as a result the sampling can start before the label arrives at the image region. Recently, the fast 3D readout sequences, including 3D gradient echo and spin echo (Günther *et al.*, 2005) and 3D stack-of-spirals (Ye, Frank, *et al.*, 2000), have gained great popularity because they offer improved SNR and reduced susceptibility artefacts compared to 2D EPI readouts. They rely on interleaving an EPI or a spiral readout, respectively, within a turbo-spin-echo readout. The benefit is a large SNR due to the larger 3D excitation volume rather than single slices. A benefit of these particular 3D sequences over other 3D sequences such as fast spin echo and gradient echo is that the time of excitation is well defined at the point of the initial 90° excitation pulse. This not only makes quantification easier, but it also enhances the effect of background suppression that is critical for these sequences. To avoid excessive blurring, the acquisition is split into four to eight segments. Even a small degree of motion between these different segments acquired several seconds apart would cause severe image artefacts were it not for background suppression.

16.4 ASL: Quantification

16.4.1 Kinetic modelling approaches

Having acquired the ASL images with appropriate sequence parameters, that is, without contamination from MT, vascular artefacts or motion as discussed above, we are left with a perfusion weighted image ΔM. To convert it to a quantitative measure of cerebral blood flow in [ml/100g/min] we need to apply some sort of kinetic modelling, because the relation of ΔM to perfusion depends on various other local and global parameters. These include arrival time, tissue relaxation parameters of both blood and tissue as well as their equilibrium magnetisation, bolus duration and inversion efficiency. The complexity of the applied model also needs to be considered in relation to the data acquired, e.g. whether multiple PLDs or only a single PLD were sampled and so forth.

16.4.1.1 Single-compartment model

The simplest and most common approach is to view the ASL signal as originating from a single compartment system, that is, the situation where the labelled blood exchanges instantaneously with the tissue compartment upon arrival to the capillary bed. The Bloch equation describing the change in MR signal over time can then be modified to

$$\frac{dM_t(t)}{dt} = \frac{M_{t,0} - M_t(t)}{T_{1t}} + CBF \cdot (M_a(t) - M_v(t)) \qquad (16.4)$$

where M_t, $M_{t,0}$, M_a and M_v are the tissue, equilibrium, arterial and venous magnetisation, respectively, and T_{1t} is the relaxation time of the tissue. Because we assume a fast exchange regime, *Mv* is equal to the tissue magnetisation corrected by differences in the water content also described by the blood-brain partition coefficient λ, such that $M_v = M_t/\lambda$. Notice the similarity of Equation 16.4 to Equation 16.1 and, indeed, when considering the water as freely diffusible, ASL quantification is similar to the early nitrous oxide method, with the main difference being that ASL quantification is done on a voxel-wise basis rather than as a whole-brain measure. Many elaborate analytical solutions have been proposed over the years (Kwong *et al.*, 1995; Calamante *et al.*, 1996) for converting ΔM into *CBF*, taking into account differences in T_1 between tissue and blood, accounting for the arterial transit time τ_a and the actual bolus duration τ_b.

Buxton's model is a general kinetic model for ASL (Buxton *et al.*, 1998), which describes ΔM based on the convolution integral in Equation 16.2, and it includes the effect of T_1 decay:

$$\Delta M(t) = 2 \cdot M_{a,0} \cdot CBF \cdot \int_0^t c(\tau) r(t-\tau) m(t-\tau) d\tau \qquad (16.5)$$

where $M_{a,0}$ is the equilibrium magnetisation in a blood-filled voxel, c(t) is the delivery function or fractional arterial input function (AIF), r(t−τ) is the residue function describing the fraction of labelled spins arriving to a voxel at time τ that still remains within the voxel at time t. The magnetisation relaxation term m(t−τ) quantifies the longitudinal magnetisation fraction of labelled spins arriving to the voxel at time τ that remains at time t. Solving this for PASL and assuming single-compartment kinetics and plug flow is done by setting

$$c(t) = \begin{cases} 0, & t < \tau_a \\ \alpha \cdot e^{-t \cdot R_{1a}}, & \tau_a \le t < \tau_d \\ 0, & t \ge \tau_d \end{cases}$$

$$r(t) = e^{-\frac{t \cdot CBF}{\lambda}} \qquad (16.6)$$

$$m(t) = e^{-t \cdot R_{1t}}$$

where R_{1a} and R_{1t} are the relaxation rates for arterial blood and tissue, respectively and α is the inversion efficiency. The trailing edge τ_d of the bolus is equal to the arterial transit time τ_a plus the bolus duration τ_b. This result in a stepwise-defined function:

$$\Delta M(t) = \begin{cases} 0, & t < \tau_a \\ \dfrac{-2 \cdot \alpha \cdot M_{a,0} \cdot CBF}{\delta R} \cdot e^{-t \cdot R_{1a}} \cdot (1 - e^{(t-\tau_a) \cdot \delta R}), & \tau_a \le t < \tau_d \\ \dfrac{-2 \cdot \alpha \cdot M_{a,0} \cdot CBF}{\delta R} \cdot e^{-\tau_d \cdot R_{1a}} \cdot (1 - e^{\tau_b \cdot \delta R}) \cdot e^{-(t-\tau_d) \cdot R_{1app}}, & t \ge \tau_d \end{cases} \qquad (16.7)$$

where $\delta R = R_{1a} - R_{1app}$ and $R_{1app} = R_{1t} + CBF/\lambda$, also called the *apparent tissue relaxation rate*. If one wants to solve it for CASL, then c(t) in Equation 16.6 simply lacks the additional T_1 decay during the bolus duration, that is $\alpha \cdot e^{-t \cdot R1a}$ becomes $\alpha \cdot e^{-\tau a \cdot R1a}$. If multiple post-labelling delays are sampled, this would be the simplest kinetic model to fit and as is evident, various parameters such as the arterial transit time τ_a, blood–tissue partition coefficient λ, $M_{a,0}$, R_{1a} and R_{1t} all need to be estimated or measured in order to obtain quantitative *CBF* values.

Nevertheless, most clinical ASL examinations today are performed using pCASL or PASL at a single PLD. The current consensus (Alsop *et al.*, 2015) is to simplify the quantification to

$$\mathrm{CBF} = \frac{6000 \cdot \lambda \cdot \Delta M \cdot e^{PLD \cdot R_{1a}}}{2 \cdot \alpha \cdot T_{1a} \cdot M_{t,0} \cdot (1 - e^{-\tau_b \cdot R_{1a}})} \left[\mathrm{ml/100g/min} \right] \text{ for pCASL} \tag{16.8}$$

and

$$\mathrm{CBF} = \frac{6000 \cdot \lambda \cdot \Delta M \cdot e^{PLD \cdot R_{1a}}}{2 \cdot \alpha \cdot \tau_b \cdot M_{t,0}} \left[\mathrm{ml/100g/min} \right] \text{ for PASL} \tag{16.9}$$

In this approach, it is assumed that the label only decays with T_1 of blood and that no blood leaves the voxel. This means that we consider it as a slow exchange system and, correspondingly, that blood stays in the vascular bed during the experimental time of an ASL experiment of a few seconds. Even if exchange of the label to the tissue occurs, the introduced errors due to differences in T_1 are expected to be small, in grey matter at least (Parkes and Tofts, 2002). By ensuring that the PLD is rather long, on the order of 2 s, one can assume that there will be negligible error due to unknown arterial transit time, for a relatively normal range of transit times (i.e. <2 s) seen in most subjects not suffering from vascular disease. If the transit times are abnormally long due to pathology, *CBF* quantification will lose accuracy. On perfusion maps this is reflected by the presence of vascular artefacts. The factor of 6000 ensures the conversion from [ml/g/s] into the usual [ml/100g/min].

16.4.1.2 Multicompartment models

Aquaporin water channels selectively transfer water molecules back and forth through the blood–brain barrier or the capillary wall. This causes a restricted exchange of the labelled water from the vasculature to the tissue, and a single compartment model may not be the best description of the system. In practice, it means that the water remains longer in the vasculature and as a result the signal decays with T_1 of blood rather than that of tissue for a longer period of time. This results in an underestimation of *CBF*. Various flavours of multicompartment approaches have been proposed to correct for the effect of restricted permeability (St Lawrence *et al.*, 2000; Zhou *et al.*, 2001; Parkes and Tofts, 2002). The system then has a permeability-surface product incorporated in a two or more compartment system, much like for dynamic contrast-enhanced perfusion imaging (Tofts and Kermode, 1991). In general, the biggest limitation of these more advanced kinetic models is that they require data at several different PLDs with a good SNR in order to fit the additional parameters. This is currently hard to achieve within clinically realistic scanning times and the approaches are therefore rarely used in clinical studies. Nevertheless, these techniques will potentially become more feasible and maybe even necessary for accurate *CBF* quantification at ultra-high field of 7T or above where the SNR is increased and the T_1 and T_2 relaxation parameters differ more between arterial and venous blood compared to 1.5T and 3T.

16.4.1.3 Model free

An alternative to kinetic modelling is to use a sampling scheme where both the tissue signal curve and the arterial input curve is acquired (Petersen *et al.*, 2006). This offers information about the arterial input function or $2 \cdot M_{a,0} \cdot c(t)$ in Equation 16.5, while having acquired the tissue signal $\Delta M(t)$ permits to calculate the tissue residue function r(t)m(t) in Equation 16.5 by means of deconvolution. This is similar to the dynamic susceptibility contrast perfusion method (Ostergaard *et al.*, 1996) where *CBF* is extracted from the peak of the resulting residue function, knowing that r(0)m(0) equals 1 by definition. A great advantage of this 'model free' approach is the fact that no prior assumptions about the underlying kinetic modelling, such as whether a single or multiple compartment system, are needed.

In ASL, the AIF can be extracted locally by considering the arterial signal on a voxel-by-voxel basis. This is achieved by interleaving two ASL experiments with acquisition at multiple PLDs, one with and one without vascular crushers. The extraction of AIF comes at the cost of a 33% increase in acquisition time due to the combination of two experiments (Petersen *et al.*, 2006; Petersen *et al.*, 2010). In the non-crushed experiment, the difference signal $\Delta M_{ncr}(t)$ consists of the tissue signal $\Delta M(t)$ plus $\Delta M_{art}(t)$, which is the arterial contribution in a given voxel. The crushed experiment, on the other hand, gives a difference signal ΔM_{cr}, which is equal to the desired tissue signal $\Delta M(t)$. The needed information $2 \cdot M_{a,0} \cdot c(t)$ therefore lays in $\Delta M_{art}(t)$, which is isolated through subtraction ($\Delta M_{ncr}(t) - \Delta M_{cr}(t)$). Because $\Delta M_{art}(t)$ originates from the arterial blood volume (aBV) and not a pure arterial voxel as expected for $2 \cdot M_{a,0} \cdot c(t)$, a scaling of the area under the curve is needed. The sequence used is a PASL sequence with QUIPSSII bolus saturation and a Look-Locker multi-PLD readout. In this sequence the bolus area is known to be $2 \cdot M_{a,0} \cdot \alpha \cdot \tau_b$ and scaling can therefore be done accordingly prior to the deconvolution. It also facilitates the assessment of the aBV on a voxel-by-voxel basis using the relationship:

$$aBV = \frac{\int_0^{\infty} \Delta M_{art}(t) \cdot e^{t/T_{1a}} dt}{2 \cdot M_{a,0} \cdot \tau_b \cdot \alpha}$$

This technique gives additional information to *CBF* in the form of aBV and ATT, which can be extracted from the multiple PLD acquisition, as we will see below.

16.4.2 Parameter estimation

For perfusion quantification, we need knowledge about other parameters than the initially acquired ΔM, such as the equilibrium magnetisation, bolus shape and arrival time, all depending on the complexity of the kinetic modelling in use. Table 16.2 highlights the usual processing steps that are used and the typical parameters assumptions used for quantification.

16.4.2.1 Blood and tissue equilibrium magnetisation and relaxation times

Regardless of the chosen labelling scheme and modelling approach, the amplitude of the equilibrium magnetisation $M_{a,0}$ is always required for absolute quantification. It can in theory be extracted from a 100% blood-filled voxel; however, due to a typically low and unknown blood fraction within the individual voxels it is hard to extract in practice. Instead, we typically measure tissue proton density, and we rely on the known differences between the water content of blood and tissue, which is the blood–brain partition coefficient λ, giving the relationship $M_{a,0} = M_{t,0}/\lambda$. The tissue proton density or equilibrium magnetisation is easy to estimate from a scan with a long repetition time and otherwise similar readout as for the ASL acquisition. The currently used blood–brain partition coefficient was established by H_2O^{15} PET (Herscovitch and Raichle, 1985) with whole brain λ = 0.9 ml/g, grey matter λ_g = 0.98 ml/g and white matter λ_w = 0.82. Typically, the single whole brain value is used, as tissue classification is not always available and it has been shown that spatial variation also exists within a single tissue type (Roberts *et al.*, 1996). Although $M_{a,0}$ ought to be a single global parameter, this method will have a spatial variation depending on $M_{t,0}$. This turns out to be a benefit, because it automatically includes correction for differences in coil sensitivity and other inhomogeneities across the brain that otherwise would need to be corrected separately. An alternative approach is to use a global $M_{t,0}$ and λ, which essentially would avoid the error due to the spatial variation in lambda. However, it will not correct the coil inhomogeneities, which differs between scanners and scan sessions and therefore is harder to control, making the former approach the preferred method of choice.

Having established $M_{a,0}$ by assuming λ and measuring $M_{t,0}$, the relaxation time of blood T_{1a} is likewise needed for any quantification. Most often, it is taken as a general value across the population with a value of 1650 ms at 3T (Alsop *et al.*, 2015). This approach does, in fact, introduce an error because it depends on the subject's current haematocrit (Varela *et al.*, 2011; Zhang *et al.*, 2013; De Vis *et al.*, 2014). Therefore, a better approach is to measure the T_1 of blood in the individual subjects using a separate scan, or if blood samples exist, to convert haematocrit into T_{1a}. The T_1 of tissue is mainly used in more advanced quantification schemes with several PLD, where information about both $M_{t,0}$ and T_{1t} automatically exists. If background suppression is used, both parameters must be extracted from a separate scan.

16.4.2.2 Arterial transit time

The actual time it takes for the blood to travel from the label region to the voxel of interest is called the *ATT* or τ_a in the formulas above. In multiple PLD acquisition it is fitted as part of the quantification, using e.g. Equation 16.7, but it can also be

TABLE 16.2 Suggested Processing Steps and Parameter Assumptions

Processing Steps	What, How and Why!
Acquired images	The basic ASL experiment requires an ASL scan with a sufficient number of label–control pairs (typical scan duration 3–4 minutes). Always export all the label–control pairs. This allows for subsequent coregistration, subtraction and outlier detection. The vendors provide some preprocessing on the console; however, it is recommended to only use this for the initial quality control as the underlying processing isn't always clear. In addition, a M_0 scan without background suppression is needed for the subsequent *CBF* quantification. It should be matched to the ASL scan, but without labelling or background suppression and with a long TR for proton density weighting.
Coregistration, subtraction and outlier detection	If there is enough signal in the images, coregistration of the time series is an option if the subject moved. However, efficient background suppression disallows this. Prospective motion correction can then be used, but this is currently not provided by the vendors and has to be implemented by the researchers. Outlier detection can on the other hand always be used looking at the individual subtraction pairs either in a supervised fashion or being automated.
CBF quantification	After the initial quality control and outlier detection, it is time to average the label-control pairs and quantify *CBF* using Equation 16.8. Each site tends to use their own packages for doing all the processing steps. Browsing the Internet will soon show packages such as the BASIL package from FSL (Chappell *et al.*, 2009) or the ASL toolbox for SPM (Wang *et al.*, 2008) but also a large range of others. It is therefore up to the researchers to judge which package will best fit into the existing processing pipelines. The following parameters should be used unless other information exists for the investigated population. Blood–brain partition coefficient λ = 0.9 ml/g, T_1 of blood 1650 ms and a labelling efficiency α = 0.85, all for pCASL at 3T.
Interpretation and analysis	In the clinic, the evaluation of perfusion images is usually qualitative, that is, the radiologist excludes artefacts and evaluates low or high flow areas according to a reference, e.g., the contralateral side. Because white matter perfusion in general is low and in particular shows low SNR in ASL, perfusion evaluation mainly concerns grey matter. In research studies, group analysis is the most common approach. Voxel-wise analysis is often used in image analysis, but it is not always beneficial for ASL. The reason is that the relative low SNR in ASL maps combined with correction for multiple comparisons requires very large groups or very large effect sizes to ensure the power of the experiment. Larger predefined regions of interest can therefore be a better option in studies where a clear hypothesis exists. In addition, global effects have to be well controlled. *CBF* is altered by the intake of substances such as caffeine and the time of the day – even the level of anxiety at the time of the scan plays a role. Therefore, analysis often targets spatial variations between groups and for this the maps can be normalised, e.g. to global perfusion, and the temporal *CBF* fluctuations can thereby be avoided.

Note: CBF = cerebral blood flow.

estimated by searching for the rising edge of the tissue curve. In the single PLD methods it is impossible to estimate ATT and one relies on acquisition at long PLDs assumed to be longer than both the bolus duration τ_b and ATT together. In such case, the error is negligible, as compared to the general errors from other sources such as in the estimates of $M_{a,0}$ and T_{1a}.

16.4.2.3 Bolus shape, dispersion and decay

The shape of the bolus or AIF has an impact on the quantification. In this case, it is mainly the multi-PLD methods that are sensitive to dispersion of the bolus, whereas the single-PLD methods are relatively insensitive, as long as the condition PLD $> \tau_a + \tau_b$ is fulfilled. Dispersion models can be integrated in the kinetic model fit (Hrabe and Lewis, 2004), and in the model free approach the shape is measured in all voxels with a large enough aBV (Petersen *et al.*, 2006).

16.5 ASL: Applications

ASL has gained great popularity as a research tool, in particular within the field of neuroscience. Recently, robust techniques such as the pCASL sequence have become generally available on most systems, either as a clinical package from the vendor or as shared sequences between research sites. This has allowed researchers to add valuable perfusion information to their studies in addition to the standard structural and functional images by adding an ASL scan a few minutes long. In the following section, we will revisit some of these applications ranging from basic neuroscience to clinical studies. It is not an exhaustive list as there are more than 2000 ASL publications registered in PubMed today (2017) and the numbers are rapidly increasing.

16.5.1 Neuroscience

The complete non-invasiveness of ASL makes it a preferable choice for monitoring perfusion changes in many neuroscience applications. Neuronal activation, which particularly increases the cerebral metabolic rate of oxygen ($CMRO_2$), is accompanied by a regional increase in *CBF* to ensure sufficient supply of oxygen to the mitochondria. The classical method for mapping neuronal activation with MRI is called *blood oxygen level dependent* (BOLD) *contrast* (Ogawa *et al.*, 1990). As the name suggests, the observed signal changes due to neuronal activation result from T_2* changes caused by alterations in blood oxygen saturation within the voxel. Changes in the oxygen saturation originate from a complex interplay of changes in $CMRO_2$, *CBF* and *CBV*, and hence the BOLD signal is hard to interpret. ASL can likewise detect neuronal activation by quantitatively measuring the regional *CBF* changes. The quantitative aspect makes functional ASL more reproducible over time and between subjects than BOLD (Detre and Wang, 2002; Wang, Aguirre, *et al.*, 2003). Also, the signal is better localised to the activated area (Duong *et al.*, 2001) as opposed to BOLD, where strong signal changes also occur in draining venous vessels further away from the activated region, although this effect is field strength dependent. Neuronal activation paradigms have many flavours, with the simplest being a block design, meaning with repeated on and off conditions of up to several minutes in length and where the activation could consist of a simple visual checkerboard or finger tapping, or far more complex cognitive paradigms. A continuous signal sampling is therefore required for the subsequent regression analysis and, similar to BOLD, this can be done for ASL by analysing subsequent control–label pairs. Contrary to BOLD, where the optimum block duration is 16 s due to drift, the quantitative nature of ASL makes a very long block duration of up to days (Wang, Aguirre, *et al.*, 2003) possible. This allow the study of different cognitive paradigms where there is interest in a certain 'state' that may vary slowly in time or the impact of a slow-acting drug.

However, there is a penalty in the temporal resolution compared to BOLD due to the required post-labelling delay and the need to collect control–label pairs. Neither BOLD nor ASL directly map the $CMRO_2$ change caused by the neuronal activation. Calibrated BOLD tackles this issue with a new imaging scheme where the two approaches are combined and $\Delta CMRO_2$ is extracted using a biophysical model of the *CBF* and BOLD signal (Hoge *et al.*, 1999; Chiarelli *et al.*, 2007; Bulte *et al.*, 2012; De Vis *et al.*, 2015a). An additional global *CBF* challenge is needed, typically a part where CO_2-enriched air is inhaled, causing the baseline *CBF* to increase while $CMRO_2$ stays more or less constant (Figure 16.6). A caveat of this approach is the fact that the setup is more complicated and harder to perform in some sensitive subjects and, therefore, efforts are made to simplify the calibration (Blockley *et al.*, 2012) to avoid the hypercapnia challenge.

ASL is therefore becoming an important tool that helps to disentangle the complex behaviour of the brain, both for gaining a better understanding of the physiological events accompanying activation but also for the cognitive neuroscience field aiming at a better understanding of how the brain processes events.

16.5.2 Clinical

Most neuronal diseases carry a perfusion aspect, such as stroke and large vessel disease, dementia and cancer. For example, Alzheimer's disease is marked by hypoperfusion (Austin *et al.*, 2011), whereas high-grade brain tumours are characterised by hyperperfusion (Dangouloff-Ros *et al.*, 2016). Perfusion images can aid in the diagnosis or follow-up of these pathologies. Moreover, the brain has the most advanced autoregulative system for maintaining sufficient perfusion, at all times, including during neuronal activation (De Vis *et al.*, 2015b). An example of this is the dilation of blood vessels to increase perfusion in response to hypercapnia (elevated CO_2 levels). ASL provides important complementary information to the standard structural MR images. The paediatric population, where ionising radiation and contrast agents should be avoided where possible, is in particular benefitting from ASL. Several studies have been performed in neonates (De Vis *et al.*, 2016) and kids (Oguz *et al.*, 2003; Wang, Licht, *et al.*, 2003; Dangouloff-Ros *et al.*, 2016) with various diseases for which there are no good alternatives available for a whole-brain perfusion assessment, especially for the

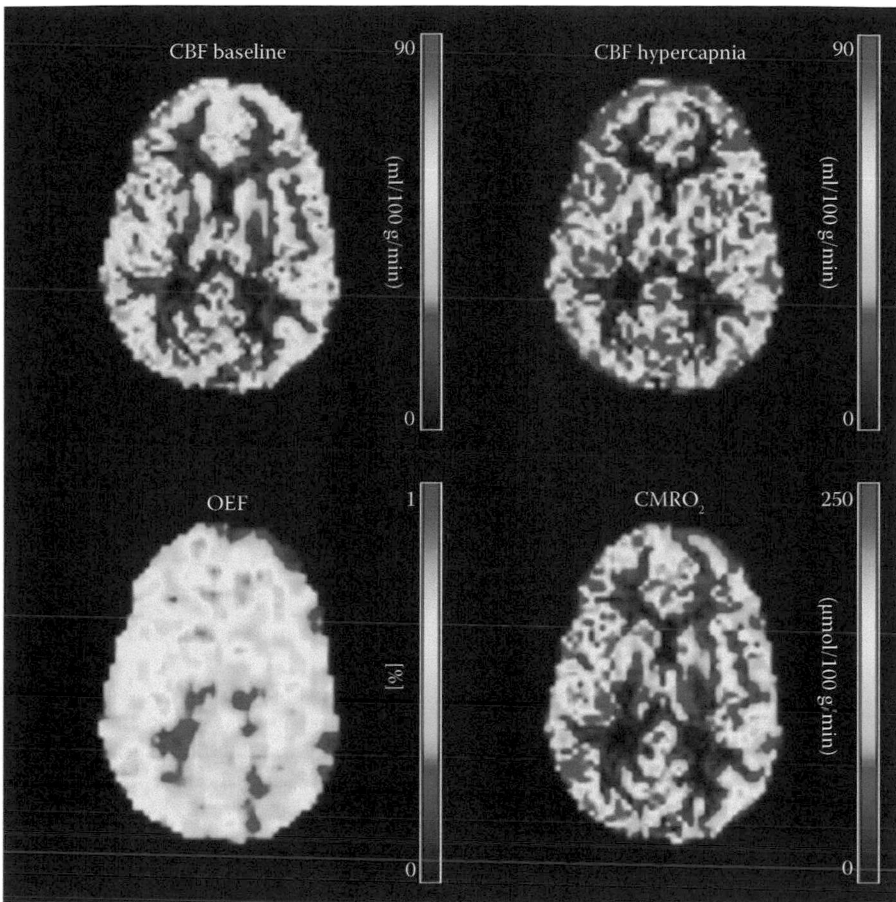

FIGURE 16.6 An example of a calibrated ASL and BOLD experiment in a healthy volunteer using a pre- and post-hypercapnia as well as hyperoxia challenge. This allows the quantification of oxygen extraction fraction (OEF) and the cerebral metabolic rate of oxygen ($CMRO_2$) in addition to *CBF*.

neonates. ASL has also been applied in ageing studies (Parkes *et al.*, 2004; De Vis *et al.*, 2015a) and in neurodegenerative diseases, where it is becoming increasingly apparent that perfusion plays a significant role (Mak *et al.*, 2012; Fernández-Seara *et al.*, 2015; Leeuwis *et al.*, 2016) and regional differences can easily be detected in group analysis between different diseases. Perfusion is an indirect marker of metabolism and indeed the observed areas of hypoperfusion in these patients agree reasonably well with fluorodeoxyglucose PET, a direct measure of metabolism (Haller *et al.*, 2016). However, an ongoing quest is how to sufficiently improve the sensitivity to allow robust discrimination between the subtle variations in perfusion patterns on an individual subject basis, which is needed for clinical diagnosis and disease monitoring (Collij *et al.*, 2016). Another area where ASL has proven versatile is in the field of cerebrovascular diseases where ASL not only provides perfusion information but also information regarding the extent of the feeding vascular territories (Hartkamp *et al.*, 2013). If in addition, a scan is performed before and after the injection of acetazolamide (Bokkers *et al.*, 2010), the cerebrovascular reserve can be assessed as well. Therefore, ASL can provide a comprehensive clinical picture of the haemodynamic status of stroke patients (Hernandez *et al.*, 2012; Wang *et al.*, 2013) and patients with large vessel disease (Figure 16.7) (Chng *et al.*, 2008; Hendrikse *et al.*, 2009; Bokkers *et al.*, 2016). For cancer, ASL can also aid in cancer staging (Yang *et al.*, 2016) and the detection of treatment response. In particular the challenge of distinguishing between radiation necrosis and recurrent tumour (Ye *et al.*, 2016) are ongoing clinical ASL research topics. Perfusion assessment using ASL is done in many other brain diseases such as epilepsy, migraine and psychiatric disorders, and the recent review paper by Haller *et al.* (2016) gives a comprehensive overview of current clinical applications of ASL in the brain.

16.6 Image quality and quantification quality

16.6.1 Pitfalls

A few things need to be kept in mind when interpreting ASL perfusion maps in the clinic or for research purposes.

First, it must be realised that the generated label has a limited lifespan lasting in the order of a few seconds, whereas a gadolinium tracer does not clear significantly within the typical

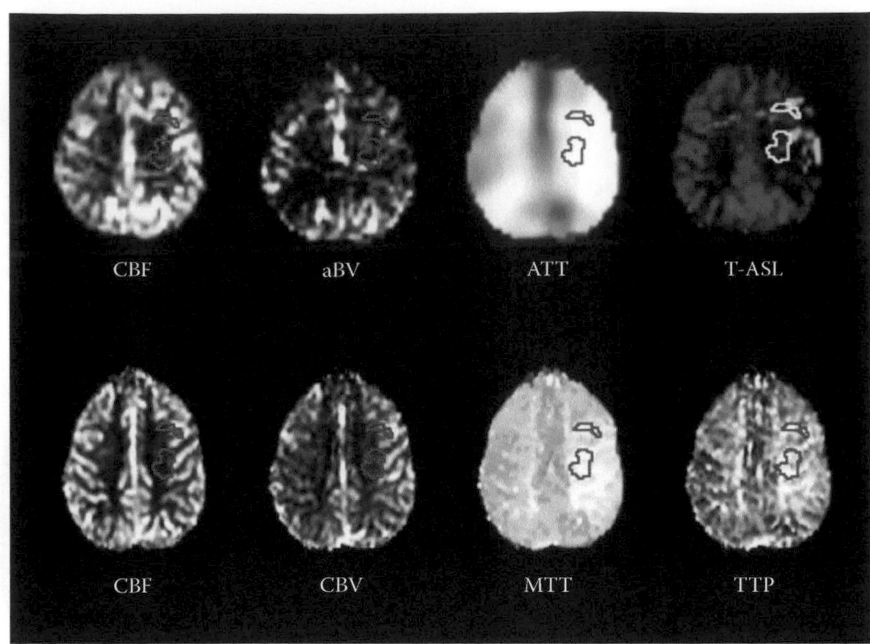

FIGURE 16.7 A stroke patient demonstrating the various haemodynamic maps available from ASL and gadolinium-based perfusion MRI. **Upper row**: From the left, ASL cerebral blood flow (*CBF*), arterial blood volume (aBV), arterial transit time (ATT) and territorial ASL (TASL) where the changes in the left middle cerebral artery perfusion territory can be seen. The red outline marks the infarct core extracted from the diffusion-weighted images. **Bottom row**: Dynamic susceptibility-weighted perfusion MRI after the injection of a gadolinium tracer. From the left, *CBF*, cerebral blood volume (*CBV*), mean transit time (MTT) and time-to-peak of the tissue curve (TTP).

experimental time of several minutes. The consequence of this will mainly affect images from patients with compromised flow due to a stenosis or an occlusion, which results in collateral perfusion via alternative routes and thereby causes very long transit times. If the ATT is longer than the single acquired PLD, then the region will appear dark or with no perfusion at all, making it impossible to quantify the flow. However, in a clinical setting the information can be combined with diffusion and structural images, usually available in such patients, to help determine if it is a very late arrival or no flow, e.g. due to an infarct. Sequences using multiple PLD are less sensitive to prolongation of ATT. However, the prolongation can be so large that ASL would not stand a chance. In intermediate cases, the signal can appear bright in the larger vasculature, suggesting that it is on its way but at the time of sampling it still had not reached the parenchyma.

Second, motion artefacts can destroy ASL perfusion maps. They are often seen as rings in the periphery of the brain with large intensities. If it is present in the final map and you are still at the scanner, repeat the scan. If offline, inspect the individual label–control pairs to see whether discarding a few pairs where the gross motion occurred potentially can save the data.

Third, if using pCASL, keep an eye on the labelling efficiency. This technique is sensitive to accurate B_0 in the label region and the vessels should be perpendicular to the label plane. Both can vary between the different target vessels, and should reduced flow appear in an entire perfusion territory, be suspicious before drawing your conclusions. Again, try to redo the positioning of the label plane and if possible the shim, too. If the observation persists, compare the perfusion map with other clinical images, particularly an angiogram, to confirm the observed phenomenon.

Last but not least, while it is common practice to use a single haematocrit and therefore T_1 of blood for quantification throughout the population (Alsop *et al.*, 2015), it does mean care should be taken when comparing different populations where the haematocrit varies between the groups. This is particularly problematic in the neonatal population where there are large variations between subjects depending on the time since birth (Varela *et al.*, 2011; De Vis *et al.*, 2014) but also in patients with sickle cell disease, for example (Václavů *et al.*, 2016). Therefore, to obtain quantitative perfusion in these populations, one must assess blood T_1 either directly via MRI or indirectly via a haematocrit measurement from the blood with subsequent conversion into T_1.

16.6.2 Quality assurance and normal perfusion values

In addition to the factors described above, image quality of ASL images also depends on the functionality of the system and the functionality of the coils. Therefore, regular checks of the system and coils are essential to ensure high quality ASL data. In addition to the weekly vendor-specific quality assurance, one should consider a stability test in a standardised phantom such as the one proposed by the FBIRN (Greve *et al.*, 2011). The introduction of up to 64 channel receive coils on clinical systems makes it harder to detect failure of the individual coils, and therefore testing the SNR of the individual receive elements over time could be considered, if the system allows it.

Even with a stable system, differences in the quantified perfusion between systems and sequences may still be observed. Scanning the same subjects using the standard vendor-provided ASL sequences not only results in differences in the global perfusion values, but they also have spatial variations from region to region (Mutsaerts *et al.*, 2015). This scenario improves by implementing near-identical sequences on the scanners with similar label module, readout and resolution (Mutsaerts *et al.*, 2015), but these sequences are not readily available on all systems. Global perfusion values could be calibrated between ASL sequences, sites and vendors using a gold standard flow phantom for ASL. However, the spatial variations caused by hardware and readout will still be an issue and, similar to other perfusion imaging modalities, the term 'normal' perfusion values hardly exist. It is therefore important to acquire a matching control group at each site, also in multicentre studies. In other words, one should determine the 'normal' perfusion value for the given setup. The perfusion values reported in the literature vary primarily between 30 and 80 ml/100 g/min for grey matter, but it is hard to pinpoint what the value should be even for a given sequence.

Finally, cerebral perfusion depends on a range of factors, including various pathologies, age and gender. Ageing is associated with a decline in *CBF* (Parkes *et al.*, 2004; Liu *et al.*, 2012), and women have been found to have a higher *CBF* and shorter ATT compared to men (Petersen *et al.*, 2010; Liu *et al.*, 2012). The influence of atrophy is of concern when interpreting data, for instance in Alzheimer's patients versus age-matched controls, or between young and old subjects. In these cases the question often arises of whether a flow reduction is simply due to increased partial volume of CSF, to general tissue atrophy or if in fact the grey and white matter perfusion is reduced as well. Methods exist for correcting for partial volume based on structural information (Asllani *et al.*, 2008; Ahlgren *et al.*, 2014); however there is no real consensus on when it should be applied, if at all! In addition, baseline *CBF* fluctuates and is easily affected by medication and substances such as caffeine, which has been found to decrease whole-brain perfusion by about 20% (Haller *et al.*, 2013).

16.7 Reproducibility

16.7.1 Validation and reproducibility studies

Both for clinical and research purposes, it is important to know how robust the method is and how well it compares to the established methods. For ASL, validation studies have been carried out in animals (Walsh *et al.*, 1994; Ewing *et al.*, 2003) using radioactive microspheres and ^{14}C-iodoantipyrine autoradiography, and in humans (Ye, Berman, *et al.*, 2000; Heijtel *et al.*, 2014, 2016; Fan *et al.*, 2016) using PET. Good correlations are observed, both at baseline and during a vascular challenge, which is promising for the quantitative usage of ASL in both clinical and research settings.

ASL has also proven reproducible in several test–retest studies of various ASL techniques (Petersen *et al.*, 2010; Gevers *et al.*, 2011; Mutsaerts *et al.*, 2014, 2015; Wu *et al.*, 2014). Both the intra- and intersubject coefficients of variation of approximately 10% and 15%, respectively, are fully in line with that of the ^{15}O-PET studies (Fan *et al.*, 2016).

16.7.2 Consensus papers

Recently, a consensus paper aimed at directing vendors, researchers as well as clinicians towards a set of standard ASL sequences was published (Alsop *et al.*, 2015). This paper makes a good reference for the minimum requirements needed for perfusion studies performed both clinically and in research. Access to similar ASL sequences with default parameters across vendors will allow a faster harmonisation of the clinical results, to the benefit for the entire field. Still, research may involve more elaborate ASL versions, which then typically must be implemented by the researcher, as is the case today.

16.8 ASL: The future

The future for ASL looks bright, the number of clinical research studies using ASL is rapidly increasing and clinical use is picking up! The reproducibility is good and consensus has been reached on how to perform the basic PASL and pCASL experiments. At the same time, a range of advanced new techniques continue to emerge.

16.8.1 New methodologies

ASL is an extremely versatile method, as we have seen above: it allows not only measurement of *CBF*, but also ATT, aBV and it can give territorial information. New approaches for improving *CBF* quantitation have been developed, such as the Hadamard encoded pCASL sequence (Teeuwisse *et al.*, 2014; von Samson-Himmelstjerna *et al.*, 2016), which facilitates time-efficient mapping of ATT and *CBF* within the same sequence scan using a Hadamard encoding matrix. Another recent approach for making time efficient collection of *CBF*, ATT and aBV information is the Turbo QUASAR sequence, which relies on the generation of a train of boluses with continuous sampling in between (Petersen *et al.*, 2013).

However, the story does not end here, and advanced sequences keep emerging from the ASL technique. By separating the blood in the sagittal sinus using a PASL labelling approach together with a series of T_2-preparation pulses (Lu and Ge, 2008), global venous oxygenation can be determined, which together with whole-brain perfusion information allows the assessment of global $CMRO_2$. This has opened an entire field for ASL-based oxygenation techniques where recent developments aim at separating the vascular compartment from the tissue on a voxel-by-voxel level in order to assess oxygen extraction

(Bolar *et al.*, 2011; Guo and Wong, 2012) or oxygen saturation of the tissue (Alderliesten *et al.*, 2016). Magnetic resonance fingerprinting has also been applied to ASL, potentially allowing estimation of more parameters from within a single scan (Su *et al.*, 2016), another new direction within ASL imaging aiming to acquire more robust perfusion estimates in shorter acquisition times.

16.8.2 Concluding remarks

The last decade has seen great progress in the ASL technique. Robust sequences such as the pCASL sequence combined with great technical advances, such as 3*T* MR scanners and 16 or more parallel receive channels, have paved the way for robust perfusion evaluation for both clinical and research purposes. Users have gained a non-invasive, accurate and repeatable imaging method for quantitative perfusion.

References

Ahlgren A, Wirestam R, Petersen ET, Ståhlberg F, Knutsson L. Partial volume correction of brain perfusion estimates using the inherent signal data of time-resolved arterial spin labeling. NMR Biomed 2014; 27: 1112–22.

Alderliesten T, De Vis JB, Lemmers PMA, van Bel F, Benders MJNL, Hendrikse J, et al. T2-prepared velocity selective labelling: a novel idea for full-brain mapping of oxygen saturation. NeuroImage 2016; 139: 65–73.

Alsop DC, Detre JA. Multisection cerebral blood flow MR imaging with continuous arterial spin labeling. Radiology 1998; 208: 410–6.

Alsop DC, Detre JA, Golay X, Günther M, Hendrikse J, Hernandez-Garcia L, et al. Recommended implementation of arterial spin-labeled perfusion MRI for clinical applications: a consensus of the ISMRM perfusion study group and the European consortium for ASL in dementia. Magn Reson Med 2015; 73: 102–16.

Asllani I, Borogovac A, Brown TR. Regression algorithm correcting for partial volume effects in arterial spin labeling MRI. Magn Reson Med 2008; 60: 1362–71.

Austin BP, Nair VA, Meier TB, Xu G, Rowley HA, Carlsson CM, et al. Effects of hypoperfusion in Alzheimer's disease. J Alzheimers Dis JAD 2011; 26 Suppl 3: 123–33.

Blockley NP, Griffeth VEM, Buxton RB. A general analysis of calibrated BOLD methodology for measuring CMRO2 responses: comparison of a new approach with existing methods. NeuroImage 2012; 60: 279–89.

Bokkers RPH, De Cocker LJ, van Osch MJP, Hartkamp NS, Hendrikse J. Selective arterial spin labeling: techniques and neurovascular applications. Top Magn Reson Imaging TMRI 2016; 25: 73–80.

Bokkers RPH, van Osch MJP, van der Worp HB, de Borst GJ, Mali WPTM, Hendrikse J. Symptomatic carotid artery stenosis: impairment of cerebral autoregulation measured at the brain tissue level with arterial spin-labeling MR imaging. Radiology 2010; 256: 201–8.

Bolar DS, Rosen BR, Sorensen AG, Adalsteinsson E. QUantitative Imaging of eXtraction of oxygen and TIssue consumption (QUIXOTIC) using venular-targeted velocity-selective spin labeling. Magn Reson Med 2011; 66: 1550–62.

Bulte DP, Kelly M, Germuska M, Xie J, Chappell MA, Okell TW, et al. Quantitative measurement of cerebral physiology using respiratory-calibrated MRI. NeuroImage 2012; 60: 582–91.

Buxton RB, Frank LR, Wong EC, Siewert B, Warach S, Edelman RR. A general kinetic model for quantitative perfusion imaging with arterial spin labeling. Magn Reson Med 1998; 40: 383–96.

Calamante F, Williams SR, van Bruggen N, Kwong KK, Turner R. A model for quantification of perfusion in pulsed labelling techniques. NMR Biomed 1996; 9: 79–83.

Chappell MA, Groves AR, Whitcher B, Woolrich MW. Variational Bayesian Inference for a nonlinear forward model. Trans Sig Proc 2009; 57: 223–36.

Chiarelli PA, Bulte DP, Gallichan D, Piechnik SK, Wise R, Jezzard P. Flow-metabolism coupling in human visual, motor, and supplementary motor areas assessed by magnetic resonance imaging. Magn Reson Med 2007; 57: 538–47.

Chng SM, Petersen ET, Zimine I, Sitoh Y-Y, Lim CCT, Golay X. Territorial arterial spin labeling in the assessment of collateral circulation: comparison with digital subtraction angiography. Stroke J Cereb Circ 2008; 39: 3248–54.

Collij LE, Heeman F, Kuijer JPA, Ossenkoppele R, Benedictus MR, Möller C, et al. Application of machine learning to arterial spin labeling in mild cognitive impairment and Alzheimer disease. Radiology 2016; 281: 865–75.

Dai W, Garcia D, de Bazelaire C, Alsop DC. Continuous flow-driven inversion for arterial spin labeling using pulsed radio frequency and gradient fields. Magn Reson Med 2008; 60: 1488–97.

Dangouloff-Ros V, Deroulers C, Foissac F, Badoual M, Shotar E, Grévent D, et al. Arterial spin labeling to predict brain tumor grading in children: correlations between histopathologic vascular density and perfusion MR imaging. Radiology 2016; 281: 553–66.

Davies NP, Jezzard P. Selective arterial spin labeling (SASL): perfusion territory mapping of selected feeding arteries tagged using two-dimensional radiofrequency pulses. Magn Reson Med 2003; 49: 1133–42.

De Vis JB, Alderliesten T, Hendrikse J, Petersen ET, Benders MJNL. Magnetic resonance imaging based noninvasive measurements of brain hemodynamics in neonates: a review. Pediatr Res 2016; 80: 641–50.

De Vis JB, Hendrikse J, Bhogal A, Adams A, Kappelle LJ, Petersen ET. Age-related changes in brain hemodynamics; a calibrated MRI study. Hum Brain Mapp. 2015; 36: 3973–87.

De Vis JB, Hendrikse J, Groenendaal F, de Vries LS, Kersbergen KJ, Benders MJNL, et al. Impact of neonate haematocrit variability on the longitudinal relaxation time of blood: implications for arterial spin labelling MRI. NeuroImage Clin 2014; 4: 517–25.

De Vis JB, Petersen ET, Bhogal A, Hartkamp NS, Klijn CJM, Kappelle LJ, et al. Calibrated MRI to evaluate cerebral hemodynamics in patients with an internal carotid artery occlusion. J Cereb Blood Flow Metab Off J Int Soc Cereb Blood Flow Metab 2015; 35: 1015–23.

Detre JA, Wang J. Technical aspects and utility of fMRI using BOLD and ASL. Clin Neurophysiol Off J Int Fed Clin Neurophysiol 2002; 113: 621–34.

Dixon WT, Du LN, Faul DD, Gado M, Rossnick S. Projection angiograms of blood labeled by adiabatic fast passage. Magn Reson Med 1986; 3: 454–62.

Dixon WT, Sardashti M, Castillo M, Stomp GP. Multiple inversion recovery reduces static tissue signal in angiograms. Magn Reson Med 1991; 18: 257–68.

Duong TQ, Kim DS, Uğurbil K, Kim SG. Localized cerebral blood flow response at submillimeter columnar resolution. Proc Natl Acad Sci U S A 2001; 98: 10904–9.

Edelman RR, Siewert B, Darby DG, Thangaraj V, Nobre AC, Mesulam MM, et al. Qualitative mapping of cerebral blood flow and functional localization with echo-planar MR imaging and signal targeting with alternating radio frequency. Radiology 1994; 192: 513–20.

Ewing JR, Wei L, Knight RA, Pawa S, Nagaraja TN, Brusca T, et al. Direct comparison of local cerebral blood flow rates measured by MRI arterial spin-tagging and quantitative autoradiography in a rat model of experimental cerebral ischemia. J Cereb Blood Flow Metab Off J Int Soc Cereb Blood Flow Metab 2003; 23: 198–209.

Fan AP, Jahanian H, Holdsworth SJ, Zaharchuk G. Comparison of cerebral blood flow measurement with [15O]-water positron emission tomography and arterial spin labeling magnetic resonance imaging: a systematic review. J Cereb Blood Flow Metab Off J Int Soc Cereb Blood Flow Metab 2016; 36: 842–61.

Fernández-Seara MA, Mengual E, Vidorreta M, Castellanos G, Irigoyen J, Erro E, et al. Resting state functional connectivity of the subthalamic nucleus in Parkinson's disease assessed using arterial spin-labeled perfusion fMRI. Hum. Brain Mapp 2015; 36: 1937–50.

Gevers S, van Osch MJ, Bokkers RPH, Kies DA, Teeuwisse WM, Majoie CB, et al. Intra- and multicenter reproducibility of pulsed, continuous and pseudo-continuous arterial spin labeling methods for measuring cerebral perfusion. J Cereb Blood Flow Metab Off J Int Soc Cereb Blood Flow Metab 2011; 31: 1706–15.

Golay X, Stuber M, Pruessmann KP, Meier D, Boesiger P. Transfer insensitive labeling technique (TILT): application to multislice functional perfusion imaging. J Magn Reson Imaging JMRI 1999; 9: 454–61.

Greve DN, Mueller BA, Liu T, Turner JA, Voyvodic J, Yetter E, et al. A novel method for quantifying scanner instability in fMRI. Magn Reson Med 2011; 65: 1053–61.

Günther M, Bock M, Schad LR. Arterial spin labeling in combination with a look-locker sampling strategy: inflow turbo-sampling EPI-FAIR (ITS-FAIR). Magn Reson Med 2001; 46: 974–84.

Günther M, Oshio K, Feinberg DA. Single-shot 3D imaging techniques improve arterial spin labeling perfusion measurements. Magn Reson Med 2005; 54: 491–8.

Guo J, Wong EC. Venous oxygenation mapping using velocity-selective excitation and arterial nulling. Magn Reson Med 2012; 68: 1458–71.

Haller S, Rodriguez C, Moser D, Toma S, Hofmeister J, Sinanaj I, et al. Acute caffeine administration impact on working memory-related brain activation and functional connectivity in the elderly: a BOLD and perfusion MRI study. Neuroscience 2013; 250: 364–71.

Haller S, Zaharchuk G, Thomas DL, Lovblad K-O, Barkhof F, Golay X. Arterial spin labeling perfusion of the brain: emerging clinical applications. Radiology 2016; 281: 337–56.

Hartkamp NS, Petersen ET, De Vis JB, Bokkers RPH, Hendrikse J. Mapping of cerebral perfusion territories using territorial arterial spin labeling: techniques and clinical application. NMR Biomed 2013; 26: 901–12.

Heijtel DFR, Mutsaerts HJMM, Bakker E, Schober P, Stevens MF, Petersen ET, et al. Accuracy and precision of pseudo-continuous arterial spin labeling perfusion during baseline and hypercapnia: a head-to-head comparison with ^{15}O H_2O positron emission tomography. NeuroImage 2014; 92: 182–92.

Heijtel DFR, Petersen ET, Mutsaerts HJMM, Bakker E, Schober P, Stevens MF, et al. Quantitative agreement between [(15)O]H2O PET and model free QUASAR MRI-derived cerebral blood flow and arterial blood volume. NMR Biomed 2016; 29: 519–26.

Helle M, Norris DG, Rüfer S, Alfke K, Jansen O, van Osch MJP. Superselective pseudocontinuous arterial spin labeling. Magn Reson Med 2010; 64: 777–86.

Hendrikse J, van der Grond J, Lu H, van Zijl PCM, Golay X. Flow territory mapping of the cerebral arteries with regional perfusion MRI. Stroke 2004; 35: 882–7.

Hendrikse J, Petersen ET, Chèze A, Chng SM, Venketasubramanian N, Golay X. Relation between cerebral perfusion territories and location of cerebral infarcts. Stroke J Cereb Circ 2009; 40: 1617–22.

Henkelman RM, Huang X, Xiang QS, Stanisz GJ, Swanson SD, Bronskill MJ. Quantitative interpretation of magnetization transfer. Magn Reson Med 1993; 29: 759–66.

Hernandez DA, Bokkers RPH, Mirasol RV, Luby M, Henning EC, Merino JG, et al. Pseudocontinuous arterial spin labeling quantifies relative cerebral blood flow in acute stroke. Stroke 2012; 43: 753–8.

Herscovitch P, Raichle ME. What is the correct value for the brain–blood partition coefficient for water? J Cereb Blood Flow Metab Off J Int Soc Cereb Blood Flow Metab 1985; 5: 65–9.

Hoge RD, Atkinson J, Gill B, Crelier GR, Marrett S, Pike GB. Linear coupling between cerebral blood flow and oxygen consumption in activated human cortex. Proc Natl Acad Sci 1999; 96: 9403–8.

Hrabe J, Lewis DP. Two analytical solutions for a model of pulsed arterial spin labeling with randomized blood arrival times. J Magn Reson San Diego Calif 1997 2004; 167: 49–55.

Kety SS, Schmidt CF. The nitrous oxide method for the quantitative determination of cerebral blood flow in man: theory, procedure and normal values. J Clin Invest 1948; 27: 476–83.

Kim SG. Quantification of relative cerebral blood flow change by flow-sensitive alternating inversion recovery (FAIR) technique: application to functional mapping. Magn Reson Med 1995; 34: 293–301.

Kwong KK, Chesler DA, Weisskoff RM, Donahue KM, Davis TL, Ostergaard L, et al. MR perfusion studies with T1-weighted echo planar imaging. Magn Reson Med 1995; 34: 878–87.

Leeuwis AE, Benedictus MR, Kuijer JPA, Binnewijzend MAA, Hooghiemstra AM, Verfaillie SCJ, et al. Lower cerebral blood flow is associated with impairment in multiple cognitive domains in Alzheimer's disease. Alzheimers Dement J Alzheimers Assoc 2017; 13(5), 531–40.

Liu Y, Zhu X, Feinberg D, Guenther M, Gregori J, Weiner MW, et al. Arterial spin labeling MRI study of age and gender effects on brain perfusion hemodynamics. Magn Reson Med 2012; 68: 912–22.

Lu H, Ge Y. Quantitative evaluation of oxygenation in venous vessels using T2-Relaxation-Under-Spin-Tagging MRI. Magn Reson Med 2008; 60: 357–63.

Mak HKF, Chan Q, Zhang Z, Petersen ET, Qiu D, Zhang L, et al. Quantitative assessment of cerebral hemodynamic parameters by QUASAR arterial spin labeling in Alzheimer's disease and cognitively normal Elderly adults at 3-tesla. J. Alzheimers Dis. JAD 2012; 31: 33–44.

Mato Abad V, García-Polo P, O'Daly O, Hernández-Tamames JA, Zelaya F. ASAP (Automatic Software for ASL Processing): a toolbox for processing Arterial Spin Labeling images. Magn Reson Imaging 2016; 34: 334–44.

Mutsaerts HJMM, Steketee RME, Heijtel DFR, Kuijer JPA, van Osch MJP, Majoie CBLM, et al. Inter-vendor reproducibility of pseudo-continuous arterial spin labeling at 3 Tesla. PLoS One 2014; 9: e104108.

Mutsaerts HJMM, van Osch MJP, Zelaya FO, Wang DJJ, Nordhøy W, Wang Y, et al. Multi-vendor reliability of arterial spin labeling perfusion MRI using a near-identical sequence: implications for multi-center studies. NeuroImage 2015; 113: 143–52.

Ogawa S, Lee TM, Kay AR, Tank DW. Brain magnetic resonance imaging with contrast dependent on blood oxygenation. Proc Natl Acad Sci U S A 1990; 87: 9868–72.

Oguz KK, Golay X, Pizzini FB, Freer CA, Winrow N, Ichord R, et al. Sickle cell disease: continuous arterial spin-labeling perfusion MR imaging in children. Radiology 2003; 227: 567–74.

Ostergaard L, Weisskoff RM, Chesler DA, Gyldensted C, Rosen BR. High resolution measurement of cerebral blood flow using intravascular tracer bolus passages. Part I: mathematical approach and statistical analysis. Magn Reson Med 1996; 36: 715–25.

Parkes LM, Rashid W, Chard DT, Tofts PS. Normal cerebral perfusion measurements using arterial spin labeling: reproducibility, stability, and age and gender effects. Magn Reson Med 2004; 51: 736–43.

Parkes LM, Tofts PS. Improved accuracy of human cerebral blood perfusion measurements using arterial spin labeling: accounting for capillary water permeability. Magn Reson Med 2002; 48: 27–41.

Petersen ET, De Vis JB, van den Berg CAT, Hendrikse J. Turbo-QUASAR: a signal-to-noise optimal arterial spin labeling and sampling strategy. ISMRM 21st Annual Meeting & Exhibition in Salt Lake City, USA, #2146. 2013.

Petersen ET, Lim T, Golay X. Model-free arterial spin labeling quantification approach for perfusion MRI. Magn Reson Med 2006; 55: 219–32.

Petersen ET, Mouridsen K, Golay X, all named co-authors of the QUASAR test-retest study. The QUASAR reproducibility study, Part II: results from a multi-center Arterial Spin Labeling test-retest study. NeuroImage 2010; 49: 104–13.

Petersen ET, Zimine I, Ho Y-CL, Golay X. Non-invasive measurement of perfusion: a critical review of arterial spin labelling techniques. Br J Radiol 2006; 79: 688–701.

Qiu D, Straka M, Zun Z, Bammer R, Moseley ME, Zaharchuk G. *CBF* measurements using multidelay pseudocontinuous and velocity-selective arterial spin labeling in patients with long arterial transit delays: comparison with xenon CT *CBF*. J Magn Reson Imaging JMRI 2012; 36: 110–9.

Roberts DA, Rizi R, Lenkinski RE, Leigh JS. Magnetic resonance imaging of the brain: blood partition coefficient for water: application to spin-tagging measurement of perfusion. J Magn Reson Imaging JMRI 1996; 6: 363–6.

Schmid S, Ghariq E, Teeuwisse WM, Webb A, van Osch MJP. Acceleration-selective arterial spin labeling. Magn Reson Med 2014; 71: 191–9.

Schmid S, Heijtel DFR, Mutsaerts HJMM, Boellaard R, Lammertsma AA, Nederveen AJ, et al. Comparison of velocity- and acceleration-selective arterial spin labeling with [15O]H2O positron emission tomography. J Cereb Blood Flow Metab Off J Int Soc Cereb Blood Flow Metab 2015; 35: 1296–303.

Schmid S, Petersen ET, Van Osch MJP. Insight into the labeling mechanism of acceleration selective arterial spin labeling. MAGMA. 2017; 30(2): 165–74.

St Lawrence KS, Frank JA, McLaughlin AC. Effect of restricted water exchange on cerebral blood flow values calculated with arterial spin tagging: a theoretical investigation. Magn Reson Med 2000; 44: 440–9.

Su P, Mao D, Liu P, Li Y, Pinho MC, Welch BG, et al. Multiparametric estimation of brain hemodynamics with MR fingerprinting ASL. Magn Reson Med 2016.

Teeuwisse WM, Schmid S, Ghariq E, Veer IM, van Osch MJP. Time-encoded pseudocontinuous arterial spin labeling: basic properties and timing strategies for human applications. Magn Reson Med 2014; 72: 1712–22.

Tofts PS, Kermode AG. Measurement of the blood-brain barrier permeability and leakage space using dynamic MR imaging. 1. Fundamental concepts. Magn Reson Med 1991; 17: 357–67.

Trampel R, Jochimsen TH, Mildner T, Norris DG, Möller HE. Efficiency of flow-driven adiabatic spin inversion under realistic experimental conditions: a computer simulation. Magn Reson Med 2004; 51: 1187–93.

Václavů L, van der Land V, Heijtel DFR, van Osch MJP, Cnossen MH, Majoie CBLM, et al. In Vivo T1 of blood measurements in children with sickle cell disease improve cerebral blood flow quantification from arterial spin-labeling MRI. AJNR Am J Neuroradiol 2016; 37: 1727–32.

Varela M, Hajnal JV, Petersen ET, Golay X, Merchant N, Larkman DJ. A method for rapid in vivo measurement of blood T1. NMR Biomed 2011; 24: 80–8.

von Samson-Himmelstjerna F, Madai VI, Sobesky J, Guenther M. Walsh-ordered hadamard time-encoded pseudocontinuous ASL (WH pCASL). Magn Reson Med 2016; 76: 1814–24.

Walsh EG, Minematsu K, Leppo J, Moore SC. Radioactive microsphere validation of a volume localized continuous saturation perfusion measurement. Magn Reson Med 1994; 31: 147–53.

Wang DJJ, Alger JR, Qiao JX, Gunther M, Pope WB, Saver JL, et al. Multi-delay multi-parametric arterial spin-labeled perfusion MRI in acute ischemic stroke - Comparison with dynamic susceptibility contrast enhanced perfusion imaging. NeuroImage Clin 2013; 3: 1–7.

Wang J, Aguirre GK, Kimberg DY, Roc AC, Li L, Detre JA. Arterial spin labeling perfusion fMRI with very low task frequency. Magn Reson Med 2003; 49: 796–802.

Wang J, Licht DJ, Jahng G-H, Liu C-S, Rubin JT, Haselgrove J, et al. Pediatric perfusion imaging using pulsed arterial spin labeling. J Magn Reson Imaging JMRI 2003; 18: 404–13.

Wang Z, Aguirre GK, Rao H, Wang J, Fernández-Seara MA, Childress AR, et al. Empirical optimization of ASL data analysis using an ASL data processing toolbox: ASLtbx. Magn Reson Imaging 2008; 26: 261–9.

Werner R, Norris DG, Alfke K, Mehdorn HM, Jansen O. Continuous artery-selective spin labeling (CASSL). Magn Reson Med 2005; 53: 1006–12.

Williams DS, Detre JA, Leigh JS, Koretsky AP. Magnetic resonance imaging of perfusion using spin inversion of arterial water. Proc Natl Acad Sci U S A 1992; 89: 212–6.

Wong EC. Vessel-encoded arterial spin-labeling using pseudocontinuous tagging. Magn Reson Med 2007; 58: 1086–91.

Wong EC, Buxton RB, Frank LR. Implementation of quantitative perfusion imaging techniques for functional brain mapping using pulsed arterial spin labeling. NMR Biomed 1997; 10: 237–49.

Wong EC, Buxton RB, Frank LR. Quantitative imaging of perfusion using a single subtraction (QUIPSS and QUIPSS II). Magn Reson Med 1998; 39: 702–8.

Wong EC, Cronin M, Wu W-C, Inglis B, Frank LR, Liu TT. Velocity-selective arterial spin labeling. Magn Reson Med 2006; 55: 1334–41.

Wu B, Lou X, Wu X, Ma L. Intra- and interscanner reliability and reproducibility of 3D whole-brain pseudo-continuous arterial spin-labeling MR perfusion at 3T. J Magn Reson Imaging JMRI 2014; 39: 402–9.

Wu W-C, Wong EC. Feasibility of velocity selective arterial spin labeling in functional MRI. J Cereb Blood Flow Metab Off J Int Soc Cereb Blood Flow Metab 2007; 27: 831–38.

Yang S, Zhao B, Wang G, Xiang J, Xu S, Liu Y, et al. Improving the grading accuracy of astrocytic neoplasms noninvasively by combining timing information with cerebral blood flow: a multi-TI arterial spin-labeling MR imaging study. AJNR Am J Neuroradiol 2016; 37: 2209–16.

Ye FQ, Berman KF, Ellmore T, Esposito G, van Horn JD, Yang Y, et al. H(2)(15)O PET validation of steady-state arterial spin tagging cerebral blood flow measurements in humans. Magn Reson Med 2000; 44: 450–6.

Ye FQ, Frank JA, Weinberger DR, McLaughlin AC. Noise reduction in 3D perfusion imaging by attenuating the static signal in arterial spin tagging (ASSIST). Magn Reson Med 2000; 44: 92–100.

Ye FQ, Mattay VS, Jezzard P, Frank JA, Weinberger DR, McLaughlin AC. Correction for vascular artifacts in cerebral blood flow values measured by using arterial spin tagging techniques. Magn Reson Med 1997; 37: 226–35.

Ye J, Bhagat SK, Li H, Luo X, Wang B, Liu L, et al. Differentiation between recurrent gliomas and radiation necrosis using arterial spin labeling perfusion imaging. Exp Ther Med 2016; 11: 2432–6.

Zhang W, Silva AC, Williams DS, Koretsky AP. NMR measurement of perfusion using arterial spin labeling without saturation of macromolecular spins. Magn Reson Med 1995; 33: 370–6.

Zhang W, Williams DS, Koretsky AP. Measurement of rat brain perfusion by NMR using spin labeling of arterial water: in vivo determination of the degree of spin labeling. Magn Reson Med 1993; 29: 416–21.

Zhang X, Petersen ET, Ghariq E, De Vis JB, Webb AG, Teeuwisse WM, et al. In vivo blood T(1) measurements at 1.5 T, 3 T, and 7 T. Magn Reson Med 2013; 70: 1082–6.

Zhou J, Wilson DA, Ulatowski JA, Traystman RJ, van Zijl PC. Two-compartment exchange model for perfusion quantification using arterial spin tagging. J Cereb Blood Flow Metab Off J Int Soc Cereb Blood Flow Metab 2001; 21: 440–55.

17

Image Analysis[1]

Contents

Siawoosh Mohammadi
University Medical Center Hamburg-Eppendorf

Martina F. Callaghan
University College London

17.1 Utility and pitfalls of quantitative MRI data

A typical aim for a neuroimaging study is to determine a marker that captures inter-individual difference in behaviour, function, dysfunction or pathology. This may be done by comparing two groups (e.g. patients and controls) or by testing for a relationship between a neuroimaging metric and some parametric descriptor (e.g. age, behavioural score, clinical outcome, etc.). A well-established neuroimaging metric is the volume of brain or spinal cord structures, which is typically measured using standard structural MRI scans (e.g. T_1-weighted Magnetic Resonance (MR) images). A change in such a volumetric neuroimaging metric, e.g. in the hippocampus (Woollett *et al.*, 2008; Bonnici *et al.*, 2012), thalamus (Keller *et al.*, 2012) and spinal cord (Freund *et al.*, 2013), can be assessed either over time or between individuals. The utility of this approach is evidenced by the fact that it has provided great insight into the plasticity of the human brain (e.g. Maguire *et al.*, 1997; Draganski *et al.*, 2004) and into pathological processes occurring in diseases such as epilepsy (Keller *et al.*, 2011), traumatic brain injury (Freund *et al.*, 2013) or spinal cord diseases (Martin *et al.*, 2016).

However, the signal intensity on conventional weighted MR images, which are generally used for morphometric analyses, are influenced by a number of different MRI properties, as well as by the scanner architecture (e.g. receive or transmit field inhomogeneities) and the specifics of the imaging protocol used (i.e. choice of TR, TE, flip angle, etc.). The resulting dependence of morphometric measures on multiple inherent physical MRI properties of the tissue is complex, making it challenging to accurately and robustly estimate the volume of anatomical regions or to interpret changes in these volume estimates. One striking example of where MRI-based volume estimates can be erroneous is the underestimation of the cortical thickness in the sensorimotor cortex. Here, the cortical contrast of T_1- and T_2-weighted images is mainly driven by the myelin compartment, which results in reduced contrast at the white matter boundary (Zilles and Amunts, 2015).

In contrast to conventional weighted MRI, quantitative MRI (qMRI) measures absolute physical parameters (e.g. a relaxation time [Koenig *et al.*, 1993] in relaxometry, an estimate of water diffusivity in diffusion MRI [Le Bihan, 2003] or the BOLD signal change in functional MRI [Logothetis, 2008]), which are corrected for instrumental biases (Lutti *et al.*, 2010; Weiskopf *et al.*, 2013). This standardised nature facilitates comparison across sites, time points (Weiskopf *et al.*, 2013) and participants, enabling multicentre trials. Furthermore, the direct relation of qMRI parameter maps to single physical properties of the tissue increases its sensitivity to microscopic tissue properties (Kleffner *et al.*, 2008; Lutti *et al.*, 2014; Yeatman *et al.*, 2014; Gomez *et al.*, 2017) as compared to standard anatomical images (e.g. T1 weighted images). For example the myelin-sensitive neuroimaging biomarkers produced from the quantitative multiparameter mapping (MPM) protocol in the brain and cervical spinal cord have provided novel insights into the underlying processes of atrophy-related pathology in spinal-cord injury (Freund *et al.*, 2013; Grabher *et al.*, 2015). In addition, the higher tissue contrast of qMRI parameter maps can not only improve the automated

[1] Edited by Nicholas G. Dowell.

estimation of anatomical structures in the brain (e.g. deep grey matter regions; Helms *et al.*, 2009; Callaghan *et al.*, 2016b) but the greater specificity can also help to identify spurious volume changes due to signal changes in standard anatomical images that are in fact driven by microstructural tissue change (e.g. as a consequence of iron accumulation in aging; Lorio *et al.*, 2016).

Despite its improved sensitivity to microscopic tissue changes, it should be borne in mind that each qMRI metric is affected by multiple underlying biological features existing at the microstructural level. As such, there can be multiple explanations for why a qMRI measure may be altered. For example, both demyelination and axonal loss lead to a reduction in the fractional anisotropy measure provided by diffusion MRI (Deppe *et al.*, 2007; Wheeler-Kingshott *et al.*, 2014), while an increase in iron concentration or an increase in myelination both manifest as a shortening of the longitudinal relaxation time (T_1) (Rooney *et al.*, 2007; Callaghan *et al.*, 2015a). Combining multiple qMRI measures can improve the interpretation of observed volumetric changes, especially if they are used as biomarkers for tissue compartments, e.g. apparent transverse relaxation time (T_2*) as a marker for iron content in ageing (Draganski *et al.*, 2011; Callaghan *et al.*, 2014; Lorio *et al.*, 2014).

Finally, qMRI measures can also be combined with biophysical models of the MR signal to extract more biologically specific quantitative histological measures of brain tissue microstructure, an idea termed *in vivo histology using MRI* or hMRI (Weiskopf *et al.*, 2015). One approach to account for this multifeature relationship is to use multivariate models incorporating the quantitative T_1, T_2* and magnetisation transfer (MT) saturation estimates as biomarkers for the myelin and iron compartments (Stüber *et al.*, 2014; Callaghan *et al.*, 2015a; Mangeat *et al.*, 2015). Alternatively, simple geometric relations describing the underlying anatomy can be used to derive multimodal biophysical models, such as the model of the MR g-ratio (describing the relative thickness of a fibre, including its myelin sheath, to that of the axon alone) that is based on biomarkers of myelin and fibre density (Stikov *et al.*, 2015) or the hollow cylinder model of the gradient echo signal decay that combines relaxometry and diffusion MRI (dMRI) (Wharton and Bowtell, 2012, 2013; Gil *et al.*, 2016). The accurate spatial correspondence during the combination of different maps is a crucial step to achieve reliable quantification of biophysical tissue properties. Given that each quantitative MRI technique comes with its own specific vulnerability to artefacts, the latter category of models, which combine data from different MRI acquisition techniques, are most susceptible to modality-specific artefacts.

In this chapter we discuss image analysis approaches that capitalise on the use of qMRI data and their combination, focussing on the potential pitfalls. It is beyond the scope of this chapter to discuss all relevant artefacts that may be encountered in specific qMRI techniques (e.g. for dMRI there are a number of books and review articles that discuss artefacts and models for this modality in great detail, e.g. Bammer *et al.*, 2005; Le Bihan *et al.*, 2006; Johansen-Berg and Behrens, 2009; Jones, 2010). Here, we will instead focus on those artefacts that we consider most important for combining multimodal qMRI data, namely *subject motion*, *spatial distortions* and *thermal noise*. Having presented methods for addressing these artefacts, we then discuss a number of approaches for analysing qMRI markers at the group level in order to determine functionally or behaviourally relevant inter-individual differences.

17.2 Processing to maximise data quality

17.2.1 Motivation

The quantification of MRI parameters typically relies on the combination of multiple images with different contrast (e.g. multiple gradient echo images acquired with different flip angles to map T_1 times). For optimal combination, any spatial misalignment due to subject motion must be minimised. It can also be necessary to combine multimodal images acquired with different MRI techniques using different readout schemes. For example, the g-ratio hMRI model requires combining 2D echo planar imaging (EPI) data (dMRI-based biomarkers for fibre density) with 3D gradient echo data (MT-based biomarkers for myelin density) on a voxel-by-voxel level. However, the different readout schemes used mean that spatial image distortions will manifest differently, having a more severe impact on EPI-based data because of the sensitivity of this technique to susceptibility-related field inhomogeneities. If uncorrected these differential effects lead to spatial mismatch that varies locally. It should also be noted that susceptibility distortions scale with the magnetic field strength, i.e. they are greater at 7T than at 3T than at 1.5T. Figure 17.1 demonstrates the misalignment that results at 3T if this artefact is not corrected appropriately, as well as the benefit of addressing the issue. If present, this misalignment strongly biases the estimated MR g-ratio map, e.g. in the genu of the corpus callosum (see crosshair in Figure 17.1).

For hMRI models, high spatial resolution is required, e.g. to better resolve finer details such as the laminar structure of the cortex, to delineate small, functionally specific nuclei or to separate functionally relevant fibres at central junctions containing complex fibre configurations (Schmahmann and Pandya, 2006; Roebroeck *et al.*, 2008; Weiskopf *et al.*, 2015). Thus, each qMRI technique should be acquired at the highest possible resolution. The role of *thermal noise* in MRI data becomes more relevant when pushing the spatial resolution to its limits. While fast-low-angle-shot (FLASH)-based estimates of MT or relaxation rates can be acquired at very high resolution (e.g. using the MPM technique [Helms *et al.*, 2008a, 2008b; Weiskopf *et al.*, 2013], 800 micron isotropic resolution can be routinely achieved at 3T [Lutti *et al.*, 2014]), it is particularly challenging to acquire *in vivo* dMRI data at voxel sizes smaller than 1.5 mm, because of the already low signal-to-noise ratio (SNR) inherent in the diffusion-weighting process and the time required to achieve this (dictating the minimum echo time [TE]). Low SNR not only can bias the qMRI estimates (for details see Section 17.2.4) but also diminish the image quality. Denoising (i.e. smoothing) is a way to use image processing to retrospectively improve image quality. However, denoising usually leads to blurring (e.g. Gaussian denoising) and thus reduces the effective image resolution. Adaptive denoising methods hold promise to reduce the noise level without reducing the effective resolution. Figure 17.2 demonstrates how

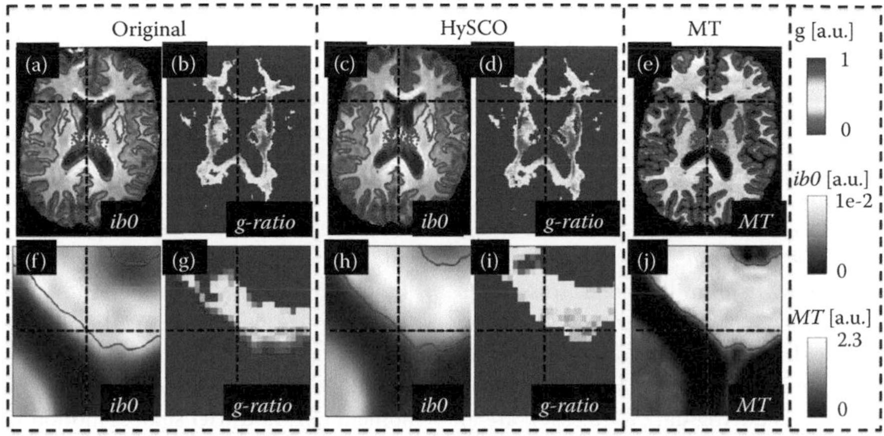

FIGURE 17.1 Susceptibility-induced geometric distortions in diffusion MRI (dMRI) data and their effects on the MR g-ratio map from a representative subject. The MR g-ratio and contrast-inverted b0 maps (ib0) from the original dMRI data (a,b,f,g) and following hyperelastic susceptibility-distortion-correction (HySCO) (c,d,h,i) were compared to a largely distortion-insensitive FLASH-based biomarker for myelin density (here we used the semi-quantitative magnetisation transfer saturation; Helms, 2008b) (e,j). The spatial mismatch between anatomical structures in dMRI and MT data (see red contours) was greatly reduced after susceptibility correction. The susceptibility-related mismatch between the uncorrected dMRI map and the MT map leads to severe, locally varying bias in the g-ratio maps (e.g., the crosshair highlights one of the voxels with an unrealistic g ≈ 1 at the edge of the genu [g]). (Reproduced from Mohammadi, S., *et al.*, *Front. Neurosci.*, 9, 441, 2015. With permission.)

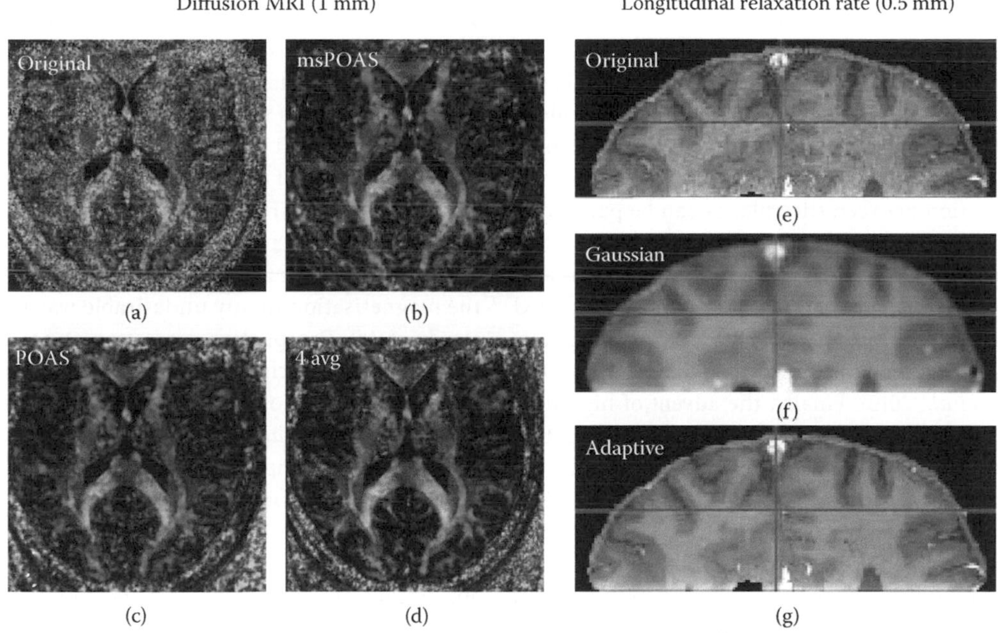

FIGURE 17.2 The effect of different adaptive smoothing algorithms on ultra-high resolution diffusion MRI (a through d) and longitudinal relaxation rate (e through g), acquired at 1 mm isotropic and at 0.5 mm, respectively. The four images on the left depict a colour-coded fractional anisotropy (FA) map (from a single-shell dataset acquired on a 7T MR system; Becker *et al.*, 2014): (a) original noisy data, (b) multishell position orientation adaptive smoothing (msPOAS) reconstruction, (c) POAS reconstruction (Becker *et al.*, 2013), (d) gold standard achieved by spending four times as long scanning and averaging the data. The three images on the right depict the original (e), the Gaussian denoised (f) and the adaptive denoised (g) longitudinal relaxation rate map acquired with a 0.5 mm isotropic resolution multiparameter mapping protocol at 3T (Tabelow *et al.*, 2016). Note that, unlike Gaussian denoising (f), adaptive denoising significantly reduced the noise level without removing structural boundaries in both qMRI modalities (g). (a–d) (Reproduced from Becker S.M.A., *et al.*, *NeuroImage*, 95, 90–105, 2014. With permission.)

adaptive denoising methods can enable ultra-high-resolution qMRI (e.g. diffusion MRI at 1 mm isotropic resolution [Becker *et al.*, 2014] and longitudinal relaxation rates at 500 micron isotropic resolution [Tabelow *et al.*, 2016]).

17.2.2 Subject motion

Subject motion is inevitable and the biggest source of image degradation in clinical MRI (Andre *et al.*, 2015). Motion occurs over multiple timescales ranging from milliseconds to seconds. It can be rigid body (e.g. slow head motion such as nodding) or non-linear (e.g. due to the ballistic motion induced by cardiac pulsation). The amplitude and type of motion, its timescale and the point at which it occurs during data acquisition are all important determinants of the degree of image degradation that will result. The impact of the motion will vary depending on whether it occurs between acquisitions (interscan motion) or during the acquisition (intrascan motion). The impact of intrascan motion will additionally depend on how many lines of *k*-space it affects and on whether it occurs during the acquisition of central or outer *k*-space lines. Motion between scans can be compensated for quite substantially through retrospective rigid body co-registration (i.e. estimating and reversing three translational and three rotational degrees of freedom). Popular open-source neuroimaging tools (SPM, FSL, etc.) use different optimisation functions to estimate the required rigid body transformation, depending on whether spatial alignment is required between different modalities or the same modality. For example, to register EPI and T_1-weighted images one could use mutual information (Studholme *et al.*, 1999), while to register images within an functional MRI (fMRI) time series one could use normalised sum-of-squares difference (Friston *et al.*, 1995). In 2D imaging, motion between slice planes can be partly corrected retrospectively by slice-wise registration. Although this is usually less stable and more time-consuming than rigid-body registration, it finds application in quantitative spinal cord MRI (Mohammadi *et al.*, 2013a; Xu *et al.*, 2013) where physiologically related involuntary motion can lead to misalignment between slices (Yiannakas *et al.*, 2012). Finally, the advent of high-density receiver arrays with highly non-uniform sensitivity profiles means that signal intensity in the images is modulated by the coil's net sensitivity profile in a position-specific manner. Rigid body co-registration does not account for this position-specific modulation. Therefore the net sensitivity profile must be separately estimated (on a participant- and position-specific basis) and removed. This prevents biases in qMRI parameter estimation (Papp *et al.*, 2016) that will otherwise be introduced even by small amplitude interscan motion.

Motion can also modulate the MRI signal, leading to bias in morphometry results (Tisdall *et al.*, 2016) and qMRI metrics (Callaghan *et al.*, 2015b). The 3D gradient echo signal is typically acquired over a timescale of minutes and will be affected by any type of motion that occurs during this period. For MRI modalities that are motion sensitive (e.g. diffusion MRI; Wedeen *et al.*, 1994; Storey *et al.*, 2007) and/or require a high temporal sampling rate for the entire brain (e.g. functional MRI), the EPI acquisition is favoured because of its speed. In single-shot 2D EPI a whole image plane can be acquired in less than 100 ms (for details see Bernstein *et al.*, 2004). The speed with which 2D EPI can be acquired affords some robustness to slow motion. However, it is much more sensitive to rapid, rotational-type motion. Such motion will introduce a linear phase variation across the tissue and thus a displacement of the data in *k*-space, which is proportional to the magnitude of the angular velocity of the rotational movement and to the first moment of the gradients applied between excitation and readout (Wedeen *et al.*, 1994). As a result, this type of motion can cause signal dropout in 2D EPI sequences that apply gradients with a high first moment (e.g. in dMRI subject motion [Storey *et al.*, 2007], table vibration [Gallichan *et al.*, 2010; Mohammadi *et al.*, 2012a] or cardiac pulsation [Skare and Andersson, 2001; Nunes *et al.*, 2005; Mohammadi *et al.*, 2013b]).

There are several retrospective intrascan motion correction approaches for 3D gradient echo acquisition techniques that aim to reverse MRI signal modulations caused by rigid-body subject motion. Some aim to estimate the motion trajectory from the complex MR signal (Atkinson *et al.*, 1997, 1999; Batchelor *et al.*, 2005; Cheng *et al.*, 2012), while others estimate it from interleaved acquisitions of so-called navigator echoes (Fu *et al.*, 1995; Welch *et al.*, 2002; Magerkurth *et al.*, 2011; Cheng *et al.*, 2012; Nöth *et al.*, 2014). While these retrospective approaches have demonstrated image quality improvements, their widespread use is limited by the fact that they require significant computational effort and do not compensate for spin history effects nor do they provide an indication of final image quality until after the scanning session, at which time the reacquisition of data would require a new imaging session and visit by the participant. Navigator echoes have the additional drawback of requiring dead time during the acquisition in which the navigator must be acquired without perturbing the magnetisation in any undesirable way (e.g. disrupting the steady state). Prospective correction approaches, using navigator echoes (Welch *et al.*, 2002; van der Kouwe *et al.*, 2006; White *et al.*, 2010) or external tracking devices (Zaitsev *et al.*, 2006; Maclaren *et al.*, 2012, 2013; Callaghan *et al.*, 2015b), can also be used to compensate for motion during data acquisition by estimating the motion in real time and using this information to update the imaging gradients and the frequency and phase of the radiofrequency field (RF). However, these approaches are still limited by their inability to compensate for the full array of motion-related effects, e.g. changes in the receive field, transmit field, B_0 shimming and non–rigid-body motion (Greitz *et al.*, 1992). An alternative approach is to exploit any redundancy in the acquired data. If the qMRI model is overdetermined, that is, more data points are available than necessary to fit the model, data-driven methods can be used to detect and remove motion-corrupted data points (e.g. for dMRI [Chang *et al.*, 2005], fMRI [Diedrichsen and Shadmehr, 2005], T_2* models [Weiskopf *et al.*, 2014] or a T_1 model [Callaghan *et al.*, 2016b]). For 2D EPI time series, motion regressors can be added to the qMRI model to estimate the contribution of these signal variations (e.g. as can be done within the general linear modelling framework of fMRI [Friston *et al.*, 1996] or dMRI [Mohammadi *et al.*, 2013b]).

17.2.3 Spatial image distortions

As noted above, 2D EPI is attractive for many applications because of its great rapidity; the speed of this approach stems from the fact that a whole plane of data is acquired in a single shot. This comes at the cost of low bandwidth along the phase-encoding direction, which means that adjacent *k*-space points in this direction are typically sampled two orders of magnitude more slowly than along the frequency-encoding direction (where the inter-echo spacing, or sampling interval along the phase-encoded direction, is in the region of 500 μs, the dwell time, or sampling interval along the frequency-encoded direction, could be 5 μs). This lower bandwidth leads to the manifestation of various artefacts in the phase encoding direction (e.g. chemical shift artefacts, Nyquist ghosting, and image distortions – for details see e.g. Bernstein *et al.*, 2004). Among these artefacts, image distortions in EPI can particularly compromise multimodal combinations of qMRI data (as exemplified in Figure 17.1). These distortions are caused by off-resonance effects (e.g. main magnetic field inhomogeneity, magnetic susceptibility variations and eddy currents). The degree to which the off-resonance effects translate into spatial distortions in the image depends on their magnitude relative to the bandwidth of the linear encoding gradients, which determines how far apart adjacent spatial points are represented in frequency space. This relationship is illustrated in Figure 17.3: the green and blue curves represent the relationship between image space and frequency space that is introduced by the phase- and frequency-encoding gradients, respectively. Note that the slope of the green curve is much lower than the slope of the blue curve and thus the green curve is more greatly perturbed by the additive off-resonance field (see red curves).

As image distortions are among the best understood and modelled phenomena in EPI, there are various processing methods that can be used to retrospectively correct these effects. Field inhomogeneities and magnetic susceptibility variations can be measured with additional MRI sequences for 'field mapping' (Jezzard and Balaban, 1995; Reber *et al.*, 1998; Hutton *et al.*, 2002) or estimated from the distorted images themselves using physical models (Chang and Fitzpatrick, 1992). Once the field inhomogeneity is known, the distortion can be reversed in an unwarping procedure (e.g. Hutton *et al.*, 2002; Markl *et al.*, 2003). While this method has proved popular, it requires accurate phase

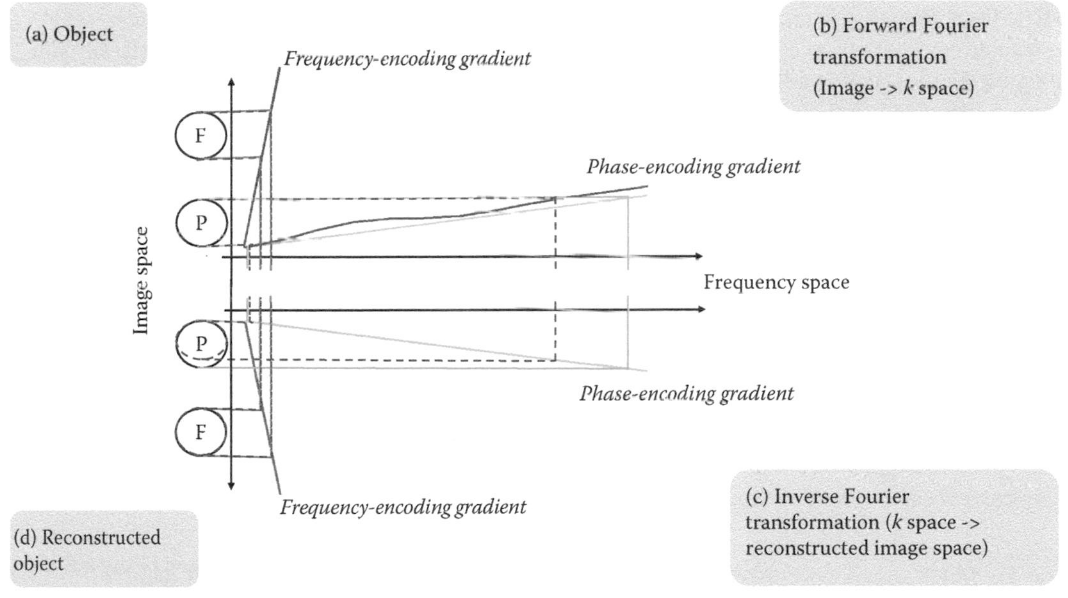

FIGURE 17.3 The linear imaging gradient produces a linear relationship between the position of the object (a) and frequency of the MRI signal (b). In an ideal situation, where this linear relationship is not perturbed, the object can be accurately reconstructed (black circles, d) by applying the inverse Fourier transform to the MR signal (c). The slope of this linear curve (relating image and *k*-space) depends on the bandwidth at which the data is sampled by the imaging encoding gradient. The much higher bandwidth of the frequency encoding gradient leads to a much greater slope for this gradient (blue line) than that of the phase encoding direction (green line), leading to differential impact of non-linear off-resonance perturbations (red curves). Consequently, the projection of the object along the frequency encoding gradient direction (black circle, F) leads to accurate reconstruction (red dashed circle, F), whereas this projection leads to distortion along the phase-encoding direction (black circle, P, distorted to red dashed circle, P). However, if the off-resonance field (and hence the shape of the perturbed imaging gradients) are known, for these distortions it can be corrected either by modifying the reconstruction itself, by using a different transform or by resampling the distorted image using the off-resonance field as a voxel-wise displacement map.

unwrapping, which can be challenging, particularly at susceptibility interfaces where the phase changes rapidly. Alternative approaches have been developed in recent years, e.g. to acquire an intermediate image that matches the contrast of one modality and the distortions of the other (Renvall *et al.*, 2016), to estimate and correct the distortion using the point spread function (Zaitsev *et al.*, 2004; In *et al.*, 2016) or to robustly estimate the field map using two EPI images with reversed phase-encoding directions (Andersson *et al.*, 2003; Holland *et al.*, 2010; Ruthotto *et al.*, 2012). These latter methods are available in open-source toolboxes (e.g. Andersson *et al.*, 2003, in FSL and Ruthotto *et al.*, 2012, 2013, in SPM-ACID). An additional component to the off-resonance fields are eddy currents. These are particularly problematic for dMRI data because of the strong diffusion-weighting gradients used, which manifest differently between images acquired with different diffusion-encoding gradient directions (Jezzard *et al.*, 1998). If the off-resonance field induced by the eddy currents is modelled up to first order, it leads to image distortions that can be described by four affine transformations (e.g. Mohammadi *et al.*, 2010); modelled up to second order it leads to image distortions that can be described by 10 transformation parameters (Rohde *et al.*, 2004), while 20 parameters are needed to describe the distortions induced by a field up to the third order (Andersson and Sotiropoulos, 2016). These transformation parameters can be combined with rigid-body co-registration to simultaneously correct for eddy current distortions and subject motion (see e.g. FSL, SPM-ACID). Even though promising new methods have succeeded in combining motion and distortion corrections (Andersson and Sotiropoulos, 2016; Andersson *et al.*, 2017), their interaction can be challenging, e.g. motion in the presence of gradient non-linearities can not only lead to varying image distortions but also to signal intensity variations (Bammer *et al.*, 2003; Mohammadi *et al.*, 2012b). While one should be mindful of this source of distortion (Schmitt *et al.*, 2012), hardware advances mean that modern human gradient systems show less severe non-linearities, greatly reducing the impact of this source of artefact.

There are only a few comparison studies published as peer-reviewed papers, making it difficult to specify the optimal method for distortion correction. However, a summary of preprocessing tools for dMRI can be found e.g. in the paper by Soares *et al.* (2013), while a recent study compared the performance of different tools for eddy current correction (Graham *et al.*, 2016).

17.2.4 Noise in MRI

While participant motion introduces noise into the MRI signal in the form of undesired signal fluctuations as discussed above, thermal noise also affects the quality of the acquired data by reducing the SNR and contrast-to-noise ratio, thereby limiting the smallest possible effect size that can be detected within the data, which is problematic when looking for subtle differences, e.g. in the prodromal stage of disease. In MRI, the term *noise* can have different meanings depending on the context. In this chapter, we will discuss the effect of thermal noise leading to variance in the detected voltage that is caused by random fluctuations in the receiver chain or the participant (i.e. due to random motion of molecules), during the acquisition of the MRI signal. The presence of noise in the acquired signal is a problem that affects not only the discriminatory power of the images (effect size as well as visual assessment for clinical evaluation) but may also interfere with further processing techniques such as segmentation and co-registration. In addition, noise can bias the observed MR signal (e.g. Rician noise bias is introduced by taking the signal magnitude) and thus affect any quantitative model estimates derived from the data.

The characterisation of noise in MRI is challenging, especially with the introduction of multiple receiver coils and parallel imaging techniques. To first order approximation, noise can be modelled as additive Gaussian-distributed complex noise that contributes to the real and imaginary parts of the acquired complex MR signal. Traditionally, magnitude images have been preferred but it is increasingly recognised, particularly with the enhanced susceptibility-based contrast at higher field strengths, that there is significant information about the microstructure of the tissue to be gleaned by additionally examining the phase of the MRI signal (Haacke *et al.*, 2004, 2015; Duyn *et al.*, 2007). For a single-channel acquisition, the magnitude reconstruction provides an image whose signal is Rician-distributed. Nowadays, multichannel receiver coils are routinely used and preferred to quadrature receiver coils because they provide higher SNR and offer the potential to reduce acquisition times and geometric distortions by using parallel imaging techniques (Deshmane *et al.*, 2012). The noise properties are influenced by any parallel imaging technique used as well as by the reconstruction filters applied. With advanced parallel imaging methods, the noise distribution becomes spatially dependent, and the signal properties require more complex modelling (Aja-Fernandez *et al.*, 2008; Robson *et al.*, 2008; Landman *et al.*, 2009; Maximov *et al.*, 2012; Veraart *et al.*, 2013). A review of the noise characteristics under these different configurations can be found in Dietrich *et al.* (2008).

The Rician noise in MR magnitude images leads to the phenomenon of noise rectification. As a consequence, as the amplitude of the true MR signal decreases, the acquired signal tends to a positive, non-zero value instead of tending to an average value of zero. This effect, termed the non-zero 'noise floor', can bias model estimates, particularly when the MR signal is of the same magnitude as the noise, i.e. in low SNR regimes. In diffusion MRI this can happen routinely because of the additional signal reduction purposefully imposed by the diffusion-weighting gradients. The noise floor can make the signal attenuation appear to be less than it is, such that the diffusion coefficient is underestimated (e.g. Dietrich *et al.*, 2001). The estimation of trace, anisotropy (Jones and Basser, 2004) and beyond-tensor diffusion properties such as the diffusion kurtosis (André *et al.*, 2014) can also be biased. These noise biases can be corrected prior to the model fitting by averaging complex data (Eichner *et al.*, 2015) or modelling its effect on magnitude data (Sijbers *et al.*, 1998; Sijbers and den Dekker, 2004; Aja-Fernandez *et al.*, 2008; André *et al.*, 2014). However, it can also be corrected at the modelling level (Basu *et al.*, 2006;

Andersson, 2008; Koay *et al.*, 2009; Landman *et al.*, 2009; Zhang *et al.*, 2012; Tabelow *et al.*, 2015; Polzehl and Tabelow, 2016). In relaxometry, particularly the T_2 (Bjarnason *et al.*, 2010; Bai *et al.*, 2014; Bouhrara *et al.*, 2015) and T_2* (Yokoo *et al.*, 2015) estimation can profit from a Rician bias correction, especially in iron-rich regions where the MR signal decays rapidly. Furthermore, high-order biophysical models (e.g. Wharton and Bowtell, 2013; Gil *et al.*, 2016; Callaghan *et al.*, 2016a) can particularly profit from the noise-bias correction, because they probe small signal changes.

The desire to maximise image resolution while minimising scan times (e.g. by using parallel imaging methods to subsample the data) means that quantitative MRI methods often suffer from low SNR. To enhance the SNR and at the same time minimise any loss in spatial resolution, edge-preserving denoising methods can be applied (Figure 17.4). Several such denoising methods (Ding *et al.*, 2005; Aja-Fernandez *et al.*, 2008; Lohmann *et al.*, 2010; Manjón *et al.*, 2010) have been suggested but it is typical for such methods that their performance (i.e. the adaptation of the smoothing kernel to the local anatomical boundaries) depends on the initial image SNR and on the magnitude of the signal change at tissue boundaries. Quantitative MRI protocols typically acquire multiple weighted volumes with varying contrast, e.g. multiple flip angles to quantify T_1 or multiple diffusion weightings. This provides redundant structural information that can be exploited by edge-preserving denoising methods if the total signal, and its variation at tissue boundaries, across the entire qMRI dataset can be used simultaneously (see Figure 17.2). Typical examples are dMRI and fMRI. These two imaging techniques also exemplify two very different demands on the SNR. In fMRI, the SNR of a single image is very high whereas the effect of interest, i.e. the BOLD effect, is of very low amplitude (0.5%–3% at 1.5T [Jezzard *et al.*, 2003] and increases by a factor of 2 for 3T [Triantafyllou *et al.*, 2011]). As a result, the temporal SNR, or more specifically the functional sensitivity, across the entire time series must be maximised. In dMRI, the image SNR is very low because of the strong diffusion-weighting gradients that are applied in addition to the imaging gradients. However, the effect size in dMRI can be quite high (particularly in white matter), allowing the diffusion parameters to be estimated from images with lower SNR. Adaptive denoising can also be used for other quantitative MRI techniques, e.g. high-resolution relaxometry (Callaghan *et al.*, 2016b; Tabelow *et al.*, 2016), by capitalising on the multicontrast nature of the acquisitions needed to quantify the MRI parameters. In Figures 17.2 and 17.4, examples of different adaptive denoising methods are demonstrated. While in Figure 17.4 the adaptive denoising of a single T_1-weighted

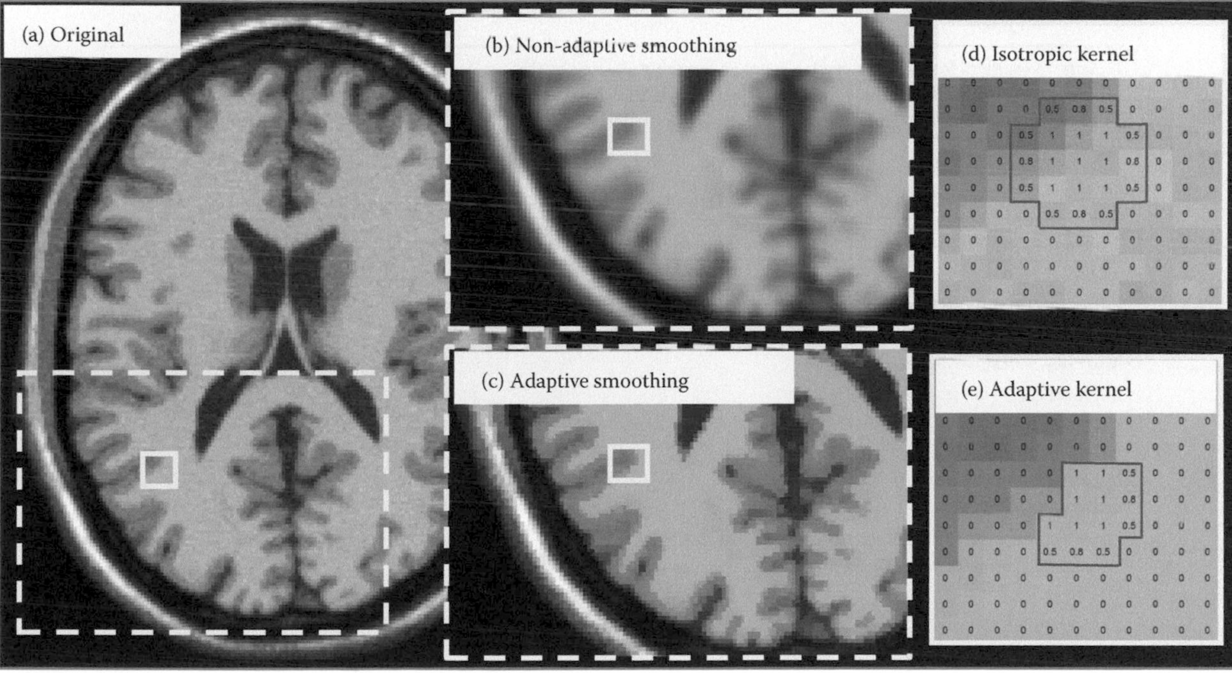

FIGURE 17.4 Extreme examples of adaptive (c) and non-adaptive (b) denoising on a simulated T_1-weighted image (Brainweb-Simulator, http://brainweb.bic.mni.mcgill.ca/brainweb/selection_normal.html) with 3% additive noise (a). In (d) and (f) the smoothing kernel (bounded by the red line) in a region that includes tissue boundaries is magnified (yellow box in (a–c)). The weights of the smoothing kernel (depicted as numbers in (d,e)) indicate how strongly that voxel contributes to the weighted mean calculated in the denoising procedure. While the weights of the adaptive smoothing kernel go to zero beyond tissue boundaries (e), the weights of the non-adaptive smoothing kernel remain isotropically distributed around the kernel centre (d). Note that this is an extreme case of adaptive smoothing to demonstrate that, in the limiting case, although tissue boundaries are preserved the structure within the tissue classes are removed. In practice, the parameters of the adaptive smoothing methods need to be tuned such that the SNR is maximised while preserving tissue features (as in Figure 17.2). (Courtesy of: Weights were calculated by Dr. Karsten Tabelow.)

image largely removed the small anatomical variation in white and grey matter (Figure 17.4c), the examples in Figure 17.2 better preserved small anatomical variation in white and grey matter. This is because the examples in Figure 17.2 used an entire dMRI (Figure 17.2a through d) or MPM (Figure 17.2e through g) dataset.

Generally, two different categories of edge-preserving denoising methods can be used: (1) methods that are based on signal models (e.g. the general linear model in fMRI [Tabelow *et al.*, 2006], the diffusion tensor model in dMRI [Arsigny *et al.*, 2006; Tabelow *et al.*, 2008] and the MPM model [Tabelow *et al.*, 2016]) and (2) methods that are independent of signal models but instead rely on symmetries within the data (e.g. on the neighbourhood information in image- and q-space; Duits and Franken, 2011; Becker *et al.*, 2012, 2014).

17.2.5 Methods for artefact correction

Artefacts manifest differently in the image depending on the acquisition technique used. There are two classes of prominently used acquisition techniques in qMRI that must be distinguished in terms of artefact: The 3D gradient echo technique is often used for relaxometry, MT and quantitative phase imaging, whereas 2D EPI is often used for dMRI and fMRI. Table 17.1 summarises selected prominent artefacts associated with 2D EPI and 3D gradient echo acquisitions. Note, however, that when it is said that, e.g., the 3D gradient echo approach is not affected by susceptibility-related distortions, it is meant that the artefact level is much smaller than in 2D EPI. In Table 17.2 we present some tools that can be used to minimise the impact of these artefacts.

TABLE 17.1 Overview of the most prominent artefacts for combining multimodal qMRI

Artefact		Subject motion			Image distortion		Noise	
Manifestation		Rigid body motion	Slice-Wise differences	Mr signal modulation	Susceptibility	Eddy currents	Bias	Image quality
2D EPI	Diffusion	! ✓	! (✓)	! (✓)	! ✓	! ✓	! ✓	! ✓
	Function	! ✓	! (✓)	! (✓)	! (✓)	×	×	×
3D GE	R1, R2	! ✓	×	! –	×	×	×	! ✓
	R2*	! ✓	×	! (✓)	×	×	! ✓	! ✓
	MT, PD	! ✓	×	! –	×	×	×	! ✓
	QSM	! ✓	×	! –	×	×	×	! ✓

Note: ! = suffers from this artefact; × = not appreciably affected by this artefact; (✓) = can be (partly) reduced via image processing; – = cannot be reduced via image processing; R1, R2 (R2*) = longitudinal, (apparent) transverse relaxation rate; MT = magnetisation transfer saturation rate; PD = proton density map; QSM = quantitative susceptibility mapping; qMRI = quantitative MRI; EPI = echo planar imaging.

TABLE 17.2 A collection of tools that can potentially reduce the artefacts highlighted in table 17.1

Tools for artefact correction	Subject motion	EPI Distortions	Noise
AFNI https://afni.nimh.nih.gov/afni	✓		
BrainVoyager http://www.brainvoyager.com/	✓		✓
ExploreDTI http://www.exploredti.com/	✓		
FSL https://fsl.fmrib.ox.ac.uk/fsl/fslwiki/	✓	✓	
MRI Denoising Software https://sites.google.com/site/pierrickcoupe/softwares/denoising-for-medical-imaging/mri-denoising/mri-denoising-software			✓
3D Slicer https://www.slicer.org/	✓	✓	✓
SPM http://www.fil.ion.ucl.ac.uk/spm/	✓	✓	
SPM-ACID www.diffusiontools.com	✓	✓	✓
TORTOISE https://science.nichd.nih.gov/confluence/display/nihpd/TORTOISE	✓	✓	
WIAS software collection for Neuroscience http://www.wias-berlin.de/software/imaging/			✓

Note: This collection might be incomplete and not all tools can handle all qMRI data appropriately. Moreover, this table was created in 2017 and new features might have been added to the tools over the years. We apologise to those developers whose tools were not mentioned.

Note that the order in which the processing approaches are presented here is not how we would recommend applying these approaches. For example, denoising and bias correction should, as far as possible, be applied on unprocessed data because interpolation due to the resampling step that is involved in distortion and motion correction can change the noise behaviour in the data. On the other hand, adaptive denoising methods for diffusion MRI or MPM data require the whole dataset to be spatially aligned, i.e. motion and eddy current distortions need to be corrected before adaptive denoising can be applied.

17.2.6 Outlook

We would like to emphasise that only a limited selection of artefacts and tools are presented, reflecting our focus on combining multimodal qMRI data. Other artefacts that are important for individual qMRI techniques (e.g. B_1^+ correction in relaxometry and magnetisation transfer imaging; Lutti *et al.*, 2010; Pohmann and Scheffler, 2013) are also of crucial importance for accurate quantification of MRI tissue properties. Furthermore, artefacts that cause higher-order effects on multimodal qMRI combination have not been covered. For example, 2D EPI data usually have broader point spread functions than 3D gradient echo data. This is caused by the appreciable amount of T_2* decay that occurs across the relatively long EPI echo train in the phase-encoded direction (Farzaneh *et al.*, 1990). This can increase partial volume effects in 2D EPI data and therefore multimodal qMRI approaches. One way to address the problem of varying effective resolution could be super-resolution approaches (Coupé *et al.*, 2013; Poot *et al.*, 2013; Ruthotto *et al.*, 2014; Setsompop *et al.*, 2015). It should also be mentioned that we focused here on tools for the brain and that there are many other important tools for other parts of the body that were not discussed here. Finally, it should be highlighted that powerful general-purpose tools exist (e.g. ANTs [http://stnava.github.io/ANTs/], ITK [https://itk.org/Doxygen/html/RegistrationPage.html] or FAIR [https://github.com/C4IR/FAIR.m]) that can be used to develop user-specific correction approaches. In fact, some of these tools are used, in part, in the tools listed in Table 17.2 (e.g. FAIR in SPM-ACID).

17.3 Processing for group analysis

Group studies can be conducted in a variety of ways. A central choice for such analyses is whether they are to be conducted in native (subject) space for each individual, e.g. by extracting a summary metric within a particular anatomical structure such as the average T_2* value in the red nucleus in the native space, or whether the analyses should be carried out in a common group space, e.g. by spatially normalising to some stereotactic space such as the commonly used Montreal Neurological Institute space. Manual delineation of specific anatomical structures are time-consuming and require expert knowledge of the anatomy of the structure of interest. An alternative is to use automated segmentation tools. This can be done in native space, e.g. to classify voxels as grey or white matter or even to identify subnuclei of the brainstem (Lambert *et al.*, 2013; Bianciardi *et al.*, 2015). Or atlas-based region-of-interest (ROI) labels can be warped from a template space into native space using the inverse deformations calculated by a non-linear spatial normalisation scheme (Klein *et al.*, 2009). The enhanced specificity of ROI analyses of qMRI data can reveal important insight about microstructural changes, e.g. the temporal pattern of microstructural changes in ageing (Lorio *et al.*, 2014; Steiger *et al.*, 2016). Group studies conducted in a stereotactic space have the advantage of direct accesses to predefined atlases in order to identify anatomical structures. In addition, there is added flexibility in that it is not necessary to have an *a priori* hypothesis about the brain area in which correlations or differences are expected; instead whole-brain analyses may be used to identify areas of interest, which can then be reported in a standardised fashion. However, to achieve anatomical correspondence across the group under investigation, these methods also rely on highly accurate spatial normalisation algorithms in order to map the individual qMRI data to the template space. When considering the quantitative nature of the data under investigation, another key aspect that warrants special attention is the need to preserve these quantitative values during this spatial normalisation process. Once in this common space, existing second-level statistical analysis frameworks (such as implemented in SPM, FreeSurfer, FSL, etc.) can be utilised for the group analysis.

17.3.1 Analysis in native space

A number of different approaches can be used to extract summary measures from qMRI data in native (subject) space. In the first instance the structure of interest, which might range in size from the entire cortex through to a specific subnucleus, needs to be defined. This may be done via an automated segmentation algorithm or by expert delineation (Keller and Roberts, 2009). Shape analysis quantifies abnormalities of form. Thereby, the term *shape* is defined as the set of geometric properties of an object that are independent of position, size and orientation (Bookstein, 2008). For example, over the course of *in utero* development, the cortex becomes increasingly convoluted. Therefore assessing the shape (degree of convolution) of the cortex can be a complementary measure to the brain size and thus help trying to better understand the consequences of preterm birth and the consequences of this key stage of development occurring *ex utero*. A variety of techniques exist to characterise the shape of the brain (e.g. outline analysis [Bookstein, 1997; Joshi *et al.*, 2013], deformation-based morphometry [Ashburner *et al.*, 1998; Ceyhan *et al.*, 2012; Joseph *et al.*, 2014] and surface-based morphometry for cortical folding patterns [Mangin *et al.*, 2004; Nordahl *et al.*, 2007]), many of which are still evolving. Texture, on the other hand, characterises the relationship between a pixel's intensity and that of its nearest neighbours, using a variety of measures. Texture analysis may be particularly well suited for lesion segmentation (Loizou *et al.*, 2015), the characterisation of brain tumours and the prediction of seizures in epilepsy (Kassner and Thornhill, 2010). However, these methods are mostly based on standard clinical MRI images (e.g. T_1- or T_2-weighted), and their application to quantitative MRI needs to be further evaluated.

17.3.1.1 ROI analyses

ROI analyses are based on the delineation of predefined anatomical areas of interest and are a commonly used strategy in qMRI (e.g. Vymazal *et al.*, 1999; Gouw *et al.*, 2008; Martin *et al.*, 2008; Schmalbrock *et al.*, 2016). It is often done manually because it requires little technical know-how and is stable. However, it is rather time-consuming, particularly if it is used for group analysis within specific anatomical regions that need to be delineated on a participant-specific basis. To improve efficiency and reliability of manual volume and surface estimation of brain structures in MRI, for example, the Cavalieri method of design-based stereology in conjunction with point counting can be used (Gundersen and Jensen, 1987; Roberts *et al.*, 2000). This method is a 100% investigator interactive technique that requires manual determination of sampling density for a given brain structure (i.e. the stereological parameters necessary to produce a reliable volume estimate) and investigator decisions on whether or not sampling probes (i.e. points) intersect the brain ROI.

17.3.1.2 Tractography-Defined ROIs

One disadvantage of manual ROI analysis is that it requires a clear hypothesis and profound anatomical knowledge to draw the ROI within the same anatomical structure for each subject. Even if all of these requisites are fulfilled, there is inter-individual variability of the structural and functional areas (Kanai and Rees, 2011), which is often difficult to capture when drawing the ROI based on a single MRI modality. For example, the different fibre pathways that are intersecting the same ROI cannot be distinguished by most qMRI techniques. One exception is diffusion MRI, which is today the sole *in vivo* tool for mapping the human structural connectome. By measuring the anisotropic diffusion of water throughout the brain, diffusion MRI reveals the course of long-range anatomical pathways using various methods under the umbrella term of *tractography* (e.g. Basser and Jones, 2002; Mori and van Zijl, 2002; Jbabdi and Johansen-Berg, 2011). The ability of tractography to localise specific white matter tracts has been used extensively as a means of defining ROIs for white matter analyses. *Tractometry*, a term coined for these techniques that consist of making tract-specific measurements, was one of the first fruitful applications of diffusion MR tractography (Berman *et al.*, 2005; Gong *et al.*, 2005; Jones *et al.*, 2005b). The idea is simply to use tractography as a means of determining ROIs (in this case tracts) in white matter from which to extract quantitative measurements. Note that there are more complex approaches for combining tract and microstructural information (for a review see Daducci *et al.*, 2016). Even though these approaches can reveal new insights about anatomy (e.g. Reich *et al.*, 2007; Yeatman *et al.*, 2014), they have to be used with caution since the dMRI-based tractography methods are inherently ill-posed (Jones *et al.*, 2013). While tractography has great utility for identifying specific tracts, given seed and target regions in an ROI, it is notoriously difficult to distinguish tracts at the subvoxel level. This fact is well known in the tractography community as the problem of distinguishing between kissing and crossing fibres, regardless of whether simplistic DTI models or more sophisticated models are used: reconstructed fibre tracts can take 'wrong turns' and produce a number of false positive and false negative connections (Jbabdi *et al.*, 2015; Neher *et al.*, 2015). Note that susceptibility tensor imaging (Liu *et al.*, 2012) is another emerging qMRI technique that might become an alternative or complementary *in vivo* tool for tractography but still needs to address severe challenges. A recent review about quantitative susceptibility imaging, including susceptibility tensor imaging, was conducted by Deistung *et al.* (2017).

17.3.2 Histogram analysis

Histogram analysis is an alternative means of summarising qMRI values over a region, which might be the entire brain, a specific tissue class or an anatomical ROI. A histogram summarises the number of voxels with a particular parameter value over the range of possible values investigated. Histogram parameters, such as mean, variance or mode, can then be compared across individuals in the group under investigation. This approach is most often used at the whole brain level, making it more robust to ROI placement and therefore the need for expert intervention (Klose *et al.*, 2013). However, such summary measures do not directly provide any information about the spatial location of the changes or differences and will have low sensitivity when these are spatially localised, since in this case they will be obscured by the greater number of unchanged voxels elsewhere in the brain. Additionally, such analyses need to consider the range over which the parameter value will be analysed, as well as the number of bins (and therefore the bin width), and may incorporate a normalisation step in order to account for inter-individual differences in overall brain size. A detailed treatment of histograms can be found in Chapter 18 of the first edition of this book (Tofts, 2003).

17.3.3 Spatial normalisation

In order to make operator-independent between-subject comparisons at a local level (in the extreme case on a voxel-by-voxel basis), qMRI data from different subjects need to be brought into alignment in the same anatomical framework. To achieve this, one needs to remove confounding factors such as extrinsic differences (e.g. head position and orientation) and intrinsic differences (e.g. brain size and shape, gyral and sulcal variation) in order to detect meaningful physiological and pathological differences between groups. While the extrinsic differences can be accounted for using rigid-body registration, the intrinsic differences require non-linear registration approaches. Numerous algorithms have emerged to non-linearly register brains to one another. As an alternative to volume-wise registration methods for the entire brain, surface-based algorithms (Davatzikos and Bryan, 1996; Drury *et al.*, 1996; Thompson and Toga, 1996;

Fischl *et al.*, 1999) have been suggested to improve the accuracy of cortical registration and thereby reduce intersubject variability. A comparison of 14 non-linear deformation algorithms applied to brain volume image registration has shown that the following registration tools give the best results according to overlap and distance measures: ART, SyN, IRTK and SPM's DARTEL (Klein *et al.*, 2009). In a follow-up study, selected volume-based registration methods were compared to surface-based methods (Klein *et al.*, 2010). The estimated non-linear deformation fields can be used to enable operator-independent analyses of neuroimaging data by either transforming ROIs from stereotactic atlases of the human brain (as available in typical neuroimaging toolboxes, e.g. SPM, FSL, FreeSurfer) or by transforming the MR images or quantitative maps themselves into the stereotactic space.

17.3.4 Analysis in stereotactic space

17.3.4.1 Voxel-Based morphometry

One of the most popular analysis approaches conducted in stereotactic space is voxel-based morphometry (VBM) of grey matter (Ashburner and Friston, 2000). This is probably because it allows a hypothesis-free analysis of group effects over the entire brain, which is particularly beneficial for the discovery of novel insights for which there is no *a priori* knowledge. In a VBM analysis, each participant's grey matter tissue probability map (produced by a segmentation routine applied in native space) is spatially normalised to template space using the estimated deformation field and modulated by the Jacobian determinants of that deformation. In other words, where the tissue has been contracted in order to map to template space, the probability is increased, and where the tissue has been expanded the probability is decreased. This has the effect of preserving the total grey-matter (GM) probability when transforming the data from native space to template space. Any residual differences in alignment across the group (see Figure 17.5) are compensated for by smoothing in template space after spatial normalisation. The degree of smoothing required is reduced when more complex models (therefore having higher numbers of parameters) are used to describe the deformation. Unfortunately, there is no empirical evidence to guide the optimal choice for the smoothing

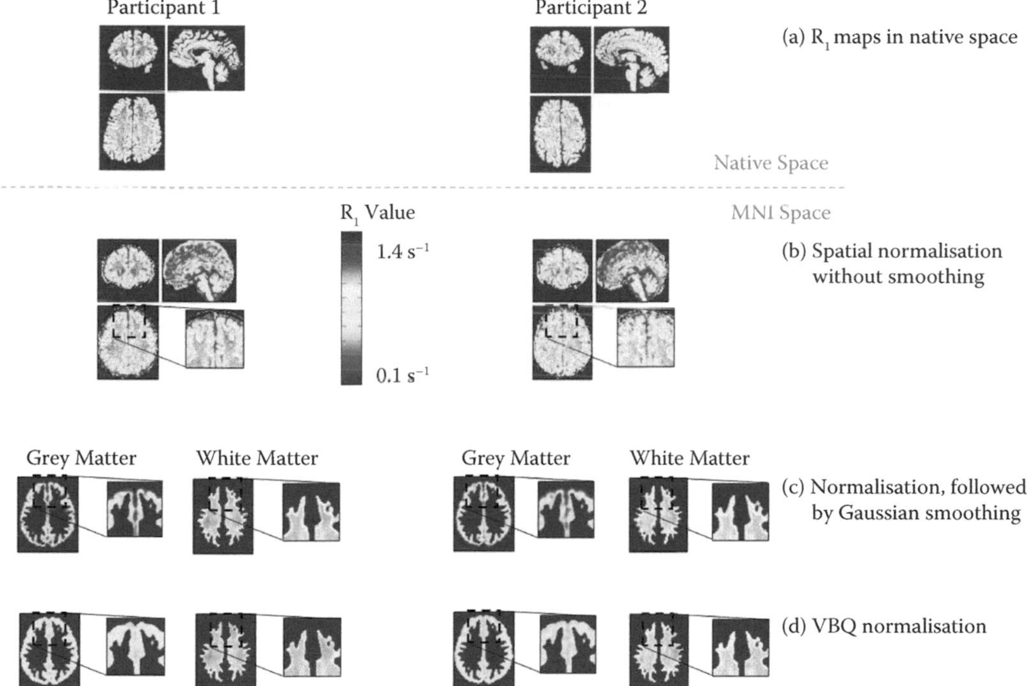

FIGURE 17.5 Illustration of the spatial normalisation and smoothing steps in VBM-style analyses in two participants. Highly parameterised non-linear deformations (e.g. the Dartel algorithm in this case; Ashburner, 2007) can largely reduce inter-individual anatomical differences (a through b). Remaining residual inter-individual differences (see zoomed box in b) can be addressed by spatially smoothing. However, even if spatial smoothing is carried out in a tissue-specific manner, it can bias qMRI values (e.g. manifesting as rapid decay of R1 values at tissue boundaries, illustrated in zoomed region in **c**). The voxel-based quantification (VBQ; Draganski *et al.*, 2011) approach has been proposed as a means of addressing this issue by taking a weighted sum of the qMRI values over the spatial extent of the smoothing kernel in native space. This approach greatly reduces smoothing-induced bias (d). The weighting is carried out in a tissue-specific manner, thereby producing a qMRI map per tissue class. It incorporates the Jacobian determinant into the weighting (e.g. voxels contracted by the transformation to Montreal Neurological Institute [MNI] space are up-weighted). The data shown here were acquired as part of an MPM protocol at 3T with 800 micron resolution. In all cases the final voxel size of the R1 maps in MNI space was 1 mm isotropic resolution. A Gaussian smoothing kernel of 4 mm full-width at half-maximum was used in (c) and (d).

kernel width. Note that spatial smoothing also has the effect of rendering the data more normally distributed, thus increasing the validity of parametric statistical tests (Friston *et al.*, 2006).

Although the spatial normalisation pipeline used for VBM is very appealing for qMRI data as well, it cannot be directly applied without modification to account for the quantitative nature of the data. For example, modulating by the Jacobian determinants would introduce erroneous quantitative values in areas estimated to undergo expansion or contraction, a clearly undesirable effect for quantitative neuroimaging biomarkers. The isotropic smoothing kernels typically adopted for such analyses would also exacerbate partial volume effects in a highly deleterious fashion. To address these issues, a tissue-specific, smoothing-compensated (T-SPOON; Lee *et al.*, 2009) approach was first introduced for voxel-based analyses of diffusion data. This approach aims to compensate for the partial-volume effects introduced by smoothing, particularly in the vicinity of edge regions.

An extension of the T-SPOON approach, termed *voxel-based quantification* (VBQ, Draganski *et al.*, 2011) has been applied to a range of different qMRI metrics (relaxometry, magnetisation transfer and diffusion metrics). This approach combines the normalisation, smoothing and compensation steps (see Figure 17.5). This approach amounts to taking a weighted average over the spatial extent of the smoothing kernel in native space, where the weights are the participant-specific tissue probabilities. This approach accounts for expansion and contraction during the normalisation procedure and additionally for partial volume effects at boundaries between neighbouring tissue classes. As a consequence, this approach produces a normalised map for each tissue type (see Figure 17.5b).

17.3.4.2 Surface-Based analysis

An alternative to voxel-based analyses is to construct a cortical surface and carry out analyses at the individual or group level directly on these surfaces. This analysis relies on the accurate delineation of the boundaries between the cortex and the white matter and pial surfaces. Once the cortex has been defined, the surface-based approach (in theory) enables sampling at arbitrary depths, while accounting for local curvature but not being limited by the acquired imaging resolution. This is an increasingly important benefit when working with higher resolution data containing greater intracortical information. Note, however, that although accounting for local curvature is important (Waehnert *et al.*, 2014), it is not standard practice today. Today, most surface-based studies of qMRI metrics either map the mean across the thickness of the cortex (Lutti *et al.*, 2014) or sample at one specific depth (Cohen-Adad, 2014). However, most of the studies still use shape descriptors (e.g. thickness, surface area, gyrification) for their analysis (Hogstrom *et al.*, 2013; Kelly *et al.*, 2013; Fairchild *et al.*, 2015).

As noted above, voxel-based analyses introduce a smoothing step to account for residual errors in spatial normalisation, which enhances partial volume effects between different tissue types. Surface-based approaches circumvent this issue by enabling smoothing to be applied tangentially along the cortical sheet rather than isotropically, which additionally avoids smoothing across gyral banks. Such smoothing could lead to inter-areal mixing within the cortex but avoids the far greater effect of mixing across tissue types. Following tessellation of these cortical surfaces, the various qMRI data from each participant can be mapped to their cortical surface for ROI or whole-brain vertex-wise analysis. By aligning the 2D manifolds, group analyses can be conducted. This might be done by aligning the data via a spherical atlas by exploiting continuity information from the entire image volume. This is achieved by matching the cortical geometry of the participants based on their individual cortical folding patterns (Fischl *et al.*, 1999). This approach has advanced our understanding of the relationship between qMRI markers of myeloarchitecture and function in basic neuroscience (e.g. Dick *et al.*, 2012; Frost and Goebel, 2012; Sereno *et al.*, 2013) and clinical research (e.g. Calabrese *et al.*, 2015).

There are multiple software packages available for surface-based analyses (e.g. see Table 17.3). Among these, FreeSurfer (http://surfer.nmr.mgh.harvard.edu/) is perhaps the most commonly used. It delineates the pial and white matter surfaces based on identifying the greatest shift in signal intensity as the transition between tissue types (Dale and Sereno, 1993; Fischl *et al.*, 1999; Fischl and Dale, 2000). This procedure has historically used T_1-weighted scans and not doing so has been reported to lead to localised segmentation errors (Dick *et al.*, 2012; Sereno *et al.*, 2013). One option to address this is to use the qMRI data to synthesise a T_1-weighted image (which is an illustration of the flexibility afforded by having qMRI data with which to synthesise arbitrary image contrasts). For the group-analysis of ultra-high resolution T1 weighted- and qMRI data (e.g. 400 μm isotropic; Bazin *et al.*, 2014), the CBS Tools software package (https://www.cbs.mpg.de/institute/software/cbs-tools) can also be used. The CBS tools not only take advantage of the higher spatial resolution but also of the multiple contrasts that are typically available in qMRI data (Tardif *et al.*, 2015).

17.3.4.3 Tract-Based spatial statistics

Tract-based spatial statistics (TBSS) is another, very extreme form of voxel-based quantification. Instead of separating the brain into white and grey matter tissue probability maps (as done in VBQ), TBSS projects volumetric data onto a white matter skeleton (i.e. a very extreme form of reducing the white matter tissue probability map) to circumvent the partial volume effect and gain statistical power from this dimensionality reduction (Smith *et al.*, 2006). The approach does not require data smoothing and addresses some of the concerns associated with VBM-style analyses (e.g. the issue of smoothing [Jones *et al.*, 2005a] or misregistration [Smith *et al.*, 2006]). It was initially created for diffusion MRI data but can also be used with other qMRI modalities (Langkammer *et al.*, 2010; Knight *et al.*, 2016). However, TBSS has limitations of its own, some of which are (1) the shape of the skeleton as well as the statistical results are rotationally variant (Edden and Jones, 2011); (2) only 10% of post-registration misalignment is reported to be corrected by the TBSS projection algorithm (Zalesky, 2011); (3) the effect of interest occurs in voxels where the local FA is highest

TABLE 17.3 A Short Summary of the Most Popular Tools That Are Used for Group Analyses of qMRI Data

Tools for Voxel-Based Group Analysis	Volume	Surface	Quasi 1D*
AFNI https://afni.nimh.nih.gov/afni	✓		
BrainVoyager http://www.brainvoyager.com/	✓	✓	✓
CBS tools http://www.nitrc.org/projects/cbs-tools/		✓	
FreeSurfer https://surfer.nmr.mgh.harvard.edu/		✓	✓
FSL https://fsl.fmrib.ox.ac.uk/fsl/fslwiki/	✓		✓
NIPY (and similar) http://nipy.org/	✓	✓	✓
SPM http://www.fil.ion.ucl.ac.uk/spm/	✓	✓	

Note: Quasi 1D* includes skeleton-based analyses as well as tractometry-type region of interest analyses. We have not included tools that were specific to the analysis of one qMRI technique only, such as tractometry for dMRI. We apologise to those developers whose tools were not mentioned.

(Van Hecke *et al.*, 2010); and (4) TBSS is not tract-specific, as it skeletonises the FA image (Bach *et al.*, 2014). More accurate registration approaches (Leming *et al.*, 2016) and other methodological improvements (Bach *et al.*, 2014) have been suggested to improve the performance of TBSS.

17.4 Outlook

MRI data acquisition has advanced enormously in recent years, providing a means to efficiently acquire high resolution data, with whole brain coverage, in reasonable scan times. The advent of ultra-high field (7T and above) will further accelerate these developments such that laminar-specific information will become more routinely available. These data will be a particular challenge for neuroimaging studies seeking to capture inter-individual difference in behaviour, function, dysfunction or pathology. Today, the most widely used tools for group analysis of qMRI data either reduce the dimensionality to simplify the registration problem (e.g. FreeSurfer reduces the problem to two dimensions, i.e. the surface, and TBSS even to one dimension, i.e. the skeleton) or keep the full three-dimensional dataset and afterwards smooth it to reduce residual spatial misalignment (VBM-style of analyses such as VBQ). However, if there were a one-to-one mapping of each anatomical point in an individual brain into the group space, then smoothing and dimension reduction could be avoided in the analyses of qMRI data, thereby preserving the spatial specificity with which the data were acquired. Therefore, a key challenge for qMRI analyses will be to further improve the methods for alignment of neuroimaging cohorts into a common group space. The qMRI approach may inherently provide some assistance in addressing this problem by virtue of the fact that rich data sets are acquired. For example, advanced neuroimaging studies use multiple contrasts, not only for parcellation of cortical areas (Glasser *et al.*, 2016) but also to drive the registration (Tardif *et al.*, 2015) to improve accuracy and ultimately to get the most complete picture of the brain's microarchitecture (Weiskopf *et al.*, 2015).

Before the full potential of hMRI can be realised, it will be necessary to combine these methodological developments (i.e. improved accuracy in spatial normalisation, artefact correction and multimodal data combination) with further advances in biophysical modelling of the MR signal. The ultimate goal of non-invasively characterising biologically relevant metrics *in vivo* will be brought within reach by combining artefact-free multimodal qMRI data with biophysical models and methods for accurate spatial integration of qMRI data into group space.

17.5 Acknowledgements

We would like to thank Lars Ruthotto for helpful insight into image registration and EPI distortions and Karsten Tabelow for his input on noise in MRI. We would also like to note that our views on qMRI have been greatly influenced by fruitful discussions with Nikolaus Weiskopf during joint years spent working at the Wellcome Trust Centre for Neuroimaging in London. SM received funding from the European Union's Horizon 2020 research and innovation programme under the Marie Sklodowska-Curie grant agreement No 658589. SM, and MFC were also supported by the ERA-Net Neuron Cofund under the BMBF No 01EW1711A.

17.6 Appendix: image registration

There are a large number of application areas that require registration procedures, e.g. art, astronomy, astrophysics, biology, chemistry, criminology, genetics, physics and imaging. It is beyond the scope of this chapter to give a full introduction to image registration (in-depth literature about image registration include work by Friston *et al.*, 2006; Modersitzki, 2009), but we will give a short introduction to the most important aspects.

A spatial transformation is applied to an image in order to change the position, orientation or shape of structures (such as the brain) in the image. Usually, by *transformation* we mean a six-parameter rigid-body transformation, which includes three translation and three rotation parameters; i.e. it moves the image (e.g. the brain) but does not change its form. The simplest transformation that includes deformation of the image is the 12-parameter affine transformation, including three translation, three rotation, three scaling and three shearing parameters. Other low-dimensional registration methods may employ a number of other forms of basis function to parameterise the warps. These include Fourier bases (Christensen and Johnson, 2001), sine and cosine transform basis functions (Christensen *et al.*, 1994; Ashburner and Friston, 1999), B-splines (Studholme *et al.*, 2000; Thévenaz and Unser, 2000) and piecewise affine basis functions. The small number of parameters will not allow every feature to be matched exactly, but it will permit the global head shape to be modelled. The rationale for adopting a low dimensional approach is that it prevents the identification of implausible transformations and allows for rapid modelling of global brain shape. While it might still be necessary to use non-parametric image transformation models to align brain images from different subjects (for more information see Friston *et al.*, 2006; Modersitzki, 2009; Sotiras *et al.*, 2013; Mangeat *et al.*, 2015), it should be treated with caution because image registration is considered an ill-posed (i.e. underdetermined) problem (Fischer and Modersitzki, 2008).

At its simplest, image registration involves estimating a mapping that best aligns a pair of images. One image is assumed to remain stationary (the *reference image*), whereas the other (the *source image*) is spatially transformed to match it. Typically this is done by examining the similarities between the source and reference, which can be measured using some function specific to the imaging modality. In order to transform the source to match the reference, it is necessary to determine the spatial transformation to map from each voxel position in the reference to the corresponding position in the source. The mapping can be thought of as a function of a set of estimated *transformation parameters*. Registration commonly involves numerical optimisation in which the source image is transformed many times, using different parameters, until some matching criterion (*distance/misfit function*) is deemed optimal (more details about the optimisation processes can be found, e.g., in work by Boyd and Vandenberghe, 2004; Nocedal and Wright, 2006; Beck, 2014). After the final parameters are estimated, the source is then *resampled* at the estimated positions.

Resampling (an interpolation process) involves determining, for each voxel in the transformed image, the corresponding intensity given the original image. Usually, this requires sampling between the centres of the original voxels, such that some form of interpolation is needed. It is beyond the scope of this chapter to go into the detail of available interpolation methods, but it should be mentioned that the optimum method for applying rigid-body transformations to MRI images with minimal interpolation artefact is to do it in Fourier space, which makes sinc interpolation (applied in the image domain) a natural candidate for resampling even though there are more efficient ways available (more details can be found, e.g., in Friston *et al.*, 2006; Modersitzki, 2009).

References

Aja-Fernandez S, Niethammer M, Kubicki M, Shenton ME, Westin C-F. Restoration of DWI data using a Rician LMMSE estimator. IEEE Trans Med Imaging 2008; 27: 1389–403.

Andersson JLR. Maximum a posteriori estimation of diffusion tensor parameters using a Rician noise model: why, how and but. NeuroImage 2008; 42: 1340–56.

Andersson JLR, Graham MS, Drobnjak I, Zhang H, Filippini N, Bastiani M. Towards a comprehensive framework for movement and distortion correction of diffusion MR images: within volume movement. Neuro Image 2017; 152: 450–66.

Andersson JLR, Skare S, Ashburner J. How to correct susceptibility distortions in spin-echo echo-planar images: application to diffusion tensor imaging. Neuro Image 2003; 20: 870–88.

Andersson JLR, Sotiropoulos SN. An integrated approach to correction for off-resonance effects and subject movement in diffusion MR imaging. NeuroImage 2016; 125: 1063–78.

André ED, Grinberg F, Farrher E, Maximov II, Shah NJ, Meyer C, et al. Influence of noise correction on intra- and inter-subject variability of quantitative metrics in diffusion kurtosis imaging. PLoS One 2014; 9: e94531.

Andre JB, Bresnahan BW, Mossa-Basha M, Hoff MN, Smith CP, Anzai Y, et al. Toward quantifying the prevalence, severity, and cost associated with patient motion during clinical MR examinations. J Am Coll Radiol JACR 2015; 12: 689–95.

Arsigny V, Fillard P, Pennec X, Ayache N. Log-Euclidean metrics for fast and simple calculus on diffusion tensors. Magn Reson Med 2006; 56: 411–21.

Ashburner J. A fast diffeomorphic image registration algorithm. NeuroImage 2007; 38: 95–113.

Ashburner J, Friston KJ. Nonlinear spatial normalization using basis functions. Hum Brain Mapp 1999; 7: 254–66.

Ashburner J, Friston KJ. Voxel-based morphometry – the methods. NeuroImage 2000; 11: 805–21.

Ashburner J, Hutton C, Frackowiak R, Johnsrude I, Price C, Friston K. Identifying global anatomical differences: deformation-based morphometry. Hum Brain Mapp 1998; 6: 348–57.

Atkinson D, Hill DL, Stoyle PN, Summers PE, Clare S, Bowtell R, et al. Automatic compensation of motion artifacts in MRI. Magn Reson Med. 1999; 41: 163–70.

Atkinson D, Hill DL, Stoyle PN, Summers PE, Keevil SF. Automatic correction of motion artifacts in magnetic resonance images using an entropy focus criterion. IEEE Trans Med Imaging 1997; 16: 903–10.

Bach M, Laun FB, Leemans A, Tax CMW, Biessels GJ, Stieltjes B, et al. Methodological considerations on tract-based spatial statistics (TBSS). NeuroImage 2014; 100: 358–69.

Bai R, Koay CG, Hutchinson E, Basser PJ. A framework for accurate determination of the T_2 distribution from multiple echo magnitude MRI images. J Magn Reson San Diego Calif 1997 2014; 244: 53–63.

Bammer R, Markl M, Barnett A, Acar B, Alley MT, Pelc NJ, et al. Analysis and generalized correction of the effect of spatial gradient field distortions in diffusion-weighted imaging. Magn Reson Med Off J Soc Magn Reson Med Soc Magn Reson Med 2003; 50: 560–9.

Bammer R, Skare S, Newbould R, Liu C, Thijs V, Ropele S, et al. Foundations of advanced magnetic resonance imaging. NeuroRx J Am Soc Exp Neurother 2005; 2: 167–96.

Basser PJ, Jones DK. Diffusion-tensor MRI: theory, experimental design and data analysis – a technical review. NMR Biomed 2002; 15: 456–67.

Basu S, Fletcher T, Whitaker R. Rician noise removal in diffusion tensor MRI. Med Image Comput Comput Assist Interv 2006; 9: 117–25.

Batchelor PG, Atkinson D, Irarrazaval P, Hill DLG, Hajnal J, Larkman D. Matrix description of general motion correction applied to multishot images. Magn Reson Med 2005; 54: 1273–80.

Bazin P-L, Weiss M, Dinse J, Schäfer A, Trampel R, Turner R. A computational framework for ultra-high resolution cortical segmentation at 7Tesla. NeuroImage 2014; 93 (Pt 2): 201–9.

Beck A. Introduction to nonlinear optimization: theory, algorithms, and applications with MATLAB. SIAM Bookst; 2014. Available from: http://bookstore.siam.org/mo19/ [cited 17 January 2017].

Becker S, Tabelow K, Mohammadi S, Weiskopf N, Polzehl J. Adaptive smoothing of multi-shell diffusion-weighted magnetic resonance data by msPOAS. Weierstrass Inst Appl Anal Stoch Prepr 1809; 2013. Available from: http://www.wias-berlin.de/publications/wias-publ/run.jsp?temp-late=abstract&type=Preprint&year=2013&number=1809 [cited 7 November 2013].

Becker SMA, Tabelow K, Mohammadi S, Weiskopf N, Polzehl J. Adaptive smoothing of multi-shell diffusion weighted magnetic resonance data by msPOAS. NeuroImage 2014; 95: 90–105.

Becker SMA, Tabelow K, Voss HU, Anwander A, Heidemann RM, Polzehl J. Position-orientation adaptive smoothing of diffusion weighted magnetic resonance data (POAS). Med Image Anal 2012; 16: 1142–55.

Berman JI, Mukherjee P, Partridge SC, Miller SP, Ferriero DM, Barkovich AJ, et al. Quantitative diffusion tensor MRI fiber tractography of sensorimotor white matter development in premature infants. NeuroImage 2005; 27: 862–71.

Bernstein MA, King KF, Zhou XJ. Handbook of MRI pulse sequences. 1st ed. Academic Press, Amsterdam; 2004.

Bianciardi M, Toschi N, Edlow BL, Eichner C, Setsompop K, Polimeni JR, et al. Toward an in vivo neuroimaging template of human Brainstem Nuclei of the ascending arousal, autonomic, and motor systems. Brain Connect 2015; 5: 597–607.

Bjarnason TA, McCreary CR, Dunn JF, Mitchell JR. Quantitative T2 analysis: the effects of noise, regularization, and multivoxel approaches. Magn Reson Med 2010; 63: 212–17.

Bonnici HM, Chadwick MJ, Kumaran D, Hassabis D, Weiskopf N, Maguire EA. Multi-voxel pattern analysis in human hippocampal subfields. Front Hum Neurosci 2012; 6: 290.

Bookstein FL. Landmark methods for forms without landmarks: morphometrics of group differences in outline shape. Med Image Anal 1997; 1: 225–43.

Bookstein FL. Morphometric tools landmark data: geometry and biology. Revised. Cambridge: Cambridge University Press; 2008.

Bouhrara M, Reiter DA, Celik H, Bonny J-M, Lukas V, Fishbein KW, et al. Incorporation of Rician noise in the analysis of biexponential transverse relaxation in cartilage using a multiple gradient echo sequence at 3 and 7 Tesla. Magn Reson Med 2015; 73: 352–66.

Boyd S, Vandenberghe L. Convex optimization – Boyd and Vandenberghe. 2004. Available from: http://stanford.edu/~boyd/cvxbook/ [cited 17 January 2017].

Calabrese M, Magliozzi R, Ciccarelli O, Geurts JJG, Reynolds R, Martin R. Exploring the origins of grey matter damage in multiple sclerosis. Nat Rev Neurosci 2015; 16: 147–58.

Callaghan M, Pine K, Tabelow K, Polzeh J, Weiskopf N, Mohammadi S. Mapping higher order components of the GRE signal decay at 7T with short TE data through adaptive smoothing. Proc Int Soc Magn Reson Med 2016; 24: 1539.

Callaghan MF, Freund P, Draganski B, Anderson E, Cappelletti M, Chowdhury R, et al. Widespread age-related differences in the human brain microstructure revealed by quantitative magnetic resonance imaging. Neurobiol Aging 2014; 35: 1862–72.

Callaghan MF, Helms G, Lutti A, Mohammadi S, Weiskopf N. A general linear relaxometry model of R1 using imaging data. Magn Reson Med 2015a; 73: 1309–14.

Callaghan MF, Josephs O, Herbst M, Zaitsev M, Todd N, Weiskopf N. An evaluation of prospective motion correction (PMC) for high resolution quantitative MRI. Front Neurosci 2015b; 9: 97.

Callaghan MF, Mohammadi S, Weiskopf N. Synthetic quantitative MRI through relaxometry modelling. NMR Biomed 2016; 29: 1729–38.

Ceyhan E, Beg MF, Ceritoğlu C, Wang L, Morris JC, Csernansky JG, et al. Metric distances between hippocampal shapes indicate different rates of change over time in nondemented and demented subjects. Curr Alzheimer Res 2012; 9: 972–81.

Chang H, Fitzpatrick JM. A technique for accurate magnetic resonance imaging in the presence of field inhomogeneities. IEEE Trans Med Imaging 1992; 11: 319–29.

Chang L-C, Jones DK, Pierpaoli C. RESTORE: robust estimation of tensors by outlier rejection. Magn Reson Med 2005; 53: 1088–95.

Cheng JY, Alley MT, Cunningham CH, Vasanawala SS, Pauly JM, Lustig M. Nonrigid motion correction in 3D using autofocusing with localized linear translations. Magn Reson Med 2012; 68: 1785–97.

Christensen GE, Johnson HJ. Consistent image registration. IEEE Trans Med Imaging 2001; 20: 568–82.

Christensen GE, Rabbitt RD, Miller MI. 3D brain mapping using a deformable neuroanatomy. Phys Med Biol 1994; 39: 609–18.

Cohen-Adad J. What can we learn from T2* maps of the cortex? NeuroImage 2014; 93 (Pt 2): 189–200.

Coupé P, Manjón JV, Chamberland M, Descoteaux M, Hiba B. Collaborative patch-based super-resolution for diffusion-weighted images. NeuroImage 2013; 83: 245–61.

Daducci A, Dal Palú A, Descoteaux M, Thiran J-P. Microstructure informed tractography: pitfalls and open challenges. Brain Imaging Methods 2016: 247.

Dale AM, Sereno MI. Improved localizadon of cortical activity by combining EEG and MEG with MRI cortical surface reconstruction: a linear approach. J Cogn Neurosci 1993; 5: 162–76.

Davatzikos C, Bryan N. Using a deformable surface model to obtain a shape representation of the cortex. IEEE Trans Med Imaging 1996; 15: 785–95.

Deistung A, Schweser F, Reichenbach JR. Overview of quantitative susceptibility mapping. NMR Biomed 2017; 30: e3569.

Deppe M, Duning T, Mohammadi S, Schwindt W, Kugel H, Knecht S, et al. Diffusion-tensor imaging at 3 T: detection of white matter alterations in neurological patients on the basis of normal values. Invest Radiol 2007; 42: 338–45.

Deshmane A, Gulani V, Griswold MA, Seiberlich N. Parallel MR imaging. J Magn Reson Imaging JMRI 2012; 36: 55–72.

Dick F, Tierney AT, Lutti A, Josephs O, Sereno MI, Weiskopf N. In vivo functional and myeloarchitectonic mapping of human primary auditory areas. J Neurosci 2012; 32: 16095–105.

Diedrichsen J, Shadmehr R. Detecting and adjusting for artifacts in fMRI time series data. NeuroImage 2005; 27: 624–34.

Dietrich O, Heiland S, Sartor K. Noise correction for the exact determination of apparent diffusion coefficients at low SNR. Magn Reson Med 2001; 45: 448–53.

Dietrich O, Raya JG, Reeder SB, Ingrisch M, Reiser MF, Schoenberg SO. Influence of multichannel combination, parallel imaging and other reconstruction techniques on MRI noise characteristics. Magn Reson Imaging 2008; 26: 754–62.

Ding Z, Gore JC, Anderson AW. Reduction of noise in diffusion tensor images using anisotropic smoothing. Magn Reson Med 2005; 53: 485–90.

Draganski B, Ashburner J, Hutton C, Kherif F, Frackowiak RSJ, Helms G, et al. Regional specificity of MRI contrast parameter changes in normal ageing revealed by voxel-based quantification (VBQ). NeuroImage 2011; 55: 1423–34.

Draganski B, Gaser C, Busch V, Schuierer G, Bogdahn U, May A. Neuroplasticity: changes in grey matter induced by training. Nature 2004; 427: 311–12.

Drury HA, Van Essen DC, Anderson CH, Lee CW, Coogan TA, Lewis JW. Computerized mappings of the cerebral cortex: a multiresolution flattening method and a surface-based coordinate system. J Cogn Neurosci. 1996; 8: 1–28.

Duits R, Franken E. Left-invariant diffusions on the space of positions and orientations and their application to crossing-preserving smoothing of HARDI images. Int J Comput Vis 2011; 92: 231–64.

Duyn JH, van Gelderen P, Li T-Q, de Zwart JA, Koretsky AP, Fukunaga M. High-field MRI of brain cortical substructure based on signal phase. Proc Natl Acad Sci U S A 2007; 104: 11796–801.

Edden RA, Jones DK. Spatial and orientational heterogeneity in the statistical sensitivity of skeleton-based analyses of diffusion tensor MR imaging data. J Neurosci Methods 2011; 201: 213–19.

Eichner C, Cauley SF, Cohen-Adad J, Möller HE, Turner R, Setsompop K, et al. Real diffusion-weighted MRI enabling true signal averaging and increased diffusion contrast. NeuroImage 2015; 122: 373–84.

Fairchild G, Toschi N, Hagan CC, Goodyer IM, Calder AJ, Passamonti L. Cortical thickness, surface area, and folding alterations in male youths with conduct disorder and varying levels of callous–unemotional traits. NeuroImage Clin 2015; 8: 253–60.

Farzaneh F, Riederer SJ, Pelc NJ. Analysis of T2 limitations and off-resonance effects on spatial resolution and artifacts in echo-planar imaging. Magn Reson Med 1990; 14: 123–39.

Fischer B, Modersitzki J. Ill-posed medicine – an introduction to image registration. Inverse Probl 2008; 24: 034008.

Fischl B, Dale AM. Measuring the thickness of the human cerebral cortex from magnetic resonance images. Proc Natl Acad Sci U S A 2000; 97: 11050–5.

Fischl B, Sereno MI, Tootell RB, Dale AM. High-resolution intersubject averaging and a coordinate system for the cortical surface. Hum Brain Mapp 1999; 8: 272–84.

Freund P, Weiskopf N, Ashburner J, Wolf K, Sutter R, Altmann DR, et al. MRI investigation of the sensorimotor cortex and the corticospinal tract after acute spinal cord injury: a prospective longitudinal study. Lancet Neurol 2013; 12: 873–81.

Friston KJ, Ashburner J, Frith CD, Poline J-B, Heather JD, Frackowiak RSJ. Spatial registration and normalization of images. Hum Brain Mapp 1995; 3: 165–89.

Friston KJ, Ashburner JT, Kiebel SJ, Nichols TE, Penny WD. Statistical parametric mapping: the analysis of functional brain images. 1st ed. London: Academic Press; 2006.

Friston KJ, Williams S, Howard R, Frackowiak RS, Turner R. Movement-related effects in fMRI time-series. Magn Reson Med 1996; 35: 346–55.

Frost MA, Goebel R. Measuring structural-functional correspondence: spatial variability of specialised brain regions after macro-anatomical alignment. NeuroImage 2012; 59: 1369–81.

Fu ZW, Wang Y, Grimm RC, Rossman PJ, Felmlee JP, Riederer SJ, et al. Orbital navigator echoes for motion measurements in magnetic resonance imaging. Magn Reson Med 1995; 34: 746–53.

Gallichan D, Scholz J, Bartsch A, Behrens TE, Robson MD, Miller KL. Addressing a systematic vibration artifact in diffusion-weighted MRI. Hum Brain Mapp 2010; 31: 193–202.

Gil R, Khabipova D, Zwiers M, Hilbert T, Kober T, Marques JP. An in vivo study of the orientation-dependent and independent components of transverse relaxation rates in white matter. NMR Biomed 2016; 29: 1780–90.

Glasser MF, Coalson TS, Robinson EC, Hacker CD, Harwell J, Yacoub E, et al. A multi-modal parcellation of human cerebral cortex. Nature 2016; 536: 171–8.

Gomez J, Barnett MA, Natu V, Mezer A, Palomero-Gallagher N, Weiner KS, et al. Microstructural proliferation in human cortex is coupled with the development of face processing. Science 2017; 355: 68–71.

Gong G, Jiang T, Zhu C, Zang Y, Wang F, Xie S, et al. Asymmetry analysis of cingulum based on scale-invariant parameterization by diffusion tensor imaging. Hum. Brain Mapp 2005; 24: 92–8.

Gouw AA, Seewann A, Vrenken H, van der Flier WM, Rozemuller JM, Barkhof F, et al. Heterogeneity of white matter hyperintensities in Alzheimer's disease: post-mortem quantitative MRI and neuropathology. Brain J. Neurol. 2008; 131: 3286–98.

Grabher P, Callaghan MF, Ashburner J, Weiskopf N, Thompson AJ, Curt A, et al. Tracking sensory system atrophy and outcome prediction in spinal cord injury. Ann Neurol 2015; 78: 751–61.

Graham MS, Drobnjak I, Zhang H. Quantitative evaluation of eddy-current and motion correction techniques for diffusion-weighted MRIs. 24th Proceedings of the International Society Magnetic resonance Medicine, Singapore, abstract: 0003, 2016.

Greitz D, Wirestam R, Franck A, Nordell B, Thomsen C, Ståhlberg F. Pulsatile brain movement and associated hydrodynamics studied by magnetic resonance phase imaging. The Monro-Kellie doctrine revisited. Neuroradiology 1992; 34: 370–80.

Gundersen HJ, Jensen EB. The efficiency of systematic sampling in stereology and its prediction. J Microsc 1987; 147: 229–63.

Haacke EM, Liu S, Buch S, Zheng W, Wu D, Ye Y. Quantitative susceptibility mapping: current status and future directions. Magn Reson Imaging 2015; 33: 1–25.

Haacke EM, Xu Y, Cheng Y-CN, Reichenbach JR. Susceptibility weighted imaging (SWI). Magn Reson Med 2004; 52: 612–18.

Helms G, Dathe H, Dechent P. Quantitative FLASH MRI at 3T using a rational approximation of the Ernst equation. Magn Reson Med 2008a; 59: 667–72.

Helms G, Dathe H, Kallenberg K, Dechent P. High-resolution maps of magnetization transfer with inherent correction for RF inhomogeneity and T1 relaxation obtained from 3D FLASH MRI. Magn Reson Med 2008b; 60: 1396–407.

Helms G, Draganski B, Frackowiak R, Ashburner J, Weiskopf N. Improved segmentation of deep brain grey matter structures using magnetization transfer (MT) parameter maps. NeuroImage 2009; 47: 194–8.

Hogstrom LJ, Westlye LT, Walhovd KB, Fjell AM. The structure of the cerebral cortex across adult life: age-related patterns of surface area, thickness, and gyrification. Cereb Cortex 2013; 23: 2521–30.

Holland D, Kuperman JM, Dale AM. Efficient correction of inhomogeneous static magnetic field-induced distortion in Echo Planar Imaging. NeuroImage 2010; 50: 175–83.

Hutton C, Bork A, Josephs O, Deichmann R, Ashburner J, Turner R. Image distortion correction in fMRI: a quantitative evaluation. NeuroImage 2002; 16: 217–40.

In M-H, Posnansky O, Speck O. PSF mapping-based correction of eddy-current induced distortions in diffusion-weighted echo-planar imaging. Magn Reson Med 2016; 75: 2055–63.

Jbabdi S, Johansen-Berg H. Tractography: where do we go from here? Brain Connect 2011; 1: 169–83.

Jbabdi S, Sotiropoulos SN, Haber SN, Van Essen DC, Behrens TE. Measuring macroscopic brain connections in vivo. Nat Neurosci 2015; 18: 1546–55.

Jezzard P, Balaban RS. Correction for geometric distortion in echo planar images from B0 field variations. Magn Reson Med 1995; 34: 65–73.

Jezzard P, Barnett AS, Pierpaoli C. Characterization of and correction for eddy current artifacts in echo planar diffusion imaging. Magn Reson Med 1998; 39: 801–12.

Jezzard P, Matthews PM, Smith SM. Functional MRI: an introduction to methods. New Ed. Oxford: Oxford University Press; 2003.

Johansen-Berg H, Behrens TEJ. Diffusion MRI from quantitative measurement to in-vivo neuroanatomy. Amsterdam: Academic Press; 2009.

Jones DK. Diffusion MRI: theory, methods, and applications. Oxford University Press, Oxford; 2010.

Jones DK, Basser PJ. 'Squashing peanuts and smashing pumpkins': how noise distorts diffusion-weighted MR data. Magn Reson Med 2004; 52: 979–93.

Jones DK, Knösche TR, Turner R. White matter integrity, fiber count, and other fallacies: the do's and don'ts of diffusion MRI. NeuroImage 2013; 73: 239–54.

Jones DK, Symms MR, Cercignani M, Howard RJ. The effect of filter size on VBM analyses of DT-MRI data. NeuroImage 2005a; 26: 546–54.

Jones DK, Travis AR, Eden G, Pierpaoli C, Basser PJ. PASTA: pointwise assessment of streamline tractography attributes. Magn Reson Med 2005b; 53: 1462–7.

Joseph J, Warton C, Jacobson SW, Jacobson JL, Molteno CD, Eicher A, et al. Three-dimensional surface deformation-based shape analysis of hippocampus and caudate nucleus in children with fetal alcohol spectrum disorders. Hum Brain Mapp 2014; 35: 659–72.

Joshi SH, Narr KL, Philips OR, Nuechterlein KH, Asarnow RF, Toga AW, et al. Statistical shape analysis of the corpus callosum in Schizophrenia. NeuroImage 2013; 64: 547–59.

Kanai R, Rees G. The structural basis of inter-individual differences in human behaviour and cognition. Nat Rev Neurosci. 2011; 12: 231–42.

Kassner A, Thornhill RE. Texture analysis: a review of neurologic MR imaging applications. Am J Neuroradiol 2010; 31: 809–16.

Keller SS, Ahrens T, Mohammadi S, Möddel G, Kugel H, Ringelstein EB, et al. Microstructural and volumetric abnormalities of the putamen in juvenile myoclonic epilepsy. Epilepsia 2011; 52: 1715–24.

Keller SS, Gerdes JS, Mohammadi S, Kellinghaus C, Kugel H, Deppe K, et al. Volume estimation of the thalamus using freesurfer and stereology: consistency between methods. Neuroinformatics 2012; 10: 341–50.

Keller SS, Roberts N. Measurement of brain volume using MRI: software, techniques, choices and prerequisites. J Anthropol Sci Riv Antropol JASS 2009; 87: 127–51.

Kelly PA, Viding E, Wallace GL, Schaer M, De Brito SA, Robustelli B, et al. Cortical thickness, surface area, and gyrification abnormalities in children exposed to maltreatment: neural markers of vulnerability? Biol Psychiatry 2013; 74: 845–52.

Kleffner I, Deppe M, Mohammadi S, Schiffbauer H, Stupp N, Lohmann H, et al. Diffusion tensor imaging demonstrates fiber impairment in Susac syndrome. Neurology 2008; 70: 1867–69.

Klein A, Andersson J, Ardekani BA, Ashburner J, Avants B, Chiang M-C, et al. Evaluation of 14 nonlinear deformation algorithms applied to human brain MRI registration. NeuroImage 2009; 46: 786–802.

Klein A, Ghosh SS, Avants B, Yeo BTT, Fischl B, Ardekani B, et al. Evaluation of volume-based and surface-based brain image registration methods. NeuroImage 2010; 51: 214–20.

Klose U, Batra M, Nägele T. Age-dependent changes in the histogram of apparent diffusion coefficients values in magnetic resonance imaging. Front Aging Neurosci 2013; 5: 78.

Knight MJ, McCann B, Tsivos D, Dillon S, Coulthard E, Kauppinen RA. Quantitative T2 mapping of white matter: applications for ageing and cognitive decline. Phys Med Biol 2016; 61: 5587–605.

Koay CG, Ozarslan E, Basser PJ. A signal transformational framework for breaking the noise floor and its applications in MRI. J Magn Reson San Diego Calif 1997 2009; 197: 108–19.

Koenig SH, Brown RD, Ugolini R. A unified view of relaxation in protein solutions and tissue, including hydration and magnetization transfer. Magn Reson Med 1993; 29: 77–83.

Lambert C, Lutti A, Helms G, Frackowiak R, Ashburner J. Multiparametric brainstem segmentation using a modified multivariate mixture of Gaussians. NeuroImage Clin. 2013; 2: 684–94.

Landman BA, Bazin P-L, Smith SA, Prince JL. Robust estimation of spatially variable noise fields. Magn Reson Med 2009; 62: 500–9.

Langkammer C, Enzinger C, Quasthoff S, Grafenauer P, Soellinger M, Fazekas F, et al. Mapping of iron deposition in conjunction with assessment of nerve fiber tract integrity in amyotrophic lateral sclerosis. J Magn Reson Imaging JMRI 2010; 31: 1339–45.

Le Bihan D. Looking into the functional architecture of the brain with diffusion MRI. Nat Rev Neurosci 2003; 4: 469–80.

Le Bihan D, Poupon C, Amadon A, Lethimonnier F. Artifacts and pitfalls in diffusion MRI. J Magn Reson Imaging JMRI 2006; 24: 478–88.

Lee JE, Chung MK, Lazar M, DuBray MB, Kim J, Bigler ED, et al. A study of diffusion tensor imaging by tissue-specific, smoothing-compensated voxel-based analysis. NeuroImage 2009; 44: 870–83.

Leming M, Steiner R, Styner M. A framework for incorporating DTI Atlas Builder registration into Tract-Based Spatial Statistics and a simulated comparison to standard TBSS. Proc SPIE Int Soc Opt Eng 2016; 9788.

Liu C, Li W, Wu B, Jiang Y, Johnson GA. 3D fiber tractography with susceptibility tensor imaging. NeuroImage 2012; 59: 1290–8.

Logothetis NK. What we can do and what we cannot do with fMRI. Nature 2008; 453: 869–78.

Lohmann G, Bohn S, Müller K, Trampel R, Turner R. Image restoration and spatial resolution in 7-tesla magnetic resonance imaging. Magn Reson Med 2010; 64: 15–22.

Loizou CP, Petroudi S, Seimenis I, Pantziaris M, Pattichis CS. Quantitative texture analysis of brain white matter lesions derived from T2-weighted MR images in MS patients with clinically isolated syndrome. J Neuroradiol J Neuroradiol 2015; 42: 99–114.

Lorio S, Kherif F, Ruef A, Melie-Garcia L, Frackowiak R, Ashburner J, et al. Neurobiological origin of spurious brain morphological changes: a quantitative MRI study. Hum Brain Mapp 2016; 37: 1801–815.

Lorio S, Lutti A, Kherif F, Ruef A, Dukart J, Chowdhury R, et al. Disentangling in vivo the effects of iron content and atrophy on the ageing human brain. NeuroImage 2014; 103: 280–9.

Lutti A, Dick F, Sereno MI, Weiskopf N. Using high-resolution quantitative mapping of R1 as an index of cortical myelination. NeuroImage 2014; 93 (Part 2): 176–88.

Lutti A, Hutton C, Finsterbusch J, Helms G, Weiskopf N. Optimization and validation of methods for mapping of the radiofrequency transmit field at 3T. Magn Reson Med 2010; 64: 229–38.

Maclaren J, Armstrong BSR, Barrows RT, Danishad KA, Ernst T, Foster CL, et al. Measurement and correction of microscopic head motion during magnetic resonance imaging of the brain. PLoS One 2012; 7: e48088.

Maclaren J, Herbst M, Speck O, Zaitsev M. Prospective motion correction in brain imaging: a review. Magn Reson Med 2013; 69: 621–36.

Magerkurth J, Volz S, Wagner M, Jurcoane A, Anti S, Seiler A, et al. Quantitative T*2-mapping based on multi-slice multiple gradient echo flash imaging: retrospective correction for subject motion effects. Magn Reson Med 2011; 66: 989–97.

Maguire EA, Frackowiak RS, Frith CD. Recalling routes around London: activation of the right hippocampus in taxi drivers. J Neurosci 1997; 17: 7103–10.

Mangeat G, Govindarajan ST, Mainero C, Cohen-Adad J. Multivariate combination of magnetization transfer, T2* and B0 orientation to study the myelo-architecture of the in vivo human cortex. NeuroImage 2015; 119: 89–102.

Mangin J-F, Rivière D, Cachia A, Duchesnay E, Cointepas Y, Papadopoulos-Orfanos D, et al. A framework to study the cortical folding patterns. NeuroImage 2004; 23 (Suppl 1): S129–38.

Manjón JV, Coupé P, Martí-Bonmatí L, Collins DL, Robles M. Adaptive non-local means denoising of MR images with spatially varying noise levels. J Magn Reson Imaging JMRI 2010; 31: 192–203.

Markl M, Bammer R, Alley MT, Elkins CJ, Draney MT, Barnett A, et al. Generalized reconstruction of phase contrast MRI: analysis and correction of the effect of gradient field distortions. Magn Reson Med 2003; 50: 791–801.

Martin AR, Aleksanderek I, Cohen-Adad J, Tarmohamed Z, Tetreault L, Smith N, et al. Translating state-of-the-art spinal cord MRI techniques to clinical use: a systematic review of clinical studies utilizing DTI, MT, MWF, MRS, and fMRI. NeuroImage Clin. 2016; 10: 192–238.

Martin WRW, Wieler M, Gee M. Midbrain iron content in early Parkinson disease: a potential biomarker of disease status. Neurology 2008; 70: 1411–17.

Maximov II, Farrher E, Grinberg F, Shah NJ. Spatially variable Rician noise in magnetic resonance imaging. Med Image Anal 2012; 16: 536–48.

Modersitzki J. FAIR: flexible algorithms for image registration. SIAM, Philadelphia; 2009.

Mohammadi S, Carey D, Dick F, Diedrichsen J, Sereno MI, Reisert M, et al. Whole-brain in-vivo measurements of the axonal G-ratio in a group of 37 healthy volunteers. Front Neurosci 2015; 9: 441.

Mohammadi S, Freund P, Feiweier T, Curt A, Weiskopf N. The impact of post-processing on spinal cord diffusion tensor imaging. NeuroImage 2013a; 70: 377–85.

Mohammadi S, Hutton C, Nagy Z, Josephs O, Weiskopf N. Retrospective correction of physiological noise in DTI using an extended tensor model and peripheral measurements. Magn Reson Med 2013b; 70: 358–69.

Mohammadi S, Möller HE, Kugel H, Müller DK, Deppe M. Correcting eddy current and motion effects by affine whole-brain registrations: evaluation of three-dimensional distortions and comparison with slicewise correction. Magn Reson Med 2010; 64: 1047–56.

Mohammadi S, Nagy Z, Hutton C, Josephs O, Weiskopf N. Correction of vibration artefacts in DTI using phase-encoding reversal (COVIPER). Proc Int Soc Magn Reson Med 2012a; 20: 1898.

Mohammadi S, Nagy Z, Möller HE, Symms MR, Carmichael DW, Josephs O, et al. The effect of local perturbation fields on human DTI: characterisation, measurement and correction. NeuroImage 2012b; 60: 562–70.

Mori S, van Zijl PCM. Fiber tracking: principles and strategies – a technical review. NMR Biomed 2002; 15: 468–80.

Neher PF, Descoteaux M, Houde J-C, Stieltjes B, Maier-Hein KH. Strengths and weaknesses of state of the art fiber tractography pipelines – a comprehensive in-vivo and phantom evaluation study using Tractometer. Med Image Anal 2015; 26: 287–305.

Nocedal J, Wright SJ. Numerical optimization. New York: Springer; 2006.

Nordahl CW, Dierker D, Mostafavi I, Schumann CM, Rivera SM, Amaral DG, et al. Cortical folding abnormalities in autism revealed by surface-based morphometry. J Neurosci 2007; 27: 11725–35.

Nöth U, Volz S, Hattingen E, Deichmann R. An improved method for retrospective motion correction in quantitative T2* mapping. NeuroImage 2014; 92: 106–19.

Nunes RG, Jezzard P, Clare S. Investigations on the efficiency of cardiac-gated methods for the acquisition of diffusion-weighted images. J Magn Reson San Diego Calif 1997 2005; 177: 102–10.

Papp D, Callaghan MF, Meyer H, Buckley C, Weiskopf N. Correction of inter-scan motion artifacts in quantitative R1 mapping by accounting for receive coil sensitivity effects. Magn Reson Med 2016; 76: 1478–85.

Pohmann R, Scheffler K. A theoretical and experimental comparison of different techniques for B_1 mapping at very high fields. NMR Biomed 2013; 26: 265–75.

Polzehl J, Tabelow K. Low SNR in Diffusion MRI Models. J Am Stat Assoc 2016; 111: 1480–90.

Poot DHJ, Jeurissen B, Bastiaensen Y, Veraart J, Van Hecke W, Parizel PM, et al. Super-resolution for multislice diffusion tensor imaging. Magn Reson Med 2013; 69: 103–13.

Reber PJ, Wong EC, Buxton RB, Frank LR. Correction of off resonance-related distortion in echo-planar imaging using EPI-based field maps. Magn Reson Med 1998; 39: 328–30.

Reich DS, Smith SA, Zackowski KM, Gordon-Lipkin EM, Jones CK, Farrell JAD, et al. Multiparametric magnetic resonance imaging analysis of the corticospinal tract in multiple sclerosis. NeuroImage 2007; 38: 271–9.

Renvall V, Witzel T, Wald LL, Polimeni JR. Automatic cortical surface reconstruction of high-resolution T1 echo planar imaging data. NeuroImage 2016; 134: 338–54.

Roberts N, Puddephat MJ, McNulty V. The benefit of stereology for quantitative radiology. Br J Radiol 2000; 73: 679–97.

Robson PM, Grant AK, Madhuranthakam AJ, Lattanzi R, Sodickson DK, McKenzie CA. Comprehensive quantification of signal-to-noise ratio and g-factor for image-based and k-space-based parallel imaging reconstructions. Magn Reson Med 2008; 60: 895–907.

Roebroeck A, Galuske R, Formisano E, Chiry O, Bratzke H, Ronen I, et al. High-resolution diffusion tensor imaging and tractography of the human optic chiasm at 9.4 T. NeuroImage 2008; 39: 157–68.

Rohde GK, Barnett AS, Basser PJ, Marenco S, Pierpaoli C. Comprehensive approach for correction of motion and distortion in diffusion-weighted MRI. Magn Reson Med 2004; 51: 103–14.

Rooney WD, Johnson G, Li X, Cohen ER, Kim S-G, Ugurbil K, et al. Magnetic field and tissue dependencies of human brain longitudinal 1H2O relaxation in vivo. Magn Reson Med 2007; 57: 308–18.

Ruthotto L, Kugel H, Olesch J, Fischer B, Modersitzki J, Burger M, et al. Diffeomorphic susceptibility artifact correction of diffusion-weighted magnetic resonance images. Phys Med Biol 2012; 57: 5715–31.

Ruthotto L, Mohammadi S, Heck C, Modersitzki J, Weiskopf N. Hyperelastic susceptibility artifact correction of DTI in SPM. In: Meinzer H-P, Deserno TM, Handels H, Tolxdorff T, eds. Bildverarbeitung für die Medizin 2013. Springer Berlin Heidelberg; 2013, pp. 344–9. Available from: http://link.springer.com/chapter/10.1007/978-3-642-36480-8_60 [cited 8 June 2013].

Ruthotto L, Mohammadi S, Weiskopf N. A new method for joint susceptibility artefact correction and super-resolution for dMRI. 2014. p. 90340P–90340P–4. Available from: https://www.spiedigitallibrary.org/conference-proceedings-of-spie/9034/1/A-new-method-for-joint-susceptibility-artefact-correction-and-super/10.1117/12.2043591.short?SSO=1

Schmahmann JD, Pandya DN. Fiber pathways of the brain. Oxford University Press, Oxford; 2006.

Schmalbrock P, Prakash RS, Schirda B, Janssen A, Yang GK, Russell M, et al. Basal ganglia iron in patients with multiple sclerosis measured with 7T quantitative susceptibility mapping correlates with inhibitory control. AJNR AmJ Neuroradiol 2016; 37: 439–46.

Schmitt F, Stehling MK, Turner R. Echo-planar imaging: theory, technique and application. Springer Science & Business Media, Berlin; 2012.

Sereno MI, Lutti A, Weiskopf N, Dick F. Mapping the human cortical surface by combining quantitative T(1) with retinotopy. Cereb Cortex N Y N 1991 2013; 23: 2261–8.

Setsompop K, Bilgic B, Nummenmaa A, Fan Q, Cauley S, Huang S, et al. SLIce Dithered Enhanced Resolution Simultaneous MultiSlice (SLIDER-SMS) for high resolution (700 um) diffusion imaging of the human brain. Proc Intl Soc Magn Reson Med 2015; 23: 0339.

Sijbers J, den Dekker AJ. Maximum likelihood estimation of signal amplitude and noise variance from MR data Magn Reson Med 2004; 51: 586–94.

Sijbers J, den Dekker AJ, Scheunders P, Van Dyck D. Maximum-likelihood estimation of Rician distribution parameters. IEEE Trans Med Imaging 1998; 17: 357–61.

Skare S, Andersson JL. On the effects of gating in diffusion imaging of the brain using single shot EPI. Magn Reson Imaging 2001; 19: 1125–8.

Smith SM, Jenkinson M, Johansen-Berg H, Rueckert D, Nichols TE, Mackay CE, et al. Tract-based spatial statistics: voxelwise analysis of multi-subject diffusion data. NeuroImage 2006; 31: 1487–505.

Soares JM, Marques P, Alves V, Sousa N. A hitchhiker's guide to diffusion tensor imaging. Front Neurosci 2013; 7: 31.

Sotiras A, Davatzikos C, Paragios N. Deformable medical image registration: a survey. IEEE Trans Med Imaging 2013; 32: 1153–90.

Steiger TK, Weiskopf N, Bunzeck N. Iron level and myelin content in the ventral striatum predict memory performance in the aging brain. J Neurosci 2016; 36: 3552–8.

Stikov N, Campbell JSW, Stroh T, Lavelée M, Frey S, Novek J, et al. In vivo histology of the myelin g-ratio with magnetic resonance imaging. NeuroImage 2015; 118: 397–405.

Storey P, Frigo FJ, Hinks RS, Mock BJ, Collick BD, Baker N, et al. Partial k-space reconstruction in single-shot diffusion-weighted echo-planar imaging. Magn Reson Med 2007; 57: 614–19.

Stüber C, Morawski M, Schäfer A, Labadie C, Wähnert M, Leuze C, et al. Myelin and iron concentration in the human brain: a quantitative study of MRI contrast. NeuroImage 2014; 93 (Part 1): 95–106.

Studholme C, Constable RT, Duncan JS. Accurate alignment of functional EPI data to anatomical MRI using a physics-based distortion model. IEEE Trans Med Imaging 2000; 19: 1115–27.

Studholme C, Hill DLG, Hawkes DJ. An overlap invariant entropy measure of 3D medical image alignment. Pattern Recognit 1999; 32: 71–86.

Tabelow K, D'Alonzo C, Polzehl J, Callaghan M, Ruthotto L, Weiskopf N, et al. How to achieve very high resolution quantitative MRI at 3T? Proceedings of the 22nd Human Brain Mapping meeting, Geneva, 2016.

Tabelow K, Polzehl J, Spokoiny V, Voss HU. Diffusion tensor imaging: structural adaptive smoothing. NeuroImage 2008; 39: 1763–73.

Tabelow K, Polzehl J, Voss HU, Spokoiny V. Analyzing fMRI experiments with structural adaptive smoothing procedures. NeuroImage 2006; 33: 55–62.

Tabelow K, Voss HU, Polzehl J. Local estimation of the noise level in MRI using structural adaptation. Med Image Anal 2015; 20: 76–86.

Tardif CL, Schäfer A, Waehnert M, Dinse J, Turner R, Bazin P-L. Multi-contrast multi-scale surface registration for improved alignment of cortical areas. NeuroImage 2015; 111: 107–22.

Thévenaz P, Unser M. Optimization of mutual information for multiresolution image registration. IEEE Trans Image Process Publ IEEE Signal Process Soc 2000; 9: 2083–99.

Thompson P, Toga AW. A surface-based technique for warping three-dimensional images of the brain. IEEE Trans Med Imaging 1996; 15: 402–17.

Tisdall MD, Reuter M, Qureshi A, Buckner RL, Fischl B, van der Kouwe AJW. Prospective motion correction with volumetric navigators (vNavs) reduces the bias and variance in brain morphometry induced by subject motion. NeuroImage 2016; 127: 11–22.

Triantafyllou C, Wald LL, Hoge RD. Echo-time and field strength dependence of BOLD reactivity in veins and parenchyma using flow-normalized hypercapnic manipulation. PLoS One 2011; 6: e24519.

Tofts PS. Quantitative MRI of the brain: measuring changes caused by disease. New York: Wiley; 2003.

van der Kouwe AJW, Benner T, Dale AM. Real-time rigid body motion correction and shimming using cloverleaf navigators. Magn Reson Med 2006; 56: 1019–32.

Van Hecke W, Leemans A, De Backer S, Jeurissen B, Parizel PM, Sijbers J. Comparing isotropic and anisotropic smoothing for voxel-based DTI analyses: a simulation study. Hum Brain Mapp 2010; 31: 98–114.

Veraart J, Sijbers J, Sunaert S, Leemans A, Jeurissen B. Weighted linear least squares estimation of diffusion MRI parameters: strengths, limitations, and pitfalls. NeuroImage 2013; 81: 335–46.

Vymazal J, Righini A, Brooks RA, Canesi M, Mariani C, Leonardi M, et al. T1 and T2 in the brain of healthy subjects, patients with Parkinson disease, and patients with multiple system atrophy: relation to iron content. Radiology 1999; 211: 489–95.

Waehnert MD, Dinse J, Weiss M, Streicher MN, Waehnert P, Geyer S, et al. Anatomically motivated modeling of cortical laminae. NeuroImage 2014; 93 (Part 2): 210–20.

Wedeen VJ, Weisskoff RM, Poncelet BP. MRI signal void due to in-plane motion is all-or-none. Magn Reson Med 1994; 32: 116–20.

Weiskopf N, Callaghan MF, Josephs O, Lutti A, Mohammadi S. Estimating the apparent transverse relaxation time (R2*) from images with different contrasts (ESTATICS) reduces motion artifacts. Brain Imaging Methods 2014; 8: 278.

Weiskopf N, Mohammadi S, Lutti A, Callaghan MF. Advances in MRI-based computational neuroanatomy: from morphometry to in-vivo histology. Curr Opin Neurol 2015; 28: 313–22.

Weiskopf N, Suckling J, Williams G, Correia MM, Inkster B, Tait R, et al. Quantitative multi-parameter mapping of R1, PD*, MT and R2* at 3T: a multi-center validation. Front. Brain Imaging Methods 2013; 7: 95.

Welch EB, Manduca A, Grimm RC, Ward HA, Jack CR. Spherical navigator echoes for full 3D rigid body motion measurement in MRI. Magn Reson Med 2002; 47: 32–41.

Wharton S, Bowtell R. Fiber orientation-dependent white matter contrast in gradient echo MRI. Proc Natl Acad Sci 2012; 109: 18559–64.

Wharton S, Bowtell R. Gradient echo based fiber orientation mapping using R2* and frequency difference measurements. NeuroImage 2013; 83: 1011–23.

Wheeler-Kingshott CA, Stroman PW, Schwab JM, Bacon M, Bosma R, Brooks J, et al. The current state-of-the-art of spinal cord imaging: applications. NeuroImage 2014; 84: 1082–93.

White N, Roddey C, Shankaranarayanan A, Han E, Rettmann D, Santos J, et al. PROMO: real-time prospective motion correction in MRI using image-based tracking. Magn Reson Med 2010; 63: 91–105.

Woollett K, Glensman J, Maguire EA. Non-spatial expertise and hippocampal gray matter volume in humans. Hippocampus 2008; 18: 981–4.

Xu J, Shimony JS, Klawiter EC, Snyder AZ, Trinkaus K, Naismith RT, et al. Improved in vivo diffusion tensor imaging of human cervical spinal cord. NeuroImage 2013; 67: 64–76.

Yeatman JD, Wandell BA, Mezer AA. Lifespan maturation and degeneration of human brain white matter. Nat Commun 2014; 5: 4932.

Yiannakas MC, Kearney H, Samson RS, Chard DT, Ciccarelli O, Miller DH, et al. Feasibility of grey matter and white matter segmentation of the upper cervical cord in vivo: a pilot study with application to magnetisation transfer measurements. NeuroImage 2012; 63: 1054–9.

Yokoo T, Yuan Q, Sénégas J, Wiethoff AJ, Pedrosa I. Quantitative R2* MRI of the liver with Rician noise models for evaluation of hepatic iron overload: simulation, phantom, and early clinical experience. J Magn Reson Imaging JMRI 2015; 42: 1544–59.

Zaitsev M, Dold C, Sakas G, Hennig J, Speck O. Magnetic resonance imaging of freely moving objects: prospective real-time motion correction using an external optical motion tracking system. NeuroImage 2006; 31: 1038–50.

Zaitsev M, Hennig J, Speck O. Point spread function mapping with parallel imaging techniques and high acceleration factors: fast, robust, and flexible method for echo-planar imaging distortion correction. Magn Reson Med 2004; 52: 1156–66.

Zalesky A. Moderating registration misalignment in voxelwise comparisons of DTI data: a performance evaluation of skeleton projection. Magn Reson Imaging 2011; 29: 111–25.

Zhang H, Schneider T, Wheeler-Kingshott CA, Alexander DC. NODDI: practical in vivo neurite orientation dispersion and density imaging of the human brain. NeuroImage 2012; 61: 1000–16.

Zilles K, Amunts K. Anatomical basis for functional specialization. In: fMRI: from nuclear spins to brain functions Uludag K, Ugurbil K, Berliner L, eds. New York: Springer; 2015, pp. 27–66.

18

Future of Quantitative MRI[1]

Contents

Mara Cercignani
University of Sussex

During the past few decades the field of quantitative MRI has continued to grow with tremendous technological advances and establishment of new methodologies. Applications in the brain have flourished, stemming from the original focus on disease, with the aim of providing a better understanding of pathophysiological mechanisms, to neurosciences in a broader sense, aiming to measure tissue features that can explain inter-individual differences in terms of performance and lifestyle.

Regardless of the specific application, the ultimate goal of quantitative MRI (qMRI) is to be able to provide non-invasive histology, as well as assessments of physiology and function, thus quantifying tissue components and characterising its pathology. However, the MRI signal is a very indirect measure of the tissue properties we are interested in, and, while some MRI contrasts may be influenced by myelin content or axonal packing, other contributors to the overall signal prevent a one-to-one association between MRI biomarkers and biological substrate. The challenges that remain to be addressed before qMRI can establish itself in clinical routine are to improve specificity and reproducibility (especially between scanners) while at the same time reducing acquisition times.

We thus expect that the future will see the development of increasingly sophisticated models of signal behaviour, as well as the implementation of faster acquisition strategies. Technological advances and the wider availability of MRI scanners with state-of-the-art hardware will be fundamental to this progress. Here we summarise some of the most recent trends in qMRI research that we expect to be central to its development in the near future.

18.1 Stronger and faster

Over the past two decades we have observed a tremendous improvement in MRI hardware, with over 50 7T human systems currently in use worldwide, while 3T has become almost the standard magnetic field for brain MRI. Ultra-high magnetic fields provide several advantages, including boosted signal-to-noise ratio (SNR) (and improved spatial resolution), frequency separation of magnetic resonance peaks, increasing the sensitivity of MR spectroscopy and of methods tuned to magnetisation transfer effects (including chemical exchange saturation transfer [CEST]), increased magnetic susceptibility, which is essential for MRI contrasts such a blood oxygenation level dependent (BOLD) and susceptibility weighted imaging (SWI). Figure 18.1 gives an example of a clinical application of quantitative susceptibility mapping at 7T (Costagli *et al.*, 2016): the use of an ultra-high magnetic field allowed the authors to identify increased susceptibility of the primary motor cortex in patients with amyotrophic lateral sclerosis (ALS), which also correlates with clinical measures of ALS progression.

In addition to enabling new methodologies, higher field strengths also present new challenges and artefacts that are typically neglected at lower fields. This is important for the purpose of quantification, because it prompts the development of new strategies for artefact corrections, which can also be used at lower field strengths, ultimately benefitting the accuracy of qMRI as a whole. (This happened, for example, with B_1 mapping techniques after the advent of 3T scanners.)

Alongside bigger magnets, hardware developments have also led to the implementation of stronger magnetic field gradients, up to 300 mTm^{-1}. While these were created for bespoke systems such as the Human Connectome Project scanner (Fan *et al.*, 2014), a handful of scanners with comparable hardware are already installed worldwide. Larger magnetic field gradients are particularly important for microstructural techniques based on diffusion imaging, as larger gradients enable much higher b values to be used (see Figure 18.2) (Fan *et al.*, 2016) and smaller structures to be resolved, but they may also offer advantages

[1] Edited by Paul S. Tofts

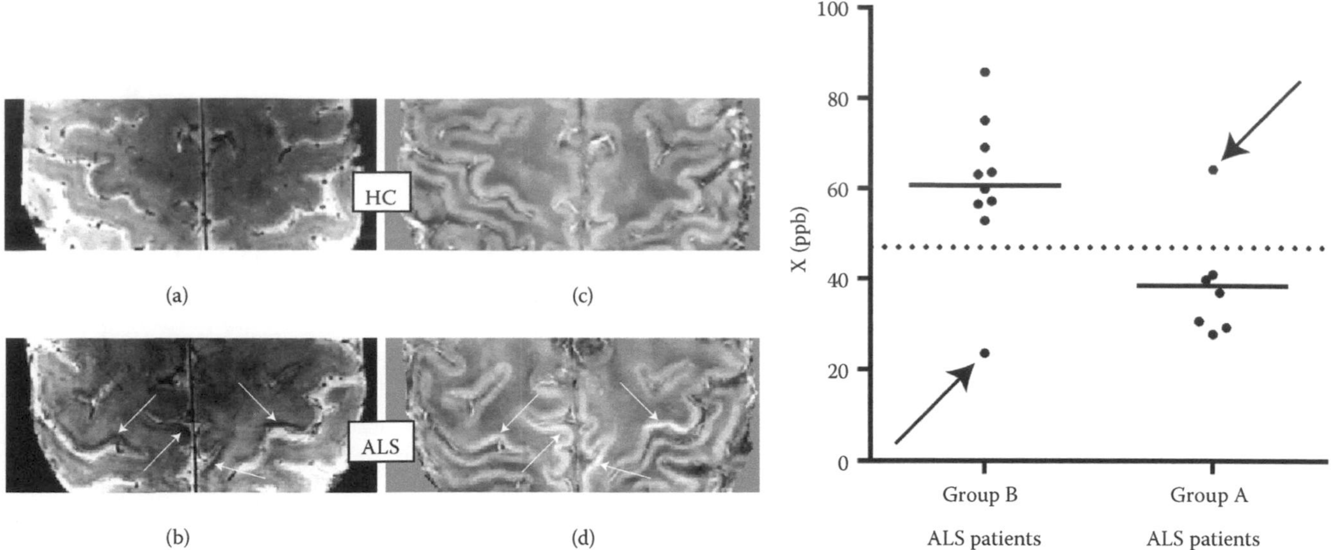

FIGURE 18.1 Increased magnetic susceptibility in the primary motor cortex (M1) of patients with amyotrophic lateral sclerosis (ALS) demonstrated at 7T. T_2*-weighted images targeting the M1 of (a) one healthy control (HC) and (b) one ALS patient; (c) and (d) the corresponding susceptibility maps. Arrows indicate the T_2* signal hypointensity and quantitative susceptibility mapping hyperintensity in patients' deep layers of the primary motor cortex. (Reproduced from *NeuroImage Clin.*, 12, Costagli, M., *et al.*, Magnetic susceptibility in the deep layers of the primary motor cortex in Amyotrophic Lateral Sclerosis, 965–969, copyright 2016, with permission from Elsevier.)

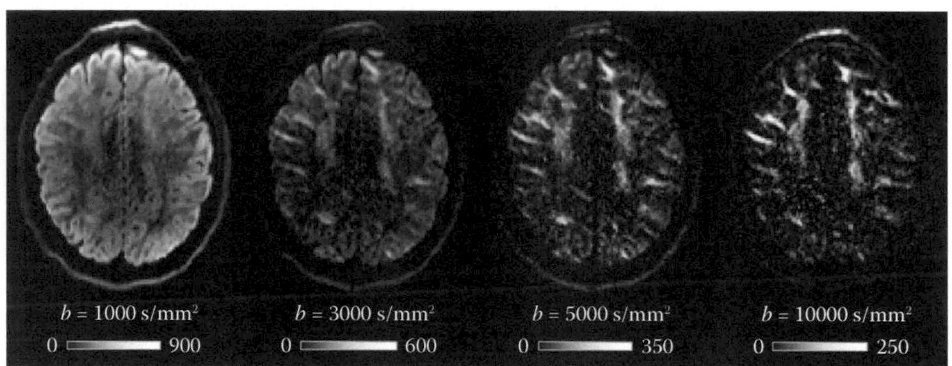

FIGURE 18.2 The advent of ultra-strong gradients (up to 300 mTm^{-1}) has made it possible to achieve diffusion b factors much higher than before. This figure shows the change in signal and contrast when sweeping the b-space from 1000 to 10,000 s/mm^2. At b = 10,000, only slow-diffusing protons are visible, and the effects of anisotropy in the white matter become very obvious. (Reproduced from *NeuroImage*, 124, Fan, Q., *et al.*, MGH-USC human Connectome Project datasets with ultra-high b-value diffusion MRI, 1108–1114, copyright 2016, with permission from Elsevier.)

for other techniques. Overall, we expect these technological improvements to strongly impact on the development of qMRI, allowing researchers to overcome constraints imposed by SNR and resolution. The increased resolution will, for example, enable a much more accurate characterisation of cortical tissue than in the past and hopefully allow its laminar structure to be resolved.

The increasing complexity of qMRI models requires the collection of large datasets, thus requiring fast acquisitions. While the improved SNR provided by better hardware can help, the development of time-efficient pulse sequences is essential to ensure the future of qMRI. Recent advances, such as multiband, aka simultaneous multislice (Moeller *et al.*, 2010), have made a real difference both in terms of achievable temporal resolution for dynamic scans and acquisition time for techniques based on diffusion MRI (dMRI). This opens new avenues for the implementation of multi-modal acquisitions.

18.2 Multiple contrasts and multimodal approaches

Multimodal MRI is a broad concept that refers to any attempt to combine information coming from more than one MRI contrast. The possible approaches thus span from simply measuring several

MRI parameters in the same individuals, to developing joint models, to using complex computational approaches to derive new measures. As the limited specificity of each single technique is acknowledged, we expect to see an increase in the development of joint-models (e.g. T_1, T_2 MT and diffusion), relying either on combined (e.g. De Santis *et al.*, 2016; Stanisz *et al.*, 1998) or separate acquisitions. Although the most informative way of combining several MRI parameters is not immediately obvious, the added value of multimodal protocols is generally recognised, and both data- and model-driven approaches have been proposed. Data-driven approaches require the use of multivariate approaches, spanning from data-reduction methods, such as principal or independent component analysis (ICA), to machine learning. The driving principle is that, if two techniques are both sensitive to a specific substrate (e.g. myelin) but also respectively affected by other factors, a method like ICA should be able to isolate their shared information (e.g. myelin concentration), assumed to reflect the feature of interest (e.g. Mangeat *et al.*, 2015). Conversely, model-based methods attempt to combine parameters extracted by existing biophysical or signal models to obtain new parameters, which are believed to be more accurate or more specific than the original ones. An example is provided by the recently introduced g-ratio framework (see Figure 18.3) (Stikov *et al.*, 2015).

Pushing the concept of multimodality even further, qMRI is increasingly associated with neurophysiological measures, such as electro-encephalography, magneto-encephalography and transcranial magnetic stimulation. These techniques nicely complement MRI by characterising brain function and connectivity with high temporal resolution. The highly detailed anatomical and microstructural information derived from MRI can then be related to physiology elucidating the mechanisms underpinning the structure-to-function relationship.

Another promising union of modalities is the combination of MRI and positron emission tomography (PET). PET can exploit ligands to specific neurobiological substrates to provide high specificity. Thanks to the spread of hybrid imagers, simultaneously acquired MRI and PET data can be easily fused. The specificity of PET can be used to validate some MRI biomarkers, or the complementarity of the two kinds of information can be exploited to build novel biomarkers.

18.3 Big data

With the increasing interest in multimodal studies, the development of sophisticated models of the MRI signal and the ability to acquire more data in shorter times, the size of typical MRI datasets are increasing quickly with time. We are thus facing new challenges in the management and the analysis of MRI data. New computational methods, including strategies to reduce the complexity of these datasets, are needed. In addition, there is great interest in relating imaging with other big data fields, particularly genetics. Overall, this trend calls for new tools able to extract meaningful information from the data and promises to attract scientists from fields other than MRI physics, such as computer science, mathematics and engineering. In parallel, the ever-increasing list of multisite initiatives (Human Connectome Project: www.humanconnectome.org; ENIGMA: http://enigma.ini.usc.edu; UK biobank: http://www.ukbiobank.ac.uk; ADNI: http://www.adni-info.org, to mention just a few) require rigorous procedures for quality assurance and data harmonisation. We therefore expect efforts to be devolved towards improving the reproducibility of quantitative MRI data by minimising sources of bias and artefacts.

18.4 Accuracy and reproducibility

This topic is covered in more detail in Chapter 3 of this book. It is however important to highlight that artefact correction and bias reduction methods are becoming increasingly popular.

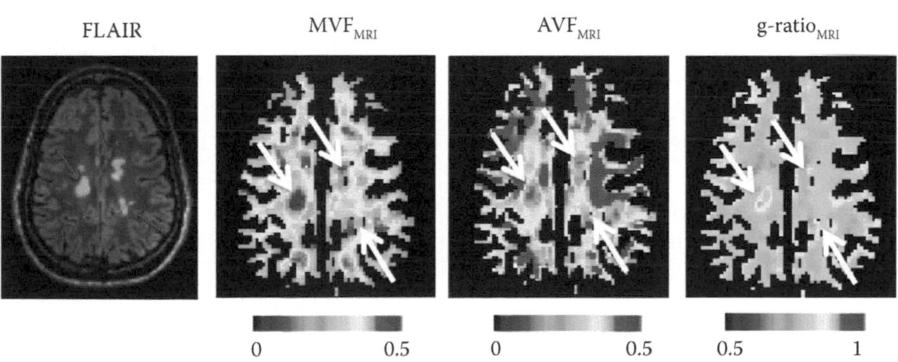

FIGURE 18.3 Imaging the aggregate g-ratio by qMRI. Combining magnetisation transfer and diffusion MRI it is possible to obtain voxel-wise estimates of the myelin volume fraction (MVF) and the axon volume fraction (AVF), which in turn can be used to estimate the g-ratio, i.e. the inner-to-outer axon diameter ratio. This quantity is related to the speed of conduction of an axon and therefore has a functional interpretation. The figure shows a case of a patient with MS, with several hyperintense lesions visible on the Fluid-attenuated inversion-recovery (FLAIR) scan. The lesions appear as hypointense on the MVF and AVF maps, suggesting the presence of both demyelination and axonal loss. The g-ratio map shows that only one of the lesions has a g-ratio greater than 0.8 – suggesting impaired (hindered) signal transmission. (Reproduced from *NeuroImage*, 118, Stikov, N., *et al.*, In vivo histology of the myelin g-ratio with magnetic resonance imaging, 397–405, copyright 2015, with permission from Elsevier.)

Noise reduction and data correction (e.g. Veraart *et al.*, 2016) will probably become standard practice as the use of quantitative MRI grows. Precision and reproducibility may be even more important, especially in the context of longitudinal and multicentre studies. Methods for data harmonisation have been proposed (e.g. Muller *et al.*, 2016). In this context, manufacturer collaboration with scientists will be essential, to ensure accurate data collection and analysis, as well the ability to implement comparable protocols on different systems.

18.5 Biological interpretation and clinical utility

Ultimately, the biggest challenge remains the biological interpretation of MRI parameters. While the number and sophistication of the available MRI parameters has grown remarkably over the past few decades, the meaning of these parameters and of their changes remains elusive. Ideally we are aiming to develop *biomarkers* that can be used as measures of physiological or pathological changes. However, beyond being quantitative and reproducible, biomarkers must change concurrently with clinical/physical status. For clinical translation, MRI techniques must be easy to implement but also evidence-based. Validation of MRI markers is challenging, because it can only be performed in animal models or post-mortem tissue. In both cases, the properties of the tissue under observation differ from the *in vivo* human condition and therefore can only be partially valid. In addition, the so-called gold standard is usually histology, which suffers from important limitations as well. The MRI community is becoming increasingly aware of the need for validation, and a series of initiatives are springing up encouraging scientists to share their data and methods.

18.6 Conclusions

The first edition of this book ended with the following paragraph: 'We are present at a true technological revolution which is exposing our inner biological workings in ever increasing detail. A few decades ago this was inconceivable; in a few decades' time the techniques will be as routine as measuring the mass of the body'. Admittedly, we have not yet reached this stage; however, the progress observed during the past 15 years suggests that the field of qMRI is still productive (this book contains three chapters dedicated to techniques that were not covered in the first edition!), and we can expect it to grow for many years to come.

References

Costagli, M., Donatelli, G., Biagi, L., Caldarazzo Ienco, E., Siciliano, G., Tosetti, M., Cosottini, M. 2016. Magnetic susceptibility in the deep layers of the primary motor cortex in amyotrophic lateral sclerosis. Neuroimage Clin 12, 965–969.

De Santis, S., Assaf, Y., Jeurissen, B., Jones, D.K., Roebroeck, A. 2016. T1 relaxometry of crossing fibres in the human brain. Neuroimage 141, 133–142.

Fan, Q., Nummenmaa, A., Witzel, T., Zanzonico, R., Keil, B., Cauley, S., Polimeni, J.R., et al. 2014. Investigating the capability to resolve complex white matter structures with high b-value diffusion magnetic resonance imaging on the MGH-USC Connectom scanner. Brain Connect 4, 718–726.

Fan, Q., Witzel, T., Nummenmaa, A., Van Dijk, K.R., Van Horn, J.D., Drews, M.K., Somerville, L.H., et al. 2016. MGH-USC Human Connectome Project datasets with ultra-high b-value diffusion MRI. Neuroimage 124, 1108–1114.

Mangeat, G., Govindarajan, S.T., Mainero, C., Cohen-Adad, J. 2015. Multivariate combination of magnetization transfer, T2* and B0 orientation to study the myelo-architecture of the in vivo human cortex. Neuroimage 119, 89–102.

Moeller, S., Yacoub, E., Olman, C.A., Auerbach, E., Strupp, J., Harel, N., Ugurbil, K. 2010. Multiband multislice GE-EPI at 7 tesla, with 16-fold acceleration using partial parallel imaging with application to high spatial and temporal whole-brain fMRI. Magn Reson Med 63, 1144–1153.

Muller, H.P., Turner, M.R., Grosskreutz, J., Abrahams, S., Bede, P., Govind, V., Prudlo, J., et al. 2016. A large-scale multicentre cerebral diffusion tensor imaging study in amyotrophic lateral sclerosis. J Neurol Neurosurg Psychiatry 87, 570–579.

Stanisz, G.J., Li, J.G., Wright, G.A., Henkelman, R.M. 1998. Water dynamics in human blood via combined measurements of T2 relaxation and diffusion in the presence of gadolinium. Magn Reson Med 39, 223–233.

Stikov, N., Campbell, J.S., Stroh, T., Lavelee, M., Frey, S., Novek, J., Nuara, S., et al. 2015. In vivo histology of the myelin g-ratio with magnetic resonance imaging. Neuroimage 118, 397–405.

Veraart, J., Novikov, D.S., Christiaens, D., Ades-Aron, B., Sijbers, J., Fieremans, E. 2016. Denoising of diffusion MRI using random matrix theory. Neuroimage 142, 394–406.

Appendix A: Greek Alphabet for Scientific Use[1]

Lower case	Upper case	
α	Α	Alpha
β	Β	Beta
γ	Γ	Gamma
δ	Δ	Delta
ε	Ε	Epsilon
ζ	Ζ	Zeta
η	Η	Eta
θ	Θ	Theta
ι	Ι	Iota
κ	Κ	Kappa
λ	Λ	Lambda
μ	Μ	Mu
ν	Ν	Nu
ξ	Ξ	Xi
ο	Ο	Omicron
π	Π	Pi
ρ	Ρ	Rho
σ	Σ	Sigma
τ	Τ	Tau
υ	Υ	Upsilon
φ	Φ	Phi
χ	Χ	Chi
ψ	Ψ	Psi
ω	Ω	Omega

[1] These are the same letters as used in modern Greek, although the pronunciation is different for a few letters (e.g. π is pronounced 'pee' in modern Greek and 'pie' here).

Appendix B: MRI Abbreviations and Acronyms

2D	two-dimensional
3D	three-dimensional
ADC	Apparent Diffusion Coefficient
AIF	Arterial Input Function
ASL	Arterial Spin Labelling
CEST	Chemical Exchange Saturation Transfer
CNS	Central Nervous System
CSF	Cerebrospinal Fluid
CV	Coefficient of Variation
DCE	Dynamic Contrast-Enhanced
DTI	Diffusion Tensor Imaging
DTPA	Diethylene Triamine Pentaacetic Acid
EPI	Echo Planar Imaging
FA	Flip Angle
FLAIR	Fluid Attenuated Inversion Recovery
FSE	Fast Spin Echo
GE	Gradient Echo
GRASE	Gradient- and spin-echo
ISMRM	International Society for Magnetic Resonance in Medicine
IR	Inversion Recovery
LL	Look-Locker
MD	Mean Diffusivity
MR	Magnetic Resonance
MRS	Magnetic Resonance Spectroscopy
MRI	Magnetic Resonance Imaging
MS	Multiple Sclerosis
MT	Magnetisation Transfer
MTR	Magnetisation Transfer Ratio
NAA	N-acetyl aspartate
PD	Proton Density
PE	Phase Encoding
QA	Quality Assurance
qMR	Quantitative Magnetic Resonance
qMRI	Quantitative Magnetic Resonance Imaging
RF	Radiofrequency
ROI	Region of interest
SAR	Specific Absorption Rate
SD	Standard Deviation
SE	Spin Echo
SEM	Standard Error of the Mean
SNR	Signal to Noise Ratio
STIR	Short TI Inversion Recovery
T	Tesla
TE	Echo Time
TI	Inversion Time
TR	Repetition Time
VFA	Variable Flip Angle
WM	White Matter

Index

Note: The letters *f* and *t* represents figure and tables, respectively.

C

D

Q

R

S

T

U

V

W

X

Y

Z

T - #0566 - 071024 - C360 - 280/210/16 - PB - 9780367781538 - Gloss Lamination